THE MATHEMATICS THAT EVERY SECONDARY SCHOOL MATH TEACHER NEEDS TO KNOW

What knowledge of mathematics do secondary school math teachers need to facilitate understanding, competency, and interest in mathematics for all of their students? This unique text and resource bridges the gap between the mathematics learned in college and the mathematics taught in secondary schools. Written in an informal, clear, and interactive learner-centered style, it is designed to help pre-service and in-service teachers gain the deep mathematical insight they need to engage their students in learning mathematics in a multifaceted way that is interesting, developmental, connected, deep, understandable, and often, surprising and entertaining.

Launch questions at the beginning of each section capture interest and involve readers in the learning of the mathematical concepts.

Student Learning Opportunities provide chances to practice problems associated with what has been learned in the chapter, to complete proofs that are mentioned but not proved, and to apply what has been learned to solving real-life problems.

Questions from the Classroom are featured in every chapter and in the Student Learning Opportunities, such as the deep "why" conceptual questions that middle or secondary school students are curious about, questions requiring analysis and correction of typical student errors and misconceptions, questions focused on counterintuitive results, and questions that contain activities and/or tasks suitable for use with secondary school students.

Highlighted themes throughout the chapters aid readers in becoming teachers who have great "MATH-N-SIGHT"

- **M** Multiple Approaches/Representations
- **A** Applications to Real Life
- **T** Technology
- **H** History
- **N** Nature of Mathematics: Reasoning and Proof
- **S** Solving Problems
- **I** Interlinking Concepts: Connections
- **G** Grade Levels
- **H** Honing of Mathematical Understanding and Skills
- **T** Typical Errors

This text is aligned with the recently released Common Core State Standards, and is ideally suited for a capstone mathematics course in a secondary mathematics certification program. It is also appropriate for any methods or mathematics course for pre- or in-service secondary mathematics teachers, and is a valuable resource for classroom teachers.

Alan Sultan is Professor, Mathematics, Queens College of the City University of New York.

Alice F. Artzt is Professor, Secondary Mathematics Education, Queens College of the City University of New York.

STUDIES IN MATHEMATICAL THINKING AND LEARNING
Alan H. Schoenfeld, Series Editor

Artzt/Armour-Thomas/Curcio• *Becoming a Reflective Mathematics Teacher: A Guide for Observation and Self-Assessment, Second Edition*

Baroody/Dowker (Eds.)• *The Development of Arithmetic Concepts and Skills: Constructing Adaptive Expertise*

Boaler• *Experiencing School Mathematics: Traditional and Reform Approaches to Teaching and Their Impact on Student Learning*

Carpenter/Fennema/Romberg (Eds.)• *Rational Numbers: An Integration of Research*

Chazan/Callis/Lehman (Eds.)• *Embracing Reason: Egalitarian Ideals and the Teaching of High School Mathematics*

Cobb/Bauersfeld (Eds.)• *The Emergence of Mathematical Meaning: Interaction in Classroom Cultures*

Cohen • *Teachers' Professional Development and the Elementary Mathematics Classroom: Bringing Understandings to Light*

Clements/Sarama/DiBiase (Eds.) • *Engaging Young Children in Mathematics: Standards for Early Childhood Mathematics Education*

English (Ed.)• *Mathematical and Analogical Reasoning of Young Learners*

English (Ed.)• *Mathematical Reasoning: Analogies, Metaphors, and Images*

Fennema/Nelson (Eds.)• *Mathematics Teachers in Transition*

Fennema/Romberg (Eds.)• *Mathematics Classrooms That Promote Understanding*

Fernandez/Yoshida • *Lesson Study: A Japanese Approach to Improving Mathematics Teaching and Learning*

Greer/Mukhopadhyay/Powell/Nelson-Barber (Eds.)• *Culturally Responsive Mathematics Education*

Kaput/Carraher/Blanton (Eds.)• *Algebra in the Early Grades*

Lajoie• *Reflections on Statistics: Learning, Teaching, and Assessment in Grades K-12*

Lehrer/Chazan (Eds.) • *Designing Learning Environments for Developing Understanding of Geometry and Space*

Ma• *Knowing and Teaching Elementary Mathematics: Teachers' Understanding of Fundamental Mathematics in China and the United States*

Martin• *Mathematics Success and Failure Among African-American Youth: The Roles of Sociohistorical Context, Community Forces, School Influence, and Individual Agency*

Martin (Ed.)• *Mathematics Teaching, Learning, and Liberation in the Lives of Black Children*

Reed• *Word Problems: Research and Curriculum Reform*

Romberg/Fennema/Carpenter (Eds.) • *Integrating Research on the Graphical Representation of Functions*

Romberg/Carpenter/Dremock (Eds.)• *Understanding Mathematics and Science Matters*

Romberg/Shafer • *The Impact of Reform Instruction on Mathematics Achievement: An Example of a Summative Evaluation of a Standards-Based Curriculum*

Sarama/Clements• *Early Childhood Mathematics Education Research: Learning Trajectories for Young Children*

Schliemann/Carraher/Brizuela (Eds.) • *Bringing Out the Algebraic Character of Arithmetic: From Children's Ideas to Classroom Practice*

Schoenfeld (Ed.)• *Mathematical Thinking and Problem Solving*

Senk/Thompson (Eds.) • *Standards-Based School Mathematics Curricula: What Are They? What Do Students Learn?*

Solomon• *Mathematical Literacy: Developing Identities of Inclusion*

Sophian• *The Origins of Mathematical Knowledge in Childhood*

Sternberg/Ben-Zeev (Eds.)• *The Nature of Mathematical Thinking*

Stylianou/Blanton/Knuth (Eds.)• *Teaching and Learning Proof Across the Grades: A K-16 Perspective*

Sultan & Artzt• *The Mathematics That Every Secondary School Mathematics Teacher Needs to Know*

Watson• *Statistical Literacy at School: Growth and Goals*

Watson/Mason• *Mathematics as a Constructive Activity: Learners Generating Examples*

Wilcox/Lanier (Eds.) • *Using Assessment to Reshape Mathematics Teaching: A Casebook for Teachers and Teacher Educators, Curriculum and Staff Development Specialists*

Wood/Nelson/Warfield (Eds.)• *Beyond Classical Pedagogy: Teaching Elementary School Mathematics*

Zaskis/Campbell (Eds.)• *Number Theory in Mathematics Education: Perspectives and Prospects*

For additional information on titles in the Studies in Mathematical Thinking and Learning Series visit www.routledge.com/education

THE MATHEMATICS THAT EVERY SECONDARY SCHOOL MATH TEACHER NEEDS TO KNOW

Alan Sultan
Queens College of the City University of New York

Alice F. Artzt
Queens College of the City University of New York

NEW YORK AND LONDON

First published 2011
by Routledge
711 Third Avenue, New York, NY 10017

Simultaneously published in the UK
by Routledge
2 Park Square, Milton Park, Abingdon, Oxon OX14 4RN

Routledge is an imprint of the Taylor & Francis Group, an informa business

© 2011 Routledge, Taylor & Francis

Typeset in Stone Sans and Stone Serif by Swales & Willis Ltd, Exeter, Devon

All rights reserved. No part of this book may be reprinted or reproduced or utilised in any form or by any electronic, mechanical, or other means, now known or hereafter invented, including photocopying and recording, or in any information storage or retrieval system, without permission in writing from the publishers.

Trademark Notice: Product or corporate names may be trademarks or registered trademarks, and are used only for identification and explanation without intent to infringe.

Library of Congress Cataloging in Publication Data
A catalog record has been requested for this book

British Library Cataloguing in Publication Data
A catalogue record for this book is available from the British Library

ISBN 13: 978-0-415-99413-2 (pbk)
ISBN 13: 978-0-203-85753-3 (ebk)

BRIEF CONTENTS

Preface/Introduction		xvii
Notes to the Reader/Professor		xxi
Acknowledgments		xxvii
CHAPTER 1	Intuition and Proof	1
CHAPTER 2	Basics of Number Theory	17
CHAPTER 3	Theory of Equations	69
CHAPTER 4	Measurement: Area and Volume	113
CHAPTER 5	The Triangle: Its Study and Consequences	159
CHAPTER 6	Building the Real Number System	215
CHAPTER 7	Building the Complex Numbers	307
CHAPTER 8	Induction, Recursion, and Fractal Dimension	357
CHAPTER 9	Functions and Modeling	397
CHAPTER 10	Geometric Transformations	451
CHAPTER 11	Trigonometry	513
CHAPTER 12	Data Analysis and Probability	599
CHAPTER 13	Introduction to Non-Euclidean Geometry	677
CHAPTER 14	Three Problems of Antiquity	705
Bibliography		721
Appendix		723
Index		729

CONTENTS

Preface/Introduction	xvii
Notes to the Reader/Professor	xxi
Acknowledgments	xxvii

CHAPTER 1 Intuition and Proof — 1

1.1 Introduction	1
1.2 Can Intuition Really Lead Us Astray?	1
1.3 Some Fundamental Methods of Proof	8
1.3.1 Direct Proof	9
1.3.2 Proof by Contradiction	10
1.3.3 Proof by Counterexample	12
1.3.4 The Finality of Proof	13

CHAPTER 2 Basics of Number Theory — 17

2.1 Introduction	17
2.2 Odd, Even, and Divisibility Relationships	17
2.3 The Divisibility Rules	25
2.4 Facts about Prime Numbers	31
2.4.1 The Prime Number Theorem	36
2.5 The Division Algorithm	38
2.6 The Greatest Common Divisor (GCD) and the Euclidean Algorithm	42
2.7 The Division Algorithm for Polynomials	48
2.8 Different Base Number Systems	51
2.9 Modular Arithmetic	56
2.9.1 Application: RSA Encryption	59
2.10 Diophantine Analysis	61

Contents

CHAPTER 3 Theory of Equations — 69

3.1 Introduction — 69
3.2 Polynomials: Modeling, Basic Rules, and the Factor Theorem — 70
3.3 Synthetic Division — 75
3.4 The Fundamental Theorem of Algebra — 80
3.5 The Rational Root Theorem and Some Consequences — 85
3.6 The Quadratic Formula — 90
3.7 Solving Higher Order Polynomials — 95
    3.7.1 The Cubic Equation — 95
    3.7.2 Cardan's Contribution — 100
    3.7.3 The Fourth Degree and Higher Equations — 101
3.8 The Role of the Graphing Calculator in Solving Equations — 103
    3.8.1 The Newton–Raphson Method — 104
    3.8.2 The Bisection Method–Unraveling the Workings of the Calculator — 108

CHAPTER 4 Measurement: Area and Volume — 113

4.1 Introduction — 113
4.2 Areas of Simple Figures and Some Surprising Consequences — 113
4.3 The Circle — 125
    4.3.1 An Informal Proof of the Area of a Circle — 126
    4.3.2 Archimedes' Proof of the Area of a Circle — 127
    4.3.3 Limits And Areas of Circles — 130
    4.3.4 Using Technology to Find the Area of a Circle — 131
    4.3.5 Computation of π — 133
    4.3.6 Finding Areas of Irregular Shapes — 139
4.4 Volume — 146
    4.4.1 Introduction to Volume — 146
    4.4.2 A Special Case: Volumes of Solids of Revolution — 150
    4.4.3 Cavalieri's Principle — 153
    4.4.4 Final Remarks — 154

CHAPTER 5 The Triangle: Its Study and Consequences — 159

5.1 Introduction — 159
5.2 The Law of Cosines and Surprising Consequences — 159
    5.2.1 Congruence — 161
5.3 The Law of Sines — 165
5.5 $\operatorname{Sin}(A + B)$ — 175
5.6 The Circle Revisited — 179
    5.6.1 Inscribed and Central Angles — 180
    5.6.2 Secants and Tangents — 183
    5.6.3 Ptolemy's Theorem — 185

5.7 Technical Issues	191
5.8 Ceva's Theorem	195
5.9 Pythagorean Triples	200
5.10 Other Interesting Results about Areas	204
5.10.1 Heron's Theorem	205
5.10.2 Pick's Theorem	206

CHAPTER 6 Building the Real Number System — 215

6.1 Introduction	215
6.2 Part 1: The Beginning Laws: An Intuitive Approach	216
6.3 Negative Numbers and Their Properties: An Intuitive Approach	220
6.4 The First Rules for Fractions	224
6.5 Rational and Irrational Numbers: Going Deeper	231
6.6 The Teacher's Level	234
6.7 The Laws of Exponents	242
6.7.1 Integral Exponents	242
6.8 Radical and Fractional Exponents	245
6.8.1 Radicals	245
6.8.2 Fractional Exponents	247
6.8.3 Irrational Exponents	250
6.9 Working with Inequalities	253
6.10 Logarithms	259
6.10.1 Rules for Logarithms	263
6.11 Solving Equations	267
6.11.1 Some Issues	268
6.11.2 Logic Behind Solving Equations	269
6.11.3 Equivalent Equations	272
6.12 Part 2: Review of Geometric Series: Preparation for Decimal Representation	277
6.13 Decimal Expansion	280
6.14 Decimal Periodicity	289
6.15 Decimals: Uniqueness of Representation	293
6.16 Countable and Uncountable Sets	297
6.16.1 Algebraic and Transcendental Numbers Revisited	303

CHAPTER 7 Building the Complex Numbers — 307

7.1 Introduction	307
7.2 The Basics	307
7.2.1 Operating on the Complex Numbers	308

7.3 Picturing Complex Numbers and Connections to Transformation Geometry	315
7.3.1 An Interesting Problem	319
7.3.2 The Magnitude of a Complex Number	323
7.4 The Polar Form of Complex Numbers and De Moivre's Theorem	325
7.4.1 Roots of Complex Numbers	330
7.5 A Closer Look at the Geometry of Complex Numbers	334
7.6 Some Connections to Roots of Polynomials	340
7.7 Euler's Amazing Identity and the Irrationality of e	343
7.8 Fractal Images	347
7.8.1 Other Ways to Generate Fractal Images	351
7.9 Logarithms of Complex Numbers and Complex Powers	352

CHAPTER 8 Induction, Recursion, and Fractal Dimension   357

8.1 Introduction	357
8.2 Recursive Relations	357
8.2.1 Solving Recursive Relations	361
8.3 Induction	372
8.3.1 Taking Induction to a Higher Level	376
8.3.2 Other Forms of Induction	378
8.4 Fractals Revisited and Fractal Dimension	387
8.4.1 The Chaos Game	389
8.4.2 Fractal Dimension	390

CHAPTER 9 Functions and Modeling   397

9.1 Introduction	397
9.2 Functions	397
9.2.1 The Historical Notion of Function	398
9.2.2 Functions Today	398
9.2.3 Functions – The More General Notion	400
9.2.4 Ways of Representing Functions	401
9.3 Modeling with Functions	406
9.3.1 Some Types of Models	407
9.3.2 Which Model Should We Use?	412
9.4 What Does Best Fit Mean?	420
9.4.1 What is Behind Finding the Line of Best Fit?	421
9.4.2 How Well Does a Function Fit the Data?	425
9.5 Finding Exponential and Power Functions That Fit Curves	427
9.5.1 How Calculators Find Exponential and Power Regressions	428
9.5.2 Things to Watch Out for in Curve Fitting	431
9.6 Fitting Data Exactly With Polynomials	433

9.7 1–1 Functions	439
9.7.1 The Rudiments	439
9.7.2 Why Are 1–1 Functions Important?	442
9.7.3 Inverse Functions in More Depth	442
9.7.4 Finding the Inverse Function	444
9.7.5 Graphing the Inverse Function	446

CHAPTER 10 Geometric Transformations — 451

10.1 Introduction	451
10.2 Transformations: The Secondary School Level	452
10.2.1 Basic Ideas	453
10.3 Bringing in the Main Tool – Functions	458
10.4 The Matrix Approach – a Higher Level	466
10.4.1 Reflections, Rotations, and Dilations	466
10.4.2 Compositions of Transformations	469
10.4.3 Reflecting about Arbitrary Lines	474
10.5 Matrix Transformations	482
10.5.1 The Basics	482
10.5.2 Matrix Transformations in More Detail – A Technical Point	485
10.6 Transforming Areas	492
10.7 Connections to Fractals	497
10.7.1 Translations	498
10.8 Transformations in Three dimensions	504
10.9 Reflecting on Reflections	507

CHAPTER 11 Trigonometry — 513

11.1 Introduction	513
11.2 Typical Applications Using Angles and Basic Trigonometric Functions	514
11.2.1 Engineering and Astronomy	514
11.2.2 Forces Acting on a Body	516
11.3 Extending Notions of Trigonometric Functions	523
11.3.1 Trigonometric Functions of Angles More than 90 Degrees	524
11.3.2 Some Useful Trigonometric Relationships	528
11.4 Radian Measure	537
11.4.1 Conversion	537
11.4.2 Areas and Arc Length in Terms of Radians	539
11.5 Graphing Trigonometric Curves	544
11.5.1 The Graphs of Sin θ and Cos θ	545
11.5.2 The Graph of $y = \text{Tan } \theta$	552
11.6 Modeling with Trigonometric Functions	555
11.7 Inverse Trigonometric Functions	560
11.8 Trigonometric Identities	567

11.9	Solutions of Cubic Equations Using Trigonometry	575
11.10	Lissajous Curves	578
11.11	Vectors	581
	11.11.1 Basic Vector Algebra	582
	11.11.2 Components of Vectors	587
	11.11.3 Using Vectors to Prove Geometric Theorems	592

CHAPTER 12 Data Analysis and Probability — 599

12.1	Introduction	599
12.2	Basic Ideas of Probability	600
	12.2.1 Different Approaches to Probability	600
	12.2.2 Issues with the Approaches to Probability	603
12.3	The Set Theoretic Approach to Probability	605
	12.3.1 Some Elementary Results in Probability	608
12.4	Elementary Counting	613
12.5	Conditional Probability and Independence	618
	12.5.1 Some Misconceptions in Probability	623
12.6	Bernoulli Trials	626
12.7	The Normal Distribution	629
12.8	Classic Problems: Counterintuitive Results in Probability	635
	12.8.1 The Birthday Problem	636
	12.8.2 The Monty Hall Problem	636
	12.8.3 The Gunfight	637
	12.8.4 Simulation	638
12.9	Fair and Unfair Games	641
	12.9.1 Games Where No Money is Involved	641
	12.9.2 Games Where Money is Involved	643
	12.9.3 The General Notion of Expectation	644
	12.9.4 The Cereal Box Problem	646
12.10	Geometric Probability	650
	12.10.1 Some Surprising Consequences	652
	12.10.2 Monte Carlo Revisited	654
12.11	Data Analysis	658
	12.11.1 Plotting Data	659
	12.11.2 Mean, Median, Mode	664
12.12	Lying with Statistics	669
	12.12.1 What Can you Do to Talk Back to Statistics?	673

CHAPTER 13 Introduction to Non-Euclidean Geometry — 677

13.1	Introduction	677
13.2	Can We Believe Our Eyes?	678
	13.2.1 What Are the Errors in the Proofs?	683

13.3 The Parallel Postulate ... 685
    13.3.1 What Can We Prove with the Parallel Postulate? ... 686
    13.3.2 What Can We Prove Without the Parallel Postulate? ... 688
13.4 Undefined Terms ... 690
13.5 Strange Geometries ... 692
    13.5.1 Hyperbolic Geometry ... 693
    13.5.2 Euclid's Axioms in the Hyperbolic World ... 695
    13.5.3 Area in Hyperbolic Space ... 700
    13.5.4 Spherical Geometry ... 702

CHAPTER 14 Three Problems of Antiquity ... 705

14.1 Introduction ... 705
14.2 Some Basic Constructions ... 705
14.3 Three Problems of Antiquity and Constructible Numbers ... 710
    14.3.1 Constructible Numbers ... 711
    14.3.2 Geometrically Constructible Numbers ... 711
    14.3.3 The Constructible Plane ... 713
    14.3.4 Solving the Three Problems of Antiquity ... 716

Bibliography ... 721
Appendix ... 723
Index ... 729

PREFACE/INTRODUCTION

What knowledge of mathematics is needed for teaching secondary school math? This question has been at the forefront of research for many years, and has yet to be fully answered. While it is widely accepted that mathematics teachers require a depth of knowledge that extends beyond what they teach, the specific details and nature of this knowledge need to be clearly delineated. Despite the fact that the research literature on this issue is in its infancy, those who have worked as mathematics teachers and as mathematics teacher educators and researchers know that excellence in teaching requires an understanding of mathematics that is quite different than that of their students. This different type of knowledge is often referred to as pedagogical content knowledge (Shulman, 1986, 1987) or knowledge of mathematics for *teaching* (Ball, 1991; Ball, Thames, & Phelps, 2008). The purpose of this book is to provide pre-service and/or in-service secondary mathematics teachers with a resource that exposes them to multiple levels and types of mathematical understanding that we believe will extend and deepen their insight into the mathematics of the secondary school curriculum in ways that will enable them to facilitate their own students' understanding, competency, and interest in mathematics.

To be more specific, mathematics teachers need to have knowledge of how to make mathematical understanding and skills accessible for all of their students. For example, experienced teachers will tell you that year after year their students have trouble understanding certain specific topics in the curriculum. In this book, we address these typical areas of difficulty and students' common misconceptions. We examine why these difficulties exist and mathematical approaches teachers can use in clarifying the concepts and procedures. In so doing, we emphasize the use of multiple ways of representing and solving problems so that you will be able to meet the needs of your students who will most definitely have diverse learning styles and abilities. The use of technology is incorporated to add to the multiple ways problems can be represented. Additionally, by its ability to dynamically represent concepts and simulate problems, technology can be used to help students solve problems through discovery, pattern recognition, and inductive reasoning. Throughout this book we examine the strengths and weaknesses of different technologies in representing mathematical ideas, as well as provide references to multiple informational and sometimes interactive websites that will be of use to you and your students.

In addition to raising students' competency and understanding of mathematics, another important part of a teacher's job is to be able to interest their students in the subject matter. All too often mathematics has been portrayed as a cold subject, devoid of human emotion. Nothing could be further from the truth! Indeed, mathematics has a very rich history and the lives of many mathematicians are quite fascinating. Although this book is not about the history of math, we have included the most interesting human stories underlying the development of mathematics that will most assuredly capture your attention as well as that of your students.

Overview of the Book

A teacher's mathematical knowledge needs to go well beyond skills and understanding of discrete mathematical topics. It needs to be comprehensive, deep, meaningful, and connected. Let us explain what we mean here by setting forth what this text contains. Throughout we try to bridge the gap between the mathematics you have learned in college and that which you will be teaching in secondary school by showing how many of the mathematical concepts you have learned in your college courses are connected and woven together and how they relate to the secondary school curriculum. This will give you a better idea of why colleges ask you to take all those advanced courses that seem to have no connection to the secondary school curriculum, but which in fact, have important connections to it.

Another unique feature of this book is that we often examine the content from an elementary school-level perspective, trace its development to a college-level perspective, and highlight the linkages between the higher-level courses you took and the courses you will most likely teach. Of course, at the root of all of these interconnections lies the nature of mathematics and proof and why proof is so important in mathematics. We believe this reflection on the big picture will enhance both your understanding of the mathematical development of some of the topics, and provide you with a unique perspective that you can offer your students.

In addition to making connections between different fields of mathematics and different grade levels of mathematics, throughout the book we highlight applications of the mathematical content to the real world. You and your students will surely be intrigued by how some of the most theoretical concepts are applied in real life. By interlinking and connecting these mathematical ideas and applications as we have described, we hope to deepen your understanding and appreciation of certain topics and help you to answer the age old question, often asked by school students "When am I ever gonna use this?"

The style and structure of this book are designed to support an instructional approach in which you will become actively involved in building your new understandings.

"Launch" questions. Each section opens with a motivational question, which we call a "launch," to capture interest and involve you in the learning of the mathematical concepts.

"Student Learning Opportunities." At the end of each section an assortment of questions provide opportunities for you to practice problems associated with what you have learned in the chapter, complete proofs that were mentioned but not proved in the chapter, and apply what you have learned to solving real-life problems.

"Questions from the Classroom." Additionally, we have included questions that you might be asked by your own students some day. These are indicated with a (C) in front of each such question. Often, these are the deep "why" conceptual questions that middle or secondary school students often ask their teachers. You will notice that some of the questions are so rich that they might be used for student projects or in-class activities to use with your future secondary school students. Other questions of this type are examples of typical student errors or misconceptions that you will need to critique and correct.

The following themes are woven throughout the content chapters so that you will be a teacher who has great "MATH-N-SIGHT."

M   **Multiple Approaches/Representations** (Knowledge of the different ways problems can be represented and solved)
A   **Applications to Real Life** (Knowledge of the role of mathematics in real life)
T   **Technology** (Knowledge of the role of technology in solving mathematical problems and developing mathematical ideas)
H   **History** (Knowledge of the human story behind the development of mathematical concepts)
N   **Nature of Mathematics: Reasoning and Proof** (Knowledge of the role of reasoning, definitions and proof)
S   **Solving Problems** (Knowledge of the different problem solving strategies, and ways of generalizing, and extending the problems)
I   **Interlinking Concepts: Connections** (Knowledge of the linkages between and among different branches of mathematics and areas outside of mathematics as well as connections between secondary school and college-level mathematical concepts)
G   **Grade Levels** (Knowledge of the grade levels in which the foundations of advanced concepts appear)
H   **Honing of Mathematical Understanding and Skills** (Experience in revisiting mathematical concepts, with opportunities for developing more mature perspectives and skills)
T   **Typical Errors** (Knowledge of the most common misconceptions students have that contribute to the most typical errors they make when doing mathematics)

To help you understand how these themes can be applied to deepen your insight into mathematics for teaching, we will ask you to respond to some questions in relation to a most famous mathematical theorem, the Pythagorean Theorem. But before we do this, we want to share the story behind why we focus on this particular theorem. Years ago we asked a group of college freshmen who were not mathematics majors to recall any theorem they remembered from their study of school mathematics. They unanimously recalled the Pythagorean Theorem and were able to state it as follows: $a^2 + b^2 = c^2$. However, the most they could tell us about the theorem was that it had something to do with a triangle. They were not even sure what the letters in the theorem stood for. We asked ourselves why their knowledge of this historic formula was so devoid of meaning and appreciation. We wondered how they were exposed to this theorem. Were they asked questions about when the theorem could be used? Did they learn anything about the history of the theorem? Did they learn about any ways the theorem could be applied? Did they do hundreds of problems using the formula? The answer to the last question was no doubt, a resounding, "Yes."

In light of this story, we want you to consider the following questions, in hopes that in the future, your students will recall more about the Pythagorean Theorem than the ones we have described. (You can consult the text to help you respond to some of the questions.)

Student Learning Opportunities

1 In what grades do students typically learn the Pythagorean Theorem? What is the prior knowledge they need to have to fully understand the meaning of the theorem? What definitions are associated with the Pythagorean Theorem? [Hint: check *Principles and Standards for*

School Mathematics (NCTM, 2000) and the March 2005 New York State Learning Standard for Mathematics at www.nysed.gov.]

2 What is the history of the theorem? When was it discovered? Who discovered it? Was it really discovered by Pythagoras?

3 How do we know that the Pythagorean Theorem is true? Was it first created through intuition or proof? Is there a proof of the theorem? If so, give one.

4 Why is the Pythagorean Theorem so famous? Why should we care about it?

5 What is surprising or mysterious about the Pythagorean Theorem? Be specific.

6 What different areas of mathematics are connected to the Pythagorean Theorem? Explain.

7 How can technology be incorporated to facilitate the learning or teaching of the Pythagorean Theorem? Give several suggestions that go beyond the mere help with computation.

8 What are some interesting Internet sites regarding the Pythagorean Theorem? Give several suggestions, including some sites that contain applets that support the geometric interpretation of the theorem.

9 What are some typical mistakes that students make when using the Pythagorean Theorem? What are some typical misconceptions that students have regarding the Pythagorean Theorem?

We believe that the book's focus on multiple perspectives on mathematical knowledge for teaching will hone your own mathematical understanding and skills and facilitate your ability to engage your students in learning mathematics in a multifaceted way that is interesting, developmental, connected, deep, clear, and often, surprising and entertaining. We hope and expect that this book will be a valuable resource for you during your career as a teacher of mathematics.

Bibliography

Ball, D.L. (1991). Teaching mathematics for understanding: What do teachers need to know about subject matter? In Mary M. Kennedy (Ed), *Teaching Academic Subjects to Diverse Learners*. (p. 63–83). New York, NY: Teachers College Press.

Ball, D.L., Thames, M.H., & Phelps, G. (2008). Content knowledge for teaching: What makes it special? *Journal of Teacher Education*, 59(5), 389–407.

National Council of Teachers of Mathematics (2000). *Principles and standards for school mathematics*, Reston, VA: The Council.

Shulman, L. (1986). Those who understand: Knowledge growth in teaching. *Educational Researcher*, 15(2), 4–14.

Shulman, L. (1987). Knowledge and teaching: Foundation of the new reform. *Harvard Educational Review*, 57, 1–22.

NOTES TO THE READER/PROFESSOR

This book has multiple uses, ranging from a very helpful resource, to a text that accompanies any methods or mathematics course for pre- or in-service secondary mathematics teachers. It was specifically written to accompany a culminating mathematics course for prospective secondary mathematics teachers. The style and structure of the book is therefore designed to support a student-centered instructional approach. Since, the method we envision and use with our own classes is quite different from the lecture approach, we explain how the book is designed to support such an interactive approach.

Before most sections there is a motivational question, which we call a "launch" that is meant to create student interest and involvement in the lesson. These questions take different forms depending on the section. For example, sometimes they are designed to create a need to learn the new material, by pointing to a void in their knowledge. Other times the questions are designed to create curiosity about why a particular relationship exists. The students can first work on this "launch" individually so that they can give the problem some thought, arrive at some preliminary ideas about the solution, and formulate questions about the solution process. After their individual work, students can then be encouraged to work with a small group of their peers to try to agree on a solution and arrive at questions for class discussion. After the students have worked in their groups, the whole class discussion can then focus on the new concepts imbedded in the launch question, which of course, is the topic of the lesson. Since in class, the development of the topics is problem-based and discussion-driven the students may not get to see a structured development of the topics, (characteristic of most lecture-style approaches) until they read the text. It is specifically for this reason that the book is written in a style which we have tried to make extremely clear and interesting for the reader. This explains the use of the informal language, humor, and historical interludes in the text. Additionally, time is taken in the text to arouse the students' appreciation of the ingenious mathematical concepts they are learning about.

At the end of each section there is an assortment of questions or "Student Learning Opportunities" which can be used for homework assignments or even class work. These questions include opportunities for students to practice problems associated with what they have learned in the chapter, complete proofs that were mentioned, but not proved in the chapter, and apply what they have learned to solve real-life problems. One of the unique features of this book is that we have also included questions that your students (future teachers) might be asked by their own secondary school students some day. We refer to these questions as "Questions from the Classroom," and we indicate them with a (C) in front of each such question. Often, these are the deep "why" conceptual questions that secondary school students are curious about. Some of these (C) questions require analysis and correction of typical student errors and misconceptions. Other (C) questions focus on counter-intuitive results. Finally, some of the (C) questions contain activities and/or tasks that are suitable for use with secondary school students. Therefore, this book should provide a valuable resource for the pre-service teachers in their own future classes.

In the Preface we have outlined themes that are woven throughout the content chapters which we have abbreviated as "MATH-N-SIGHT." To help students reflect upon and organize their learning, we believe it will be worthwhile if throughout their learning, you request that they identify which theme is being addressed.

Finally, since the material in this book is more extensive than can be addressed in one course, we include a short description of each of the chapters, so that you can select the ones which you believe will be most suitable for your course. You will note that throughout the book we incorporate the five content strands: Numbers and Operations, Algebra, Geometry, Measurement, and Data Analysis and Probability, which are the underlying concepts of the secondary school curriculum. The descriptions follow.

CHAPTER SUMMARIES

Chapter 1: Intuition and Proof

In this opening chapter we discuss a variety of problems where the solution seems clear but the "obvious" solution is wrong. This leads to a discussion of why we need proof and some of the different methods of proof. We include examples that relate to later chapters and the type of proofs used in the secondary school curriculum.

Chapter 2: Basics of Number Theory

We begin this chapter with the basic definitions in Number Theory that relate to the secondary school curriculum: even, odd, divisible by, and so on and then show that by using proper definitions, one can prove elusive relationships very easily. We discuss the different tests for divisibility, why they work, and the Euclidean Algorithm. We apply these divisibility results to **UPC** codes, **RSA** encryption, prime numbers, computer design, recreational problems, different systems of numeration, Diophantine Equations, and how these concepts and applications relate to topics in the secondary school mathematics curriculum.

Chapter 3: Theory of Equations

As this is a major part of the secondary school curriculum, in this chapter we include a thorough treatment of polynomials and issues related to their use. First we discuss relations between roots, factors, rational and irrational numbers and show in a very understandable way why synthetic division works. We provide applications to modeling, the Fundamental Theorem of Algebra, polynomiography, and methods that calculators use to find square roots and maxima and minima. We link these concepts to the solution of polynomial equations of higher order and carefully go back and forth between the technological and theoretical issues. We emphasize why the technology is an important adjunct to the mathematics, and conversely, why the mathematics is an important adjunct to the technology. This chapter, like most other chapters, is filled with historical vignettes bringing the principals we discuss to life by sharing some of their more interesting stories.

Chapter 4: Measurement: Area and Volume

In this chapter we provide a thorough analysis of area and volume. We begin by deriving several area formulas and then discuss the issues involved in defining area and volume. From a few basic assumptions we quickly derive the Pythagorean Theorem and its converse. We turn to the circle and then give Archimedes' remarkable proof that the area of a circle is πr^2. We use the technology and several simple theoretical arguments to show that the ratio of the circumference of a circle to its diameter is constant, and indeed end up proving this quite by accident. This chapter is filled with history and links between the theoretical concepts and technology. We then discuss some of the volume issues from a higher (calculus) level with Cavalieri's principle playing a central role.

Chapter 5: The Triangle: Its Study and Consequences

Many fascinating facts about triangles are discussed in this chapter. We begin by including basic derivations of secondary school results, and then take the unusual approach of using the Law of Sines and the Law of Cosines to corroborate the congruence and similarity theorems one normally learns in secondary school. From these basic rules and laws about triangles we move to circles and use our triangle laws to develop basic geometric results about circles, and present some rather surprising proofs of the formulas for the sine of the sum and difference of angles. We discuss Pythagorean triples, Pick's theorem, Ceva's theorem, Heron's theorem, and Ptolemy's theorem. Along the way we find such interesting surprises as how to use some of these ideas to corroborate that the square root of two is irrational! This chapter makes many important connections, tying together the concepts of area, trigonometry, circles, number theory, and many of the main theorems about them in ways that are not available in most secondary school texts.

Chapter 6: Building the Real Number System

This chapter develops the number system. Because of its length we have divided this summary into two parts.

Part 1: We start with the basic commutative, associative, and distributive laws, discuss why they are true, and then develop the rules for the real number system. For example, we address such commonly asked questions as: Why is it true that a negative times a negative is a positive? Why do we invert and multiply when we divide fractions? Why can't we divide by zero? Why is it true that the square root of the product is the product of the square roots? Why do the rules of exponents hold even when the exponents are irrational? We then give a thorough discussion of solutions of equations and the typical errors students make and ways to avoid them. Following are discussions of inequalities in which we derive the rules needed to solve typical secondary school problems, all the while including the use of technology as a way to corroborate important issues. This first section ends with the topic of logarithmic and exponential equations and a most fascinating discussion of how it was determined that the shroud of Turin (supposedly worn by Jesus) was a fake and how one determines time of death as applications.

Part 2: In this section, decimals and their representations are discussed in great depth. We address such questions as: How many way are there of representing decimals? How can one predict the size of the periodic part of a fraction? Why does the method of long division work for finding the decimal expansion of a number? We end with the notions of countability and uncountability and talk a little about the issues of algebraic and transcendental numbers, their links to the other material discussed in this and prior chapters, and why they were studied. We prepare the students for later discussions of surprising mathematical results including proving the impossibility of trisecting an angle with compass and straightedge and other results in that genre.

Chapter 7: Building the Complex Numbers

Similar to how we developed the real numbers in chapter 6, in this chapter we develop the complex numbers and show why the same rules that work for the real numbers work for the complex numbers. We discuss the history of complex numbers, why they ultimately were studied, why they originally were ignored, and some of the remarkable results that come from their study. We address such realistic applications of complex numbers as how they can be used in the design of shock absorbers, and how they can be used to solve a treasure hunt. We also give applications of complex numbers that are quite powerful. Here the students discuss in detail some of the main results involving complex numbers like DeMovire's theorem, and Euler's result. Connections are highlighted between many of the concepts we discussed in the previous chapters as well as how complex numbers relate to transformation geometry. We end by showing how fractals relate to the previously discussed concepts.

Chapter 8: Induction, Recursion, and Fractal Dimension

This chapter focuses on the important concept of recursion and the related topic of induction. The emphasis is on real applications as well as on modeling. Many non- routine examples as well as routine examples of induction are given. We discuss interesting links between this material and that of previous chapters. For example, we show fractals are formed by recursive routines. We also investigate the dimension of a fractal and interesting issues involving their perimeter and area. The Chaos game is presented, whose results are quite surprising to all. Links to arithmetic and geometric sequences are also made.

Chapter 9: Functions and Modeling

In this chapter we discuss functions and modeling in great depth. We begin by carefully examining what constitutes a function and then turn to modeling with functions, using only real or realistic examples. We discuss curve of best fit, what it means, derive the formula for line of best fit using simple calculus and solving simultaneous equations, and discuss some of the issues with regression. We discuss how to decide which types of curves we should try to use to fit data and then examine exact fits to certain types of data. Finally, we talk about ways of representing functions, 1-1 functions, their importance, inverse functions, and connect these concepts to transformations which are discussed fully in the next chapter

Chapter 10: Geometric Transformations

The notions of transformation geometry from both an elementary point of view and a matrix point of view are discussed in this chapter. Here your students will get to see how the matrices they study in college (and which are now studied in secondary school) are used in graphics programs, and how the results of many different transformations lead to animation. We highlight the many interesting relationships between reflections, rotations, translations, and so on, and derive formulas for different kinds of transformations which explain many of the concepts they will teach in secondary school. Using a very concrete approach we discuss the uses of composition of functions and deal with issues of how transformations affect figures and their areas. Again, we link many different areas of mathematics that they teach in a very cohesive way. We finish by examining how fractals can be generated by combining the transformations discussed in this chapter providing another perspective on this interesting topic. Finally we will show how certain important laws of optics arise from the discussions we gave as well as how certain difficult to solve (but easy to state) geometric problems were solved using the notions of geometric transformations

Chapter 11: Trigonometry

In this chapter we discuss the basic trigonometric functions. The applications of these simple trigonometric concepts are plentiful and powerful. We highlight many of them right from the start, by discussing the application of trigonometric functions to physics, astronomy, engineering, problems from everyday life, and other areas of mathematics. Radian measure, referred to as a "dimensionless" measure, is introduced and beyond its role in trigonometry, its value for scientists and mathematicians is pointed out. As well, students are encouraged to integrate ideas of transformation geometry to sketch the graphs of trigonometric curves. We emphasize the role of technology pointing out the benefits and pitfalls of depending on the graphing calculator or computer for graphing trigonometric functions. We discuss the trigonometric identities that are part of the secondary school curriculum and highlight them as multiple forms of representation which have uses in transformation geometry and our current technology. We even link the study of trigonometry to the solution of polynomial equations, thereby connecting two very diverse areas together. We end this chapter with a fascinating study of how geometric proofs can be done using vectors. In fact, we point out how otherwise very difficult proofs can be made rather simple with such an approach.

Chapter 12: Data Analysis and Probability

We begin this chapter with a discussion of the basic concepts of probability and the notion of likelihood and follow these with a discussion of some of the issues with the definitions used in probability. Once we establish the classical approach to probability, we discuss the basic laws of probability and conditional probability and illustrate them with several examples. The counting arguments we develop and their applications lead us naturally to the normal distribution. This is followed by some exceptionally interesting counter- intuitive results in probability as well as a discussion of common misconceptions, and fair and unfair games. Geometric probability follows giving us a rather interesting way to view some probabilistic outcomes. After this we get into some basic ideas of statistics that are used in the secondary schools such as organizing data through the use of histograms, stem and leaf plots, and box and whisker plots. We conclude with an interesting section on how it is possible to be fooled by media that lie with statistics.

Chapter 13: Introduction to Non-Euclidean Geometry

In this chapter we discuss Euclidean and Non-Euclidean geometries, and their interesting origins. We begin by discussing some fallacies that can arise in geometry through misleading diagrams or flawed logic, allowing us to emphasize the importance of checking every step in a proof carefully and critically examining what your visual intuition leads you to believe. We show how this careful approach to proof and the assumption that the parallel postulate did not hold led to Non-Euclidean geometry. The characters responsible for this whole development are introduced and some interesting historical vignettes are given.

Chapter 14: Three Problems of Antiquity

In this chapter we provide a detailed discussion of geometric constructions and the famous Three Problems of Antiquity: squaring the circle, doubling the cube, and trisecting an angle. The solutions of these problems integrate many of the concepts developed in this text and are presented in a clear manner which is accessible to the student. We highlight how problems which stumped mathematicians for thousands of years and which seemed to have nothing to do with polynomials, were solved using them. It is quite an interesting story!

ACKNOWLEDGEMENTS

Behind this book are many people who have made contributions without which it could never have been written. These people range from supportive family members, to insightful students, to encouraging colleagues, to helpful and wise experts in the field. While it is not possible to thank all of these people we will make our best effort to mention most of them.

The idea of writing this book originated with our work together in teaching a culminating mathematics class for prospective secondary math teachers. The feedback we got from our students throughout the process was invaluable. Although all of our students provided feedback on each draft of the text, certain students deserve special mention. Maria Leon Chu and Madeline Leno greatly enhanced the quality of this book through the extensive time and effort they devoted to proofreading and critiquing various sections.

Special thanks go to Marti Wayland of Baylor High School, who read through the first five chapters of this book and made substantive suggestions which greatly improved the exposition. Our deep appreciation goes to Naomi Silverman, senior editor at Routledge, for her consistent support, encouragement, patience, and advice from the beginning till the end of the book-writing process. We would also like to thank Mary Sanders, Swales & Willis Limited, and RefineCatch Limited who assisted in the production process.

Our deep gratitude goes to Alan Schoenfeld, who has always encouraged us to move forward with this text. Having the support and feedback of such a giant in the field of mathematics education was essential to the book's development.

Finally, we want to thank our devoted spouses, Ann and Russ, and our loving children, Jason, Michele, Kurt, Julie, Ricky, Allison, and Greg, who were always understanding and supportive of the time required to work with our students and compose the book.

CHAPTER 1

INTUITION AND PROOF

1.1 Introduction

The history of mathematics is replete with examples where observation and intuition led mathematicians to correct conclusions. However, there are just as many cases where it led to incorrect conclusions. For example, for many years mathematicians believed that there was only one kind of geometry–Euclidean. That proved to be false. They also believed that negative numbers had no meaning. Yet you know from your studies that negative numbers are essential in real-life applications.

As a secondary school student, you were probably only given the correct final results, like a negative number multiplied by a negative number is a positive number, and not made aware of the bumpy path it took for results like this to be discovered. This most likely left you with the false impression that mathematics evolved in a systematic way in which mathematicians created only correct results. To get a true understanding of the work of mathematicians, and the need for proof, it is important for you to experiment with your own intuitions, to see where they lead, and then to experience the same failures and sense of accomplishment that mathematicians experienced when they obtained the correct results. Through this, it should become clear that, when doing any level of mathematics, the roads to correct solutions are rarely straight, can be quite different, and take patience and persistence to explore.

We begin this process by exposing you to some of the instances in history where intuition led mathematicians astray and give you a chance to test your own intuition on these problems. Hopefully, by the end of this chapter, you will understand why proof is so important. These types of situations are what account for the present-day rigor that is part of today's mathematics curriculum. In the second section of this chapter you will experience the variety of methods that you can use to either prove that your own mathematical observations or intuitions are correct, or possibly even incorrect!

1.2 Can Intuition Really Lead Us Astray?

LAUNCH

Evaluate the expression $N^2 + N + 41$ for integer values of N from 1 through 5. Do you believe that this expression represents a prime number for all positive integers, N? Justify your answer.

2 Intuition and Proof

Most people who see this problem for the first time easily verify that the expression is prime for each of $N = 1, 2, 3, 4, 5$. After checking $N = 6, 7, 8, \ldots, 20$ and seeing that we still get prime answers, our intuition starts to kick in and tells us "Maybe this really is prime for all values of N." How can we be sure? How many cases must we take before we know with certainty? We will return to this "launch" question later in the chapter to resolve our dilemma. But first we will turn to some historical examples that exemplify the process of mathematical discovery.

As our first example we tell an interesting story about the famous mathematician Pierre Fermat (1601–1665). Until the day he died, he believed that the numbers $N = 2^{(2^k)} + 1$ were prime for all positive integers k. For example, if k is 1, we get $N = 5$ which is surely prime. You should test his hypothesis out for the cases $k = 2, 3,$ and 4 to see that it still holds. If you continue on and test it for $k = 5$, you will get $N = 4,294,967,297$. Is this a prime number? Fermat's intuition was always right on target and he had proved many illustrious and deep theorems. Who could doubt the great Fermat? For approximately 100 years this number's primality remained unresolved until Leonhard Euler (1707–1783), who some said did mathematics as effortlessly as men breathed, proved, using an ingenious argument, that

$$4,294,967,297 = 641 \times 6,700,417.$$

That is, Euler showed that Fermat was wrong. The value of N was not prime when $k = 5$. Things got worse. The value of N obtained when $k = 6$ was also not prime, having a factor of $274,177$, and it was subsequently shown that *none* of the values of N when $k = 7, 8, 9 \ldots 27$ were prime. Fermat couldn't have been more wrong. Of course, if Fermat could make mistakes, how much more suspicious should we be of our own mathematical beliefs?

Even Euler, who was considered one of the greatest mathematicians who ever lived and whose original works comprise more than 70 volumes, most of which are considered seminal, made his mistakes. For example, he conjectured that one cannot find four different positive integers a, b, c, and d, which make

$$a^4 + b^4 + c^4 = d^4.$$

This statement was believed by many to be true for more than 245 years, yet no one could prove it was true or prove it was false—until 1987. Then Noam Elkies of Harvard University discovered that, if we let $a = 2,682,440$, $b = 15,365,639$, $c = 18,796,760$, and $d = 20,615,673$, we have

$$2,682,440^4 + 15,365,639^4 + 18,796,760^4 = 20,615,673^4$$

Not only that, he also showed that there were *infinitely many* sets of numbers that worked, all of them huge—quite a surprising result indeed!

Now, join us in trying to find the sum, S, of the series

$$S = a + ar + ar^2 + \ldots.$$

We multiply both sides by r to get

$$rS = ar + ar^2 + \ldots$$

and then subtract this equation from the previous to get

$$S - rS = a.$$

We now factor out S to get

$(1-r)S = a$

and we finally divide by $1-r$ to get

$$S = \frac{a}{1-r}.$$

Is this correct? Check it out by applying it to the series $1 + 2 + 4 + 8 \ldots$. Here $a = 1$ and $r = 2$. According to our "proof" the sum of the series is $S = \frac{1}{1-2} = -1$. Yet all the terms are positive! Something is very wrong here. What is it? [Hint: Not all infinite series have finite sums. So when you say "Let S equal the sum of an infinite series," you had better be sure that the series has a finite sum. If it doesn't, you have no business manipulating it as we did above.]

The next example is a good one to share with your future students who have learned the Pythagorean Theorem. It is quite visual and concerns the staircase in Figure 1.1. Below, in Figure 1.1(a) is shown a line AB.

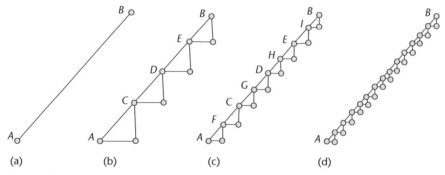

Figure 1.1

We begin by constructing in Figure 1.1(b), a staircase pattern. We then refine that staircase pattern in Figure 1.1(c) and then refine it again in Figure 1.1(d). Our eyes tell us that the smaller the vertical and horizontal segments are, the closer the length of the staircase comes to the length of our line. (By length of the staircase, we mean the sum of the lengths of the vertical and horizontal segments.) So we are led to conclude that, if we continue this, then the limit of the lengths of the staircases constructed is the length of the line AB. Do you believe it?

Well, if you do, then you have to accept that the hypotenuse of any right triangle is the sum of the lengths of the legs, and not believe the Pythagorean Theorem.

Let's see why. Begin with the right triangle shown below in Figure 1.2 with legs 3 and 4 and hypotenuse 5. Call the hypotenuse AB.

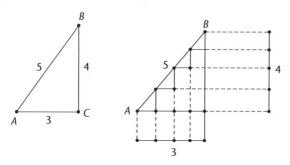

Figure 1.2

4 Intuition and Proof

Draw one of the staircases on the hypotenuse as shown. Move all the vertical segments on the staircase to the right as shown and all the horizontal segments on the staircase down as shown in Figure 1.2. Then it is clear that the sum of the lengths of the vertical segments is 4, the length of the vertical leg. The sum of the lengths of the horizontal segments is 3, the length of the horizontal leg. This same thing works regardless of which staircase pattern we use. Thus, the length of the staircase, regardless of which staircase we take, is 7 and not 5, the length of the hypotenuse.

Our intuition really fooled us here. Our point is that our eyes deceived us. What they showed us was false. We cannot trust what we see. We need proof that what we think we see is correct.

We know you'll want to try this next example with your own students some day. Start by drawing a circle and then pick two points on the circumference as shown in Figure 1.3(a). Draw the chord connecting them. It divides the circle into two regions as indicated in Figure 1.3(a) below. Next, draw another circle and put 3 points on the circumference and connect each pair. What is the maximum number of regions into which the circle is divided? The picture in Figure 1.3(b) below shows 4 regions.

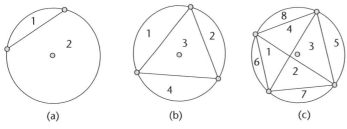

(a)　　　　　(b)　　　　　(c)

Figure 1.3

Now draw another circle and put 4 points on the circumference and connect each pair. What is the maximum number of regions into which the circle can be cut? The answer is 8 as the picture in Figure 1.3(c) above shows. Now, guess the answers to the maximum number of regions the circle is divided into, with 5 points put on the circumference. Did you guess 16? Yes! And did you draw it? Yes! And were you correct? Yes!

Now finish the following sentence: If we put n points on a circle and connect them, the maximum number of regions the circle is broken into is _____." We hope that you guessed 2^{n-1}. Now go through the process one last time for 6 points. Be sure to check your answers by drawing a picture. Did you get 2^5 regions? Well, if you did, then you didn't draw the picture correctly. For, when $n = 6$, we get 31 regions as a maximum and that is not 2^{n-1} since $2^{6-1} = 32$. Our pattern broke down.

These examples are just a few of many similar examples, and are meant to show us the danger of accepting or making statements without proof.

Let us give one final example. This next one is easy. In Figure 1.4 below, which is longer, a (the shorter base of the top trapezoid) or b (the longer base of the bottom trapezoid)?

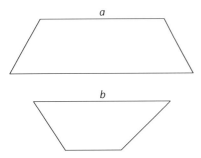

Figure 1.4

Did you say *b*? Well if you did, then you have good eyesight. What? You didn't? Well, that is just the point! Measure them with a ruler! *b* is a bit longer than *a*.

Are you beginning to distrust your intuition? your vision? your reasoning? If you said, "Yes," then you are beginning to think like a mathematician!

Student Learning Opportunities

Many of the problems that follow are intended to get you to think more critically about situations. Some of them are tricky and will be discussed further later on in the book. For now, see how you fare on them.

1 Suppose you were offered the choice of buying an item at a 30% discount, and then adding on 10% sales tax, or adding the sales tax first and then taking the 30% discount. What does your intuition tell you is the better deal? Prove it. Was your intuition correct?

2 The cylindrical cup on the left in Figure 1.5 has a radius of 3 inches and a height of 5 inches. The cylindrical glass on the right has a radius of 2 inches and height of 11.25 inches. You are very thirsty and want to buy some lemonade at a state fair. Both cups are filled to the brim with lemonade. The cup on the left sells for $2 and that on the right for $3. What does your intuition tell you is the better buy? How can you tell if your intuition was correct? Find out which is really the better buy. Was your intuition correct?

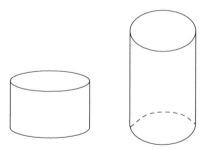

Figure 1.5

3 In the following Figure 1.6, which circle is bigger, the one in the center on the left, or the one in the center on the right? Estimate how much bigger the radius of the larger one is and then measure the radii to see how accurate you were. Was your estimate correct? In this case, where do intuition and proof come into play?

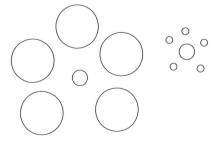

Figure 1.6

6   Intuition and Proof

4 **(C)** A student (Lucky Larry) says to you that "The fraction $\frac{19}{95}$ is $\frac{1}{5}$. You just cancel the 9's." Larry is right that the answer is $\frac{1}{5}$. When you tell him that his method is wrong, he says, "How much is $\frac{16}{64}$?" You say, "$\frac{1}{4}$" to which he responds, "Yes!!! Cancel the 6's!" What would you tell him and how would you convince him that he is using an incorrect method? Explain how intuition and proof are involved in the above scenario.

5 There are only four fractions where the numerator and denominator consists of two digits, and you can cancel as in Exercise 4 and get the right answer. See if you can find the other two.

6 **(C)** Jack, a student of yours comes rushing up to you one day very excited and says: "I'll finally be able to buy the car that I've always wanted." You ask him if he suddenly came into some money and he replies: "No, but I've been playing the lottery for the past 8 years and I've lost every time. I know that I am now surely due for a win." How do you respond to Jack? Is he really due for a win? If the probability of Jack winning is 1/1000 and he plays 10,000 games or 100,000 or even 1,000,000 games, isn't he guaranteed to win? How did intuition play a role in Jack's reasoning?

7 **(C)** Some students in your class come to you bewildered by some strange things one of the tricksters in the math club told them. How do you respond to the following proofs they were shown? Explain how intuition and proof came into play in both of these examples.

(a) A glass half empty is a glass half full. In symbols,

$$\frac{1}{2} \text{ glass empty} = \frac{1}{2} \text{ glass full}.$$

Multiply both sides by 2 to get 1 glass empty = 1 glass full.

(b) $\frac{1}{4}$ of a dollar = 25 cents. Taking the square root of both sides we get $\frac{1}{2}$ dollar = 5 cents.

8 **(C)** A student is asked to solve the following equation: $\frac{x-5}{x+1} = \frac{x-5}{x+3}$. The student smiles smugly and says, "There is no solution. Cross multiply to get $(x-5)(x+3) = (x-5)(x+1)$. Divide both sides by $x-5$ and I get $x+3 = x+1$. Subtracting x from both sides, I get $1 = 3$, which is impossible. So there is no solution." Is he right? Explain the role of intuition and proof in this situation.

9 **(C)** The math tricksters are at it again and have just shown your very bright student Maria a very convincing proof that 1 = 2. Maria comes to you very disturbed that she can't figure out what the flaw is. Here is the proof they gave: Start with the statement $a = b$. Multiply both sides by b to get $ab = b^2$. Subtract a^2 from both sides to get $ab - a^2 = b^2 - a^2$. Factor the left and right sides of the equation to get $a(b-a) = (b-a)(b+a)$. Now divide both sides by $b-a$ to get $a = b+a$. Finally, let $a = b = 1$ in this final result to get the statement that 1 = 2. How can you help Maria understand what the problem is with this proof?

10 **(C)** The tables have been turned and now Maria gets back at the math tricksters (see previous problem) by bewildering them with the following proof that 2 > 3 : Begin with the statement that $\frac{1}{4} > \frac{1}{8}$. Rewrite this as $(0.5)^2 > (0.5)^3$. Take the logarithm of both sides to the base 10 which we abbreviate as "log," to get $\log(0.5)^2 > \log(0.5)^3$. Now use the property of logarithms that allows you to pull out exponents. That leaves you with $2\log(0.5) > 3\log(0.5)$. Finally, divide by $\log(0.5)$ to get 2 > 3. How can you help the math tricksters see the flaw in Maria's proof? Explain the role of intuition and proof in this situation.

11 (C) You asked your class to compute $(-8)^{\frac{1}{3}}$ and much to your surprise they came up with two different methods of doing it. The first: $(-8)^{\frac{1}{3}} = \sqrt[3]{-8} = -2$. The second: $(-8)^{\frac{1}{3}} = (-8)^{\frac{2}{6}} = \sqrt[6]{(-8)^2} = \sqrt[6]{64} = 2$. How would you help your students understand which of these statements is wrong and why it is incorrect? Explain the role of intuition and proof in this situation.

12 (C) Lucky Larry's first cousin (see Student Learning Opportunity 4) solves the equation $-x^2 + x + 6 = 4$ by factoring first to get $(3 - x)(x + 2) = 4$. From this he concludes that either $(3 - x) = 4$ or $(x + 2) = 4$. He then solves each equation and gets $x = -1$ and $x = 2$ and both solutions check. Is this a valid way to solve quadratics? If not, why not? Explain the role of intuition and proof in this situation. An interesting question (which we are not asking you to answer), is: When will this procedure give you the correct answers?

13 Consider the quadratic equation:

$$\frac{(x-1)(x-2)}{2} + \frac{(x-2)(x-3)}{2} - (x-1)(x-3) = 1.$$

You can check that $x = 1, 2,$ and $x = 3$ solve this equation. But a quadratic equation only has at most two different solutions. What is wrong here? Explain the role of intuition and proof in this situation.

14 (C) A student, Hannah, sees the fraction $\frac{a^2-b^2}{a-b}$ and is asked to reduce it to lowest terms. She says "Cancel the a on the bottom with the a^2 on the top, to get a on the top, and cancel the b on the bottom with the b^2 on the top to get b on the top and finally, note that when you divide two negatives (the one on the top between a^2 and b^2 and the one on the bottom between a and b) we get a positive. So the answer is $a + b$." Now, we know the answer is $a + b$ but what was done is nonsense. Respond to the following questions, and then describe how asking Hannah these questions would help her to see the flaws in her procedures.

(a) Would the same thing work for $\frac{a^3+b^3}{a+b}$ to get $a^2 + b^2$? How can you check it?
(b) When can you "cancel" a term in the numerator and a term in the denominator? What are you really doing when you are "cancelling?"
(c) Can you "cancel" zeros? That is, is $\frac{0}{0} = 1$? Explain.

15 A 17 foot ladder is leaning against a wall. The base of the ladder is 8 feet from the wall, and the top of the ladder is 15 feet above the floor. The top of the ladder begins to slide down the wall. The ladder is sliding down the wall at the rate of 1 foot per second. Is the base of the ladder also sliding away from the wall at that rate? What does your intuition tell you? Can you prove it?

16 (a) Let S be the infinite series

$$S = 1 - 1 + 1 - 1 \ldots . \tag{1.1}$$

Multiply both sides by -1. This will give us

$$S = 1 - 1 + 1 - 1 \ldots \tag{1.2}$$

$$-S = -1 + 1 - 1 + 1 \ldots . \tag{1.3}$$

Subtract equation (1.3) from equation (1.2) to get

$2S = 1.$

Hence,

$$S = \frac{1}{2}. \tag{1.4}$$

How can this be since all the terms are integers?

(b) Let us manipulate the series in part (a) differently. First write $S = (-1 + 1) + (-1 + 1) \ldots$ so that $S = 0$. Now write $S = 1 + (-1 + 1) + (-1 + 1) \ldots$ so that now $S = 1$. How could S be both 0 and 1? This argument was once used to prove the existence of God. If we can turn nothing (0 was considered as representing nothing) into something (namely 1), then there is a God! What is the flaw in the reasoning?

(c) Explain the role of intuition and proof in this situation.

17 Consider the circle with center at (1, 0) and radius 1. Pick any point P on the circle, and then pick a point Q on the y-axis such that $OP = OQ$. (See Figure 1.7 below.) Draw QP and let it intersect the x-axis at R. As P approaches O it is clear that R moves to the right. True or false: The point R approaches a finite number. If so, which number? Try verifying your guess by hand or with some software. Try proving your guess is correct.

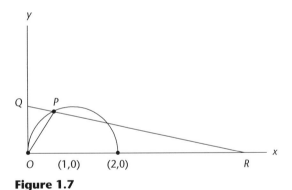

Figure 1.7

1.3 Some Fundamental Methods of Proof

Make a convincing argument for or against the following statement: The sum of the squares of any 3 consecutive integers is divisible by 14. Does your argument constitute a proof? Why or why not?

After responding to the launch question, you are probably beginning to question whether or not you really supplied a "proof" for your answer. In this section we will resolve your question by reviewing some of the different proof techniques that every mathematics student should be familiar with and which relate in one form or another to the secondary school curriculum, not to mention all of mathematics. The types of proofs we will describe are: direct proof; proof by contradiction, and proof by counterexample. In each section we first describe the method of proof and then follow it with several examples. (Please note that there is one more method of proof that is extremely important in mathematics, and that is proof by induction. However, since that method is a bit more abstract and not as prevalent in the secondary school curriculum, yet of critical importance, we have relegated it to a later part of the book, Chapter 8.)

1.3.1 Direct Proof

The first type of proof we speak about is one that you have used throughout school and is called **direct proof**. Here we prove something directly from known facts. We simply string the known results together or perform correct mathematical manipulations to come out with our conclusion. We give a few examples.

The first example of a direct proof is of a result that is used quite often in mathematics. In fact, there is a famous legend that is related to this result, of a mathematical genius Karl Freidrich Gauss who was asked by his elementary school teacher to sum the first 100 numbers to keep him busy and out of trouble. Much to his teacher's surprise, after only a few minutes he came up with his solution. How could he possibly have done it so quickly? Essentially, he used the method used to prove Theorem 1.1 below which is an illustration of a direct proof.

Theorem 1.1 *The sum of the first n integers is $\frac{n(n+1)}{2}$. That is,*

$$1 + 2 + 3 + \ldots + n = \frac{n(n+1)}{2}.$$

Proof. We call the given sum, S. Thus,

$$S = 1 + 2 + 3 + \ldots + n. \tag{1.5}$$

Now rewrite the sum starting at the last term and going to the first. This yields

$$S = n + (n-1) + (n-2) + \ldots + 1. \tag{1.6}$$

We now add the two series equation (1.5) and equation (1.6) for S term by term to get

$$2S = (n+1) + (n+1) + \ldots + (n+1). \tag{1.7}$$

In equation (1.7) we have n terms, all equal to $n+1$. So the sum on the right side of the equation (1.7) is $n(n+1)$. Thus,

$$2S = n(n+1) \tag{1.8}$$

and dividing equation (1.8) by 2, we get $S = \frac{n(n+1)}{2}$. ∎

Thus, if we were asked, as little Gauss was, to find the sum of the integers from 1 to 100, the sum would be $\frac{100(101)}{2}$, or 5050. And that is how Gauss did it!

This method exemplifies a direct proof since we used what we knew about how to represent the sum of n integers algebraically in several ways and then logically combined and manipulated our representations in a way that resulted in our theorem.

Our next example of a direct proof involves a theorem from geometry that most secondary school students learn. We need only recall one fact from geometry: If two sides of a triangle are equal, then the angles opposite them are equal.

Theorem 1.2 *An angle inscribed in a semicircle is a right angle.*

Proof. Begin with angle ABC inscribed in the semicircle as shown in Figure 1.8 below. (Recall that an inscribed angle is one whose vertex is on the circle.)

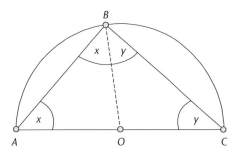

Figure 1.8

Now draw radius OB. Since all radii are equal, $OB = OA = OC$. Hence angle OAB = angle OBA since triangle AOB has two equal sides. We call both these angles x. Similarly, angle OCB = angle OBC since triangle OBC is isosceles, and we call these angles y as marked in the diagram. Now, we know that the sum of the angles of a triangle is 180 degrees. So summing the angles of triangle ABC, we get, $A + B + C = 180$ or $x + (x + y) + y = 180$, which simplifies to $2x + 2y = 180$. Dividing both sides of this equation by two, we get that $x + y = 90$. But $x + y$ is angle B, and our goal was to show that angle B, the inscribed angle, was 90 degrees. So we are done. ∎

This method exemplifies a direct proof since we used what we knew about the radii of a circle, the sum of the angles of a triangle, and the angles of an isosceles triangle to create an algebraic equation that represented the relationships of the angles. We then correctly manipulated our equation to arrive at our result. Indeed, it is a thing of beauty!

1.3.2 Proof by Contradiction

The direct proofs are so elegant, you might be asking yourself why we need to learn different methods of proof. This is certainly a valid question. Consider trying to prove, as Euclid did, that $\sqrt{2}$ is irrational. In ancient times, the Greeks believed that all numbers were rational and the fact that $\sqrt{2}$ did not fit the criteria for a rational number was of great concern to them. But, how do we prove it? A direct proof in this case is not obvious. Therefore, in the true spirit of problem solving, when one method does not present itself easily, one must think of another approach, in this case, proof by contradiction! In a **proof by contradiction**, sometimes known as an **indirect proof**, we are, as in all proofs, trying to prove that a statement is true. Given that we know that a statement must either be true or false, we do a tricky thing and assume that what we are trying to prove is false! We then show that this leads to a contradiction of something we know is true. Since the assumption that the theorem was false led to a contradiction of something we know is true, the original statement COULD NOT have been false. Thus, our original statement must have been true. Got that? Observe how clever this method is!

Let us now examine Euclid's most famous proof by contradiction that $\sqrt{2}$ is irrational. It is a masterpiece of simplicity. We present it here. We just recall that a rational number is one that is a ratio of two integers which, of course, may be written in lowest terms. We assume that the reader accepts the fact that if the square of an integer is even, the integer itself must be even. We discuss these issues more in Chapter 2.

Theorem 1.3 $\sqrt{2}$ *is irrational.*

Proof. Suppose not. Then $\sqrt{2}$ is rational. That is,

$$\sqrt{2} = \frac{p}{q} \tag{1.9}$$

where p and q are integers and $\frac{p}{q}$ is in lowest terms. What that means is that p and q have no common factor.

Now since $\sqrt{2} = \frac{p}{q}$, we may square both sides of this equation to get $2 = \frac{p^2}{q^2}$. Next we multiply both sides by q^2 to get

$$p^2 = 2q^2. \tag{1.10}$$

This tells us that p^2, being twice the number, q^2, must be even, hence p *is even*.

Since every even number is twice some number, and p is even, we may write $p = 2k$ and substitute into equation (1.10) to get $4k^2 = 2q^2$. Divide both sides by 2 to get

$$2k^2 = q^2. \tag{1.11}$$

Since equation (1.11) says that q^2 is twice some number, q^2 is even and hence q *is even*.

We have shown that both p and q were even. This contradicts that $\frac{p}{q}$ was in lowest terms.

Our assumption that $\sqrt{2}$ was rational led to a contradiction of something we knew was true. Thus, $\sqrt{2}$ is irrational. ∎

Isn't this proof amazing?

Before continuing, let us say a few words about how to prove "If–then" statements by contradiction. The statement "If A then B" is telling you to assume that A is true, and to show that B follows from it. Now, either B follows from it, or it doesn't. In a proof by contradiction, we assume that B *doesn't* follow and show this leads to a contradiction. So, if we want to prove the statement "If n is odd, then n^2 is odd" by contradiction, we assume that n is odd and n^2 is *not* odd and proceed from there to find a contradiction. If we wanted to prove the statement "If n is divisible by 2, then n^3 is divisible by 8" by contradiction, we begin by assuming that n is divisible by 2 and n^3 is *not* divisible by 8. Finally, if we are trying to prove that "If $a + b < 6$, then either *a* or *b* is less than 3." We assume that $a + b < 6$ but it is false that *a* or *b* is less than 3." If it is false that *a* or *b* is less than 3, then *both* *a* and *b* must be greater than or equal to 3. (For more on the logic of this, see Student Learning Opportunity 9.) We are now ready to proceed with some other proofs by contradiction.

> **Theorem 1.4** *If a and b are real numbers and $a + b < 6$, then either a or b is less than 3.*

Proof. Suppose $a + b < 6$ and it is not true that either *a* or *b* is less than 3. Then both *a* and *b* are greater than or equal to three. That is, $a \geq 3$ and $b \geq 3$. Adding these two inequalities yields $a + b \geq 6$, and this contradicts what we were given, namely, that $a + b < 6$. Since our supposition that it is not the case that either *a* or *b* was less than 3 led to a contradiction of something we knew, that supposition must be wrong. So *a* or *b must be* less than 3. ∎

> **Theorem 1.5** *If the coordinates of a quadrilateral are A: (0, 0), B: (1, 3), C: (−3, 5), and D: (−1, 2), then ABCD is not a parallelogram.*

Proof. Suppose *ABCD* is a parallelogram. Then *AB* must be parallel to *CD*. Thus, *AB* and *CD* must have the same slope. But the slope of *AB* is $\frac{3-0}{1-0} = 3$ and the slope of $CD = \frac{2-5}{-1--3} = \frac{-3}{2}$. They are not the same. So *ABCD* cannot be a parallelogram. ∎

Although proof by contradiction seems strange at first, we all experience it first hand in the courtrooms. Suppose that a lawyer is trying to prove that Mr. Smith didn't kill his wife. A proof by contradiction would proceed as follows: "Suppose that Mr. Smith did kill his wife. Then he had to be there with her when she was killed. But he was at a party at that time (here is our contradiction) and this was verified by 73 different guests. So Mr. Smith didn't kill his wife."

1.3.3 Proof by Counterexample

A **conjecture** is a statement of a relationship that one believes is true based on evidence or intuition, or both, but not yet proven. In Section 1.2, we discussed Fermat's conjecture that the numbers $N = 2^{(2^k)} + 1$ are prime for all positive integers *k*. Euler had a suspicion that Fermat's conjecture was false. But how could he show that? Certainly, if he could find one example where it did not work, then he would have shown that Fermat's statement wasn't true. An example which shows that a statement is false is called a **counterexample**. Euler found that, when $k = 5$, we get $N = 4,294,967,297$, which is NOT a prime. So Euler found a counterexample to Fermat's conjecture.

Suppose we are given the statement "The square of any odd number is even" and we want to disprove it. We need only produce one counterexample. In this case the odd number 3 is a counterexample. The square of 3 is 9 and 9 is not even. Therefore, the statement that "The square of any odd number is even," is false.

So, when we believe that a conjecture is not true, we have a method to prove it is false. All we have to do is find a counterexample.

In the launch in the beginning of the chapter we asked if $N^2 + N + 41$ is prime for all positive integers *N*. After taking several examples, our intuition told us it might be! However, if we can find one counterexample to this, then we have our answer. Try $N = 41$ and you can then see that $(41)^2 + (41) + 41$ is divisible by 41. So it is not true that $N^2 + N + 41$ is always prime.

1.3.4 The Finality of Proof

In mathematics when one produces a correct proof, one never has to question its truth again. That is what mathematical proof is all about. It is about finality and knowing for sure that, for eternity, something is true or not true. That is very different from proof in the sciences, where, for the most part, theories are constructed based on evidence. In the sciences one rarely can prove the theory, but one accepts it because it explains the physical phenomena. However, it is always subject to change. If new evidence surfaces, the whole theory might change. In mathematics, it is very different since theories are proven, and with a correct proof they are true forever!

Student Learning Opportunities

1 (a) Give a direct proof that the exterior angle of a triangle is the sum of the two remote interior angles. That is, $w = x + y$. [Hint: The sum of the angles of a triangle is 180° as is the sum of z and w. See Figure 1.9 below.]

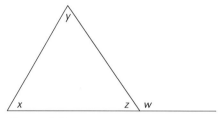

Figure 1.9

   (b) As a corollary of (a), deduce that the exterior angle of a triangle is greater than either of the remote interior angles.
   (c) As another corollary, prove by contradiction that, from a point outside a line, there can only be one perpendicular drawn to that line.

2 Give a direct proof that the figure with coordinates $(5, 0)$, $(3, 3)$, $(-5, 0)$, and $(-3, -3)$ is a parallelogram.

3 Give a direct proof that $1 + 3 + 5 \ldots + (2n - 1) = n^2$ by showing that the sum is $(1 + 2 + \ldots + 2n) - (2 + 4 + 6 + \ldots + 2n) = (1 + 2 + \ldots + 2n) - 2(1 + 2 + 3 + \ldots + n)$ and then using Theorem 1.1.

4 Using the facts from trigonometry that $\sin^2 \theta + \cos^2 \theta = 1$, and that $\cos 2\theta = \cos^2 \theta - \sin^2 \theta$, give a direct proof that $\cos 2\theta = 1 - 2\sin^2 \theta$, and hence that $\sin^2 \theta = \dfrac{1 + \cos 2\theta}{2}$ for any angle θ.

5 (C) Students are asked to expand the expression $(a + b)^3$. They do the computations and get $a^3 + 3a^2b + 3ab^2 + b^3$. How do they arrive at this expression? Is this a proof that $(a + b)^3 = a^3 + 3a^2b + 3ab^2 + b^3$? If so, what kind of a proof is it? Why?

6 Give a direct proof that $\dfrac{1}{1\cdot 2} + \dfrac{1}{2\cdot 3} + \ldots + \dfrac{1}{(n-1)\cdot n} = \dfrac{n-1}{n}$ [Hint: $\dfrac{1}{1\cdot 2} = 1 - \dfrac{1}{2}$, $\dfrac{1}{2\cdot 3} = \dfrac{1}{2} - \dfrac{1}{3}$, $\dfrac{1}{3\cdot 4} = \dfrac{1}{3} - \dfrac{1}{4}$ etc.]

7 Give a direct proof that, when n is even, $1^2 - 2^2 + 3^2 - \ldots + (-1)^n(n-1)^2 + (-1)^{n+1}n^2 = \dfrac{-n(n+1)}{2}$. [Hint: Rewrite this as $(1^2 - 2^2) + (3^2 - 4^2) + \ldots + ((n-1)^2 - n^2)$ and then factor to get $(1-2)(1+2) + (3-4)(3+4) + \ldots + (n-1-n)(n-1+n)$.]

8 Here is an interesting pattern:

$$1 + \boxed{2} + 1 = \boxed{2}^2$$
$$1 + 2 + \boxed{3} + 2 + 1 = \boxed{3}^2$$
$$1 + 2 + 3 + \boxed{4} + 3 + 2 + 1 = \boxed{4}^2$$

and so on. This pattern continues where the sum is always the middle number squared. Using Theorem 1.1 see if you can explain the pattern. [Hint: The typical expression on the left can be written as $1 + 2 + 3 + \ldots + \boxed{n} + n - 1 + n - 2 + \ldots + 1$.]

9 (C) Students often have trouble with proofs by contradiction. They don't understand why when you negate an "if–then" statement, you assume the "if" part and negate the "then" part. Show, using logic tables, that the negation of $(p \to q)$ is equivalent to $(p \land \sim q)$. Then explain how this equivalence is used as the basis for a proof by contradiction.

10 Give a proof by contradiction that, if $3n + 5$ is even, then n must be odd.

11 Give a proof by contradiction that, if $x + y < 12$, then either $x < 6$ or $y < 6$.

12 Give a proof by contradiction to show that, if two lines l and m are cut by a transversal, in such a way that the alternate interior angles, x and y are equal, then the lines are parallel. [Hint: If the lines aren't parallel, then they meet at some point P as shown in the second picture of Figure 1.10 below. Now use Student Learning Opportunity 1 part (b).]

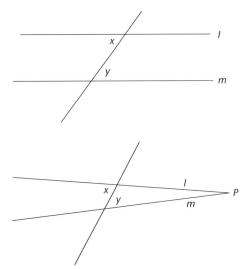

Figure 1.10

13 True or False: When you give a proof by contradiction, you must contradict something that is given. Explain.

14 (C) A student asks if you can use the same method to prove $\sqrt{3}$ is irrational as you used to show $\sqrt{2}$ is irrational. How do you guide the student to see the differences and similarities in the proofs?

15 Give a proof by contradiction that $4 + \sqrt{3}$ is irrational. (You may need to use the fact that the difference of rational numbers is rational.)

16 Give a proof by contradiction that there cannot be a quadrilateral whose consecutive sides are $AB = 2$, $BC = 3$, $CD = 5$, and $DA = 12$. [Hint: Draw diagonal AC cutting the quadrilateral into two triangles. Using the fact that the shortest distance between two points is a straight line, show that the length of AC is less than 5. Now work with the other triangle. The shortest distance from A to D should be 12.]

17 Prove or disprove the following statement: "$3^n > n + 2$ for each positive integer n." Explain what method you used.

18 What method of proof was used to disprove Euler's conjecture that there are no positive integers a, b, c, and d which make $a^4 + b^4 + c^4 = d^4$?

19 Find a counterexample to the statement "The smallest natural number, n, such that the sum of the first n natural numbers is greater than 1000 is $n = 50$."

20 (C) Your students are convinced that the following statement is true: $\frac{x^2 - 1}{x - 1} = x + 1$. Are they correct? Give a proof for why this is or is not correct. What type of proof did you give?

21 (C) Your students are convinced that the following statement is true: If $a < b$ then $a^2 < b^2$. Are they correct? Give a proof for why this is or is not correct. What type of proof did you give?

22 (C) Your students have proven that, when they add 2 consecutive integers, they always get an odd number. Now they have begun to investigate what happens when then add 3 consecutive integers. Some have decided that the sum is always divisible by 3. Others have decided that the sum is always divisible by 6. Prove or disprove each of your students' conjectures. Which method or methods did you use?

23 Prove or disprove: If three consecutive integers are multiplied together, and the second, in order of size, is added to the product, the result is always a perfect cube.

24 (C) A student asks, " Since we can always use direct proof, why do we need to know proof by contradiction and proof by counterexample." What is your reply?

25 (C) A student asks, "What happens when you try to prove that $\sqrt{4}$ is irrational in a manner similar to the way we proved that $\sqrt{2}$ was irrational? Won't that same proof show that $\sqrt{4}$ is irrational?" How do you explain to your student that the same method won't work? Where does the proof break down?

CHAPTER 2

BASICS OF NUMBER THEORY

2.1 Introduction

Throughout the school curriculum, an emphasis is placed on numbers, their properties, and relationships between and among them. In this chapter we present some of the basics of number theory that middle school and secondary school teachers should be aware of. Those who took courses in number theory will likely find much of this material familiar, but will enjoy revisiting some of the important and elegant results which often even secondary school students can prove. The results provide insight into what proof is all about and the important role of good definitions. Throughout the chapter we also intersperse interesting applications that range from recreational areas such as numerical curiosities and tricks, to such serious practical applications as the workings of a computer and high-level security systems.

2.2 Odd, Even, and Divisibility Relationships

 LAUNCH

Find five odd numbers whose sum is 100.

After exploring the launch question, you may be somewhat frustrated. Were you able to find any sets of five odd numbers that met the conditions? How many sets of numbers did you try? You might suspect that there are no such five integers. Did you notice any patterns in the sums that you found? Were they all odd? It might have hit you that "Wait, the sum of five odd numbers must be odd, so this sum cannot be 100." Ahhhh! The light came on! The solution of the problem depends on realizing that the sum of two odd numbers is always even and the sum of an even number and an odd number is odd. But how do you know that the sum of two odd numbers is always even? Just because you try sum after sum of two odd numbers and you get an even number does not always mean the sum of ALL pairs of odd numbers is even. Is there a simple way to show this? Sure! We will now see how.

So then, what is an even number? We know numbers like, 2, 4, 6, and so on, are even. But how can we define an even number so that, if we want to prove things about them, we can? Think about it for a few minutes before proceeding and see what you come up with.

There are a few ways to proceed, but a particularly elegant and simple way is to realize that every even number can be represented as twice another integer. For example, $2 = 2 \cdot 1$, $4 = 2 \cdot 2$, and so on. That leads us to our definition of even number. An **even** number is any integer that can be written as double an integer. That is, N is even if $N = 2k$ for some integer k. It follows that, if an even number is a number of the form $2k$, then an odd number, being one more than an even number, is also easy to define. That is, an integer N is **odd** if $N = 2k + 1$ for some integer k.

One of the first patterns that middle school children observe is that the sum of two even numbers always results in an even number. Similarly, they notice that the sum of two odd numbers is even. What about the product of even and odd numbers? Try multiplying a few pairs of integers. What patterns do you observe? While middle school children can draw conclusions based on these observations, they typically believe they have enough evidence to accept these relationships as facts. It is the teacher's responsibility to let students know that observations alone do not ensure that the relationships will hold for all cases. Having a higher level of understanding of the concepts and proofs that underlie the relationships is essential. Even if the proofs are too sophisticated for middle or secondary school students, teachers can make them accessible to their students in a more informal way, but only if they themselves have insights into the proofs. So, here we begin the process with our first theorems interspersed with examples, tricks, and applications that involve number concepts.

> **Theorem 2.1**
> *(a) The sum of two even numbers is even.*
> *(b) The sum of two odd numbers is even.*
> *(c) The product of two odd numbers is odd.*
> *(d) If an even number is multiplied by any integer, the result is even.*

Proof. We won't give the proof of all of these, as they are worthwhile tasks for you. But we will give the proof of parts (a) and (c):

(a) Suppose M and N are even numbers. Then, by the *definition* of even number, each of these is twice some integer. Thus $M = 2k$ and $N = 2l$ for some integers k and l. We need to show that the sum of these numbers is even, and that means that the sum must also be shown to be twice some integer. We do that as follows:

$$M + N = 2k + 2l$$
$$= 2(k + l)$$

and we are done. We have shown that $M + N$ equals twice the integer $k + l$. How simple it was to prove using the proper definitions!

(c) Suppose M and N are odd integers. Then by the *definition* of odd number, each of these is one more than an even number. That is, $M = 2k + 1$ and $N = 2l + 1$ for some integers k and l. We need to show that MN is odd. That is, it is of the form $2m + 1$ for some integer m. But

$$MN = (2k + 1)(2l + 1)$$
$$= 4kl + 2l + 2k + 1$$

$$= 2(2kl + l + k) + 1$$
$$= 2m + 1$$

where $m = 2kl + l + k$. Thus the product of two odd numbers is odd. ∎

We can solve some surprisingly difficult problems by just considering when the numbers involved are odd or even. Here are some on the secondary school level. The first came from a secondary school contest. No calculators were allowed, and the students had about 2 minutes to solve the problem. See if you can do it before looking at the solution.

> **Example 2.2** *Of the following pairs (x, y), only one of them does not satisfy the equation $187x - 104y = 41$. Which one is it?*
> *Here are the pairs:* (107, 192), (211, 379), (314, 565), (419, 753), (523, 940).

Solution. A quick solution would run as follows: $104y$ is even (Why?) Add $104y$ to both sides of the equation $187x - 104y = 41$ to get $187x = 104y + 41$. The right side of this new equation, being a sum of an even number and an odd number, is odd. Thus, the left side of this new equation, $187x$, must be odd. This eliminates the pair whose x coordinate is 314 since 187 times 314 is even. Thus, (314, 565) doesn't work.

This next example shows how concepts of odd and even can be used to figure out tricks.

> **Example 2.3** *Tell a friend to take a dime and nickel and put one coin in one hand and the other coin in the other hand. You can turn your back while he does this. Tell him to multiply the value of the coin in his right hand by 8 and the value of the coin in his left hand by 3 and tell you the sum. If he tells you an even number, the dime is in his left hand. If he tells you an odd number, the dime is in his right hand. Explain this trick.*

Solution. The trick here is to realize that the dime has an even number of cents and that the nickel has an odd number of cents. Let R be the value of the coin he has in his right hand in cents, and L be the value of the coin he has in his left hand in cents. You are asking him to compute $8R + 3L$. Now $8R$ is always even and $3L$ will be odd or even depending on whether L is odd or even. If L is even, that is, if the dime is in his left hand, the sum $8R + 3L$ will be even. If L is odd, that is, if the coin in his left hand is the nickel, the sum will be odd.

Concepts of odd or even rest on the question of whether or not a number is divisible by 2. We can now extend this notion to examine numbers that are divisible by numbers other than 2. For example, one of the topics we emphasize in schools today is pattern recognition. So, for instance, if we list the numbers 3, 6, 9, and so on, we see that each number is a multiple of 3, or put another way, each number in the list is divisible by 3. What does it mean for a number to be divisible by 3? What does it mean for a number to be divisible by 4, and so on? We are guided by the definition of an even number. A number N is divisible by 3 if $N = 3k$ for some integer k. A number, N, is divisible by 4 if $N = 4k$ for some integer k, and so on. Thus, a number N is **divisible** by a, an integer, if $N = ak$ for some (unique) integer k.

There are several other ways of saying that N is divisible by a. One is that N has a factor of a. (Recall that, when we write a number as a product, each number in the product is called a **factor**.)

Another is that N is a **multiple** of a, or that a **divides** N, or that a is a **divisor** of N. Thus, if we know that an integer N = 11k for some integer k, right away we know that N is divisible by 11 or, said another way, N is a multiple of 11, or 11 is a factor of N, or 11 divides N, or 11 is a divisor of N.

> **Example 2.4** Show that, for any integer k, $(2k + 1)^2 - (2k - 1)^2$ is divisible by 8.

Solution. If we square the expressions in parentheses and simplify, we get

$(2k + 1)^2 - (2k - 1)^2$

$= (4k^2 + 4k + 1) - (4k^2 - 4k + 1)$

$= 8k.$

Clearly, this result is divisible by 8.

> **Example 2.5** A man buys apples at 3 cents a piece and oranges at 6 cents a piece, and hands the salesperson a 5 dollar bill. His change is $4.12. Did he receive the right change?

Solution. At first glance, this problem seems impossible to answer, but it can be answered. If the man bought x apples and y oranges (both positive integers), his cost would be $3x + 6y$ cents. Since he received change of $4.12 cents, his cost must have been 88 cents. That is, $3x + 6y = 88$. But, since one can factor out 3 from the left side of the equation, $3x + 6y$ is divisible by 3. But the right side, 88, isn't. This means that $3x + 6y$ can't be 88, and hence he couldn't have paid 88 cents for the fruit. Thus, his change could not have been correct.

Isn't it nice how the simple divisibility facts help us in solving this problem?

Let us give one last example from geometry.

> **Example 2.6** Recall that a regular polygon is one whose sides are equal and whose angles are equal. Thus, an equilateral triangle is a regular polygon, as is a square. If we take 4 squares and arrange them around a point, we can do it so that there is no space left between them. If we take 6 equilateral triangles that are congruent, we can arrange them around a point so that no space is left between them. (See Figure 2.1 below.)
>
>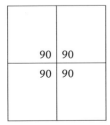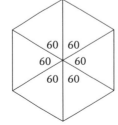
>
> **Figure 2.1**
>
> Show that there are only three regular polygons for which we can do this.

Solution. We need to recall a fact from geometry, namely, each interior angle of a regular polygon is $\frac{180(n-2)}{n}$ where n is the number of sides. If k of these polygons are put together so that there is no space left over at the center, then the sum of the angles at the center must be 360 degrees. That is, $\frac{k \cdot 180(n-2)}{n} = 360$. If we divide by 180, we get $\frac{kn-2k}{n} = 2$. Multiplying both sides by n we get

$$kn - 2k = 2n$$

and subtracting $2n$ from both sides and adding $2k$ to both sides we get

$$kn - 2n = 2k.$$

Finally, factoring out n from the left side and dividing by $k-2$ we get

$$n = \frac{2k}{k-2} = 2 + \frac{4}{k-2}. \tag{2.1}$$

Since the left side of equation (2.1) is an integer, so is the right side. Thus, $\frac{4}{k-2}$ is an integer and so $k-2$ must divide 4. Since $k-2$ divides 4, $k-2$ must be 1, 2, or 4, and therefore $k = 3$, 4 or 6. Substituting these values into equation (2.1) we get that $n = 6$, 4, or 3, respectively. Thus, the only regular polygons that will accomplish this are hexagons ($n = 6$), squares ($n = 4$) and equilateral triangles ($n = 3$).

Let's move on to some other results involving concepts of divisibility that are interesting and useful.

Theorem 2.7 *If M and N are each divisible by a, then so are M + N and M − N. This generalizes: The sum and/or difference of a collection of numbers, each divisible by a, is divisible by a.*

The proof of each is almost identical to the proof of Theorem 2.1 (a) so we leave it to you as a simple but instructive exercise.

Let us illustrate this theorem with a simple example.

Example 2.8 *Show that the only positive integer n that divides both an integer a and a + 1 is 1.*

Solution. Most people don't know where to start with this one. But, if n divides both a and $a + 1$, then n divides their difference $(a + 1) - a$, or 1. But the only positive integer that divides 1 is 1. Thus, $n = 1$.

Theorem 2.9 *If M is divisible by a, and N is any integer, then MN is divisible by a.*

Proof. We need to show that $MN = ak$ for some integer k. But, M is divisible by a, so for some integer m,

$$M = am.$$

22 Basics of Number Theory

This is what it means for M to be divisible by a. Multiplying both sides of this equation by N, we get that

$MN = amN$.

Thus, $MN = ak$ where k is mN and therefore MN is divisible by a. ∎

Most middle school students know the following rule: A number is even if the units digit of the number is 0, 2, 4, 6, or 8. There are rules to tell if a number is divisible by 3, 4, 5, 6, 7, 8, 9, and 11, some of which are quite easy to remember. To develop the rules for divisibility and their proofs, we need some notation. If a number has tens digit t and units digit u, then the value of the number is $10t + u$. Since the number 36 has tens digit 3 and units digit 6, the value of the number is $10(3) + 6$. Similarly, if a three digit number has hundreds digit h, tens digit t, and units digit u, then the value of the number is $100h + 10t + u$, and so on.

The following is a typical secondary school problem, which can be solved by trial and error with some analysis, or by algebra. We take the algebraic approach.

> **Example 2.10** *If the digits of a two digit number are reversed, the resulting number is 9 more than the original number. The sum of the original number and the number with the digits reversed is 55. Find the original number.*

Solution. We let t be the tens digit and u be the units digit of the original number. Then the value of the original number is $10t + u$. When the digits are reversed, t becomes the units digit, and u the tens digit. Thus, the new number will have value $10u + t$. From the given information, when we reverse the digits the resulting number is 9 more than the original number. That is, $10u + t = 9 + 10t + u$. Subtracting $10t + u$ from both sides we have that $9u - 9t = 9$ and when dividing by 9 we get

$$u - t = 1. \tag{2.2}$$

From the information that the sum of the original number and the number with the digits reversed is 55, we have $10t + u + 10u + t = 55$ which simplifies to

$$11u + 11t = 55. \tag{2.3}$$

Dividing this by 11 we get that

$$u + t = 5. \tag{2.4}$$

Adding equations (2.2) and (2.4) we get

$2u = 6$,

hence $u = 3$. Substituting $u = 3$ into equation (2.4) and solving for t we get $t = 2$, and so the number we are seeking is 23. And indeed, we can see that our answer works out. That is, $32 - 23 = 9$ and $32 + 23 = 55$!

Just as a note, it is always important to check back after doing a problem to see if your answer works out. This is as much a part of the problem-solving process as solving the problem, since it is possible that computational errors were made or the algebraic representation of the problem was incorrect. Also, even if all the mathematical work was correct, it is possible that answers that don't work were introduced. (See Chapter 6, Section 11 for more on this.) Teachers need to set the example that good problem solving involves reflecting on one's solutions.

Here is an even more interesting problem which is the basis of many number tricks. (See Example 2.13 in the next section for one of them.)

Example 2.11 *Show that, if we take any three digit number and scramble the digits, then subtract the smaller of the two numbers from the larger one, the result will always be divisible by 9.*

Solution. Just to understand what this is saying, let us take a few examples. Suppose the original number is 921 and the scrambled number is 291. Then the difference is $921 - 291 = 630$, which is divisible by 9. If the scrambled number were 129 the difference would be $921 - 129 = 792$, which is also divisible by 9.

Now let us give the idea of the proof in general. We begin by assuming that the hundreds digit of the number is h, the tens digit is t, and the units digit is u. Let us scramble the digits, making, say, u, the hundreds digit, h, the tens digit, and t, the units digit. Then our original number has value $100h + 10t + u$ and the number with the digits scrambled is $100u + 10h + t$. Let us assume that the original number is the larger one. Then when we subtract, we get $100h + 10t + u - (100u + 10h + t) = 90h + 9t - 99u = 9(10h + t - 11u)$. Since this difference is $9k$ where $k = 10h + t - 11u$, we see the difference of the numbers is divisible by 9. A similar proof works for any scrambling of the digits. Since this works for all of the permutations of the digits, we are done!

Student Learning Opportunities

1 (a) Give a direct proof that the sum of two odd integers is even.
  (b) Give a direct proof that the sum of any odd integer and any even integer is odd.
  (c) Give a direct proof that multiplying any integer by an even integer gives an even integer.
  (d) Give a proof by contradiction that if n^2 is odd, then n is odd.

2 (C) A student asks if 0 is odd, even, or neither. How do you respond? How do you explain your answer?

3 (C) A student gives the following proof of Theorem 2.1 part (a): Suppose M and N are even integers. Then by the definition of even number, each of these is twice some integer. Thus $M = 2k$ and $N = 2k$. We need to show that the sum of these numbers is even. But, that is easy:

$$M + N = 2k + 2k$$
$$= 2(k + k)$$

so the sum is even.
Criticize this proof. Show how to do it correctly.

4 (a) Show that if a number is divisible by a, then it is divisible by any factor of a.
  (b) Prove that If M and N are each divisible by a, then so is $M + N$.

24 Basics of Number Theory

(c) Prove that If M and N are each divisible by a, then so is $M - N$.

(d) Prove that If M and N are each divisible by a, then so is M^p for any positive integer power p. Is it true if p is a negative integer power? Explain.

5 Show that $3k^2 + 3k$ is always divisible by 6 for any integer k.

6 Prove that the square of the product of 3 consecutive integers is always divisible by 12.

7 (C) Your students have been investigating the square roots of the following square numbers: 4, 16, 36, 64, etc. and have come up with the conjecture that, if N^2 is even, where N is an integer, then so is N. Are they correct? How can they prove or disprove their hypothesis? (Note that we used this property in the proof that $\sqrt{2}$ was irrational in Chapter 1.)

8 Show that the sum of the cubes of 3 consecutive integers is divisible by 9. [Hint: The identity $(a + b)^3 = a^3 + 3a^2b + 3ab^2 + b^3$ helps.]

9 (C) One of your curious students noted the following interesting relationship: When she added 2 consecutive odd integers, the sum was divisible by 2. When she added 3 consecutive odd integers, the sum was divisible by 3. She made the conjecture that the sum of n consecutive odd numbers is always divisible by n. Is she correct? How would you help her prove or disprove her conjecture?

10 The product of 66 integers is 1. Can their sum be zero? Explain.

11 Show that there are no positive integers such that $ab(a - b) = 703345$. [Hint: Consider the cases when one of the numbers is even, when both are even, and when both are odd.]

12 If $a + 2$ is divisible by 3, show that $8 + 7a$ is also.

13 (C) The integers 1 to 10 are written along a straight line. You dare your students to insert "+" signs and "−" signs in between them so that their sum is zero. Will they be able to do it? Why or why not?

14 In Example (2.11) we assumed that the larger number was $100h + 10t + u$. What if that were the smaller number and $(100u + 10h + t)$ were the larger. Would the conclusion that the difference is divisible by 9 still hold? Explain.

15 Find all two digit numbers such that the sum of the digits added to the product of the digits gives the number

16 (C) You ask your students to do the following: Select any three digit number. Form all possible two digit numbers that can be formed from the digits of this three digit number. Sum these two digit numbers and divide the result by the sum of the digits of the original number. What do you get? Now start over with a different three digit number. What result do you get now? Do you notice anything? Start again with another 3 digit number. Explain your results algebraically.

17 (C) Here is an interesting mind reading trick that you can do with your students that is based on place value representation of numbers. Have each of your students think of a card. An Ace is worth 1 and deuce 2, and so on. A Jack is worth 11, a Queen 12, and a King 13. A club is worth 5, a Diamond 6, a Spade 7, and a Heart 8. Let each take the numerical value of his or her card, (for example, Jack is numerically worth 11) and add the next consecutive number.

(So, in the case of a Jack, the next consecutive number would be 12.) Then have everyone multiply the sum by 5 and add the suit value. (Thus, if a student thought of a Club, his suit value would be 5.) Ask one of your students to tell you the result he or she got. Subtract 5 from whatever the student tells you. The last digit of the remaining number is the suit value, and the first digit (or the first two digits in the case of a 3 digit number) gives you the card the student chose. Thus, if you end up with 127, the student chose the Queen of Spades. Call on several other students and demonstrate repeatedly that you can "read their minds." Ask your students to figure out how you do it. Ask one of them to play the trick on the class and then explain how it works. How does the trick work?

18 (C) Here is a mind-boggling trick you can play with your students. Ask a particular student to think of the price of an item he recently bought that cost more than 10 dollars. Ask that student to write this price in large numbers on an index card. Then while you turn your back to the class and cover your eyes, ask that student to show the price written on the card to the class without your seeing it. Then ask the class to do the following computations: Write down the price without the decimal. (Thus 16.95 is written as 1695.) Take the first two digits of the number (in this case 16) and add the next consecutive number (in this case 17). Take the sum of these two numbers (in this case 33) and multiply by 5 (here, 165.) Tell the students to place a zero to the right of the number to make a four digit number. (Here, 1650.) Now use the particular student you are working with to pick any number between 10 and 99, and have that student tell the class the number out loud. Have everyone in the class add it to the last number they got. (So, if the original number was 35, they would now all have 1685.) Finally, ask the students to add the number formed by the last two digits of the original number they wrote down (in this case 95) to the number they now have (1685) and tell you the result (1780). To find the student's original number, subtract 50 plus the number he told you (35). The number you get (in this case 1695) tells you the price the student paid for the item. Ask your students to figure out what you did. Explain why this works.

2.3 The Divisibility Rules

**LAUNCH**

What is the smallest positive integer composed of only even digits that is divisible by 9? Justify your answer.

Did you spend a great deal of time trying to come up with an answer to the launch question? How many integers did you try? Did you use any divisibility rules to help you arrive at your solution? What were they? If you have not solved this problem as yet, continue reading the chapter and return to it later, after you have more "tools" to work with regarding divisibility.

Most middle school and secondary school students know the rule that, if a number is divisible by 2, then its last digit is divisible by 2. The divisibility rules for other numbers are less well known

26 Basics of Number Theory

to middle school students, and their proofs are not given. But, as a teacher, you should be aware of the proofs so that they are not a mystery to you and you can make sense of them to your students.

There are several divisibility rules that we present here and, although we will prove these only for three digit numbers, the proofs extend to numbers with any number of digits. Remember, we are only giving the proofs for three digit numbers now.

1. Divisibility by 2: *If the final digit of a number, N, is divisible by 2, then so is N divisible by 2. Conversely, if the number, N, is divisible by 2, so is its final digit.*

Proof of 1: (The first part.) Let N be a 3 digit number and suppose that its final digit is divisible by 2. Then N can be written as $100h + 10t + u$. Clearly, $100h + 10t$ is divisible by 2 since we can factor out a 2. So we have

$$N = \underbrace{100h + 10t}_{\text{divisible by 2}} + \underbrace{u}_{\text{divisible by 2 by assumption}}.$$

Now N is the sum of two numbers divisible by 2. So N must be divisible by 2 by Theorem 2.7.

To prove the converse, rewrite $N = 100h + 10t + u$ as $N - (100h + 10t) = u$. Now we are assuming that N is divisible by 2 and, since $100h + 10t$ is also clearly divisible by 2, their difference, u, is divisible by 2 by Theorem 2.7 with $a = 2$.

2. Divisibility by 3: *If the sum of the digits of a number, N, is divisible by 3, then the number, N, is also divisible by 3. Conversely, if the number is divisible by 3, so is the sum of the digits.*

To illustrate, let us investigate this property with the number 231. We sum the digits to get $2 + 3 + 1 = 6$. Since 6 is divisible by 3, we know that 231 is as well. And we can verify this since $231/3 = 77$. Conversely, if we take the number 69, which we know is divisible by 3, we see that the sum of the digits, $6 + 9 = 15$ is also divisible by 3. So if we want to determine if a large number such as 1235492 is divisible by 3, we just sum the digits: $1 + 2 + 3 + 5 + 4 + 9 + 2 = 26$. Since 26 is not divisible by 3, the original number is not divisible by 3.

Proof of 2. Let the number be $N = 100h + 10t + u$. We are going to show that, if the sum of the digits, $h + t + u$, is divisible by 3, then the number N is. Rewrite N as $N = 99h + h + 9t + t + u$ or just as

$$N = \underbrace{(99h + 9t)}_{\text{divisible by 3}} + \underbrace{(h + t + u)}_{\text{divisible by 3 by assumption}}. \tag{2.5}$$

The expression in the first parentheses of equation (2.5) is divisible by 3, since we can write it as $3(33h + 3t)$. Furthermore, we are assuming the expression in the second parentheses, the sum of the digits, is also divisible by 3. So N, being the sum of two parenthetical expressions each divisible by 3, is divisible by 3. (Theorem 2.7.)

To prove the converse we just rewrite $N = (99h + 9t) + (h + t + u)$ as

$$\underbrace{N}_{\text{assuming is divisible by 3}} - \underbrace{(99h + 9t)}_{\text{is divisible by 3}} = (h + t + u)$$

and then argue that, since we are assuming that N is divisible by 3, and since clearly $(99h + 9t)$ is divisible by 3, the difference, which is $(h + t + u)$, is divisible by 3. Of course, $h + t + u$ is the sum of the digits. Thus, if N is divisible by 3, so is the sum of the digits.

3. Divisibility by 4: *A number, N, is divisible by* 4 *if the 2 digit number formed by the tens digit and units digit is divisible by* 4 *and conversely.*

Let us illustrate. To see if the number 1235492 is divisible by 4, you look at the number formed by the last two digits, 92. Since 92 is divisible by 4, so is the number 1235492.

Proof of 3: We prove if the two digit number formed by the tens digit and units digit of a number is divisible by 4, then so is the number. We prove this for a 4 digit number whose thousands digit is a, whose hundreds digit is b, whose tens digit is c, and whose units digit is d. Then

$$N = 1000a + 100b + 10c + d = \underbrace{(1000a + 100b)}_{\text{divisible by 4}} + \underbrace{(10c + d)}_{\text{divisible by 4 by assumption}}.$$

The number in the first parentheses is divisible by 4 automatically since we can factor out 4 from it, and the number in the second parentheses is the number formed from the last two digits of N which we are assuming is divisible by 4. We have written N as the sum of two parenthetical phrases, each of which is divisible by 4. So N is also divisible by 4 by Theorem 2.7.

The converse is left as an exercise.

4. Divisibility by 5: *A number, N, is divisible by* 5 *if the final digit is divisible by* 5 *and conversely.*

The proof is similar to the proof of the rule for divisibility by 2 and we leave it to the reader.

5. Divisibility by 6: *A number, N, is divisible by* 6 *if it is divisible by both* 2 *and* 3.

Proof of 5: If the number is divisible by 2, then when we factor it, it has a factor of 2. Similarly, since it is divisible by 3, when we factor, it will have a factor of 3. Thus, when we factor the number, it will have a factor of 2 and a factor of 3. Thus $N = 2 \cdot 3 \cdot k$. This tells us $N = 6k$ and N is divisible by 6.

The test for divisibility by 7 is complicated and not used much, so we omit it.

6. Divisibility by 8: *A number, N, is divisible by* 8 *if the number formed by the last* 3 *digits is divisible by eight, and conversely.*

Proof of 6: The proof is similar to the proof of divisibility by 4. We leave it as a Student Learning Opportunity.

As an illustration, 12345678 is not divisible by 8 because the number formed by the last three digits, 678 is not divisible by 8.

7. Divisibility by 9: *A number is divisible by* 9 *if the sum of the digits is divisible by* 9 *and conversely.*

Proof: The proof is virtually identical to the proof of divisibility by 3 and your students will very likely be curious about why this is true. It is left as a Student Learning Opportunity at the end of this chapter.

8. Divisibility by 11: *A number, N is divisible by* 11 *if the sum of the digits in the odd positions minus the sum of the digits in the even positions is divisible by* 11.

To illustrate, the number 12345674 is divisible by 11 since $(1 + 3 + 5 + 7) - (2 + 4 + 6 + 4) = 0$ which is divisible by 11. So the original number is divisible by 11. Use your calculator and check it!

28 Basics of Number Theory

There are some fascinating examples based on divisibility by 9 which we will show now.

> **Example 2.12** *A number, c, consists of N 1's and only N 1's. What is the smallest number, c, like that, that will be divisible by 9?*

Solution. This is easier than you might think. To be divisible by 9, the sum of the digits must be divisible by 9. Since N consists only of 1's, we will need 9 ones. Thus our number is 111111111.

> **Example 2.13** *Try this next trick with a friend or with your students. Tell them to do the following: Take a number and scramble the digits. Subtract the smaller number from the larger number and obtain the result. Now cross out any NONZERO digit in the result and sum the remaining digits. If they tell you the sum of the digits they got, you will be able to tell them the digit they crossed out. For example, if they told you the sum was 7, you could tell them the digit they crossed out was 2. If they told you the sum of the digits was 12, you could tell them the digit they crossed out was 6. How do you do it?*

Solution. The solution is based on Example 2.11. In that example we said that, if you take any 3 digit number and scramble the digits, the difference between the larger and the smaller number will be divisible by 9. Although we only showed it for a 3 digit number, it is true for any size number and the proof is essentially the same. Knowing that the difference of the number and the scrambled number is divisible by 9 means that the sum of the digits of the resulting number must be divisible by 9. So if you crossed out a nonzero digit and the sum of the remaining digits is 12, then what you crossed out has to be a 6, since the sum of all the digits in the remaining number had to be a multiple of 9, and 18 is the next multiple of 9. Similarly, if the sum of the digits was 7, then since the next multiple of 9 closest to 7 is 9, you must have crossed out a 2. When you try this on your friends, they will be amazed with your powers.

We have talked about divisibility, and while these results seem to just be theoretical, that is hardly the case. We now talk about a practical application of some of the things we have been doing. When you go to the supermarket, you notice that each item you buy has its own UPC label. A typical label looks something like that in Figure 2.2.

Figure 2.2

This label identifies the item. The first 6 digits of this code represent the manufacturer and the next 6 digits describe the item. Each manufacturer has its own code. The label is read by a scanner, which then identifies the manufacturer and the item and then finds the price of the item. The label we have shown here is the label from a small box of Dole Raisins. A typical UPC label has 12 digits as you see in the picture, counting the 0 in the beginning and the 9 at the end.

Now suppose that the scanner at the checkout counter scanned the label incorrectly and reports that you bought a box of detergent instead of a box of raisins. So, rather than being charged say 20 cents for the box, you are charged $2.50. This would not be a good thing. So, each UPC code has what is known as a check digit to alert us if it made a mistake so that the item should be re-scanned. The check digit is always the last digit. In this case it is the last digit 9.

What the scanner does is add all the digits in the odd numbered places to get $0 + 5 + 0 + 0 + 2 + 4 = 11$ and multiplies this sum by 3 to get 33. It adds to this all the numbers in the even numbered positions, but ignores the check digit. That is, it adds to this $7 + 7 + 0 + 3 + 1 = 18$. So, the total so far is $33 + 18 = 51$. The check digit is always chosen so that, when added to the total, the resulting sum is divisible by 10. Of course, $51 + 9 = 60$ and that is divisible by 10. If, when the machine adds the check digit to its reading, it doesn't come out with a multiple of 10, it alerts the checker that the item needs to be re-scanned.

Sometimes a machine cannot scan a label and the checker has to key in by hand what he or she sees as the UPC code. Of course, he or she can make a mistake. The most common mistakes are keying in a wrong digit, or switching two adjacent digits, say by typing in 57 instead of say, 75. This system will always catch a mistake if one digit is entered incorrectly, and will, in most cases, find an error if two adjacent numbers are switched.

Why will this system catch an error if a single digit is keyed in wrong? Well, if an odd digit is incorrect, say the 5 is keyed in as 8 (or read by the scanner as 8), then when it is multiplied by 3 the new sum (counting the check digit) differs from the correct sum by a multiple of 3 (in this case 3 times 3). And no single digit multiple of 3 can bring us back up to the next multiple of 10. Similarly, if the error is made in an even position, the new sum (including the check digit) differs from the true sum by a single digit, and thus cannot be divisible by 10. So, in summary, this system will allow us to detect many errors.

Unfortunately, this system will not allow us to correct *all* errors. For example, if the first two digits in the UPC code were "16" instead of "05" and the digits were read as "61" instead, then we would have an error and our check digit, 9, at the end would still work, giving us a multiple of 10, as you should verify. So, what we are saying is, you may actually be charged for detergent as the machine may not detect the error. Our advice: (a) watch as your items are scanned, and (b) be grateful that we have a mechanism that will correct many errors.

In a similar manner, zip codes have a built-in error detection scheme based on divisibility by 10, but postal money orders have a check digit scheme based on divisibility by 7. Even Avis Rent-A-Car uses a divisibility by 7 scheme to identify rental cars. You can find out more about this on the Internet.

As we have said, the methods that we have discussed for detecting errors in zip codes, UPC bar codes, car rental codes, UPS codes, and so on, are not foolproof. There are much more sophisticated methods used, depending on the application, that can not only detect errors, but correct errors. Such technology is used in CD players in their anti-skip capability. Thus, the laser in the CD may at first skip a note, but it is detected and immediately corrected

Student Learning Opportunities

1 Factor the following numbers completely by using the divisibility rules from this section. Explain how you used the different rules.

(a) 111
(b) 297
(c) 255
(d) 18,144

30 Basics of Number Theory

2 Without converting to decimals, what is the least positive integer x for which $\dfrac{1}{640} = \dfrac{x}{10^y}$ given that y is a positive integer also.

3 (C) A student is curious about the test for divisibility by 9 and asks you to prove it for any 3 digit number. How do you do it?

4 (C) A student claims that if the final digit of any number N is divisible by 5, then so is N. How can you prove this is so?

5 If the number 412 is added to 3b2 and the result is divisible by 9, tell what the value of b is.

6 (C) A student claims to have made a discovery that, if you take any two odd numbers, m and n, then the difference of their squares is divisible by 8. She shows the example $9^2 - 5^2 = 56$ which is divisible by 8 and claims it is always true. Is she correct? Prove or disprove this.

7 Investigate on the Internet the test for divisibility by 7. Explain what makes it complex.

8 Prove the test for divisibility by 8.

9 State a test for divisibility by 10. Prove it works.

10 What is the smallest positive integer composed of only even digits that is divisible by 9? Justify your answer.

11 Show that, if a number, N, is divisible by a and b, and a and b have no common factor other than 1, then N is divisible by ab.

12 Suppose that x and y are integers and that $2x + 3y$ is a multiple of 17. Show that $9x + 5y$ is also a multiple of 17. [Hint: Start with $17x + 17y$.]

13 How many numbers less than 1000 are divisible by either 5 or 7? Justify your answer.

14 Are there single digit values for a and b that make the number 4324a5b4 divisible by both 4 and 9? If so, what are they? If not, why not?

15 The UPC codes for several items are given below. Find the check digits which have been replaced by question marks.
  (a) Wise Potato Chips 20 ounces size: 04126228563?
  (b) Wesson Canola Oil 48 ounce size: 02700069048?
  (c) Del Monte Fruit Cocktail in light syrup, 15 ounce size: 02400016707?

16 In each of the following UPC codes, a digit is missing from the full UPC code and is replaced by a question mark. As usual, the last digit is the check digit. Find the missing digit.
  (a) Silk Soy Milk, Chocolate. 8.25 ounce size: 02529?600850
  (b) Diamond Crystal Salt 16 ounces size: 0136?0000100
  (c) Bon Ami Cleanser 14 ounces size: 01?500044151

17 (C) After playing around with her calculator, a student notices that the following numbers are all divisible by 11: (a) 123,123; (b) 742,742; (c) 685,685. She is convinced that the number abc, abc where a, b, c are single digit natural numbers will always be divisible by 11. Prove or disprove her conjecture. (Here abc does not mean the product of a, b, and c, rather the digits in the representation of the number.)

18 Let $a_2 = 1001$, $a_3 = 1001001$, $a_4 = 1001001001$, and so on where the subscript of a represents the number of 1's. Show that every single a_n is factorable if n is divisible by 3, or if n is even. (In fact, regardless of what n is, a_n is factorable, but this is harder to prove.)

2.4 Facts about Prime Numbers

The number $N = 49{,}725$ represents the ages of a group of teenagers multiplied together. How many teenagers are there and what are their ages? Explain how you got your answer.

We hope that you learned a lot about factoring integers as you engaged in trying to solve the launch problem. How did you figure out the number of teenagers that were needed? Did your result include any prime numbers? During your solution process, we hope you developed some intuitive ideas about some properties of prime numbers that you are curious to learn more about. We will explore some of these properties in this section.

Essential to this topic is the concept of prime. We say that an integer, N, *greater than 1*, is **prime**, if the only way to factor N with positive factors is $1 \cdot N$. For example, since the only way to factor 2 is $1 \cdot 2$, 2 is a prime. (We consider the factorization $2 \cdot 1$, where the factors are the same but the order is different to be the same factorization.) Similarly, 3 is a prime. Notice that a prime number is a number *greater than 1*. The reason for this is somewhat technical. It makes the statements and proofs of our theorems much simpler and avoids having to make many qualifying statements.

An integer greater than one is called **composite** if it is not prime. What that means is that the integer can be factored into two or more smaller primes. Thus 9 is composite because it can be factored into $3 \cdot 3$. Similarly, 14 is composite because it can be factored into $2 \cdot 7$. In finding the greatest common divisor of a set of numbers, we often have to factor the numbers completely down to primes. We will talk more about this later in the chapter. For example, $36 = 4 \cdot 9 = 2 \cdot 2 \cdot 3 \cdot 3$. While it certainly seems obvious that every composite number can be factored into primes, we really need to be sure. The next theorem tells us that is true and essentially follows the calculations that we did above.

Theorem 2.14 *Every composite number N can be factored into primes.*

Proof. If N is composite, then it can be factored into two smaller numbers, a and b. If both are prime, then we are done. If not, then each composite factor can be factored into smaller numbers. If these smaller numbers are all primes, we are done. If not, each composite factor can be factored further into smaller numbers. The key word in this proof is "smaller." We cannot continue to factor indefinitely, since each time we factor we get smaller factors, and *there are only a finite number of* *smaller whole numbers less than N*. Thus the process must end, and when it does, it does so because we can find no smaller factors. At that point, each remaining number in the factorization of N is prime. ∎

> **Corollary 2.15** *Every integer $N > 1$ is either prime, or can be factored into primes.*

Proof. Either the number is prime, or it is composite. If it is prime, we are done. If it is composite, it can be factored into primes by the theorem. ∎

This theorem seems pretty mundane. "So what?" you may think. But, it is the fact that every number can be factored into primes, and that this can be a very difficult thing to do when the number is large, that has major applications. In fact, it is this fact that is the basis of our national security. Many of our country's secrets are encrypted (as are your credit card numbers when you order online) using a scheme that can only be broken if the prime factors of certain large numbers are found. The problem is, these numbers are huge (consisting of several hundred digits) and finding the prime factors, even with our super computers, can take decades. So, for now, or until someone finds a fast way of factoring numbers into primes, we are safe. This encryption scheme is an interesting application of prime numbers and we will have more to say about it later in the chapter.

The next theorem is one we will also use.

> **Theorem 2.16** *If a PRIME number p divides a product ab, then the prime number p must divide a or b.*

You might be thinking this result is obvious. We just factor a and b into primes, and if p divides this product of primes, it must be one of them. The issue really is that there may be more than one way to factor a number into primes, and one of these ways may not involve the prime p. This is a subtle issue, and once we resolve it at the end of the section, we will give the proof of this theorem.

Notice the word "prime" in the theorem. The result is not true if the word "prime" is omitted. For example, 18 can be factored into $2 \cdot 9$, and the composite number 6 divides $2 \cdot 9$. But the composite 6, does not divide either 2 or 9.

Theorem 2.16 is often used in proofs. So, for example, if we know that 3 divides some number $(p^2 + 1)(q - 2)$ and we know it does not divide the first number $p^2 + 1$, then it must divide the second number, $q - 2$, since 3 is prime. We will use this idea later on in the book in the proof of the rational root theorem and its applications. [See Chapter 3, Section 5.]

If we start listing the primes numbers in order, we have: 2, 3, 5, 7, 11, 13, 17, 19, and so on and it appears that there are no particularly large gaps (differences) between consecutive primes. For example, 2 and 3 differ by 1; 3 and 5 differ by two, as do 5 and 7; 7 and 11 differ by 4. It is natural to ask the question, how large can the gap between consecutive primes get? Can there be a gap of at least 10,000 between consecutive primes? Put another way, can we find 10,000 consecutive integers which are composite, or must a prime occur somewhere in this list of 10,000 consecutive numbers? The answer, is, surprisingly, that we CAN find 10,000 consecutive numbers which are composite. In fact, we can even show them to you. They are $(10,001)! + 2, (10,001)! + 3, (10,001)! + 4, \ldots, (10,001)! + 10,001$. (Recall that $10,001!$ is the product of all the integers from 1 to 10,001.) The key to proving that these are all composite is that *$10,001!$ is divisible by each of the numbers*

2,3,4, and so on up to 10,001. Thus, the first number, (10,001)! + 2, is the sum of two numbers each of which is divisible by 2 hence is divisible by 2. The second number in the set, (10,001)! + 3, is the sum of two numbers, each of which is divisible by 3, and so it is divisible by three. Similarly, the next number in the set is the sum of two numbers, each of which is divisible by 4. Hence it is divisible by 4. Continuing in this manner, we see that each of the 10,000 numbers in this set is composite. There is nothing special about 10,000 here. In fact, we have the following:

Theorem 2.17 *If N is any positive integer, we can find a string of N consecutive composite numbers.*

Proof. Consider the N numbers, $(N+1)! + 2$, $(N+1)! + 3$, $(N+1)! + 4$, ... $(N+1)! + (N+1)$. Realizing that $(N+1)!$ is divisible by all integers from 2 to $N+1$ inclusive, and using the same kind of argument as above, we see that each is composite. That is, the first is composite because it is the sum of 2 numbers divisible by 2. The second is composite because it is the sum of two numbers divisible by three, and so on. ■

Thus, we can find a million, or a billion or even a trillion consecutive numbers in a row with no primes in sight. This seems to indicate the primes may be becoming scarcer and scarcer, and it might be that there are only a finite number of primes. And, even if there were infinitely many primes, how on earth would one go about proving it? Well, there are infinitely many primes, as we know, and Euclid proved it in the following way. This proof certainly counts as one of the most efficient, ingenious, and elegant proofs in all of mathematics. We should all see it.

Theorem 2.18 *There are infinitely many primes.*

Proof. Using proof by contradiction, suppose it is not the case that there are infinitely many primes. Then there would be a finite number of primes which we can call $p_1, p_2, p_3 \ldots p_L$ where p_L represents the last prime. Now, form the following number:

$$N = p_1 p_2 \ldots p_L + 1. \tag{2.6}$$

By corollary 2.15, this number N is either prime or can be factored into primes, and in this latter case would have a prime factor, p. N can't be prime because it is bigger than p_L and p_L was the largest prime. So N must be factorable into primes and have a prime factor of p. But p must be one of the primes occurring in the product $p_1 p_2 \ldots p_L$ since this is supposedly the list of *all* primes. Thus $p_1 p_2 \ldots p_L$ is divisible by p.

Now, since N is divisible by p and since $p_1 p_2 \ldots p_L$ is also divisible by p, their difference, $N - p_1 p_2 \ldots p_L$, is divisible by p. But their difference is 1, by equation (2.6). Thus 1 is divisible by p. How can this be, since the prime p is bigger than 1?

Our assumption that there were finitely many primes led us to the contradiction that p must divide 1. Thus our original assumption that there was a finite number of primes was false and there must be infinitely many primes. ■

Are you smiling? You should be. You have to admit, this is one beautiful proof!

We now return to a proof of Theorem 2.16 that we promised. But first we have to prove something related. This is a "structural theorem." Our goal is to show that, when we factor a number into primes, the factorization is unique.

34 Basics of Number Theory

> **Lemma 2.19** *If there is a smallest number, N, that can be factored into primes in two different ways, then any primes in one factorization of N will not occur in the other factorization of N.*

Proof. We give a proof by contradiction. Remember that we are letting N represent the *smallest* number that can be factored into primes in two different ways. Suppose that two **different** ways we may factor N are $N = p_1 p_2 \ldots p_n$ and $N = q_1 q_2 \ldots q_k$ and suppose that these two factorizations of N have a prime factor, say p_1 in common. Then we can rearrange the primes in the factorizations of N so that p_1, comes first. That is, we can assume that $p_1 = q_1$. Thus, $N = p_1 p_2 \ldots p_n$ and $N = p_1 q_2 \ldots q_k$. Divide each of these equations by p_1. This yields $\frac{N}{p_1} = p_2 \ldots p_n$ and $\frac{N}{p_1} = q_2 \ldots q_k$. What these two equations say is that the number $\frac{N}{p_1}$ can be factored in *two different ways*, $p_2 \ldots p_n$ and $q_2 \ldots q_k$. But this number is smaller than N, and this contradicts the fact that N was the *smallest* number that could be factored in different ways. This contradiction which arose from assuming there were two different factorizations of N with a common prime factor, shows that, if there is a smallest number that can be factored into primes in two different ways, they cannot have a common factor. ∎

> **Theorem 2.20** *Any natural number greater than 1 can be factored into primes in only one way.*

Proof. Again, using proof by contradiction, suppose it is not the case. Then there is some natural number that cannot be factored in only one way. Hence, there must be a smallest natural number that cannot be factored into primes in only one way. Call it N. Then by the previous lemma, N has two different factorizations: $N = p_1 p_2 \ldots p_n$ and $N = q_1 q_2 \ldots q_k$ and none of the p's and q's are the same. Thus $p_1 \neq q_1$. Suppose $p_1 < q_1$. (If the reverse is true we do a similar argument.) Our plan is to construct a number P smaller than N with two different factorizations, and this will contradict the fact that N is the smallest such number. Here is our candidate for P:

$$P = (q_1 - p_1) q_2 \ldots q_k. \tag{2.7}$$

We first observe that $(q_1 - p_1) < q_1$. We now multiply both sides of this inequality by $q_2 \ldots q_k$, to get

$$(q_1 - p_1) q_2 \ldots q_k < q_1 q_2 \ldots q_k. \tag{2.8}$$

But the left side of inequality (2.8), is P and the right side is N. Thus

$$P < N.$$

We have shown that $P < N$. Now we will show that P has two different factorizations.

The first factorization of P is obtained from equation (2.7). The q's are all primes and none of them are p_1, but $q_1 - p_1$ may not be prime and may have p_1 as a factor. Let us see what happens if $q_1 - p_1$ has a factor of p_1. If it does,

$$q_1 - p_1 = k p_1$$

for some k, and solving for q_1 we get

$$q_1 = k p_1 + p_1 = p_1(k + 1).$$

This says that q_1 is a multiple of the prime p_1. But this cannot be since q_1 is a prime and has no positive factors other than 1 and itself. Thus,

the factorization of P given in equation (2.7) does not contain p_1. (2.9)

Now we return to find another factorization of P that DOES contain a factor of p_1. This coupled with (2.9) will provide us with the two factorizations of P and will give us the contradiction we seek. We start with equation (2.7):

$$
\begin{aligned}
P &= (q_1 - p_1)q_2 \ldots q_k \\
&= q_1 q_2 \ldots q_k - p_1 q_2 \ldots q_k \quad \text{(Distributive Law)} \\
&= N - p_1 q_2 \ldots q_k \quad \text{(Since } q_1 q_2 \ldots q_k \text{ is one of the ways of factoring N)} \\
&= p_1 p_2 \ldots p_n - p_1 q_2 \ldots q_k \quad \text{(Since } p_1 p_2 \ldots p_n \text{ is another way of factoring } N\text{)} \\
&= p_1(p_2 \ldots p_n - q_2 \ldots q_k) \quad \text{(Factoring out } p_1\text{)}.
\end{aligned}
$$
(2.10)

This last factorization provides us with another factorization of P which DOES contain the factor p_1.

So, let us summarize. We took the smallest number N that did not have a unique factorization, and produced a smaller number P that did not have a unique factorization. One factorization of P, the one in equation (2.10) had a factor of p_1 in it, the other, the one in equation (2.7) didn't as we showed and highlighted in (2.9). This contradicted the fact that N was the *smallest* number that did not have a unique factorization. This contradiction arose from assuming that there was some number that didn't have a unique factorization, and hence that there was a smallest one. Thus, this assumption was wrong, and this tells us that all natural numbers greater than 1 have a unique factorization into primes. ∎

We just want to make one last comment on this theorem. A prime, like 2 is already considered "factored into primes." Now let us give a rather unexpected consequence of this:

Example 2.21 *Show that* $\log_2 3$ *is irrational.*

Solution. We do this by contradiction. Suppose that $\log_2 3$ is rational and that $\log_2 3 = \frac{a}{b}$ where a and b are positive integers. Then $2^{\frac{a}{b}} = 3$. (See Chapter 6 for a review of logarithms.) Now raise both sides of the equation to the bth power to get $2^a = 3^b$. Call the common value of these two numbers, N. Thus $N = 2^a = 3^b$. Now what? Well, we are done! We have that the positive integer N has been written in two different ways as a product of primes. In the first factorization it is a product of 2's. In the second it is a product of 3's. This contradicts Theorem 2.20, so the assumption that $\log_2 3$ is rational cannot be true. Therefore, $\log_2 3$ is irrational.

Isn't this neat? It is so logical!

We can now give the proof of Theorem 2.16 that "If a prime p divides a product ab, then either p divides a or p divides b."

Proof. If p divides ab, then $pk = ab$ for some integer k by the definition of divisor of ab. Factor both sides of this equation into primes. Since there is only one way to factor a number into primes, and p occurs on the left side of the equation as a factor, p must also occur on the right side of the equation as a factor. That is, p had to arise as a factor of either a or b. And we are done.

When a and b are positive numbers that have no prime factors in common, we say that a and b are **relatively prime**. Thus $8 = 2^3$ and $27 = 3^3$ are relatively prime, since they have no common prime factor. The same is true of 18 and 35 since $18 = 2 \cdot 3^2$, and $35 = 5 \cdot 7$. When a rational number $\frac{a}{b}$ is in lowest terms, then a and b must be relatively prime.

One other useful result we will need is:

Theorem 2.22 *If a and b are relatively prime, and if a divides kb for some some integer k, then a must divide k.*

Proof. If a divides kb, then all prime factors of a divide kb. But since a and b have no prime factors in common, being relatively prime, all prime factors of a must divide k. And, if all the prime factors of a divide k, then k contains all prime factors of a and hence is a multiple of a. That is, a divides k. ∎

2.4.1 The Prime Number Theorem

Knowing that there are infinitely many primes, and that there can be very large gaps between a prime and the next prime, led mathematicians to wonder about the distribution of primes. How many primes roughly are there less than some number N? The mathematician Gauss, at the age of 14 (Yes, 14!!!!) studied this problem, and came up with a conjecture about the number of primes. He said, let $\pi(N)$ be the number of primes less than or equal to N. (Here $\pi(N)$ is a function of N, it has nothing to do with the number π. But this is pretty standard notation for this.) Since there are 4 primes less than or equal to 7, and they are, 2, 3, 5, and 7, $\pi(7) = 4$. Similarly, $\pi(13) = 6$, since there are six primes less than or equal to 13 and these are: 2, 3, 5, 7, 11, 13.

What Gauss conjectured was that the ratio $\frac{\pi(N)}{N}$, which represents the fraction of primes less than or equal to N, is roughly $\frac{1}{\ln N}$ when N is large, where $\ln N$ is the natural logarithm of N. Yes, $\frac{1}{\ln N}$!!!! Your first reaction might be disbelief. How does a 14 year old come up with the estimate $\frac{1}{\ln N}$ for LARGE N, when he doesn't even have a computer, and more so, why should the natural logarithm have anything to do with prime numbers? It is just amazing!

Let us examine the ratios of $\frac{\pi(N)}{N}$ and $\frac{1}{\ln N}$ for some specific values. The values were obtained by computer. When N is 1 million, $\frac{\pi(N)}{N}$ is 0.0784 and $\frac{1}{\ln N} = 0.0723$. When N is 10 million, $\frac{\pi(N)}{N} = 0.0664$ and $\frac{1}{\ln N} = 0.0620$. (So, approximately 6% of the numbers less than or equal to 10 million are prime.) When N is 100 million, $\frac{\pi(N)}{N} = 0.0576$ and $\frac{1}{\ln N} = 0.0542$, and so on. Gauss seemed to be on the right track. But Gauss was a genius (as you might have guessed). Who among us could have ever guessed that rule? Of course, for Gauss, this is just a conjecture. He did not prove this conjecture was true. It took another 100 years for his conjecture to be proven and the proof required some very sophisticated mathematics.

Student Learning Opportunities

1 What is the largest prime factor of 14,300,000? Justify your answer.

2 (C) A student asks you to explain why the only even prime number is 2. Show how you could prove it by contradiction.

3 **(C)** A student asks whether 1 is prime or composite? How would you explain the answer to this question?

4 The numbers 2 and 3 are consecutive integers, which are both prime. Show that no other pair of consecutive integers is prime.

5 If $n = 2^3$, then there are only 4 factors of n and they are $2^0, 2^1, 2^2$, and 2^3. Similarly if $n = 2^3 3^4$, the factors are of the form $2^a 3^b$ where $0 \leq a \leq 3$ and $0 \leq b \leq 4$. Thus there are 20 factors of n. (Why?) Show that, if a number N is factored into primes, say $N = p_1^{n_1} \cdot p_2^{n_2} \cdot \ldots \cdot p_k^{n_k}$, then the number of factors of N is $(n_1 + 1)(n_2 + 1) \ldots (n_k + 1)$.

6 **(C)** A student asks what is the best way to show if a number N is prime. What do you say? At first, students think one has to try to divide N by all primes less than N and if none divide N evenly, then N is prime. But, a better way is to show by contradiction that, if $N = pq$, then one of p and q must be less than or equal to $\sqrt{N}$. Thus, when trying to determine if a number is prime, they need only check for prime divisors less than or equal to $\sqrt{N}$. Show how you would prove this result and help your students see how to use it to show that 143 is not prime and that 569 is prime.

7 Suppose x and y are both integers. Find a solution of $(2x + y)(5x + 3y) = 7$.

8 If $3^{x-1} 5^{y+2} = 7^z 11^t$, where the exponents are nonnegative integers, how do we know that there are no solutions other than $x = 1$, $y = -2$, $z = 0$, and $t = 0$?

9 Apply your knowledge of prime numbers to answer the launch question: The number $N = 49{,}725$ represents the ages of a group of teenagers multiplied together. How many teenagers are there and what are their ages? Explain.

10 How many distinct ordered pairs (x, y), where x and y are positive integers, are there that make $x^4 y^4 - 10 x^2 y^2 + 9 = 0$? Explain.

11 What is the largest prime factor of $29! + 30!$? Explain. [Hint: Factor out $29!$.]

12 Find a number n such that $n + 2, \ldots, n + 2007$ is composite. Justify your answer.

13 Show that, if n is composite, then it is not possible for $(n - 1)! + 1$ to be divisible by n. [Hint: By contradiction: If $(n - 1)! + 1 = kn$, where n is composite, then $(n - 1)! + 1 = kab$, where a and b are factors of n. Rewrite this as $kab - (n - 1)! = 1$. Since a and b are less than n, they each divide $(n - 1)!$. Finish it.]

14 Here is another proof that $\sqrt{2}$ is not rational: Write $\sqrt{2} = \frac{p}{q}$ where $\frac{p}{q}$ is in lowest terms. Then square and multiply both sides by q^2 to get $2q^2 = p^2$. Now p^2 and q^2 being squares, have each of their prime factors raised to even powers. But the extra two on the left side of $2q^2 = p^2$ means that 2 occurs to an odd power on the left side. The fact that 2 appears an odd number of times on the left and an even number of times on the right gives us our contradiction. So, $\sqrt{2}$ is irrational.

(a) Give a similar proof to show that $\sqrt{3}$ is irrational. Then show that $\sqrt{N}$ is irrational when N is an integer that is not a perfect square.

(b) If we tried to give a similar proof to show that $\sqrt{4}$ is irrational, where would the proof break down?

38 Basics of Number Theory

15 Show that $\log_5 7$ is irrational. Is $\log_b a$ irrational if a and b are primes with $a \neq b$? What if $a = b$?

16 (C) A student asks if there is an integer $N > 1$ such that the square root, cube root, and fourth root of N are all integers, and if so, what is the smallest one? How do you respond? Justify your answer and show how you could help him find such a value for N.

17 If we tried to show, as in Example 2.21, that $\log_2 8$ is irrational (it isn't!), where would the proof break down? What is $\log_2 8$ equal to?

18 Find $\pi(10)$. Then compute $\frac{\pi(10)}{10}$ and compare it to $\frac{1}{\ln 10}$ using your calculator.

2.5 The Division Algorithm

LAUNCH

A magician is in possession of a piece of paper on which there is written an integer. He tells you that this integer is being divided by the number 23 and, if you guess what the remainder is, you will win a trip to Las Vegas! He allows you 10 guesses to figure out the remainder. Do you have a good chance of winning the trip? Explain.

Chances are that, when you first read the launch problem, you thought maybe some information was missing. Hopefully, after some exploration and examination of various integer divisions, you began to get an inkling of some of the qualities of their quotients and remainders. Maybe you have even developed some intuitive ideas about the relationship between divisors and remainders in integer division. We will investigate this further now in our discussion of the division algorithm.

Suppose that we divide $N = 28$ by 4. It goes in 7 times, or put another way, the quotient is 7 and the remainder is 0. When 29 is divided by 4, the quotient is again 7, but the remainder is 1. When 30 is divided by 4, the quotient is again 7 and the remainder is 2. When 31 is divided by 4, the quotient is again 7 and the remainder is 3. Then everything begins to repeat. 32 divided by 4 gives a quotient of 8 and leaves a remainder of 0 and so on. As we increase N by 1 each time, the quotients get bigger and the remainders cycle, 0, 1, 2, 3, 0, 1, 2, 3. When we divide an integer by 4, there are only 4 possible remainders, and they are, 0, 1, 2 or 3. In a similar manner, when we divide a number by 5, there are 5 possible remainders, 0, 1, 2, 3, and 4. In general, if a number is divided by a positive integer b, there can only be b remainders, and they are 0, 1, 2, ... $b - 1$. We learned this in elementary school: When a positive number N is divided by a positive number b, there is a quotient q and a remainder of r. Furthermore, if we multiply the quotient by the divisor and add the remainder, we get N. That is, $N = bq + r$. We can get an intuitive picture of why this is true by looking on the real number line. There you see b, $2b$, $3b$, and so on. We can imagine the space between 0 and b to represent a segment of length b, and the space between b and $2b$ to represent a segment of length b and so on. (Each segment includes the left endpoint but not the right endpoint.) These are back to back and cover the whole number line. It follows that any number N is either an endpoint of one of these segments or lies inside one of these segments. What that is essentially saying is that every number N is either a multiple of b, or lies between two multiples of b. If the left part of that segment is the largest multiple of b less than or equal to N, that means that the *difference* between N and bq is some nonnegative integer, r, less than b. See Figure 2.3 below.

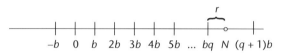

Figure 2.3

From the picture, we can see that $N = bq + r$. Of course, this diagram alone is not a proof of that fact but is probably convincing to most secondary school students. The real proof is not much different from this intuitive explanation. In fact, the picture we drew and our observations drive the proof. In it, we consider *differences* between N and multiples of b. Here is the real proof:

> **Theorem 2.23** (Division Algorithm) *If an integer N is divided by a positive integer b, then there is always some integer q_0 and some remainder r where $0 \leq r < b$ such that $N = bq_0 + r$. Furthermore, q_0 and r are unique.*

Proof. We give the proof of the case when N and b are both positive, as this makes things just a bit simpler. The theorem is still true when N is negative and b is positive.

So, suppose that N and b are both positive. Consider the set, S, of numbers of the form $N - bq$, where $q = 0, 1, 2, \ldots$. This set clearly has nonnegative integers since, for example, N is in it. (Just take $q = 0$.) Now every set of nonnegative integers has a smallest element. Let the smallest element of this set, S, occur when $q = q_0$ and call this element r. Thus

$$N - bq_0 = r. \tag{2.11}$$

Since r is the smallest nonnegative integer in this set by choice, $r \geq 0$.

We will show that r must be less than b. We will do this by showing that, if $r \geq b$, then we can find a smaller nonnegative member of S than that shown in equation 2.11, which will give us our contradiction. Suppose then that $r \geq b$ then $r - b \geq 0$. Consider $N - (q_0 + 1)b$, which is smaller than $N - q_0 b$. Here is the proof that $N - (q_0 + 1)b$ is nonnegative

$$N - (q_0 + 1)b$$
$$= (N - q_0 b) - b$$
$$= r - b \geq 0. \text{ (Since we are assuming } r \geq b.) \tag{2.12}$$

Since $N - (q_0 + 1)b$ is nonnegative, and hence a member of S, and since this number is smaller than the smallest element, $(N - q_0 b)$, of S, as we have shown, we have our contradiction. Since this contradiction arose from the assumption that $r \geq b$, it must follow that $r < b$.

To prove the uniqueness of q and r, suppose that

$$N = bq_0 + r_1 \tag{2.13}$$

and that

$$N = bq_1 + r_2. \qquad (2.14)$$

Our goal is to show that $q_0 = q_1$ and that $r_1 = r_2$. Subtracting equation (2.13) from equation (2.14), we get that $0 = b(q_1 - q_0) + r_2 - r_1$, which implies that

$$-b(q_1 - q_0) = r_2 - r_1. \qquad (2.15)$$

Taking the absolute values of both sides of equation (2.15) we get

$$b\,|(q_1 - q_0)| = |r_2 - r_1|. \qquad (2.16)$$

Now, the left side of equation (2.16) is a nonzero multiple of b, if $q_0 \neq q_1$, and thus must be greater than, or equal to, b. Since both r_1 and r_2 are between 0 and b, it follows that $|r_2 - r_1|$ is less than b, since this absolute value is the distance between the points. (See Figure 2.4 below.)

Figure 2.4

So, the left side of equation (2.16) is greater than, or equal to, b if $q_0 \neq q_1$ and the right side of equation (2.16) is less than b. This is impossible. So $q_0 = q_1$. Substituting this into equation (2.16), it follows that $|r_2 - r_1| = 0$, or that $r_1 = r_2$. ∎

Notice how much was involved in writing the proof of something that geometrically, using the number line, seemed obvious!

In elementary school, before students learn about rational numbers, they express all of their solutions to division problems in terms of quotients and remainders. Thus, when 16 is divided by 3, the quotient is 5 and the remainder is 1. When they advance to the study of rational numbers they suddenly relinquish all discussion of remainders and express the answer to a division problem, like 16 divided by 3 as $5\frac{1}{3}$. What this means in terms of the Division Algorithm is that, instead of $N = bq + r$, they would write instead, $\frac{N}{b} = q + \frac{r}{b}$. It is important that you as a teacher be aware of this extension from the integers to the rationals.

Theorem 2.23 is a fundamental result about division of integers and has widespread use both in and outside of mathematics. As just one example, when computers process data, every piece of data is changed into strings of digits consisting of 0's and 1's. Numbers are stored using their binary representation, which we will talk more about later. However, to get the binary representation of numbers, we need to use the division algorithm. Thus, numerical computations done on computers use the division algorithm in some implicit way! This is neat!

Let us illustrate this theorem.

Example 2.24 *Suppose we divide each of the numbers $N = 32$ and $M = -32$ by $b = 6$. Find q and r in each case.*

Solution. If we divide $N = 32$ by $b = 6$, we get a quotient, q, of 5 and a remainder, r, of 2. Notice that $N = bq + r$ and that r is between 0 and 6, as the theorem says it should be. When we divide $M = -32$ by 6, you may think that the quotient, q, is -5. But if q were, in fact, -5, then the only way $N = bq + r$ would be if $r = -2$, and that contradicts the fact that the remainder is between 0 and b. So, instead, we take the quotient to be -6, and then r would be 4, and now $M = -32 = 6(-6) + 4 = bq + r$ where r is between 0 and 6. This is consistent with the proof we gave. We always find the largest multiple of b less than or equal to N when using the formula $N = bq + r$, and in the case when $N = -32$, that largest multiple of 6 less than or equal to -32 is $6(-6)$. This is very surprising to students and, at first, seems quite strange.

> **Example 2.25** *Suppose that $N = 4q + 1$. Can we say that the remainder when N is divided by 4 is 1?*

Solution. Yes. According to the theorem, there is only one b and one r less than 4 that makes $N = bq + r$. Since $N = 4k + 1$ says that $b = 4$ and $r = 1$ "works" this must be our unique pair of numbers. So, the remainder when N is divided by 4 *must be* 1.

Student Learning Opportunities

1 Find the quotient and remainder when each of the following numbers is divided by 5 :
   (a) 17
   (b) -17
   (c) 33
   (d) -33

2 (C) Suppose that you wished to find the quotient and remainder when 17, 589 is divided by 834. Typically, the calculators students use in secondary schools express non-integer division results using decimals. Your students ask you how they could find the quotient and remainder using such calculators. What do you say?

3 If a natural number a is divided by a natural number b, the quotient is c and the remainder is d. When c is divided by b', the quotient is c' and the remainder is d'. What is the remainder when a is divided by bb'?

4 Use the division algorithm to show that any number N can be written as either $N = 3k$, $N = 3k + 1$, or $N = 3k + 2$. Use this to show that the product of any three consecutive integers must be divisible by 3. (In fact, it must be divisible by 6. Why?)

5 (C) One of your very insightful students checks the squares of several integers and notices that every time the square is divided by 3, it leaves a remainder of 0 or 1. Several other students corroborate this with other examples. How can you use the results of the previous Student Learning Opportunity to show that this is true? Is it true that, if the square of an integer is divided by 4, the remainder can only be 0 or 1? How do you know?

6 (C) A student asks whether there could be any integers that are neither odd or even. How would you prove to your student that every integer must either be odd or even? [Hint: When we defined an even number, we said it was of the form $2m$, and an odd number is of the form $2m + 1$. It is theoretically possible with this definition that a number is neither odd nor even. That is, there might be numbers that are not picked up by this definition. Using the division algorithm with divisor 2, show that every integer must be odd or even. As a result

of this, it follows that consecutive integers have "opposite parity." That is, if one is odd, the other is even.]

7 A pair of primes that differ by two is called a twin pair of primes. For example, the pair of numbers 3, 5 is a twin pair.

(a) Find two more twin pairs of primes.

(b) It is unknown if there are infinitely many twin primes. (That is certainly a hard problem to work with if you have the time and inclination.) Let us try a simpler problem. A set of 3 primes, for example, 3, 5, 7 is a prime triple if the differences between the first and second and the second and the third are both two. Using the following hint, show that the set of prime triples is finite: Call the primes in the prime triple, p, $p+2$, and $p+4$. When p is divided by 3 it leaves a remainder of 0, 1, or 2. That is, $p = 3k$, $3k+1$, or $3k+2$. If $p = 3k$, then p divisible by 3. Since the only prime divisible by 3 is 3, we get the triple, 3, 5, 7. Now, show that, if $p = 3k+1$, then $p+2$ is not prime, and that if $p = 3k+2$, $p+4$ is not prime. Thus there is only one prime triple.

(c) Using the same method as in part (a), show that the only prime number p such that p and $8p^2 + 1$ are prime is $p = 3$.

2.6 The Greatest Common Divisor (GCD) and the Euclidean Algorithm

**LAUNCH**

Find the greatest common divisor of 20 and 35. What method did you use to find the answer? Now find the greatest common divisor of 16, 807 and 14, 406. If you were able to find it, did you use the same method you used in the first problem? Why or why not?

We imagine that you probably had a lot of difficulty finding the greatest common divisor of the large numbers presented in the launch question. You might have used a "factoring tree" quite easily with the first pair of small numbers, but it is much more difficult to do with the pair of large numbers. The purpose of this section is to introduce you to an algorithm that will help you find the GCD rather easily and will give you other insights about numbers and their greatest common divisors.

One of the fundamental topics stressed throughout the middle and secondary school curriculum is the **greatest common divisor** or **greatest common factor** of two numbers a and b. We denote this by gcd (a, b). This is the largest number that divides both a and b. So, for example, gcd $(6, 8)$ is 2 and gcd $(10, 15) = 5$.

The greatest common divisor is not only useful in mathematics. It has found uses in developing secure codes that even the National Security agency can't break, and hence is useful for our own national security. It has been used in developing certain musical rhythms and also in neutron accelerators as well as in computer design and so on. [An interesting article detailing some of this is "The Euclidean Algorithm Generates Traditional Musical Rhythms", by Godfried Toussaint (2005).]

In the book *Number Theory in Science and Communication*, by M.R. Shroeder (1988), we find the following: "An interesting and most surprising application of the greatest common divisor occurs in human perception of pitch: the brain, when confronted with harmonically related frequencies, will perceive the GCD [greatest common divisor] of these frequencies as the pitch." (page 5)

When the greatest common divisor of two numbers a and b is 1, then a and b have no prime factors in common and so they are relatively prime. Thus, another way to define the expression "a and b are relatively prime" is to say that $\gcd(a, b) = 1$. So, 8 and 15 are relatively prime since $\gcd(8, 15) = 1$ as are the numbers 14 and 17 since $\gcd(14, 17) = 1$. Of course, $\gcd(a, b) = \gcd(b, a)$. We also observe that, if a is positive,

$\gcd(a, a) = a$

$\gcd(a, 1) = 1$ and

$\gcd(a, 0) = a$.

Make sure you can explain why each of these statements is true.

Finding the greatest common divisor of two numbers when the two numbers are factored into primes is simple. For each *common* prime in the factorizations, we take the lowest power of the prime we see and multiply the results. Thus, if we wanted to find the greatest common divisor of the numbers $M = 2^6 \cdot 3^9 \cdot 7$ and $N = 2^8 \cdot 3^4 \cdot 11$, we would notice that the common primes in the factorizations are 2 and 3, and that the lowest power of 2 we see is 2^6, while the lowest power of 3 we see is 3^4. Thus, $\gcd(M, N) = 2^6 \cdot 3^4$. We use this idea in algebra when we factor expressions. Thus, if we have $3a^3b^2$ and $6ab^3$ and we want to find the greatest common factor of these two expressions (which is synonymous with greatest common divisor), we treat the a and b as if they are primes. Thus, $\gcd(3a^3b^2, 6ab^3) = 3ab^2$, where we have factored out the smallest power of each common "prime." We tell our students that, in any factoring problem, we always factor out the greatest common divisor of the terms first. Thus, if we have to factor $3a^3b^2 + 6ab^3$, we factor out $3ab^2$ and we get $3ab^2(a^2 + 2b)$. In a similar manner, if we want to factor $x^4 - 9x^2$, we factor out the gcd first, which is x^2 and we get $x^2(x^2 - 9) = x^2(x - 3)(x + 3)$.

Finding the gcd of numbers by factoring into primes is easy when the numbers are small or are already in factored form. For example, if we wanted to find the $\gcd(24, 18)$ we get 6 very quickly. Imagine though trying to find the gcd of 4562 and 2460 or numbers much larger than these. Factoring would be cumbersome and time consuming. In practice, it is not done this way since in practical applications the numbers are usually extremely large, making it inefficient and difficult to factor even with the help of computers. Instead, there is a better method known as the Euclidean Algorithm to find the gcd of two numbers. It shows us how to transform the gcd of two numbers into the gcd of two numbers which are at best, smaller. Continued application of this yields the gcd of the two numbers.

Theorem 2.26 (*Version 1 of Euclidean Algorithm*) *If a and b are integers, then $\gcd(a, b) = \gcd(b, a - b)$.*

Proof. We will show that the set of divisors of a and b coincides with the set of divisors of b and $a - b$. Thus, the largest number which divides a and b is also the largest number that divides b and $a - b$. That is, $\gcd(a, b) = \gcd(b, a - b)$. For example, $\gcd(15, 6) = \gcd(6, 9) = 3$.

Now let h be any divisor of a and b. Then, since h is a divisor of a and b, it divides $a - b$ by Theorem 2.7 . **Thus any divisor of a and b is a divisor of b and $a - b$.**

Now we show the reverse. Suppose that h is any divisor of b and $a - b$. Then h certainly divides b, and by Theorem 2.7, h divides the sum $(a - b) + b$ or just a. That is, **any divisor of b and $a - b$ divides both a and b. Thus h is a divisor of a and b.**

We have shown in the bold statements that each divisor, h, of a and b is a divisor of b and $a - b$ and conversely. Thus, the divisors of a and b and b and $a - b$ are the same. So, the greatest common divisor of a and b is the greatest common divisor of b and $a - b$. ∎

This algorithm is very easy to program in a computer and is very quick and efficient. Here is an algebraic example of how to use this theorem.

Example 2.27 *Show that for any integer n, $\gcd(n + 1, n) = 1$. (This is Example 2.8 redone differently.)*

Solution. $\gcd(n + 1, n) = \gcd(n, n + 1 - n) = \gcd(n, 1) = 1$. Observe the factoring method is useless here. In a similar manner we can show that $\gcd(2n + 1, n) = 1$ for any integer n.

Here is another version of the Euclidean Algorithm which one sees more frequently.

Theorem 2.28 *(Euclidean Algorithm version 2). Suppose that $a > b$ and that, when we divide a by b, we get a remainder of r. Then $\gcd(a, b) = \gcd(b, r)$.*

Proof. The proof is almost the same as the proof of Theorem 2.26. We know that $a = bq + r$ for some q. Looking at the right side of this equation, we see that any divisor of b and r must divide the left side, a (Theorem 2.7). Thus, **the divisors of b and r are divisors of a (and b)**. From the relationship $a - bq = r$, we observe by looking at the left side of this equation that any divisor of a and b divides the right side, r and of course, b (again by Theorem 2.7). Thus, **the divisors of a and b are divisors of r and b**. Two bolded statements together show us that the divisors of a and b are the same as those of b and r.

We have shown that the divisors of a and b are the same as those of b and r. It follows that $\gcd(a, b) = \gcd(b, r)$. ∎

You might think of Theorem 2.28 as the "new and improved" version of Theorem 2.26. This new Algorithm is very efficient, and computers always employ it when finding $\gcd(a, b)$ by repeatedly applying this theorem. In fact, when books refer to "the" Euclidean Algorithm, they mean repeated application of version 2 of the Euclidean Algorithm, and we will refer to it likewise. Notice, we always find the greatest common divisor of the smaller number and the *remainder* when the larger number is divided by the smaller number. We repeat this over and over until we finally get to $\gcd(g, 0)$ at which point we know that g is the greatest common divisor of a and b. A way to express this mathematically, calling the remainders at each stage, r_1, r_2, and so on is, $\gcd(a, b) = \gcd(b, r_1) = \gcd(r_1, r_2) = \ldots \gcd(g, 0) = g$. To clarify how to use this version of the Euclidean Algorithm, let us revisit some examples we mentioned earlier.

Example 2.29 *Find (a) $\gcd(24, 18)$ and then find (b) $\gcd(4562, 2460)$.*

Solution. (a) 24 and 18 are both easy to factor. $24 = 2^3 \cdot 3$ and $18 = 2 \cdot 3^2$, so by our factoring method of taking the lowest power of each common factor and multiplying them together, we get gcd (24, 18) = 2 · 3 = 6. Had we done this using version 2 of the Euclidean Algorithm, the steps would have been gcd (24, 18) = gcd (18, 6) (since the remainder when we divide 24 by 18 is 6) and gcd (18, 6) = gcd (6, 0) since the remainder when 18 is divided by 6 is 0. But now we are done since we know that gcd (6, 0) = 6.

Let us show, in Figure 2.5, how this would look by long division.

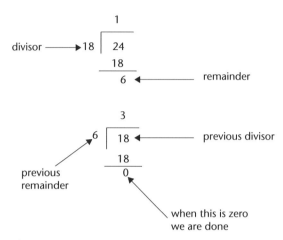

Figure 2.5

In our first step, we divide the larger number 24 by the smaller number 18. We look only at the remainder, 6. That becomes our new divisor, and we try to divide the most recent divisor, 18, by the remainder 6. The method is always, "Divide the most recent divisor by the remainder until you get a remainder of 0 in which case your latest divisor is the gcd."

(b) This is more difficult to do by the factoring method. Let us do this by version 2 of the Euclidean Algorithm: gcd (4562, 2460) = gcd (2460, 2102) = gcd(2102, 358) = gcd (358, 312) = gcd (312, 46) = gcd (46, 36) = gcd (36, 10) = gcd (10, 6) = gcd (6, 4) = gcd (4, 2) = gcd (2, 0) = 2.

There is a useful result that follows from Theorem 2.7, which is not obvious, and which we will use later on when we study Diophantine equations. We will also use its corollary, which gives us neater proofs of some theorems. They really are quite important in the study of number theory.

> **Theorem 2.30** *If g is the greatest common divisor of a and b, then g = ma + nb for some integers m and n.*

Proof. Consider the set of numbers of the form $ma + nb$ where m and n are integers. Then this set has some positive elements. (See if you can tell why.) Let P be the set consisting of positive numbers of the form $ma + nb$. Suppose s is the smallest element of this set. We claim that s is the gcd of a and b. Since s is in the set and must be of this form too,

$$s = m_0 a + n_0 b \tag{2.17}$$

for some m_0 and n_0. By Theorem 2.7, any divisor of a and b divides $m_0 a + n_0 b$ or just s. In particular, g, the greatest common divisor, divides s. This implies that

$$g \leq s. \tag{2.18}$$

46 Basics of Number Theory

We will now show that s divides both a and b and, as a divisor of both a and b, it will be less than or equal to g, the greatest common divisor of a and b. That is,

$$s \leq g. \tag{2.19}$$

From (2.18) and (2.19) it will follow that $s = g$. And, since $s = m_0 a + n_0 b$, it will follow that $g = m_0 a + n_0 b$ since $s = g$. Of course, this is what we wanted to show. So, now we proceed to show that s divides a. A similar proof will show that s divides b.

Proof that s divides a: Suppose s doesn't divide a. Then by the Division Algorithm,

$$a = sq + r \tag{2.20}$$

for some q where r is positive and

$$r \text{ is less than } s. \tag{2.21}$$

But $a = sq + r$ means $a = (m_0 a + n_0 b)q + r$ by equation (2.17) or upon simplifying, that

$$r = a(1 - m_0) + (-n_0 q)b. \tag{2.22}$$

Thus, r, being positive, and having the right form (a multiple of a added to a multiple of b) is in P, and r being less than s by (2.21), contradicts the fact that s was the smallest positive element in the set.

Our contradiction arose from assuming that s didn't divide a. Thus s divides a. Similarly, s divides b and therefore being a divisor of both a and b, $s \leq g$, the greatest common divisor. ∎

The next result will be used when we study Diophantine Equations later in the chapter.

> **Corollary 2.31** *If a and b are relatively prime, then there exist integers m and n such that $ma + nb = 1$.*

Proof. Take g in Theorem 2.30 to be 1. ∎

To give an example, if $a = 6$ and $b = 13$, we can write $(-2)6 + 1(13) = 1$. If $a = 23$ and $b = 5$, we have $2(23) - 9(5) = 1$.

Not only is the greatest common divisor useful in algebra and elsewhere, but so is the least common multiple (lcm). For example, students encounter this concept in elementary school when they wish to add fractions with unlike denominators. By the **least common multiple** of two positive integers N_1 and N_2, we mean the smallest positive integer that is a multiple of both N_1 and N_2. Thus, the least common multiple of 2 and 3 is 6 since 6 is the smallest positive integer which is a multiple of both of these.

When N_1 and N_2 are factored into primes, finding the least common multiple, of N_1 and N_2 is easy: We look at all primes that occur in the prime factorization of these numbers and take the *highest power* of each of these primes we see. Then we multiply them. So if $N_1 = 2^2 \cdot 3 \cdot 7$ and $N_2 = 3^2 \cdot 5 \cdot 11$, the least common multiple of N_1 and N_2 is $2^2 \cdot 3^2 \cdot 5 \cdot 7 \cdot 11$. We denote the least common multiple of N_1 and N_2 by lcm (N_1, N_2). Similarly, in algebra if we want to find the least common multiple of two algebraic expressions, we factor each completely, and considering each variable factor as a prime, we take the highest power of each "prime" we see. So, to find the lcm of $3a^3 b^2$ and $4ab^3 c$, we get $12a^3 b^3 c$.

In algebra we use the least common multiple when finding the least common denominator. Thus, if we want to add

$$\frac{4}{3x^5y^2z} + \frac{7}{6xy^3}$$

we would get the least common denominator first which is $18x^5y^3z$, and that is lcm $(3x^5y^2z, 6xy^3)$. Then we convert each fraction to an equivalent fraction with that denominator by multiplying the numerator and denominator of each fraction by an appropriate quantity to build the denominator to the lcm. Similarly, if we want to add

$$\frac{2x}{3(x-1)^3(x+2)} + \frac{4x-3}{4(x-1)^2(x+2)^4}$$

our common denominator would be $12(x-1)^3(x+2)^4$, which is the lcm of $3(x-1)^3(x+2)$ and $4(x-1)^2(x+2)^4$.

As we have mentioned, factoring large numbers is difficult, so one might expect that finding the lcm of two numbers requires a separate algorithm from the Euclidean Algorithm. Actually, that is not true. Once we find the gcd of two numbers, it is easy to find the lcm of the two numbers. We use this result:

Theorem 2.32 *If N_1 and N_2 are two natural numbers, then* lcm $(N_1, N_2) \cdot$ gcd $(N_1, N_2) = N_1 N_2$. *Thus to find* lcm (N_1, N_2), *we simply multiply $N_1 N_2$ and divide by* gcd (N_1, N_2), *which we can find by the Euclidean Algorithm.*

We illustrate this by an example.

Example 2.33 *Verify theorem (2.32) for (a) $N_1 = 24$ and $N_2 = 45$ and (b) for the algebraic expression $N_1 = 3ab^3$ and $N_2 = 4a^2b^2c$.*

Solution. (a) $N_1 = 2^3 \cdot 3$ and $N_2 = 3^2 \cdot 5$. Now gcd $(N_1, N_2) = 3$ and lcm $(N_1, N_2) = 2^3 3^2 5$. From this we have

gcd $(N_1, N_2) \cdot$ lcm $(N_1, N_2) = 2^3 3^3 5 = N_1 N_2$.

(b): We have gcd $(N_1, N_2) = ab^2$ and lcm $(N_1, N_2) = 12a^2b^3c$. Thus gcd $(N_1, N_2) \cdot$ lcm $(N_1, N_2) = (ab^2)(12a^2b^3c) = 12a^3b^5c = N_1 N_2$

We leave the proof of Theorem (2.32) to you as it is an instructive Student Learning Opportunity.

Student Learning Opportunities

1 Find gcd $(2^4 \cdot 5^6 \cdot 7^{44}, 2 \cdot 5^3 \cdot 7)$.

2 Find gcd $(3a^2b, 6ab^3)$.

48 Basics of Number Theory

3 Find
  (a) gcd (234, 342) and lcm (234, 342).
  (b) gcd (6156, 7255) and lcm (6156, 7255).
  (c) gcd (42650, 36540) and lcm (42650, 36540).

4 Find
  (a) gcd $(4ab^3, 12a^5bc)$ and lcm $(4ab^3, 12a^5bc)$.
  (b) gcd $(2x^2y, 3z)$ and lcm $(2x^2y, 3z)$.
  (c) gcd $(5(x+3)^2(x-1), 6(x+2)(x-1)^4)$ and lcm $(5(x+3)^2(x-1), 6(x+2)(x-1)^4)$.

5 (C) Your student says the greatest common divisor of two positive integers is greater than the least common multiple. Why do think the student is asking this question? How would you respond?

6 Show that gcd$(3n+1, n) = 1$ and, in fact, that gcd$(an+1, n) = 1$ when a is a positive integer.

7 Show that the greatest common divisor of $2n+13$ and $n+7$ is 1. As a consequence of this, show that, for any positive integer, n, $\dfrac{n+7}{2n+13}$ is always in lowest terms.

8 Use Theorem 2.30 to give a quick proof of Theorem 2.16. Here is the outline. Since p is prime, its only divisors are p and 1. If p does not divide a, then gcd$(a, p) = 1$. Thus there are integers m and n such that $ma + np = 1$. Multiplying by b we get that $mab + npb = b$. Use the fact that p divides ab and this equation to show p divides b. We prefer the proof we gave in this chapter, since it very directly addresses the issues connected to prime factorization, while in our opinion Theorem 2.30 is somewhat of an indirect approach to the problem.

9 Prove Theorem 2.32.

2.7 The Division Algorithm for Polynomials

LAUNCH

Using your graphing calculator, enter and graph the function $f(x) = (4x^2 + 13x + 8)/(x - 3)$. Do you see some type of a curve with an asymptote at $x = 3$? Now, zoom out several times until you see what appears to be a line. Has our curve become a line? What do you think that line represents?

If you are bewildered by how zooming out on the calculator seemed to turn a curve into a straight line, then you will be interested in reading this section. We will describe how the division algorithm for polynomials can unravel the mystery of what you have seen on your calculator.

In secondary school, students learn how to divide polynomials. Given that algebra can be thought of as a generalization of arithmetic, it is only natural to examine if, and how, the division algorithm for integers portrayed in the previous section can be extended to polynomials. We proved in Theorem 2.23 that, for any integers a and b, there are integers q and r such that $a = bq + r$ and $0 \leq r < b$. Put another way, when we divide a by b, we get a quotient q and a remainder r, and the remainder that we get is strictly less than the divisor b. There is, in fact an analog of this for polynomials.

If $a(x)$ and $b(x)$ are polynomials, then there exists a polynomial $q(x)$ such that $a(x) = b(x)q(x) + r(x)$ and the DEGREE of $r(x)$ < DEGREE of $b(x)$. The proof is essentially the same as the method for dividing polynomials that you learned in secondary school, but with a smattering of induction. We review that method of dividing polynomials here.

Suppose that we wanted to divide the polynomial $b(x) = 4x^3 + 3x^2 + 7$ by the polynomial $a(x) = x^2 + 3$. We set this up as

$$x^2 + 3 \overline{\smash{)}\, 4x^3 + 3x^2 + 0x + 7}$$

Notice that we wrote $0x$ to leave a place for any x terms that may occur in the process. Our first step is to divide the lead term $4x^3$ in $b(x)$ (the dividend) by the lead term x^2 in $a(x)$ (the divisor.) Since $\frac{4x^3}{x^2} = 4x$, we put a $4x$ on top, and then multiply the divisor, $x^2 + 3$, by what we just put on top to get $4x^3 + 12x$. That goes on line 2 as shown below. We then subtract line 2 from line 1 to get line 3. Now we start all over again. We divide the lead term in line 3, $3x^2$, by the lead term in the divisor, x^2 to get 3. That goes on top. We multiply the divisor by *what we just put on top*, namely, 3. We get $3x^2 + 9$. That goes on line 4. We subtract line 4 from line 3 to get line 5. We are done, since the degree of the polynomial on line 5 is less than the degree of the divisor.

```
              4x     +3
x² + 3 | 4x³  +3x²   +0x    +7      (Line 1)
         4x³         +12x           (Line 2)
         ─────────────────────
                3x²  −12x   +7      (Line 3)
                3x²         +9      (Line 4)
         ─────────────────────
                     −12x   −2      (Line 5)
```

Thus, when $4x^3 + 3x^2 + 7$ is divided by $x^2 + 3$ we get a quotient of $q(x) = 4x + 3$ and a remainder of $r(x) = -12x - 2$ and we can verify by multiplication that $4x^3 + 3x^2 + 7 = (x^2 + 3)(4x + 3) + (-12x - 2)$. That is, that $a(x) = b(x)q(x) + r(x)$.

Another way of writing $a(x) = b(x)q(x) + r(x)$ is $\frac{a(x)}{b(x)} = q(x) + \frac{r(x)}{b(x)}$ as we see when we divide both sides of the equation by $b(x)$. Thus in the previous example, we can write $\frac{4x^3 + 3x^2 + 7}{x^2 + 3} = 4x + 3 + \frac{-12x - 2}{x^2 + 3}$. There is something to be noticed about this. As x gets larger and larger in absolute value, the second fraction on the right, $\frac{-12x - 2}{x^2 + 3}$ gets smaller and smaller and has a limit of zero. What this is saying is that, when x is large in absolute value, the quotient $\frac{4x^3 + 3x^2 + 7}{x^2 + 3}$ is approximately $4x + 3$. Graphically, this means that the graph of $\frac{4x^3 + 3x^2 + 7}{x^2 + 3}$ is **asymptotic to** (that is, approaches) the line $4x + 3$ when one moves far out to the right or left. We can see this by graphing both curves on the same set of axes (Figure 2.6). Notice how the curve, plotted with dark ink, gets closer and closer to the line $y = 4x + 3$ plotted in lighter ink.

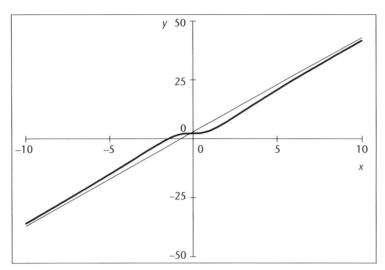

Figure 2.6

We will study consequences of this division algorithm for polynomials in the next chapter.

Student Learning Opportunities

1 (C) A student asks, "When we divide $5x^2 - 6x + 8$ by $x + 1$, is the quotient $5x - 11$ or is it $5x - 11 + 19/(x + 1)$?" How do you respond?

2 Perform the following divisions:

(a) $\dfrac{x^3 - x^2 + 3x - 2}{x - 1}$

(b) $\dfrac{x^3 - 8}{x^2 - 1}$

(c) $\dfrac{16x^4 - 2x^2 + 3x - 2}{2x - 1}$

(d) $\dfrac{x^5 + x + 1}{x^2 + x + 3}$

3 (C) You ask your students to perform the following division problem: $\dfrac{2x^2 + 7x + 5}{x + 1}$. They put their work on the board and you discover that some have done it by factoring and others have used the division algorithm. They ask you which method is better. How do you respond?

4 In each of the following problems, find the line that the function is asymptotic to as the absolute value of x goes to infinity. Graph both the function and its asymptote on the same set of axes using your calculator.

(a) $f(x) = \dfrac{x^3 - 3x + 2}{x^2 - 4}$

(b) $f(x) = \dfrac{3x^2 + 5x}{x - 2}$

(c) $f(x) = \dfrac{x^4 - 3x^2 + 3}{x^3 - 4}$

5 In each of the following, the given function is asymptotic to a curve when $|x|$ is large. Find the curve in each case and graph both the function and the curve on the same set of axes using your calculator.

(a) $f(x) = \dfrac{4x^4 + 3x - 1}{2x - 3}$

(b) $f(x) = \dfrac{3x^3 - 5x + 1}{x - 4}$

6 What is the remainder when $x^{100} - 1$ is divided by $x^2 - 3x + 2$? [Hint: The remainder is of the form $ax + b$. Use the division algorithm, and take convenient values of x to find a and b.]

2.8 Different Base Number Systems

LAUNCH

Tanica, a very bright student, claims that she can show you that she can represent the number 35 by using only 1's and 0's and that it is really equal to 100011. Can you explain what Tanica is talking about? [Hint: You know that 35 really means $3 \times 10 + 5 \times 1$ or $3 \times 10^1 + 5 \times 10^0$ in base 10.]

We hope that this problem got you thinking about how numbers can be represented in multiple ways by using different bases. What you might not realize is that the ability to do this is at the root of some of the most important advances in technology.

In this section we examine one of the most ground-breaking applications of the division algorithm: the representation of numbers in different bases. The development of computers hinged on the ability to represent numbers using only 1's and 0's (on and off switches). This was achieved by representing numbers in base 2. (See later for definition.) Also, representing numbers in base 8 and 16 are critical in the design and working of any computer. Writing a number in base 2 is part of the reason that arithmetic can be done so quickly on a computer. When numbers are represented in base two, the addition of the numbers is trivial and proceeds at lightning speed. The applications of representing numbers in different bases are numerous, so for those who find applications motivational, there is no shortage of examples. However, in addition, and just as importantly, to really understand the base 10 concepts we use on a daily basis, yet rarely think about, it is most informative to examine how numbers can be represented using other bases. Just as it is helpful to study the grammar of a new language to better understand one's first language, it is helpful to study different base number systems to better understand base 10, our number system.

The number 3245 is a short way of representing the number $3 \times 1000 + 2 \times 100 + 4 \times 10 + 5$. When this is written in exponential notation, we have $3245 = 3 \times 10^3 + 2 \times 10^2 + 4 \times 10 + 5 \times 10^0$. This is called the base 10 representation of the number 3245. And, of course, each digit in the representation of that number is less than 10. If we were to replace each 10 by say, 5, we would get a completely different number. That number $3 \times 5^3 + 2 \times 5^2 + 4 \times 5 + 5 \times 5^0$ is really the number 450. That representation of the number 450 is called the base 5 representation of 450, since the base used is 5. This base 5 representation of 450 is denoted by $(450)_5$. In general, when a positive integer is written as a sum of powers of a positive integer b where the coefficients of each power of b are less than b, we say that we have written the number in base b. Thus, the number $a_n(b)^n + a_{n-1}(b)^{n-1} + \ldots + a_0$, where all the $a_i's$ are less than b, is the base b representation of some number N. We will often write $(a_n a_{n-1} \ldots a_0)_b$ to abbreviate this. Notice that the exponent of b in the beginning, b^n, is one less than the number of digits in the representation of the number. To get a sense of this, let us give some examples.

> **Example 2.34** *What is the value of each of the following numbers?* (a) $(1222)_3$ (b) $(345)_6$ (c) $(43,216)_8$

Solution. In part (a) we are given the base 3 representation of a number. Our representation has 4 digits, so we begin with a power of 3 one less than the number of digits. That is, with 3^3. Our number really is the number $1 \times 3^3 + 2 \times 3^2 + 2 \times 3^1 + 2 \times 3^0$ or just 53. So we could write $53 = (1222)_3$. In part (b) we are given the base 6 representation of a number. Since the representation given has 3 digits, our number begins with a power of 6, one less than the number of digits. Thus $(345)_6 = 3 \times 6^2 + 4 \times 6^1 + 5 \times 6^0$ or just, 137. So $137 = (345)_6$. In (c), we are given the base eight representation of a number. The representation has 5 digits, so we begin with 8^4. Our number $(43,216)_8 = 4 \times 8^4 + 3 \times 8^3 + 2 \times 8^2 + 1 \times 8^1 + 6 \times 8^0 = 18,062$.

We will now discuss how to convert a number from our ordinary system to base b. Let's say we want to convert a number N to base 3. Then we know it will look like $a_n(3)^n + a_{n-1}(3)^{n-1} + \ldots + a_1(3^1) + a_0$ after the conversion. How do we find a_0? Well, if we factor out a 3 from all but the last term, we see that we can write $a_n(3)^n + a_{n-1}(3)^{n-1} + \ldots + a_1(3^1) + a_0$ as

$3p + a_0$ where $0 \leq a_0 < 3$.

(Here $p = a_n(3)^{n-1} + a_{n-1}(3)^{n-2} + \ldots + a_1$.) Said in terms that we are more familiar with, when we divide the number N by 3, we get a quotient of p and a remainder of a_0. So to find a_0, we divide the original number by 3 and a_0 is our remainder.

Now let us look at the quotient p. Since $p = a_n(3)^{n-1} + a_{n-1}(3)^{n-2} + \ldots a_2(3) + a_1$, we see by factoring out 3 from all but the last term, that it is of the form $3q + a_1$. (Here $q = a_n(3)^{n-2} + a_{n-1}(3)^{n-3} + \ldots + a_2$.) Said in more familiar terms, when p is divided by 3, it leaves a quotient of q and a remainder of a_1. So, to find a_1, we divide our original quotient, p, by 3. The remainder is a_1. Now we can see how the method works. To find a_2, we divide q, our latest quotient by 3 and our remainder is a_2, and so on.

Thus, the algorithm (rule) that allows us to find the base b representation of a number N is to divide N by b. Take the quotient and divide by b again. Take the resulting quotient and divide by b again. At each stage, put the remainders on the side. The remainders we generate will be the digits of the base b representation of the number, *but from last to first*. So we just reverse the order of the remainders generated. Let us illustrate this by a numerical example.

> **Example 2.35** *Find the base 3 representation of the number* 53.

Solution. Here are the steps: we divide 53 by 3. The quotient is 17 the remainder 2. Put the two on the side. Now divide our previous quotient, 17 by 3. We get a quotient of 5 and a remainder of 2. Put the 2 on the side. Next, we divide our previous quotient, 5, by 3. The quotient is 1 and the remainder 2. Put the two on the side. Finally, divide our last quotient 1 by 3. The quotient is 0 and the remainder is 1. When the quotient is 0, we stop. Put the remainder on the side. All the work is shown in Figure 2.7 below which makes it clearer.

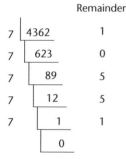

Figure 2.7

We look at our remainders from the *bottom up*. We get 1222, which is the base 3 representation of 53, which tells us that 53 is $1(3)^3 + 2(3)^2 + 2(3)^1 + 2(3)^0$.

Example 2.36 *Find the base 7 representation of the number* 4362.

Solution. Here are the steps (Figure 2.8):

```
         Remainder
7 | 4362    1
7 |  623    0
7 |   89    5
7 |   12    5
7 |    1    1
       0
```

Figure 2.8

Thus $(15, 501)_7 = 4362$. We can check this. $(15, 501)_7 = 1 \times 7^4 + 5 \times 7^3 + 5 \times 7^2 + 0 \times 7 + 1 \times 7^0 = 4362$.

We said that one of the major applications of representing numbers in different bases, especially base 2 is that computers use this to represent numbers and do arithmetic rapidly. In **base 2 representation**, also called **binary** representation, the only digits used are digits less than 2, that is, 0 and 1. Why is base 2 representation the important one? The answer is simple. A computer's memory consists of a large number of electrical switches and switches can only take on two positions, on and off. Thus, if we want these on-off switches to represent a number we seem to have no choice except to use base 2. In this representation, a 1 means "switch on" and a zero "switch off." Since the number 8, in base 2 is $(1000)_2$, it can be represented by 4 switches, called bits, where the first is on and the other three are off. This is an overly simplistic description of what actually goes on inside the computer, but is the essence of it all. The on and off "switches" are simply parts of the memory magnetized into positive and negative charges.

We said that adding in base 2 is very fast. That is because all the digits are 0 and 1. The only rules for addition are that $0 + 0 = 0$, $0 + 1 = 1$, and $1 + 1 = 0$, but we must "carry" a 1 over to the next column. Thus to give a very simple application, if we want to add $8 + 9$, we write both in binary. $8 = (1000)_2$, $9 = (1001)_2$. Our addition is shown in Figure 2.9 below. Starting from right to left, we have $0 + 1 = 1$, $0 + 0 = 0$, $0 + 0 = 0$, and then $1 + 1 = 0$ but we carry a 1 to the next column and then add it to what is there, which is nothing, giving us a 1.

```
8  ———▶  1000
9  ———▶  1001
         ————
         10001
```

Figure 2.9

Thus our sum is $(10001)_2$, which you can check is 17.

There are some very sophisticated card tricks that are based on base 3 representation of numbers, and other tricks that are based on base 2 representation of numbers. Here is one often played on middle school and secondary school students.

Example 2.37 *A student is asked to pick a number between* 1 *and* 31 *but not to tell you the number. You then show the student the 5 cards shown in Figure 2.10 below.*

```
16 17 18 19      8  9 10 11      4  5  6  7
20 21 22 23     12 13 14 15     12 13 14 15
24 25 26 27     24 25 26 27     20 21 22 23
28 29 30 31     28 29 30 31     28 29 30 31
   Card 1          Card 2          Card 3

 2  3  6  7      1  3  5  7
10 11 14 15      9 11 13 15
18 19 22 23     17 19 21 23
26 27 30 31     25 27 29 31
   Card 4          Card 5
```

Figure 2.10

You ask the student to point to each card that has his number and you immediately tell him what number he chose. So if he chose cards 1 and 2, you instantly tell him his number is 24. If he tells you cards 1, 3, and 5, you instantly tell him his number is 21. How does this card trick work?

Solution. Each card is worth a certain amount (which you can write on the back of the card if you wish). The first card is worth 16, the second 8 the third, 4, the fourth, 2, and the fifth, 1. We keep a running total as we progress. Any time the first card is chosen, we add 16, and any time the second card is chosen we add 8. When the third card is chosen, we add 4. When the fourth card is chosen, we add 2 and, when the fifth card is chosen, we add 1. So, if a person picks cards 1 and 2, his number is $16 + 8$ or 24. If he picks cards 1, 3, and 5, his number is $16 + 4 + 1$ or 21.

This card trick is based on binary representation of numbers. Given a 5 digit binary number whose binary digits are *a*, *b*, *c*, *d*, *e* working from left to right, the value of that number is $a \times 2^4 + b \times 2^3 + c \times 2^2 + d \times 2^1 + e \times 2^0$. That is, when you write one of these numbers from 1 to 31 in binary, you are decomposing the number of 16's it has, and then how many additional 8's it has and then how many additional 4's it has, and so on. On card one, we have all the numbers from 1 to 31 whose a digit is 1. All these numbers thus have one 2^4 or 16 in them, which is why we write 16 on the back of the first card. On card, 2, we have all the numbers from 1 to 31 whose b digit is 1. If their b digit is 1 then they have one additional 2^3 or 8 in them. That is why we write an 8 on the back of that card. On card three, we have all the numbers from 1 to 31 that have an additional 4 in them. That is, their c digit in the binary representation is 1. If the d digit is 1, they have an additional 2 in them, and so on. So, if a person picks only cards 1 and 2, he is telling you the binary representation of the number is 11000 or just $1 \times 16 + 1 \times 8 + 0 \times 4 + 0 \times 2 + 0 \times 1$. That is, his number is 24. If the person says his number is only on cards 1, 3 and 5, then he is telling you the binary representation of his number is 10,101, and this is worth $1 \times 16 + 1 \times 4 + 1$ or 21. The numbers on the backs of the cards are always added up to give you his number.

Here is a list of the binary representation of numbers from 1 to 31 to make this clearer. Check that all the numbers that are on the first card have their a digit equal to 1, all the numbers on the second card have their b digit equal to 1, and so on.

Number	1	2	3	4	5	6	7	8
Binary representation	1	10	11	100	101	110	111	1000
Number	9	10	11	12	13	14	15	16
Binary representation	1001	1010	1011	1100	1101	1110	1111	10,000
Number	17	18	19	20	21	22	23	24
Binary representation	10,001	10,010	10,011	10,100	10,101	10,110	11,111	11,000
Number	25	26	27	28	29	30	31	
Binary representation	11,001	11,010	11,011	11,100	11,101	11,110	111,111	

Student Learning Opportunities

1 Find the base 10 representation of the number $(356)_5$.

2 Find the base 7 representation of the number 456.

3 Find the base 8 representation of 223.

4 Find the binary (base 2) representation of 15.

5 (C) One of your students asks if it is possible to have a negative number for a base. Can you?

6 What is the minimum number of weights needed to weigh all integer quantities from 1 to 80 on a standard two pan balance where the weights may only be put on the left pan? What does this problem have to do with base number systems?

7 In what base b will the number $(111)_b = 73_{10}$?

56 Basics of Number Theory

2.9 Modular Arithmetic

 **LAUNCH**

On Saturday, March 29th, a litter of puppies was born. I was told that I could not take one of the puppies home with me until at least 56 complete days had passed. What is the very first day and date that I could take the puppy home with me? Could I get it in time to give to my sister for her birthday on June 6th? Explain how you figured out the answer.

Surely, you could have done this problem by looking at a calendar and counting off 56 days. If you did it this way, it must have been quite tedious. If you found a short cut to doing this problem, then you probably have some inkling into concepts of modular arithmetic. In fact, that is the focus of this next section, which will extend the study of remainders by examining the basics of modular arithmetic, whose applications are significant and effect us on a daily basis. We begin by examining how modular arithmetic is applied in the security of data (like your credit card) when you buy online. Security systems are fundamental to our country's well being, and most are based on modular arithmetic. We start with a typical middle school problem, which is recreational in nature

Example 2.38 *You give your students the following table:*

A	B	C	D	E	F	G	H
1	2	3	4	5	6	7	8
9	10	11	12	13	14	15	16
17	18	19	20	21	22	23	24
25	etc						

and you ask them to determine the column in which the number 283 lies.

Solution. Students play with this and soon realize the pattern here. Each row has a group of 8 consecutive numbers in it. So, to find which column 283 lies in, you divide by 8, and your remainder will tell you what column you are in. If your remainder is 1, you are in column *A*. If your remainder is 2, you are in column *B*, and so on. When 283 is divided by 8, the remainder is 3. Thus 283 lies in column *C*.

Example 2.39 *On September 4th I bought an insurance policy. That was a Monday. The policy would not be activated until 45 complete days had passed. I wanted to know what day and date that would be. What is the answer?*

Solution. We need only realize that every 7 days, we are at a Monday. So, if we divide 45 by 7, we get a remainder of 3. Thus, the day the policy takes effect is 3 days from Monday, or on Thursday, October 19th.

Both Examples (2.38) and (2.39) make use of of modular or "clock" systems. In such a system, the remainders are the important thing. They are called clock systems, since they model clocks. So, for example, if it is 2 o'clock now and we want to know what time it will be 50 hours from now, we simply realize that every 12 hours, we are at the same time (neglecting AM or PM). So, we simply divide 50 by 12, to get 4 (groups of 12) and get a remainder of 2. The remainder tells us how many hours after our start time it was. So, 50 hours from now, it will be 4 o'clock.

Now that we understand the concept, we get to the abstract mathematical analysis. Suppose that a and b are integers and that m is positive. We say that a **is congruent to** b mod m, if a and b have the same remainder when divided by m. We write $a \equiv b \bmod m$ and read this as "a is congruent to b mod m." Thus, $12 \equiv 19 \bmod 7$, since both leave a remainder of 5 when divided by 7. There is another way of telling if 2 numbers have the same remainder when divided by m without performing the divisions. That result is useful and is given in this next theorem:

Theorem 2.40 $a \equiv b \bmod m$ if and only if $a - b$ is divisible by m.

An "if and only if" proof is an argument that goes both ways. So, to prove this theorem, we have to prove two things. (1) If $a \equiv b \bmod m$, then $a - b$ is divisible by m and (2) If $a - b$ is divisible by m, then $a \equiv b \bmod m$. The first statement is indicated by the arrow $\Longrightarrow$, while the second is indicated by the arrow $\Longleftarrow$.

Proof. ($\Longrightarrow$) : We are assuming that $a \equiv b \bmod m$ and we want to show that $a - b$ is divisible by m. Since $a \equiv b \bmod m$, a and b have the same remainder, r, when divided by m. That means, by the division algorithm, that $a = pm + r$, and $b = qm + r$. Clearly, if we subtract these two equations and factor out m, we get $a - b = (p - q)m$. This says that $a - b$ is divisible by m, which is what we wanted to prove. ∎

Proof. ($\Longleftarrow$) Now we are assuming that $a - b$ is divisible by m and we want to show that a and b have the same remainder when divided by m. Suppose that, when a and b are divided by m, they leave remainders r_1 and r_2, respectively. What this means, by the Division Algorithm, is that

$$a = pm + r_1$$

and

$$b = qm + r_2$$

where both r_1 and r_2 are less than m and nonnegative. If we compute $a - b$ we get,

$$a - b = (p - q)m + r_2 - r_1.$$

Since we are given that $a - b$ is divisible by m, $a - b = km$ and this last equation can be written as

$$km = (p - q)m + r_2 - r_1,$$

or

$$km - (p - q)m = r_2 - r_1.$$

The left side is a multiple of m since we can write it as

$$m[k - (p - q)] = r_2 - r_1. \tag{2.23}$$

58 Basics of Number Theory

But the right side of equation (2.23) is the difference of two nonnegative numbers less than m and so must have absolute value less than m. So, the right side can't be a multiple of m unless it is zero. That is, r_1 must be equal to r_2 and we have shown that the remainder is the same when both numbers, a and b, are divided by m. ∎

Thus, to see if 43 and 75 are congruent mod 6, we need only subtract them to get 32, and since 32 is not divisible by 6, they are not congruent mod 6. That is, they have different remainders when divided by 6. Here are some relationships that are true when working with mods.

> **Theorem 2.41** *If $a \equiv b \bmod m$ and $c \equiv d \bmod m$, then*
> *(a) $a + c \equiv b + d \bmod m$.*
> *(b) $a - c \equiv b - d \bmod m$.*
> *(c) $ap \equiv bp \bmod m$ for any integer p.*
> *(d) $ac \equiv bd \bmod m$.*
> *(e) $a^n \equiv b^n \bmod m$ for any positive integer n.*

Proof. We will prove some parts leaving the rest for you in the Student Learning Opportunities for your own enjoyment and to get better at doing proofs.

(a) By Theorem 2.40 we only have to show that the difference $(a + c) - (b + d)$ is divisible by m. But, from the given facts that $a \equiv b \bmod m$ and $c \equiv d \bmod m$, we have, again using Theorem 2.40, that $a - b$ is divisible by m and $c - d$ divisible by m. So their sum, $(a - b) + (c - d)$, must be divisible by m. But this sum simplifies to $(a + c) - (b + d)$. So $(a + c) - (b + d)$ must be divisible by m. We have proved what we set out to prove. Part (b) is proved similarly.

(c) See Student Learning Opportunity 8.

(d) This result might seem a bit surprising at first. Since we are given that $a \equiv b \bmod m$ and $c \equiv d \bmod m$, we know that both $a - b$ and $c - d$ are divisible by m. Thus $c(a - b) + b(c - d)$ must be divisible by m by theorems 2.7 and 2.9 But this last result simplifies to $ac - bd$. So $ac - bd$ is divisible by m and it follows from theorem 2.40 that $ac \equiv bd \bmod m$.

(e) See Student Learning Opportunity 8. ∎

One important observation to make is that, if a number n is divisible by an integer k, then $n \equiv 0 \bmod k$. For example, 6 is divisible by 3 so $6 \equiv 0 \bmod 3$.

> **Example 2.42** *Show that, if n is an integer, then $n^2 + 1$ is never divisible by 3.*

Solution. At first glance, this result seems rather surprising and difficult to prove, but observe how clear it is using mods. When a number n is divided by 3, there are only 3 possible remainders, 0, 1, or 2. That is $n \equiv 0, 1,$ or $2 \bmod 3$. Suppose that $n \equiv 0 \bmod 3$. Then $n^2 \equiv 0 \bmod 3$ and $n^2 + 1 \equiv 1 \bmod 3$. When $n \equiv 1 \bmod 3$, $n^2 + 1 \equiv 2 \bmod 3$. Finally, when $n \equiv 2 \bmod 3$, $n^2 \equiv 2^2 \bmod 3$ or equivalently $n^2 \equiv 1 \bmod 3$ again making $n^2 + 1 \equiv 2 \bmod 3$. To summarize, $n^2 + 1 \equiv 1$ or $2 \bmod 3$, and this means it is never divisible by 3, since if a number is divisible by 3 it must be congruent to $0 \bmod 3$.

> **Example 2.43** *What are the last two digits of the number 3^{25} when this number is expanded?*

Solution. The last two digits can be obtained by dividing the number by 100 and seeing what the remainder is. (Thus, if we divide 1235 by 100, the remainder is 35, the last two digits.) That is, we are interested in 3^{25} mod 100. We observe that $3^4 \equiv 81 \bmod 100$ and squaring both sides we get that $3^8 \equiv 81^2 \bmod 100 = 6561 \bmod 10 \equiv 61 \bmod 100$, and cubing both sides of this last congruence we get $3^{24} \equiv 61^3 \bmod 100 = 226\,981 \bmod 100 \equiv 81 \bmod 100$. Now multiply both sides of this by 3 to get $3^{25} \equiv 243 \bmod 100 \equiv 43 \bmod 100$. So, the last two digits of 3^{25} are 43. Can you imagine the work on this without mods? If you are thinking, well, "I could have done it on my calculator," then we suggest you try it. The calculator will not be able to calculate this number since it is just too large. The calculator will round the answer and lose some of the digits.

Certain rules for mods seem pretty obvious. For example, $a + b \equiv b + a \bmod m$. But certain things that, hold for real numbers do not hold for mods. As one example, we know that, if the product of two real numbers is 0, then one of them must be 0. The analogous result for mods is not true. For example $2 \cdot 3$ is 6, which is congruent to 0 mod 6. But neither 2 nor 3 is congruent to 0 mod 6.

2.9.1 Application: RSA Encryption

A rather sophisticated and very important application of modular arithmetic is RSA encryption. Suppose that we want to send information to another party, but we want it to be safe from people who might be interested in that information. (For example, when you use your credit card to purchase something online from, say, the fictitious website mathiestuff.com, you need to make sure that no outsiders can access it.) The way this is done is by RSA encryption. RSA stands for the discoverers of this method, Ron Rivest, Adi Shamir, and Leon Adelman. This method is extremely secure and was only recently discovered in 1977.

In RSA encryption each digit of the credit card number is changed or encrypted. So, the number that we send looks nothing like the original number. When the encrypted message is sent, the receiver has to have a "key" to decrypt the message you sent and get back your original credit card number. The method is simple to implement, and almost impossible to compromise. Discovering the key that decrypts the code requires the factoring of huge numbers, which even the most sophisticated computers cannot do in real time.

In what follows, we summarize how this encryption and decryption method works. We will not give the full details of why the method works, but describe how the ideas used in this section can be used to accomplish the goals. Certainly, this is one place you can answer your students' question, "Where do we ever use this stuff?" rather emphatically!

We will deal with the case of the company mathiestuff that uses RSA encryption software. We would like to order some materials for our classroom using our credit card. How is it secured?

1. RSA software picks two primes p and q and a number e which is relatively prime to the number $\varphi = (p-1)(q-1)$. Usually p and q are taken to be huge primes though in our illustrative example below, we will choose them small. The letter e is used to represent "encryption exponent."

2. The software finds a number d less than $n = pq$ such that $ed \equiv 1 \bmod \varphi$. This can always be done, and such a number can be found by the Euclidean Algorithm. d is the "decryption" exponent.

3. The credit card number, c, is raised to the encryption power e. The result mod pq is sent. We call this resulting message, s, for sent. To get the original credit card number c back, we raise

60   Basics of Number Theory

the sent message, s to the decryption exponent, d, and take the result mod pq. We will get back c. Let us illustrate this with some small numbers.

> **Example 2.44** *Let us imagine that our credit card has only one number, 9. We want to encrypt it. So we pick primes, say 5 and 7, and then pick a number e relatively prime to $\varphi = (5-1)(7-1)$ or 24. Let us choose e to be 5. Now we find a number d such that $ed \equiv 1 \bmod 24$. (Notice that 24 is φ.) Such a number is $d = 5$, since $ed = 25 \equiv 1 \bmod 24$. Now we take our original message, 9 and raise it to the e power and compute the result* mod 35. *(35 is pq.) We know that $9^5 \equiv 4$ mod 35. So, 4 is the message sent. Now to find the original message, we raise the sent message, 4 to the decryption exponent 5, and compute the result* mod 35. *We get 4^5 which is $\equiv 9$ mod 35. Thus our original message is recovered.*

To break this code, we would need to be able to find p and q. That requires factoring pq. If p and q are large, say, with more than 200 digits each, then even with our supercomputers, factoring pq is a gargantuan task, which is not easily done, and can take months to do. Of course, companies that use RSA encryption keep changing the large primes p and q (some daily) to make it all but impossible to crack the code.

All kinds of messages in the world are sent via RSA encryption. Messages with words are transformed into numbers by using a different number to represent each different letter of the alphabet as well as for periods, commas, and spaces. The message is sent as a number (which in turn is turned into binary) and the result decrypted back into words. RSA encryption is quite an amazing algorithm, and uses nothing more than modular arithmetic, which is often presented in secondary schools.

There is a wonderful website where you can play with encrypting messages and decrypting them. One of our Student Learning Opportunities will refer you to that website:

http://www.profactor.at/~wstoec/rsa.html.

The website makes the encryption and decryption computations painless and it is fun to play with.

Student Learning Opportunities

1 (C) Your students claim that, if $ab \equiv 0 \bmod m$, then because of the zero property (if a and b are real numbers and $ab = 0$, then either $a = 0$ or $b = 0$ or both a and $b = 0$), it stands to reason that either $a \equiv 0 \bmod m$ or $b \equiv 0 \bmod m$. How do you respond?

2 Compute
  (a) 7^{35} mod 3
  (b) 84^{99} mod 85
  (c) 25^{13} mod 7

3 What are the last two digits of 7^{9999}? Explain.

4 What are the last three digits of 10^{30}. Explain. What are the last 3 digits of 9^{30}?

5 (C) Your students ask you why they can't just use their calculator to find the last two or three digits of numbers raised to large powers. What is it about the calculator that makes using it ineffective in problems like this?

6 Using mods, give a proof that, if $3 + a$ and $25 - b$ are divisible by 11, so is $a + b$.

7 Is $232^{554} + 554^{232} + 5$ divisible by 7? How do you know?

8 Prove parts (c) and (e) of Theorem 2.41.

9 Show by example that, if $ab \equiv ac \bmod m$, then it is NOT necessarily true that $b \equiv c \bmod m$. Thus, while we can add, subtract and multiply with mods, division requires care.

10 Show that, if $ab \equiv ac \bmod m$ and a and m are relatively prime, THEN it follows that $b \equiv c \bmod m$. (Compare with Student Learning Opportunity 9.)

11 Using mods, show that the square of any natural number can only be congruent to either 0 or 1 mod 3.

12 Using the previous exercise, show that the equation $a^2 - 6b^2 = 8$ has no solutions if a and b are integers.

13 Prove that, if p and $p + 2$ are both odd primes, and $p > 3$, then $p + 1$ is divisible by 6. [Hint: What p can be congruent to mod 3?]

14 Prove that, for any natural number, n, $n^5 - n$ is congruent to 0 mod 10. [Hint: You need to show that it is divisible by 2 and 5. Showing divisibility by 2 is the easier part. To show divisibility by 5, ask yourself what n can be congruent to mod 5 and work from there.]

15 What are the only possible remainders when the square of an odd number is divided by 4? Using your answer, show that the sum of the squares of 3 odd numbers cannot be a perfect square.

16 If x and y are integers and neither of them is divisible by 5, show that $x^4 + 4y^4$ will be divisible by 5. [Hint: Consider all that y can be congruent to, and for each y decide what possible values the x's can take on.]

17 Suppose that $2x + 3y$ is a multiple of 17. Using mods show that $9x + 5y$ is also a multiple of 17.

18 Using the website mentioned right before the Student Learning Opportunities, encrypt the message "6" using primes 11 and 13. Then decrypt it and show it works. Afterwards, use different primes and show that this method works with your new set of primes also.

2.10 Diophantine Analysis

**LAUNCH**

How many solutions are there to the linear equation, $3x + 6y = 4$? Give three examples of ordered pairs (x, y) that are solutions to this equation. Do any of your ordered pairs consist of x and y values that are both integers? Can you find such a solution? Why or why not?

If you decided that it was impossible to find an integer solution to the equation $3x + 6y = 4$, you might already have an idea about why this was the case. You might also be wondering if there is a general method to immediately tell whether a linear equation has any integer solutions. This next section, will indeed satisfy your curiosity as it will focus on an area of algebra, solving linear equations that has an interesting relation to the number concepts developed in this chapter.

An equation like $2x + 1 = 5$ has only one solution, namely $x = 2$. An equation like $x + y = 11$, has infinitely many solutions, like $x = 2$, $y = 9$, $x = 3$, $y = 8$, $x = 4.2$, $y = 6.8$, and so on. All the solutions of this equation can be pictured. They lie on the line which results when we graph $x + y = 11$. That line is graphed in Figure 2.11 below.

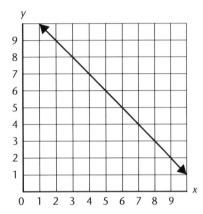

Figure 2.11

Consider the following problem.

Example 2.45 *A man purchases 14 cents worth of stamps consisting of 4 cent stamps and 5 cent stamps. How many of each did he buy?*

Solution. It does not take a lot of thought to figure out that he had to buy one 4 cent stamp and two 5 cent stamps. Yet if we wanted to, we could have set up an equation to model this situation as follows: If x is the number of 4 cents stamps purchased, then the cost of these stamps is $4x$ and, if y is the number of 5 cent stamps purchased, then the cost of these stamps is $5y$. The total expenditure on stamps is $4x + 5y$ and this must be 14. So

$$4x + 5y = 14. \tag{2.24}$$

Now, had we blindly written this equation, we could have said, "Oh, this equation has infinitely many solutions, so there must be many ways of purchasing the stamps to make 14 cents." But, at once we realize that this equation is different from the $x + y = 11$ equation above in that this is a practical problem. The number of each type of stamp can only have nonnegative values. Furthermore, they must be integers. In addition, once x exceeds 4, the cost of the stamps is already more than 14. So this limits us further. The point is that x can only take on integer values from 0 to 3 and y can only take on integer values from 0 to 2. Letting $x = 0, 1, 2, 3$, and solving for y in each case using equation (2.24) above, we see that the only value of x that makes y integral is $x = 1$, and in this case, $y = 2$. So, there is only one solution to this practical problem.

A **Diophantine equation** is an equation whose solutions *we require to be integers*. (This was extensively studied by the mathematician Diophantus.) They need not be linear as eqution (2.24) above. They can be quadratic, cubic, or anything else. Thus, $x^2 = y^3 + 1$ is a Diophantine equation provided we require our solutions to be integers. Furthermore, we are not even requiring that the solutions be positive integers. They can be any integers, although in a specific problem only positive integers may make sense. Diophantine equations can have any number of solutions from 0 to infinity. Let us consider a few of these. We will only consider linear Diophantine equations.

Example 2.46 *Find all integer values of x and y that satisfy*

$$2x + 4y = 7.$$

Solution. On careful analysis, it is easy to see that there are no integral solutions to this equation; for, if x and y are integers, then $2x$ is divisible by 2, $4y$ is divisible by 2, hence $2x + 4y$ is divisible by 2. Thus their sum can never be 7 since 7 is not divisible by 2. So, this equation has no solution. This example illustrates the general principle that, if the greatest common divisor of a and b does not divide c, then the Diophantine equation $ax + by = c$ has no solution.

At the opposite extreme we have:

Example 2.47 *Solve the Diophantine equation $3x + 4y = 7$.*

Solution. We must remember that, when we use the word Diophantine, we are requiring that our solutions be integers. It almost jumps out at us that $x = 1$ and $y = 1$ is one solution. But are there more? Actually, in this case there are infinitely many integer solutions, and they are $x = 1 + 4t$ and $y = 1 - 3t$ for ANY integer t. (We will explain later where this came from.) We could try different values of t and see that this works, but it is so much easier to substitute these into the original equation and see that it works. Here are the steps.

$$3x + 4y = 3(1 + 4t) + 4(1 - 3t)$$
$$= 3 + 12t + 4 - 12t$$
$$= 7.$$

Done!

The astute reader may have noticed that our general solution above $x = 1 + 4t$ and $y = 1 - 3t$ consisted of two parts— our initial solution, $x = 1$, $y = 1$, and multiples of t that were the coefficients of the equations but in reverse order. The solution for x involved the coefficient of y, and the solution for y involved the coefficient of x in the original equation but with opposite sign. Is it always true that, if we can find one integral solution to a linear Diophantine equation, that we can find infinitely many integral solutions and that they are of this form? The answer is, "Yes." Let's examine one other example before giving the general result.

Example 2.48 *Consider the equation $3x - 4y = 8$. One integer solution is $x = 4$ and $y = 1$. Show that this Diophantine equation has infinitely many integer solutions.*

Solution. Guided by what we did above, we try $x = 4 - 4t$ and $y = 1 - 3t$ where t is any integer. We substitute into the equation and see that

$$3x - 4y = 3(4 - 4t) - 4(1 - 3t)$$
$$= 12 - 12t - 4 + 12t$$
$$= 8.$$

So it works. You should now be able to show that the solutions of any linear Diophantine equation are obtained in this way and you will be asked to do that in Student Learning Opportunity 1. We state this as a theorem.

Theorem 2.49 *If (x_0, y_0) is a solution of the Diophantine equation $ax + by = c$, where a, b and c are integers, then $x = x_0 + bt$, $y = y_0 - at$ are also integer solutions of this equation for any integer t.*

Note: One can easily get insight into this theorem by remembering something that is taught in secondary school. Students are taught to plot lines by first finding a point (x_0, y_0) on a line and then using the slope to find another point. Slope is $\frac{\text{rise}}{\text{run}}$. Let us illustrate. If we want to graph a line passing through the point $A = (1, 2)$ with slope $\frac{3}{5}$, starting at $A = (1, 2)$, we rise 3 and move over 5 to the right and we will get another point, B on the line. Now, from that point, we again rise 3 and move over 5 to the right and we will get another point, C, on the line. (See Figure 2.12 below.)

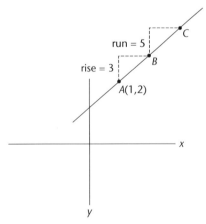

Figure 2.12

We can rise as many times as we want, say *t* times, as long as we run *t* times, and we will get new points on the line. That is, points on the line are given by $x = 1 + 5t$ (the original *x* plus *t* runs of 5) and $y = 2 + 3t$ (the original *y* plus *t* rises of 3). Now getting back to our Diophantine equation $ax + by = c$, the slope is $\frac{-a}{b}$. Starting at the point (x_0, y_0) on the line, we run *t* times a quantity *b* and rise $-a$ times to yield a new point on the line. That new point is $x = x_0 + bt$ and $y = y_0 - at$. This is essentially why the theorem holds.

So, we know how to generate infinitely many integer solutions if we have one. But how do we even know if we have one solution? After all, if we have no solutions, then we are wasting our time looking. The following theorem gives us our answer.

> **Theorem 2.50** *If a and b are relatively prime integers, then $ax + by = c$ where c is an integer, always has integral solutions.*

Proof. By Corollary 2.31, we can find integers x_0 and y_0 such that $ax_0 + by_0 = 1$. If we multiply both sides of this equation by c, we get that $cax_0 + cby_0 = c$ or, put another way, that $a(cx_0) + b(cy_0) = c$. Thus, the integers cx_0 and cy_0 both solve the given equation. ∎

If a and b are not relatively prime, then $ax + by = c$ will have a solution only if $\gcd(a, b)$ divides c. We leave that as a Student Learning Opportunity. We now turn to the question of how we find a particular solution of $ax + by = c$. There are two approaches to this, which are essentially the same. One is with modular arithmetic. Let us give the mod free approach first.

> **Example 2.51** *Find a particular integral solution of $6x + 5y = 13$.*

Solution. We begin by solving for y in terms of x. We get

$$y = \frac{13 - 6x}{5} = 2\frac{3}{5} - (1\frac{1}{5})x.$$

We now separate off the integer part of each term on the right leading to

$$y = 2 + \frac{3}{5} - (1 + \frac{1}{5})x$$
$$y = 2 + \frac{3}{5} - x - \frac{1}{5}x \quad \text{or}$$
$$y = 2 - x + \frac{3 - x}{5}.$$

Now x needs to be an integer. This implies that the term $2 - x$, which occurs on the right of the above equation, is an integer. This means that the only way y on the left will be an integer is if $\frac{3-x}{5}$ on the right is an integer. We can try different integer values for x (between 0 and 4) and see which makes $\frac{3-x}{5}$ an integer, but it is obvious that $x = 3$ does the job. Substituting this into our original equation, we see that $y = -1$ is a solution. So $(3, -1)$ is a solution. Now we can find infinitely many other solutions as we did above by letting $x = 3 + 5t$ and $y = -1 - 6t$ for any integer value of t.

This method that we used above always works, but we can make it much shorter than we did above. What we did above was just for illustration. Starting with $y = \frac{13-6x}{5}$, we divide the numerator by 5 and consider only the remainders. When 13 is divided by 5, a 3 is left over. When $6x$ is divided by 5, there is $1x$ left over, but we keep the negative sign. Thus, the remaining expression is $3 - x$. This must be divisible by 5. And now we proceed as before.

Let's take another example.

Example 2.52 *Solve*

$$5x - 3y = 7 \tag{2.25}$$

for integer values of x and y.

Solution. Solving for y we get

$$y = \frac{5x - 7}{3}.$$

Now, when $5x$ is divided by 3, $2x$ is left over. When 7 is divided by 3, 1 is left over, but we keep the negative sign. So, the left over is $2x - 1$, which must be divisible by 3. Trying $x = 0, 1, 2$, we see that $x = 2$ works. Substituting into equation (2.25), we see that $y = 1$. Now we can generate infinitely many solutions: $x = 2 - 3t$ and $y = 1 - 5t$.

Now let us present the mod approach to this same problem. We can work with either mod 3 or mod 5, since both of these are coefficients of the variables. Let us work with mod 3 since what we did above was essentially working with divisibility by 3. We first observe that any multiple of 3 is $\equiv 0 \bmod 3$, thus $3y$ is congruent to 0 mod 3. Also, $5x \equiv 2x \bmod 3$. Thus, $5x - 3y \equiv (2x - 0) \bmod 3$, or just $2x \bmod 3$. Similarly, the right side of equation (2.25), 7, is $\equiv 1 \bmod 3$. Thus, when we "mod" both sides of equation (2.25) by 3, equation (2.25) becomes,

$$2x \equiv 1 \bmod 3. \tag{2.26}$$

Now we can just substitute numbers in for x, say 0, 1, and 2, and we see right away that $x = 2$ solves the mod equation (2.26). Thus, one solution is $x = 2$, just as we got before. Now we just substitute into equation (2.25) and get $y = 1$.

Let us do one last example.

Example 2.53 *Find integer solutions to the equation $13x - 7y = 9$.*

Solution. In order to eliminate y, we "mod" out everything mod 7, realizing that $13x \equiv 6x \bmod 7$ and $7y \equiv$ to 0 mod 7 and $9 \equiv 2 \bmod 7$ and we get

$$6x \equiv 2 \bmod 7.$$

Now we only have to use values of x from 0 to 6 to find a solution. We see that $x = 5$ works. So, when we substitute this into our original equation, we get $y = 8$. Hence all solutions are $x = 5 - 7t$, $y = 8 - 13t$.

There is a fine point that we have left out. We said that, if we could find one solution of a linear Diophantine equation, $ax + by = c$, then we could find infinitely many others, as we showed above. But we never showed that the solutions we generated by the above method represent ALL the integral solutions. We do that next.

Basics of Number Theory

Theorem 2.54 *If (x_0, y_0) is a solution of the Diophantine equation $ax + by = c$, where a and b are relatively prime, and c is also an integer, then all solutions of this equation are of the form $x = x_0 + bt$, $y = y_0 - at$ where t takes on all integer values.*

Proof. We will show that, if (x_1, y_1) is any integral solution of $ax + by = c$, then $x_1 = x_0 + bt$, $y_1 = y_0 - at$ for some t. That is, x and y are of the desired form. Now, since (x_1, y_1) satisfies $ax + by = c$,

$$ax_1 + by_1 = c. \tag{2.27}$$

Also, since (x_0, y_0) is a solution of $ax + by = c$,

$$ax_0 + by_0 = c. \tag{2.28}$$

Subtracting equation (2.28) from equation (2.27) we get $a(x_1 - x_0) + b(y_1 - y_0) = 0$, which implies that $a(x_1 - x_0) = -b(y_1 - y_0))$. This last equation can be rewritten as:

$$(x_1 - x_0) = \frac{b(y_0 - y_1)}{a}. \tag{2.29}$$

Now the left side of equation (2.29) is an integer being the difference of integers, so the right side must also be an integer. Since a and b have no common factors, $y_0 - y_1$ must be divisible by a, for the a's to divide out and give us an integer. This means that $(y_0 - y_1) = at$ for some t. The terms can be rearranged to $y_1 = y_0 - at$. Substituting this into equation (2.29) we get $(x_1 - x_0) = \frac{b}{a}(y_0 - (y_0 - at)) = bt$ which, when rearranged, gives us, $x_1 = x_0 + bt$ which is what we wanted to show. ∎

Student Learning Opportunities

1 Prove Theorem 2.49. Just remember that saying (x_0, y_0) is a solution of $ax + by = c$ means that $ax_0 + by_0 = c$ already. Is it true that $x = x_0 - bt$, $y = y_0 + at$ will also be solutions? Explain.

2 In Example 2.47 let $t = -1$, then $t = 2$, then $t = 3$. Show that in each case we get a solution of $3x + 4y = 7$.

3 (C) When we solved $6x + 5y = 13$ in Example (2.51), we said that we need only try values of x between 0 and 4 to see which make $\frac{3-x}{5}$ an integer. Your students ask you the following questions about this. How do you respond?

(a) Why should we consider only values of x between 0 and 4? Why not 5, 6, and so on?

(b) How come, if we are given the Diophantine equation, $3x + 17y = 29$, it is better to mod out by 3 than by 17?

4 (C) Make up two different examples of linear Diophantine equations that you can give your students that have no solutions. How did you create these equations?

5 (C) Make up two different examples of linear Diophantine equations that you can give your students that have an infinite number of solutions. How did you create these questions?

6 (C) You ask your students to model the following situation algebraically and then graph their solution: "You roll two dice and the sum of the numbers on each die is 7." Your students draw the line $x + y = 7$, where x represents the roll on the first die and y the roll on the second die. How do you respond?

7 Show that, if a, b, and c, are integers and if $\gcd(a, b)$ does not divide c, then $ax + by = c$ has no solutions.

8 Solve each of the following Diophantine equations. Be sure to first check that these have solutions before you waste time trying to find them.

(a) $4x + 5y = 12$
(b) $5x - 10y = 7$
(c) $3x - 7y = 1$
(d) $2x - 6y = 1$
(e) $9x + 7y = 5$
(f) $3x + 6y = 9$.

9 Suppose that $\gcd(a, b) = d$, and that d divides c. Assuming that the equation $ax + by = c$ has one solution, (x_0, y_0), find all other solutions.

10 You have an unlimited supply of 5 cent stamps and 7 cent stamps, and want to make a total of 89 cents worth of postage.

(a) Set up a Diophantine equation that will help you to solve this.
(b) Solve the equation from part (a) and list all of the ways we can make 89 cents using only 3 cent and 5 cent stamps.

CHAPTER 3

THEORY OF EQUATIONS

3.1 Introduction

A great percentage of the middle and secondary school curriculum is centered on polynomials: adding them, multiplying them, and most importantly, finding solutions (or roots) of polynomial equations. In fact, the butt of many jokes aimed at pointing out the uselessness of learning secondary school mathematics concerns the famous, or perhaps infamous, quadratic formula used to find the roots of quadratic equations. If one looks under roots of polynomials on the Internet, one is quite surprised to see the thousands of articles written on this topic, mostly in applied journals. So, why the keen interest in roots of polynomials? For starters, polynomials are in widespread use in modeling applications in real life, and finding roots of polynomials is the source of important mathematical problems. In this chapter we will begin by reviewing methods of finding roots of quadratic equations and then branch into the intriguing techniques and findings involved in solving higher order equations. We concentrate only on polynomials, finding their roots, and examining their many applications. They have been used in such areas as the study of vibrations, electrical systems, genetics, chemical reactions, quantum mechanics, mechanical stress, economics, geometry, statistics, and error correcting codes used in scanners, cd players, and the like. In fact, we will use them later in the book to solve certain recurrence relations, which have quite a few other significant practical applications. In the process we will see that the calculator will not be able to do all that we want it to which is why we need the results of this chapter and why there are so many articles written on this subject.

As usual, we start off simple, but quickly find ourselves discussing sophisticated concepts. As we study roots of polynomials, we will get to meet some of the interesting characters responsible for the development of this subject matter and get some sense of the relevant historical issues. We will also see the mathematics behind how a calculator finds roots of polynomials and how it computes functions like square roots. We will also show how the material from this chapter can be used in the design of the calculator. We hope you enjoy the journey.

3.2 Polynomials: Modeling, Basic Rules, and the Factor Theorem

 **LAUNCH**

A container manufacturer has just received a large order for metal boxes that must be able to hold 50 cubic inches. He plans to make these boxes out of rectangular pieces of metal 8 inches by 10 inches by cutting out squares from the corners and folding up the sides. He needs to know what size square he should cut out to achieve his goal. How would you use the given information to solve this problem?

We are assuming that, in planning to solve this problem, you immediately employed the use of your algebraic skills. Before we review this problem, we would like to point out that what you have just engaged in is the process of mathematical modeling where you attempted to model the essence of the problem by using mathematical concepts. In this case you most likely used a polynomial to model the problem. In fact, it was a cubic polynomial. Mathematical modeling is big business these days, and consultants are highly sought after to solve problems using such techniques. The general approach is to begin by finding a simple model for the problem, using polynomials if possible, since they are usually easy to work with. If the polynomial model does not fit the situation, you try to use other functions. We will have more to say about this in Chapter 9.

Getting back to our launch problem, let us see how algebra can be used to model the situation. To help us visualize the situation, we use the helpful problem-solving strategy of drawing a diagram (see Figure 3.1 below). Of course, if you use this in your classroom, then cutting out the squares from an 8" × 10" piece of paper would demonstrate this very clearly.

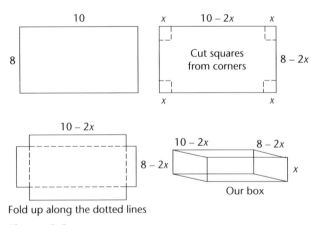

Figure 3.1

Notice that we have let x represent our unknown, the side of the square to be cut out in inches. Then the dimensions of the box it forms have length: $(10 - 2x)$, width: $(8 - 2x)$, and height x. The volume of the box will therefore be length times width times height, or $(10 - 2x)(8 - 2x)\,x$. Since the manufacturer wants the volume of this box to be 50, we want to solve the equation $x(8 - 2x)(10 - 2x) = 50$. We hope that this is the equation you arrived at as well. If we simplify the expression on the left, we get the equation $80x - 36x^2 + 4x^3 = 50$. We notice that the left side of this equation is a polynomial. We are now interested in the solution. If we graph the curve $y = 80x - 36x^2 + 4x^3$ and restrict ourselves to x between 0 and 4 which are the only values of x which are physically possible in this problem situation, we get the following picture (Figure 3.2).

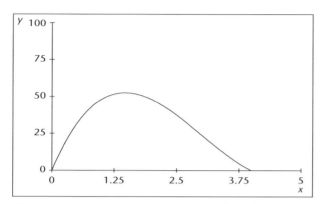

Figure 3.2 The graph of $y = 80x - 36x^2 + 4x^3$ for x between 0 and 4

We can see that, if we want to make $y = 50$, we need to take x to be somewhere between 1.25 and 2. (Try to find the solutions using your calculator!)

Before getting into a deep discussion of finding roots of polynomials, we review the definition of a polynomial. This is probably the most misunderstood word in secondary school mathematics. A **polynomial** in x, denoted by $p(x)$, consists of one or more terms of the form cx^n where c is a constant, and n is a nonnegative integer. For example $p(x) = 7$ is a polynomial, since this can be written as $7x^0$. (It is also a monomial, but that doesn't stop it from being a polynomial!) Each of the following are also polynomials: $p(x) = 3x + 2$, $p(x) = 7x^2 + \pi x - \sqrt{2}$. But $\frac{1}{x} = x^{-1}$ is not a polynomial since it has a negative exponent and the exponents in polynomials must be nonnegative integers. Also $\sqrt{x} = x^{\frac{1}{2}}$ is not a polynomial, since the exponent is fractional. When a value $x = c$ makes $p(x) = 0$, we say that $x = c$ is a **root** or a **zero** of $p(x)$. Thus, zeroes of the polynomial $p(x) = x^2 - 5x + 6$ are $x = 2$ and $x = 3$, since both $p(2) = 0$ and $p(3) = 0$. We can talk about the roots of any function, $f(x)$, regardless of whether or not it is a polynomial. These are simply the numbers that make $f(x) = 0$.

In the previous chapter we discussed the division algorithm for polynomials which explained that, if we had a polynomial, $a(x)$, of degree n, and we divided it by a polynomial $b(x)$ of smaller degree, then there would be a quotient $q(x)$ and a remainder $r(x)$ such that $a(x) = b(x)q(x) + r(x)$, where the degree of $r(x)$ is smaller than the degree of $b(x)$. We gave some examples to illustrate the method of long division that would be used to find $q(x)$ and $r(x)$.

In this section we concentrate on the specific case when the divisor is a polynomial of the form $x - c$. As we shall soon see, this is a particularly important case to consider because it is related to finding the roots of polynomials. Thus, in this case, $a(x) = (x - c)q(x) + r(x)$. Since our divisor is of degree one and our remainder must be of degree less than 1, it has to be of degree 0. Thus, it must be a constant. So, in this case we will simply write $a(x) = (x - c)q(x) + r$. Our first theorem is based on this idea and is a standard one in precalculus courses.

Theorem 3.1 *When a polynomial $p(x)$ is divided by $x - c$, the remainder, r, that you get, is $p(c)$.*

Note: It is important that the divisor be written in the form $x - c$.

Proof. When we divide $p(x)$ by $x - c$, we get a quotient $q(x)$ and a remainder r and $p(x) = (x - c)q(x) + r$. Now replace x by c and we get that $p(c) = (0)q(c) + r = r$. That is, the remainder r, when we divide $p(x)$ by $x - c$, is $p(c)$. ∎

Thus, if we divide the polynomial $p(x) = x^3 - 3x^2 + 4x - 8$ by $x - c = x - 1$, the remainder will be $p(1)$ or -6 since here, c is 1. If we divided the same polynomial by $x + 2$, which can be written as $x - (-2)$, the remainder will be $p(-2)$ or -36 since here, $c = -2$. Let us illustrate this second result in long division form. (A review of long division occurs in Section 2.7.)

> **Example 3.2** Show, using long division that, when we divide $p(x) = x^3 - 3x^2 + 4x - 8$ by $x + 2$, we get a remainder of -36.

Solution. Here is the long division:

$$
\begin{array}{r}
x^2 \quad -5x \quad +14 \\
x+2 \overline{\smash{)}\; x^3 \quad -3x^2 \quad +4x \quad -8 } \\
\underline{x^3 \quad +2x^2 } \\
-5x^2 \quad +4x \\
\underline{-5x^2 \quad -10x } \\
14x \quad -8 \\
\underline{14x \quad +28 } \\
-36
\end{array}
$$

(Line a)
(Line b)
(Line c)
(Line d)
(Line e)
(Line f)

A corollary of Theorem 3.1, known as the **Factor Theorem**, is:

> **Corollary 3.3** If $p(x)$ is a polynomial and if $p(c) = 0$, then $x - c$ is a factor of $p(x)$.

Proof. Since $p(c)$ is 0, we have by the previous theorem, that the remainder when $p(x)$ is divided by $x - c$, that is, r, is zero. Thus, our division algorithm statement, $p(x) = (x - c)q(x) + r$, now reads $p(x) = (x - c)q(x)$. That is, $x - c$ is a factor of $p(x)$. We are done. To find $q(x)$, the other factor, we simply divide $p(x)$ by $(x - c)$. After all, $q(x)$ is the quotient! ∎

To illustrate, suppose that $p(x) = x^2 - 4$. Since $p(2) = 0$, $x - 2$ is a factor of $p(x)$. In a similar manner, since $p(-2) = 0$, $(x - -2)$, that is, $x + 2$ is a factor of $p(x)$.

Let us illustrate this with another example.

> **Example 3.4** Find the roots of the polynomials
>
> (a) $p(x) = x^3 - 2x^2 - 5x + 6$
>
> (b) $q(x) = x^3 - 2x^2 + 6x + 5$
>
> without the use of a calculator.

Solution. (a) By inspection, we see that $x = 1$ is a root of the first equation, since $p(1) = 0$. Thus, $x - 1$ is a factor of $p(x)$. Now divide $p(x)$ by $x - 1$ using long division and you find that the other factor is $x^2 - x - 6$. So $p(x) = (x - 1)(x^2 - x - 6) = (x - 1)(x - 3)(x + 2)$. It follows that $p(x) = 0$ when $x = 1$, $x = 3$, and $x = -2$.

(b) Again, $x = 1$ makes $q(x) = 0$, and again $x - 1$ is a factor of $q(x)$. By long division we see that the other factor of $q(x)$ is $x^2 - x - 5$. So $q(x) = (x - 1)(x^2 - x - 5)$. Now $q(x) = 0$ when $x = 1$ or $x^2 - x - 5 = 0$, and this latter is zero when $x = \frac{1 \pm \sqrt{21}}{2}$ by the quadratic formula.

There are some interesting factoring results that can be obtained by Corollary 3.3. We illustrate some of them.

Example 3.5 *It is a common fact taught in secondary school that the expressions $x^n - b^n$ are always divisible by $x - b$. Show how this follows from our Corollary 3.3. More specifically, show that*

$$x^n - b^n = (x - b)(x^{n-1} + x^{n-2}b + x^{n-3}b^2 + \ldots + xb^{n-2} + b^{n-1})$$

Solution. Let $p(x) = x^n - b^n$. Since $p(b) = b^n - b^n = 0$, by the above Corollary 3.3, $x - b$ is a factor of $p(x)$. We can find the other factor by long division, or by synthetic division. (See the next section for a relatively complete discussion of synthetic division.) In fact, the other factor is $(x^{n-1} + x^{n-2}b + x^{n-3}b^2 + \ldots + xb^{n-2} + b^{n-1}.)$ Thus, $p(x) = (x - b)(x^{n-1} + x^{n-2}b + x^{n-3}b^2 + \ldots xb^{n-2} + b^{n-1})$, which is what we were trying to prove.

Let us show what this says in two special cases, $n = 3$ and $n = 4$. When $n = 3$, we have

$$x^3 - b^3 = (x - b)\left(x^2 + xb + b^2\right) \tag{3.1}$$

and, when $n = 4$, we have

$$x^4 - b^4 = (x - b)(x^3 + x^2b + xb^2 + b^3). \tag{3.2}$$

You should verify that, if you multiply the expressions on the right side of the equality in both equations (3.1) and (3.2), we get the left sides of these equations, respectively.

Student Learning Opportunities

1 Find the remainder when $3x^2 - 4x + 1$ is divided by $x - 3$.

2 Find the remainder when $x^{44} + 3x^{23} - 2$ is divided by $x + 1$.

3 Find all roots of the following equations by first observing that, in each case, there is a simple number, either 0, 1, or 2 that satisfies each equation.

(a) $x^2 - x + x^3 - 1 = 0$.
(b) $x^3 - 8x + 7 = 0$.
(c) $x^3 - 5x^2 + 6x = 0$.
(d) $2x^3 - 11x^2 + 17x - 6 = 0$.

4 The polynomial $p(x)$ has the property that $p(2) = p(3) = p(-1) = 0$. Find two such polynomials. Find a third such polynomial that also makes $p(4) = 14$.

5 Show that $a^3 + b^3$ can be factored into $(a+b)(a^2 - ab + b^2)$. Show that $(a^5 + b^5) = (a+b)(a^4 - a^3b + a^2b^2 - ab^3 + b^4)$. Generalize to finding factors of $a^n + b^n$ when n is a positive odd integer.

6 Factor each of the following completely: You may need to use the results of the previous problem.
 (a) $x^4 - 1$
 (b) $y^3 + 8$
 (c) $a^6 - b^6$
 (d) $8x^3 + 27y^3$
 (e) $16x^6 - 81y^6$

7 Find all real solutions of the polynomial equations below by factoring.
 (a) $x^4 - 2x^3 + 3x^2 = 0$
 (b) $2x^3 - x^2 - 18x + 9 = 0$
 (c) $x^6 - 2x^3 = -1$

8 For which values of m is $x - 1$ a factor of $x^3 + m^2x^2 + 3mx + 1$?

9 (C) Show that the polynomial $p(x) = x^5 + b^5$ always has a root and use the root to factor $p(x)$.

10 If two factors of the polynomial $2x^3 - hx + k = 0$ are $x + 2$ and $x - 1$, what are the values of h and k?

11 Find 4 different factors all in terms of x and y, that when multiplied equal $8^{2x} - 27^{2y}$. [Hint: First write this as $a^2 - b^2$ and factor. Then each factor will be the sum or difference of two cubes.]

12 What is the sum of the prime factors of $2^{16} - 1$?

13 If the polynomial $p_1(x) = ax^2 + bx + c$ has roots r and s, show that the polynomial that has roots $\frac{1}{r}$ and $\frac{1}{s}$ is $p_2(x) = cx^2 + bx + a$.

14 (C) After doing the previous problem, one of your students asks if it is true that if we have a cubic polynomial with roots r, s, and t, then a polynomial that has roots $\frac{1}{r}$, $\frac{1}{s}$, and $\frac{1}{t}$ is just the polynomial with the coefficients reversed? How do you respond? Justify your answer.

15 Model the following problem using polynomials: A grain silo consists of a main part which is a cylinder, topped by a hemispherical roof. Suppose the height of the cylindrical portion is to be 50 feet and the volume of the silo, including the hemisphere on top is 20, 000 cubic feet. What is the radius of the cylindrical portion? (The volume of a cylinder is given by $V = \pi r^2 h$ and the volume of a sphere is $V = \frac{4}{3}\pi r^3$. For more information on volume, see Chapter 4.)

16 The United States Post Office will not accept a box whose girth (distance around) plus length is more than 108 inches. Suppose that we want to build a container with a square base whose volume is as large as possible and whose girth is precisely the maximum 108 inches. Model this situation by letting x be the length of the side of the square base, and expressing the volume in terms of x. Then use your calculator to estimate the dimensions of the box.

17 If p is an odd number greater than 1, show that $(p-1)^{\left(\frac{p-1}{2}\right)} - 1$ is divisible by $p - 2$. [Hint: Let $p = 2n + 1$.]

18 Show that $x - a$ is a factor of $x^2(a - b) + a^2(b - x) + b^2(x - a)$. [Hint: Call the given expression $p(x)$.]

19 Show that $x - c$ is a factor of $(x - b)^3 + (b - c)^3 + (c - x)^3$.

20 When the polynomial $p(x) = 2x^3 + ax^2 + b$ is divided by $x - 1$, the remainder is 1, but when it is divided by $x + 1$, the remainder is -1. If possible, find the values of a and b that will make this true, or prove that it is impossible.

3.3 Synthetic Division

LAUNCH

Examine the following two displays. What does the first display tell you about what happens when $2x^3 - 3x^2 + 4x - 1$ is divided by $x + 1$. What is the quotient? What is the remainder? How is the second display similar? Using the information in the first display, explain what the numbers in lines 1, 2, and 3 represent in the second display.

```
              2x²    −5x    +9
      ┌─────────────────────────
x + 1 │  2x³   −3x²   +4x    −1        (Line a)
         2x³   +2x²                    (Line b)
      ─────────────────────────
                −5x²   +4x              (Line c)
                −5x²   −5x              (Line d)
              ─────────────────
                        9x    −1       (Line e)
                        9x    +9       (Line f)
                      ──────────
                              −10
```

```
 −1 │  2   −3    4    −1              (Line 1)
           −2    5    −9              (Line 2)
       ─────────────────
        2   −5    9   −10             (Line 3)
```

After responding to the launch question and examining the two figures above, you might be getting the idea that, at times, there is a short cut to the usual division algorithm used for polynomials. Guess what? You are right!. When a polynomial is divided by $x - c$, the division can be done very rapidly by a method known as synthetic division. The purpose of this section is to provide you with a brief review and explanation of that method, which is usually shown in a precalculus course.

76 Theory of Equations

Let us go back to the long division $x+2 \overline{\smash{\big)}\, x^3 - 3x^2 + 4x - 8}$ that we did in the last section. There we found that the quotient was $1x^2 - 5x + 14$ and the remainder was -36.

If we write the divisor $x+2$ in the form $x-c$, we see that $c = -2$. The following shortcut is known as synthetic division, which we illustrate on this example. We begin by writing down the coefficients of the dividend, writing down a coefficient of 0 for any power of x that is missing. On the side, we write the number c. We have bolded $c = -2$ so that you can see it emphasized, as we will use it many times. Thus, we have

$$\underline{-2\,\big|}\ \ 1\quad -3\quad 4\quad -8 \quad \text{(Line 1)}$$

Now we bring down the lead coefficient, 1, in the dividend, to line 3, as shown below. We then multiply it by the $c = -2$ we put aside. The product, -2, goes on line 2 under the -3 from the first line as shown below.

$$\begin{array}{r|rrrr}
\mathbf{-2} & 1 & -3 & 4 & -8 \\
& \downarrow & -2 & & \\
\hline
& 1 & & &
\end{array}
\quad
\begin{array}{l}
\text{(Line 1)} \\
\text{(Line 2)} \\
\text{(Line 3)}
\end{array}$$

We now add the numbers in the second column, -3 and -2 to get -5 and put this on line 3 to get:

$$\begin{array}{r|rrrr}
\mathbf{-2} & 1 & -3 & 4 & -8 \\
& \downarrow & -2 & & \\
\hline
& 1 & -5 & &
\end{array}
\quad
\begin{array}{l}
\text{(Line 1)} \\
\text{(Line 2)} \\
\text{(Line 3)}
\end{array}$$

We now multiply the -5 on line 3 by the -2 we put aside to get 10. We put that on line 2 under the 4 from line 1 and then add those two numbers to give us 14 which is put on line 3. That yields:

$$\begin{array}{r|rrrr}
\mathbf{-2} & 1 & -3 & 4 & -8 \\
& \downarrow & -2 & 10 & \\
\hline
& 1 & -5 & 14 &
\end{array}
\quad
\begin{array}{l}
\text{(Line 1)} \\
\text{(Line 2)} \\
\text{(Line 3)}
\end{array}$$

Finally, we multiply the 14 we just put on line **3** by the -2 to give us -28, put it below the -8 on line 1 and then add to give us:

$$\begin{array}{r|rrrr}
\mathbf{-2} & 1 & -3 & 4 & -8 \\
& \downarrow & -2 & 10 & -28 \\
\hline
& 1 & -5 & 14 & -36
\end{array}
\quad
\begin{array}{l}
\text{(Line 1)} \\
\text{(Line 2)} \\
\text{(Line 3)}
\end{array}$$

The rule is "bring down the leading coefficient and then successively multiply the latest entry you put on line 3 by -2, or, in the general case, c, and add the result to the next number on line 1 until you are done." The coefficients of our quotient are given on line 3. The variable begins with 1 power less than the dividend. So, in this case, our quotient is $1x^2 - 5x + 14$ and the last number on line 3, which is -36, is our remainder when we divide by $x + 2$. In this example we wrote down every step. But we can do it all in one step very quickly. Let us give another example just to make sure it is clear. In this example, some powers are missing in the dividend.

Example 3.6 *Divide $x^3 + 2x - 27$ by $x - 4$ using synthetic division.*

Solution. Our starting synthetic division tableau is

$$
\begin{array}{c|cccc}
4 & 1 & 0 & 2 & -27 \\
  &   &   &   &   \\
\hline
  & 1 &   &   &
\end{array}
\quad \begin{array}{l}\text{(Line 1)} \\ \text{(Line 2)} \\ \text{(Line 3)}\end{array}
$$

Notice that, since the polynomial we started with was missing an x^2 term, we had to put a 0 in for the missing term. Now we follow the algorithm. We multiply whatever new number we put on line 3 by the bolded number, 4, in the corner, and add the result to the number on line one in the next column and continue till we get to the end.

$$
\begin{array}{c|cccc}
4 & 1 & 0 & 2 & -27 \\
  &   & 4 & 16 & 72 \\
\hline
  & 1 & 4 & 18 & 45
\end{array}
\quad \begin{array}{l}\text{(Line 1)} \\ \text{(Line 2)} \\ \text{(Line 3)}\end{array}
$$

Our quotient is $1x^2 + 4x + 18$ and our remainder is 45. You can check the division is correct by computing $(x - 4)(1x^2 + 4x + 18) + 45$ and showing the result is $x^3 + 2x - 27$.

The simplicity of the method of Synthetic Division is surely to be appreciated. But you must be wondering why it works? We explain this with our first Example 3.2 above. Using long division, we had,

$$
\begin{array}{r}
x^2 \quad -5x \quad +14 \\
x+2 \enclose{longdiv}{x^3 \quad -3x^2 \quad +4x \quad -8} \\
\underline{x^3 \quad +2x^2 } \\
-5x^2 \quad +4x \\
\underline{-5x^2 \quad +10x } \\
14x \quad -8 \\
\underline{14x \quad +28} \\
-36
\end{array}
\quad \begin{array}{l}\text{(Line a)} \\ \text{(Line b)} \\ \text{(Line c)} \\ \text{(Line d)} \\ \text{(Line e)} \\ \text{(Line f)}\end{array}
$$

Notice the redundancy in the division above. Each time we subtract, we subtract the lead term of the previous line. Thus, on line b we subtract x^3 from the x^3 on line a. On line *d* we subtract $-5x^2$ from the same term on line c and so on. Since the result of this subtraction is 0, we replace all the lead terms in lines b, d, and f, by 0. This yields

$$
\begin{array}{r}
x^2 \quad -5x \quad +14 \\
x+2 \enclose{longdiv}{x^3 \quad -3x^2 \quad +4x \quad -8} \\
+2x^2 \\
\underline{} \\
-5x^2 \quad +4x \\
-10x \\
\underline{} \\
14x \quad -8 \\
+28 \\
\underline{} \\
-36
\end{array}
\quad \begin{array}{l}\text{(Line a)} \\ \text{(Line b)} \\ \text{(Line c)} \\ \text{(Line d)} \\ \text{(Line e)} \\ \text{(Line f)}\end{array}
$$

78 Theory of Equations

Furthermore, there really is no need to bring down the next term each time we subtract. We realize that we are subtracting the $-10x$ on line d from the $4x$ on line a. We are subtracting the 28 on line f, from the -8 on line a. So let's not bring things from line 1 down as we go along, but let us keep everything lined up. (That is, put x's under x's and x^2's under x^2's and so on.) Our division now looks like:

$$
\begin{array}{r|rrrr}
 & x^2 & -5x & +14 & \\
x+2 \,\big)\, & 1x^3 & -3x^2 & +4x & -8 \quad \text{(Line a)} \\
 & & +2x^2 & & \quad \text{(Line b)} \\ \hline
 & & -5x^2 & & \quad \text{(Line c)} \\
 & & & -10x & \quad \text{(Line d)} \\ \hline
 & & & 14x & \quad \text{(Line e)} \\
 & & & & +28 \quad \text{(Line f)} \\ \hline
 & & & & -36
\end{array}
$$

Now observe the remaining coefficients of the terms left on lines b, d, and f. We see that each is generated by multiplying the lead coefficient in the previous line (indicated in bold) by 2, the constant term in the divisor. The arrows show us the flow. Thus, the 2 in line b (the coefficient of x^2) is the result of multiplying the lead coefficient, 1, in line a, by 2, the constant term in the divisor. The -10 in line d is the lead coefficient, -5, in line c, multiplied by 2, the constant term in the divisor, and the remaining lead coefficient in line f, 28 is the lead coefficient, 14, in line e, multiplied by 2, the constant term in the divisor. So, all terms are multiplied by the constant term in the divisor before they are being subtracted. But subtraction is the same as addition of the negative. Thus, another way of saying what we said above is that all these remaining terms on lines, b, d, and f, are being multiplied by -2, the opposite of the constant term in the divisor, and then being **added** to the next term in the dividend! So, to remember this, we suppress the x in the divisor and change the constant term in the divisor to -2, and then think of adding. Our result now looks like:

$$
\begin{array}{r|rrrr}
 & x^2 & -5x & +14 & \\
-2 \,\big)\, & 1x^3 & -3x^2 & +4x & -8 \quad \text{(Line a)} \\
 & & \boxed{-2x^2} & & \quad \text{(Line b)} \\ \hline
 & & -5x^2 & & \quad \text{(Line c)} \\
 & & & \boxed{+10x} & \quad \text{(Line d)} \\ \hline
 & & & 14x & \quad \text{(Line e)} \\
 & & & & \boxed{-28} \quad \text{(Line f)} \\ \hline
 & & & & -36
\end{array}
$$

Notice we have changed the signs of those terms we subtracted. Thus, the $2x^2$ changed to $-2x^2$, the $-10x$ changed to $+10x$, and so on. We have also boxed some terms. Since these will be added to their "like term" partners in line a, we might as well put them right under their like term partners in line b. The bolded terms are simply the result of the like term addition, so they should go on line *c* once we have moved the boxed terms to line b. That is, let's just collapse the table above and bring the boxed terms up to line b and the bolded terms to line c. This gives us

$$-2 \,\big|\, \begin{array}{cccc} 1x^3 & -3x^2 & +4x & -8 \\ & -2x^2 & +10x & -28 \\ \hline 1x^3 & -5x^2 & 14x & -36 \end{array} \quad \begin{array}{l} \text{(Line a)} \\ \text{(Line b)} \\ \text{(Line c)} \end{array}$$

Now we just suppress the x's, and we have

$$-2 \,\big|\, \begin{array}{cccc} 1 & -3 & +4 & -8 \\ & -2 & 10 & -28 \\ \hline 1 & -5 & 14 & -36 \end{array} \quad \begin{array}{l} \text{(Line a)} \\ \text{(Line b)} \\ \text{(Line c)} \end{array}$$

And this, folks, is synthetic division!

We will give a second, and perhaps nicer, proof of synthetic division in the next section after we discuss the Fundamental Theorem of Algebra.

Note: Although we have shown synthetic division when the divisor is $x - c$ where c is real, the same works even if c is a complex number. Since we are waiting until another chapter to discuss the complex numbers, we will just accept this for now.

Student Learning Opportunities

1 Use synthetic division to find the quotient and remainder when $x^3 - 3x^2 + 2x - 3$ is divided by $x - 1$.

2 Use synthetic division to find the quotient and remainder when $x^4 - 2$ is divided by $x - 2$.

3 Use synthetic division to find the quotient and remainder when $2x^3 - 7x + 3$ is divided by $x - 3$.

4 (C) Your student complains that he finds the method of synthetic division very confusing and difficult to remember. He requests that you allow him to do division of polynomials the way he originally learned it, because that's the way that makes the most sense to him. How do you respond?

5 Use synthetic division to find the quotient and remainder when $4x^3 - 2x + 4$ is divided by $2x - 3$. [Hint: $\dfrac{4x^3 - 2x + 4}{2x - 3} = \dfrac{2x^3 - x + 2}{x - 3/2}$.]

6 What are the quotient and remainder when $x^5 - mx + 2$ is divided by $x - 1$?

7 When a polynomial $p(x)$ with all odd powers is divided by $x - 2$, the remainder is 4. What is the remainder when the polynomial is divided by $x^2 - 4$? [Hint: Do you know $p(2)$? How about $p(-2)$? Now, by the division algorithm, $p(x) = (x^2 - 4)q(x) + r(x)$ where $r(x)$ is of degree < 2. That is, $r(x) = ax + b$.] Take it from there.]

8 Use synthetic division to show that when $x^n - b^n$ is divided by $x - b$ where n is a positive integer, then the other factor is $x^{n-1} + x^{n-2}b + x^{n-3}b^2 + \ldots + xb^{n-2} + b^{n-1}$.

3.4 The Fundamental Theorem of Algebra

LAUNCH

1. State the number of solutions in each of the following equations:
    (a) $x^2 - 3x + 2 = 0$
    (b) $x^3 - 5x^2 + 6x = 0$
2. How many solutions do you think a 4th degree polynomial has? An nth degree polynomial?
3. Find the solutions of each of the following equations:
    (a) $x^2 - 2x + 1 = 0$
    (b) $x^4 = 4x - 3$
4. Did the number of solutions you found for 3(a) and 3(b) support your conjecture in question 2? What seems to be the problem?

If, after having done the launch question, you are somewhat confused regarding the number of solutions to an nth degree polynomial, then you will be interested in reading this next section.

Historically, there was great interest in knowing how to solve polynomial equations. This led to the development of the quadratic formula and other formulas for calculating the roots of cubic equations and fourth degree equations which we present later. A reference for some of this background is: http://www.thalesandfriends.org/gr/images/marina/crimes/eng/Equations.doc. Experience showed that linear equations, that is equations of the form $ax + b = c$, have only one solution, namely, $x = \frac{c-b}{a}$. Quadratic equations, that is equations of the form $ax^2 + bx + c = 0$, have two different solutions most of the time and they can be found by the quadratic formula. But, we know that sometimes the quadratic formula leads to only one solution. For example, if one used the quadratic formula on the equation $x^2 - 2x + 1 = 0$ one would find that only $x = 1$ is a solution. If we tried to solve $x^2 - 2x + 1 = 0$ by factoring, we would find that it factors into $(x - 1)^2 = 0$. Although $x = 1$ is the only solution to this last equation, we say that it has **multiplicity** 2 since the exponent that the factor $x - 1$ is raised to is 2. So, if we count the multiplicity of a root, it appears that every quadratic equation has 2 roots.

The polynomial $p(x) = (x - 1)^3(x + 2)^2(x - 3)$ has 3 roots. They are 1, -2, and 3. Looking at the exponents of the factors, we see that the root $x = 1$ has multiplicity 3, the root $x = -2$ has multiplicity 2, and the root $x = 3$ has multiplicity 1. Of course, if $p(x)$ has such a factorization, then $p(x)$ has to be of degree 6 to begin with. If we sum the multiplicities of each root, we get $3 + 2 + 1 = 6$, the same as the degree of the equation. This is always the case, as we shall see.

One of the questions that arose historically is, "Does every polynomial in x, say $p(x)$, have a root?" (Or, in other words, is there always a value of x that makes $p(x) = 0$?) Furthermore, if a polynomial has degree n, how many roots does such a polynomial have? If we restricted ourselves to just real numbers, then it is not true that every polynomial has a zero which is a real number. For example, for the polynomial $p(x) = x^2 + 1$, there is no real value of x that makes $p(x)$ equal to zero. But, if we allow complex solutions, then this polynomial has two zeroes, i and $-i$. (See Chapter 8 for a complete discussion of complex numbers.) If we allowed complex solutions, then how many zeroes would a polynomial have, counting multiplicities? The mathematical genius Gauss, proved the following theorem.

> **Theorem 3.7** (a) *(Fundamental Theorem of Algebra) If one allows complex numbers as roots, then every single polynomial of degree n > 0 has a zero* (b) *Furthermore, if c is a zero of p(x), then (x−c) is a factor of the polynomial. Finally,* (c) *Every nonzero polynomial of degree n > 0 has n roots if one counts multiplicities.*

Part (a) of the theorem is one of the most remarkable results in mathematics and that is what is called the Fundamental Theorem of Algebra. It says that, if we adjoin the complex numbers to our system of real numbers, you have all you need to find roots of every single polynomial equation. You might think that this is talking only about polynomials with coefficients that are real numbers. It is not. It is true even if the coefficients are complex numbers! Parts (b) and (c) are consequences of part (a) and again there is no restriction on c being a real number.

Proof. (a) Since all proofs of this are extremely sophisticated and use the calculus of complex valued functions, we are omitting it. However, given that it can be proven true, we will use it to prove part (c). (One can find a proof in the book, *Complex Variables and Applications* by Churchill and Brown (2004).)

Part (b) of the theorem says that each zero of a polynomial provides a factor of the polynomial. That is, if $p(c) = 0$, then $x - c$ is a factor of the polynomial. Again, c can be complex. So, for the polynomial $p(x) = x^2 + 1$, since $x = i$ is a root, we know that $x - i$ is a factor. Indeed, $p(x) = x^2 + 1 = (x - i)(x + i)$. We can prove this theorem exactly the way we did in Section 2 using the Division Algorithm, which is also true for polynomials even if the coefficients are complex numbers. We outline another proof of this in the Student Learning Opportunities that does not use the division algorithm at all and gives us some further insight into this.

The proof of part (c) of the theorem follows from part (a). By part (a), any polynomial $p(x)$ has a zero, $x = c_1$. By part (b), $x - c_1$ is a factor of $p(x)$. This means that $p(x) = (x - c_1)q(x)$ where $q(x)$ is a polynomial of degree one less than $p(x)$. But, by part (a), $q(x)$ also has a root, $x = c_2$. So it too can be factored into $(x - c_2)r(x)$. But $r(x)$ is also a polynomial and it too has a root $x = c_3$ and can be factored. So, we continue finding roots and factoring, each time getting a polynomial of one degree smaller until we are left with a constant, c, which obviously has no zero. Here is how it can be represented.

$$p(x) = (x - c_1)q(x)$$
$$= (x - c_1)(x - c_2)r(x)$$
$$= (x - c_1)(x - c_2)(x - c_3)s(x)$$
$$= \ldots\ldots\ldots\ldots$$
$$= c(x - c_1)(x - c_2)\ldots(x - c_n).$$

Thus, every polynomial of degree n has n linear factors (and some factors may be repeated). ∎

It follows from this theorem that every nth degree polynomial has n roots counting multiplicity. So every 5th degree polynomial will have 5 roots counting multiplicity. Every 6th degree polynomial will have 6 roots counting multiplicity, and so on.

If you have seen this result before, you probably thought that it was true only for polynomials with real coefficients. It is not. It is true for all polynomials of degree n. Thus, the polynomial $(2 + i)x^2 - (3 - i)x + 6$, being of second degree, also has 2 roots. Notice, we can't use the graphing calculator to solve polynomials in general, since our graphs only provide us with real roots, and polynomials may have complex solutions. Furthermore, and this may surprise you, imaginary numbers have very real and important applications in the real world. Thus, complex roots of polynomials are essential to study.

A relatively new field related to finding roots of polynomials, called **polynomiography**, represents a beautiful fusion between mathematics and art. Very striking pictures are drawn using approximate roots of polynomials. The designs are so pretty that some of them have been used in Iranian carpets. Here is a picture of one done in black and white (Figure 3.3). [Special thanks to Professor Kalantari of Rugers University for permission to use this image. You can also visit:http://www.polynomiography.com, where you can find a great deal of information on this topic and see these stunning pictures in color.]

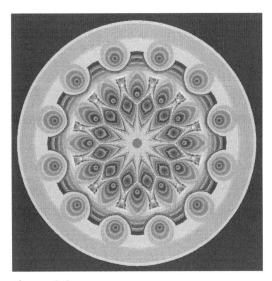

Figure 3.3

There is an interesting result related to Theorem 3.7 and that is:

Theorem 3.8 *If a polynomial, p(x) appears to be of degree n and takes on the value zero for n + 1 different numbers, then the polynomial must be 0 for all numbers, and hence is p(x) = 0. Thus, this polynomial really has degree 0.*

What Theorem 3.8 is saying is that, if a polynomial which appears to be of second degree has three roots, the polynomial must be identically 0. If a polynomial which appears to be of degree 3 has 4 roots, it must be identically 0, and so on.

Proof. We give a nice short proof by contradiction. If $p(x) \neq 0$, then it has n roots by part (c) of Theorem 3.7. But, we are given that the polynomial has $n + 1$ roots. This contradicts what is given in the statement of part (c) of the theorem. Since our contradiction arose from assuming that $p(x) \neq 0$, it must be that $p(x) = 0$. ∎

We can now resolve an exercise we gave in Chapter 1 :

Example 3.9 *Consider the "quadratic" equation:*

$$\frac{(x-1)(x-2)}{2} + \frac{(x-2)(x-3)}{2} - (x-1)(x-3) = 1.$$

You can check that $x = 1, 2,$ and $x = 3$ are solutions of this equation. But a quadratic equation only has, at most, two different solutions. What is wrong here?

Solution. Since the polynomial

$$p(x) = \frac{(x-1)(x-2)}{2} + \frac{(x-2)(x-3)}{2} - (x-1)(x-3) - 1 \qquad (3.3)$$

obtained by subtracting 1 from both sides of the equation, has 3 roots, $x = 1, 2,$ and 3, and the $p(x)$ appears to be quadratic, it must be that $p(x)$ is identically 0. If you expand the left side of (3.3) and simplify, you will see that, indeed, $p(x) = 0$.

A corollary of Theorem 3.8 is:

Corollary 3.10 *If two polynomials of degree n take on the same values for $n + 1$ different values of x, then the two polynomials must be the same.*

Proof. Suppose we have two polynomials $p(x)$ and $q(x)$ both of degree n. And suppose that

$p(c_1) = q(c_1),$

$p(c_2) = q(c_2),$

$\ldots,$

$p(c_{n+1}) = q(c_{n+1}).$

Now form a new polynomial $h(x) = p(x) - q(x)$. Since $p(c_1) = q(c_1)$, we have that $h(c_1) = p(c_1) - q(c_1) = 0$. Since $p(c_2) = q(c_2)$, we have that $h(c_2) = 0$, and so on. It follows that the polynomial $h(x)$ takes on the value 0 for each of the $n + 1$ numbers, $c_1, c_2, \ldots, c_{n+1}$. Thus, by Theorem 3.8, $h(x) = 0$. Saying $h(x) = 0$ means that $p(x) = q(x)$ for all x. ∎

A special case of this is:

Corollary 3.11 *If two polynomials are equal for all values of x, then they must be the same.*

Thus, if you had that $ax^2 + bx + c = 3x^2 + 4x + 5$ for all x, then it must be that $a = 3, b = 4,$ and $c = 5$. There is no other polynomial that has this property. This last corollary is the basis of the method of equating coefficients when you find partial fractions in calculus, but can also be used to explain the method of synthetic division as we will now see.

We will concentrate on a simple example, which generalizes to polynomials of any degree.

84 Theory of Equations

Suppose that we wanted to divide the cubic polynomial

$$ax^3 + bx^2 + cx + d$$

by $x - h$. From the division algorithm, we know that there will be a quotient of degree 2, $ex^2 + fx + g$, and a remainder of degree 0, r, which is a constant. So

$$ax^3 + bx^2 + cx + d = (ex^2 + fx + g)(x - h) + r$$

Our goal is to solve for e, f, g, and r. If we expand the right side of this last equation, we get

$$ax^3 + bx^2 + cx + d = ex^3 + (f - he)x^2 + (g - hf)x + r - hg$$

If we equate coefficients, we get $a = e$, $b = f - he$, $c = g - hf$ and $d = r - hg$. Solving for e, f, and g, we get $e = a$, $f = b + he$, $g = c + hf$ and $r = d + hg$. If we put these equations in a table, we get the table below, which is precisely the table we would get had we used synthetic division, and which explains why the method of synthetic division works.

h	a	b	c	d
		he	hf	hg
	e	f	g	r

Student Learning Opportunities

1 (C) A student says the equation $p(x) = x^5 - 1$ has only one root, $x = 1$, and proves it to you by showing you the graph of $x^5 - 1$ on the calculator and pointing out to you that it crosses the $x-$ axis only once. So that is the only root. The student questions the Fundamental Theorem of Algebra. What misconception does the student have here?

2 Find all roots of $p(x) = x^5 - x$.

3 Find all solutions, real and complex, of the equation $x^3 - 8 = 0$.

4 Suppose that we have the polynomial $p(x) = ax^2 + bx + c$ and that $p(-1) = p(2) = p(3) = 0$. Find $a + b + c$.

5 (C) A student asks "If $ax^3 + bx^2 + cx + d = 3x^3 + 4x^2 - 3x - 1$ for all values of x, is it necessarily true that $a = 3$, $b = 4$, $c = -3$ and $d = -1$?" How do you respond and what explanation would you give? The student continues, "What if the polynomials on the left and right side of the equation only agree for 4 values of x. Is it still true that $a = 3$, $b = 4$, $c = -3$, and $d = -1$?" Now what is your answer? Justify it.

6 (C) We know from the quadratic formula that the roots of $x^2 + 2x - 2$, are $r = -1 - \sqrt{3}$, and $s = -1 + \sqrt{3}$. Thus, we can factor $x^2 + 2x - 2$ into $(x - r)(x - s)$. Show that, when you multiply $(x - r)(x - s)$ with these values of r and s, you actually do get $x^2 + 2x - 2$.

7 Here is a proof of part (b) of Theorem 3.7 that does not use the division algorithm. We will take a specific case, the polynomial $p(x) = x^3 + 3x^2 + 2x + 1$ and prove it for that. Suppose r

is a zero of $p(x)$. Then $p(r) = 0$ by definition of a zero. Now

$$p(x) = p(x) - p(r) \quad \text{(Since } p(r) = 0\text{)}$$
$$= x^3 + 3x^2 + 2x + 1 - (r^3 + 3r^2 + 2r + 1) \quad \text{(and by regrouping terms here)}$$
$$= (x^3 - r^3) + 3(x^2 - r^2) + 2(x - r).$$

Now each of the terms on the right in parentheses has a factor of $x - r$, so we can write

$$p(x) = (\mathbf{x - r})(x^2 + rx + r^2) + 3(\mathbf{x - r})(x + r) + 2(\mathbf{x - r})$$
$$= (\mathbf{x - r})[x^2 + rx + r^2 + 3(x + r) + 2].$$

Thus, we see that $x - r$ is a factor of $p(x) = 0$. The proof for a general polynomial is essentially the same and you should convince yourself by going through the steps that, if r is a zero of $p(x) = ax^3 + bx^2 + cx + d$, then $x - r$ is a factor of this expression. This proof assumes that $x^n - r^n$ has a factor of $x - r$, which we know it has.

8 If $\dfrac{24x^2 + 72x + 3m}{6} = (ax + b)^2$ for all x, then find a, b, and m.

3.5 The Rational Root Theorem and Some Consequences

**LAUNCH**

State whether the following are rational or irrational and justify your answer.

1. $\sqrt{3} + \sqrt{5}$
2. 3^π
3. $\sqrt{11 - 6\sqrt{2}} + \sqrt{11 + 6\sqrt{2}}$

How sure do you feel about your answers to the launch questions? After reading this section, you will want to revisit your responses to see if you were indeed correct, or if in fact, anyone really knows the answers.

As we alluded to earlier, the study of polynomial equations allows us to investigate some very interesting mathematical questions. For example, we have seen in chapter 1 that $\sqrt{2}$ is irrational. A similar proof shows that $\sqrt{3}$ is irrational and in fact, $\sqrt{N}$ is irrational when N is a positive integer which is not a perfect square. What about numbers like $\sqrt{2} + \sqrt{3}$, or $\sqrt[6]{7}$ or $\sqrt{3 - 2\sqrt{2}} + \sqrt{3 + 2\sqrt{2}}$? Are these also irrational? You might think, "Sure. $\sqrt{2} + \sqrt{3}$ is irrational because the sum of two irrational numbers is irrational." Well, that is false, as the following example shows: $1 + \sqrt{2}$ is irrational and so is $1 - \sqrt{2}$. Yet their sum is 2 which is rational. In fact, the third number we presented, $\sqrt{3 - 2\sqrt{2}} + \sqrt{3 + 2\sqrt{2}}$, looks pretty irrational to most people, but in fact, it is rational!

86 Theory of Equations

Certain numbers "look" like they should be irrational, like 2^π and π^π but no one knows if these are rational or not. Today's best mathematicians, with all the computer technology available, have not determined the nature of these numbers. And what about numbers like $\sin 1°$ or $\log_2 3$? Are these irrational?

The purpose of this section is to treat a large number of these expressions from a single and rather elegant point of view which uses a theorem about the roots of polynomials: the Rational Root Theorem. This is taught in many secondary school precalculus courses. Our goal in this section is to present the theoretical background for some sharp mathematical observations that have been used to solve some very difficult problems in mathematics and give us a powerful arsenal of useful information as well.

> **Theorem 3.12** (Rational Root Theorem) *If a polynomial with integer coefficients*
>
> $$p(x) = a_n x^n + a_{n-1} x^{n-1} + a_{n-2} x^{n-2} + \ldots + a_0$$
>
> *has a rational root $\frac{a}{b}$ where $\frac{a}{b}$ is in lowest terms, then the numerator a must divide the constant term a_0 and the denominator b must divide the lead coefficient a_n.*

Proof. We give the proof for the specific polynomial $p(x) = 3x^3 + 2x + 5$ since it will make it easier for you to follow. Afterwards, we give the general proof. Now, saying that a/b is a root of $p(x)$ means that $p(a/b) = 0$. Substituting $x = a/b$ into $p(x) = 0$ yields

$$3(a/b)^3 + 2(a/b) + 5 = 0$$

or just

$$3\frac{a^3}{b^3} + 2\frac{a}{b} + 5 = 0.$$

Multiplying both sides by b^3 we get

$$3a^3 + 2ab^2 + 5b^3 = 0. \tag{3.4}$$

Now, if we subtract $5b^3$ from both sides of the equation we get

$$3a^3 + 2ab^2 = -5b^3$$

or just

$$a(3a^2 + 2b^2) = -5b^3. \tag{3.5}$$

Thus a is a divisor of the left side of equation (3.5). So it must divide the right side of equation (3.5) also. That is, it must divide $-5b^3$. Now, a/b is in lowest terms. So, a and b have no common factors. Therefore, since a divides $-5b^3$, it must be that a divides 5, since it can't divide b^3 by Theorem 2.22 of Chapter 2. In summary, a divides the constant term, 5, of $p(x)$.

Now we use a similar method to make the conclusion we want about b, namely, that it divides 3. We subtract from both sides all the terms of equation (3.4) that have b in them. This yields

$$3a^3 = -2ab^2 - 5b^3. \tag{3.6}$$

This shows that b is a divisor of the right side of equation (3.6) and hence must divide the left side of equation (3.6,) which is $3a^3$. Since b has no common factor with a, b must divide 3, which is the lead coefficient of $p(x)$.

In summary, we have shown that any rational roots $\frac{a}{b}$ of this polynomial $p(x) = 3x^3 + 2x + 5$ have the property that a divides 5 and b divides 3. (Thus, the only possible rational roots of this equation are $\pm\frac{5}{1}$, $\pm\frac{5}{3}$, $\pm\frac{1}{1}$ and $\pm\frac{1}{3}$, and in fact, -1 works.)

The general proof follows the same idea. If a/b is a root of $p(x) = a_n x^n + a_{n-1} x^{n-1} + \ldots + a_0 = 0$, then

$$p(a/b) = a_n(a/b)^n + a_{n-1}(a/b)^{n-1} + a_{n-2}(a/b)^{n-2} + \ldots + a_0 = 0. \tag{3.7}$$

Multiplying both sides of equation (3.7) by b^n and simplifying, we get

$$a_n a^n + a_{n-1} a^{n-1} b + \ldots + a_0 b^n = 0.$$

We subtract the last term $a_0 b^n$ from both sides and we get

$$a_n a^n + a_{n-1} a^{n-1} + \ldots + a_1 a = -a_0 b^n \tag{3.8}$$

and, since a can be factored out of the left side of equation (3.8), the left side is divisible by a. Thus the right side, $-a_0 b^n$ is also divisible by a. Since a and b have no common factor, the only way a can divide the right side is if a divides a_0, the constant term. You can finish the proof mimicking what we did earlier to show that b divides a_n. ∎

Let us now illustrate how this theorem can help us find the rational roots of a polynomial equation.

Example 3.13 *What are the possible rational roots of $p(x) = 2x^3 + 3x - 5$ and which, if any, are actual roots of $p(x)$?*

Solution. Any rational root a/b of $p(x)$ has the property that a must divide the constant term 5 and that b must divide the lead coefficient, 2. Thus, $a = \pm 1, \pm 5$, and $b = \pm 1, \pm 2$. It follows that $a/b = \pm\frac{1}{1}, \pm\frac{1}{2}, \pm\frac{5}{1}$, and $\pm\frac{5}{2}$. If we compute $p(1)$, we get zero, but the value of p at any of the other possible rational roots is not zero. So the only rational root of $p(x)$ is $x = 1$.

Now we give an example which is more in line with what we have set out to do, which is, to discover whether certain numbers are rational or irrational.

Example 3.14 *(a) Show that the only rational roots the equation $p(x) = x^2 - 2 = 0$ can have are $x = \pm 1$ and $x = \pm 2$. (b) Show that none of these are roots. (c) Show that $\sqrt{2}$ is a root of this equation. (d) Give another proof using (a), (b) and (c) that $\sqrt{2}$ is irrational.*

Solution. (a) By the rational root theorem, if a/b is any root of the equation $x^2 - 2 = 0$, then a must divide 2 and so must be either ± 2 or ± 1. Also b must divide 1, meaning b must be ± 1. Thus, a/b must be either ± 2 or ± 1.

(b) Substituting each of these values into $p(x)$, we see that $p(x)$ is not zero for any of these values. Thus *$p(x) = 0$ has no rational roots*.

(c). It is clear that $p(\sqrt{2}) = 0$. (d) Since there are no rational roots, $\sqrt{2}$, which is a root of $p(x)$, cannot be rational.

Can you see how powerful this technique is in proving that a number is irrational? Let us try another example and show that $\sqrt[3]{7}$ is irrational. Since $\sqrt[3]{7}$ satisfies the polynomial $p(x) = x^3 - 7 = 0$, and the only rational roots possible for this equation are ± 7 and ± 1, none of which work, $\sqrt[3]{7}$ is irrational. Similarly, we can show $\sqrt[4]{15}$ is irrational, or even $\sqrt[n]{A}$ where A is an integer that is not a perfect nth power. Even more elaborate numbers like $\sqrt[3]{1 + \sqrt{2}}$ can be shown to be irrational in a similar manner. For example, if we let $x = \sqrt[3]{1 + \sqrt{2}}$, and cube both sides, we get $x^3 = 1 + \sqrt{2}$ and subtracting 1 from both sides and squaring, we get that $(x^3 - 1)^2 = 2$ or that $x^6 - 2x^3 - 1 = 0$. The only rational roots of this are ± 1 by the rational root theorem and none of them work. So, this equation has no rational roots. But $\sqrt[3]{1 + \sqrt{2}}$ is a root of this equation, so it must be irrational. What a nice tool the rational root theorem is!

We have just seen that several irrational numbers can be obtained as roots of polynomials. For example, $\sqrt{2}$ is a root of the polynomial $p(x) = x^2 - 2$, and $\sqrt[3]{1 + \sqrt{2}}$ is a root of $p(x) = x^6 - 2x^3 - 1$. It is a natural question to ask if *all* irrational numbers are roots of polynomials with integer coefficients. For a while, many people believed that. But, to allow for the possibility that this was not true, mathematicians defined the term algebraic number.

An **algebraic number** is a number that is the root of a polynomial with integral coefficients. Thus, $\sqrt{2}$ and $\sqrt[3]{1 + \sqrt{2}}$ and $\sqrt[4]{15}$ are algebraic as we saw in the last two paragraphs. A number which is not algebraic is called **transcendental**. Thus, a transcendental number is a number that is not a root of any polynomial with *integral* coefficients (though it can be a root of a polynomial whose coefficients are not integers).

As we have pointed out, the prevailing thought was that all irrational numbers were algebraic, and thus transcendental numbers did not exist! This was wrong. It took many years for the discovery of the first transcendental number. One number was discovered by the mathematician Louiseville in around 1851 and it is the number 0.110001000000000000000010000000000000... where the number 1 occurs in only the factorial positions. That is, in the 1!, 2!, 3! and so on positions (in the 1st, 2nd, 6th, 24th, etc. position). Proving that this number is transcendental is quite involved. We refer the reader to the Internet for several variations on proofs of this or to the book, *Numbers, Rational and Irrational* (1961) by Ivan Niven. It took until 1873 until the mathematician Hermite proved that the number e so prevalent in the study of calculus, was transcendental, and then another 9 years before the mathematician Lindemann proved that π was transcendental. Thus e and π, though irrational, are not roots of polynomials with integral coefficients. Historically, finding numbers that are transcendental was slow. This might lead you to believe that very few exist. But in mathematics, things are not always what they seem.

In 1887, George Cantor surprised the mathematical world when he proved that there were *infinitely many* transcendental numbers, and in fact, there were more of them than rational numbers! Indeed, "almost all" irrational numbers are transcendental. What a surprise! But even though there are so many transcendental numbers, proving that a number is transcendental seems to be extremely difficult. In fact, it took until 1999 just to prove that numbers like $e^{\pi\sqrt{2}}$ and $e^{\pi\sqrt{3}}$ and so on are transcendental. For more information about transcendental numbers, see Chapter 6 Page 304.

Defining and finding algebraic and transcendental numbers seemed to just be a game intellectuals played. But it turned out that these notions held the key to problems that had baffled mathematicians for thousands of years. Some of the problems that were solved by studying algebraic numbers were the problem of squaring the circle, duplicating the cube, and trisecting an angle, using only an unmarked ruler and compass. These problems of antiquity seem to be irrelevant in today's world. But they were puzzles that could not be solved by even the best minds for over 2000 years. We discuss these problems in Chapter 14.

We have shown how the rational root theorem could be used to prove that certain numbers are irrational. We can also use this theorem to show that $\cos n°$ is irrational for all rational values of θ such that $0 < \theta < 90°$, except for cos 60 degrees. Before showing this, we will state the following result will be proven in Chapter 8. (See Example 8.24.) That result is, that for any rational angle θ between 0 and 90 degrees, the quantity $2\cos\theta$ satisfies an equation of the form

$$1x^n + a_{n-1}x^{n-1} + \ldots + a_0 = 0 \tag{3.9}$$

where the coefficients are integers. (See the corollary to Example 8.24.)

We now use this result.

Example 3.15 *Show that $\cos\theta°$ is irrational for all rational angles θ where $0° < \theta < 90°$ except for cos 60 degrees.*

Solution. By the rational root theorem, any rational solution of equation (3.9) is of the form a/b, where a divides a_0 and b divides 1. Of course, if b divides 1, b is either 1 or -1 and the fraction a/b is an integer. Thus, the only rational roots of equation (3.9) are integers. Now, $x = 2\cos\theta$ is a root of equation (3.9). Thus, if $x = 2\cos\theta$ is rational, it must be an integer. Since $2\cos\theta$ is strictly between 0 and 2 (that is, $0 < 2\cos\theta < 2$) when θ is strictly between 0 and 90 degrees, it follows that the only integer $x = 2\cos\theta$ can be is 1. And that happens when $\cos\theta = 1/2$, which happens when $\theta = 60°$. Thus, the only rational root of this equation occurs when $\theta = 60°$.

Student Learning Opportunities

1 Finish the proof of Theorem 3.12.

2 Set up a polynomial that each of the following numbers is a root of, and then use the rational root theorem to show that each of these is irrational.

(a) $\sqrt[4]{13}$
(b) $5 + \sqrt{2}$
(c) $\sqrt[9]{2}$
(d) $\sqrt[2]{2} + \sqrt[2]{7}$
(e) $\sqrt{2} + \sqrt[3]{3}$

3 Show that $\sqrt{3 - 2\sqrt{2}} + \sqrt{3 + 2\sqrt{2}}$ is rational by observing that $3 - 2\sqrt{2}$ is the square of $\left(\sqrt{2} - 1\right)$ and making a similar observation for $3 + \sqrt{2}$.

4 (C) A student asks you whether an irrational number raised to an irrational power can be rational? Can it? Explain.

5 Use the rational root theorem to find all the roots of the following equations

(a) $x^3 - 4x^2 + 3 = 0$
(b) $4x^3 - x^2 + 5 = 0$
(c) $2x^3 + 6x^2 = 8$

(d) $4x^3 + 4x^2 = x + 1$
(e) $x^2(4x + 8) - 11x = 15$
(f) $x^3 - 2x^2 = 1 - 2x$

6 (C) Your students understand that, since the number $\frac{2}{3}$ satisfies the equation $3x - 2 = 0$, which has integral coefficients, $\frac{2}{3}$ must be algebraic by definition of algebraic. But they have the following questions and need help in answering and then proving their answers. How would you explain the answers to these questions? Use variables in part (a).

(a) Are all rational numbers algebraic?
(b) Are all transcendental numbers irrational?

3.6 The Quadratic Formula

LAUNCH

1. Solve the following quadratic equations by hand and show all of your work.
   (a) $3x^2 - 4x + 1 = 0$
   (b) $3x^2 + 2x + 1 = 0$
2. What method did you use to solve 1(a)? 1(b)?
3. Where did the method you used for 1(b) come from? How do you know that it gives you the correct results?
4. Could you have used the method you used for 1(b) to find the solutions for 1(a)? If so, do it and check that you arrive at the correct solution.

After having done the launch question you are well aware that this next section concerns the quadratic formula, which you are surely familiar with from your secondary school studies. We hope that you appreciate the power of this formula and that, at the same time, if you don't know already, you are curious about where the formula comes from. You might also be wondering if we have such formulas for solving all polynomial equations. How nice that would be! While we do have formulas to find roots of cubic equations and fourth degree equations (some of which you will see later), it was proved by the mathematicians Abel and Ruffini (see Theorem 3.23) that there is no formula that will give us solutions to equations of 5th degree or higher. Some of the best minds worked on this problem but with no success. The theorem was a triumph and the solution was unexpected. It used group theory to prove the result.

In this section we concentrate on solving quadratic equations. We know from secondary school that the equation $y^2 = a$ is very easy to solve: $y = \pm\sqrt{a}$. Thus, if $y^2 = 7$, $y = \pm\sqrt{7}$. If we can somehow reduce a quadratic equation $ax^2 + bx + c = 0$ to the form $y^2 = a$, then solving it would be easy. The method that is often taught in secondary school is the method of completing the square. This method has applications to many different areas in mathematics other than solving quadratics.

For example, it can be used to find the center and radius of circles that are not in the "right" form. It can be used to find key information about ellipses, parabolas, and hyperbolas (some of which find applications in astronomy). It can also be used to solve some rather complicated integrals in calculus that occur in the sciences. So we spend some time on it now.

What does it mean to **complete a square**? What it means is that you start with an expression of the form $1x^2 + bx$, and try to determine what must be added to this expression to make it the square of a binomial. What we must add is $(\frac{b}{2})^2$. That is, we add the square of half the coefficient of b. To see that this is correct, we simply check that $1x^2 + bx + (\frac{b}{2})^2 = (x + \frac{b}{2})^2$, and is therefore a perfect square. Thus, if one asks what must be added to $1x^2 + 5x$ to make it a perfect square, the answer is $(\frac{5}{2})^2$ or $\frac{25}{4}$. Now we can verify that $1x^2 + 5x + \frac{25}{4}$ is the square of $(x + \frac{5}{2})^2$. To complete the square of $y^2 - 6y$, we add 9 (half of −6 all squared.) It is easy to check that $y^2 - 6y + 9$ is $(y - 3)^2$. Let us now illustrate a typical secondary school problem where a quadratic equation is solved by completing the square. Notice that this method requires that a, the coefficient of x^2, be equal to 1.

Example 3.16 *Using the method of completing the square, solve the equation* $x^2 + 6x + 1 = 0$.

Solution. We subtract 1 from each side of the equation to get $x^2 + 6x = -1$. We complete the square on the left side by adding 9. Of course, to keep the equation balanced, we need to do the same to the right hand side. Our equation becomes: $x^2 + 6x + 9 = -1 + 9$. This is the same as $(x + 3)^2 = 8$. Thinking of $(x + 3)$ as y, this tells us we have $y^2 = 8$ and hence $y = \pm\sqrt{8}$. Replacing y by $x + 3$ we have, $x + 3 = \pm\sqrt{8}$. So $x = -3 \pm \sqrt{8}$.

Example 3.17 *Use the method of completing the square to solve the equation* $3x^2 + 4x - 2 = 0$

Solution. We add 2 to both sides to get

$$3x^2 + 4x = 2.$$

To use the method of completing the square, we need the coefficient of x^2 to be 1. So we divide the equation by 3 to get

$$x^2 + \frac{4}{3}x = \frac{2}{3}.$$

We add $[\frac{1}{2}(\frac{4}{3})]^2 = \frac{4}{9}$ to both sides of the equation to get

$$x^2 + \frac{4}{3}x + \frac{4}{9} = \frac{2}{3} + \frac{4}{9}$$

which just becomes

$$\left(x + \frac{2}{3}\right)^2 = \frac{10}{9}.$$

From this we get that

$$\left(x + \frac{2}{3}\right) = \pm\sqrt{\frac{10}{9}},$$

and so,

$$x = -\frac{2}{3} \pm \sqrt{\frac{10}{9}}.$$

It is exactly in this way that we derive the quadratic formula. Here it is for completeness.

Example 3.18 *Derive the quadratic formula.*

Solution. We start with the equation $ax^2 + bx + c = 0$ where $a > 0$. We then subtract c from both sides to get

$$ax^2 + bx = -c.$$

Since we need the coefficient of x^2 to be 1, we divide both sides by a to get

$$x^2 + \frac{b}{a}x = -\frac{c}{a}.$$

We complete the square on the left side by adding $(\frac{1}{2} \cdot \frac{b}{a})^2$ or just $\frac{b^2}{4a^2}$. We get

$$x^2 + \frac{b}{a}x + \frac{b^2}{4a^2} = -\frac{c}{a} + \frac{b^2}{4a^2}. \tag{3.10}$$

Now the left side of equation (3.10) is a perfect square, the square of $(x + \frac{b}{2a})$. Thus, we have

$$\left(x + \frac{b}{2a}\right)^2 = -\frac{c}{a} + \frac{b^2}{4a^2}$$

which can be rewritten as

$$\left(x + \frac{b}{2a}\right)^2 = -\frac{4ac}{4a^2} + \frac{b^2}{4a^2}.$$

Combining the two fractions on the right, we have

$$\left(x + \frac{b}{2a}\right)^2 = \frac{b^2 - 4ac}{4a^2}.$$

Thus,

$$\left(x + \frac{b}{2a}\right) = \pm\sqrt{\frac{b^2 - 4ac}{4a^2}}$$

and this can be rewritten as

$$\left(x + \frac{b}{2a}\right) = \pm\frac{\sqrt{b^2 - 4ac}}{2a}.$$

Subtracting $\frac{b}{2a}$ from both sides we get

$$x = -\frac{b}{2a} \pm \frac{\sqrt{b^2 - 4ac}}{2a} = \frac{-b \pm \sqrt{b^2 - 4ac}}{2a}$$

and we are done!

The quadratic formula holds even if the coefficients a, b, and c, are complex numbers, but, of course, then quantities like $\sqrt{b^2 - 4ac}$ would lead to taking square roots of imaginary numbers. What on earth does this mean? We will talk about this later when we discuss complex numbers in depth. Let us mention that, once we define what this means, the quadratic formula *will* hold for all quadratic equations, even if the coefficients are complex.

Given the pressure of completing a crowded curriculum and preparing students and preparing students for standardized exams, many teachers ponder the value of sharing this proof with their students. However, there may be students who are curious about where the quadratic formula came from and after having done several numerical examples with completing the squares this proof should not be difficult for them to follow. We offer another proof of the quadratic formula in the Student Learning Opportunities, which is much simpler. Although that proof is simpler, the method of completing the square occurs in several places in the secondary school curriculum, relating to conic sections and their transformations, which is why we addressed it here.

Student Learning Opportunities

1 Here is another way to derive the quadratic formula without getting bogged down in a lot of fractions. This might be more useful for a fraction-phobic classroom. Begin with $ax^2 + bx + c = 0$ and multiply both sides of this equation by $4a$ to get $4a^2x^2 + 4abx + 4ac = 0$. Now, subtract $4ac$ from both sides and add b^2 to both sides to get $4a^2x^2 + 4abx + b^2 = b^2 - 4ac$. Observe that the left side is a perfect square. Take it from there.

2 (C) For the quadratic equation $ax^2 + bx + c = 0$, where a, b, and c are integers, the quantity $b^2 - 4ac$ is called the discriminant. In secondary school the following rule is taught: If the discriminant is 0, there is only one root of the quadratic equation $ax^2 + bx + c = 0$. If the discriminant is positive and a perfect square, then the two roots are real and rational and unequal. If the discriminant is positive and not a perfect square, the roots of the quadratic equation are irrational and unequal, and if the discriminant is negative, the roots are imaginary. Describe how you would justify these rules to your students. What would you tell them if they asked whether the rules were still true if b is irrational?

3 Solve the following quadratic equations by completing the square.
   (a) $x^2 - 6x = -8$
   (b) $y^2 - 7y + 6 = 0$
   (c) $z(z - 1) + 1 = 0$
   (d) $3z^2 - 2z + 1 = 0$

4 (C) One of your students was asked to solve $x^2 - 8x - 25 = 0$ by completing the square. Her work appears below. She notices that neither of her solutions work and concludes this quadratic equation has no answers. Comment on her work and on her conclusions. If she is not correct, how would you help her to modify her work so that she gets a correct answer?

$$x^2 - 8x - 25 = 0$$
$$x^2 - 8x = 25$$
$$x^2 - 8x + 16 = 25$$
$$(x-4)^2 = 25$$
$$x - 4 = \pm\sqrt{25}$$
$$x = 4 + 5 \quad \text{or} \quad x = 4 - 5$$
$$x = 9 \quad \text{or} x = -1$$

5. Find the values of k for which the roots of $w^2 - kw + 6 = 0$ are equal. What are the roots for this value of k?

6. Solve for x : $\sqrt{2}x^2 - 5x + \sqrt{8} = 0$.

7. Find all solutions of $\sqrt{x+10} + \dfrac{5}{\sqrt{x+10}} = 6$.

8. Solve for n in terms of S : $S = \dfrac{n(n+1)}{2}$.

9. Solve for r in terms of A and h. $A = \pi r^2 + 2\pi r h$.

10. The length of a rectangle is 4 feet more than the width. The area is 22 square feet. Find the width.

11. (C) Using the solutions from the quadratic formula, how would you explain to your students why the sum of the roots of a quadratic equation $ax^2 + bx + c = 0$ is $\dfrac{-b}{2a}$ and that the product of the roots is $\dfrac{c}{a}$? Using this fact, how would you demonstrate how to find the sum and product of the roots of $2x^2 - 3x - 1 = 0$? Show how you would justify that this answer was correct by finding the actual roots and adding them and multiplying them to check that the answer is correct.

12. In the previous problem you showed that the sum of the roots for the quadratic $ax^2 + bx + c = 0$ is $\dfrac{-b}{a}$ and that the product of the roots is $\dfrac{c}{a}$. We now wish to generalize this to cubic equations. Suppose that you have the cubic equation $ax^3 + bx^2 + cx + d = 0$. Dividing by a this becomes $x^3 + \dfrac{b}{a}x^2 + \dfrac{c}{a}x + \dfrac{d}{a} = 0$. Suppose the roots of this cubic are r, s, and t. Then by Theorem 3.8 this polynomial can be factored into $(x-r)(x-s)(x-t) = 0$. Expand this product and equate coefficients (Corollary 3.11) to conclude that $r + s + t$, the sum of the root is $\dfrac{-b}{a}$ and $rs + st + rt$, the sum of the roots taken two at a time is $\dfrac{c}{a}$ and that the product of the roots is $-\dfrac{d}{a}$.

13. Suppose that the two roots of the equation $x^2 + px + q = 0$ differ by 1. Show that $p^2 - 4q = 1$.

14. Prove or disprove: If a, b, and c are odd integers, the roots of $ax^2 + bx + c = 0$ cannot be rational.

3.7 Solving Higher Order Polynomials

LAUNCH

1. Solve the cubic equation $x^3 + 7x = -48$.
2. How many real solutions are there? How can we find the remaining solutions?
3. Is there a formula, like the quadratic formula, that can be used to find the solutions of this equation?

According to the Fundamental Theorem of Algebra we know that a cubic equation should have 3 solutions. If you are wondering if there is a formula for finding these solutions, similar to the quadratic formula for quadratic equations, you will be interested in reading this section.

3.7.1 The Cubic Equation

The secondary school curriculum focuses primarily on solving linear and quadratic equations. The history of solving polynomial equations of higher order is rich with surprises and contributions to important mathematics. In this section we examine polynomial equations of higher degree. There is more here than meets the eye. You will see how the ideas in this section brought about some strange results, which led to the subsequent development and understanding of complex numbers. You will learn in Chapter 7, that complex numbers have major applications in many fields. Some of the equations that we encounter in this section are complicated. Bear with them, for they will bear fruit.

One would think that, to be able to solve the equation $x^3 + bx^2 + cx + d = 0$, all one would have to do is complete the cube in some way similar to the way we completed the square. Unfortunately, when we cube something like $(x + p)$, we get $x^3 + 3px^2 + 3p^2x + p^3$, which tells us immediately that the coefficient of the x^2 term is $3p$ and the coefficient of the x term is $3p^2$. So, if we have an equation like $x^3 + x^2 + 5x + 1$, looking at the coefficient of the x^2 term and the x term, we get that $3p = 1$ and $3p^2 = 5$. The first equation tells us that $p = 1/3$. But, if this is substituted into the second equation, $3p^2 = 5$, we get an untrue statement. Thus, there seems to be no hope of completing the cube. This was certainly noticed by the many people who tried, in vain, to solve the cubic equation.

The first progress in solving the cubic was made by the mathematician Scipione del Ferro (1465–1526). He didn't solve the general cubic but instead, solved what is called the "depressed cubic"

$$x^3 + px = q \text{ where } p \text{ and } q \text{ are positive.} \tag{3.11}$$

Depressed cubic simply meant the x^2 term was missing. Later we will show that this equation has only one real solution.

What Ferro said is that this equation has only one real solution and it is of the form $x = u + v$ where

$$3uv + p = 0 \tag{3.12}$$

96 Theory of Equations

and
$$u^3 + v^3 = q. \tag{3.13}$$

Furthermore, one can always find such a u and v. If you are wondering where Ferro got these equations, you are not alone. Here lies his brilliance. Now Ferro was right, but how he figured that out was anyone's guess. The Polish mathematician Mark Kac, made a distinction between the ordinary mathematical genius and the magician mathematician. He essentially said that the ordinary genius is one whose mind is so much better than ours and one who can see things we can't. But once we are presented with what he sees, we can understand how his mind worked. In contrast, the magician genius, which Ferro might be considered by some, is one whose mind works in the dark. Even after we see what they have done, we have no clue how they ever thought of it.

To see that $x = u + v$ solves the equation $x^3 + px = q$ under conditions of equations (3.12) and (3.13), we substitute $x = u + v$ into the left side of this equation to get

$$(u + v)^3 + p(u + v). \tag{3.14}$$

We will show this is equal to q, the right hand side of our equation (3.13), and thus, $x = u + v$ solves our equation. Now we know that $(u + v)^3 = u^3 + v^3 + 3uv(u + v)$. (Just expand and check.) Thus when expression (3.14) is expanded, we get

$$(u + v)^3 + p(u + v)$$

$$= (u^3 + v^3) + 3uv(u + v) + p(u + v), \text{ which by factoring out } u + v \text{ yields}$$

$$= (u^3 + v^3) + (u + v)(3uv + p). \text{ Now using equations (3.13) and (3.12), respectively, we see this}$$

$$= q + 0$$

$$= q.$$

So, we have shown that, *if* we can find u and v that satisfy equations (3.12) and (3.13), then we have solved our cubic equation. Of course, there is the issue of how to find such u and v.

Ferro then set to the task of showing that we can always solve equations (3.12) and (3.13). Let us assume that there are such u and v and try to find them. From equation (3.12) we get $v = -\frac{p}{3u}$. When this is substituted into equation (3.13), we get

$$u^3 - \frac{p^3}{27u^3} = q \tag{3.15}$$

and when we multiply both sides of equation (3.15) by u^3, we get $u^6 - \frac{p^3}{27} = qu^3$. Getting all terms over to one side, we get

$$u^6 - qu^3 - \frac{p^3}{27} = 0. \tag{3.16}$$

If we can solve this for u, then we can find v from $v = -\frac{p}{3u}$, and then we can find x since $x = u + v$. Equation (3.16) looks intimidating. But have nor fear, it looks worse than it is. If in equation (3.16) we make the substitution $z = u^3$ then equation (3.16) becomes $z^2 - qz - \frac{p^3}{27} = 0$, which is a quadratic in z! So, we can use the quadratic formula, to get

$$z = \frac{q \pm \sqrt{q^2 + \frac{4p^3}{27}}}{2} \quad \text{(Verify this!)}$$

Since $z = u^3$, we get from this that

$$u^3 = \frac{q \pm \sqrt{q^2 + \frac{4p^3}{27}}}{2}$$

hence that

$$u = \sqrt[3]{\frac{q \pm \sqrt{q^2 + \frac{4p^3}{27}}}{2}}. \qquad (3.17)$$

Now from equation (3.13), $v = \sqrt[3]{q - u^3}$ so substituting equation (3.17) in this we get, after finding a common denominator and distributing the negative sign, that

$$v = \sqrt[3]{\frac{q \mp \sqrt{q^2 + \frac{4p^3}{27}}}{2}}. \qquad (3.18)$$

(Again, verify it!) Notice that the cube root in equation (3.17) has plus/minus, while that in equation (3.18) has minus/plus. Thus, if we take the plus sign in one radical, we must take the opposite sign in the other radical.

Finally, since $x = u + v$, we have

$$x = \sqrt[3]{\frac{q \pm \sqrt{q^2 + \frac{4p^3}{27}}}{2}} + \sqrt[3]{\frac{q \mp \sqrt{q^2 + \frac{4p^3}{27}}}{2}} \qquad (3.19)$$

and we have solved our cubic equation!

Now it looks like we have two solutions for x, but it can be verified (and is an exercise in algebraic manipulation) that the two solutions are really the same. Thus, our official (unique) real solution to the cubic equation (3.11) is

$$\boxed{x = \sqrt[3]{\frac{q + \sqrt{q^2 + \frac{4p^3}{27}}}{2}} + \sqrt[3]{\frac{q - \sqrt{q^2 + \frac{4p^3}{27}}}{2}}.} \qquad (3.20)$$

Since this formula is so complex, it is understandable that it is not included in the secondary school curriculum. Let's now see how we can solve a cubic equation.

Example 3.19 *Let us apply formula (3.20) to solve the equation $x^3 + 6x = 20$.*

Solution. Here $p = 6$ and $q = 20$. Substituting into equation (3.20) we get

$$x = \sqrt[3]{\frac{20 + \sqrt{20^2 + \frac{4(6)^3}{27}}}{2}} + \sqrt[3]{\frac{20 - \sqrt{20^2 + \frac{4(6)^3}{27}}}{2}}. \qquad (3.21)$$

Now the program with which this chapter was written is capable of doing these kinds of computations, (as are your hand held calculators) and when we asked it to evaluate this numerically, the program gave us $x = 2$. Indeed, $x = 2$ does work as we see by substituting $x = 2$ into $x^3 + 6x - 20 = 0$. Neat! Of course, the solution in the form of equation (3.21) is intimidating, but it does the job.

We said earlier that a cubic equation has 3 solutions counting multiplicity. What are the other two? Our equation is really $x^3 + 6x - 20 = 0$. To find the other two solutions, we need only divide $x^3 + 6x - 20$ by $x - 2$ to get the other factor, which is a quadratic and then solve the resulting quadratic equation. We leave this to you to solve. You will discover that $x = 2$ is the only real solution.

We will have you practice more problems like this in the Student Learning Opportunities. But, for now, we thought you would be interested in knowing that all equations $x^3 + px = q$ or equivalently $x^3 + px - q = 0$ where p and q are positive, have only one *real* solution and why that is. We need to recall from calculus, the Intermediate Value Theorem:

> **Theorem 3.20** (*Intermediate Value Theorem*)) *If $f(x)$ is continuous on $[a, b]$ and $f(a)$ and $f(b)$ have opposite signs, then there is a value, c, strictly between a and b where $f(c) = 0$.*

What this is saying is that, if the graph of $f(x)$ is below the x-axis at one endpoint of $[a, b]$ and above the x-axis at the other endpoint of $[a, b]$, (see Figure 3.4 below) then it must cross the x-axis between a and b, which seems intuitively clear if the function is continuous.

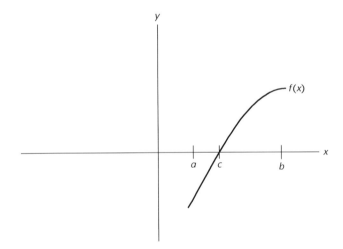

Figure 3.4

Now, let us apply this theorem to show that $f(x) = x^3 + px - q$ has a real root, and has only one real root. We observe that $f(0) = -q$, which is negative since q was taken to be positive. Also, if $x = N$, where N is a very large positive number, then $x^3 + px$ will be a very large positive number and will be bigger than q if N is large enough. So $f(N)$ will be positive. Since $f(0) < 0$ and $f(N) > 0$, $f(x)$ must cross the x-axis somewhere between 0 and N. That is, $f(x)$ has a real root.

Now, we show that $f(x)$ has only one root. Recall from calculus that, if the derivative of f is positive, then the function f must be increasing. Since $f'(x) = 3x^2 + p$ and since $3x^2$ is always nonnegative, and p is positive, $f'(x)$ is positive. Thus, our function is always increasing. What this means is that, once it crosses the x-axis, it keeps going up. And so it can't cross the x-axis again. That is, it has no other real root.

The above proof that $f(x) = x^3 + px - q$ has only one real root was straightforward. But when you realize that Cartesian coordinates, functions, and calculus hadn't yet been discovered when

Ferro did his work, you can appreciate Ferro's realization about this equation having only one real root.

Since mathematicians of his time didn't believe negative numbers had any meaning, which by today's standards, is almost incomprehensible, in his proof Ferro assumed p was nonnegative. But, in fact, there is nothing wrong with assuming that p is negative in equation (3.11). Ferro's formula gives us a solution in this case (though in this case there may be more than one real solution.)

Once we have one real solution of the cubic, $x = r$, we can find the other two solutions by dividing $f(x)$ by $x - r$. This reduces the equation to a quadratic equation whose solutions we can find by the quadratic formula. Let us illustrate by example.

Example 3.21 *Suppose we have the equation*

$$x^3 - 4x = 15. \qquad (3.22)$$

Solve for x using Ferro's formula.

Solution. Here $p = -4$ and $q = 15$. Substituting in Ferro's formula, we get

$$x = \sqrt[3]{\frac{15 + \sqrt{15^2 + \frac{4(-4)^3}{27}}}{2}} + \sqrt[3]{\frac{15 - \sqrt{15^2 + \frac{4(-4)^3}{27}}}{2}},$$

and using a calculator or computer software to compute, we get that $x = 3$, which we can easily verify by letting $x = 3$ in equation (3.22). Now, if we rewrite equation (3.22) as $x^3 - 4x - 15 = 0$ and divide $x^3 - 4x - 15$ by $x - 3$, say, using synthetic division, we find that the other factor is $x^2 + 3x + 5$ and thus $x^3 - 4x - 15 = 0$ factors into $(x - 3)(x^2 + 3x + 5) = 0$. To find the other two roots, we need only set the other factor $x^2 + 3x + 5$ to 0, and solve by the quadratic formula. The other solutions are: $-\frac{3}{2} \pm \frac{1}{2}i\sqrt{11}$.

While Ferro's formula seemed to work well, there was something strange about the results that will become evident in the next example and which led to the development of imaginary numbers.

Let us consider the equation, $x^3 - 15x = 4$, one of whose solutions is $x = 4$. Not only is one solution 4, but, if we graph the equation $f(x) = x^3 - 15x - 4$, we get (Figure 3.5):

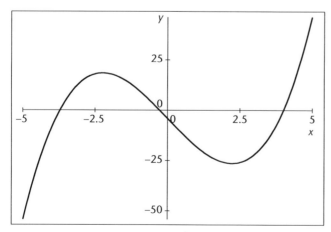

Figure 3.5 The graph of $fx = x^3 - 15x - 4$

and clearly ALL 3 solutions of $f(x) = 0$ are real, since the graph crosses the axis 3 times. If we use Ferro's formula, with $p = -15$ and $q = 4$ we get

$$\sqrt[3]{\frac{4 + \sqrt{4^2 + \frac{4(-15)^3}{27}}}{2}} + \sqrt[3]{\frac{4 - \sqrt{4^2 + \frac{4(-15)^3}{27}}}{2}}$$

which simplifies to $\sqrt[3]{\frac{4+\sqrt{-484}}{2}} + \sqrt[3]{\frac{4-\sqrt{-484}}{2}}$. Look! All the solutions are real, yet square roots of negative numbers are appearing in the algebraic solution! It was this kind of mystery that made mathematicians look much more carefully at square roots of negative numbers and develop the set of complex numbers. Contrary to what students are taught, the imaginary number i was not developed to solve the equation $x^2 = -1$; rather it was developed to explain the kind of situation that was occurring here when square roots of negatives appeared in equations whose solutions were obviously real! (See the chapter on imaginary numbers for more on this.)

It was the Italian mathematician Bombelli who discovered that, if we multiply the square roots of negative numbers as we do with square roots of positive numbers, we could explain much of what was happening. That is, if we treat the seemingly meaningless $\sqrt{-484}$ as if it satisfied the relationship $\sqrt{-484} \cdot \sqrt{-484} = -484$, much could be explained. Thus, he established that these imaginary numbers should be given status as bonafide numbers. Alas, this was the "birth" of complex numbers.

It will follow from work we do in Chapter 8 that $\sqrt[3]{\frac{4+\sqrt{-484}}{2}} + \sqrt[3]{\frac{4-\sqrt{-484}}{2}}$ reduces to the real number 4.

There is much more to be said here about complex roots of polynomials, which we will be in a better position to address in the chapter on imaginary numbers. For now, we continue to investigate how to solve cubic equations.

3.7.2 Cardan's Contribution

Girolamo Cardano (1501–1576), known as Cardan, made a key step in solving the general cubic equation of the form.

$$x^3 + bx^2 + cx + d = 0. \tag{3.23}$$

He came to the interesting realization that, if you make the substitution $x = y - \frac{b}{3}$ in equation (3.23), you get

$$\left(y - \frac{b}{3}\right)^3 + b\left(y - \frac{b}{3}\right)^2 + c\left(y - \frac{b}{3}\right) + d = 0$$

which simplifies to

$$y^3 + \left(c - \frac{1}{3}b^2\right)y = -d + \frac{1}{3}bc - \frac{2}{27}b^3. \tag{3.24}$$

(Do the algebra if you have the patience.) This equation is of the form $y^3 + py = q$ where $p = c - \frac{1}{3}b^2$ and $q = -d + \frac{1}{3}bc - \frac{2}{27}b^3$. And thus, we can solve this for y using Ferro's formula, and then, since

$x = y - \frac{b}{3}$, we can find out what the original solution to x is in equation (3.23). Thus, we have learned how to solve all cubic equations of the form of equation (3.23)!!!! Let us give one example to illustrate how this works.

Example 3.22 *Solve the equation* $x^3 - 15x^2 + 81x - 175 = 0$.

Solution. We make the substitution $x = y - \frac{(b)}{3} = y - \frac{(-15)}{3} = y + 5$, but instead of substituting in the original equation, which would yield an even more complex cubic equation, we make use of the fact that the reduced equation will be of the form $y^3 + py = q$ where

$$p = c - \frac{1}{3}b^2 \text{ and} \tag{3.25}$$

$$q = -d + \frac{1}{3}bc - \frac{2}{27}b^3. \tag{3.26}$$

Using the values $b = -15$, $c = 81$, $d = -175$, and substituting into equations (3.25) and (3.26), we get $p = 81 - \frac{1}{3}(-15)^2 = 6$, $q = -(-175) + \frac{1}{3}(-15)(81) - \frac{2}{27}(-15)^3 = 20$, so our reduced equation is

$$y^3 + 6y = 20.$$

We solved this earlier by Ferro's formula to get $y = 2$ (see Example 3.19) and thus, $x = y + 5 = 7$, which we can verify solves our equation. We leave the task of finding the other two solutions to you.

Cardan was a rather interesting character. Morris Kline in his book, *Mathematics for the Non Mathematician* (1967), tells us that Cardan suffered from many illnesses which seemed to prompt him to become a physician. In fact, he became quite a celebrated physician as well as a professor of medicine. Yet, with all his fame, Kline tells us about Cardan, "He was aggressive, high tempered, disagreeable and even vindictive as if anxious to make the world suffer for his early deprivations. Because illness continued to harass him and prevented him from an enjoying life, he gambled daily for many years. This experience undoubtedly helped him to write a now famous book, *On Games of Chance*, which treats the probabilities in gambling. He even gives advice on how to cheat, which was also gleaned from experience." (p.119)

3.7.3 The Fourth Degree and Higher Equations

Once the solution of cubics was found, the quest continued to find solutions of 4th degree, 5th degree, 6th degree polynomials, and so on. It was not long before the general 4th degree equation $ax^4 + bx^3 + cx^2 + dx + e = 0$ was solved. The formula is very complex and makes the formula for the solution of cubic equations look like child's play. Since the development of this formula is so complicated, we will not discuss it here. The difficulty of the formula makes it too difficult for secondary school students to learn, but their teachers are encouraged the visit the website http://www.karlscalculus.org/quartic.html to learn some of the details of this method. As difficult as this formula is, it is certainly usable in computer software programs that solve these 4th degree equations.

Mathematicians' intellectual curiosity led them to see if they could find formulas that would solve higher degree equations regardless of their complexity. Solving the 5th degree polynomial turned out to be much harder than people expected. Different methods were tried, but no one succeeded, because it turned out, as later proved by the mathematicians Abel and Ruffini, that there are no formulas that will solve these equations. This was a big surprise. The theorem follows.

Theorem 3.23 (*Abel–Ruffini*). *One cannot find a formula similar to the formulas for solving quadratic, cubic and quartic equations that will solve all 5th degree equations and higher. That is, one cannot find a formula that will solve all of these equations by using radicals.*

The proof of this theorem involved one of the first applications of group theory and is a very sophisticated result. One can contact the website http://en.wikipedia.org/wiki/Abel-Ruffini_theorem to learn more about this. This last theorem is not saying polynomials of higher degree can't be solved. They can be by a variety of methods, some of which we will discuss. But there is no FORMULA, like the quadratic formula that can be used to solve these equations. Thus, until the 20th century, when graphing calculators came on the scene, solving polynomials of degree 5 or higher was difficult, and in many cases, impossible. Even with these calculators, the solutions are not always exact. There is however, one very good method for solving these equations known as the Newton–Raphson method, which we discuss later on in this chapter. Without a computer, however, the method can be all but impossible. In fact, technology plays an important role in the solutions of equations, which we will discuss in the next section.

Student Learning Opportunities

1 Show that the two solutions in equation (3.19) are the same.

2 Find the other two solutions in Example 3.19.

3 Use the rational root theorem to find the real solutions of the following equations. Then use the formulas in this section to get the real solutions. Using your calculator show that the solutions are the same.
  (a) $x^3 + 4x = 5$
  (b) $x^3 + 2x = 12$
  (c) $x^3 - 4x = 3$

4 Use Cardan's idea and then synthetic division to reduce the following equations to depressed cubics (when necessary). Then solve the equations using equation (3.20)
  (a) $x^3 + 6x^2 - 9x - 14 = 0$
  (b) $x^3 - 4x^2 + 7x = 4$
  (c) $x(x^2 - 2) = 4$

3.8 The Role of the Graphing Calculator in Solving Equations

LAUNCH

1. Solve the following equations by entering them in your graphing calculator and finding where they cross the x-axis:
   (a) $y = x^3 - x^2 - 2x$
   (b) $y = x^3 + 199.99x^2 - 1.5 \times 10^5 x + 1500$
2. How many roots were you able to see in each case? How many roots should you see?
3. If there was a problem seeing the roots in equation 1(b), what do you think was the cause?

While it is wonderful that we now have graphing calculators to help us find solutions to higher order equations, you can see from your experience doing the launch that technology must be used with insight and skill. Let us begin by reviewing how to find real solutions using the graphing calculator and what to do in cases where the roots are not visible in the window.

When we wish to find real solutions of equations, we simply put the equation in the form $f(x) = 0$, graph it, and find where it crosses the x-axis. Let us illustrate this for a cubic equation and find its roots. The same idea works regardless of the degree of the equation.

Example 3.24 *Solve the equation $2x^3 = 7x - 1$ to 3 place decimal accuracy.*

Solution. We write the equation as $2x^3 - 7x + 1 = 0$, and plot the function $f(x) = 2x^3 - 7x + 1$. Our goal is to find the zeroes or roots of $f(x)$, which on the graph is where $f(x)$ crosses the x-axis. We see that, in this case, it crosses the x-axis in 3 places which are the solutions of $f(x) = 0$. That is, these crossings are the solutions of $2x^3 - 7x + 1 = 0$. Here is the graph of $f(x)$ (Figure 3.6):

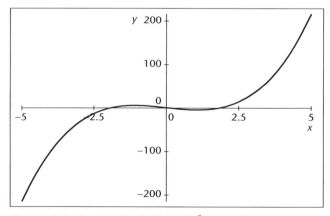

Figure 3.6 The graph of $f(x) = 2x^3 - 7x + 1$

We can zoom in on the graph near the points where it crosses the x-axis to get a better idea of the solutions, or we use the root solving capability that most graphing calculators have to find the zeroes of the function. We find that the three zeroes of $f(x)$ are $x \approx -1.9385\ldots$, $x \approx 0.14371\ldots$ and $x \approx 1.7948\ldots$. How easy it is today if we have an accurate graph of $f(x)$! The only real difficulty occurs when the roots are very far from the origin or close to the origin, or very close together, or very far apart, because when we zoom out or in, we just can't see them. This is exactly what happened in the second launch equation above. That is, when we wanted to solve $y = x^3 + 199.99x^2 - 1.5 \times 10^5 x + 1500$, we tried graphing it on a standard calculator used in secondary school, in a standard window (x and y both go from -10 to 10), but we could see very little. Zooming out we could still not see a root. In fact, we would have to zoom out three times before we could see two of the roots, but then our picture would look like two vertical lines, which we know is not correct. The third root is not within sight. Our polynomial happens to factor into $(x - 0.01)(x + 500)(x - 300) = 0$, which is not obvious but which tells us the roots right away namely, $x = 0.01$, $x = -500$ and $x = 300$. (This factoring we did is not magic. We just made up the problem beginning with the factors.) Here is the graph of our function using a more sophisticated computer program (Figure 3.7). (The numbers on the y-axis are powers of 10. Thus, 1e + 9 means 1.0×10^9.)

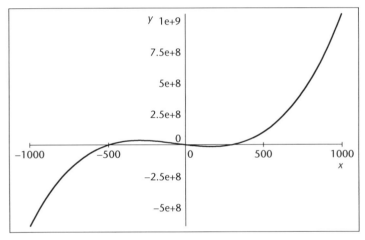

Figure 3.7

3.8.1 The Newton–Raphson Method

Thus far, we have talked about finding solutions of polynomial equations of the form $f(x) = 0$ and some of their applications. Since any equation can be put in the form $f(x) = 0$, as we shall see, *the methods we will present can be used to solve any equation that occurs in any field.* So, if we wanted to say solve $\sin x = \cos 4x - 2^x$, we simply bring all the terms to one side of the equation and, instead, solve $\sin x - \cos 4x + 2^x = 0$. The solution of this equivalent equation will give us the solution of our original equation. No wonder why finding roots of equations is so important! It can be used everywhere for any kind of application! More importantly, what we will do in this section generalizes and gives us a tool to solve complex systems of equations that occur in practice, which makes it very practical even in very sophisticated applications. So let's discuss this special technique called the Newton–Raphson method.

In calculus we often were asked to find the equation of a tangent line to a curve at a point. Amazingly, this equation be can be adapted and used to quickly find solutions to *all* equations and thus, has many applications in the sciences. Let us now show how this method can be used to get quick solutions of equations.

We need to recall how to find the equation of a tangent line to a curve $y = f(x)$ at a point (a, b) on the curve. First, we compute $f'(a)$, the derivative of $f(x)$ evaluated at a, as this gives the slope of the tangent to the curve $f(x)$ at a. Then the equation of the tangent line is

$$y - b = f'(a)(x - a). \tag{3.27}$$

But since (a, b) is a point on the curve $y = f(x)$, $b = f(a)$. Thus, the above equation can be written as

$$y - f(a) = f'(a)(x - a). \tag{3.28}$$

Let us give a numerical example as a review.

Example 3.25 *Find the equation of the tangent line to the curve $f(x) = 4x^2 - 3x + 1$ at the point $(a, b) = (1, 2)$.*

Solution. The slope of the curve at the point $(1, 2)$ is $f'(1)$. Since $f'(x) = 8x - 3$, $f'(1) = 5$ and the equation of the tangent line is, by (3.27) $y - 2 = 5(x - 1)$.

This brings us to the Newton–Raphson method. To understand it fully, we need to be able to figure out where a tangent line to the curve hits the x-axis. This is simple. Using equation (3.28) we can easily find where the tangent line hits the x-axis by setting $y = 0$. This yields

$$-f(a) = f'(a)(x - a). \tag{3.29}$$

Dividing both sides of (3.29) by $f'(a)$ and then adding a to both sides of the resulting equation, we get that the x coordinate of the point where the tangent line hits the x-axis is

$$x = a - \frac{f(a)}{f'(a)}. \tag{3.30}$$

Now refer to Figure 3.8 below.

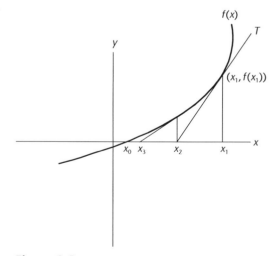

Figure 3.8

The goal is to find the zero of the function shown; that is, where the function $f(x)$ crosses the x-axis. This occurs at x_0. We observe that if we draw the tangent line T at a nearby point with coordinates $(x_1, f(x_1))$, the point, x_2, where the tangent line T crosses the x-axis, is closer to x_0 than x_1 is. (In the Student Learning Opportunities, we will show that this is not always true, but for now it suffices.) Furthermore, by equation (3.30), using x_1 for a, we get

$$x_2 = x_1 - \frac{f(x_1)}{f'(x_1)}. \tag{3.31}$$

Now we draw the tangent line to the curve at $(x_2, f(x_2))$ and look at where it crosses the x-axis. We see that it crosses at x_3 which is even closer to x_0 than x_2 is. Furthermore,

$$x_3 = x_2 - \frac{f(x_2)}{f'(x_2)}.$$

We continue generating x's in this way and get closer and closer to our zeroes of f (assuming we don't run into one of the problems described in the Student Learning Opportunities given later). This series of calculations can be done quite easily on a calculator, as you will see from the next numerical example.

Example 3.26 *Beginning with the number $x_1 = 1$, find the zero of the function $f(x) = 2x^3 - 7x + 1$ generated by the Newton–Raphson method.*

Solution. We show how we can do this very quickly on the TI calculator by doing the following steps. Begin by keying the function $2x^3 - 7x + 1$ in for Y_1. Put its derivative, $6x^2 - 7$ in Y_2. Since we are starting with $x = 1$, we begin by storing 1 in the variable X. We do that by typing $\boxed{1}\,\boxed{\text{STO}}\,\boxed{\text{x}}$ and then we press $\boxed{\text{Enter}}$. Next, key in $X - \frac{Y_1(X)}{Y_2(X)}\,\boxed{\text{STO}}\,X$ and keep pressing enter. This yields the following values for X which we call x_2, x_3, and so on.

$x_2 = -3$

$x_3 = -2.3191489$

$x_4 = -2.0139411$

$x_5 = -1.9424509$

$x_6 = -1.9385486$

$x_7 = -1.9385372$

$x_8 = -1.9385372$

$x_9 = -1.9385372$

We are at our solution. It rounds to what we got before, -1.9385 in Example 3.24.

What is nice about the Newton–Raphson method is that if it works, it works very fast. More specifically, when x_n has 3 digits of our solution correct, then x_{n+1} will have approximately 6 digits correct, and x_{n+2} approximately 12 digits correct. Thus, we quickly zero in on an accurate solution.

We will use the Newton–Raphson method again later in the book (Chapter 8) when we discuss fractals. That is yet another area where it is useful, only there we work with complex numbers.

Years back, before the days of graphing calculators, a former student who had become an electrical engineer called the first author of this book with a problem that came up in his job that was baffling him. He had this difficult equation that he needed to solve, but didn't have a clue how to proceed. I asked him if he tried the Newton–Raphson method and his response, "The what method?" told me I needed to explain it to him, which I did. He took out his scientific calculator and began computing. The next day he called me all excited because he had found the solution. I am sure he never forgot that.

In the article *"Getting to Real Time Load Flow,"* by Regina Llopsis-Rivas in *Electric Perspectives* (an Internet journal) Jan-Feb 2003 issue, we see the following quote:

"But today non-linear equations solving algorithms based on the Newton–Raphson method are used industry-wide to analyze the behavior of electrical power systems."

So this method is important, very important!

(For the full article, see http://findarticles.com/p/articles/mi_qa3650/is_200301/ai_n9168698/)

We will end this section by surprising you with how some calculators use an algorithm based on the Newton–Raphson method, to compute square roots very rapidly and accurately. We illustrate this with the next example.

Example 3.27 *Many calculators compute $\sqrt{N}$ using the following scheme: It picks say x_1, and successively generates new terms according to the formula*

$$x_{n+1} = \frac{1}{2}\left(x_n + \frac{N}{x_n}\right).$$

Show how this arises from the Newton–Raphson method and discuss how efficient it is.

Solution. We apply the Newton–Raphson method to the function $f(x) = x^2 - N$ starting with a positive value of x_1 to find a zero of this function. Of course, the zero is $\sqrt{N}$. Since $f'(x) = 2x$, the Newton–Raphson method tells us that

$$\begin{aligned}
x_{n+1} &= x_n - \frac{f(x_n)}{f'(x_n)} \\
&= x_n - \frac{x_n^2 - N}{2x_n} \\
&= \frac{2x_n^2 - (x_n^2 - N)}{2x_n} \\
&= \frac{x_n^2 + N}{2x_n} \\
&= \frac{1}{2}\left(\frac{x_n^2 + N}{x_n}\right) \\
&= \frac{1}{2}\left(x_n + \frac{N}{x_n}\right)
\end{aligned}$$

and we are done.

This method is very fast and very accurate, even if the calculator starts far away from the square root of the number and always converges to the square root if the initial guess is positive. For example, suppose that we wanted to compute $\sqrt{10}$ and we start with our initial guess of $x_1 = 100$, which is way off. Here is what the Newton–Raphson method gives us. Of course, all this is done with lightning speed on the computer:

$x_2 = 50.05$

$x_3 = 25.1249001$

$x_4 = 12.76145582$

$x_5 = 6.772532736$

$x_6 = 4.124542607$

$x_7 = 3.274526934$

$x_8 = 3.164201587$

$x_9 = 3.16227766$

$x_{10} = 3.16227766$

and then the subsequent values of x_i are all the same as x_{10}. So we are done. If we take as our initial guess N, things go faster, and we get our solution at x_5.

The algorithm for the computation of a square root that we presented goes back long before Newton, and was known to the ancient Greeks. Of course, how they got it is anyone's guess!

3.8.2 The Bisection Method – Unraveling the Workings of the Calculator

Students today use calculators in their mathematics classes on a daily basis. How the calculator gets its results is usually a mystery to them and their teachers. The purpose of this section is to reveal what some calculators do when they find the zeros of an equation. Different calculators may have different methods, but we concentrate on the widely used TI series calculator and what it does. The calculator's method is really not sophisticated at all. It uses a technique called the bisection method.

We begin with an interval containing a solution of $f(x) = 0$. This is an interval we can pick and one where $f(x)$ at the left endpoint of the interval and the right endpoint of the interval have opposite signs. The bisection method keeps cutting the interval in half and generates a smaller interval (one half the size) one of whose endpoints is now the latest midpoint and which contains the root.

Here is the **bisection method**: (1) Begin with an interval $[a, b]$ where $f(a) \cdot f(b) < 0$. (This is just another way of saying that $f(a)$ and $f(b)$ have opposite signs.) Let m be the midpoint of the interval. If $f(a) \cdot f(m) < 0$, that is, $f(a)$ and $f(m)$ have opposite signs, then our new and smaller interval containing a root is the interval $[a, m]$. If $f(b) \cdot f(m) < 0$, then our new interval containing the root is $[b, m]$. What is important to realize is that *the calculator is using the Intermediate Value Theorem to find the smaller interval containing the root.* Here we see yet another place where what we study in school can be applied.

Let us illustrate how the bisection method works with a specific example.

Example 3.28 *Solve $f(x) = 2x^3 - 7x + 1 = 0$.*

Solution. Every polynomial is continuous everywhere. A quick computation shows that $f(-2)$ is negative and that $f(0)$ is positive. That is, f is negative at the left endpoint and positive at the right endpoint. So, by the Intermediate Value Theorem, there is a root between -2 and 0. Let us take the midpoint of this interval, -1. If we compute $f(-1)$ we find that it is positive. Thus, $f(-2)f(-1) < 0$. So the new interval containing our root is $[-2, -1]$. If we look at $f(x)$ at the midpoint of the interval $[-2, -1]$ which is -1.5, we see that it is positive. Thus, $f(-2)f(-1.5) < 0$. So, we can reduce our interval containing a root of our equation to $[-2, -1.5]$. Again, we bisect the interval $[-1, -1.5]$ to get -1.75 where f is positive. Since $f(-2)f(-1.75) < 0$, our new interval containing a root is $[-2, -1.75]$. The midpoint is -1.875 where f is still positive, so our new interval containing the root is $[-2, -1.875]$. The midpoint is -1.9375 where f is still positive. So the new interval containing our solution is $[-2, -1.9375]$. The midpoint of this new interval is -1.96875 where the value of $f(x)$ is NEGATIVE. So our new interval is not $[-2, -1.96875]$ (since f is negative at both endpoints), but $[-1.96875, -1.9375]$ (where f at the two endpoints has opposite signs). The midpoint of this interval is -1.953125 where f is negative. So our new interval containing the solution is $[-1.953125, -1.9375]$, and so on.

Here is a table that summarizes the work. a will always stand for "left endpoint of the interval containing the root," and b for "right endpoint of the interval containing the root," and m for "midpoint of that interval."

a	b	m	$f(a) \cdot f(m)$	New interval containing root
-2	0	-1	negative	$[-2, -1]$
-2	-1	-1.5	negative	$[-2, -1.5]$
-2	-1.5	-1.75	negative	$[-2.1.75]$
-2	-1.75	-1.875	negative	$[-2. -1.875]$
-2	-1.875	-1.9375	negative	$[-2, -1.9375]$
-2	-1.9375	-1.96875	positive	$[-1.96875, -1.9375]$
-1.96875	-1.9375	-1.953125	negative	$[-1.953125, -1.9375]$

The calculator does these computations very rapidly leading to the solution $x \approx -1.9384$. Although compared to the Newton–Raphson method, the bisection method is relatively slow, it does work all the time for *any* continuous function on an interval containing a single root and so is an excellent method for finding solutions.

Although we have used the Intermediate Value Theorem to explain the bisection method, it has many theoretical consequences, one of which we will use later on in the book when we discuss the theory behind radicals. That result is,

Theorem 3.29 *Every positive number has a square root. Furthermore, there is only one positive square root.*

Proof. We give the proof for the number 7. The proof is similar for any other positive number we choose. Form the function $f(x) = x^2 - 7$. Now $f(0)$ is negative and $f(10)$ is positive. Thus, there is a root of $f(x)$ between 0 and 10. That root satisfies $x^2 - 7 = 0$ or just $x^2 = 7$. That is, there is a number whose square is 7, and thus 7 has a square root.

To show that there is only one positive square root, suppose that there are at least two positive square roots, and that x and y are two of them. Then by definition of square root of 7 (a number whose square is 7,) $x^2 = 7$ and $y^2 = 7$. Thus $x^2 = y^2$. Hence $x^2 - y^2 = 0$ from which it follows that $(x - y)(x + y) = 0$. This means that either $x - y = 0$ or $x + y = 0$. Since x and y are both assumed to be positive, $x + y > 0$. So the second equation, $x + y = 0$, can't hold. Thus, $x - y = 0$ and it follows that $x = y$. We have shown that any two positive square roots of 7 must be the same. Thus, there is only one positive square root of 7. The proof is similar for any positive number. ■

For a continuation of the study of solutions of equations, the reader should go to Section 6.11, where we discuss extraneous solutions and what types of operations on equations can get us into trouble.

Student Learning Opportunities

1 Use your calculator to find all real roots of the following equations:
  (a) $x^3 - 3x^2 - 5 = 0$
  (b) $x^2 + 1 = -x$
  (c) $\sin x = (2/3)x$
  (d) $4^x = 2x + 1$
  (e) $\log x = x - 5$.

2 Use the bisection method to find each of the zeroes of the function below in the indicated interval.
  (a) $f(x) = x^2 + x - 1$ on $[-1, 1]$
  (b) $f(x) = 2x^3 - x - 3$ on $[0, 2]$
  (c) $f(x) = \sin x - x/2$ on $[1, 3]$
  (d) $f(x) = 2^x - x - 1$ on $[0, 2]$.

3 Use the Newton–Raphson method and your calculator to find each root of the functions in Question 2 (above) taking the right endpoint as your initial guess. Do you prefer doing this by hand or using the root-finding capabilities of the calculator? (You need to recall that the derivative of $\sin x$ is $\cos x$ and that the derivative of 2^x is $2^x \ln 2$.)

4 Suppose one used the Newton–Raphson method with the function $f(x) = \sqrt[3]{x}$ with initial guess 1 to find the roots of $f(x)$. Obviously, the only zero of $f(x)$ is $x = 0$. Show that Newton's method fails to converge when $x = 1$, or any number not equal to 0.

5 (C) Your student tried to use the Newton–Raphson method to find the roots of the function $f(x) = 2x^3 - 6x^2 + 6x - 1$ by beginning with an initial guess of 1 and was unsuccessful. Why? What is special about the tangent line to $f(x)$ at $x = 1$ that causes this behavior? What can you suggest to your student to help find the roots using this method?

6 (C) Your student tried to use the Newton–Raphson method to find the roots of the function $f(x) = x^3 - 2x + 2$ by beginning with an initial guess of 1. She was unsuccessful and asked you for help. Why was she unsuccessful? Explain using tangent lines what is causing this behavior. What can you suggest to her to help her find the roots of $f(x)$?

7 **(C)** Your students are familiar with the algorithm for finding the square root of a number. They ask you if there is a similar algorithm for finding the cube root of a number? The fourth root of a number? The fifth root of a number? Modify example 3.27 to find such formulas.

8 Show that every positive number has a unique positive fourth root.

9 The torque in foot pounds of a certain engine is approximated by

$$T = 0.8x^3 - 18x^2 + 71x + 112 \quad \text{for } 1 \leq x \leq 5$$

where x is the number of revolutions per minute. Using any method discussed in this section, find the approximate values of x that make the torque 140 foot pounds.

10 A rectangular sheet of metal 8 inches by 15 inches is to be used to construct a box by cutting out squares from the corners and folding up the sides. If the volume of the resulting box is 80 cubic inches, what size square must be cut out to accomplish this? Use a method presented in this section.

CHAPTER 4

MEASUREMENT: AREA AND VOLUME

4.1 Introduction

Starting in elementary school, children learn about such important concepts in measurement as area and volume. These are measures that are used in our lives on a daily basis. For example, we buy carpeting according to square footage and this is a measure of area. We buy paint according to the area of the surface that must be covered. We buy milk by the quart, which is a measure of volume, and we build tanks to hold gallons of oil, another measure of volume. These are just a few of many, many applications of area and volume that are used in various fields on a day-to-day basis. In this chapter we take a closer look at these concepts by linking basic geometry, algebra, trigonometry, probability, and the rudiments of calculus together with modern technology.

Since we want to stress some extremely interesting approaches and relationships involved in area and volume, we will avoid a strict formal approach and we will assume that you accept certain facts such as: congruent figures have congruent areas; parallel lines are everywhere equidistant; and regular polygons with an increasing number of sides inscribed in a fixed circle of radius r have areas approaching the area of a circle. Afterwards, we discuss some of the issues with this mostly informal approach and the need for axiomatizing certain relationships.

4.2 Areas of Simple Figures and Some Surprising Consequences

LAUNCH

1. Show, using a picture, that the area of a rectangle with sides 2 inches wide and 3 inches long has an area of 6 square units. [Hint: divide the rectangle into square inches.]
2. In a similar manner, show, using pictures, that the area of a rectangle with sides $\frac{1}{2}$ of a unit and $\frac{1}{3}$ of a unit is $\frac{1}{6}$ of a square unit.

We imagine that the first launch question was quite simple for you, as you probably have done similar problems in elementary school. But, did you ever wonder why we define area as we do? Is the formula for finding the area of a rectangle a theorem or a definition? Have you ever seen a proof of the formula, or do you believe that, by breaking up the rectangle as you have done that in actuality, you have just proven it? The second problem was probably much more challenging and likely, less familiar to you. Do you think that using an area model is a helpful way to justify how we multiply fractions? Would this method work if you were to use improper fractions as well? We hope that your curiosity has been piqued by these questions, which will all be addressed in the section that follows. Let us begin by examining how we measure area.

As you well know, we measure area in square units. But what is a square unit? A **square unit** is exactly what is sounds like. It is a square, whose sides are all 1 unit, as shown in Figure 4.1 below.

Figure 4.1

Thus, a square foot, is a square 1 foot by 1 foot, and a square yard, is a square 1 yard by 1 yard. Carpeting and flooring are often sold by the square foot or square yard, as are many other materials used in construction.

If the length of a rectangle is 3 units and the width is 5 (of the same) units, then the area of the rectangle is 15 square units, as is easy to see. We simply break the rectangle into 15 square units by drawing horizontal lines 1 unit apart and vertical lines 1 unit apart as shown below.

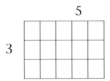

We see that the area is 15 square units. Similarly, if one side of a rectangle is 4 units and the other, 8, then we can divide the rectangle into 32 square units. So, we see that the area of this 4 by 8 rectangle is 32 square units. It seems clear then that, to find the area of a rectangle whose length and width are whole numbers, we just multiply the length by the width, and that counts the number of square units.

This method of multiplying also works when the sides are fractional. For example, suppose that the length of a rectangle is $\frac{2}{3}$ of a unit and that the width is $\frac{3}{5}$ of a unit. Figure 4.2 below shows a unit square and a darkened rectangle with dimensions $\frac{2}{3}$ of the unit and $\frac{3}{5}$ of the unit. We can see from the figure that the square unit is broken into 15 congruent rectangles, each of which is $\frac{1}{15}$ of the square unit. We see that our rectangle with dimensions $\frac{2}{3}$ of a unit and $\frac{3}{5}$ of a unit takes up $\frac{6}{15}$ of the square unit.

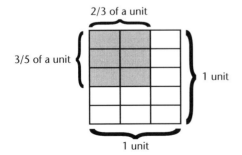

Figure 4.2

Thus, the area of a rectangle with dimensions $\frac{2}{3}$ by $\frac{3}{5}$ of a unit is $\frac{6}{15}$ of a square unit; again, length times width.

You may be thinking, "So what is the big deal? The area of a rectangle is length times width. That is the formula for the area of a rectangle!" Would it surprise you to know that we cannot prove that the area of a rectangle is length times width? It is a definition that arose from examples like the one above.

To show that we need a definition, consider the following problem: What would the area be of a rectangle whose width is $\sqrt{2}$ units long and whose length is $\sqrt{3}$ units long? Both $\sqrt{2}$ and $\sqrt{3}$ are irrational. If we write out their decimal equivalents, they will go on forever (i.e. $\sqrt{2} = 1.4142\ldots$ and $\sqrt{3} = 1.7321\ldots$). How can we divide this rectangle with sides $\sqrt{2}$ and $\sqrt{3}$ into squares whose sides are 1 unit, or even $\frac{1}{5}$ of a unit, or $\frac{1}{10}$ of a unit? Of course, the answer is, we can't. So, how do we know that the area is $\sqrt{2} \cdot \sqrt{3}$? The answer to why the area of a rectangle is DEFINED as length times width, is so that it will be consistent with those examples where we *can* divide the rectangles up into unit squares.

This business of defining area troubles many people. Area is the amount of space taken up by a figure. How can we define what this is? It is no different from defining a foot and then measuring length with a ruler that represents a foot. A measurement of 1 foot is an object created by human minds. We could just as well have defined a measurement of length to be the distance from the tip of your nose to your bellybutton, called that a "bod," and then measured how many bods there are in, say, a mile. The definitions we use for area, length, temperature, and so on, are totally constructed by human beings. By establishing standard measures, we are able to make sense of what we observe.

Having made the definition of the area of a rectangle as length times width, we can now easily derive the formulas for areas of other figures. Yes, we did say **derive**. It is quite remarkable that we can go from one figure to the next and find their areas, all from the area of a rectangle. What is especially nice is that, in doing so, we will see a direct interplay between algebra and geometry. The first few results are routine and will be gone through quickly, but soon some surprising results will emerge.

Theorem 4.1 *The area, A, of a right triangle with legs a and b, is given by* $A = \frac{1}{2}ab$.

Proof. Start with right triangle *ABD* and observe that it is half of a rectangle *ABCD* with sides *a* and *b* (Figure 4.3).

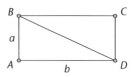

Figure 4.3

Since the area of the rectangle is *ab*, the area of the triangle is $\frac{1}{2}ab$. ∎

Given a triangle, *ABC*, the custom is to denote the side opposite angle *A* by *a*, and the side opposite angle *B* by *b*, and the side opposite side *C* by *c*. By an **altitude** of a triangle, we mean a line drawn from a vertex, perpendicular to the opposite side, extended if necessary. Below, in

Figure 4.4, we see a triangle *ABC* and notice that, to draw the altitude to side *b*, we need to extend it.

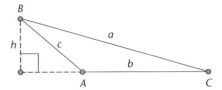

Figure 4.4

In geometry, the term "corresponding" is frequently used, especially in congruence theorems. In that context, corresponding parts are parts that match when one triangle is placed upon another so that all parts fit exactly. In the following theorem, the term "corresponding" refers to the specific base to which the height is drawn.

Theorem 4.2 *The area of any triangle, one of whose bases is b and whose corresponding height is h, is $\frac{1}{2}bh$.*

Proof. We give the proof for a triangle whose altitude is inside the triangle. In the Student Learning Opportunities, you will prove the formula for the case where the altitude falls outside of the triangle. Suppose *ABC* is a triangle with altitude *BD* drawn to base *AC* = *b*, dividing *AC* into segments *x* and *y*, as shown in Figure 4.5 below.

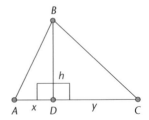

Figure 4.5

This altitude divides the triangle into two right triangles, *ADB* and *CDB*. The area of *ABD* is $\frac{1}{2}xh$, by the previous theorem, and similarly, the area of triangle *DBC* is $\frac{1}{2}yh$. The area of triangle *ABC* is the sum of these areas. Thus, the area of triangle *ABC* is $\frac{1}{2}xh + \frac{1}{2}yh = \frac{1}{2}(x+y)h = \frac{1}{2}bh$. ∎

Note that any side of a triangle may be taken as the base, and the height is the altitude drawn to that base. No matter which side is considered to be the base, *b*, if we draw *h*, the altitude to that side, and compute $\frac{1}{2}bh$, we will get the area. Thus, we have three possible ways to get the area depending on which base we use.

Theorem 4.3 *The area of a parallelogram, with base b and height h, is bh.*

Proof. We begin with parallelogram $ABCD$ and draw diagonal BD. This diagonal divides the parallelogram into two congruent triangles ABD and CDB as shown in Figure 4.6 below. (Why?)

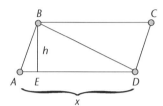

Figure 4.6

We draw the altitude BE of triangle ABD and call it h, while we call the base, AD, of the parallelogram, x. Now, the area of triangle ABD is $\frac{1}{2}AD \cdot BE = \frac{1}{2}xh$. Thus, the area of the parallelogram being the sum of the areas of the two congruent triangles is $\frac{1}{2}xh + \frac{1}{2}xh$ which, of course, is xh or just base times height. ∎

Theorem 4.4 *The area of a trapezoid is $\frac{1}{2}h(b_1 + b_2)$, where b_1 is the length of the shorter base and b_2 is the length of the longer base.*

Proof. Start with trapezoid $ABCD$ and draw altitudes, BE and FD as shown in Figure 4.7 below, and diagonal BD.

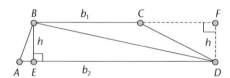

Figure 4.7

Then the area of triangle ABD is $\frac{1}{2}b_2 h$ and the area of triangle CBD is $\frac{1}{2}b_1 h$. (Both triangles have the same height because parallel lines are everywhere equidistant.) The area of the trapezoid is the sum of the areas of these two triangles and thus, is $\frac{1}{2}b_1 h + \frac{1}{2}b_2 h = \frac{1}{2}h(b_1 + b_2)$. ∎

Although we haven't done much with areas, we are already in a position to get some impressive results. Here is one—the well known Pythagorean Theorem.

Theorem 4.5 *In a right triangle with legs a and b and hypotenuse c, $a^2 + b^2 = c^2$.*

Proof. We begin with right triangle ABC with right angle at C. Place a triangle BED congruent to ABC (and with right angle at D) in such a way that CBD is a straight line. Then draw AE. (See Figure 4.8 below.)

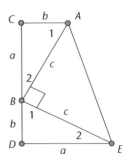

Figure 4.8

Since angles C and D are right angles, AC and DE are perpendicular to the same line CBD and, hence, are parallel. That makes figure $ACDE$ a trapezoid.

Furthermore, since angles 1 and 2 in triangle ABC add up to 90 degrees and angle CBD is 180 degrees, angle ABE is also a right angle. (Why?) Thus, we have three right triangles in the figure.

Now, by the previous theorem, the area of the trapezoid is one half the height times the sum of the bases. The height of the trapezoid is CD or just $(a + b)$. The parallel bases are AC and DE. Thus,

$$\text{Area of trapezoid } ACDE = \frac{1}{2}CD(AC + DE) = \frac{1}{2}(a+b)(a+b). \tag{4.1}$$

Now, we know that the area of the trapezoid is the sum of the areas of the 3 right triangles, ACB, EBD, and ABE, and by Theorem 4.1 we have that

$$\text{Area of triangle } ACB = \frac{1}{2}ab \tag{4.2}$$

$$\text{Area of triangle } EBD = \frac{1}{2}ab \tag{4.3}$$

and

$$\text{Area of triangle } ABE = \frac{1}{2}c^2. \tag{4.4}$$

Since the area of the trapezoid is equal to the sum of the areas of the triangles, by using equation (4.1)–equation (4.4) we have that

$$\underbrace{\frac{1}{2}(a+b)(a+b)}_{\text{Area of Trapezoid } ACDE} = \underbrace{\frac{1}{2}ab + \frac{1}{2}ab + \frac{1}{2}c^2}_{\text{Sum of the areas of the triangles}} \tag{4.5}$$

Upon multiplying equation (4.5) by 2 we have

$$(a+b)(a+b) = ab + ab + c^2,$$

which simplifies to

$$a^2 + 2ab + b^2 = 2ab + c^2.$$

Subtracting $2ab$ from each side we get

$a^2 + b^2 = c^2$

and we are done. How nice! ∎

In the Student Learning Opportunities we outline yet another proof of the Pythagorean Theorem using areas. It is interesting and surprising that we can actually prove the Pythagorean Theorem using *areas* of triangles and trapezoids. What is also interesting, from a historical point of view, is that this proof just given was not done by a mathematician, rather, by the 20th president of the United States, James Garfield!

We just proved the Pythagorean Theorem: In a right triangle with legs a and b and hypotenuse c, $a^2 + b^2 = c^2$. The converse of this theorem is also true and its proof is rarely found in secondary school textbooks. We now give a proof of this for your reference. What is unusual about this proof is that it uses the Pythagorean Theorem to prove the converse of the Pythagorean Theorem. Since it is uncommon in mathematics for a theorem to be used to prove its converse, this proof is somewhat special.

> **Theorem 4.6** *In a triangle ABC with sides a, b, and c, if $a^2 + b^2 = c^2$, the triangle is a right triangle and angle C is the right angle.*

Proof. We begin with triangle ABC which we do not know is right. Starting at C, we draw a line CD perpendicular to BC of length AC, and then draw BD. (See Figure 4.9 below. Notice the right angle on the right side only.)

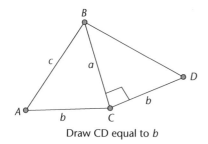

Draw CD equal to *b*

Figure 4.9

Our goal is to show that triangles ABC and CBD are congruent, since their corresponding parts will be congruent (That is, they will have the same measure.) It will follow that angle BCA is a right angle, which is what we want to prove.

Now, by construction, $AC = DC$, and of course BC is common to both triangles. Thus, we have two sides of one triangle equal to two sides of the other triangle. If we can show the third sides are equal (that is, that $c = BD$), then the two triangles will be congruent and we will be done. So, we proceed to show that $c = BD$.

By construction, triangle CBD is a right triangle, so we can apply the Pythagorean Theorem to THAT triangle to get

$(BC)^2 + (CD)^2 = (BD)^2.$

But $BC = a$ and $CD = b$ by construction, so the above equation becomes

$a^2 + b^2 = (BD)^2$.

Now, using the fact that we were given $a^2 + b^2 = c^2$ in triangle ABC, we substitute for $a^2 + b^2$ into the above equation, to get

$c^2 = (BD)^2$

and hence $c = BD$ and we are done.

The two triangles are now congruent since 3 sides of one triangle are congruent to three sides of the other. Thus, angle BCA must be a right angle, because in congruent triangles corresponding parts are congruent. With a few hints, this proof could be given to some astute secondary school students. ∎

It is easy to go from the area of a triangle to the area of a regular polygon (one whose sides all have the same length and whose angles all have the same measure) by breaking the polygon into triangles and summing the areas of the triangles. We leave that for you in the Student Learning Opportunities. But we need to first review the formula for the area of a polygon for reference.

It can be shown that every regular polygon can be inscribed in a circle. If we draw a perpendicular line from the center of the circle to any side of the inscribed polygon, that line is called an apothem. The figure below shows an apothem for a square and for a pentagon, both inscribed in a circle of radius r. It is a fact, and you will prove this in the Student Learning Opportunities, that the area of a regular polygon is $\frac{1}{2}ap$ where a is the apothem and p is the perimeter of the polygon. It is also a fact, and we will use this later, that as the number of sides of the inscribed regular polygons increases, the lengths of the apothems of the polygons approach the radius of the circle. (See Figure 4.10.)

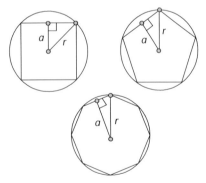

Figure 4.10

Student Learning Opportunities

1 (C) A student asks you to justify, using pictures, that the area of a rectangle with sides $\frac{1}{3}$ of a unit and $\frac{1}{4}$ of a unit is $\frac{1}{12}$ of a square unit. Show the diagram and explain how you would demonstrate it. Do the same for a rectangle with sides, $\frac{3}{2}$ and $\frac{1}{3}$.

2 (C) A student asks how you can prove that the shortest straight line distance from a point to a line is the perpendicular distance from that point to the line. How would you show this using the Pythagorean Theorem?

3 **(C)** Your students are familiar with how to prove the formula for the area of a triangle when the altitude falls within the triangle. But they are curious how to prove this formula when the altitude drawn to the base is outside the triangle. How would you help them do it?

4 If the base of a triangle is increased by 10% and the altitude is decreased by 10%, by what percentage is the area changed and is it increased or decreased? Explain.

5 Find the length and width of a rectangle if, when the length of a rectangle is increased by 2 and its width is decreased by 2, its area stays the same, while if the length is increased by 2 and the width is decreased by 1, we also get the same area.

6 One side of a triangle is 5 and the altitude to that side is 4. Another side of the triangle is 3. Can you tell what the length of the altitude to that side of the triangle is? If not, why not? If so, show what it is.

7 **(C)** Your students come across the following formula for the area of an equilateral triangle: $A = \dfrac{s^2\sqrt{3}}{4}$, where s is the length of the side. They ask you why it is true. How do you help them derive the formula for themselves? [Hint: When you draw an altitude, it cuts the base in half.]

8 Find the area enclosed by Figure 4.11 below.

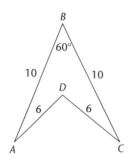

Figure 4.11

9 Prove that the area of a regular polygon is $\dfrac{1}{2}ap$, where a is the apothem and p is the perimeter.

10 **(C)** One of your students asks if it is ever the case that the numerical area of a rectangle with integer sides is equal to its numerical perimeter. How do you reply? How many such rectangles are there?

11 Imagine that, on a fictitious planet of Zor, a strange sort of geometry exists. Make believe that on this planet of Zor, the area of a rectangle is defined to be the length plus the width. And suppose that, on Zor, they assume that congruent figures have the same area, and that the rules for triangles being congruent are the same on Zor as in the Euclidean plane.

(a) Find the area of a rectangle with length 3 and width 4 on Zor.
(b) Derive the formula for the area of a right triangle on Zor.

(c) Show that one gets two different formulas for the area of a triangle on Zor, depending on whether the altitude to the base is inside the triangle or outside.

(d) Show that, on Zor, parallelograms other than rectangles do not have well defined areas. Do this by splitting the parallelogram into two triangles in two different ways as shown in Figure 4.12 below and then by using part (c).

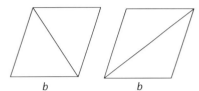

Figure 4.12

12 In Figure 4.13 below, line EF is parallel to line CB. Which triangle has greater area, triangle CGB, or triangle CFB, or is it impossible to tell? Explain.

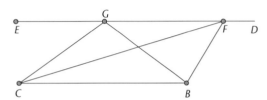

Figure 4.13

13 A rectangle has length 7 inches and width 9 inches. There is a border of $\frac{1}{2}$ inch around the rectangle. Guess what percentage the area of the border is to the entire rectangle plus the border, and then check if your guess is right. Are you surprised?

14 In quadrilateral ABCD, AB = 3, BC = 4, CD = 12, and DA = 13. Angle B is a right angle. Find the area of the quadrilateral.

15 (C) Your students have asked to see a proof of the Pythagorean Theorem that they could easily understand. You decide to give them a visual, hands-on method of proving the theorem. You begin by giving them cut-outs of four congruent right triangles with legs of length a and b and ask them to arrange them so that they form a large square with sides of length $a + b$, as depicted in Figure 4.14 below. Answer the following that you will be asking your own students, and show how the answers lead to a proof of the Pythagorean Theorem.

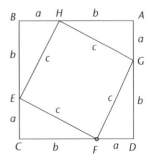

Figure 4.14

(a) Explain why *EHGF* must be a square.

(b) The area of the square *ABCD* is the area of the 4 triangles plus the area of the square *EHGF*. Find all these areas and set up an equation expressing this relationship. What have you found? How does this yield yet another proof of the Pythagorean Theorem?

16 **(C)** One of your astute students notices that the Pythagorean Theorem can be interpreted as follows: If we draw squares on the three sides of a right triangle, the sum of the areas of the squares on the legs of the triangle is the area of the square drawn on the hypotenuse. She asks if a similar relationship holds if we draw regular pentagons or regular hexagons on all three sides of the triangle. How do you respond to this student and how do you justify your answer?

17 A ship is located at *A*, 10 miles south of a ship located at *B*. The ship at *B* is going to travel east at a rate of 2 miles per hour while the ship at *A* can travel at 5 miles per hour. He wishes to meet the ship traveling from *B* at some point *C*, only he needs to know where *C* is so that he can set his course. (See Figure 4.15 below.)

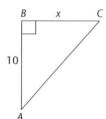

Figure 4.15

Find the value of *x* by using the formula: time traveled $= \dfrac{\text{distance}}{\text{rate}}$ for both ships and setting the times equal to each other.

18 In 2005, the Pythagorean Theorem was a deciding factor in a case before the New York State Court of Appeals. A man named Robbins was convicted of selling drugs within 1000 feet of a school. In the appeal, his lawyers argued that the man wasn't actually within the required distance when caught and so should not get the stiffer penalty that school proximity calls for. Here are the details from a "Math Trek" article by Ivars Peterson (mathland/mathtrek_11_27_06.html):

"The arrest occurred on the corner of Eighth Avenue and 40th Street in Manhattan. The nearest school, Holy Cross, is on 43rd Street between Eighth and Ninth Avenues. Law enforcement officials applied the Pythagorean Theorem to calculate the straight-line distance between the two points. They measured the distance up Eighth Avenue (764 feet) and the distance to the church along 43rd Street (490 feet). Using the data to find the length of the hypotenuse, (*x*) feet. Robbins' lawyers contended that the school is more than 1000 feet away from the arrest site, because the shortest (as the crow flies) route is blocked by buildings. They said the distance should be measured as a person would walk the route. However, the seven-member Court of Appeals unanimously upheld the conviction, asserting that the distance in such cases should be measured 'as the crow flies.'"

Find the value of the hypotenuse *x* and explain why the lawyers argued the way they did.

19 In this section we defined the area of a rectangle to be length times width. From this, it follows that the area of a square with side *x* is x^2 since every square is a rectangle. Suppose we decided to go the other way, namely, define only the area of a square with side x to be x^2. Using this formula, we can show that the area of a rectangle with length l and width w, is lw. Finish the details of the following proof.

Begin with rectangle $ABCD$, where $AB = w$ and $BC = l$. (See Figure 4.16.) Extend AB to E so that $BE = l$, and extend BC to F so that CF is w, and draw the lines shown to form the square $AEGI$ with side $(l + w)$ and area $(l + w)^2$. Then note that the area of $AEGI$ is the sum of the areas of square $EBCH$ and square $DCFI$ and the two congruent rectangles $ABCD$ and $CFGH$. Take it from there.

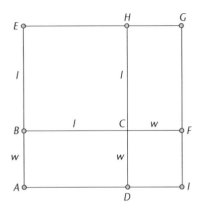

Figure 4.16

20 In Figure 4.17 below, AE, BF, and CD are medians in triangle ABC.

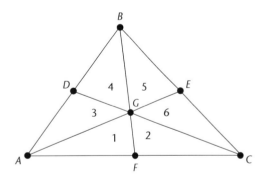

Figure 4.17

(a) Show that triangles 1 and 2 have the same area.
(b) Ditto for triangles 3 and 4.
(c) Ditto for triangles 5 and 6.
(d) Also show that the sum of areas 1, 3, and 4 is the sum of areas 2, 5, and 6. Why does it follow that the area of triangle AGB = the area of triangle BGC?
(e) Now, why does it follow that area of triangle 3 = area of triangle 6?
(f) Show that all areas of all triangles 1 through 6 are the same.
(g) Then show that the ratio of BG to GF is 2:1. (This is one part of the famous result that the medians meet at a point $\frac{2}{3}$ of the way from any vertex.)

21 Begin with a square each of whose sides is $2s$. Connect the midpoints of the sides. Show that the resulting quadrilateral is a square and find its area in terms of s.

22 Pick a point inside an equilateral triangle, and draw the lines representing the distances to each side. Call these distances, h_1, h_2 and h_3. Prove, using areas of triangles, that $h_1 + h_2 + h_3 = h$, where h is the altitude of the equilateral triangle.

23 In Figure 4.18 below

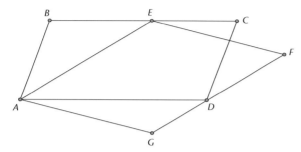

Figure 4.18

ABCD is a parallelogram. *E* is any point on *BC*. Show that the area of parallelogram *ABCD* is the same as the area of *EFGA*. [Hint: Draw *ED*.]

4.3 The Circle

 **LAUNCH**

1 The equatorial diameter of the earth is 7926 miles. Calculate the distance around the earth at the equator, using 3.14 as an approximation of π.
2 What formula did you use for your calculation?
3 Where did this formula come from? Is it a theorem or a definition?
4 Where does π come from? In terms of a circle, what does it represent? What is so extraordinary about it?

We are sure that you had no difficulty figuring out the distance around the earth at the equator. The formula you used to do it, that you learned many years ago, has been used by others since before the third century BC. Since having tried to answer the launch questions, are you now wondering where and how this formula originated? Are you now curious about the meaning of π? We hope so, since the history is fascinating! The section that you are about to read will reveal many interesting stories about circles and their features.

As you must be aware, the study of the circle is a major part of the middle and secondary school curriculum. Therefore, as a future teacher, we're sure you will agree that it is important for you to know and appreciate the history and meaning of all the formulas and amazing relationships regarding the circle that you will be teaching your students. We will begin with a discussion of the circumference of a circle and then examine its area.

The circumference of a circle is $2\pi r$. How do we know that? The answer may surprise you. Several thousand years ago it was discovered that the ratio of the circumference, C, to the diameter, d, of a circle, appeared to be the same no matter what the size of the circle. This ratio was a bit over 3. This was a discovery verified repeatedly by experimentation. There was no formal proof of it. Thus, this was an accepted fact about the nature of the circumference of a circle. It was an axiom based on observation. This may disturb those of you who need to see proof. We will give several corroborations of this relationship soon, in an attempt to convince you that this is true.

So, based on observation, we are accepting that the ratio of the circumference to the diameter is always the same. Why not give this ratio a name? An English mathematician, William Oughtred, in a book written in 1647, *Clavis Mathematicae*, felt it would be good to have a symbol to represent the ratio of the **p**eriphery of a circle (the English word for circumference) to its diameter. And since mathematicians are in the habit of using Greek letters to represent mathematical objects, he used the letter π. It stood to remind us of where it came from—**p**eriphery. So, π was defined as the ratio of the circumference of a circle to its diameter, which was observed to be the same for each circle. That is, $\pi = \frac{C}{d}$ by *definition*. Thus, the statement $C = (\pi)d = (\pi)2r = 2\pi r$, followed from a definition based on observations. We now move to the area of a circle.

4.3.1 An Informal Proof of the Area of a Circle

To go from the area of a polygon to the area of a circle is a bit sophisticated, since polygons have straight sides and circles have curvature. On the middle school level, once you have introduced the area of a rectangle, the following "proof" convinces many that the area of a circle is πr^2. This proof, originally done by the Greeks, is over 2000 years old. We begin with a circle, which we cut into an even number of sectors. (See Figure 4.19 below.)

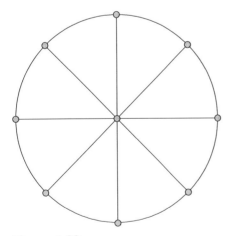

Figure 4.19

Next, we cut out the sectors and arrange them along a line as shown in Figure 4.20 below:

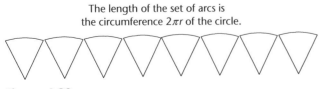

Figure 4.20

Next, we cut this string of sectors in half and fit the bottom half of "teeth" into the top half like the teeth in your mouth (well, if you are an alligator!). (See Figure 4.21.) We get the following figure:

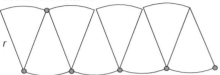

The length of the arcs on top is half the circumference of the circle, or πr. The same is true for the bottom.

Figure 4.21

The area of this figure is the area of the circle regardless of the number of sectors into which we cut the circle. The more sectors, the narrower the sectors are, and the closer the above figure approximates a rectangle with length πr and height r. Here is the picture we got by dividing the circle into 18 sectors each with central angle 30 degrees (obtained by using a commercial graphics program). One can already see that Figure 4.22 almost looks like a rectangle.

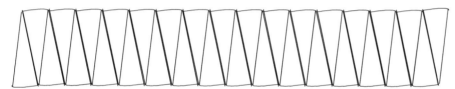

Figure 4.22

Since these figures approach a rectangle with length πr and height r, and the area of a rectangle is base times height= (πr) times r, the areas of these figures approach πr^2. But all the areas are the same, the area of a circle. Thus, the area of a circle must be πr^2!

4.3.2 Archimedes' Proof of the Area of a Circle

We now take a journey through genius. Earlier, we gave a plausible argument that the area of a circle was πr^2. Actually, Archimedes is responsible for this formula. Archimedes had an amazing mind, and many of his proofs were extremely clever. His proof that the area of a circle is πr^2 is no different. Again, he used the observed fact that the circumference of a circle is $2\pi r$.

Before we proceed, observe that, as we inscribe polygons of more and more sides in a circle, as shown in Figure 4.23 below,

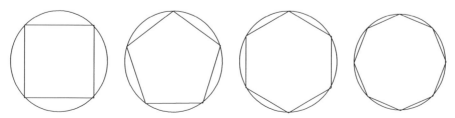

Figure 4.23

the areas of the polygons approach the area of the circle. Archimedes relied on this observation in his proof.

Here is Archimedes' proof for the area of the circle. The proof astounds one for its simplicity.

Theorem 4.7 *The area of a circle is πr^2.*

Proof. Begin with a circle of radius r and call its area A. Draw a right triangle, one of whose legs is r and whose other leg is the circumference of the circle, namely $2\pi r$. (See Figure 4.24 below.)

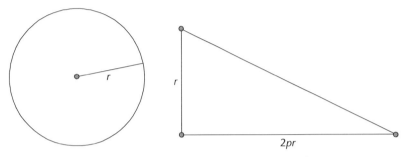

Figure 4.24

Call the area of the triangle A_T. We will show (actually, Archimedes will show!) that the area of the circle is the same as the area of the triangle. Now, the area of the triangle is $\frac{1}{2}$(base)$\times$(height) = $\frac{1}{2}(2\pi r)(r) = \pi r^2$. So, if we can show that the area of the circle is equal to the area of the triangle, we will have shown that the area of the circle is πr^2. Remember now, A is the area of the circle, and A_T the area of the triangle. Our proof will be a proof by contradiction. Suppose that $A \neq A_T$. Then there are two cases to consider.

Case 1. Suppose that $A > A_T$.

In this case we *inscribe* a many-sided regular polygon in the circle as shown below, making the circumference of the circle larger than the perimeter of the polygon. Then the area of the polygon, A_P, can be made to differ from the area of the circle by less than $A - A_T$. (See Figure 4.25 below and realize that, since the areas of the polygons are approaching the area of the circle, the difference between them can be made as small as we want. In particular, it can be made less than $A - A_T$.)

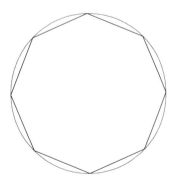

Figure 4.25

Then our polygon has the property that

$$A - A_P < A - A_T.$$

Add $-A$ to both sides to get

$$-A_P < -A_T$$

and then multiply both sides by -1 to get

$$A_P > A_T. \tag{4.6}$$

But,

$$\begin{aligned}
A_P &= \text{The area of the polygon} \\
&= \frac{1}{2} a \cdot p \quad \text{(Where } p \text{ is the perimeter of the polygon.)} \\
&< \frac{1}{2} a \cdot c \quad \text{(Where } c \text{ is the circumference of the circle.)} \\
&= \frac{1}{2} a \cdot 2\pi r \\
&= A_T \quad \text{(The area of a triangle is 1/2 its base times height.)}
\end{aligned}$$

Putting this string of equalities and inequalities together, we have

$$A_P < A_T. \tag{4.7}$$

Comparing inequalities (4.6) and (4.7) we see we have a contradiction. So, this case can't hold.

Case 2. Suppose that $A < A_T$.

In this case we *circumscribe* a many-sided polygon in the circle as shown below, making the circumference of the circle smaller than the perimeter of the polygon. Then, since the areas of the circumscribed polygons get close to the area of the circle, we can make the difference $A_P - A$ as small as we want by taking a polygon with a sufficiently large number of sides. (See Figure 4.26 below.) Thus, we can make $A_P - A < A_T - A$.

Polygon with *n* sides circumscribing circle

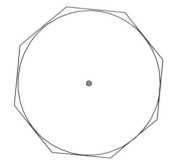

Figure 4.26

Since our polygon has the property that

$$A_P - A < A_T - A$$

we can add A to both sides to get

$$A_P < A_T. \tag{4.8}$$

But,

A_P = The area of the polygon

$= \frac{1}{2} a \cdot p$ (Where p is the perimeter of the polygon.)

$> \frac{1}{2} a \cdot c$ (Where c is the circumference of the circle.)

$= \frac{1}{2} a \cdot 2\pi r$

$= A_T$ (The area of a triangle is 1/2 its base times height.)

Putting this string of inequalities together, we have

$A_P > A_T.$ (4.9)

Comparing inequalities (4.8) and (4.9) we have a contradiction. Thus, this case cannot hold.

Since the cases that $A > A_T$ and $A < A_T$ both led to contradictions, there is only one possibility left, namely, $A = A_T$. Put another way, the area of a circle is πr^2. ■

How did Archimedes think of this? This is a stunning proof. His construction of a right triangle with base $2\pi r$ seems to come out of nowhere. This is just one of the many, many things that Archimedes did by using absolutely ingenious arguments.

You must still be mystified by how Archimedes thought of this proof, and rightly so. Of course, we will never know, but here is a picture that gives us insight into what he might have thought. (See Figure 4.27.) Imagine the circle being composed of lots of very thin circular strips. We show a few below. Now cut each strip and straighten them out, and align them all to the left. They form a right triangle whose height is r and whose base is $2\pi r$. Was this what he was thinking?

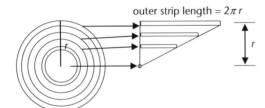

Figure 4.27

4.3.3 Limits And Areas of Circles

Here is an alternate "proof" of the fact that the area of a circle is πr^2. In some ways, it is more natural than Archimedes' proof.

Let P_n be a regular polygon with n sides inscribed in a circle. Let a_n be the apothem of P_n, let p_n be its perimeter, and let A_n be its area. The lower case "p" stands for perimeter. Figure 4.28 shows polygons of increasing numbers of sides inscribed in a circle.

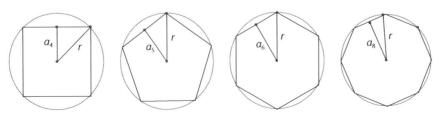

Figure 4.28

In the first figure we show P_4, a regular polygon with 4 sides inscribed in the circle. Its area is A_4 and its perimeter is p_4. Next, we have P_5, then P_6, and then P_8 with respective areas A_5, A_6, and A_8 and perimeters p_5, p_6, and p_8.

Now, if you inscribe regular polygons with an increasing number of sides within a circle, the perimeters of the polygons approach the circumference of the circle and the areas of the polygons approach the area of the circle. The pictures tell us this and indeed this is what Archimedes believed as well. Using calculus notation, we can write this as, $\lim_{n\to\infty} p_n = 2\pi r$ and $\lim_{n\to\infty} A_n = A$, where A is the area of the circle. Also, as we see from the picture, the apothems of the polygons approach r, the radius of the circle. In symbols, $\lim_{n\to\infty} a_n = r$. Now the proof of the area of a circle is simple:

As you showed in a previous Student Learning Opportunity, the area of a regular polygon is $\frac{1}{2}ap$, where a is the apothem of the polygon and p is the perimeter. Thus, $A_n = \frac{1}{2}a_n p_n$ and the area of the circle, $A = \lim_{n\to\infty} A_n = \lim_{n\to\infty} \frac{1}{2} a_n p_n = \frac{1}{2}\left(\lim_{n\to\infty} a_n\right) \cdot \left(\lim_{n\to\infty} p_n\right) = \frac{1}{2}(r)(2\pi r) = \pi r^2$. Here we used the fact that the limit of the product was the product of the limits. Calculus has given us an edge.

Although we are using the concept of limits, this is easy to present to secondary school students without the word limit. Students have an intuitive feel for this concept from the picture, since they can see the areas of the polygons approaching the areas of the circle.

4.3.4 Using Technology to Find the Area of a Circle

We are spending a lot of time on the circle because its study is so rich with connections. In both proofs we gave for the area of the circle, we used the fact that the circumference of a circle is $2\pi r$. If you are anything like we are, then accepting the fact that the circumference of a circle is $2\pi r$ or, put another way, that the ratio of the circumference of a circle to the diameter of a circle is always the same, is difficult. We really would like to see this from another point of view. In this section we give that point of view. Only we add two new ingredients to the mix—trigonometry and technology. What we show now is a method which secondary school teachers can present to their students.

We begin by inscribing in a circle a regular polygon of n sides which we call P_n. We denote its perimeter by p_n. We divide the polygon into n congruent isosceles triangles as shown in the diagram below. Let us focus on one triangle which is shown on the right of Figure 4.29.

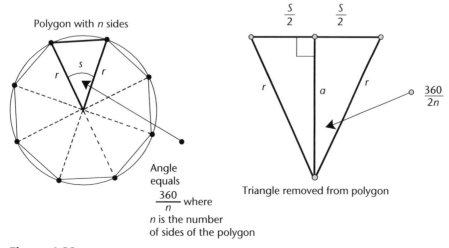

Figure 4.29

The central angle is $\frac{360}{n}$. Draw the altitude of that triangle (which is the apothem of the polygon). This divides the triangle into two congruent triangles and the central angle of each sub-triangle is $\frac{1}{2}$ of $\frac{360}{n}$, or $\frac{360}{2n}$ as shown in the picture above. Now the side, s, of the polygon may be obtained from trigonometry. From the triangle on the right, $\sin(\frac{360}{2n}) = \frac{\text{opposite}}{\text{hypotenuse}} = \frac{\frac{s}{2}}{r} = \frac{s}{2r}$. Multiplying both sides of this equation by $2r$ we get,

$$s = 2r \sin\left(\frac{360}{2n}\right).$$

Since there are n sides to the polygon, the perimeter of the polygon is $p_n = ns$ or just

$$n\left(2r \sin\left(\frac{360}{2n}\right)\right)$$

which we write as

$$p_n = 2r\, n\, \sin\left(\frac{360}{2n}\right).$$

Now as the number of sides of the polygons get greater, the perimeters of the polygons approach the circumference, C of the circle. That is,

$$C = \lim_{n \to \infty} p_n$$
$$= \lim_{n \to \infty} 2r\, n\, \sin\left(\frac{360}{2n}\right)$$
$$= 2r \lim_{n \to \infty} n \sin\left(\frac{360}{2n}\right). \tag{4.10}$$

We will compute this limit in equation (4.10) soon, but we must stop to notice something remarkable that results from equation (4.10). Since the length C of the circumference of the circle is finite, and $2r$ is finite, the limit, $\lim_{n \to \infty} n \sin(\frac{360}{2n})$, from equation (4.10), *must* exist. Let $k = \lim_{n \to \infty} n \sin\left(\frac{360}{2n}\right)$. Since the definition of k as a limit depends only on n and not on the radius r of the circle, *k must be the same regardless of what the radius of the circle is.* Now equation (4.10) tells us that $C = 2r \lim_{n \to \infty} n \sin(\frac{360}{2n}) = 2rk$. Thus, we see that the circumference of a circle is a constant, k, times the diameter, $2r$, regardless of the size of the circle! It appears that, by using the concept of limit, we have a proof of something that was always just accepted for thousands of years, namely, the ratio of the circumference of a circle to its diameter is always the same, the number we called k. However, our proof depends on our believing that the circumferences of the inscribed polygons approach the circumference of the circle as the number of sides gets larger and larger. If this is true, then we have indeed proved that the ratio of the circumference of a circle to the diameter is a constant, provided we believe the circumference of a circle is finite. We need only show that $k = \pi$ and we will have established that the circumference of a circle is $2\pi r$.

We are now ready to find k. That is where the technology comes in. Using your calculator, compute $n \sin\left(\frac{360}{2n}\right)$ for larger and larger values of n. (Make sure your calculator is in degree mode.) We give a table here.

n	$n \sin\left(\frac{360}{2n}\right)$
50	3.13952
500	3.14157...
5000	3.14159.
10000	3.14159....

As we can see as n gets larger and larger, $n \sin\left(\frac{360}{n}\right)$ stabilizes around 3.14159, which is approximately π. Thus, this table makes it plausible that $\lim\limits_{n \to \infty} n \sin\left(\frac{360}{2n}\right)$, or k, is π, and therefore the circumference of the circle is $C = 2\pi r$ according to (4.10). Are you all aglow now?

One final note: There are several calculus proofs purporting to show that the area of a circle is πr^2. All of them that we know of use the fact that the derivative of $\sin x$ is $\cos x$ in one form or another. To prove we need to use the fact that if θ is in radians, then $\lim\limits_{\theta \to 0} \frac{\sin \theta}{\theta} = 1$. This fact, however, uses the fact that the area of a sector of a circle is $\frac{1}{2}r^2\theta$, which in turn uses the fact that the area of a circle is πr^2. So our point is that all calculus proofs that we know of that prove that the area of a circle is πr^2 use (indirectly) the same fact that we are trying to prove, namely, that the area of the circle is πr^2! Thus, all these proofs are circular! (No pun intended.) This is not to say that the calculus proofs are bad. Hardly. They just corroborate what we know. It is always good to see something from a different point of view.

4.3.5 Computation of π

Archimedes' Computation of π

Archimedes did some magnificent mathematics. His mind was always churning. It is said that he carried with him a tablet of sand on which he would draw diagrams whenever he got an idea (his version of the modern day laptop!). One of the tasks he set for himself was an estimate of the value of π. He essentially did this by inscribing polygons with more and more sides in the circle of radius 1, and found the perimeters of the resulting figures. As we have observed, their perimeters will approach the circumference of the circle which is $2\pi(1)$ or just 2π. So these perimeters will provide us with an estimate of 2π and thus dividing by 2, we get an estimate of π.

We begin by proving a theorem we will need to continue.

> **Theorem 4.8** *If a regular polygon of n sides, inscribed in a circle with radius 1, has side s, and we double the number of sides, the side of the new polygon has length t where $t = \sqrt{2 - \sqrt{4 - s^2}}$.*

As a special case of this theorem, below we see a picture of a 4-sided regular polygon (a square *ABCD*) inscribed in circle of radius 1, together with an 8-sided regular polygon (an octagon). The square in Figure 4.30 has sides all of length s, the octagon has sides all of length t. The theorem relates the length of t to that of s, namely, $t = \sqrt{2 - \sqrt{4 - s^2}}$.

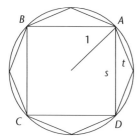

Figure 4.30

Now to the proof.

Proof. The square roots seem to indicate that the Pythagorean Theorem might play a part and indeed it does. Let us start with an *n*-sided regular polygon and suppose that *s* is the length of any side of our *n*-sided polygon as shown in Figure 4.31 below where we show *only one side*, *AD*, of our *n*-sided polygon.

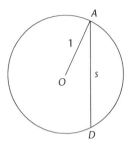

Figure 4.31

We now draw in *OD* giving us Figure 4.32.

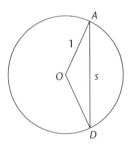

Figure 4.32

Now, since radii *OA* and *OD* are equal, triangle *OAD* is isosceles, and we can draw the altitude, *OP*, to the base and extend it to *B*. Since the altitude of an isosceles triangle bisects the base (a well known fact from geometry), our picture now looks like that in Figure 4.33.

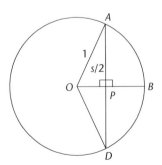

Figure 4.33

Finally, we draw *AB* and *BD*, which are the sides of a 2*n*-sided regular polygon. Let $AB = t$. So our figure now looks like Figure 4.34

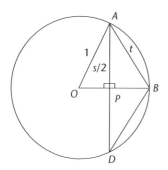

Figure 4.34

and it is starting to look complicated. So, let us just pull out of the picture what we need, namely, triangles *OAP* and *BAP* and let us call $OP = x$ and $PB = y$. This yields Figure 4.35.

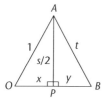

Figure 4.35

Now, using the Pythagorean Theorem on triangle *OAP* we get,

$$x^2 + \left(\frac{s}{2}\right)^2 = 1. \tag{4.11}$$

Using the Pythagorean Theorem on triangle *BAP* we get

$$y^2 + \left(\frac{s}{2}\right)^2 = t^2 \tag{4.12}$$

and if we subtract equation (4.11) from equation (4.12) we get

$$y^2 - x^2 = t^2 - 1. \tag{4.13}$$

We observe from the picture that $x + y = OB$, which is the radius of the circle which is 1. So, $y = 1 - x$. Substituting this into equation (4.13) we get

$$(1 - x)^2 - x^2 = t^2 - 1 \tag{4.14}$$

which, upon squaring and simplifying, gives us

$$1 - 2x = t^2 - 1.$$

We solve for *t* to get

$$t = \sqrt{2 - 2x}. \tag{4.15}$$

We are almost there. We need to do a little algebra. From equation (4.11) we get, when we solve for x, that $x = \sqrt{1 - \left(\frac{s}{2}\right)^2} = \sqrt{1 - \frac{s^2}{4}} = \sqrt{\frac{4-s^2}{4}}$ or just

$$x = \frac{\sqrt{4-s^2}}{2}.$$

Substituting this into equation (4.15) we get

$$t = \sqrt{2 - 2x} = \sqrt{2 - \sqrt{4-s^2}} \tag{4.16}$$

and we are done. ∎

Corollary 4.9 *If we start with a regular polygon of n sides inscribed in a circle of radius 1, then the perimeter of the 2n-sided polygon is just $2n\sqrt{2 - \sqrt{4-s^2}}$, where s is the side of the n-sided polygon.*

Proof: Since each side of the 2n-sided polygon has length $\sqrt{2 - \sqrt{4-s^2}}$, the perimeter of that polygon is

$$2n\sqrt{2 - \sqrt{4-s^2}}.$$

Now we tie it all together. The perimeters of the 2n-sided polygons, $2n\sqrt{2 - \sqrt{4-s^2}}$, approach the circumference of our circle, which is 2π as the number of sides gets large. Let us begin with a square ($n = 4$) inscribed in a circle of radius 1. Thus, the diameter of the circle is 2. Here is the picture, Figure 4.36:

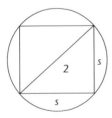

Figure 4.36

and using the Pythagorean Theorem we can see that the side s must be $\sqrt{2}$. (Verify!) Thus, by equation (4.16), the side of the octagon, the polygon with twice the number of sides, must be

$$t = \sqrt{2 - \sqrt{4-s^2}} = \sqrt{2 - \sqrt{4-2}} = \sqrt{2 - \sqrt{2}}.$$

And this is now our new value of the side, s, of the inscribed polygon. We now double the number of sides again. We get a 16-sided polygon. Again, by equation (4.16), our new side, t, is

$$t = \sqrt{2 - \sqrt{4-s^2}} = \sqrt{2 - \sqrt{4 - (\sqrt{2-\sqrt{2}})^2}} = \sqrt{2 - \sqrt{2 + \sqrt{2}}}.$$

We can again double the number of sides to get a 32-sided polygon with side = $\sqrt{2 - \sqrt{2 + \sqrt{2 + \sqrt{2}}}}$, and so on.

Let us stop at this 32-sided polygon. Since the side of it has length $\sqrt{2-\sqrt{2+\sqrt{2+\sqrt{2}}}}$, its perimeter will be $32\sqrt{2-\sqrt{2+\sqrt{2+\sqrt{2}}}}$, which our calculator tells us is 6.2731. But this is approximately the circumference of the circle, which we know is 2π. So our estimate for π now is $6.2731/2 \approx 3.1366 \approx 3.14$.

Archimedes started with a hexagon, not a square, as we did. He then computed the perimeter of a 12-, 24-, 48-, and 96-sided polygon. He stopped there. Now, given that there were no calculators, and notations for decimal representation of numbers had not yet been invented, he had to compute each of the monstrous square roots by hand. He might have very well used the algorithm that we presented in Chapter 3 page 107, since it has been around for thousands of years. Also, algebraic notation had not been invented in Archimedes' time. So he had to do all these algebraic manipulations in his head or with the aid of geometry. You can't help but be astonished by what he did.

The Monte Carlo Method for Estimating π

A method that is often used in industrial problems as well as theoretical analyses is something called the Monte Carlo Method. This is a statistically based method for determining certain quantities that may otherwise be very difficult to compute. It is widely used and has numerous applications. It also can be used to approximate π. That it gives you an accurate value of π from just random data is mind boggling. Thus, this section links geometry and probability and in the course of doing it, also uses some analytic geometry.

Suppose we want to compute π. We know the area of the circle is πr^2, for we have proved it. Thus, if we take a circle of radius 1, its area will be π. Now imagine a quarter of circle of radius 1 placed inside a square with side 1, shown in Figure 4.37 below.

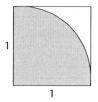

Figure 4.37

Imagine throwing darts at the above figure. Now imagine that, although you are not a particularly skilled dart thrower, at least you can hit a picture when you are close enough. If these are truly random throws, then the probability that a dart ends up in the shaded portion is

$$\frac{\text{Area of the quarter circle}}{\text{Area of the square}} = \frac{\frac{\pi}{4}}{1} = \frac{\pi}{4}.$$

To estimate this probability (or equivalently, to estimate $\frac{\pi}{4}$), we throw darts randomly at the board and compute

$$\frac{\text{The number of darts that hit the shaded area (including the boundary)}}{\text{The total number of darts hitting the square (including the boundary)}}.$$

If we do this for a large number of throws, we should get an estimate of $\frac{\pi}{4}$ and thus π.

Now, we need a large number of throws, and they must be random. So, we do what is called a simulation. We have the computer generate pairs of numbers (x, y), where both x and y are between 0 and 1. To generate these points, we use a random number generator. This generates random points (more or less) and we can easily decide whether or not the points generated are in the quarter of a circle or not by realizing that the equation of the circle is $x^2 + y^2 = 1$. Thus, the point is in (or on) the circle if $x^2 + y^2 \leq 1$.

Here is a summary of the procedure. We will be calling D the number of points we generate that lie in the circle and T the total number of points that we have generated. (D stands for the number of darts that lie in the quarter circle, and T for the total number of darts thrown.)

1 Generate points (x, y) randomly.
2 Determine if the point is in or on the circle. If it is, increase the count of D by 1.
3 Compute the ratio $\frac{D}{T}$. This is our estimate of $\frac{\pi}{4}$. To find the estimate of π, just multiply by 4.

Here is a program that was used to do this on the TI series calculator.

1: $0 \to T : 0 \to D$	(Initialize the values of the total number of darts thrown, and those that hit the circle.)
2: FOR $(I, 1, 1000)$	(We are about to generate 1000 sets of random numbers, (x, y).)
3: rand $\to x$	(Generate 1 random number for x.)
4: rand $\to y$	(Generate 1 random number for y.)
5: $T + 1 \to T$	(Each time we throw a dart, we increase the count by of T by 1.)
6: If $x^2 + y^2 \leq 1 : D + 1 \to D$	(If the dart is in the circle, increase the count of D by 1.)
7: END	(This signals the end of the generation of our pairs of numbers.)
8: "Our estimate for pi is"	(We are telling the machine to write on the screen the words, "Our estimate for pi is".)
9: Display $\frac{4D}{T}$	(The machine displays our estimate of π.)

We actually ran this program and got the following: $\pi \approx 3.1$. This is both good and bad. We had to generate 1000 points to get to just 3.1. If we want a better estimate, we need to generate many more points. But, with the speed of computers today, this is hardly an issue.

You can key in the program and run it for several thousand more trials if you wish. (Just change the number 1000 in step 2 to 100,000 for example.) See what you get. Also, bring a book along with you while you are waiting. The program takes a long time to run on the TI calculator.

Although you may be getting the impression that the Monte Carlo Method is inefficient, with the speed of modern computers, this happens to be a viable method. See, for example, the website: http://polymer.bu.edu/java/java/montepi/montepiapplet.html, where you generate estimates of π at high speed.

We have said that Monte Carlo Methods have many applications. Here are some which we found on the Internet: (1) radiation transport, (2) operations research, (3) design of nuclear reactors, (4) the study of molecular dynamics, (5) the study of long chain coiling polymers, (6) global illumination computations which produce photorealistic images of virtual 3D models with applications in video games, (7) architectural design, (8) computer generated films with applications to special effects in cinema, (9) business and economics, (10) the evaluation of some very difficult integrals that occur in applications. The list goes on and on. In fact, there is a journal called *International Journal of Monte Carlo Methods* which is devoted purely to applications of the method.

4.3.6 Finding Areas of Irregular Shapes

Historically, while finding the areas of polygons was not that difficult, finding areas of irregularly shaped figures was quite a challenge. It took over 1000 years to go from one to the other. What follows are some methods that have been developed over many years.

One method of approximating the area of an irregular figure is by putting it on a grid and counting the number of squares inside and on the boundary of the figure. For example, suppose we wanted to estimate the area of Figure 4.38 below.

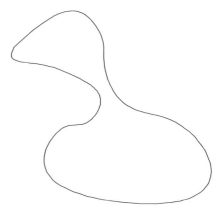

Figure 4.38

We can put in on a grid as shown in Figure 4.39 below,

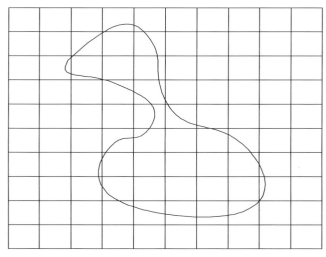

Figure 4.39

and count the number of squares that are inside the figure or which cut the boundary of the figure. The squares have to be in standard units. For example, if we are on a map, and we are measuring the area of New York state, a possible standard unit for the length of a square might be a 100 miles.

In the above figure we count 38. So we estimate the figure to have 38 square units (where a square has some kind of standard measure). To get a finer approximation to the area, we can subdivide the squares in the grid further and further into halves, quarters, and so on. This is tedious but will give a better approximation to the area of the figure.

One of the many triumphs of the calculus is that we can find exact areas of certain irregular objects. Of course, you learned how to do this in calculus. If the graph of $f(x)$ was above the x-axis as in Figure 4.40 below

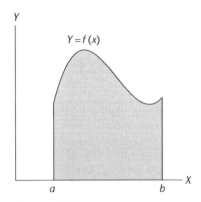

Figure 4.40

and you wanted to find the area under the curve from $x = a$ to $x = b$ (the shaded area), you simply computed $\int_a^b f(x)dx$.

As you recall, $\int_a^b f(x)dx$ is computed by finding an antiderivative $F(x)$ of $f(x)$ and evaluating $F(b) - F(a)$. This result is fundamental and, in fact, is called the **Fundamental Theorem of Integral Calculus (FTIC)**. What a remarkable formula to find the area! So simple, and so unexpected! What on earth, after all, do antiderivatives have to do with area? Of course, the notion of integral goes far beyond areas under curves. No scientist can do his or her job today without calculus and integrals. They are as fundamental to the scientist as having lights are to the everyday person.

We assume that the reader remembers the following basic formulas of integration. In what follows, c and p are *constants*.

$$\int c\,dx = cx + k \text{ where } k \text{ is constant} \tag{4.17}$$

$$\int cx^p dx = \frac{cx^{p+1}}{p+1} + k \text{ where } k \text{ is constant and } p \neq -1 \tag{4.18}$$

$$\int (f(x) + g(x))dx = \int f(x)\,dx + \int g(x)dx \tag{4.19}$$

Thus, the integral $\int 5\,dx = 5x + k$ by equation (4.17). $\int 7x^5 dx = \dfrac{7x^6}{6} + k$ by equation (4.18) and $\int (3x + 5) = \frac{3x^2}{2} + 5x + k$ since equation (4.19) says that the integral of the sum is the sum of the integrals.

Example 4.10 *Find the area under the curve $f(x) = 2x^2 + 3$ from $x = 1$ to $x = 2$.*

Solution. The graph of $f(x)$ is shown below (Figure 4.41).

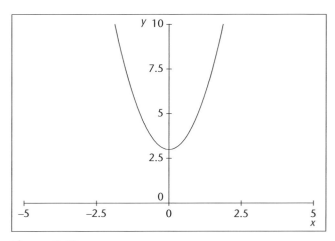

Figure 4.41

Since the curve is above the x-axis, from $x = 1$ to $x = 2$, we have that the area under the curve is $\int_1^2 (2x^2 + 3)dx = (\frac{2x^3}{3} + 3x)|_1^2 = \frac{2(2)^3}{3} + 3(2) - (\frac{2(1)^3}{3} + 3(1))$ or just $\frac{23}{3}$.

Now that we have reviewed how to compute an integral, let's talk a bit about what an integral is, because we will need to use that when we get to volumes of certain solids.

When we find the area under the curve, we first approximate it by rectangles. To be more specific, we begin by breaking the interval from a to b into n equal parts of length Δx (see Figure 4.42 below) and labeling the division points as follows: call $a = x_0$, and then the successive division points are labeled x_1, x_2, and so on up to b which we call x_n. Notice that x_1 is the right endpoint of the first subinterval that $[a, b]$ is broken into, x_2 is the right endpoint of the second subinterval that $[a, b]$ is broken into, and so on. We draw lines from the x-axis to the curve at these points, and form rectangles as show in the diagram below. We have drawn only the first few rectangles and have shaded the first.

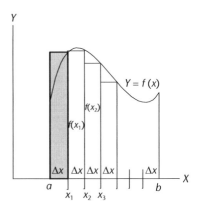

Figure 4.42

The height of the first rectangle is $f(x_1)$ and the width is Δx, so the area of the first rectangle is $f(x_1)\Delta x$. Similarly, the area of the second rectangle is $f(x_2)\Delta x$. The sum of the areas of the rectangles is

$$f(x_1)\Delta x + f(x_2)\Delta x + f(x_3)\Delta x + \ldots + f(x_n)\Delta x \tag{4.20}$$

which we can write in more compact form as:

$$\sum_{1}^{n} f(x_i)\Delta x. \tag{4.21}$$

Keep in mind that this complicated looking expression in display (4.21) simply means "the sum of the areas of the rectangles" and the letter n simply means the number of rectangles in the picture.

Now, we observed in calculus that, as we increased the number, n, of rectangles in the picture, the sum of the areas of the rectangles more and more closely approximated the area under the curve. (Actually, we are using our intuition again. That is why, when you look in calculus books, you will see the area under a curve *defined* as the limit of the sum of the areas of the rectangles. It is defined that way because it looks like it is true!) To get a good dynamic picture of how this works, go to the applet at http://cs.jsu.edu/~leathrum/Mathlets/riemann.html.

Thus, the area under the curve is the limit of these sums of the areas of the rectangles as $n \to \infty$ or in symbols we say that the:

$$\text{area under the curve from } a \text{ to } b = \lim_{n\to\infty} \sum_{1}^{n} f(x_i)\Delta x$$

when the curve $f(x)$ is above the x-axis. Since we know that this area can also be computed by $\int_a^b f(x)dx$, we have

$$\int_a^b f(x)dx = \lim_{n\to\infty} \sum_{1}^{n} f(x_i)\Delta x. \tag{4.22}$$

Now, while we originally were motivated to study the right side of equation (4.22) to find the area under a curve, equation (4.22) is telling us much more. It is saying, that if in an application we can express a quantity as a limit of a sum like that on the right of equation (4.22), then we know that the quantity we seek can be computed by doing an integral. To get the integral which the right side of equation (4.22) represents, we simply replace the x_i on the right side of equation (4.22) by x and the Δx by dx. The quantities a and b are the left and right hand limits of the interval, which is being partitioned by the points x_i.

Example 4.11 *In an application, the following computation needs to be done:* $\lim_{n\to\infty} \sum_{1}^{n} (x_i)^2 \Delta x$ *where the x_is partition $[2, 3]$. What is this limit equal to?*

Solution. We notice immediately that this looks just like equation (4.22) where $f(x_i) = x_i^2$. So this limit is nothing more than $\int_2^3 x^2 dx$. We simply replace the x_i in the summation by x and the Δx by dx. We know how to evaluate this integral: $\int_2^3 x^2 dx = \frac{x^3}{3}\big|_2^3 = \frac{(3)^3}{3} - \frac{2^3}{3} = \frac{19}{3}$.

Example 4.12 *When computing the volume of a solid, a budding mathematician realizes that the volume can be expressed as* $\lim_{n\to\infty} \sum_1^n 2\pi x_i(x_i^3 + 1)\Delta x$ *where the x_i's partition $[1, 2]$, but is not sure how to continue. What is the volume of the solid?*

Solution. Once again, we notice that this looks just like equation (4.22) where $f(x_i) = 2\pi x_i(x_i^3 + 1)$. So, this limit is nothing more than $\int_1^2 2\pi x(x^3 + 1)dx$ obtained by replacing each x_i by x and Δx by dx. We, of course, can pull out the 2π to get $\int_1^2 2\pi x(x^3 + 1)dx = 2\pi \int_1^2 x(x^3 + 1)dx = 2\pi \int_1^2 (x^4 + x)\, dx = 2\pi \frac{77}{10}$. (Verify!)

The key in using the integral in applications is to express some quantity we seek by something that looks like the right side of equation (4.22) and then to quickly realize it is an integral. We will soon use this meaning of integrals in the study of volumes.

Student Learning Opportunities

1 Circle 1 is circumscribed about a square of side 6 and circle 2 is inscribed in the square. What is the ratio of the area of circle 1 to circle 2?

2 The diameter AB of a circle with center O is 6. C is a point on the circle such that angle BOC is 60 degrees. Find the length of the chord AC.

3 Using the fact that the length of an arc subtended by a central angle of $n°$ in a circle of radius r is $2\pi r \cdot \dfrac{n}{360}$, solve the following: An arc of 60 degrees on a circle has the same length as an arc of 45 degrees in another circle. What is the ratio of the areas of the circles, smaller to larger?

4 In the square below (Figure 4.43) with side 9 inches, one places a circle with radius 3 inches. Find the shaded area. Notice the upper right corner of the picture is not shaded.

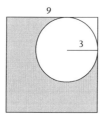

Figure 4.43

5 The odometer on a car measures the distance traveled by multiplying the circumference of a tire by the number of revolutions. Thus, if you change the tire size, and no adjustment is made in the odometer to account for the new tire size, then if you travel the same distance, with smaller tires and larger tires, the odometer will read differently. Suppose that a 450 mile trip is made with 15″ radius tires, and that the same trip is made a second time with a different size tire and no adjustment in the odometer is made. If the odometer reads 440 miles the second time, determine if the radius of the new tire is smaller or larger than 15 inches, and then find the radius of the new tire.

6 (C) A student asks you to explain how we know that the apothem has the same length no matter which side of the regular polygon we draw it to. How do you explain it?

144 Measurement: Area and Volume

7 (C) A student is curious to know why when you inscribe a regular hexagon in a circle, the length of each side of the hexagon is the same as the radius of the circle. How would you show the student why this is true? [Hint: Divide the hexagon into 6 triangles by drawing radii to the vertices of the hexagon.]

8 (C) Here is a great activity to do with your students that will intrigue them. You will need lots of string, rulers, and several pairs of scissors. Give your students 4 equal strings of length 12 inches. Ask them to form the first string into a circle and find its circumference and area. Ask them to cut the second string in half and form 2 circles from the resulting strings. Have them find the total circumference and area of these two circles. Ask them to next cut the third string into 3 equal parts and form 3 circles and again calculate the total circumference and area. Finally, ask them to cut the fourth string into 4 equal parts and form 4 circles and find the total circumference and area. Guide them in showing that, in each case, the sum of the circumferences of the smaller circles is the same, but the sum of the areas of the small circles differs drastically from one case to the next. Ask them if they think this makes sense and to explain why. Without using string, what is your answer to this last question?

9 (C) (Continuation of previous problem.) After having done the string activity with your students, their curiosity has been piqued and they ask you what would happen in a general case, where you cut a string of length 12 into n equal parts and formed circles with them. They ask if it is still true that the sum of the circumferences of the circles is the same and also ask what proportion of the first circle the sum of the areas of the smaller circles is. How do you respond? Justify your answer.

10 (C) This is a wonderful problem and calculator activity that will further convince you and your students that the area of a circle is πr^2. We gave a proof that the area of a circle was πr^2 using the fact that the circumference of a circle was $2\pi r$. But, if we agree to use the calculator, then we need not even use the fact that $C = 2\pi r$. Here is how we can show that the area is πr^2 with the calculator.

First, it is often shown in secondary school that the area, A, of a triangle with sides a, b, and c, is $A = \frac{1}{2}ab \sin C$ where a and b are the sides of the triangle, and C is the angle between the sides a and b. Now consider a polygon of n sides inscribed in a circle. It can be divided into n congruent isosceles triangles each with vertex angle $\theta = \frac{360}{n}$ degrees and each with area $\frac{1}{2}r^2 \sin \theta$. (See Figure 4.44 below where we have shown one of the triangles.)

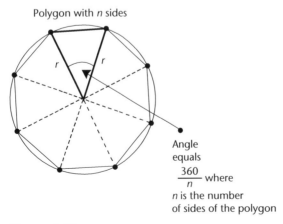

Figure 4.44

(a) Show the area, A_n of the polygon with n sides is $A_n = n(\frac{1}{2}r^2 \sin \frac{360}{n}) = r^2(\frac{1}{2}n \sin \frac{360}{n})$. To find the area of a polygon with 50 sides, we just substitute $n = 50$ in the above expression. To find the area with 500 sides we just substitute $n = 500$ in the above expression. Now, take out your calculator and make a table for $\frac{1}{2}n \sin \frac{360}{n}$ for larger and larger values of n and show that this quantity approaches π. Thus, the area of a circle is πr^2. This is pretty exciting to see and will easily convince your secondary school students that the area of a circle is πr^2.

(b) This question is appropriate for you and your calculus students. Use the expression $A_n = r^2(\frac{1}{2}n \sin \frac{360}{n})$ obtained in part (a) to find the limit of A_n as n approaches infinity. Only this time, do it without a calculator. You will first need to write the previous expression in radian form which yields $A_n = r^2(\frac{1}{2}n \sin \frac{2\pi}{n})$, and then you will have to use L'Hopital's rule since the limit you get is indeterminate. (In calculus, all the formulas for derivatives assume radian measure.) This is yet another way of getting that the area of a circle is πr^2.

11 In Figure 4.45 below we have right triangle ABC inscribed in a circle with $AC = 6$, $BC = 8$ and $AB = 10$. We construct semicircles on AC and CB. Find the sum of the areas of regions X and Y, the shaded crescents.

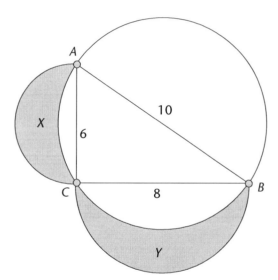

Figure 4.45

12 In your own words, describe the similarities and differences of the four proofs of the area of the circle given in this section.

13 What is the area under the curve $f(x) = x^3$ and above the x-axis from $x = 1$ to $x = 3$?

14 Find the area enclosed by the curves $f(x) = x^2$ and $g(x) = 2x$.

4.4 Volume

 **LAUNCH**

1. Take two pieces of 8" by 11" paper out of your notebook. Take one piece of paper and curl it so that the 11" sides touch one another and it forms a tall, thin cylindrical shape. Take the other piece of paper and curl it so that the 8" sides touch one another and it forms a shorter and fatter cylindrical shape.
2. If you were to fill each of these cylinders with popcorn, which one do you think would hold more popcorn? Or, do you think they would hold the same amount of popcorn?
3. What type of measurement would we have to use to figure out the answer to the above question?
4. What is the formula you would use? Do you know where this formula came from? Is it a definition or a theorem?

If you are now curious about which cylinder would hold more popcorn, and you are curious about what formula to use and where the formula came from, then you will enjoy reading this section of the text. It discusses how to find the volume of various three dimensional objects and describes where the formulas come from. Starting in middle school, students learn some elementary concepts about volume, and some even learn some of the formulas. But rarely do they learn where these formulas come from and how they are related to other formulas for area that they have already learned. As a future teacher, you will surely want to know more about how you find the volumes of various solid shapes and how you can teach these formulas to your own students by relating them to formulas about area that they already know. You might also want to return to the launch question and figure out what the actual volumes are. You can verify your results by actually pouring popcorn into each of your cylinders. In fact, this is a great activity you can do with your own students one day!

4.4.1 Introduction to Volume

So now we will turn to the important topic of how to measure the volumes of solids. To benefit the most from this reading, make sure you have read the last subsection, "Finding Areas of Irregular Shapes" of the previous section.

Since we live in a three dimensional world, and deal with three dimensional figures all the time, it is only fitting that the study of volume is a critical area of focus in the secondary school curriculum. Our first goal is to derive a general formula for the volume of a solid with known cross-sectional area. To accomplish this, all we need is the definition of volume for very simple kinds of solids: A **simple solid** is a solid that is formed by a curve enclosing an area B, moved along a line perpendicular to the curve a distance of h. h is called the height of the solid. (See Figure 4.46 below.) Notice that all cross sections parallel to the base are congruent and hence have the same area.

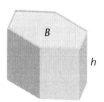

Figure 4.46

We define the **volume of a simple solid** to be Bh, where B is the area of the base and h is the height of the solid. Thus, if we have a cube with side x, as shown in Figure 4.47,

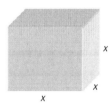

Figure 4.47

the volume = (area of base) × height = $x^2 \cdot x = x^3$. And, if we have a right circular cylinder (a can, shaped as in Figure 4.48 below)

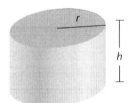

Figure 4.48

the volume of the cylinder = (Area of the base) × height = $\pi r^2 \times h$, the formula we usually teach in middle school.

Now let us take a general solid whose picture is shown below in Figure 4.49.

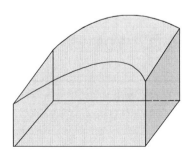

Figure 4.49

Our goal is to find a formula for its volume. We place the solid above the x-axis and let it span from $x = a$ to $x = b$ as shown in Figure 4.50.

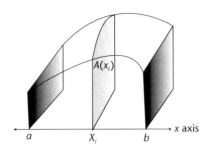

Figure 4.50

We first divide $[a, b]$ into equal parts of length Δx as we did earlier in the plane. We then imagine planes cutting the solid perpendicular to the x-axis at each of the points, x_1, x_2, x_3, and so on. (These planes are also parallel to the y-axis.) We have shown, in the picture above, one such cross section resulting from cutting the solid at x_i by a plane perpendicular to the x-axis. We call its area $A(x_i)$. Of course, when such planes are close together, they divide the solid into *thin* slabs that are approximately simple solids. We show one slab in Figure 4.51 below. Its volume is *approximately* $A(x_i)\Delta x$ (the area of the base times the height).

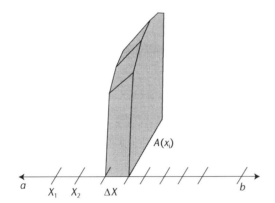

Figure 4.51

Note: The slab is not a simple solid, since the bases are not congruent, but this estimate is not far off if the slab is very thin; that is, if Δx is small.

Suppose that the cross-sectional areas at x_1, x_2, x_3, and so on, are given by $A(x_1), A(x_2), A(x_3)$, and so on. Then the volume of the first slab is approximately equal to $A(x_1)\Delta x$. The volume of the second slab is approximately $A(x_2)\Delta x$, and so on. The sum of the volumes of the slabs is approximately.

$$A(x_1)\Delta x + A(x_2)\Delta x + \ldots + A(x_n)\Delta x$$

which can, of course, be written in shorthand as

$$\sum_{i=1}^{n} A(x_i)\Delta x.$$

Now, as *n*, the number of slabs goes to infinity, the sum of the volumes of the slabs gets closer and closer to the volume of the solid. That is, our approximations get better and better and we finally have that

$$\text{the volume of the solid} = \lim_{n \to \infty} \sum_{1}^{n} A(x_i)\Delta x.$$

But this we recognize! This, by equation (4.22), is an integral! In fact, it is the integral $\int_{a}^{b} A(x)dx$. Thus, we have established the next theorem

Theorem 4.13 *Suppose that S is a solid and that the cross-sectional area of S at a distance x from the origin is given by A(x) where the cross section is perpendicular to the x-axis, and parallel to the y-axis. Then the volume of S is $\int_{a}^{b} A(x)dx$.*

Let us apply this.

Example 4.14 *The base of a solid, S, is the region, R, under the curve $f(x) = \sqrt{x}$ from $x = 1$ to $x = 5$. (a) Cross sections of S perpendicular to the x-axis (and parallel to the y-axis) at a distance x away from the origin are all squares. Find the volume of the S. (b) Suppose instead that cross sections perpendicular to the x-axis and parallel to the y-axis, are semicircles. Find the volume of S now.*

Solution. (a) First we draw *R* (Figure 4.52)

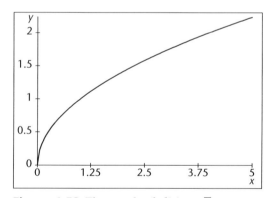

Figure 4.52 The graph of $f(x) = \sqrt{x}$

and then we attempt to draw *S*, the solid we are talking about. (See Figure 4.53 below.)

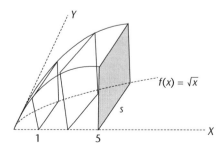

Figure 4.53

The cross-sectional areas at a distance x away from the origin are squares. At a distance x from the origin, the side, s, of the square is just the y coordinate of the curve, namely $f(x) = \sqrt{x}$. Thus, the cross-sectional area $A(x)$ of a typical such cross section at a distance x from the origin is given by $A(x) = s^2 = (f(x))^2$. Thus, by Theorem 4.13, the volume of the solid equals

$$\int_1^5 A(x)dx$$

$$= \int_1^5 (f(x))^2 dx$$

$$= \int_1^5 (\sqrt{x})^2 dx$$

$$= \int_1^5 x\,dx = \frac{x^2}{2}\Big|_1^5 = 12 \text{ cubic units.}$$

(b) Our solid now looks something like the following figure, Figure 4.54.

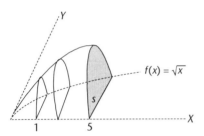

Figure 4.54

We see at once that a typical diameter, s, of our semicircular cross section is given by $s = f(x)$. Hence the radius, r, of a typical cross section is $r = \frac{s}{2} = \frac{f(x)}{2}$. Since the area of a semicircle is $\frac{1}{2}\pi r^2$, we have, using Theorem 4.13, that the volume of our solid is:

$$\int_1^5 A(x)dx$$

$$= \int_1^5 \frac{1}{2}\pi r^2 dx$$

$$= \int_1^5 \frac{1}{2}\pi \left(\frac{\sqrt{x}}{2}\right)^2 dx$$

$$= \int_1^5 \pi \frac{x}{8} dx = \frac{\pi x^2}{16}\Big|_1^5 = \frac{3}{2}\pi \text{ cubic units.}$$

4.4.2 A Special Case: Volumes of Solids of Revolution

Suppose that we have a region in the xy plane and spin the region about, say, the x-axis or the y-axis. Such a solid is known as a solid of revolution. Below, in Figure 4.55, you see such a region (the region bounded by $f(x)$, the x-axis and the lines $x = a$ and $x = b$) and the result of spinning it around the x-axis.

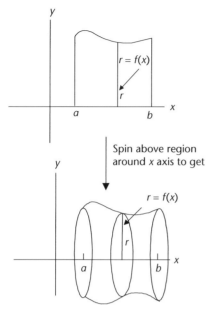

Figure 4.55

When we spin such a region around the x-axis, we see that our cross sections are now circular, and that the radius of a typical cross section is $r = f(x)$. Using Theorem 4.13 we get the following:

Theorem 4.15 *If the region bounded by the curve $y = f(x)$, $x = a$, $x = b$ and the x-axis is spun around the x-axis to get a solid of revolution, then the volume of the resulting solid of revolution is given by*
$$V = \pi \int_a^b (f(x))^2 \, dx.$$

Proof. The cross-sectional areas are circular with areas πr^2 where $r = f(x)$, as the above figure shows. Thus, by Theorem 4.13, the volume of our solid is

$$V = \int_a^b A(x)\,dx$$

$$= \int_a^b \pi r^2 \, dx$$

$$= \int_a^b \pi (f(x))^2 \, dx.$$

∎

Volume of a Cone

We can apply this theorem to find the volume of a cone with base radius r and height h, by considering it as a solid of revolution. We simply revolve the triangle shown in Figure 4.56 below about the x-axis.

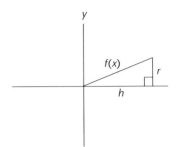

Figure 4.56

to get Figure 4.57.

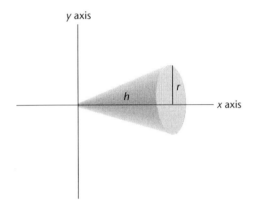

Figure 4.57

Now, we only need to find the equation of the line segment, which is the hypotenuse of the right triangle being spun around the x-axis. But the equation of a line is $y = mx + b$, where m is the slope and b is the y intercept. In this case, $b = 0$ since the line crosses the y-axis at the origin. The slope from the picture is the rise over the run, or $\frac{r}{h}$. Thus, the equation of the line is $y = f(x) = \frac{r}{h}x$. Now we apply Theorem 4.15 to get that the volume

$$V = \pi \int_a^b (f(x))^2 \, dx = \pi \int_0^h \left(\frac{r}{h}x\right)^2 dx = \int_0^h \pi \frac{r^2 x^2}{h^2} dx = \frac{\pi r^2}{h^2} \int_0^h x^2 dx = \frac{\pi r^2}{h^2} \frac{x^3}{3}\Big|_0^h$$

$$= \frac{\pi r^2}{h^2} \frac{h^3}{3} - \frac{\pi r^2}{h^2} \frac{0^3}{3} = \frac{\pi r^2 h}{3}.$$

Surely in the past you must have wondered where the $\frac{1}{3}$ in the formula for the volume of a cone came from. (We know we did.) Now you know, although you see it from a very sophisticated point of view.

Volume of a Sphere

In a similar manner, if we wanted to find the volume of a sphere with radius R, we need only think of the sphere as the result of spinning the semicircle whose equation is $y = f(x) = \sqrt{R^2 - x^2}$ around the x-axis. See Figure 4.58 below:

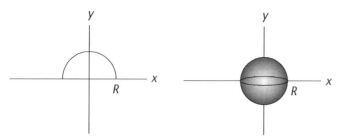

Figure 4.58 $f(x) = \sqrt{(r^2 - x^2)}$ spun around the x-axis

That yields that the volume, $V = \pi \int_{-R}^{R} \left(\sqrt{(R^2 - x^2)}\right)^2 dx$. We leave it to the reader to verify that this is $\frac{4}{3}\pi R^3$.

Archimedes is usually credited with this formula for the volume of a sphere, although his proof most unexpectedly used principles of mechanics! The reader can find his proof in *The Works of Archimedes with the Method of Archimedes* (Heath, 1953), or one can visit the following website: http://www.cut-the-knot.org/pythagoras/Archimedes.shtml.

Archimedes also established that the *surface area* of a sphere is $4\pi r^2$ by inscribing a sphere inside a cylinder and arguing cleverly. He was so exceedingly proud of this proof that, on his tombstone, is engraved a picture of a sphere inscribed in a cylinder.

Speaking of tombstones, we can't leave Archimedes without talking a bit about his death. Archimedes was intently studying diagrams he had drawn in the dirt when a Roman soldier from the army that had conquered the city where Archimedes lived came upon him. The soldier ordered Archimedes to get up and follow him to Marcellus the consul of Rome. Archimedes ignored him, which enraged the soldier, who purportedly then messed up his diagrams. As it is written, Archimedes protested and told the soldier that he needed to finish what he was working on. The soldier was so furious by what he interpreted as insolence, that he drew his sword and stabbed Archimedes to death.

It is hard to know what the real story is, since there are different accounts of this incident. But, whatever the truth, his death was exceedingly tragic. It is also written that Marcellus was so upset that Archimedes had been killed against his specific orders, that he commanded that this soldier be killed as well. Those who thought the history of mathematics was devoid of human emotion might reconsider their views after knowing this story.

4.4.3 Cavalieri's Principle

In a lighter vein, let us now turn to another beautiful result about volume called Cavalieri's principle.

Cavalieri's principle. If two solids of the same width are placed next to each other, say, on a table, in such a way that, at the same distance x from the origin, the cross-sectional area of the first solid is equal to the corresponding cross-sectional area of the second solid, even though they may be shaped very differently, the volumes of the solids are the same. For example, examine Figure 4.59 below.

154 Measurement: Area and Volume

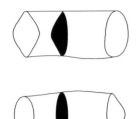

Figure 4.59

Cavalieri's principle is saying that, if the shaded areas are the same for *each* x between a and b, then the volume of the two solids shown are the same.

We have all we need to prove this, and it is done in only a few sentences. If we let $A(x)$ be the area of the cross section of the first solid at a distance x from the origin and let $B(x)$ be the corresponding area of the cross section for the second solid, then we know that $A(x) = B(x)$ for each x between a and b. So, of course, it follows that

$$\int_a^b A(x)dx = \int_a^b B(x)dx. \tag{4.23}$$

But the integral on the left side of equation (4.23) is the volume of the first solid and the integral on the right side of equation (4.23) is the volume of the second solid, as we have seen in the beginning of this section. Hence, the volumes of the two solids are the same.

Cavalieri's principle is quite abstract, and so you might be wondering whether it has any applications in real life. Well, we recently did an Internet search for Cavalieri's principle and the following articles came up unexpectedly "Volume estimation of multicellular colon carcinoma spheroids using Cavalieri's principle" by J. Bauer and others. The authors described how they use Cavalieri's principle in cancer studies. In a journal of pathology were the articles, "Application of the Cavalieri principle and vertical sections method to the lung: estimation of volume and pleural surface area," and "Estimation of Breast Prosthesis Volume by the Cavalieri principle." This was followed by articles applying Cavalieri's principle to MRI images, to dermatology, and neuropharmacology. Who would have thought?

Similar to many other mathematicians who discovered abstract relationships, Cavalieri had no idea if his principle would ever have any applications in real life. It is important to realize that we don't always know when and where the mathematical ideas will be applied. But if we aren't willing to wait for the application and concentrate on the development of these abstract concepts and relationships, the applications may never happen.

4.4.4 Final Remarks

In the chapter on properties of numbers and theory of equations, we had certain definitions which we used to prove our results. There was no question of the truth of the findings we got, since they followed from definitions only, and no other assumptions. Geometry is a very different kind of field. There, we have diagrams and we have to use our eyes. But what we see, may not be correct. In geometry, we really have to axiomatize what it is that we believe, and then the theorems we prove will be true provided the assumptions we make are true. For example, our eyes told us that, as we inscribed regular polygons with more and more sides in a circle, the areas of these polygons approached the area of the circle. We accepted that. And then it followed from that, that the area of the circle is πr^2. If someone finds an example where this is not true, then our theory has to be redone, and it might not be that in all circles the area is πr^2. But we have lots of corroboration that the area of a circle is πr^2, and so we believe that what our eyes were telling us was true.

In classical geometry, people believed that a figure could be moved in space and that its physical properties would not change. Einstein has shown that this may not be true. And so, new versions of geometry were developed, and relationships that were formerly proven by moving figures in space, became axioms in many cases; that is, relationships we accepted without proof. To deal with some of these issues, modern geometers now approach the topic of congruence from a function point of view. But in geometry we always have to make certain assumptions that are consistent with what we see.

People who are interested in applications don't worry too much about these technicalities. For them, geometry is a model of the real world, and if the results we prove "work" in the real world, then that is enough to accept them. To be stubborn, and not accept them, when all indications are that the model works well, would be foolish. We would miss all the applications.

Student Learning Opportunities

1 (C) Your students have accepted the formula for the volume of a simple solid. They are now curious to know how you can show that if you have the following solid shown in Figure 4.60

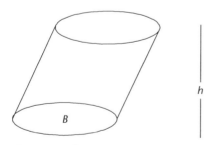

Figure 4.60

where each cross section has area B and height is h, the volume of the solid is Bh. How do you show it?

2 Many times, strokes are caused by a buildup of plaque in the arteries. Imagine the plaque buildup in an artery to be a region between two concentric circular cylinders of length L, and assume that the inner radius is 0.3 cm and the outer radius is 0.307 cm. Estimate the volume of the artery blocked by plaque. (See Figure 4.61.)

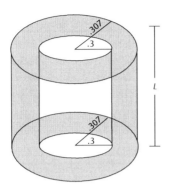

Figure 4.61

3 The French physiologist, Jean Poisseuille, discovered the law that the volume of blood flowing through an artery per minute is given by $V = kR^4$ where k is a constant, and R is the inner radius. When a person's arteries are blocked, a procedure called angioplasty is done. Here a balloon is inserted into the artery and the artery is expanded. Suppose that, under angioplasty, an artery has its radius increased 5%. Estimate the change in the volume of blood flow that results.

4 (C) A student asks why can't you use equation (4.18) to find $\int x^x dx$. What is your answer?

5 Let R be the region bounded by $y = x^2$, the x-axis, the lines $x = 1$ and $x = 5$.

(a) Suppose S is a solid whose base is R and whose cross sections perpendicular to the x-axis are semicircles. What is the volume of the solid, S?

(b) Suppose S is a solid whose base is R and whose cross sections perpendicular to the x-axis are equilateral triangles. What is the volume of the solid, S?

6 (C) A student asks how you show that the volume of a sphere with radius R is $\frac{4}{3}\pi R^3$. How do you explain it by using integrals?

7 If a plane perpendicular to a diameter at a distance a from the center of a sphere chops the sphere into two parts, the smaller part is called a spherical cap. Find the volume of a spherical cap in terms of a and R where R is the radius of the sphere.

8 (C) A student asks you how to show that the volume of a pyramid with square base whose area is B is $\frac{1}{3}Bh$. Using Figure 4.62 and the hint below, fill in the details.

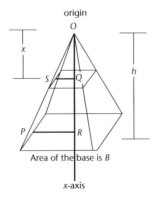

Figure 4.62

[Hint: Put the pyramid so that its apex is at the origin. We have also shown a typical cross section at a distance x from the origin (the square with the letters S and Q). Observe that triangles SOQ and POR are similar. Thus, $\frac{SQ}{PR} = \frac{x}{h}$. But SQ is half the side of the square containing SQ, and PR is half the side of the square containing PR. Thus, the sides of the squares containing SQ and PR are double SQ and PR, and the ratio of the areas of the squares containing these lines is $\frac{(2SQ)^2}{(2PR)^2}$ or just $\frac{(SQ)^2}{(PR)^2}$. Recalling that the area of the square base is B and calling the area of the cross section at distance x from the origin, $A(x)$, we have $\frac{A(x)}{B} = \frac{(SQ)^2}{(PR)^2} = \left(\frac{x}{h}\right)^2$. Thus, $A(x) = \frac{Bx^2}{h^2}$. Now finish it. The key step in the above proof was to show that $A(x) = \frac{Bx^2}{h^2}$. This is true for any pyramid with vertical axis, regardless of the shape of the base.]

CHAPTER 5

THE TRIANGLE: ITS STUDY AND CONSEQUENCES

5.1 Introduction

If you ask adults what theorem they remember from their study of mathematics, they will most probably say, the Pythagorean Theorem. Why should this theorem, usually studied in secondary school, make such a lasting impression? As will be demonstrated in this chapter, this one theorem concerning the relationship of the sides of a right triangle can be extended to the study of (a) all types of triangles, (b) relationships concerning circles, (c) key trigonometric relationships, and (d) concepts of area. It is really quite amazing!

To begin, we need only remember a few basic definitions: In a right triangle with acute angle, A,

$$\sin A = \frac{\text{side opposite } A}{\text{hypotenuse}}$$

$$\cos A = \frac{\text{side adjacent to } A}{\text{hypotenuse}}$$

$$\tan A = \frac{\text{side opposite to } A}{\text{side adjacent to } A}.$$

Also, we easily see that

$$\tan A = \frac{\sin A}{\cos A}.$$

We begin by discussing how the Pythagorean Theorem can be extended to generate the Law of Cosines and then follow it with the study of the Law of Sines, similarity, and relationships within a circle.

5.2 The Law of Cosines and Surprising Consequences

 **LAUNCH**

Draw a large triangle on a clean sheet of paper. Then measure the length of each of the sides of the triangle you have drawn. Using these same three lengths, try to draw another triangle that is NOT congruent to the first one you drew. Could you do it? Why or why not?

We hope that the launch question helped you to recall some of the work you did in secondary school regarding congruent triangles. You probably remember the theorem that one of the ways to prove that two triangles are congruent is to show that the three sides of one are congruent to the three sides of another (often represented as SSS = SSS). But, if a student asked you if this was an axiom or a theorem, would you know what to say? Don't feel badly if you wouldn't, since many secondary school textbooks have listed it in different ways. It will probably surprise you to know that it can indeed by proven, by applying the Law of Cosines. In this section, we will demonstrate how this and other similar congruence results can be shown.

The Pythagorean Theorem tells us that, in a right triangle with legs a and b and hypotenuse c, $a^2 + b^2 = c^2$. What happens if the triangle is not a right triangle? The following theorem answers this.

Theorem 5.1 *(Law of Cosines): In any triangle ABC,*

$$c^2 = a^2 + b^2 - 2ab \cos C.$$

Proof. Notice that "$c^2 = a^2 + b^2$" is part of the theorem. How interesting! It seems very likely that we will be using the Pythagorean Theorem in this proof. We begin with triangle ABC shown below and draw altitude AD. We will prove the theorem in the case when the altitude is inside the triangle, (that is when the triangle has all its angles less than 90°) (Figure 5.1). In Student Learning Opportunity 4 you will prove it for the case when the altitude is outside of the triangle.

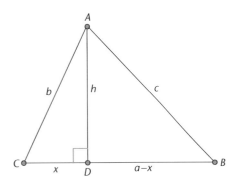

Figure 5.1

Using the Pythagorean Theorem on triangle ABD we have that $(a - x)^2 + h^2 = c^2$ which, when expanded, gives us

$$a^2 - 2ax + x^2 + h^2 = c^2. \tag{5.1}$$

Using the Pythagorean Theorem on triangle ACD we have

$$x^2 + h^2 = b^2. \tag{5.2}$$

Substituting equation (5.2) in equation (5.1) we have

$$a^2 - 2ax + b^2 = c^2. \tag{5.3}$$

Now, from triangle ACD, $\cos C = \frac{x}{b}$ from which it follows that $x = b \cos C$. Substituting this in equation (5.3) for x we get

$$a^2 - 2ab \cos C + b^2 = c^2 \qquad (5.4)$$

which can be rearranged to read

$$c^2 = a^2 + b^2 - 2ab \cos C \qquad (5.5)$$

and we are done. ∎

There are two other versions of the law of cosines:

$$a^2 = b^2 + c^2 - 2bc \cos A \quad \text{and} \qquad (5.6)$$
$$b^2 = a^2 + c^2 - 2ac \cos B \qquad (5.7)$$

and they are proved in exactly the same way, only we draw altitudes to the other sides of the triangle. You will prove one of the versions in the Student Learning Opportunities.

Taking the theorems further, if we solve for $\cos C$ in equation (5.5) we get that

$$\cos C = \frac{a^2 + b^2 - c^2}{2ab}. \qquad (5.8)$$

Similarly, we can solve for $\cos A$ and $\cos B$ in equations (5.6) and (5.7), respectively, to get

$$\cos A = \frac{b^2 + c^2 - a^2}{2bc} \qquad (5.9)$$

and

$$\cos B = \frac{a^2 + c^2 - b^2}{2ac}. \qquad (5.10)$$

What equations (5.9), (5.10), and (5.8) tell us, respectively, is, if we know the lengths a, b, and c of the three sides of a triangle ABC, then we immediately know $\cos A$, $\cos B$, and $\cos C$ and hence angles A, B, and C. This brings us to the topic of congruence.

5.2.1 Congruence

Recall that, in geometry, two triangles ABC and DEF are congruent if their sides and angles can be matched precisely. That is, $AB = DE$, $BC = EF$, $AC = DF$, and $\angle A = \angle D$, $\angle B = \angle E$, and $\angle C = \angle F$. (See Figure 5.2 below.)

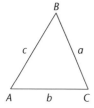 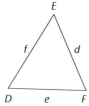

Figure 5.2

When we write $AB = DE$, **it will mean that the lengths of the sides** AB **and** DE **are the same, and when we write** $\angle A = \angle D$ **it means that angles** A **and** D **have the same measure.** Under this correspondence, angles A and D are called corresponding angles, as are the angles B and E, and C and F. Sides AB and DE are called corresponding sides, as are the sides BC and EF, and the sides AC and DF. By definition of congruent triangles, corresponding parts have the same measure, which means corresponding sides have the same length and corresponding angles have the same degree measure. Notice that the order in which we write the letters tells us the angle correspondence and side correspondence. Had we written that triangle ACB was congruent to FDE, then it would mean that $\angle A = \angle F$, $\angle C = \angle D$, and $\angle B = \angle E$, and that $AC = FD$, $CB = DE$, and $BA = EF$.

The first result we talk about is something we are all familiar with: If three sides of one triangle have the same lengths as three sides of another triangle, then the triangles are congruent. That is, all their corresponding parts match! This is quite remarkable since we have said nothing about the angles of these triangles. Yet, this follows immediately from the Law of Cosines.

> **Theorem 5.2** *(SSS = SSS) If the three sides of triangle ABC are equal to the three sides of triangle DEF, then the triangles ABC and DEF are congruent.*

Proof. Let us assume that the sides that match are a and d, b and e, and c and f. (Refer to Figure 5.2.) So $a = d$, $b = e$, and $c = f$. From equation (5.8) we have that

$$\cos C = \frac{a^2 + b^2 - c^2}{2ab} \tag{5.11}$$

and using the same law in triangle DEF with the corresponding sides, we have

$$\cos F = \frac{d^2 + e^2 - f^2}{2de}. \tag{5.12}$$

Since $a = d$, $b = e$, and $c = f$, we can substitute them in equation (5.11) to get

$$\cos C = \frac{d^2 + e^2 - f^2}{2de} \tag{5.13}$$

and we see from equations (5.12) and (5.13) that

$$\cos C = \cos F. \tag{5.14}$$

It follows that $\angle C = \angle F$.

In a similar manner, using the other versions of the Law of Cosines, equations (5.9) and (5.10), we can show that $\angle A = \angle D$ and $\angle B = \angle E$.

Thus, if three sides of one triangle are equal to three sides of another triangle, then the angles match, and so the triangles are congruent. ∎

Note: In the proof of Theorem 5.2 (refer to equation (5.14)), we used the fact that, if $\cos C = \cos F$, then $\angle C = \angle F$. While you may have accepted this, much more is involved in this statement than meets the eye. For now, we will use this fact continually and ask you to accept it. But, we will examine the reason behind it in a later section on technical issues. We now turn to a corollary of our SSS congruence theorem.

> **Corollary 5.3** *(HL = HL) Two right triangles are congruent if the hypotenuse and leg of one triangle are equal to the hypotenuse and leg of the other triangle.*

Proof. Below, in Figure 5.3 we see two right triangles where the hypotenuse and leg of one have the same length as the hypotenuse and leg of the other.

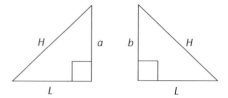

Figure 5.3

By the Pythagorean Theorem,

$$a^2 + L^2 = H^2 \tag{5.15}$$

and

$$b^2 + L^2 = H^2. \tag{5.16}$$

From equations (5.15) and (5.16) we have

$$a^2 + L^2 = b^2 + L^2.$$

Subtracting L^2 from both sides, we get that

$$a^2 = b^2$$

and therefore $a = b$. Thus, the lengths of three sides of one triangle are equal to the lengths of three sides of the other triangle and the two triangles are congruent by SSS = SSS. ∎

> **Corollary 5.4** *(SAS = SAS) If two sides and the included angle of one triangle are equal to two sides and the included angle of another triangle, then the triangles are congruent.*

Proof. Suppose we have triangles ABC and DEF and suppose that $a = d$, $b = e$, and $\angle C = \angle F$. (See Figure 5.4 below.)

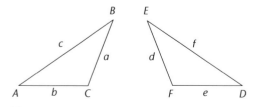

Figure 5.4

Then, by the Law of Cosines, applied to triangle *ABC*,

$$c^2 = a^2 + b^2 - 2ab \cos C. \tag{5.17}$$

By the Law of Cosines, applied to triangle *DEF*,

$$f^2 = d^2 + e^2 - 2de \cos F. \tag{5.18}$$

But we know that $a = d$, $b = e$, and $\angle C = \angle F$, and if we substitute these into equation (5.18) we get

$$f^2 = a^2 + b^2 - 2ab \cos C. \tag{5.19}$$

Since the right hand sides of equations (5.17) and (5.19) are the same, so are the left sides. That is,

$$c^2 = f^2.$$

From this, we get that $c = f$.

Thus, the three sides of the first triangle are equal to the three sides of the second triangle, and so the triangles are congruent by Theorem 5.2. ∎

Suppose we have one triangle and we only know the measures of two of its angles and one side. Would we be able to use the Law of Cosines to determine information about the other two sides? Well, the answer is, "No." Since the Law of Cosines requires us to know two sides and one angle, we do not have enough information to use it. To find the missing information about the triangle, we need another law which we will discuss in the next section: the Law of Sines.

Student Learning Opportunities

1 Given that the sides of a triangle are $a = 3$, $b = 5$, and $c = 7$, find all three angles.

2 If the sides of a parallelogram are 3 and 4, and the angle between them is 30 degrees, how long is each diagonal?

3 A surveyor needs to estimate the distance across a lake from point *A* to point *B*. Standing at point *C*, 4.6 miles from *A* and 7.3 miles from *B*, he measures the angle shown below in Figure 5.5 to be 80 degrees. Estimate the distance *AB*.

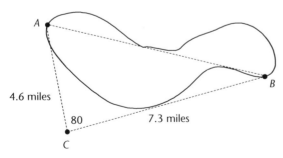

Figure 5.5

4 We will point out in the chapter on trigonometry that $\cos(180° − x) = −\cos x$. Use this fact to prove the Law of Cosines when the altitude *AD* in Figure 5.1 is outside of the triangle.

5 (C) A student asks how you prove the other 2 versions of the Law of Cosines found in equations (5.9) and (5.10). How do you do it?

6 Your students are intrigued by how the Pythagorean Theorem was used to prove the Law of Cosines. They wonder, because of the similar structure of the theorems, if one can go in reverse. That is, can one use the Law of Cosines to prove that, if $c^2 = a^2 + b^2$ holds in a triangle, then the triangle is right? What is your answer and how do you show it?

7 (C) A student asks how you can prove that, if two angles and any side of one triangle are equal to two angles and the corresponding side of another triangle, the triangles are congruent (i.e., AAS = AAS). How do you prove it?

5.3 The Law of Sines

LAUNCH

Give an example (draw it) where two angles and a side of one triangle are equal to two angles and a side of another triangle, but the triangles are not congruent. [Hint: make sure the sides are not corresponding.]

We hope that you were able to construct two different shaped triangles and that if you hadn't realized it before, you realize now, the importance of always specifying that corresponding parts be congruent in congruence proofs. You most likely remember the theorem, which says that one of the ways to prove that two triangles are congruent is to show that they have two angles and a corresponding side that are congruent. What you probably did not know is why this is true. In this section you will be surprised to see how instrumental the Law of Sines can be in verifying the proof of this relationship.

Theorem 5.5 *(Law of Sines): In any triangle ABC*
$$\frac{a}{\sin A} = \frac{b}{\sin B} = \frac{c}{\sin C}.$$

Proof. We may use Figure 5.1 (from earlier in the chapter), which we copy here for convenience.

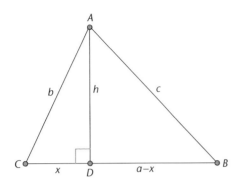

Figure 5.1

In triangle ADB, $\sin B = \frac{h}{c}$. Hence,

$$h = c \sin B. \tag{5.20}$$

In triangle ADC, $\sin C = \frac{h}{b}$. Thus,

$$h = b \sin C. \tag{5.21}$$

Setting the two expressions equal for h in equations (5.20) and (5.21) we have,

$$c \sin B = b \sin C, \tag{5.22}$$

and dividing both sides of equation (5.22) by $\sin B \sin C$ we get

$$\frac{c}{\sin C} = \frac{b}{\sin B}. \tag{5.23}$$

That is the first half of our theorem. In the Student Learning Opportunities you will draw a different altitude and show that

$$\frac{a}{\sin A} = \frac{c}{\sin C}. \tag{5.24}$$

And the two relationships, equations (5.23) and (5.24) together tell us that

$$\frac{a}{\sin A} = \frac{b}{\sin B} = \frac{c}{\sin C}.$$

∎

Since this law involves several angles, we can now find the missing parts of a triangle in which we are given two angles and a side or two sides and an angle.

Corollary 5.6 *(ASA = ASA) If two angles and the included side of one triangle are equal to two angles and the included side of another triangle, the triangles are congruent.*

Proof. Suppose that we have two triangles *ABC* and *DEF* from Figure 5.2 earlier in the chapter, which we copy here,

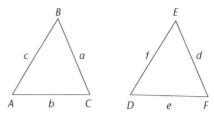

Figure 5.2

and we are given that $\angle A = \angle D$, $\angle B = \angle E$, and $c = f$. Since $\angle A = \angle D$ and $\angle B = \angle E$, we also have $\angle C = \angle F$, because the sum of the angles of a triangle is 180 degrees. Using the Law of Sines in triangle *ABC*, we get that

$$\frac{a}{\sin A} = \frac{c}{\sin C}. \tag{5.25}$$

Using the Law of Sines in triangle *DEF*, we have

$$\frac{d}{\sin D} = \frac{f}{\sin F}. \tag{5.26}$$

Since $\angle A = \angle D$, $\angle C = \angle F$, and $c = f$, we can substitute these values into equation (5.25) to get

$$\frac{a}{\sin D} = \frac{f}{\sin F} \tag{5.27}$$

and since the right sides of equations (5.26) and (5.27) are the same, we see that

$$\frac{a}{\sin D} = \frac{d}{\sin D}. \tag{5.28}$$

Multiplying both sides of equation (5.28) by sin *D*, we get that $a = d$. Since we were given $c = f$ and we showed that $a = d$ and we were given that $\angle B = \angle E$ (see figure above), by Corollary 5.4. (SAS = SAS) we have that the two triangles *ABC* and *DEF* are congruent. ∎

Student Learning Opportunities

1 (C) Your students tell you that they are confused about when they should use the Law of Cosines and when they should use the Law of Sines, when trying to find missing parts of a triangle. What do you tell them?

2 In triangle *ABC*, $\angle A = 37°$, $\angle B = 64°$, and $c = 12$. Find the lengths of all three sides.

3 In triangle *ABC*, $AC = 56$, $AB = 80$, $\angle C = 64°$. Find $\angle B$ to the nearest degree.

168 The Triangle: Its Study and Consequences

4 Find ∠ACD to the nearest degree in Figure 5.6 below.

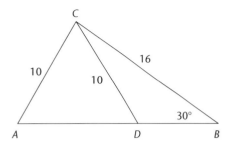

Figure 5.6

5 A satellite orbiting the earth is being tracked. The observation stations are 300 miles apart in two different towns A and B. When the satellite is visible from both towns, the angles of elevation of the satellite are recorded and found to be 63 and 72 degrees, respectively, as shown in Figure 5.7 below.

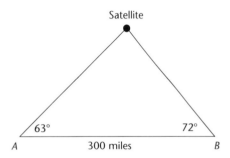

Figure 5.7

How far is the satellite from station A? from station B?

6 (C) The proof of the Law of Sines we gave was for acute triangles only. A student is curious to know how the proof would have to be modified to show that it is still true if h is outside the triangle. How would you show it? [Hint: It is a fact that $\sin(180° - x) = \sin x$.]

7 The following interesting result also follows from the Law of Sines: If ABC is any triangle and BD is the angle bisector of angle B, then $\frac{AB}{BC} = \frac{AD}{DC}$. (The angle bisector divides the opposite side into segments having the same ratio as the sides.) Prove this. You will need the fact that $\sin(180 - p)° = \sin p$. [Hint: Use Figure 5.8 below.]

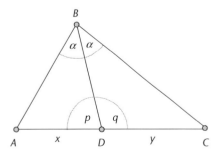

Figure 5.8

8 In triangle BAC, AC is the shortest side. Angle bisector BD is drawn to AC, dividing it into segments AD and BD. If AC = 14 and the ratio of the sides of the triangle is 2 : 3 : 4, how long is the shorter of the segments AD and BD? Explain.

9 (C) Your students want to know if there is a way to find the area of a triangle without knowing its altitude. Show them that there is, by proving that the area of a triangle ABC is $\dfrac{a^2 \sin B \sin C}{2 \sin A}$. [Hint: The area of a triangle is $\dfrac{1}{2}$ base × height. Take the base to be a.]

5.4 Similarity

LAUNCH

1. Using a ruler, a pen, and a piece of paper, draw a triangle. Label the vertices, A, B, and C. Starting at vertex A, extend side AB its own length to point D (side AD will now be twice as long as side AB). Now, from vertex A, extend side AC its own length to point E. (Side AE will now be twice as long as side AC.)
2. Measure the length of side BC. Based on what you have measured, what do you predict is the length of side DE? Are you correct?
3. What can you say about the shapes of triangles ABC and AED? Why do you believe this is true?

After having done the launch question, you are probably beginning to recall some of the basic properties of similar triangles. Did you ever question how the similarity theorems you learned in secondary school could be proven? In this section we will surprise you with how they can be done.

We have used the Law of Sines and the Law of Cosines to derive all the congruence theorems that are taught in geometry. But now let us turn to another set of results that are also critically important, those that deal with similarity. Applications of similarity range from the mundane to the surprising. For example, similarity is used on a daily basis by engineers who use scale drawings to create a model of a building that is going to be constructed. When you take a picture of a person, the picture you get is similar to the person. Similarity is also used (surprisingly) in radiation therapy for cancer patients for accuracy in focusing the beam. To find out more about this, visit the website: http://www.learner.org/resources/series167.html?pop=yes&vodid=533776&pid=1800# and watch the video on similar triangles.

We will now show how the Law of Sines and the Law of Cosines can be used to develop the main results about similarity. The versatility of these laws is quite remarkable. Recall that two triangles, ABC and DEF are called **similar** if they have the same shape, but not necessarily the same size. This means that one is a scaled version of the other. The formal definition of similarity between triangles ABC and DEF is that the angles of triangle ABC are congruent to those of triangle

DEF, and that the sides opposite the corresponding congruent angles are proportional. When one writes that two triangles are similar, the order in which the letters are written reveals which angles are congruent. Thus, when we write that triangle *ABC* is similar to triangle *DEF*, it follows that $\angle A \simeq \angle D$, $\angle B \simeq \angle E$, and $\angle C \simeq \angle F$. Saying that the sides of the triangles are proportional means that there is a number k such that $a = kd$, $b = ke$, and $c = kf$ where a, b, c, and d, e, f, are respectively, the corresponding sides of the triangles. See Figure 5.9 below.

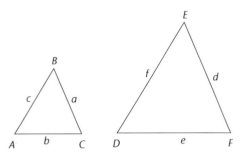

Figure 5.9

Another way of expressing that the sides of two triangles *ABC* and *DEF* are proportional is to write $\dfrac{a}{d} = \dfrac{b}{e} = \dfrac{c}{f} = k$.

Here is our first theorem on similarity.

Theorem 5.7 *If the corresponding sides of two triangles are proportional, then the corresponding angles of the triangle are equal. Hence, the two triangles are similar.*

Proof. Suppose that our triangles are *ABC* and *DEF*, as in Figure 5.9, and suppose that we are given that $a = kd$, $b = ke$, and $c = kf$. Using equation (5.8) we have that, in triangle *ABC*

$$\cos C = \frac{a^2 + b^2 - c^2}{2ab}. \tag{5.29}$$

In triangle *DEF* we have the similar result that

$$\cos F = \frac{d^2 + e^2 - f^2}{2de}. \tag{5.30}$$

But we know that $a = kd$, $b = ke$, and $c = kf$. Substituting these in equation (5.29) we have

$$\cos C = \frac{(kd)^2 + (ke)^2 - (kf)^2}{2kdke}$$

$$= \frac{k^2(d^2 + e^2 - f^2)}{2k^2 de}$$

$$= \frac{d^2 + e^2 - f^2}{2de}$$

$$= \cos F. \quad \text{[From equation (5.30).]}$$

Since $\cos C = \cos F$, $C = F$.

In a similar manner we can show that the other angles are equal. Thus, since the corresponding sides were in proportion and, as a result of this, we showed the corresponding angles were equal, the triangles *ABC* and *DEF* are similar. ∎

Now we prove the converse.

Theorem 5.8 *If the three angles of one triangle are equal to three angles of another triangle, then the corresponding sides of the triangles are in proportion. Hence, the two triangles are similar.*

Proof. Suppose that we have triangles *ABC* and *DEF* where angle $\angle A = \angle D$, $\angle B = \angle E$ and $\angle C = \angle F$. (See the figure from the previous theorem.) Then by the Law of Sines, applied to triangle *ABC* we have

$$\frac{a}{\sin A} = \frac{b}{\sin B}$$

which can be written as

$$\frac{a}{b} = \frac{\sin A}{\sin B}. \tag{5.31}$$

Using the Law of Sines in triangle *DEF*, we have in a similar manner that

$$\frac{d}{e} = \frac{\sin D}{\sin E} \tag{5.32}$$

But since $\angle A = \angle D$ and $\angle B = \angle E$, we can substitute $\angle A$ and $\angle B$ for $\angle D$ and $\angle E$ in equation (5.32) and we get

$$\frac{d}{e} = \frac{\sin A}{\sin B}. \tag{5.33}$$

Since the right sides of equations (5.31) and (5.33) are the same, the left sides are also, so we have that

$$\frac{a}{b} = \frac{d}{e}.$$

Therefore, $ae = bd$ by cross multiplying, and dividing both sides by *de* we get that

$$\frac{a}{d} = \frac{b}{e}.$$

In the Student Learning Opportunities you will show in a similar manner that $\frac{a}{d} = \frac{c}{f}$. So, with your work and ours, we get

$$\frac{a}{d} = \frac{b}{e} = \frac{c}{f}$$

which says the sides are in proportion. ∎

The following result is probably the most familiar to you.

Corollary 5.9 (AA = AA) *If two angles of one triangle are equal to two angles of another triangle, the triangles are similar.*

Proof. The third angles of the triangles will also be equal, since the sum of the angles of a triangle is 180°. The result now follows from Theorem 5.8. ∎

What Theorems (5.7) and (5.8) are saying is that to show that two triangles are similar, we need to show that *either* the corresponding sides are in proportion *or* that the corresponding angles are equal. One automatically implies the other.

There is one other result about similar triangles which is useful, but less well known.

Theorem 5.10 *If two sides of one triangle are proportional to two sides of another triangle, and the angle between the proportional sides of these triangles is the same, then the two triangles are similar.*

Proof. We may suppose that the sides that are in proportion are b and e, and c and f, and that the angle A between sides b and c, is equal to the angle D that is between e and f. Saying that the sides b, e, c, and f are in proportion means that

$$\frac{b}{e} = \frac{c}{f} = k. \tag{5.34}$$

To prove the triangles are similar, we will show that the third sides are also in proportion. That is, we will show $\frac{a}{d}$ is also k. Once we have that all three sides are in proportion, we know by Theorem 5.7 that the triangles are similar.

Now, using the Law of Cosines in triangle *ABC* we have that

$$a^2 = b^2 + c^2 - 2bc \cos A. \tag{5.35}$$

Using the Law of Cosines in triangle *DEF* we have

$$d^2 = e^2 + f^2 - 2ef \cos D. \tag{5.36}$$

From equation (5.34) we have $b = ke$ and $c = kf$, and we were given that $\angle A = \angle D$. Replacing b by ke, c by kf, and A by D in equation (5.35) we get

$$a^2 = (ke)^2 + (kf)^2 - 2(ke)(kf) \cos D$$
$$= k^2(e^2 + f^2 - 2ef \cos D)$$
$$= k^2 d^2. \quad \text{[Using (5.36).]}$$

This string of equalities shows that $a^2 = k^2 d^2$. Hence, $a = kd$. It follows that $\frac{a}{d}$ is also k and, using equation (5.34), we see that

$$\frac{a}{d} = \frac{b}{e} = \frac{c}{f} = k.$$

So, we have shown that all three sides are in proportion, and hence, by Theorem 5.7, triangles *ABC* and *DEF* are similar. ∎

Student Learning Opportunities

1 If triangle *ABC* is similar to triangle *DEF*, and the following facts about the sides are given, find the remaining sides and angles to the nearest degree.
   (a) $AB = 6$, $DE = 18$, $EF = 12$, $CA = 8$
   (b) $BC = 12$, $EF = 6$, $DF = 18$, $DE = 8$

2 (C) A student wants to know if, in similar triangles, corresponding altitudes are in the same ratio as corresponding sides. Are they? If so, how can you show it?

3 (C) A student wants to know if you are given two similar triangles, does it mean that any pair of their corresponding medians are in the same proportion as the sides? How do you respond? What is your explanation?

4 (C) A student wants to know if it is true that the ratio of areas of similar triangles is the same as the ratio of the corresponding sides. Is it? If not, what is true about the ratio of the areas of similar triangles and the ratio of corresponding sides and how can you show it? Take two specific similar right triangles, like the 3–4–5 right triangle and the 6–8–10 right triangle, and answer these questions before answering the general question.

5 Using Theorem 5.10, show that the length of the line segment connecting the midpoints of two sides of a triangle is $\frac{1}{2}$ the length of the third side.

6 (C) Your students notice that, when you draw the line segment connecting the midpoints of two sides of a triangle, it appears to always be parallel to the third side. They want to know if this is always true, and if it is, how can it be proven. How do you respond and how do you prove it? [Hint: Show that you have a pair of corresponding angles equal.]

7 (C) You have encouraged your students to use some dynamic geometric software to make some conjectures about the shape of the quadrilateral created by connecting the midpoints of the adjacent sides of any quadrilateral. (See Figure 5.10 below.) They have been surprised to see that, regardless of the size or shape of the exterior quadrilateral, it seems that the interior quadrilateral they have created always looks like a parallelogram. They want to know if it really is, and if it is, how it can be proven. How would you prove it? [Hint: Use the results from Student Learning Opportunity 6 for your proof that the students are correct.]

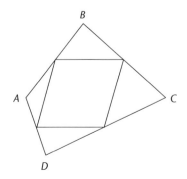

Figure 5.10

8 Finish the proof of Theorem 5.8 by proving $\dfrac{a}{d} = \dfrac{c}{f}$.

9 (C) You have had your students use some dynamic geometric software to create a right triangle and draw an altitude to the hypotenuse. After dragging the points of the right triangle, they have noticed that the two smaller triangles that are formed within the larger right triangle appear to always be similar to each other, and more surprisingly, seem to always be similar to the big triangle. They want to know if this is always true, and if it is, how can it be proven. How do you prove it?

10 Give another proof of the Pythagorean Theorem using similar triangles. [Hint: Refer to Figure 5.11 below. Show that $(AC)^2 = AB \cdot AD$ and that $(BC)^2 = AB \cdot DB$. Now add the equations. You might want to use the result of the previous Student Learning Opportunity.]

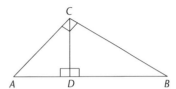

Figure 5.11

11 Find the value of x in Figure 5.12 if DE is parallel to AC.

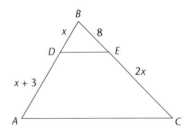

Figure 5.12

12 Two poles of heights 5 feet and 10 feet are separated by a distance of 20 feet. A wire is placed tautly from the top of each pole to the bottom of the other pole and they overlap at a point P shown in Figure 5.13 below.

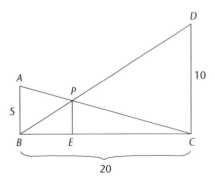

Figure 5.13

(a) How high above the ground is *P*? [Hint: Triangles *ABC* and *PEC* are similar, as are triangles *BPE* and *BDC*. Let *BE* = *x*, *EC* = 20 − *x*, and *PE* = *h*. From the first set of similar triangles, $\frac{5}{20} = \frac{h}{20-x}$. Find a similar relationship for the second pair of triangles and work from there.]

(b) Do the same problem as in (a), only now assume that the distance between the poles is 100 feet instead of 20 feet. Show that the overlap of the wires is at the same height above the ground as it was in part (*a*). Does this surprise you? Why or why not?

(c) Suppose the distance between the poles is *d*. Show that the distance *h*, that *P* is above ground, does not depend on *d*.

(d) Generalize the solution. Suppose that the poles are at heights *a* and *b* and that the distance between them is *d*. Show that
$$h = \frac{ab}{a+b}.$$

5.5 Sin(*A* + *B*)

LAUNCH

1 Most students believe that sin(*A* + *B*) = sin(*A*) + sin(*B*).

   a. Is this always true? If you said "Yes," then justify why. If you said "No," then support your answer by giving a counterexample.
   b. Is this ever true? If you said "Yes," then give one example when it is true and state how often you think it is true. If you said "No," then justify why you believe it is never true.

After having responded to the launch question, you are now most likely curious about the behaviors of the sine of the sum of two angles. You must be wondering if the trigonometric relationships share the same distributive property that algebraic relationships have. Wouldn't it be nice if they did. In this section we will pursue this question further to get the right relationships in a manner that will most likely surprise you.

We have used the concept of area to prove the Pythagorean Theorem and applied that theorem to prove the Law of Sines and the Law of Cosines, which gave us all the main theorems about congruence and similarity. We now take the concept of area and use it to prove theorems in trigonometry. It is hard not to appreciate how powerful this concept of area is.

It is a common misconception among secondary school students that sin(*A* + *B*) = sin(*A*) + sin(*B*). This is not true, which is easy to see by just taking the counterexample *A* = 30° and *B* = 60°. Using a calculator, or the well-known values of $\sin 30° = \frac{1}{2}$, $\sin 60° = \frac{\sqrt{3}}{2}$ and $\sin 90° = 1$, we see that sin(*A* + *B*) = 1 but, $\sin A + \sin B = \frac{1}{2} + \frac{\sqrt{3}}{2} \approx 1.366$. They are quite far apart. However, it is true that sin(*A* + *B*) = sin *A* cos *B* + cos *B* sin *A*, and you can verify it for the angles given above. Of course, a proof of this relationship is needed, which we will give now using areas. Our proof is

176   The Triangle: Its Study and Consequences

only valid for triangles with interior altitudes. But the theorem is true in general and will follow from results we set forth in the trigonometry chapter. First, we need a preliminary result which is commonly taught in secondary school.

Theorem 5.11 *The area of a triangle is $\frac{1}{2}ab\sin C$, where a and b are two sides of a triangle and C is the included angle.*

Proof. Using Figure 5.14 below with altitude h and base b.

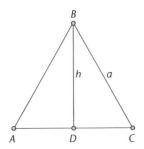

Figure 5.14

We have that the area of the triangle is

$$\frac{1}{2}bh. \tag{5.37}$$

But from right triangle BDC, $\sin C = \dfrac{\text{opposite}}{\text{hypotenuse}} = \dfrac{h}{a}$, so $h = a\sin C$. Substituting for h in (5.37) we get that the area of a triangle is $\frac{1}{2}ab\sin C$. ∎

Theorem 5.12 *If α and β are angles, then $\sin(\alpha + \beta) = \sin\alpha\cos\beta + \cos\alpha\sin\beta$.*

Proof. Place angles α and β next to each other, as shown in Figure 5.15 below, so that their common side, BH, is vertical, and together they form a larger angle $\alpha + \beta$, which we call B.

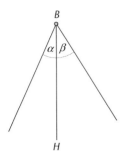

Figure 5.15

Mark off a point E at distance of 1 from B, along BH, and, through that point E, draw a line perpendicular to BH intersecting the sides of the large angle B at A and C. (See Figure 5.16 below.)

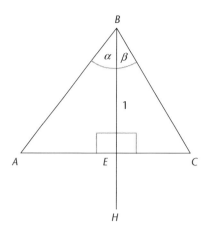

Figure 5.16

Now observe from the picture that, in right triangle AEB, $\cos\alpha = \dfrac{1}{AB}$. Cross multiplying and solving for AB we get:

$$AB = \frac{1}{\cos\alpha}. \tag{5.38}$$

Similarly, in right triangle CEB, $\cos\beta = \dfrac{1}{BC}$, so

$$BC = \frac{1}{\cos\beta}. \tag{5.39}$$

Also, from triangle AEB, we have

$$\tan\alpha = \frac{AE}{1} = AE. \tag{5.40}$$

Similarly, from triangle CEB we have

$$\tan\beta = \frac{EC}{1} = EC. \tag{5.41}$$

Now, we know that

the area of triangle ABC = the area of triangle AEB + the area of triangle CEB, (5.42)

and, from Theorem 5.11,

the area of triangle $ABC = \dfrac{1}{2} AB \cdot BC \cdot \sin(\angle ABC)$. (5.43)

Also, the areas of right triangles AEB and CEB are each $\dfrac{1}{2}$ base times height. Thus,

the area of triangle $AEB = \dfrac{1}{2} AE \cdot 1$

and

the area of triangle $CEB = \dfrac{1}{2} EC \cdot 1$. (5.44)

Substituting equations (5.43) and (5.44) into equation (5.42), we get

$$\frac{1}{2}AB \cdot BC \cdot \sin(\angle ABC) = \frac{1}{2}AE \cdot 1 + \frac{1}{2}EC \cdot 1. \tag{5.45}$$

We have one final step. Using the relationships that $AB = \frac{1}{\cos \alpha}$, $BC = \frac{1}{\cos \beta}$, $AE = \tan \alpha$, and $EC = \tan \beta$ from equations (5.38)–(5.41) and substituting into (5.45) we get

$$\frac{1}{2} \cdot \frac{1}{\cos \alpha} \cdot \frac{1}{\cos \beta} \cdot \sin(\alpha + \beta) = \frac{1}{2}\tan \alpha + \frac{1}{2}\tan \beta$$

which can be written as

$$\frac{1}{2} \cdot \frac{1}{\cos \alpha} \cdot \frac{1}{\cos \beta} \cdot \sin(\alpha + \beta) = \frac{1}{2}\frac{\sin \alpha}{\cos \alpha} + \frac{1}{2}\frac{\sin \beta}{\cos \beta} \tag{5.46}$$

since the tangent of an angle is the sine over the cosine. We now multiply both sides of equation (5.46) by $2\cos\alpha \cos \beta$ and simplify and we get

$$\sin(\alpha + \beta) = \sin \alpha \cos \beta + \cos \alpha \sin \beta. \text{ (Verify!)} \tag{5.47}$$

∎

Notice how this theorem ties together algebraic concepts with the geometric concept of area, and the trigonometric concepts of sines, cosines and tangents. What a nice interplay of concepts! Later in this chapter, we will get this result in a different and quite unexpected manner. (See Ptolemy's theorem.)

Theorem 5.12 is attributed to the Persian mathematician and astronomer Abul Wafa and dates back to the 10th century.

Student Learning Opportunities

1 (C) One of your very clever students asks: "If the two adjacent sides of a triangle are 6 and 8 and I vary the angle between them, I will get different triangles and their areas will be different. What would the measure of the included angle have to be to make a triangle of largest area?" How would you respond and does this same angle work regardless of the lengths of the adjacent sides? [Hint: Use Theorem 5.11.]

2 Verify that $\sin(30° + 60°) = \sin 30° \cos 60° + \cos 60° \sin 30°$, using the known values presented in this section.

3 A rectangle $ABCD$ has sides 3 and 6. If diagonal AC is split into three equal parts by points E and F, find the area of triangle BEF.

4 We have shown that the area of a triangle is $\frac{1}{2}ab\sin C$. We can also compute the area in two other ways: $\frac{1}{2}bc\sin A$ and $\frac{1}{2}ac\sin B$. Of course the area that we get is the same no matter

which formula we use. Show that, from this observation, we can get another proof of the Law of Sines.

5 (C) Despite learning the formula for sin(A + B) several of your students still maintain that $\sin 2\theta = \sin\theta + \sin\theta$. How can you prove to them graphically that they are incorrect and how would you prove to them that, in fact, $\sin 2\theta = 2\sin\theta\cos\theta$?

6 Using the fact that the exact values of $\sin 45°$ and $\cos 45°$ are $\frac{\sqrt{2}}{2}$, and that the exact values of $\sin 30°$ and $\cos 30°$ are, respectively, $\frac{1}{2}$ and $\frac{\sqrt{3}}{2}$, find the exact value of $\sin 75°$.

5.6 The Circle Revisited

LAUNCH

In a circle, with center O draw two adjacent central angle of 120 degrees. Let the intersection of the sides of these angles with the circle be A, B, and C as shown in Figure 5.17 below:

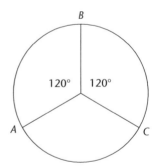

Figure 5.17

Connect A to B, B to C, and C to A, and you should now have an equilateral triangle. Pick any point P on the circle and draw line segments PA, PB, and PC. Do you notice any relationship between the lengths of the two shorter segments and the length of the longer segment? What do you notice? Pick another point P' and do the same thing. Do you notice the same relationship? What is it? Do you think this will always happen? Why or why not?

After having completed the launch, you are probably beginning to realize that the circle is a most fascinating figure, especially when you begin to inscribe other geometric Figures within it. In this section, you will learn more about the most interesting relationships that exist within a circle. You will revisit the launch problem at the end of the section in the Student Learning Opportunities, after you become familiar with Ptolemy's theorem.

5.6.1 Inscribed and Central Angles

We hope you have appreciated seeing how we have used the Law of Sines and the Law of Cosines to develop all the congruence theorems, similarity theorems, and related laws of trigonometry that are part of the secondary school curriculum. Recall that the Law of Cosines essentially depended on the Pythagorean Theorem which, in turn, depended on the concept of area. Thus, it seems that area is the driving concept in the secondary school curriculum. This is why our chapter on areas preceded this one.

Let us now investigate the circle and see what relationships we can prove.

Other than congruence and similarity, the main theorems in a geometry course are those concerning circles, their chords, their tangents, and their secants. We now prove some of the main theorems about these, and guide you through many of the others in the Student Learning Opportunities. These tasks will not only review the theorems, but show, yet again, how all the concepts connect. After this, we will continue to investigate further applications of the Pythagorean Theorem and the concept of area.

To begin, we need to recall a fact from geometry. **A central angle** is one whose vertex is at the center of the circle. Thus, in the picture below, θ is a central angle and arc AB (denoted by $\overset{\frown}{AB}$) is the **arc subtended by the central angle**. We define the measure of the arc subtended by a central angle of θ, to be θ also. That is, each central angle has the same measure as the arc subtended by it and vice versa. This is a definition. (See Figure 5.18.)

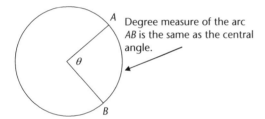

Figure 5.18

When you think about this definition, it makes sense. We know that a complete rotation is 360 degrees. So, if we drew adjacent central angles each of 1 degree, we would need 360 of them to fill the circle. But this divides the circle into 360 parts. Thus, the number of central angles of 1 degree and the number of parts of the circle are both 360. Therefore, it seems reasonable that each arc associated with a 1 degree central angle should be called a 1 degree arc. It follows from this that any arc will have the same number of degrees as its central angle.

Here is our first theorem. Recall that an **inscribed angle** is one whose vertex is on the circle and whose sides are chords of the circle, as is shown in Figure 5.19 below.

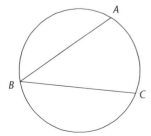

Figure 5.19 Angle ABC is an inscribed angle

Theorem 5.13 *An angle inscribed in a circle is measured by $\frac{1}{2}$ of its degree arc.*

Proof. We give one half of the proof leaving the other half to you. We begin with the central angle *AOB*. By definition, this has the same degree measure as $\overset{\frown}{AB}$. Pick a point *P* on the circle. We suppose that our picture is as given below in Figure 5.20. (This simplifies the first part of the proof. There is another picture as you will see.)

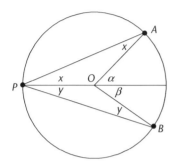

Figure 5.20

Because the radii of a circle are equal, triangles *AOP* and *BOP* are isosceles, and their base angles are equal, as indicated in the diagram above. Since the exterior angle of a triangle is the sum of the remote interior angles (Chapter 1, Section 2, Student Learning Opportunity 1) we have that $\alpha = 2x$ and in the same way, $\beta = 2y$. Thus, $\alpha + \beta = 2x + 2y$. This means that

$$x + y = \frac{\alpha + \beta}{2}. \tag{5.48}$$

But $\alpha + \beta$ is the measure of arc *AB* and $x + y$ is the measure of the inscribed angle *P*. Thus equation (5.48) says that

$$\angle P = \frac{1}{2} \angle AOB. \tag{5.49}$$

You might think the proof is complete. It isn't. There is another possible picture (see Figure 5.21 below). In the Student Learning Opportunities, you will be asked to prove the theorem for this case.

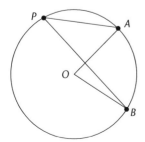

Figure 5.21

182 The Triangle: Its Study and Consequences

> **Corollary 5.14** *If an inscribed angle and a central angle intercept the same arc, the central angle is twice the inscribed angle.*

Proof. In the diagram above, $\angle P$ is measured by $\frac{1}{2}\stackrel{\frown}{AB}$, while angle O is measured by $\stackrel{\frown}{AB}$. So, the measure of the central angle is twice the measure of the inscribed angle. ∎

There is a very surprising corollary of Theorem 5.13—an extended version of the Law of Sines!

> **Corollary 5.15** *Given any triangle ABC, $\dfrac{a}{\sin A} = \dfrac{b}{\sin B} = \dfrac{c}{\sin C} = d$, where d is the diameter of the circumscribed circle.*

Proof. According to the previous corollary, the central angle intercepting the same arc as an angle of the triangle whose vertex lies on the circle will have twice the measure of the inscribed angle. Look at Figure 5.22 below, which shows an acute triangle and its circumscribed circle.

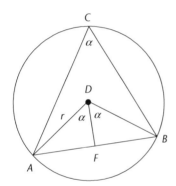

Figure 5.22

If the triangle is acute, then the center of the circle is always inside the triangle and we can draw altitude DF to isosceles triangle ADB and it will bisect angle ADB as well as the base. Now, in triangle ADF, $\sin\alpha = \dfrac{AF}{r} = \dfrac{AF}{\frac{d}{2}}$ where d is the diameter of the circle. Inverting the $\dfrac{d}{2}$ and multiplying, we get that $\sin\alpha = \dfrac{2AF}{d} = \dfrac{AB}{d}$. Rewriting this as

$$\frac{AB}{\sin\alpha} = d \tag{5.50}$$

and realizing that $\sin\alpha = \sin C$ and AB is side c in triangle ABC, equation (5.50) becomes

$$\frac{c}{\sin C} = d. \tag{5.51}$$

That is, the ratio of c to $\sin C$ is the diameter. Since, there was nothing special about angle C, a similar proof shows that $\dfrac{a}{\sin A}$ and $\dfrac{b}{\sin B}$ are both equal to d. So, $\dfrac{a}{\sin A} = \dfrac{b}{\sin B} = \dfrac{c}{\sin C} = d$ and this gives us not only the law of sines, but tells us exactly what the common ratios are, the diameter of the circumscribed circle! ∎

There is yet another corollary of this, which may seem obscure now, but will be put to good use later in Section 5.6.3.

> **Corollary 5.16** *In a circle with diameter 1, if we have an inscribed angle of measure α, intercepting arc AB, then the length of the chord joining the points A and B has length $\sin \alpha$. That is, $\sin \alpha = AB$.*

Proof. We simply let $d = 1$ in equation (5.50) and the result follows immediately by cross multiplying. ∎

5.6.2 Secants and Tangents

In this section we deal with some of the main theorems concerning tangents and secants. Recall that a **secant line** to a circle is a line drawn from an external point which cuts through the circle and stops at the other side. We have drawn a picture of a secant in Figure 5.23. AP is the secant.

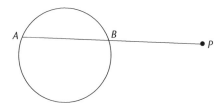

Figure 5.23

> **Theorem 5.17** *If two secants with lengths s_1 and s_2 are drawn to a circle from an external point, and the parts of them which are external to the circle are e_1 and e_2, then $s_1 e_1 = s_2 e_2$.*

Proof. Suppose that the secants hit the circle at points A, B, C and D as shown below in Figure 5.24.

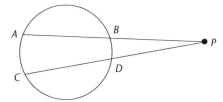

Figure 5.24

We are calling $AP = s_1$, $CP = s_2$, $BP = e_1$, and $DP = e_2$. Draw CB and AD as shown in Figure 5.25 below.

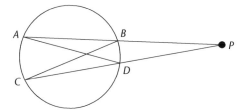

Figure 5.25

184 The Triangle: Its Study and Consequences

Then we have that $\angle A = \angle C$, since both angles A and C are inscribed angles and both are half the degree measure of arc BD. Of course, angle P is common to both triangles ADP and CBP. Thus, by Corollary 5.9 (AA = AA) triangles ADP and CBP are similar. It follows that the corresponding sides are in proportion. Thus,

$$\frac{AP}{DP} = \frac{CP}{BP}$$

Cross multiplying, we get

$$AP \cdot BP = CP \cdot DP$$

which says nothing more than

$$s_1 e_1 = s_2 e_2.$$

∎

Theorem 5.18 *If a tangent line with length t and a secant line with length s and external segment e are both drawn to a circle from an external point, then $t^2 = se$.*

Proof. Although we could give a purely geometric proof of this, we prefer instead to show you a different approach, which ties together concepts you learned in calculus with geometry.

Imagine a group of additional secants with lengths s_1, s_2, and so on being drawn to the circle, and that these secants approach the tangent line. (See Figure 5.26 below.) The bold parts of the diagram are the original secant line with length s and external segment of length e, and the original tangent whose length is t.

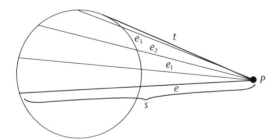

Figure 5.26

Now we know from the previous theorem that $se = s_1 e_1 = s_2 e_2$, and so on. This says that the sequence of numbers, $\{s_n e_n\}$ is constant and every term is equal to the constant se. Now we know that the limit of a constant sequence of numbers is the constant. That is

$$\lim_{n \to \infty} s_n e_n = se. \tag{5.52}$$

Furthermore, since the secant lines approach the tangent line, we see that the lengths of the secant lines approach the length of the tangent, and the lengths of the external segments of the secant lines also approach the length t of the tangent line. In symbols: $\lim_{n \to \infty} s_n = t$ and $\lim_{n \to \infty} e_n = t$. Now, using equation (5.52) we have

$$se = \lim_{n\to\infty} s_n e_n$$
$$= \lim_{n\to\infty} s_n \cdot \lim_{n\to\infty} e_n$$
$$= t \cdot t$$
$$= t^2$$

and we are done. ∎

5.6.3 Ptolemy's Theorem

One result relating to circles that is done in some secondary school courses is Ptolemy's theorem, which we present next. Ptolemy lived in the second century and was one of Greece's most influential astronomers. He propounded the theory that the earth was the center of the solar system, which was believed until about 1543, when Copernicus showed otherwise. Ptolemy is credited with making the first tables of sines, cosines, and tangents and applying these to problems in astronomy. The proof of Ptolemy's theorem uses similar triangles. Its consequences are quite unexpected and powerful and relate directly to the secondary school curriculum.

Here is Ptolemy's theorem.

Theorem 5.19 *Suppose that quadrilateral PQRS is inscribed in a circle as shown in Figure 5.27 below:*

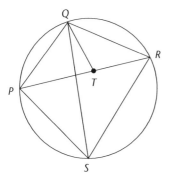

Figure 5.27

If we multiply the lengths of the opposite sides of the quadrilateral and sum the results, we get the product of the diagonals. That is,

$QR \cdot PS + PQ \cdot SR = QS \cdot PR.$

Proof. Here is a plan for the proof. We will pick a point T on diagonal PR such that $\angle RQT = \angle SQP$. We will then show that triangle RQT is similar to SQP, and triangle RQS is similar to triangle TQP and thereby establish proportions that will lead to our theorem.

So, begin by picking a point T on diagonal PR such that $\angle RQT = \angle SQP$. Thus, by construction, we have one angle of triangle RQT equal to one angle of triangle PQS.

Since $\angle QRP$ and $\angle QSP$ subtend the same arc, PQ, it follows that $\angle QRP = \angle QSP$. This gives us a second pair of equal angles in triangles RQT and SQP.

Since two angles of triangle RQT are equal to two angles of triangle SQP, triangles RQT and SQP are similar by Corollary 5.9 (AA = AA). It follows that $\dfrac{QR}{TR} = \dfrac{QS}{PS}$. Cross-multiplying, we get

$$QS \cdot TR = QR \cdot PS. \tag{5.53}$$

Now we show that triangle RQS is similar to triangle TQP. We know that, by construction $\angle RQT = \angle PQS$. If we add $\angle SQT$ to both of these angles, we get that $\angle PQT = \angle RQS$, providing us with one pair of equal angles in triangles, RQS and TQP. Also, since $\angle QPR$ and $\angle QSR$ both subtend arc QR, we have that $\angle QPR = \angle QSR$. So now triangles RQS and TQP are similar by AA = AA. So their sides are in proportion. That is, $\dfrac{QP}{PT} = \dfrac{QS}{SR}$. Cross multiplying we have

$$PT \cdot QS = QP \cdot SR. \tag{5.54}$$

Now, if we add equations (5.53) and (5.54) we get

$$QS \cdot TR + PT \cdot QS = QR \cdot PS + QP \cdot SR$$

and if we factor out QS we get

$$QS \cdot (PT + TR) = QR \cdot PS + PQ \cdot SR$$

and since $PT + TR = PR$, this simplifies to

$$QS \cdot PR = QR \cdot PS + PQ \cdot SR.$$

∎

The tendency is to say "So what?" Let's see how you feel after the next example.

Example 5.20 *Show using Ptolemy's theorem that* $\sin(\alpha + \beta) = \sin \alpha \cos \beta + \cos \alpha \sin \beta$.

Solution. Work in a circle of diameter 1 shown in Figure 5.28 below.

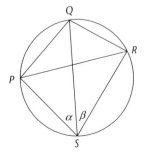

Figure 5.28

Here we have drawn a quadrilateral, one of whose sides, QS, is a diameter. Thus, $QS = 1$. Now since QPS is inscribed in a semicircle, it is a right angle. (It is measured by half its intercepted arc which is 180 degrees.) So

$$\cos \alpha = \frac{\text{adjacent}}{\text{hypotenuse}} = \frac{PS}{QS} = \frac{PS}{1} = PS \tag{5.55}$$

Similarly,

$$\cos \beta = SR, \tag{5.56}$$

$$\sin \alpha = PQ, \text{ and} \tag{5.57}$$

$$\sin \beta = QR. \tag{5.58}$$

(Verify!) Since the diameter of the circle is 1, we have by Corollary 5.16 that

$$PR = \sin(\alpha + \beta). \tag{5.59}$$

Now we are ready to proceed. By Ptolemy's theorem

$$PQ \cdot SR + PS \cdot QR = PR \cdot QS. \tag{5.60}$$

Now we know that $QS = 1$, and substituting the values for PQ, SR, PS, QR, and PR, obtained in equations (5.56)–(5.59) in (5.60) we get

$$\sin \alpha \cdot \cos \beta + \cos \alpha \cdot \sin \beta = \sin(\alpha + \beta) \cdot 1$$

which was our goal! So we have seen yet another surprising and corroborating proof of the formula for $\sin(\alpha + \beta)$.

We can get other trigonometric identities from Ptolemy's theorem, some of which we leave for the Student Learning Opportunities.

Student Learning Opportunities

1 Triangle ABC is inscribed in a circle. The smaller arcs, $\widehat{AB}$, $\widehat{BC}$, and $\widehat{CA}$ are, respectively, $2x$, $3x$, and $5x$ degrees. What are the angles of the triangle?

2 (C) Your students want you to explain why an angle inscribed in a semicircle is a right angle. Show how you would explain this by using theorems from this chapter.

3 (C) You have encouraged your students to use some dynamic geometric software to examine what happens if you create an isosceles triangle and draw an altitude from the vertex angle to the base. They have noticed that, regardless of the size and shape of the isosceles triangle they make, the altitude seems to always bisect the vertex angle as well as the base. Your students want to know if this is always true and, if so, how can you prove it. What is your response and how do you prove it? (Note that this relationship was used in Theorem 5.15.)

4 (C) Through the use of dynamic geometric software, your students have become convinced that every triangle ABC can be inscribed in a circle. (That circle is called the circumscribed circle.) They ask you how you can construct the circumscribed circle for any triangle. How do you do it? [Hint: Proceed as follows. Draw the perpendicular bisectors of two sides, say AB and BC of the triangle. They will intersect at some point P. Prove that AP = PB using congruent triangles. Now, in a similar manner PB = PC. Finish it.]

5 (C) Through the use of dynamic geometric software, your students have noticed that, whenever they inscribe a right triangle in a circle, the hypotenuse always seems to be the diameter of the circle. They want to know if this is always the case, and if so, how to prove it. How will you do it? They then want to know why the median to the hypotenuse of a right triangle has length half the hypotenuse. How can you show it, based on what you have already done?

6 Use Ptolemy's theorem to show that, if a rectangle with legs a and b and diagonal c is inscribed in a circle, then $a^2 + b^2 = c^2$. Thus, we have yet another proof of the Pythagorean Theorem.

7 Use Ptolemy's theorem to prove that $\sin(\alpha - \beta) = \sin\alpha\cos\beta - \cos\alpha\sin\beta$. Use Figure 5.29 below where we take the diameter $AD = 1$, and use Corollary 5.16. [Hint: $AB = \cos\alpha$, $BC = \sin(\alpha - \beta)$ by Corollary 5.14.]

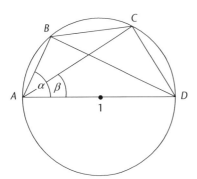

Figure 5.29

8 Prove the following theorem, which answers the launch question: If an equilateral triangle ABC is inscribed in a circle, and P is any point on the circle, then the shorter two of the three segments, PA, PB, and PC adds up to the third. [Hint: Call the side of the equilateral triangle s, and connect P to the three vertices of the triangle to form a quadrilateral. Then use Ptolemy's theorem.]

Note: We have been using Ptolemy's Law to derive geometric and trigonometric results. In fact, there is a very famous and exceedingly useful law in optics that says that, when light traveling with velocity v_1 in a medium, say air, enters a medium, say water, entering at an angle of θ_1 relative to the vertical, the light is refracted, that is, bent, and travels at an angle θ_2 to the vertical. Furthermore, its velocity in the medium it enters changes to v_2. The relationship is known as Snell's Law and says that

$$\frac{\sin\theta_1}{\sin\theta_2} = \frac{v_1}{v_2}.$$

A big surprise: Ptolemy's theorem can be used to show that Snell's Law as well as another principle known as Fermat's principle, which states that light travels in such a way so as to minimize its travel time from point *A* to point *B*, lead to the same geometry of refraction.

9 Find the area of the pentagon shown in Figure 5.30 below. Angle *A* is a right angle.

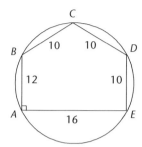

Figure 5.30

10 (C) You have encouraged your students to use some dynamic geometric software to make a conjecture regarding the angle a tangent line drawn to a circle makes with a radius drawn to that tangent line at the point of tangency. They have all discovered that these two lines always seem to be perpendicular to one another. They want to know why this happens. How can you help them to discover this? [Hint: Draw the radius of the circle to the point of tangency. So, its length is *r*, the radius of the circle. Then draw a line to any other point *P* on the tangent line and show that its length is more than *r*. How does this show it?]

11 (C) Your students have been using dynamic geometric software to investigate the relationship of two tangent lines drawn to a circle from a common external point. They have made the conjecture that the two tangent lines have the same length and now they need some guidance on how to prove it. How can you help them?

12 (C) You have encouraged your students to use some dynamic geometric software to make some conjectures. They have observed over and over that, if a line is drawn perpendicular to a chord, *AB*, it bisects the chord. They ask you for a proof. How do you prove it? (Drawing radii to the endpoints of the chord will help.)

13 Suppose we have two circles with the same center and that the area of the shaded region between them is 25π. A chord is drawn in the larger circle, which is tangent to the smaller circle. (See Figure 5.31 below.)

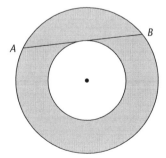

Figure 5.31

What is the length of the chord? [Hint: Draw a radius to the point of tangency and another to *B*.]

14 (C) Using dynamic geometric software, your students have noticed that, if they draw two chords that are the same distance from the center of a circle, they always have the same length. They want to know how to prove it. How can you prove it? [Hint: Draw the lines that give the distance from the center. Also, draw radii to the one endpoint of each chord.]

15 (C) Using dynamic geometric software, your students have noticed that, if in one circle they draw two chords that have the same length, then they are the same distance from the center of the circle. They want to know how to prove it. Show them.

16 (C) Using dynamic geometric software, your students have noticed that, if in one circle they draw chords of equal length, they always seem to subtend minor arcs that have the same angle measure. They want to know if this is always the case, and if it is, how can it be proven. What do you say and how can you prove it?

17 Show that, if a circle of radius r is inscribed in a triangle with perimeter P and area A, then $\frac{P}{A} = \frac{2}{r}$. [Hint: Draw radii to the sides and connect the center of the circle to the vertices. Then sum the areas of the 3 triangles formed. Each has height r.]

18 (C) Your students have learned that, if two chords intersect within a circle, then the product of the segments of one chord is equal to the product of the segments of the other chord. That is, in the following diagram, $ab = cd$. Some of your more curious students want to know how that is proven. How do you do it? [Hint draw the dotted lines as shown in Figure 5.32 below, and try to get similar triangles.]

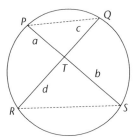

Figure 5.32

19 Prove Theorem 5.13 for figure 5.21.

20 Using Figure 5.33 below, find the missing piece of information in parts (a)–(d).

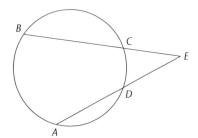

Figure 5.33

(a) $BE = 12$, $AE = 9$, $DE = 1$, $CE = ?$
(b) $BC = 5$, $CE = 4$, $DE = 9$, $AD = ?$

(c) BC = 6, AD = 4, DE = 2, CE = ?
(d) CE = 4, DE = 3, AD = 4, BC = ?

5.7 Technical Issues

LAUNCH

Historically, the cosine of an acute angle A in a right triangle was defined as the ratio of the side adjacent to A to the hypotenuse. Unless this ratio is the *same* for all right triangles having an acute angle equal to A, this definition is unusable. But, how can we be sure that the cosine of an angle is the same, regardless of the right triangle in which it occurs?

We hope that, after thinking about this, you realized that similar triangles are used to show this. The usual argument given is: "If A is an angle in two right triangles, then the triangles must be similar, since they both have angle A and a right angle. Since they are similar, their sides are in proportion. That means that the ratio of the adjacent side to the hypotenuse is the same in both triangles." That is, the cosine of A is independent of the right triangle in which it resides. Our goal in this chapter was to prove the theorems about similarity using the fact that sines and cosines are well defined, and we cannot use theorems about similar triangles to assume that the sine and cosine are well defined, or else we would be engaging in what is known as circular reasoning (using theorems about similar triangles to prove the same theorems about similar triangles).

So, to avoid this circular reasoning, we will now give an independent proof of the fact that the sine of an angle is independent of the triangle in which it resides. Surprisingly, we can use areas of triangles to prove this, which we will now do. But first we need the following preliminary theorem.

Theorem 5.21 *Suppose that, in a given triangle ABC, a line is drawn from B to AC intersecting AC at D. Then, the ratio of the areas of triangle ABD to triangle CDB is the same as the ratio of AD to DC.*

Proof. We refer to Figure 5.34 below.

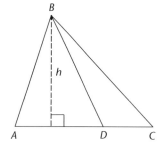

Figure 5.34

We see that both triangles ABD and DBC have the same height h. Thus,

$$\frac{\text{Area of triangle } ABD}{\text{Area of triangle } DBC} = \frac{\frac{1}{2} AD \cdot h}{\frac{1}{2} DC \cdot h} = \frac{AD}{DC}.$$

∎

Corollary 5.22 *Given a right triangle ABC with right angle at C, draw a line DE parallel to BC, where D is any point on AC and E is where this line intersects AB. Then $\frac{AC}{AB} = \frac{AD}{AE}$. That is, cosA is the same whether we take the ratio of the opposite side to the hypotenuse in right triangle AED or in right triangle ABC.*

Proof. We begin with our picture, Figure 5.35.

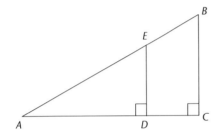

Figure 5.35

First draw EC, dividing the triangle into triangles I and II as shown in Figure 5.36 below:

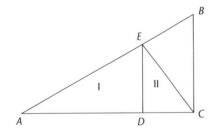

Figure 5.36

Now, from Theorem 5.21, applied to triangle AEC we have

$$\frac{\text{Area of I}}{\text{Area of II}} = \frac{AD}{DC}. \tag{5.61}$$

Now, draw DB, yielding Figure 5.37.

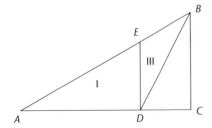

Figure 5.37

Again, by Theorem 5.21, only applying it to triangle ABD, we have

$$\frac{\text{Area of I}}{\text{Area of III}} = \frac{AE}{EB}. \tag{5.62}$$

Finally, we note that both triangles II and III have base ED and height DC (since ED is parallel to BC and hence are the same distance from each other everywhere). Thus, the areas of II and III are the same. Replacing Area III by Area II in equation (5.62) we get that

$$\frac{\text{Area of I}}{\text{Area of II}} = \frac{AE}{EB}. \tag{5.63}$$

Comparing equations (5.61) and equation (5.63), we see that

$$\frac{AD}{DC} = \frac{AE}{EB}. \tag{5.64}$$

We will now add the number 1 to both sides of equation (5.64). (You will soon see what this accomplishes.) We get

$$\frac{AD}{DC} + 1 = \frac{AE}{EB} + 1.$$

Combining each side into a single fraction, we get

$$\frac{AD + DC}{DC} = \frac{AE + EB}{EB}. \tag{5.65}$$

Dividing equation (5.65) by equation (5.64), we get that

$$\frac{\frac{AD+DC}{DC}}{\frac{AD}{DC}} = \frac{\frac{AE+EB}{EB}}{\frac{AE}{EB}}$$

which simplifies to

$$\frac{AD + DC}{AD} = \frac{AE + EB}{AE}$$

and, since $AD + DC = AC$ and $AE + EB = AB$, this is just

$$\frac{AC}{AD} = \frac{AB}{AE}. \tag{5.66}$$

From this proportion, it follows that

$$\frac{AC}{AB} = \frac{AD}{AE} \tag{5.67}$$

and we are done. ∎

The importance of this theorem cannot be underestimated. It says that, if A is any angle, cos A is unique without the assumption of similar triangles. So, if we have two right triangles, ABC and

AED, each containing an acute angle *A*, then we can just overlap them together so that we get Figure 5.38 as shown below:

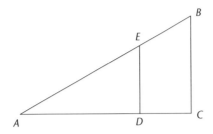

Figure 5.38

and then from equation (5.67) we see that the cosine of *A* is independent of the triangle we are using.

We can use a similar proof to show that sin *A* is independent of the triangle in which it resides, or we can prove it using trigonometric identities. (See Student Learning Opportunity 1.) Now that we know that sin *A* and cos *A* are well defined, and don't change depending on which triangle we are working in, we can use these facts freely.

There is one other result we need for our work to be complete.

Theorem 5.23 *As an acute angle A in a right triangle increases, its cosine decreases. As angle A increases, so does its sine.*

Proof. Since cos A is independent of the triangle in which it resides, we will consider only triangles with hypotenuse = 1. Such a triangle will look as shown in Figure 5.39.

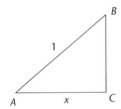

Figure 5.39

From the Pythagorean Theorem, $x^2 + (BC)^2 = 1$. It follows that $x = \sqrt{1 - (BC)^2}$. From this relationship we see that, as *BC* increases, *x* decreases. Now, if *A* increases, *BC* also increases, and hence *x* decreases. But, from the triangle, $x = \cos A$. So, as angle *A* gets bigger, cos *A* gets smaller.

The proof of the second part is similar. ∎

We used the following theorem throughout this chapter (see, for example, Theorem 5.2). Now it can be justified.

Corollary 5.24 *If A and B are acute angles and if* cos *A* = cos *B, then* ∠*A* = ∠*B*.

Proof. From the theorem, as an angle increases, its cosine decreases. Hence, it is not possible for two different acute angles to have the same cosine. Thus, it must be that ∠*A* = ∠*B*. ∎

Student Learning Opportunities

1 Draw a right triangle ABC with right angle C. Starting with the Pythagorean Theorem and dividing both sides by c^2, show that $\sin^2 A + \cos^2 A = 1$. From this, show that $\sin A = \sqrt{1 - \cos^2 A}$. (We don't use the $\pm$ square root since the cosine of an angle in a triangle can't be negative. It is the ratio of the lengths of two sides.) Use this to show that $\sin A$ is independent of the triangle in which it resides.

2 (C) Your students have been investigating triangles using dynamic geometric software. Under your direction, they have noticed that, if they create any triangle ABC, and they draw a line parallel to AC, intersecting AB at D and BC at E, then $\frac{AD}{DB} = \frac{CE}{EB}$. They want to know how to prove this. How do you respond?

3 Show that, if A and B are acute angles, and if $\sin A = \sin B$, it follows that $\angle A = \angle B$.

4 In the proof of Corollary 5.22 we said that equation (5.67) follows from equation (5.66). Show it.

5.8 Ceva's Theorem

LAUNCH

1 On a piece of standard loose-leaf paper, draw a triangle at the top half of the paper. Using a ruler, or by folding the segments, locate the midpoints of each segment and then connect these midpoints to the opposite vertices. You should have just drawn three medians. What do you notice? Do the medians intersect? Do they all intersect at the same point? Do you think this will this always happen?

2 On the bottom half of your paper, draw an entirely different type of triangle. As before, draw the three medians. Did the same thing happen? Do you think this will always happen? Explain.

Now that you have completed the launch question, we hope you are marveling at the mysteries of the triangle. The truth is that there are many interesting results related to the triangle that you will be reading more about in this section. For example, in addition to the medians meeting at a point, the three altitudes meet at a point, and the three angle bisectors meet at a point, although the points at which they meet are usually different. All of these theorems, as well as many others, follow from one remarkable result called Ceva's Theorem (after the mathematician Giovanni Ceva (1647–1674) who we know little about, except that he was a professor of mathematics in Mantua, Italy and published one of the first works in mathematical economics).

We begin with a definition. A **cevian** is a line that emanates from a vertex to the opposite side of a triangle (or its extension). Thus, altitudes, medians and angle bisectors are all cevians. In this section we will once again use areas of triangles in a completely unexpected way.

But first we need an interesting lemma about fractions. (The word "lemma" is used to refer to a preliminary result that is used to prove a more important result.)

Lemma 5.25 *If $\frac{a}{b} = \frac{c}{d}$, then $\frac{a}{b} = \frac{c}{d} = \frac{a-c}{b-d}$. Thus, if we subtract numerators and denominators of two equal fractions, we get an equivalent fraction.*

To illustrate: $\frac{2}{8} = \frac{6}{24} = \frac{2-6}{8-24}$, which is true!

Here is the proof:

Proof. Let t be the common value of the fractions $\frac{a}{b}$ and $\frac{c}{d}$. So, $\frac{a}{b} = t$ and $\frac{c}{d} = t$. Multiplying these equations by b and d, respectively, we get $a = bt$ and $c = dt$. Using these values in the fraction $\frac{a-c}{b-d}$, we get $\frac{a-c}{b-d} = \frac{bt-dt}{b-d} = \frac{t(b-d)}{b-d} = t$. Thus, $\frac{a-c}{b-d}$ has the same value, t, as the other fractions have. So all three fractions are equivalent. ∎

Theorem 5.26 *(Ceva's Theorem): In triangle ABC, if AE, BF, and CD are cevians that meet at G inside the triangle, then $\frac{BD}{DA} \cdot \frac{AF}{FC} \cdot \frac{CE}{EB} = 1$. (See Figure 5.40.)*

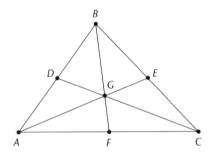

Figure 5.40

Proof. Before getting into the heart of proof, we just remind you of Theorem 5.21 which says that if two triangles have the same height, then the ratio of their areas is the same as the ratio of their bases.

Now, since triangles *ABF* and *CBF* have the same height,

$$\frac{\text{area }(ABF)}{\text{area }(CBF)} = \frac{AF}{FC}. \tag{5.68}$$

Similarly, since triangles *AGF* and *FGC* have the same height

$$\frac{\text{area }(AGF)}{\text{area }(CGF)} = \frac{AF}{FC}. \tag{5.69}$$

From equations (5.68) and (5.69),

$$\frac{\text{area } (ABF)}{\text{area } (CBF)} = \frac{\text{area } (AGF)}{\text{area } (CGF)} = \frac{AF}{FC} \tag{5.70}$$

and so, by the previous lemma, with a, b, c, and d replaced by the appropriate numerators and denominators of equation (5.70) we have

$$\frac{\text{area } (ABF) - \text{area } (AGF)}{\text{area } (CBF) - \text{area } (CGF)} = \frac{AF}{FC}. \tag{5.71}$$

But area (ABF) − area (AGF) = area(AGB) and area (CBF) − area (CGF) = area(CGB). (Look at the figure to confirm!) Substituting into equation (5.71) we have

$$\frac{\text{area}(AGB)}{\text{area}(CGB)} = \frac{AF}{FC}. \tag{5.72}$$

We now use the cevian AE in a similar manner to get that

$$\frac{CE}{EB} = \frac{\text{area}(AGC)}{\text{area}(AGB)}. \tag{5.73}$$

And then again use cevian CD in a similar manner to get

$$\frac{BD}{DA} = \frac{\text{area}(CGB)}{\text{area}(AGC)}. \tag{5.74}$$

Using equations (5.72), (5.73), and (5.74) and multiplying we get

$$\frac{AF}{FC} \cdot \frac{CE}{EB} \cdot \frac{BD}{DA} = \frac{\text{area}(AGB)}{\text{area}(CGB)} \cdot \frac{\text{area}(AGC)}{\text{area}(AGB)} \cdot \frac{\text{area}(CGB)}{\text{area}(AGC)} = 1 \text{ [dividing common areas]}.$$

∎

The converse of Theorem 5.26 is also true.

Theorem 5.27 *If in triangle ABC we have three cevians, AE, BF, and CD, and $\frac{AF}{FC} \cdot \frac{CE}{EB} \cdot \frac{BD}{DA} = 1$, then the cevians AE, BF and CD meet at a point.*

Proof. Suppose that AE and DC meet at G. Draw BG and let it intersect AC at F' as shown in Figure 5.41.

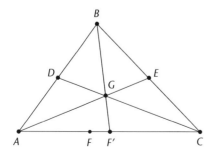

Figure 5.41

Then by Ceva's Theorem,

$$\frac{AF'}{F'C} \cdot \frac{CE}{EB} \cdot \frac{BD}{DA} = 1. \tag{5.75}$$

But we are given that

$$\frac{AF}{FC} \cdot \frac{CE}{EB} \cdot \frac{BD}{DA} = 1. \tag{5.76}$$

From equations (5.75) and (5.76) we have

$$\frac{AF'}{F'C} \cdot \frac{CE}{BE} \cdot \frac{BD}{DA} = \frac{AF}{FC} \cdot \frac{CE}{BE} \cdot \frac{BD}{DA}. \tag{5.77}$$

Dividing both sides of equation (5.77) by $\frac{CE}{BE} \cdot \frac{BD}{DA}$, we get

$$\frac{AF'}{F'C} = \frac{AF}{FC}.$$

From which it follows that $F = F'$. Thus, AF' really is AF and the three cevians go through the same point, G. ∎

Student Learning Opportunities

1 **(C)** Using their dynamic geometric software, your students have been investigating what happens in a triangle when they construct medians from all three vertices. No matter how they drag their figures and change the sizes and shapes of their triangles, it always seems to be that the medians meet at one point and that the six triangles formed by the medians have the same area (which they have programmed their software to calculate). They ask you if these relationships can be proven, and if so, how it is done. Use Ceva's Theorem to prove this.

2 Prove that, if a circle is inscribed in a triangle (see Figure 5.42 below)

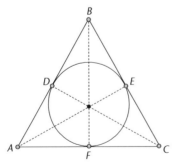

Figure 5.42

then the Cevians drawn from each vertex to the points where the circle is tangent meet in a point. (You will need to use the fact that the tangents drawn to a circle from an external point are equal. That is, $BD = BE$, and so on.) This point where they meet is known as the Gergonne point.

3 We only proved Ceva's Theorem for the case where the Cevians met inside of the triangle. But the Cevians can meet outside the triangle. Consider Figure 5.43 where we begin with triangle *ABC* and draw Cevians *AE*, *CD*, and *BF*. Go through the proof of Theorem 5.27 line by line to see if the proof of the theorem works in this case.

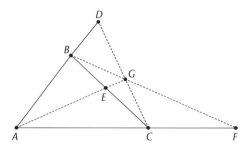

Figure 5.43

4 (C) Your students have been using their dynamic geometric software to explore what happens when you create a triangle and draw the angle bisectors from each of the three vertices. They notice that the three angle bisectors always meet at a point. They ask you how to prove that this will always happen. How can you prove it to your students, using Ceva's Theorem? [Hint: Student Learning Opportunity 7 from Section 5.3 may help.]

5 (C) Your students have continued their exploration of triangles using their dynamic geometric software and have now drawn a triangle and constructed the altitudes from all three of the sides. Much to their surprise they notice that, no matter the size or shape of their triangle, the altitudes meet at one point. They are eager to know if this always happens and why. How does Ceva's Theorem help you to prove it? [Hint: In Figure 5.44 $AD = AB \cos(\angle BAC)$. obtain similar relationships for the other segments.]

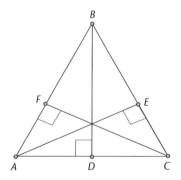

Figure 5.44

6 In the proof of the converse of Ceva's Theorem we made the statement that, if
$$\frac{AF'}{F'C} = \frac{AF}{FC}$$
then $F = F'$. Verify that this is true.

5.9 Pythagorean Triples

LAUNCH

1. Pick any two positive integers *m* and *n* where *m* is greater than *n*. Now compute *a*, *b*, and *c*, where $a = m^2 - n^2$, $b = 2mn$ and $c = m^2 + n^2$. Examine your values for *a*, *b*, and *c*. Specifically, check if $a^2 + b^2 = c^2$. What does this mean? What did you just find?
2. Pick two different positive integers *m* and *n*, where *m* is greater than *n*. Follow the directions above again. Did the same relationship exist between *a*, *b*, and *c*?
3. Do you think this will always work? What does this mean?

Now that you have completed this launch question, you may be thinking that you have found a way to generate three numbers that can serve as sides of a right triangle. Pretty cool, isn't it? We will investigate this further in this section.

Thus far, we have demonstrated how the Pythagorean Theorem and its consequences can be used to develop some very important relationships. In this section we wish to study the Pythagorean Theorem a bit further.

As you know, there are many "special" right triangles that are included in the secondary school curriculum. For example, there is the 3–4–5 right triangle, the 5–12–13 right triangle, and so on. Such sets of 3 positive integers which can serve as the sides of the same right triangle are called **Pythagorean triples.**

You may be thinking that the method you used in the launch question can be used to generate all Pythagorean triples. Surprisingly, that is true. In this section, we will show how this can be done, and in the process we will connect the material we studied in Chapter 2 on divisibility with geometry and algebra. The connections alone make the journey worthwhile.

Here is our first theorem:

> **Theorem 5.28** *Suppose we have a triangle ABC with sides a, b, and c, and suppose that there are positive integers m and n such that $a = m^2 - n^2$, $b = 2mn$, and $c = m^2 + n^2$. Then, the triangle ABC will automatically be a right triangle.*

Proof. All we have to do is show that $a^2 + b^2 = c^2$. Since $a = m^2 - n^2$,

$$a^2 = m^4 - 2m^2n^2 + n^4 \tag{5.78}$$

as you can easily verify by multiplying *a* by itself.

Since $b = 2mn$,

$$b^2 = 4m^2n^2. \tag{5.79}$$

And, since $c = m^2 + n^2$,

$$c^2 = m^4 + 2m^2n^2 + n^4. \tag{5.80}$$

However, notice that, if you add equations (5.78) and (5.79), you get equation (5.80). Thus, $a^2 + b^2 = c^2$ and, by Theorem 46 from the previous chapter, the triangle, ABC is a right triangle, regardless of the values of m and n. ∎

What is surprising is that the converse of Theorem 5.28 is true; namely, if we start with a right triangle with legs a and b, and hypotenuse c, where a, b, and c have no common factors (other than 1), then there MUST be positive integers m and n such that $a = m^2 - n^2$, $b = 2mn$, and $c = m^2 + n^2$. This result will be the main theorem of this section.

We reiterate, we are assuming from the outset that a, b, and c have no common factors. What this means is that the greatest common divisor of a, b, and c is 1.

We first observe that, if a number is squared, then all its prime factors are raised to even powers. Furthermore, if all the prime factors of a number (when it is factored completely into primes) are raised to even powers, then the number is a perfect square.

Let us give a numerical example to demonstrate. If $N = 2^3 5^6$, then $N^2 = 2^6 5^{12}$ and all exponents are even. Conversely, if $P = 3^6 7^4$, then $P = T^2$, where $T = 3^3 7^2$. That is, if all powers of the primes in the factorization of a number are even, then the number is a square. Armed with this fact, we can prove our first lemma.

Lemma 5.29 *If s and t are positive integers with no common factors and st is a perfect square, then both s and t are perfect squares.*

Proof. Let us factor st into primes. Since it is a square, all primes in the factorization are raised to even powers. Since s and t have NO COMMON FACTORS, each prime raised to its power goes with EITHER s or with t. You cannot have a prime going with s and with t because that would mean that s and t will have a common factor. Since all the powers of the primes are even, those that go with s have even powers and all the primes that go with t also have even powers. Thus, s and t are squares. ∎

Lemma 5.30 *If a, b, and c are positive integers with no common factor and if $a^2 + b^2 = c^2$, then one of a or b is even, and the other is odd.*

Proof. If both a and b are even, then c^2 being the sum of two even numbers is even, and hence c is even. That means that each of a, b, and c are even. This contradicts that they have no common factor. So a and b cannot both be even.

If both a and b are odd, then so are their squares. And since $c^2 = a^2 + b^2$, c^2 must be even, being the sum of two odd numbers. Hence c must be even. Since a and b are odd and we now know that c is even, we can write $a = 2k + 1$ and $b = 2l + 1$ and $c = 2r$ where k, l, and r are integers. Substituting this into $a^2 + b^2 = c^2$, we get that

$$(2k+1)^2 + (2l+1)^2 = (2r)^2,$$

or, after squaring and simplifying, that

$$4k^2 + 4k + 4l^2 + +4l + 2 = 4r^2,$$

which in turn can be written as

$$4r^2 - (4k^2 + 4k + 4l^2 + 4l) = 2.$$

Now, since 4 can be factored out of the left side of this equation, the left side is divisible by 4. But the right side isn't. This can't be. Thus, it can't be that both a and b are odd.

We have dispensed of the case where both a and b are even and where a and b are odd. The only case left is that one of a and b must be even and the other odd. ∎

Lemma 5.31 *If a is odd and c is odd, and if a and c have no common factor (other than 1), then $\frac{c+a}{2}$ and $\frac{c-a}{2}$ have no common factor (other than 1).*

Proof. First, we know that $\frac{c+a}{2}$ and $\frac{c-a}{2}$ are integers, since $c+a$ and $c-a$ are even. Now, if $\frac{c+a}{2}$ and $\frac{c-a}{2}$ had a common factor, say d, then, since d divides both, $\frac{c+a}{2}$ and $\frac{c-a}{2}$, d must divide their sum and difference. That is, d must divide $\frac{c+a}{2} + \frac{c-a}{2} = c$ and d must divide $\frac{c+a}{2} - \frac{c-a}{2} = a$. But c and a have no common factors other than 1. So d must be 1. ∎

We are now ready to prove the main theorem about generating our Pythagorean triples.

Theorem 5.32 *Given that a, b, and c are positive integers with no common factors, and a is odd and b is even, then, if $a^2 + b^2 = c^2$, there are positive integers m and n such that*

$$a = m^2 + n^2$$

$$b = 2mn$$

$$c = m^2 + n^2.$$

Proof. Write $a^2 + b^2 = c^2$ as

$$b^2 = c^2 - a^2$$

Dividing both sides by 4 we have

$$\frac{b^2}{4} = \frac{c^2 - a^2}{4},$$

which in turn can be written as

$$\left(\frac{b}{2}\right)^2 = \frac{(c+a)}{2} \cdot \frac{(c-a)}{2}. \tag{5.81}$$

So, on the left, we have the square of an integer, and on the right, we have the product of two integers $\frac{(c+a)}{2}$ and $\frac{(c-a)}{2}$ with no common factor other than 1. So, by Lemma 5.30 we have that both

$\frac{(c+a)}{2}$ and $\frac{(c-a)}{2}$ are perfect squares. That is,

$$\frac{(c+a)}{2} = m^2 \text{ and} \tag{5.82}$$

$$\frac{(c-a)}{2} = n^2. \tag{5.83}$$

Substituting these values from equations (5.82) and (5.83), into equation (5.81) we get

$$\left(\frac{b}{2}\right)^2 = \frac{(c+a)}{2} \cdot \frac{(c-a)}{2} = m^2 n^2,$$

from which it follows that that $\frac{b}{2} = mn$. Hence, $b = 2mn$. Also, subtracting equation (5.83) from equation (5.82), we get $a = m^2 - n^2$, and adding equations (5.83) and (5.82) we get $c = m^2 + n^2$.

Thus, we have shown that a, b, and c satisfy the formulas we gave in the theorem. ∎

This theorem is telling us that all right triangles are generated in the same way using different values for m and n. This is a rather interesting result, don't you think?

As we hope you are aware, throughout this text we have been trying to point out the connections that exist between different areas of the mathematical content studied in secondary school. We now take the opportunity to make connections between number theory and geometry by showing a second proof that $\sqrt{2}$ is irrational (though this is not as elegant as the proof in Chapter 1 which required minimal knowledge).

Begin with a right triangle whose legs are 1 and 1, and suppose that the hypotenuse, which is $\sqrt{2}$, is rational and equal to $\frac{p}{q}$, where p and q are positive integers with *no common factor*. Then the sides of our triangle are 1, 1, and $\frac{p}{q}$ and, if we multiply all the sides by q, we get a similar right triangle with legs $a = q$, $b = q$, and $c = p$. But, by Lemma 5.30, one of the legs, say a, of the right triangle has to be odd and the other, b, has to be even. But a and b are the same! How could this be? We have a contradiction!

Our contradiction arose from assuming that $\sqrt{2}$ was rational. Thus, $\sqrt{2}$ is irrational.

In the next section, we return to the area of a triangle, and get some exciting results.

Student Learning Opportunities

1 What is the Pythagorean triple generated by $m = 2$ and $n = 1$?

2 Generate a Pythagorean triple which gives us a value of 12 for one of the legs of the triangle. Is there more than one triple with this property?

3 Find the sides of a right triangle that has all integer sides, and for which one leg is 9. Can you find a Pythagorean triple one of whose sides is k where k is any odd integer? Explain.

4 Find the sides of a right triangle with all integer sides where the hypotenuse is 25.

5 Find a Pythagorean triple where the even leg is the smallest side.

6 (C) In trying to generate Pythagorean triples, some of your students have noticed that, if you start with the Pythagorean triple, 3, 4, 5 and multiply each number by 2, you get 6, 8, 10, which is another Pythagorean triple. If they multiply each number in the original triple by 3,

they get 9, 12, 15, which is again a Pythagorean triple. They are wondering if it is always true that, if each entry in a Pythagorean triple is multiplied by a positive integer k, the result will be a Pythagorean triple. How do you respond and how do you prove it?

7 (C) One of your students has made the observation that, in all of the Pythagorean triples, she has seen, at least one member is divisible by 3. For example, in the triple, 6, 8, 10, the first member 6 is divisible by 3. In the triple 5, 12, 13, the second member 12 is divisible by 3. She has made a conjecture that every Pythagorean triple has a member that is divisible by 3. Is it true, and if so, how do you prove it? [Hint: What can m and n be congruent to mod 3?]

5.10 Other Interesting Results about Areas

LAUNCH

In Figure 5.45

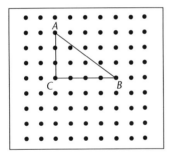

Figure 5.45

the sides of the triangle ABC, where C is the right angle and B is the vertex at the base of the triangle, are of length $a = 3$ units, $b = 4$ units, and $c = 5$ units. Let's see how many different ways you can calculate the area of this triangle.

1. First use the formula $A = \frac{1}{2} b \times h$. What did you get?
2. Now, use the formula $A = \frac{1}{2} ab \sin C$. What did you get?
3. Now, examine the grid, noting that the triangle is one half of a rectangle whose sides are 3 units and 4 units. What did you get for the area?
4. Next, you will do something that is probably unfamiliar to you.
   a. Figure out the perimeter of the triangle and let s equal 1/2 the perimeter.
   b. Now, use the strange formula $A = \sqrt{s(s-a)(s-b)(s-c)}$. Did you get the same area as before?
5. Next, you will find the area by counting only boundary and interior points on the dot diagram.

(a) Count the number of dots that lie on the boundary of the triangle and let it be, B.
(b) Count the number of dots that lie on the interior of the triangle and let it be, I.
(c) Now use the strange formula $A = I + \dfrac{B}{2} - 1$. Did you get the same area as before?

6 Compare and contrast the different formulas you used to calculate the area of the triangle. When would you use one over another?

Having completed the launch question, you are probably still wondering about the formulas you used in parts 4 and 5. Where did they come from? This section will explain.

5.10.1 Heron's Theorem

Thus far, when finding the area of a triangle, we have used either $A = \frac{1}{2}b \times h$ or $\frac{1}{2}ab\sin C$. While these are most useful formulas, they necessitate knowing either the height or an angle of the triangle. But, what if this information is not available? It would be most helpful if we could derive a formula for the area of a triangle that only requires information about the sides. Actually, given the relationships we have established thus far, we can do just that. We will use the Pythagorean Theorem, the fact that we can factor $m^2 - n^2$ into $(m-n)(m+n)$, and the formula for the area of a triangle, $\frac{1}{2}ab\sin C$. Let us proceed.

Theorem 5.33 *(Heron's formula) The area, A, of a triangle with sides a, b, and c is given by*

$$A = \sqrt{s(s-a)(s-b)(s-c)}$$

where s is half the perimeter of the triangle; that is, where $s = \frac{a+b+c}{2}$.

Proof. We know from Theorem 5.11 that the area of a triangle is $A = \dfrac{1}{2}ab\sin C$. Let us square both sides to get $A^2 = \dfrac{1}{4}a^2b^2\sin^2 C$. Since $\sin^2 C$ is $1 - \cos^2 C$, we can write this last statement as

$$A^2 = \frac{1}{4}a^2b^2(1 - \cos^2 C)$$

$$= \frac{1}{4}a^2b^2(1 + \cos C)(1 - \cos C). \tag{5.84}$$

Now, from equation (5.29) we have that $\cos C = \dfrac{a^2 + b^2 - c^2}{2ab}$. Thus, $1 + \cos C = 1 + \dfrac{a^2 + b^2 - c^2}{2ab} = \dfrac{a^2 + 2ab + b^2 - c^2}{2ab} = \dfrac{(a+b)^2 - c^2}{2ab} = \dfrac{(a+b+c)(a+b-c)}{2ab}$. Similarly, $1 - \cos C = \dfrac{(c+a-b)(c-a+b)}{2ab}$.

Substituting these expressions for $1 + \cos C$ and $1 - \cos C$ into equation (5.84) we get

$$A^2 = \frac{1}{4}a^2b^2(1 + \cos C)(1 - \cos C)$$

$$= \frac{1}{4}a^2b^2 \frac{(a+b+c)(a+b-c)}{2ab} \cdot \frac{(c+a-b)(c-a+b)}{2ab}$$

$$= \frac{(a+b+c)(a+b-c)(c+a-b)(c-a+b)}{16}$$

$$= \frac{(a+b+c)}{2} \cdot \frac{(a+b-c)}{2} \cdot \frac{(c+a-b)}{2} \cdot \frac{(c-a+b)}{2}. \tag{5.85}$$

Since $s = \frac{a+b+c}{2}$, our first factor in equation (5.85) is s. Since $s - c = \frac{a+b+c}{2} - c = \frac{a+b-c}{2}$, our second factor in equation (5.85) is $s - c$. In a similar manner, the third factor in equation (5.85) is $s - b$, and the fourth factor in equation (5.85) is $s - a$. Thus, our previous string of equalities now simplifies to

$$A^2 = s(s-c)(s-b)(s-a).$$

Taking the square root of both sides and rearranging the terms, we have

$$A = \sqrt{s(s-a)(s-b)(s-c)}.$$

∎

5.10.2 Pick's Theorem

We end this chapter with one last result, Pick's theorem. Pick's theorem is concerned with finding the area of a polygon whose vertices are at **lattice points** of the xy plane. (Lattice points are points both of whose coordinates are integers.) In around 1899 Georg Pick discovered a remarkable theorem showing how to do this that depends on nothing more than the number of lattice points on the boundary of the polygon and the number of interior lattice points of the polygon. How surprising!

We can model a portion of the xy plane using dot paper as shown in Figure 5.46 below. Each dot represents a lattice point where the distance between consecutive horizontal dots is 1, and the distance between consecutive vertical dots is 1. (Geoboards that allow students to easily form different polygons, are usually used as a manipulative to help students develop area concepts, and also develop Pick's theorem. The following website is an applet that models the geoboard: http://standards.nctm.org/document/eexamples/chap4/4.2/part2.htm#applet).

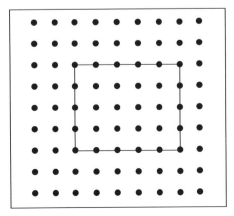

Figure 5.46

Consider the rectangle shown in the figure above. Finding its area is easy. We just count the number of 1 unit squares contained in the figure giving us an area of 20 square units. But, what if the figure was a triangle like the one shown below in Figure 5.47? How would you find its area? In this case the polygon cannot be divided into easily countable square units.

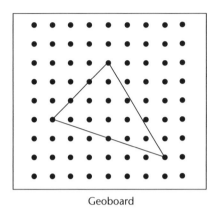
Geoboard

Figure 5.47

One could enclose the whole figure by a rectangle as shown below, and find the area of the rectangle and then subtract the areas of the individual right triangles labeled in Figure 5.48 below.

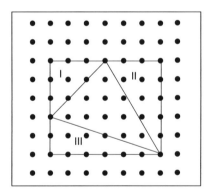

Figure 5.48

Let's do it. The area of the circumscribing rectangle is 6×5 or 30. The area of right triangle I is $\frac{3 \times 3}{2}$, of right triangle II is, $\frac{3 \times 5}{2}$, and of right triangle III, $\frac{2 \times 6}{2}$. Thus, the area of the middle triangle is

$$6 \times 5 - \left(\frac{3 \times 3}{2} + \frac{3 \times 5}{2} + \frac{2 \times 6}{2} \right) = 12.$$

Suppose we have a very complicated polygon and want to find its area. Proceeding as above would be difficult and tedious. This is where Pick's formula comes to the rescue. Here is the theorem. We will break the proof up into several small parts.

Theorem 5.34 *The area of a polygon whose coordinates are lattice points is given by the following simple formula*

$$\text{Area} = I + \frac{B}{2} - 1$$

where I is the number of lattice points inside the polygon and B is the number of lattice points on the boundary of the polygon.

Let us check the previous example using this theorem. Using the figure above, we can see that there are 10 points inside the triangle and 6 points on the boundary. Thus, the area should be $10 + 6/2 - 1$ or 12, which is exactly what we got before. Look how much simpler this was! Let us now move to the proof of this remarkable theorem.

We begin by proving Pick's Theorem for a rectangle whose sides are parallel to the x and y axes, respectively, and whose corners are at lattice points. We give the proof with a numerical example first, and then extend it to the general case. So, let us focus on the rectangle in Figure 5.49 below.

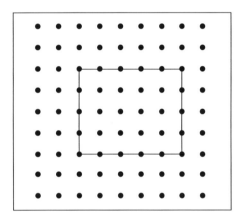

Figure 5.49

We notice that there are 6 dots on the side which represents the length, and that makes the side of the rectangle one less, or 5, as we can see. Similarly, there are 5 dots along the side which represent the width of the rectangle, so that length is one less or 4. Thus, if a side of the rectangle has a lattice points, the length of that side is $a - 1$. So a rectangle with a lattice points on one side and b lattice points on the adjacent side has area $(a - 1)(b - 1)$.

Now let's count the interior points. There are $(5 - 2)$ or 3 rows of interior points, and $(6 - 2)$ or 4 columns of interior points. Thus, the number of interior points is $(5 - 2)(6 - 2)$ or 12. In a similar manner, if the number of lattice points on two consecutive sides of a rectangle are a and b, then the number of interior points is $(a - 2)(b - 2)$.

We now turn to the number of boundary points of the rectangle. This is the number of boundary points on the left edge of the rectangle, 5, plus the number of boundary points on the right edge (also 5), plus the number of lattice points on the top edge not already counted $(6 - 2)$, plus the number of lattice points on the bottom edge not already counted $(6 - 2)$, for a total of $5 + 5 + 6 - 2 + 6 - 2 = 18$. Similarly, if the left side of the rectangle has a lattice points and the top side has b lattice points, then the number of boundary points on the rectangle is $a + a + b - 2 + b - 2$ or just $2a + 2b - 4$. We summarize these observations in Theorem 5.35.

Theorem 5.35 *If the vertices of a rectangle are at lattice points and there are a lattice points along the width and b lattice points along the length, then the area of the rectangle, A_R, is $(a-1)(b-1)$, the number of interior points, I_R, is $(a-2)(b-2)$ and the number of boundary points, B_R, is $2a + 2b - 4$.*

Corollary 5.36 *Pick's Theorem holds for rectangles.*

Proof. The area, A_R, of the rectangle by Theorem 5.35 is given by $A_R = (a-1)(b-1)$, which simplifies to

$$A_R = ab - a - b + 1. \tag{5.86}$$

Now, let us compute $I_R + \frac{1}{2}B_R - 1$. By Theorem 5.35, the number of interior points of the rectangle is $(a-2)(b-2)$ and the number of boundary points is $2a + 2b - 4$. Thus,

$$I_R + \frac{1}{2}B_R - 1$$
$$= (a-2)(b-2) + \frac{1}{2}(2a + 2b - 4) - 1$$
$$= ab - a - b + 1 \,. \quad \text{(Simplifying.)} \tag{5.87}$$

Comparing equations (5.86), and (5.87), we see that they are the same. So, Pick's Theorem works for this rectangle. ∎

We are now ready to verify Pick's Theorem for a right triangle. (Can you see where we are going with this?)

Look at the right triangle in Figure 5.50 below.

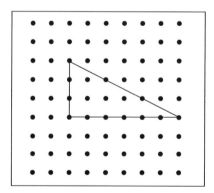

Figure 5.50

If the vertical leg has a lattice points and the horizontal leg has b lattice points (here a is 4 and b is 7), then the number of lattice points on the two legs combined will be $a + b - 1$, since before we counted the lattice point at the right angle twice, once for the vertical leg and once for the horizontal leg. Thus, we have to subtract one from the count, so as to only count the corner lattice point once.

Now, suppose that the hypotenuse of the right triangle has k points on the boundary. We have already counted 2 of them when we added the number of lattice points on the legs. So, the additional number of lattice points on the hypotenuse that we haven't yet counted is $k - 2$, and the total number, B, of boundary points will be $a + b - 1 + k - 2$ or just

$$B = a + b + k - 3. \tag{5.88}$$

Let I be the number of interior points of the triangle. We want to show that the area of the triangle, A_T, is given by Pick's theorem. What this means is that the area is

$$A_T = I + \frac{B}{2} - 1$$

which by equation (5.88) amounts to showing that the area is

$$I + \frac{a + b + k - 3}{2} - 1 \quad \text{or just} \quad I + \frac{1}{2}a + \frac{1}{2}b + \frac{1}{2}k - \frac{5}{2}. \tag{5.89}$$

Now imagine the right triangle as half a rectangle as shown in Figure 5.51 below.

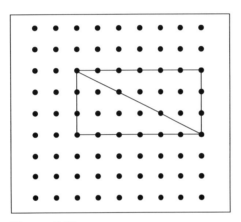

Figure 5.51

Then each of the two triangles has the same number of interior points, which is I, and the $k - 2$ vertices on the diagonal now become interior points, so the total number of interior points in the rectangle is $2I + k - 2$. The number of lattice points on the boundary is $2a + 2b - 4$ (Theorem 5.35). Thus, by Pick's theorem for the rectangle, Corollary 5.36, the area of the rectangle is

$$A_R = 2I + k - 2 + \frac{(2a + 2b - 4)}{2} - 1$$

which simplifies to

$$A_R = 2I + k + a + b - 5.$$

It follows that the area of the triangle is half of this or

$$A_T = I + \frac{1}{2}a + \frac{1}{2}b + \frac{1}{2}k - \frac{5}{2}$$

and this is exactly equation (5.89), which is what we were trying to show. We have proved Corollary 5.37

Corollary 5.37 *Pick's theorem holds for right triangles whose sides are parallel to the x- and y-axes.*

We now come to the hardest part of the proof, which is to prove Pick's Theorem for arbitrary triangles. But we seem to know what to do. We simply enclose the triangle in a rectangle whose sides are parallel to the axis, and then proceed as we did in the beginning of the section. The idea is simple, the algebra is messy, and we need to be careful when counting boundary points, interior points, and so forth.

So referring to Figure 5.52 below,

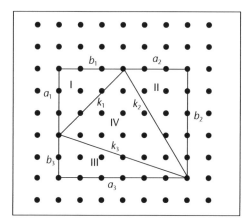

Figure 5.52

we call the number of lattice points on the legs and hypotenuse of triangle I, a_1, b_1, and k_1, and use similar definitions for triangles II, III, and IV. We let I_1, I_2, I_3, and I_4 be the number of interior points of triangles I, II, III, and IV, respectively.

Now we know that each of the $k_1 - 2$ points on the hypotenuse of the first triangle are interior points of the rectangle and the same is true for the other two triangles. Thus, the number of interior points of the rectangle which we denote by I_R is $I_1 + I_2 + I_3 + I_4 + k_1 - 2 + k_2 - 2 + k_3 - 2$ or just

$$I_R = I_1 + I_2 + I_3 + I_4 + k_1 + k_2 + k_3 - 6. \tag{5.90}$$

The number of boundary lattice points on the rectangle which we denote by B_R is

$$\underbrace{(a_1 + b_3 - 1)}_{\text{Lattice points on left edge}} + \underbrace{(b_2)}_{\text{Lattice points on right edge}}$$

$$+ \underbrace{(b_1 + a_2 - 3)}_{\text{Not already counted lattice points on upper edge}} + \underbrace{(a_3 - 2)}_{\text{Not already counted lattice points on lower edge}}.$$

(See if you can explain the -1 in the first parentheses and the -3 in the third parentheses.) This yields

$$B_R = a_1 + a_2 + b_1 + a_3 + b_2 + b_3 - 6. \tag{5.91}$$

Let us denote the area of the rectangle by A_R. Applying Pick's theorem to the rectangle and using equations (5.90) and (5.91) we have

$$A_R = I_R + \frac{B_R}{2} - 1$$

$$= I_1 + I_2 + I_3 + I_4 + k_1 + k_2 + k_3 - 6 + \frac{(a_1 + a_2 + b_1 + a_3 + b_2 + b_3 - 6)}{2} - 1. \tag{5.92}$$

Then, from our previous work (see equation 5.89), the area of right triangle I is $I_1 + \frac{1}{2}a_1 + \frac{1}{2}b_1 + \frac{1}{2}k_1 - \frac{5}{2}$, and there are similar expressions for the areas of the other two right triangles. So the sum of the areas of the three right triangles which we call S_T is $(I_1 + \frac{1}{2}a_1 + \frac{1}{2}b_1 + \frac{1}{2}k_1 - \frac{5}{2}) + (I_2 + \frac{1}{2}a_2 + \frac{1}{2}b_2 + \frac{1}{2}k_2 - \frac{5}{2}) + (I_3 + \frac{1}{2}a_3 + \frac{1}{2}b_3 + \frac{1}{2}k_3 - \frac{5}{2})$
which simplifies to

$$S_T = I_1 + I_2 + I_3 + \frac{(a_1 + a_2 + b_1 + a_3 + b_2 + b_3 + k_1 + k_2 + k_3)}{2} - \frac{15}{2}. \tag{5.93}$$

Now, if we denote the area of the middle triangle by A_T, we have that the area is the difference between the area of the rectangle and the sum of the areas of the three triangles, which in symbols is:

$$A_T = A_R - S_T. \tag{5.94}$$

If we replace A_R in equation (5.94) by equation (5.92) and S_T in equation (5.94) by equation (5.93), and do the algebra, we get.

$$A_T = \left(\underbrace{I_1 + I_2 + I_3 + I_4 + k_1 + k_2 + k_3 - 6 + \frac{(a_1 + a_2 + b_1 + a_3 + b_2 + b_3 - 6)}{2} - 1}_{\text{Area of Rectangle}} \right.$$
$$\left. - \underbrace{(I_1 + I_2 + I_3 + \frac{(a_1 + a_2 + b_1 + a_3 + b_2 + b_3 + k_1 + k_2 + k_3)}{2} - \frac{15}{2})}_{\text{sum of the areas of the right triangles}} \right)$$

which simplifies to $A_T = I_4 + \frac{k_1 + k_2 + k_3}{2} - \frac{5}{2}$, which can be written as

$$A_T = I_4 + \frac{(k_1 + k_2 + k_3 - 3)}{2} - 1. \tag{5.95}$$

We have only one last step: The number of boundary points for the middle triangle, is $(k_1 + k_2 + k_3 - 3)$ where the -3 is for the three vertices of the middle triangle which were double counted. Letting $B_4 = k_1 + k_2 + k_3 - 3$ and substituting into the above equation we get

$$A_T = I_4 + \frac{B_4}{2} - 1,$$

which is exactly what we were trying to prove.

Now, to prove Pick's Theorem for polygons, we can break the polygons up into triangles, or do an induction proof. (See Chapter 8 for a review of induction and for the proof of the remainder of this theorem.) For now, we accept it as true.

Pick's Theorem is elegant. It is easy to state, though as you can see, it is quite another thing to prove.

Student Learning Opportunities

1 (C) Your students have learned that there are multiple strategies for solving a problem and some methods come easier to them than others. Given the problem of finding the area of the triangle in Figure 5.53 below,

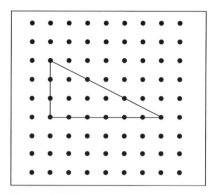

Figure 5.53

do the following:

(a) List at least four different formulas they could use to find the area.
(b) Calculate the area using each of the formulas you have listed.
(c) Which of the formulas might be helpful for your more visual learners? Explain.

2 Find the area of a triangle whose sides are 10, 12, and 15.

3 Find to the nearest integer, the area of a quadrilateral ABCD if AB = 5, BC = 6, CD = 7, and DA = 8, if ∡B = 100 degrees.

4 In Figure 5.54 below, the circles of radii 8, 10, and 12 are tangent to one another. Find the area of the region between the circles to the nearest tenth.

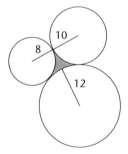

Figure 5.54

5 A person is to pay a one time tax of 10 dollars per square meter on the area of his backyard. The shape and dimensions of the parts of the backyard are given in Figure 5.55 below.

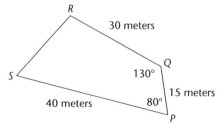

Figure 5.55

Estimate the area of the land to the nearest square foot and the tax paid on this land. [Hint: Draw RP and find its length as well as the measure of angle RPQ.]

6 Find the areas of each of the following figures (Figures 5.56, 5.57, 5.58) using Pick's Theorem. Verify your answer without using Pick's Theorem.

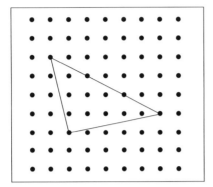

Figure 5.56

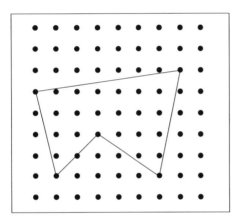

Figure 5.57

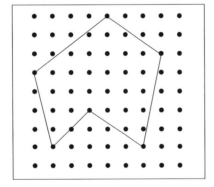

Figure 5.58

CHAPTER 6

BUILDING THE REAL NUMBER SYSTEM

6.1 Introduction

Have you ever wondered where the number system we use today came from? How did it come about? Most secondary school students have never given this question a second thought and don't realize that it was created by human beings over the course of thousands of years. The understanding of the **natural numbers** (1, 2, 3, ...) seems to come quite "naturally" to all children at very young ages. In fact, they are so basic that the mathematician Kronecker once said, "God made the natural numbers; all else was the work of man." (Bell, 1986, p. 477.) Indeed, the number 0, the negative numbers, fractions, and finally irrational and complex numbers were all human creations, as were the rules for working with them. The evolution of today's number system is most interesting and will be the subject of this chapter. We will begin with the rudiments of numbers and progress to some rather sophisticated properties and critical theorems about real numbers and their representations.

The first part of this chapter investigates the reasons for the different methods we use to perform operations on numbers. For example, we will address such questions as: "Why, when we multiply a negative number by a negative number, do we get a positive number?" "Why, when we divide fractions, do we invert and multiply?" We will discuss these informally at first, leaving the more formal aspects to a later part of the chapter.

We will describe the kinds of observations that led to the discovery of the commutative, associative, and distributive laws for whole numbers. We will then extend these rules to negative integers. This will require *defining* rules for addition and multiplication of signed integers. These rules will be motivated by practical applications. We will then show that, with these rules, the commutative, associative, and distributive laws hold for integers. We then discuss and extend the definitions of addition and multiplication to rational numbers. Afterwards, we prove again that, with these definitions, the commutative, associative, and distributive laws hold. Finally, we extend the laws to all the real numbers, which will entail using limits. Once we have these rules, we will turn to the topics of exponents and radicals and develop their laws. We will follow this by a study of logarithms, and solving equations, and inequalities.

In the second part of this chapter, we will discuss decimal representations of numbers and prove some of the basic theorems concerning decimal expansions of real numbers. We will also discuss the fascinating topic of cardinality, which we will then link to the study of algebraic and transcendental numbers we started in Chapter 3. This seems like a long and drawn out process. But there are no shortcuts to this. Although the ultimate formation of the number system as we know it today was done by mathematicians, its creation hinged on years of observations made by people in the course of their lives. It is in this spirit that we tell the story of the development of the number system. Join us on this interesting historic journey.

6.2 Part 1: The Beginning Laws: An Intuitive Approach

 **LAUNCH**

A friend of yours challenges you to show how the distributive law, $a(b+c) = a \cdot b + a \cdot c$, might be applied in real life. You tell him that you use it all the time to do quick mental multiplication. Show how you use the distributive law to calculate $7 \cdot 28$ quickly in your head instead of using the standard algorithm for multiplication.

After having done the launch problem, you may now have a suspicion that, throughout your life, you have been using the fundamental laws for numbers without even realizing it. After all, when did you last question the fact that $4 + 5 = 5 + 4$? These laws are so natural to us that we don't even think about them. In fact, surely we have used the laws long before we learned they were laws. After reading this section, we hope you will have gained a better understanding of why this is the case.

Throughout your mathematical studies, you have probably been exposed to the commutative, associative, and distributive laws many times. That is because they play an important role in the foundations of mathematics. Actually, since these laws are deeply rooted in our observations, we readily accept them. For example, think of the statement that $2 + 3$ is the same as $3 + 2$. Historically, addition meant combining. So, if you have two sticks and combine them with three sticks (say by putting them all in a bag), whether you place 2 sticks in the bag first and then place another 3 in afterwards, it is clearly the same as if you reverse the process. You will still have the same 5 sticks in the bag. The figure below is a very elementary way of visualizing what you did, where the symbol "I" represents a stick. Combining simply means moving things together, so that they are next to each other.

$$\underbrace{\text{II} + \text{III}}_{\text{combining 2 and 3}} = \text{IIIII}$$

$$\underbrace{\text{III} + \text{II}}_{\text{combining 3 and 2}} = \text{IIIII}$$

This idea holds true for all examples we construct like this and seems to point to the fact that, for any natural numbers a and b, $a + b = b + a$. Since no one has ever found an exception to this, and our intuition about this is so strong, we accept it as a rule for natural numbers and you know it as the commutative law of addition.

We can use similar examples to verify the associative law. In this case we use parentheses to mean, "consider as a unit." To illustrate the associative law, consider two different ways of combining sticks, which we represent by $(2 + 3) + 4$ and $2 + (3 + 4)$. First, examine the meaning of $(2 + 3) + 4$

$$\underbrace{\underbrace{(\text{II} + \text{III})}_{2+3} + \text{IIII}}_{(2+3)+4} = (\text{IIIII}) + \text{IIII} = \underbrace{\text{IIIIIIIII}}_{\text{result}}$$

Next, examine $2 + (3 + 4)$

$$\underbrace{\text{II} + \underbrace{(\text{III} + \text{IIII})}_{3+4}}_{2+(3+4)} = \text{II} + (\text{IIIIIII}) = \underbrace{\text{IIIIIIIII}}_{\text{result}}$$

We get the same result in both cases. In repeated examples we observe similar results. Thus, we accept that, for any three natural numbers a, b and c, $(a + b) + c = a + (b + c)$.

The distributive law for natural numbers can also be illustrated with pictures. To see, for example, that $2(3 + 4) = 2 \cdot 3 + 2 \cdot 4$, we need only draw the following picture where we use the fact that multiplication means repeated addition. That is, $2(3 + 4)$ means adding $(3 + 4)$ to itself, twice; that is, $(3 + 4) + (3 + 4)$.

$$\underbrace{(\text{III} + \text{IIII}) + (\text{III IIII})}_{2(3+4)} \overset{\text{upon rearranging}}{=} \underbrace{(\text{III} + \text{III})}_{2 \cdot 3\ +} + \underbrace{(\text{IIII} + \text{IIII})}_{2 \cdot 4}$$

Using similar pictures, we repeatedly verified this relationship and thus accepted the rule that, for natural numbers a, b, and c, $a(b + c) = a \cdot b + a \cdot c$, otherwise known as the distributive property.

Similar pictures can be used to illustrate the commutative laws of multiplication and associative laws of multiplication for natural numbers. We ask you to do this for specific cases of $2(3) = 3(2)$ and $2 \cdot (3 \cdot 4) = (2 \cdot 3) \cdot 4$, respectively, in the Student Learning Opportunities.

When we add the number 0 to the natural numbers, we get the **whole numbers**, and the rules still hold. If we combine any number of objects with nothing, where nothing is represented by the symbol 0, we get the same number of objects. That is, $a + 0 = 0 + a = a$. Again, the commutative, associative, and distributive laws still seemed to hold with 0 added to the natural numbers. Thus, for whole numbers, the following laws were postulated (accepted without question):

1. $a + b = b + a$ **Commutative Law of Addition**
2. $(a + b) + c = a + (b + c)$ **Associative Law of Addition**
3. $a(b + c) = a \cdot b + a \cdot c$ **Distributive Law**
4. $ab = ba$ **Commutative Law of Multiplication**
5. $(ab)c = a(bc)$ **Associative Law of Multiplication**
6. $a + 0 = 0 + a = a$ **Zero Property**

We would like to comment on the use of parentheses in rules 2 and 5 in particular. Recall $(a + b)$ means, "Consider $a + b$ as a single number." Why do we put parentheses in rule 2 in the first place?

The answer is more subtle than you might think. Addition is what is called a **binary operation**, meaning that you can only perform the operation of addition on two numbers at a time. You have been doing this all your life, though you probably never thought of it. For example, think about how you would add the following set of numbers:

$$\begin{array}{r} 13 \\ +26 \\ +22. \\ \hline \end{array}$$

You probably would first add the 3 to the 6 in the last column, then you would take the result, 9, and add it to 2 to get 11. Then you would carry. At each step of the way, you would add only two numbers at a time. That is the only way our brain can process addition.

Thus, when we insert the parentheses on the left side of rule 2, we are indeed adding only two numbers, the single number $a + b$, (expressed by putting parentheses around $a + b$), to the single number c. The right side is similarly saying that we are adding the single number a to the single number $b + c$. If we had just written $a + b + c$, it really would make no sense, since we can only add two numbers at a time. If we just wrote $a + b + c$, it could be interpreted in two different ways. We can see it as $(a + b) + c$ or $a + (b + c)$. Rule 2 is saying that no matter how you interpret it you will get the same answer. So, based on the associative law, if you are adding three numbers, you don't need to include the parentheses and writing the sum as $a + b + c$ is fine. If you are adding more than two numbers, this is also true and that law is known as the **generalized associative law**. We will say more about this in a later section.

The same remarks we made for rule 2 hold for rule 5. Multiplication is also a binary operation. You can only multiply two numbers at a time. No matter how you interpret the multiplication abc, whether it be $(ab)c$ or $a(bc)$, rule 5 says you get the same result. So, again, you can omit the parentheses.

Rule 3, the distributive law, is quite important since it is the basis of so many of the algebraic manipulations we do. Eventually, we will show that it works for all real numbers. We use this when we multiply $x(x + 3)$ to get $x^2 + 3x$. We also use this in the "FOIL" method that is often taught in secondary school to multiply binomials. You will establish this in the Student Learning Opportunities. Even when we solve more complicated equations like $x(x + 3) + 2(x - 4) = -2$, we need the distributive law to expand and break up the parentheses so that we can proceed to solve the equation. It is probably not a misstatement to say that the distributive law is the most useful and important law in elementary algebra.

We have agreed to accept the commutative, associative, and distributive laws for natural numbers, and also for whole numbers since not only has it been repeatedly verified, but it also defies our intuition not to accept them. Of course, if someone comes along one day and finds a counterexample to any of these laws, we are in trouble and would have to start all over again. It is not likely to happen as we see in the next paragraph.

You may be wondering if one can rigorously *prove* that these laws are true. That way, we wouldn't have to just accept them. The surprising answer is, "Yes," and a mathematician, Guisseppe Peano (1852–1932) did prove them by just assuming more basic facts about numbers, together with the principle of mathematical induction. He and others developed the entire real number system from scratch and proved all the laws that we accept. Thus, they put this area of mathematics on a solid foundation. It is a beautiful piece of work, but beyond the scope of this book, since it requires a very detailed analysis that would take at least a semester to do completely. So, we will stick to what people observed, and continue to develop the number system intuitively, just as human beings did. Ultimately, we can rest assured that mathematicians have proven these rules.

Student Learning Opportunities

1 (C) Your students who are more visual learners would benefit from seeing a picture that illustrates that 2(3) = 3(2). What picture can you draw?

2 (C) Your students think that the following relationship is quite obvious and don't really see a conceptual difference in the expressions on either side of the following equation: $2 \cdot (3 \cdot 4) = (2 \cdot 3) \cdot 4$. Draw a picture that would help them clarify this issue.

3 (C) When using the distributive property, your students often forget to distribute completely. You decide that a visual representation might help improve their understanding and use of the distributive law. What picture can you draw to illustrate this and how would you use the picture to illustrate the law? [Hint: Begin with the rectangle shown in Figure 6.1 below.

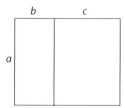

Figure 6.1

Finish it.]

4 Draw a picture similar to the picture from Student Learning Opportunity 3 and show geometrically that $a(b + c + d) = ab + ac + ad$ for positive numbers, a, b, c, and d.

5 (C) You have decided that it would be helpful for all of your students to see a geometric representation of the algebraic relationship that for positive numbers a and b, $(a+b)^2 = a^2 + 2ab + b^2$. How do you do it? [Hint: Start with the square below with side $a + b$]. (See Figure 6.2)

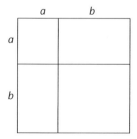

Figure 6.2

After explaining the result geometrically, prove it using the laws presented in this section.

6 Using only the laws presented in this section, do the following problems. (Although we assumed in this section that the numbers were whole numbers, the laws hold for all real numbers and you may assume that.)

(a) Prove the "FOIL" method that is taught in secondary school:
$(a + b)(c + d) = ac + ad + bc + bd$.

(b) Apply this method together with your knowledge of rules for exponents to show that $(x^2 + 1)(yz + 3) = x^2 yz + 3x^2 + yz + 3$.

(c) Show that, for whole numbers a, b, c, and d, $a(b + c + d) = ab + ac + ad$.

(d) Show in a step by step manner why $(a + b)(a^2 + ab + b^2) = a^3 + b^3 + 2ab^2 + 2a^2 b$.

6.3 Negative Numbers and Their Properties: An Intuitive Approach

LAUNCH

One of the typical questions that teacher candidates are asked when going on a job interview for a mathematics teaching position is how to explain to students that a negative number times a negative number is a positive number. What would you say, if you were asked this question on an interview?

As a mathematics major, you have used the rule that a negative number times a negative number is a positive number many times over in doing your arithmetic and algebraic computations. But, have you ever tried to figure out why this rule is used or how you could make sense of it to yourself, or anyone else for that matter? Hopefully, in answering the launch question, you have come up with an adequate explanation. In fact, there can be many explanations for this rule, some of which you will learn about in this next section.

Negative numbers were created by human beings to express opposite situations. For example, if 3 represented a gain of 3, then negative 3 represented a loss of three. So, if you combine (add) a gain of 3 and a loss of three, the net result is no gain or loss. In symbols, $(3) + (-3) = 0$. A statement like $(-4) + 3 = -1$ can be interpreted as "A loss of 4 combined with a gain of three results in a loss of 1" which makes perfect sense. And $(-4) + (-3) = -7$ simply says that "A loss of 4 combined with an additional loss of 3 results in a loss of seven."

If we want to abbreviate our last example, we can omit the plus sign and the parentheses and simply write $-4 - 3 = -7$ and read it as "A loss of three followed by a loss of four, results in a loss of seven," which is what we usually do in algebra.

In summary, mathematicians created the concept of negative numbers to allow us to express opposite situations, and the following rule holds.

7. For every whole number a, there is, by creation,

   a "number," denoted by $-a$ such that $a + (-a) = 0$.

The "number" $-a$ that we refer to above is called the **additive inverse** of a, and we have our seventh law, which is known as the **additive inverse property**. We put the word "number" in quotes just to emphasize that this is a creation. We think of it as a number, since we are going to use it in our computations.

If we wanted an additive inverse of 0, it would be a number which, when added to 0, gives 0. Since $0 + 0 = 0$, the additive inverse of 0 is taken to be 0. Put another way, $-0 = 0$. Notice in rule 7 we said there was a "number," which is an additive inverse. Does this mean that there could be more than one? As we will see later, the answer is, "No."

Having created the negative numbers, we can now extend our number system. If to the whole numbers we adjoin the negatives of whole numbers, we get the set of **integers**. Thus, the set of integers includes 0, ± 1, ± 2, and so on.

In order to apply negative numbers to real problems, we must decide on rules for computing with them, so this will ultimately make sense. Certainly, if we add two negatives, we should get a negative (since, for example, the sum of two losses is a loss). And, if we add a positive and negative, then the gain represented by the positive and loss represented by the negative must produce either a gain or a loss depending on whether the gain is greater or the loss is greater. Since you are familiar with these rules for addition of signed numbers, we need not discuss them further. But, for the sake of our future work, we give the definition of the sum of two negatives. If x and y are whole numbers then

$$-x + -y \text{ is defined to be } -(x + y). \tag{6.1}$$

So, for example, the sum of -3 and -4 is by definition, $-(3 + 4)$ or -7.

Let us turn to multiplication of signed numbers. In elementary school we learn that, when we multiply two numbers, we are performing repeated addition. Thus, 2(3) means that we add 3 to itself 2 times. In a similar manner, we can extend this idea to multiplying a positive number by a negative number. Thus, for consistency, $3(-4)$ should mean, adding negative 4 to itself 3 times. In more practical terms, this means having 3 losses of 4 resulting in a net loss of 12. In symbols: $3(-4) = -12$. Thus, at least in the business sense of gains and losses, we will have to define a positive times a negative to be a negative. And, if we are lucky, after we have constructed all our rules for computing with signed numbers, we will find that the commutative law holds and then it will follow that $(-4)(3)$ will also be -12. That is, a negative times a positive should also be a negative. At this point in our development though, we have not confirmed that the commutative law holds for signed numbers. We are just starting to operate with them. So, here are our formal definitions of multiplying a positive times a negative and a negative times a positive, based on what we have seen in practical applications: If x and y are whole numbers, then

$$(x)(-y) \text{ is defined to be } -(xy) \tag{6.2}$$

$$(-y)(x) \text{ is defined to be } -(xy). \tag{6.3}$$

Notice, we defined the product in both cases to be the same. Thus, we have built commutativity into our definition of multiplication.

How should a negative times a negative be defined so that it reflects what we see in real life? Actually, it has been defined to be a positive, but how can we make sense of this? Imagine the following scenario: If each week you lose three pounds, you can represent this loss by (-3). If this has been going on for several weeks and continues, then 4 weeks from now your weight will decrease by 12 pounds. This loss of weight can be computed as follows: $4(-3) = -12$. In this computation, 4 means 4 weeks in the future. Therefore, 4 weeks in the past, the opposite situation would be represented by -4, and 4 weeks ago your weight was 12 pounds *more* than your weight is now. So, if $4(-3)$ means what your change in weight will be 4 weeks in the future, then $(-4)(-3)$

would represent your weight change four weeks in the past. And since that weight change was positive 12, we should probably define $(-4)(-3) = 12$.

This is not a proof, but simply a way of saying that, if we are going to represent opposite real-life situations by using negative numbers, we are forced to accept the rule that a negative times a negative is a positive for reasons of consistency in *applications*. Our formal definition of multiplication of a negative times a negative is given by:

$$(-x)(-y) = xy \tag{6.4}$$

where x and y are whole numbers. Thus, by definition, $(-3)(-4) = 3 \cdot 4$ or 12. Notice that, by definition, $(-y)(-x)$ is yx which we know is xy since x and y are whole numbers. Thus, both $(-x)(-y) = xy$ and $(-y)(-x) = xy$. Thus, our definition automatically guarantees commutativity of multiplication of negative numbers.

So, now that we have definitions (6.1), (6.2), (6.3), and (6.4) for addition and multiplication, we have to establish that the commutative, associative, and distributive laws hold. They do, but since the proofs are somewhat tedious, we just demonstrate a special case of one of them, the distributive law, to give you a flavor of what is involved in the proofs and ask you to verify some other cases in the Student Learning Opportunities.

Example 6.1 *Prove the distributive law for a negative number times the sum of two positive numbers.*

Solution. Suppose a, b, and c are positive natural numbers, then $(-a)(b+c)$ is the product of a negative and two positive whole numbers. By (6.3), this expression by definition is equal to $-[a(b+c)]$, where we have replaced y by a and x by $b+c$. This, in turn, equals $-[ab+ac]$, since the distributive law holds for natural numbers and a, b, and c are natural numbers. And now, by equation (6.1) with x replaced by ab and y replaced by ac, we have, $-[ab+ac] = -(ab) + -(ac)$. Finally, by equation (6.3), this can be written as $(-a)(b) + (-a)(c)$.

In summary, we have shown that $(-a)(b+c) = (-a)(b) + (-a)(c)$ when a, b, and c are natural numbers. Thus, the distributive law holds in this case.

This was only one case. We have to deal with the cases $(-a)(-b + -c)$, and $(-a)(b + -c)$, and $(a)(-b+c)$, and so on. Since this is quite tedious, we will not do it here. Instead, we will just accept these rules and be grateful that mathematicians took the time to prove them. We now have the following theorems.

Theorem 6.2 *Rules (1) – (7) (found on pages 217 and 220), hold for the integers.*

Secondary school students have no trouble accepting rules (1)–(7) for integers. That a positive times a negative is a negative is also easily accepted using the repeated addition concept we presented earlier. However, accepting that a negative times a negative is a positive troubles them.

Another way to convince yourself or a student of this is to create an argument in which we examine a specific case. Let us consider the product $(-4)(-3)$. Assuming that we accept that 0 times anything is 0, we have that

$-3 \cdot 0 = 0.$

Rewrite this as

$$-3(4 + (-4)) = 0. \tag{6.5}$$

Now if we believe the distributive law holds, then we can distribute in equation (6.5) to get

$$(-3)(4) + (-3)(-4) = 0. \tag{6.6}$$

But we already have accepted that $(-3)(4) = 4(-3) = -12$ so equation (6.6) becomes,

$$-12 + (-3)(-4) = 0.$$

From this, it follows that $(-3)(-4) = 12$. (We are adding something to negative 12 and getting zero. So that something, $(-3)(-4)$, *must be* +12.)

Now that we have given a variety of reasons for why a negative times a negative must be a positive, you might be more comfortable about accepting it. But, we should mention that some mathematicians bitterly fought the existence of negative numbers. They didn't believe in their existence and many really did not understand them. As late as the nineteenth century, author F. Busset, in his handbook of mathematics describes negative numbers as "the roof of aberration of human reason." An eighteenth century mathematician, Francis Maseres, describes negative numbers as those which "darken...equations and make dark of the things which are in their nature excessively obvious." Given how accepted negative numbers are today, this type of perception is hard to comprehend.

We have mentioned that negative numbers were created to express opposite situations, and -3 represented the opposite of what 3 meant. What then would $-(-3)$ mean? It would mean the opposite of -3. Since -3 and 3 are opposites, the opposite of -3 is 3. That is, $-(-3) = 3$.

We have defined the rules for addition and multiplication of signed numbers. We have not defined what **subtraction of signed numbers** means. We define $a - b$ to mean $a + (-b)$. The definition we give makes sense from a practical standpoint since, when you lose money, you are adding a loss to your finances. Thus, any theorems about subtraction can be proven by turning them into addition problems. To illustrate the definition of subtraction, $2 - 3$ is defined as $2 + (-3) = -1$ and $3 - (-4) = 3 + -(-4) = 3 + 4$.

This last example shows why, when you subtract a negative, you are adding a positive. A more intuitive way of explaining this is to approach it from a practical standpoint. If a negative is thought of as a debt, then subtracting a negative means, "taking away" a debt. And, when you have a debt removed, you have gained. Thus, subtracting a negative is equivalent to adding a positive.

We have been using the negative sign in two different contexts. One is for subtraction, the other is to represent the opposite of a number. The context makes it clear which is which. Most calculators have separate buttons for subtracting and taking the negative of a number. All we really have to keep in mind is that subtraction means addition of the opposite. This is a definition!

Student Learning Opportunities

1 (C) How might you explain to a student that $a(0) = 0$ for a an integer?

2 Using equations 6.1 and 6.4, show that $(-a)(-b + -c) = (-a)(-b) + (-a)(-c)$ where a, b, and c are natural numbers.

3 Assuming the distributive law for addition holds, show that $a(b - c) = ab - ac$.

4 (C) Your students want you to give another practical example, like the one given in the text about losing weight over a period of weeks, to illustrate why a negative times a negative should be a positive and why a positive times a negative should be a negative. What example can you give?

5 Suppose that a, b, and c are natural numbers. Show that $(-a)(bc) = (-ab)(c)$.

6 Suppose that a and b are natural numbers. We define $-a + b$ to be $-(a - b)$ if $a > b$ and $b - a$ otherwise. Use these definitions to add $-4 + 3$ and $-3 + 4$.

7 Show that, if a, b, and c are natural numbers, then $-a + (-b + -c) = (-a + -b) + (-c)$.

6.4 The First Rules for Fractions

LAUNCH

It is a well known fact that fractions is one of the most confusing topics for elementary mathematics students. In fact, secondary school mathematics teachers claim that their students have extremely weak skills when it comes to fractions, and that this deficiency creates a major stumbling block when they are trying to learn algebra. One of the reasons they have such difficulty is that the rules don't make any sense to them. How would you explain to your students who are having trouble learning and accepting the rules for operating on fractions, why, when dividing fractions, you invert and multiply? Specifically, how would you explain why, $\dfrac{1/2}{3/5} = 1/2 \cdot 5/3$?

We hope that you were able to come up with an explanation for the rule for the division of fractions that would be helpful to secondary school students. If not, don't despair, since after you have read this section, you should have a clearer idea of how this rule came about, and why it works. You might even come up with more ideas for how to make this and other rules for operating with fractions more understandable for you as well as your future students.

In this section we examine fractions, starting with only those that are rational numbers whose numerator and denominator are positive. Later, we will expand our study to fractions that are the quotients of any two numbers.

Fractions are natural quantities that people are faced with during the course of their lives. For example, we often break things into parts and need to be able to describe what we see. More specifically, we are first taught that $\frac{1}{3}$ of a quantity is that which results from dividing that something up into 3 equal parts and taking one of them. Thus, $\frac{1}{3}$ of 6 is 2.

We have in our minds what a picture of $\frac{1}{3}$ of a loaf cake might be. Namely (Figure 6.3),

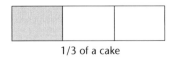

1/3 of a cake

Figure 6.3

It is from pictures like these that the first rules for working with fractions arose. Since we are developing the concept of fraction, we can define addition and multiplication any way we see fit, and so we define it based on what we observe. So, for example, if we wanted to add $\frac{2}{7} + \frac{3}{7}$, it is clear from Figure 6.4 below that the answer is $\frac{5}{7}$.

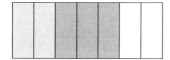

Figure 6.4

We are just adding 2 pieces of cake that have the same size, $\frac{1}{7}$, and then another 3 to it, also with size $\frac{1}{7}$. So, all together we have five pieces of cake that size, or $\frac{5}{7}$ of a cake. Thus, the rule is: to add fractions with a common denominator, we just add the numerators and keep the denominator. This follows from what the picture shows us. In symbols that rule is:

$$\frac{a}{c} + \frac{b}{c} = \frac{a+b}{c}. \tag{6.7}$$

We emphasize that this is a rule that we accept, which tells us *what we observe* in the case when a, b, and c are positive integers. Since $\frac{6}{6}$ of a cake is 1 whole cake, just as $\frac{3}{3}$ of a cake is a whole cake, our observations lead to another rule: For any nonzero number k,

$$\frac{k}{k} = 1.$$

Let us turn to multiplication of fractions. Suppose that we want to take $\frac{2}{3}$ of $\frac{4}{5}$. When we take $\frac{2}{3}$ of something, we divide it into 3 equal parts and take two of them. Thus, to compute $\frac{2}{3}$ of $\frac{4}{5}$, we divide the $\frac{4}{5}$ into 3 equal parts, and take two of those parts. Here is the picture (Figure 6.5).

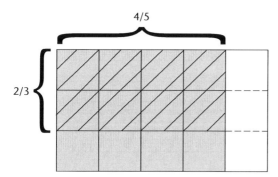

Figure 6.5

The 4 vertical strips going top to bottom of the large rectangle represent $\frac{4}{5}$ of the large rectangle. We want $\frac{2}{3}$ of this. So, we divide the $\frac{4}{5}$ portion into 3 equal parts by horizontal lines and then take two of them. This is represented by the cross hatched area. The overlap between the shaded area and cross hatched area represents $\frac{2}{3}$ of $\frac{4}{5}$. This is clearly $\frac{8}{15}$ of the cake as we can see, since the cake is divided into 15 equal parts by the horizontal and vertical lines, and we have 8 of them in this part.

It appears that, to get the result, $\frac{8}{15}$, we needed to only multiply the numerators and denominators of the fractions. We get the same sense with other similar examples. Each time we take a fraction of a fraction, it seems as if all we do is multiply the numerators and denominators of the fractions involved to get the correct answer. So, we define an operation on fractions that accomplishes this and we call this operation, **multiplication of fractions**. The *definition* of multiplication of fractions is:

$$\frac{a}{b} \cdot \frac{c}{d} = \frac{ac}{bd}$$

(where b and d are not zero).

Again, this is a rule based on pictures and observations. It explains what we are taught in elementary school namely, that the word "of" in this case means "times." The reason it means times is because taking a fraction **of** a fraction seems to be done by multiplying the numerators and denominators of the fractions, and thus it follows our definition of "times" for fractions.

When we come to adding fractions with different denominators, we have to be a bit more careful. Let's say we wanted to add $\frac{1}{3}$ of a cake to $\frac{1}{2}$ of the same size cake. Since, at this point, we only know how to add fractions that have the same denominator, we must cut both of the pieces into pieces of the same size. Hence, we have the idea of a common denominator. Finding a common denominator has the effect of cutting the pieces into the same size.

Let us illustrate. Look at Figure 6.6 below. First, we consider $\frac{1}{3}$ (the shaded section in part (a) of the figure below) and then $\frac{1}{2}$ (the shaded part of figure (b)). We divide each third in figure (a) into 2 equal parts, and each half in figure (b) into three equal parts. We have thus divided each cake into 6 equal parts of the same size. The figures also tell us right away that $\frac{1}{3}$ is the same as $\frac{2}{6}$ and that $\frac{1}{2}$ is the same as $\frac{3}{6}$.

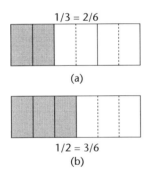

Figure 6.6

Now that our cakes have been broken into pieces of the same size, we can proceed: $\frac{1}{3} + \frac{1}{2} = \frac{2}{6} + \frac{3}{6} = \frac{5}{6}$. Thus, the idea of getting a common denominator is based on cutting objects into pieces of the same size, so that it is easy to tell how many we have.

The preceding analysis led us to the conclusion that $\frac{1}{3}$ was equivalent to $\frac{2}{6}$ and that $\frac{1}{2}$ was equivalent to $\frac{3}{6}$. In short, our analysis illustrated the fact that we can multiply the numerator and denominator of a fraction by the same quantity and get an equivalent fraction (or the same size piece of cake.) This *observation* is called the Golden Rule of Fractions.

Golden Rule of Fractions: The numerator and denominator of a fraction can both be multiplied by the same nonzero quantity, and we will get an equivalent fraction. In symbols, the Golden Rule says that, if a, b, and k are positive numbers and $b \neq 0$, then

$$\frac{a}{b} = \frac{ak}{bk} \quad \text{if} \quad k \neq 0. \tag{6.8}$$

Again, we accept this because all examples in our experience show us this is true. This at least is what happened historically.

Given our example of adding $\frac{1}{2}$ and $\frac{1}{3}$, we can explain the rule for addition of fractions that we learned in elementary school. We have to get a common denominator. Thus, to add $\frac{a}{b} + \frac{c}{d}$, where a, b, c, d are whole numbers and $b, d \neq 0$, we may use as a common denominator, bd. We use the Golden Rule to multiply numerator and denominator of the first fraction by d, and the numerator and denominator of the second fraction by b, and the sum becomes

$$\frac{a}{b} + \frac{c}{d} = \frac{ad}{bd} + \frac{cb}{bd} = \frac{ad+bc}{bd}.$$

Thus, we can define the sum of rational numbers in general as follows: $\frac{a}{b} + \frac{c}{d} = \frac{ad+bc}{bd}$. Remember, this is a *definition* that is based on what we observe for quotients of natural numbers.

Eventually, we will show that these rules for addition and multiplication hold for all fractions, even when the numerators and denominators irrational. In this last sentence we seem to be saying that, after this section, fractions will mean quotients of numbers, regardless of what the numbers are, as opposed to rational numbers which will be quotients of integers. Indeed, this will be the case. We will prove many of these laws from laws involving limits.

But first let us return to the Golden rule for a minute. If we read the symbolic representation of the Golden Rule, equation (6.8) in *reverse*, that is, from right to left, it tells us that, if the numerator and denominator of a fraction have a common *factor k*, then the common *factor* can be "cancelled" to give us an equivalent fraction. (We will use the word "cancel" to mean dividing out common factors from the numerator and denominator.) Thus, the justification for cancelation is the Golden Rule (but read from right to left). That is why in algebra we cannot simplify the fraction $\frac{a+b}{a}$ by just cancelling a in the numerator and denominator to get $\frac{1+b}{1}$, since a is not a *factor* of the numerator. That is also why, when we simplify an expression like $\frac{x^2-9}{x-3}$, we *must factor* first before we divide. Thus, $\frac{x^2-9}{x-3} = \frac{(x-3)(x+3)}{(x-3)\cdot 1} = \frac{(x-3)(x+3)}{(x-3)\cdot 1} = \frac{(x+3)}{1}$. No doubt, one of your (future) students would have gotten the same answer by dividing the x^2 in the numerator by the x in the denominator to get x, and then dividing -9 in the numerator with -3 in the denominator to get $+3$. Of course, this makes no mathematical sense. It is pure luck that it worked in this case.

The rule for division of fractions that we learned in elementary school, which is to "invert and multiply," can be explained in several ways. Here is one: Suppose we wanted to divide $\frac{1}{3}$ by $\frac{2}{5}$. This is:

$$\frac{\frac{1}{3}}{\frac{2}{5}}.$$

If the Golden Rule is to be true for all fractions, then we can multiply the numerator and denominator of this complex fraction by the same quantity, $\dfrac{5}{2}$. This yields

$$\frac{\frac{1}{3}}{\frac{2}{5}} = \frac{\frac{1}{3} \cdot \frac{5}{2}}{\frac{2}{5} \cdot \frac{5}{2}} = \frac{\frac{1}{3} \cdot \frac{5}{2}}{1} = \frac{1}{3} \cdot \frac{5}{2}.$$

Thus, for consistency with our other rules, in particular, the Golden rule, we *have to* define division of fractions by the rule "invert and multiply." A second way to explain this is that division and multiplication are opposite in the sense that, when we perform the division $\frac{15}{3}$ to get 5, we check by multiplying 3 by 5 to get 15. For consistency then, if

$$\frac{\frac{1}{3}}{\frac{2}{5}} = x$$

then $\frac{2}{5}x$ has to give $\frac{1}{3}$ by cross multiplying. What must x then be? Answer, $\frac{5}{6}$, since $\frac{2}{5} \times \frac{5}{6} = \frac{1}{3}$. But $\frac{5}{6}$ is the result of inverting and multiplying! There are other ways to justify this definition. We point out some of them in the Student Learning Opportunities.

We summarize our **definitions for working with fractions** where a, b, and c are positive integers.

$$F1 : \frac{a}{c} + \frac{b}{c} = \frac{a+b}{c}$$

$$F2 : \frac{a}{b} \cdot \frac{c}{d} = \frac{ac}{bd}$$

$$F3 : \frac{a}{b} + \frac{c}{d} = \frac{ad+bc}{bd}$$

$$F4 : \frac{\frac{a}{b}}{\frac{c}{d}} = \frac{a}{b} \cdot \frac{d}{c}.$$

Now that we have rules for working with fractions, we can ask if the set of fractions satisfies the commutative, associative, and distributive laws. They do and it is not difficult to show.

> **Theorem 6.3** *The commutative, associative, and distributive laws hold for fractions.*

Proof. We won't give the proof in all cases, but just show in a few cases how they follow from the definitions $F1-F4$ above. Let us verify the commutative law of addition. By definition, $\frac{a}{b} + \frac{c}{d} = \frac{ad+bc}{bd}$. Furthermore, by definition, $\frac{c}{d} + \frac{a}{b} = \frac{cb+da}{db}$. But, since the numerator and denominator consist of positive integers (and we already know that, for these numbers, the commutative, associative, and distributive laws hold) (Section 1), we have

$$\frac{a}{b} + \frac{c}{d} = \frac{ad+bc}{bd} \quad \text{[By definition of addition of fractions, } F1.\text{]}$$

$$= \frac{bc+ad}{bd} \quad \text{[Commutative Law for addition of whole numbers.]}$$

$$= \frac{cb+da}{db} \quad \text{[Commutative Law for multiplication of whole numbers.]}$$

$$= \frac{c}{d} + \frac{a}{b} \quad \text{[Definition of addition of fractions again.].}$$

Let us give one more proof, that multiplication of positive rational numbers is associative. We begin with three fractions, $\frac{a}{b}, \frac{c}{d},$ and $\frac{e}{f}$. We wish to show that $\left(\frac{a}{b} \cdot \frac{c}{d}\right) \cdot \frac{e}{f} = \frac{a}{b} \cdot \left(\frac{c}{d} \cdot \frac{e}{f}\right)$. Here is how it goes:

$$\left(\frac{a}{b} \cdot \frac{c}{d}\right) \cdot \frac{e}{f} = \left(\frac{ac}{bd}\right) \cdot \frac{e}{f} \quad \text{[Definition of multiplication of fractions, } F2.\text{]}$$

$$= \frac{(ac)e}{(bd)f} \quad \text{[Ditto.]}$$

$$= \frac{a(ce)}{b(df)} \quad \text{[Associative Law for whole numbers.]}$$

$$= \frac{a}{b} \cdot \left(\frac{ce}{df}\right) \quad \text{[Definition of multiplication of fractions, } F2.\text{]}$$

$$= \frac{a}{b} \cdot \left(\frac{c}{d} \cdot \frac{e}{f}\right) \quad \text{[Ditto.].}$$

■

In a very similar manner, we can prove the rest of the commutative, associative, and distributive laws for positive rational numbers. You will do some of this in the Student Learning Opportunities.

So far in this section, the word fraction has meant quotients of natural numbers. This is not standard terminology as we have pointed out. When the word fraction is used in mathematics, the numerator and denominator can be any type, including irrational numbers. A rational number, by contrast, is the quotient of *integers*, where the denominator is not zero. Do the rules $F1$–$F4$ hold for rational numbers? Since we now are allowing negative integers for the numerator and denominator, in this definition, we really are creating a new entity. So we have to define what we mean by addition and multiplication and division of rational numbers. We define them by rules $(F1)$–$(F4)$ as we did earlier. Now, if we wanted to prove Theorem 6.3 for rational numbers, the proof would be identical, since the definitions are identical and we have already pointed out that the commutative, associative, and distributive laws held for all integers (and $\frac{k}{k} = 1$ for all integers.) Thus, we have:

Theorem 6.4 *Rules (1)–(7), the commutative, associative, and distributive laws hold for all rational numbers.*

We emphasize that now we are allowing negative numbers in the numerators and denominators. Before leaving this section, we should say a few words about why, when we deal with fractions, we don't allow denominators of zero. There are many reasons for this, the primary one being that it leads to inconsistencies which are not acceptable in a mathematical structure.

Here is an elementary explanation: Multiplication was originally defined as repeated addition. Division can similarly be thought of as repeated subtraction. Thus, 15 divided by 3 can be thought of as the number of groups of 3 that we can remove (subtract) from a group of 15 before we have nothing left. Of course, the answer is 5. Now what would 15 divided by 0 mean? Answer: How many groups of nothing can we take away from 15 till we end up with nothing? Of course this has no answer, since no matter how many times we subtract 0 we will never be left with nothing. So, we don't divide by 0.

As was demonstrated in Chapter 1, much can go awry if you try to divide by zero. The following was an Student Learning Opportunity in Chapter 1: Find the flaw in the following proof that 1 = 2: Start with the statement $a = b$. Multiply both sides by b to get $ab = b^2$. Subtract a^2 from both sides to get $ab - a^2 = b^2 - a^2$. Factor the left and right sides of the equation to get $a(b - a) = (b - a)(b + a)$.

Now divide both sides by $b - a$ to get $a = b + a$. Now, if we let $a = b = 1$ we get the statement that $1 = 2$. The error here was that we divided both sides by $a - b$ which was zero. That is what led to the false statement that $1 = 2$. The literature is replete with similar kinds of examples where false conclusions result from trying to divide by zero. However, the main reason we can't divide by zero is that doing so would cause inconsistencies in the system, causing breakdowns in the development of new results. Therefore, division by 0 is banned.

Student Learning Opportunities

1. (C) You are distressed to see that some of your high school students still have difficulty adding fractions with unlike denominators. They insist on adding both the numerators and denominators to get their answers. For example, given the following example, they would do as follows: $\frac{4}{5} + \frac{1}{2} = \frac{5}{7}$. How would you use diagrams to help them see that their answer and their procedure makes no sense and that finding a common denominator is a necessity?

2. (C) Your student has done the following work and is satisfied since he got the correct result. Comment on your student's work and correct it. $\frac{x^2 - 25}{x - 5} = \frac{x^2}{x} + \frac{-25}{-5} = x + 5$.

3. We mentioned that division is repeated subtraction, just as multiplication is repeated addition. Use this idea to justify each of the following:

    (a) $1 \div \frac{1}{3} = 3$

    (b) $4 \div \frac{2}{3} = 6$

    (c) $\frac{6}{5} \div \frac{3}{5} = 2$

    (d) $\frac{16}{7} \div \frac{4}{7} = 4$

4. Based on (c) and (d) of the previous example and similar examples, it seems that $\frac{a}{b} \div \frac{c}{b}$ is equivalent to $\frac{a}{c}$. Accept this as true. Using this, give another proof of the invert and multiply rule. [Hint: Convert all fractions to a common denominator.]

5. Prove that multiplication of rational numbers is commutative.

6. Prove that the associative and distributive laws hold for rational numbers.

7. Use the rules from this section to show that $\frac{a}{b} \cdot \frac{b}{a} = 1$.

8. Using the idea of division being repeated subtraction, explain why $\frac{b}{1}$ is b for any positive integer b.

9. Use the laws of this section to show that, if a and b are positive integers, $b \cdot \frac{a}{b} = a$.

10. When one solves the equation $ax = b$ for x, one divides both sides by a. Using the laws from this section, show why the left side becomes x. That is, justify it using the rules from this section. (Here $a \neq 0$.)

11. (C) Your students ask you if integers are rational. What do you say?

6.5 Rational and Irrational Numbers: Going Deeper

LAUNCH

In elementary school, children are introduced to the number line, a one-dimensional picture of a line in which the integers are shown as specially marked points evenly spaced on the line. If we wish to fill up this line with other numbers that are not integers, how would we do it? Specifically, answer the following questions:

1. Where would the rational numbers fall on the line? How many rational numbers (points) would be found between the point 0 and the point 1/2?
2. Where would we locate the point representing $\sqrt{2}$?
3. How many irrational numbers would be on the line? How would these numbers be spaced? Evenly or unevenly?

Did you ever stop to appreciate how beautiful the number line is as a representation of real numbers? Actually, the number line was invented by John Wallis, an English Mathematician who lived from 1616–1703. He must have realized that, since there is an infinite number of real numbers, and there is an infinite number of points on a line, the correspondence of points to real numbers is perfect! But, interesting questions arise when you begin to think about where all of the numbers would appear on the line. This section will describe features of the rational and irrational numbers that will give you a better picture of the density of the number line and the quantity and distribution of all of the different types of real numbers.

The Greeks believed that all numbers were rational. That is, anything that could be measured, necessarily had a length $\frac{p}{q}$, where p and q are integers, and $q \neq 0$. As we all know now, they couldn't have been further from the truth.

The Pythagoreans, the group that gets credit for discovering the irrational numbers, was a secret society formed by the mathematician Pythagoras. The sect was very strict, lived in caves, had many rituals, and studied mathematics as part of their attempt to understand the universe. They were sworn to secrecy and many of their discoveries remained untold. Although the Pythagorean Theorem was attributed to the master of their sect, Pythagoras, it was known long before Pythagoras was born. Perhaps the reason the Pythagorean Theorem was attributed to this group was that they may have given its first deductive proof. But, as is common in history, it is hard to know exactly what happened over a span of a few thousand years, especially when many of the books which might have contained the correct history have been destroyed.

As you may have guessed from the previous paragraph, the discovery of irrational numbers hinged on the Pythagorean Theorem. That is, consider a right triangle where each leg is one. We see, using the Pythagorean Theorem that the length of the hypotenuse is $\sqrt{2}$. The surprise, of course, was that $\sqrt{2}$ is irrational as we have already shown in Chapter 1. There are infinitely many irrational numbers, as we have seen in earlier chapters, and in a somewhat surprising result, there are far more irrational numbers than rational numbers. (For more on this, see Section 6.16.)

232 Building the Real Number System

The Pythagoreans called irrational numbers "alogon" which means "unutterable." It is written that they were so shocked by this discovery, that anyone who dared mention it in public was put to death. There is a well known story that the person who discovered irrational numbers, Hippasus of Metapontum, "perished at sea." Whether this happened as a result of the leak or because of rough seas or illness, we don't know. But most accounts seem to indicate that he was drowned as a result of this discovery. One thing is clear though, this discovery was a major upset in the mathematical world. Indeed, many of the proofs in geometry that depended on the idea that all numbers were rational had to be corrected.

So now that we know that irrational numbers exist, we can join the set of rational numbers with the set of irrational numbers to form (their union) a set called the **real numbers**. Thus, by definition, every real number will be either rational or irrational.

Because of the launch question, you might be wondering about the spread of the irrational numbers on the number line. Are they evenly distributed or not? Are they scarce or everywhere? The next two results show that rationals and irrationals are everywhere. However, since in this book, we are not developing the real number system completely, we have to make use of some facts about real numbers that are intuitive. Here are the facts we accept: (a) The fraction $\frac{1}{n}$ can be made as small as we want by taking n large. (Thus, if n is one million, this fraction is $\frac{1}{1,000,000}$ which is small.) (b) Between any two numbers that differ by 1, there lies some integer. That is, for any number a, there is always some integer k, that satisfies $a < k \leq a + 1$. For example, if $a = 3.5$, this last statement says that between 3.5 and 4.5 there is an integer k, specifically the integer 4. If $a = 2$, the above statement says that there is an integer k that satisfies $2 < k \leq 3$. Obviously, that integer k is 3.

> **Theorem 6.5**
>
> (1) *Between every two real numbers there is a rational number.*
> (2) *Between every two rational numbers there is an irrational number.*

Proof. (1) Suppose x and y are any two real numbers, and that $x < y$. This implies that $y - x > 0$. Since $\frac{1}{n}$ can be made as small as we want, there is some positive number n that makes $\frac{1}{n} < y - x$. Let us take the smallest such n. Since the numbers nx and $nx + 1$ differ by 1, we know there is some number k that satisfies $nx < k \leq nx + 1$. Divide this inequality by n to get $x < \frac{k}{n} \leq x + \frac{1}{n}$. But, since we know that $\frac{1}{n} < y - x$ (notice the strict inequality), the previous inequality can be written as $x < \frac{k}{n} < x + (y - x)$, or just as $x < \frac{k}{n} < y$.

We have found that between *any* two real numbers x and y there is a rational number $\frac{k}{n}$, so we have proven the first part of the theorem.

Proof. (2) Take x and y rational and suppose that $x < y$. Multiply both sides of this inequality by $\sqrt{2}$ to get $\sqrt{2}x < \sqrt{2}y$. Now, by part (1) of the theorem, there is a rational number k between the two real numbers $\sqrt{2}x$ and $\sqrt{2}y$. That is, there is a rational number k such that $\sqrt{2}x < k < \sqrt{2}y$. Divide this inequality by $\sqrt{2}$. to get $x < \frac{k}{\sqrt{2}} < y$. And since (Student Learning Opportunity 7) a rational number divided by an irrational number is irrational, we have found an irrational number, $\frac{k}{\sqrt{2}}$, between x and y. ∎

Let us take this further. Suppose that r is any real number. Then, between the real numbers r and $r + 1$, there is a rational number, r_1 by part (1) of the theorem. Similarly, there is a rational number, r_2 between r and $r + \frac{1}{2}$ and a rational number r_3 between r and $r + \frac{1}{3}$, and so on. Since the numbers $r + 1, r + \frac{1}{2}, r + \frac{1}{3}$, and so on get closer and closer to r, the numbers r_1, r_2, r_3, and so on get closer and closer to r. We have established the following (critical!) theorem.

> **Theorem 6.6** *For any real number, r, one can find a sequence $r_1, r_2, r_3 \ldots$ of rationals converging to r. (In the language of calculus, we are saying that $r = \lim_{n \to \infty} r_n$.)*

Now we know from calculus that the limit of the sum is the sum of the limits. There are similar statement for the limit of the difference, product, and quotient (provided in the quotient, the limit of the denominator is not 0). The first statement that the limit of the sum is the sum of the limits can be expressed more formally in this case as $\lim_{n \to \infty} (a_n + b_n) = \lim_{n \to \infty} a_n + \lim_{n \to \infty} b_n$ with similar expressions for $\lim_{n \to \infty} (a_n - b_n)$, $\lim_{n \to \infty} (a_n b_n)$, and $\lim_{n \to \infty} (a_n/b_n)$ provided $\lim_{n \to \infty} b_n \neq 0$ in the last statement. Here, all the limits are assumed to exist and be finite.

In the beginning of this chapter, we said that we would accept the associative laws, commutative laws, and distributive laws for all whole numbers and then pointed out how, once the rules for multiplying negatives were established, we could extend these rules to negative numbers and eventually rational numbers which we have done. Using the theorems from this section, we can now extend the rules to *all* real numbers.

We illustrate how this is done with one example.

> **Example 6.7** *Show that, for any real numbers a and b, $a + b = b + a$ assuming that the commutative law holds only for rational numbers.*

Solution. Pick a sequence of *rational* numbers a_n converging to a, and a sequence of *rational* numbers b_n converging to b. Then

$$a = \lim_{n \to \infty} a_n \quad \text{and} \tag{6.9}$$

$$b = \lim_{n \to \infty} b_n. \tag{6.10}$$

Now

$$\begin{aligned}
a + b &= \lim_{n \to \infty} a_n + \lim_{n \to \infty} b_n && \text{[Using equations (6.9) and (6.10)]} \\
&= \lim_{n \to \infty} (a_n + b_n) && \text{[The limit of the sum is the sum of the limits from calculus.]} \\
&= \lim_{n \to \infty} (b_n + a_n) && \text{[Addition of rational numbers is commutative.]} \\
&= \lim_{n \to \infty} b_n + \lim_{n \to \infty} a_n && \text{[The limit of the sum is the sum of the limits again]} \\
&= b + a && \text{[From equations (6.9) and (6.10)].}
\end{aligned}$$

The proofs of all the other rules are similar. Thus, with the notion of limit, we can fill all the gaps and move from rationals to all real numbers. So, we finally have:

> **Theorem 6.8** *Rules (1–7) (found on pages 217 and 220), hold for all real numbers.*

Student Learning Opportunities

1. Assuming that $ab = ba$ and $a(b+c) = ab + ac$ hold for rational numbers, show, using Theorem 6.6 that they hold for all real numbers.

2. Prove, using Theorem 6.6, that multiplication is associative for all real numbers, assuming that it is associative for all rational numbers.

3. Show that, if a sequence $\{r_n\}$ of rational numbers converges to a then the sequence $\{-r_n\}$ converges to $-a$. Then show that rule 7, that $a + (-a) = 0$ holds for all real numbers a.

4. (C) Your students ask you what the last rational number is that comes right before 3. How would you explain to them that there is no "last rational number" that comes right before 3? (In their proofs, we have often heard students say things like, "Well, let's take the last rational number before 3" in an argument.)

5. Show that the sum, difference, product, and quotient of any two rational numbers are rational.

6. Show $\sqrt{2} + \sqrt{3}$ is irrational. (Assume it is rational and square both sides.)

7. Show, using a proof by contradiction, that the sum, product, difference, and quotient of a rational number and an irrational number are irrational.

8. Show that the product of two irrational numbers can be rational.

9. (C) Your students find it very hard to believe that there are really an infinite number of (a) irrational numbers and (b) rational numbers between any two real numbers. How would you prove to them that this is true?

10. Show that $\sqrt{\dfrac{1}{2}}$ and $\sqrt{\dfrac{2}{3}}$ are both irrational.

11. How many points are there with rational coordinates in the region of the plane bounded by the line $x + y = 6$, the x-axis, and the y-axis?

6.6 The Teacher's Level

LAUNCH

1. If a student asks you to *prove* that a negative times a negative is a positive, how would you proceed?
2. A teacher explains the rule this way: If you show a movie of someone walking backwards in reverse, then it looks like the person is moving forwards. Is this a proof? Comment on it.

Experienced mathematics teachers will attest to the fact that secondary school students often want to make sense out of the rules for multiplying signed numbers. In this section we provide proofs of the validity of the critical theorems underlying those questions that are used on a regular basis in algebra. Understanding these proofs should help you explain the areas that students find so confusing.

In previous sections we gave intuitive and rigorous arguments to support basic theorems about rules about operations with signed numbers and working with fractions. In this section we will actually *prove* all of these theorems; however, we will have to assume something, and what we assume is the validity of the commutative, associative, and distributive laws. Thus, if we assume that the operations of addition and multiplication satisfy rules (1)–(7) found on pages 217 and 220, then it *must* follow that a negative times a negative is a positive, and that a positive times a negative is a negative, and that any number times 0 is 0, and so on.

We will now begin an abstract development of the critical theorems you use regularly in algebra. Below are the only assumptions we make.

For *all real numbers*, a, b, and c

1. $a + b = b + a$ **Commutative Law of Addition**
2. $(a + b) + c = a + (b + c)$ **Associative Law of Addition**
3. $a(b + c) = a \cdot b + a \cdot c$ **Distributive Law**
4. $ab = ba$ **Commutative Law of Multiplication**
5. $(ab)c = a(bc)$ **Associative Law of Multiplication**
6. There is a number 0 that has the property that

   $a + 0 = 0 + a = a$ **Zero Property**
7. For each a, there exists a (unique) number $-a$,

   such that $a + (-a) = 0$. **Additive Inverse Property.**

We assume nothing else. (Actually, we also assume that the sum and product of two real numbers is a real number too, but we don't explicitly state that since it seems so obvious.)

The first theorem is essential.

Theorem 6.9 *There can only be one additive inverse of a number, x.*

Proof. Our strategy for this proof is to begin by assuming that there are two numbers that are additive inverses of a given number and then argue that they must be equal.

Suppose there were two additive inverses of x and that they are a and b. Then, by definition of additive inverse,

$$x + a = 0 \tag{6.11}$$

and

$$x + b = 0. \tag{6.12}$$

Now,

$$a = a + 0 \quad \text{(Rule 6 above)}$$
$$= a + (x + b) \quad \text{(By equation (6.12))}$$
$$= (a + x) + b \quad \text{(Associative law)}$$
$$= (x + a) + b \quad \text{(Commutative law)}$$
$$= 0 + b \quad \text{(By equation (6.11))}$$
$$= b \quad \text{(Zero property)}$$

∎

We have shown that any two additive inverses a and b of x are the same. Thus, the additive inverse is unique.

We now show how, with acceptance of rules (1)–(7), we can derive some of the usual rules of algebra. These proofs show *step by step* what is really happening in some of the typical algebraic manipulations that students do in their algebra work.

> **Theorem 6.10**
> (a) The equation $x + x = x$ has only one solution, namely $x = 0$.
> (b) If a represents any number, then $a(0) = 0$.
> (c) $(-a)(b) = -(ab)$. In particular, a negative times a positive is a negative.
> (d) $(-a)(-b) = ab$. In particular, a negative times a negative is a positive.

One would think that, because we are proving this for all real numbers, we will need limits for parts (c) and (d). However, we don't need them, since Rules (1–7) on page 235 will do.

Proof. (a) Start with $x + x = x$ and rewrite this as $(x + x) = x$. Add $-x$ to both sides to get

$$(x + x) + (-x) = x + (-x).$$

Use the Associative Law to rewrite this as

$$x + (x + (-x)) = x + (-x).$$

Use rule 7 above to rewrite this as

$$x + 0 = 0.$$

Finally, use the rule 6 above to rewrite this as

$$x = 0.$$

Thus, if $x + x = 0$, we have that $x = 0$.

(b) Since $0 + 0 = 0$ by rule 6 above, with $a = 0$, we have,

$$a(0) = a(0 + 0).$$

Distributing, we get

$$a(0) = a(0) + a(0).$$

Now, calling $a(0) = x$, this becomes

$$x + x = x.$$

And now by part (a),

$$x = 0. \tag{6.13}$$

But $x = a(0)$. Thus, equation (6.13) says, $a(0) = 0$.

(c) We now know that $a(0) = 0$. Rewrite this as $a((-b) + b) = 0$. Distribute to get

$$a(-b) + ab = 0. \tag{6.14}$$

Equation (6.14) says that $a(-b)$ is an additive inverse of ab, since they sum to zero. But there is only one additive inverse of ab and that is $-(ab)$. Thus,

$$a(-b) = -(ab). \tag{6.15}$$

(d) We already know that any number times 0 is 0. Thus, $(-a)(0) = 0$. Rewrite this as $(-a)(-b + b) = 0$. Distributing we get,

$$(-a)(-b) + (-a)b = 0. \tag{6.16}$$

By equation (6.15), equation (6.16) reduces to

$$(-a)(-b) + (-(ab)) = 0. \tag{6.17}$$

Now, equation (6.17) says that $(-a)(-b)$ is an additive inverse of $-(ab)$. But so is ab. Since the additive inverse of $(-a)(-b)$ is unique, $(-a)(-b) = ab$.

An alternative way of showing this is to start with equation (6.17). Now we add ab to both sides and follow along as in the proof of (a) or (c) to get $(-a)(-b) = ab$. You should work this through and see it happen. ∎

Notice we have just proved theorems about addition and additive inverses. We have said nothing about subtraction. We define subtraction the way it is done in secondary school. Namely $a - b$ is defined to be $a + (-b)$. Here is another set of rules that one uses in secondary school.

Theorem 6.11

(1) $-(-a) = a$,

(2) $-(a - b) = b - a$

Proof. The first part is easier than it looks. We know that

$$-a + -(-a) = 0 \tag{6.18}$$

for we are just adding $-a$ to its additive inverse. Now equation (6.18) says that $-(-a)$ is an additive inverse of $-a$. But so is a. Since there is only one additive inverse of a number, $-(-a) = a$.

The proof of the second part is similar, but requires more detail. See if you can fill in the reasons for each step below:

$(a - b) + (b - a)$

$= (a + -b) + (b + -a)$

$= ((a + -b) + b) + -a$

$= (a + (-b + b)) + -a$

$= (a + (0)) + -a$

$= (a) + -a$

$= 0.$

Since $(a - b) + (b - a) = 0$, $(b - a)$ is an additive inverse of $(a - b)$. But there is only one additive inverse of $(a - b)$ and that is, $-(a - b)$. So $-(a - b) = b - a$. ∎

As you well know, in algebra it is helpful, or sometimes essential, to be able to rearrange terms. Often students have difficulty accepting the legitimacy of claims such as $a - b + c$, is the same as $-b + c + a$. We will now show why this can be done. Similar arguments will show that, when you have a group of terms separated by plusses and/or minuses, they can be rearranged as long as the signs are unchanged.

We know that $a - b + c$ means $a + -b + c$. And we have pointed out in an earlier section that, by the associative law, it makes no difference if you interpret $a + -b + c$ as $(a + -b) + c$ or $a + (-b + c)$. The meaning is the same. So we can drop the parentheses and just write $a - b + c$. Now, by the commutative law, $a + -b + c = -b + a + c$. So we have succeeded in showing that $a - b + c$ is the same as $-b + a + c$.

In a similar manner, $abcdef$ has the same value as $decbfa$. Again, we use rules (1)–(7) to prove this and state this as a theorem for reference, since it is such a useful result. We will develop it more in the Student Learning Opportunities.

> **Theorem 6.12**
>
> (a) *In an algebraic expression consisting of terms added and subtracted, we may rearrange the terms, as long as we keep the signs intact.*
> (b) *In a product, the terms may be rearranged and we will get the same product.*

Proof. Prior to the theorem, we have indicated how this is done when we have three terms. The proof of this result in general, when there are many terms, is somewhat tricky and uses induction. We don't include it here, but we will give you a feel for how involved the proof is by demonstrating it for 4 terms. We use the expression $(a + b) + (c + d)$. Now, suppose we wanted to show that this was the same as $(d + b) + (c + a)$. Here are the steps.

$(a + b) + (c + d) = (c + d) + (a + b)$ [Commutative law]

$= c + (d + (a + b))$ [Associative law]

$= c + (d + (b + a))$ [Commutative law]

$= c + ((d + b) + a)$ [Associative law]

$= c + (a + (d + b))$ [Commutative Law]

$= (c + a) + (d + b)$ [Associative Law]

$= (d + b) + (c + a)$ [Commutative Law].

It is this kind of proof that shows that, by using a combination of the commutative and associative laws we can rearrange any sum and always arrive at an equivalent expression. That is why we can discard the parentheses in any sum and write an expression like $(a + b) + (c + d)$ as just $a + b + c + d$. No matter how you interpret this sum, the result is always the same! ∎

Part (b) of the theorem provides justification for why an expression like $(-3x^2y^3)(4x^4y)$ is equal to $-12x^6y^4$. Although secondary school students are required to perform these rearrangements in their study of algebra, they often don't feel confident in doing it and don't understand the reason it is allowed. That is, the standard procedures for representing the product is to place the number first, followed by all of the $x's$, followed by all of the $y's$. In our example above we have that $(-3)(4)x^2x^4y^3y$ and this readily yields the result $-12x^6y^4$ once the rules for exponents are employed.

There are some other properties of real numbers, which are important and which we postulate.

8. There is a number 1 with the property that

   $a \cdot 1 = 1 \cdot a = a$ for any real number a.

This property is known as the **multiplicative identity property**. (You multiply a number by 1 and you get the identical number.)

Another property that we accept is the following.

9. For each nonzero number a there is a (unique) number

   denoted by a^{-1} such that $a \cdot a^{-1} = a^{-1} \cdot a = 1$.

The number a^{-1} is called the **multiplicative inverse** of a.

Until now we have not needed to use rules (8) and (9), but we will need them now to continue extending the properties of real numbers.

Rule 8 can be used to explain many algebraic processes. For example, when we solve quadratic equations, we sometimes factor and then set each factor equal to zero. Why do we do that? The following theorem tells us why.

Theorem 6.13 *If a and b are real numbers and $ab = 0$, then $a = 0$ or $b = 0$.*

Proof. Either $a = 0$ or it isn't. If $a = 0$, then we are done. If it is not, then a^{-1} exists by rule 8. Now, multiply both sides of the equation $ab = 0$ by a^{-1} to get

$a^{-1}(ab) = a^{-1}(0)$.

Since we have proven that $a^{-1}(0)$ is 0, this simplifies to

$a^{-1}(ab) = 0$. (6.19)

Using the associative law we get

$$(a^{-1}a)b = 0$$

or just

$$1 \cdot b = 0. \tag{6.20}$$

But $1 \cdot b = b$ by rule 9. Thus, equation (6.20) becomes $b = 0$. A similar proof shows that, if $b \neq 0$, then a must be 0. ∎

Now let's examine how this theorem plays a part in solving quadratic equations. If we want to solve $x^2 - 5x + 6 = 0$, we factor the left side and get $(x - 2)(x - 3) = 0$. Thinking of $x - 2$ as a and $x - 3$ as b, we have $ab = 0$. Thus, either a or b must be 0. That is, either $x - 2$ or $x - 3$ must be 0. This principle extends. If we have a product of any number of expressions, which is equal to zero, then one of the factors is 0. This is often used in solving higher degree equations. For example, if we wish to solve $x^3 = x$, then we bring everything over to one side of the equation to get $x^3 - x = 0$, which factors into $x(x - 1)(x + 1) = 0$. This means that either $x = 0$, $x + 1 = 0$, or $x - 1 = 0$, which tells us that either $x = 0$, $x = 1$, or $x = -1$.

Students often make the following mistake when trying to solve quadratic equations. Say they want to solve $x^2 - 2x = 4$. They factor both sides to get $x(x - 2) = 1 \cdot 4$ and then conclude that $x = 1$ and $x - 2 = 4$ and therefore the solutions are $x = 1$ and $x = 6$. Of course, if they check their answers, they will see this is not correct. When we have a product like $x(x - 2) = 4$, we can make no conclusion about what either of the factors is since the number 4 can be written as a product in many ways. (It could be $1 \cdot 4$ or $2 \cdot 2$ or $6 \cdot \dfrac{2}{3}$ and so on.) So this method is completely wrong. But, if the product of two factors is *zero*, then we *can* make a conclusion, and that conclusion is given by the above theorem: Either one or the other factor is zero.

Ancient civilizations had methods for solving linear equations and for solving certain quadratic equations, but the method of solving by factoring took a very long time to evolve. Part of that may be that the concept of 0 as a number came relatively late in the history of mathematics.

In the next few sections we will discuss other aspects of the real number system. More specifically, we will examine decimal representation of numbers, which represented a major step forward for humankind. But first we need to review the concept of geometric series.

Student Learning Opportunities

1 (C) One of your students claims that $x + x = x^2$ and $2x + 3y = 6xy$. How do you help the student? What are correct statements? What laws or definitions substantiate the correct statements?

2 (C) If a student asks, "How do you know that $2x + 3y$ is the same as $3y + 2x$," how do you answer?

3 (C) A student wants to know what the justification is behind the statement from algebra, "When you add like terms to like terms, you will get a term of the same type." (For example, $3x^2 + 2x^2 = 5x^2$.) How do you answer?

4 Using the laws for real numbers, show why $a - b = -b + a$.

5 Using the laws for real numbers, show in detail why, if $y + x = 0$, then y must be $-x$.

6 Using the laws for real numbers, show in detail that $a - b - c = -c - b + a$.

7 Using *only* rules 1–7 and Theorem 6.10, give a detailed proof that $(a + b) + (a + b) = 2a + 2b$.

8 Using *only* rules 1–7 and Theorem 6.10, give a detailed proof that $(a + b) + a = 2a + b$.

9 One law that is used repeatedly in algebra is the following: If $a + b = a + c$, then $b = c$. Justify this law using whichever of rules 1–7 you need. Thus, when someone sees the equation $x + y = x + 3$, one can eliminate the x's to get $y = 3$.

10 Using the fact that we don't need parentheses when adding numbers, we can prove the following surprising result: The commutative law of addition, rule 1, did not have to be given as a postulate since it automatically follows from the other rules! Prove this. [Hint: Start with $(a + b) + (a + b) = 2a + 2b = (a + a) + (b + b)$. Rewrite this as

$$a + b + a + b = a + a + b + b. \qquad (6.21)$$

Finish it by adding the appropriate quantities to both sides of equation (6.21).]

11 Using the laws for real numbers, show in detail that $x - y - z = -z + x - y$.

12 Multiply the following numbers in your head: $(245)(342)(4341)(3533)(5235)(0)(4566)(3004)$. Explain how you did it. What rule(s) did you use?

13 (C) After learning how to solve quadratic equations by factoring, one of your students does the following work and is confused why her solution does not check:

$x^2 + 3x = 10$

$x(x + 3) = 10$

$\quad x = 5 \quad \text{and} \quad x + 3 = 2$

$\quad x = 5 \quad \text{and} \quad x = -1$.

How can you help your student? What is incorrect about this work? How can you use the zero property to solve this equation properly?

14 Show that $(-1) \cdot x = -x$. [Hint: $0 \cdot x = 0$. Write 0 as $1 + -1$.]

15 Show that $-(x + y) = -x - y$.

16 Show that $b \cdot (-a) = -ab$.

17 (C) How would you justify, in detail, the following algebraic manipulation to a student: $(x + y) - 3(x - 2y) = (x + y) - (3x - 6y)$? How would you continue to justify that this is the same as $x + y - 3x + 6y$ and that this is the same as $-2x + 7y$?

18 If a and b are integers, then $a + b \equiv b + a \bmod m$. That is, addition is commutative mod m. Which of the other rules 1–7 are true mod m? Verify those that are true.

19 Prove that the multiplicative inverse of a number is unique.

20 Show that, if a sequence $\{r_n\}$ of rational numbers converges to a, then the sequence $\{\frac{1}{r_n}\}$ converges to $\frac{1}{a}$. Then show that $a \cdot \frac{1}{a} = 1$ for any real number a. It follows that $a^{-1} = \frac{1}{a}$.

21 Let us consider the set, S, of all numbers of the form $a + b\sqrt{2}$ where a and b are integers and that we add and subtract numbers of this form the same way we did in algebra. Will the commutative, associative, and distributive laws hold for this set of numbers? How do you know?

6.7 The Laws of Exponents

**LAUNCH**

One of the typical job interview questions for a mathematics teaching position is to describe how you would explain to students why the expression 3^0 is equal to 1. How would you respond?

The secondary school curriculum requires that students have facility using the laws of exponents. In order for this to occur, they must have a basic understanding of the fundamental rules and their meanings. This section will clarify these rules and their associated theorems.

6.7.1 Integral Exponents

Algebra is a shorthand. We observe that $2 + 3 = 3 + 2$ and $5 + 7 = 7 + 5$, and so on. In order to express our observations concisely, we can use the shorthand $a + b = b + a$. This is simple, clean, and captures the whole essence of the concept that addition is commutative regardless of the numbers. The same is true for all the other laws we have given—the associative, distributive laws, and so on. When it comes to exponents, we also use shorthand. If n is a positive integer, we abbreviate

$$\underbrace{a \cdot a \cdot a \cdot \ldots \cdot a}_{n \text{ times}}$$

as a^n. Using this notation, we can establish the following laws of exponents.

Theorem 6.14 *For positive integers m and n,*

(E1) $a^m \cdot a^n = a^{m+n}$

(E2) $\dfrac{a^m}{a^n} = a^{m-n}$

(E3) $(a^m)^n = a^{mn}$

(E4) $\left(\dfrac{a}{b}\right)^n = \dfrac{a^n}{b^n}$

(E5) $(ab)^n = a^n b^n$.

We refer to these rules for exponents as rules (E1)–(E5). Since we will be referring to these rules often, we suggest you jot them down for easy access.

The first law follows immediately from the definition of a raised to a power:

$$a^m \cdot a^n = \underbrace{a \cdot a \cdot a \cdot \ldots \cdot a}_{m \text{ times}} \cdot \underbrace{a \cdot a \cdot a \cdot \ldots \cdot a}_{n \text{ times}}$$

and we see that, when we multiply the two terms on the left, we have a string of $m + n$ a's on the right. So we see that $a^m \cdot a^n = a^{m+n}$.

When first working with exponents, it is wise for students to represent a few examples in expanded form such as $a^2 a^3 = (a \cdot a)(a \cdot a \cdot a) = a^5$. This way it becomes very clear why the rules hold.

Rule $(E2)$ is explained by dividing. Most of the time rule $(E2)$ is first taught assuming that $m > n$ so that negative exponents don't have to be addressed. Thus, we may show the student some specific examples like $\dfrac{a^5}{a^3}$

$$\frac{a^5}{a^3} = \frac{a \cdot a \cdot a \cdot a \cdot a}{a \cdot a \cdot a}.$$

Now, we divide three of the five a's in the numerator with the three a's in the denominator and we end up with $a \cdot a$ or just a^2 in the numerator and 1 in the denominator. A few examples like this will clearly demonstrate why the second rule for exponents holds.

Rule $(E3)$ can be explained by expanding some simple expressions. For example: $(a^2)^3 = a^2 \cdot a^2 \cdot a^2 =$ (by rule $(E1)$) $= a^{2+2+2} = a^{3(2)} = a^6$. After a few examples, one discovers the rule that, when you "power twice," you multiply the exponents.

Rule (E4) Follows immediately since

$$\left(\frac{a}{b}\right)^n = \overbrace{\frac{a}{b} \cdot \frac{a}{b} \cdot \ldots \cdot \frac{a}{b}}^{n \text{ times}} = \frac{a^n}{b^n}$$

by the rule that, when we multiply fractions, we multiply numerators and denominators.

Thus, rules $(E1)$–$(E5)$ follow almost directly from the definition of raising a variable to a power. We leave the proof of rule $(E5)$ for the Student Learning Opportunities.

Typically, students confuse the different laws of exponents. That is why it is important to do such things as compare rules $(E1)$ and $(E3)$. That is, it is useful to compare values of expressions such as $a^4 \cdot a^3$ and $(a^4)^3$ and then point out that the first is a^7 while the second is a^{12}, and why this is so.

At this point, it is only natural to wonder if rules $(E1)$–$(E5)$ can be extended to negative and fractional exponents. First, we must ask the question, "What must the definition of a raised to a negative exponent or fractional exponent be for rules $(E1)$–$(E5)$ to hold in all cases?"

If we want rule (E2) to be true all the time, then it must be true when $m = n$. In particular, $\dfrac{a^m}{a^m}$ must be $a^{m-m} = a^0$. And since any (nonzero) number, a^m, divided by itself is 1, for consistency we must DEFINE a^0 to be 1 when $a \neq 0$.

If you check most algebra books, you will see the statement $a^0 = 1$, $a \neq 0$. Often one asks, "What happens if $a = 0$ in the above definition of a^0?" Well, we get 0^0. So, what does 0^0 equal? Some people feel that it should be defined to be 1 for consistency. But actually if you define it to be 1, you get a different inconsistency: We know that 0 raised to any power is 0. So if you define it to be 1, you run into the problem that, on the one hand, $0^0 = 1$ and, on the other hand, $0^0 = 0$. We run into a similar problem if we define 0^0 to be 0, since then you have the inconsistency that

$a^0 = 1$. Mathematicians hotly debated this issue of defining 0^0, and the final decision was made to leave it alone. It is undefined. It is like division by 0.

Continuing in this way, if we want $(E2)$ to be true in all cases, we want it to be true when $m = 0$. In particular, we want $\dfrac{a^0}{a^n}$ to be equal to a^{0-n} or a^{-n}. But we have agreed that $a^0 = 1$ (when $a \neq 0$). So $\dfrac{a^0}{a^n} = a^{-n}$ simply becomes $\dfrac{1}{a^n} = a^{-n}$ when $a \neq 0$. It was this desire for rule $(E2)$ to hold that motivated the *definition* of a^{-n} as $\dfrac{1}{a^n}$. It was fortunate that all of the rules $(E1)-(E5)$ hold with this definition as we shall see.

Let us illustrate an example that involves both positive and negative exponents and shows that rule $(E3)$ holds.

Example 6.15 *Show that* $(a^{-2})^{-3} = a^6$.

Solution. We transform the negative exponents into positive exponents, since we know by Theorem 6.14 that rules $(E1)-(E5)$ hold for positive exponents. Now,

$$(a^{-2})^{-3} = \left(\dfrac{1}{a^2}\right)^{-3} \quad \text{[Definition of negative exponent.]}$$

$$= \dfrac{1}{\left(\dfrac{1}{a^2}\right)^3} \quad \text{[Ditto.]}$$

$$= \dfrac{1}{\left(\dfrac{1}{a^6}\right)} \quad \text{[Theorem 6.14 part }(E4).\text{]}$$

$$= 1 \cdot \dfrac{a^6}{1} \quad \text{[Invert and multiply.]}$$

$$= a^6.$$

In an identical manner, we can show that $(a^{-m})^{-n} = a^{mn}$ where $-m$ and $-n$ are negative exponents.

Similarly, one can prove all the other laws, but many cases must be taken. We will ask you to prove some of the other laws in the Student Learning Opportunities. For now, we simply state the results of all this as a theorem.

Theorem 6.16 *Rules* $(E1)-(E5)$ *hold if the exponents are any integers.*

Student Learning Opportunities

1 (C) Your students are very confused by all of the algebraic rules they have learned and claim the following: $(2x^2y^4)(5x^3y) = 10x^6y^4$ and $(3x^5y^3)(4x^3y^2) = 12x^{15}y^6$. How do you help them? Which law are they confused about?

2 (C) Using only rules (E1)–(E5) for positive exponents, and the definition of a negative exponent (that is, without using Theorem 6.16), how would you help your students understand that each of the following are true.

(a) $\left(a^{-2}\right)^3 = a^{-6}$

(b) $\dfrac{a^{-3}}{b^{-4}} = \dfrac{b^4}{a^3}$

(c) $a^{-3} \cdot b^{-3} = (ab)^{-3}$

(d) $\dfrac{a^5}{a^{-3}} = a^8$

(e) $\left(a^{-6}\right)^{-2} = a^{12}$

3 Assuming that m and n are positive integers, and using only Theorem 6.14 and the definition of a number raised to a negative exponent, show that

(a) $\left(\dfrac{a}{b}\right)^{-n} = \left(\dfrac{b}{a}\right)^n$

(b) $(a^{-m})^{-n} = a^{mn}$

(c) $\left(\dfrac{a^{-n}}{a^m}\right) = a^{-n-m}$

(d) $a^{-m} a^{-n} = a^{-m-n}$

6.8 Radical and Fractional Exponents

LAUNCH

If a student asked you why $25^{\frac{1}{2}}$ is defined to be $\sqrt{25}$, what would you say? If a student asked you "What does $2^{\sqrt{3}}$ mean?," how would you respond?

A student claims that $\sqrt{25} = \pm 5$. Is that student correct? Explain.

If you were able to respond correctly to the launch question, you understand that there are some subtle rules that you must be aware of when dealing with problems involving radical and fractional exponents. The purpose of this section is to clarify these rules so that any areas of confusion you may have had will be resolved.

6.8.1 Radicals

We saw in Chapter 3 a theorem that the equation $x^2 = a$ has a solution for each positive a. In fact, there are two solutions. But only one of them is positive. In algebra, a **square root** of a positive number, a, is any number b which when multiplied by itself gives a. Thus, a square root of 9 is 3, since 3 multiplied by itself is 9. Another square root of 9 is -3 since -3 multiplied by itself is 9. Thus, there are two square roots of 9.

The positive square root of 9 is denoted by $\sqrt{9}$. Stop! Notice the words, "The positive square root." Many people think that $\sqrt{9}$ is ± 3. It is not. On the secondary school level, the use of the square root symbol means the positive square root. If we wanted to talk about the other square root of 9, we would denote it by $-\sqrt{9}$. This quantity is -3. The confusion here seems to come from the fact that the equation $x^2 = 9$ has two solutions, ± 3, or put another way, $\pm\sqrt{9}$. Yes, the *equation* has two solutions, but the symbol $\sqrt{9}$ by itself means the positive square root. Some books call the positive square root the **principal square root** of a.

As we observed, every positive number has two square roots. The positive square root of a is denoted by $\sqrt{a}$. Of course, there is only one square root of 0 and that is 0.

We will now present theorems that support our work on radicals. Here is our first theorem. We don't really need the words "If a is nonnegative" since the symbol $\sqrt{a}$ in secondary school already requires that a be nonnegative. We include it for emphasis.

Theorem 6.17

(a) *If a is nonnegative then, $\sqrt{a} \cdot \sqrt{a} = a$.*
(b) *If a and b are nonnegative, then $\sqrt{a} \cdot \sqrt{b} = \sqrt{ab}$.*
(c) *If a and b are nonnegative, then $\dfrac{\sqrt{a}}{\sqrt{b}} = \sqrt{\dfrac{a}{b}}$.*

Proof. (a) This first proof is much simpler than one might think. We defined $\sqrt{a}$ to be that nonnegative number, b, which when multiplied by itself gives a. Thus, by definition, $\sqrt{a}$ multiplied by itself must be a. That is, $\sqrt{a} \cdot \sqrt{a} = a$.

(b) Let us compute $\left(\sqrt{a} \cdot \sqrt{b}\right)^2$. This is $\left(\sqrt{a} \cdot \sqrt{b}\right) \cdot \left(\sqrt{a} \cdot \sqrt{b}\right) = \left(\sqrt{a} \cdot \sqrt{a}\right) \cdot \left(\sqrt{b} \cdot \sqrt{b}\right) = ab$. Here we have used part (a).

We have shown that $\sqrt{a} \cdot \sqrt{b}$ when multiplied by itself gives us ab. Thus $\left(\sqrt{a} \cdot \sqrt{b}\right)$ is one of the square roots of ab. Since $\left(\sqrt{a} \cdot \sqrt{b}\right)$ is nonnegative, and there is only one nonnegative square root of ab which we denote by $\sqrt{ab}$, it must be the case that $\sqrt{ab} = \sqrt{a} \cdot \sqrt{b}$.

(c) We leave this proof to you as it is very instructive. ∎

In the same way as we show that every nonnegative number has a square root, using the Intermediate Value Theorem from Chapter 3, we can show that every real number a has a unique cube root (that is, something which when multiplied by itself three times gives a.) We simply form the function $f(x) = x^3 - a$ for that specific a, and show that $f(x)$ takes on positive values for some x and negative values for other x. By the Intermediate Value Theorem, that means that the graph must cross the x-axis. Crossing the x-axis means $f(x) = 0$ and hence that $x^3 = a$ for some x. That shows that there is a cube root of a. To show that there is only one cube root of a, we need to show that the function never crosses the x-axis again. That is, $f(x)$ is never 0 again. This follows since the function is increasing. So, once it crosses the x-axis, it never crosses it again. (Refer back to Chapter 3, the last section, for a review of this.) We denote this cube root by $\sqrt[3]{a}$. There are similar definitions for 4th roots, 5th roots, and so on. Thus, an nth root of a number a is a number which when multiplied by itself n times gives us a. When n is odd, there is only one nth root. When n

is even and $a > 0$, there are two nth roots, and the positive one (or **principal nth root of a**) is denoted by $\sqrt[n]{a}$. In general, when we say the nth root of a, we will mean the positive nth root when n is even. The analog of Theorem 6.17 is Theorem 6.18.

Theorem 6.18 *If n is a positive integer*

(a) $(\sqrt[n]{a})^n = a$
(b) $\sqrt[n]{a} \cdot \sqrt[n]{b} = \sqrt[n]{ab}$
(c) $\dfrac{\sqrt[n]{a}}{\sqrt[n]{b}} = \sqrt[n]{\dfrac{a}{b}}$.

When n is even, we assume that a and b are nonnegative.

Proof. The proof is entirely analogous to the proof of the previous theorem and we leave it to you. ∎

6.8.2 Fractional Exponents

Raising a number to a fractional exponent means nothing until we give it some meaning. We begin with the definition of $a^{\frac{1}{2}}$, which we learned in secondary school means $\sqrt{a}$. Where did that come from? Well, if we want to be able to apply rule ($E1$) in all cases, then it must be true that $a^{\frac{1}{2}} \cdot a^{\frac{1}{2}} = a^{\frac{1}{2}+\frac{1}{2}} = a^1 = a$. Thus, $a^{\frac{1}{2}}$ multiplied by itself will have to give you a. By definition of square roots, this tells us that, if $a^{\frac{1}{2}}$ means anything at all, it must be a square root of a. We defined $a^{\frac{1}{2}}$ to be $\sqrt{a}$, but, since we cannot take the square roots of negative numbers and get real numbers, we restrict our definition to $a \geq 0$ on the secondary school level.

Similarly, if we want rule ($E1$) to be true for fractional exponents, then $a^{\frac{1}{3}} \cdot a^{\frac{1}{3}} \cdot a^{\frac{1}{3}} = a^{\frac{1}{3}+\frac{1}{3}+\frac{1}{3}} = a^1$. So $a^{\frac{1}{3}}$ multiplied by itself 3 times will give us a. That is, $a^{\frac{1}{3}}$ is a cube root of a. But, there is only one cube root of a. Thus, we define $a^{\frac{1}{3}} = \sqrt[3]{a}$ for consistency.

Similarly, we define $a^{\frac{1}{4}} = \sqrt[4]{a}$, and so on. In general, we define $a^{\frac{1}{n}} = \sqrt[n]{a}$ when n is a positive integer. But, if we are going to apply the rules ($E1$) – ($E5$) unconditionally, we are forced to require that $a \geq 0$. While it is true that we can define cube roots of negative numbers, and fifth roots of negative numbers and so on, we will NOT be able to make a statement like $a^{\frac{1}{2}} \cdot a^{\frac{1}{3}} = a^{\frac{1}{2}+\frac{1}{3}}$ unless $a \geq 0$, since $a^{\frac{1}{2}}$ is not defined unless a is nonnegative. That is why, in textbooks where they ask you to simplify radicals, you will often see the words, "Assume that all the variables under consideration are nonnegative."

One other thing. We denote the positive square root of a by $\sqrt{a}$. When it is convenient, as it will be later on in some proofs, we will denote this also as $\sqrt[2]{a}$.

Having defined $a^{\frac{1}{n}}$, what would be an appropriate definition for $a^{\frac{m}{n}}$? Well, if we want rule ($E3$) to apply, then $a^{\frac{m}{n}}$ must be $(a^{\frac{1}{n}})^m$. But $a^{\frac{1}{n}} = \sqrt[n]{a}$. So, one way to define $a^{\frac{m}{n}}$ is to define it as $(a^{\frac{1}{n}})^m$ which is $(\sqrt[n]{a})^m$. Of course, for this definition to make sense when n is even, we must require that $a \geq 0$. Note that, although $a^{\frac{1}{n}}$ and $a^{\frac{m}{n}}$ have been defined based on consistency, we still do not know whether rules ($E1$)–($E5$) will hold with these definitions. After we establish some theorems familiar to most secondary school students, we will begin the process of showing that these rules do hold.

> **Theorem 6.19** If $a \geq 0$, and m and n are positive integers, then $(\sqrt[n]{a})^m = \sqrt[n]{a^m}$. That is
> $$\left(a^{\frac{1}{n}}\right)^m = (a^m)^{\frac{1}{n}}.$$

Proof. We know that

$$a^m = a^{\frac{mn}{n}}$$
$$= \left(\sqrt[n]{a}\right)^{mn} \quad [\text{Definition of } a^{\frac{mn}{n}}]$$
$$= \left(\left(\sqrt[n]{a}\right)^m\right)^n \quad [\text{Rule } (E3) \text{ with } \sqrt[n]{a} \text{ in place of } a]$$
$$= \left(a^{\frac{m}{n}}\right)^n \quad [\text{Definition of } a^{\frac{m}{n}}].$$

This string of equalities read from the bottom up, tells us that

$$\left(a^{\frac{m}{n}}\right)^n = a^m.$$

In words, if we take the number $\left(a^{\frac{m}{n}}\right)$ and raise it to the n th power, we will get a^m. Thus, $\left(a^{\frac{m}{n}}\right)$ being nonnegative, must be the (principal) nth root of a^m. We display this.

$$a^{\frac{m}{n}} = \sqrt[n]{a^m}. \tag{6.22}$$

But, by the definition of $a^{\frac{m}{n}}$, we have

$$a^{\frac{m}{n}} = \left(\sqrt[n]{a}\right)^m. \tag{6.23}$$

Comparing equations (6.22) and (6.23) we see that

$$\sqrt[n]{a^m} = \left(\sqrt[n]{a}\right)^m. \tag{6.24}$$

If a were negative and n were odd, then there is nothing wrong with the string of inequalities that we have given, and thus equation (6.24) would be true in this case. We run into a problem when n is even and $a < 0$. Thus, to avoid this problem, we will agree that, when a negative number is raised to a fractional exponent, the exponent must be in lowest terms with an odd denominator. ∎

A Student Learning Opportunity that we presented in Chapter 1 will show why we need these restrictions. There, we computed the value $(-8)^{\frac{1}{3}}$ two ways. We computed $(-8)^{\frac{1}{3}}$ as $\sqrt[3]{-8}$, which is the definition of $(-8)^{\frac{1}{3}}$ and we got -2 as we should have. If we take the same $(-8)^{\frac{1}{3}}$ and rewrite it as $(-8)^{\frac{2}{6}}$ and then attempt to apply Theorem 6.19, we get $\sqrt[6]{(-8)^2}$, which is $+2$, not negative 2. We cannot apply Theorem 6.19 in this case since $a < 0$, or explained another way, since the fraction $\frac{2}{6}$ is not in lowest terms, and we agreed that when a negative number is raised to a fractional power, the fraction must be in lowest terms. This explains why, in mathematics books, you will often see the following statement, "We define $a^{\frac{m}{n}} = \sqrt[n]{a^m}$ for fractions $\frac{m}{n}$ that are in lowest terms." This is required when $a < 0$. But, when $a \geq 0$, we have seen in Theorem 6.19 that we do not have to have this concern.

So, do rules $(E1)$–$(E5)$ hold for fractional exponents? Well, the theorem that this is the case for fractional exponents when the bases being raised to powers are nonnegative has a proof which is similar in nature to the way we do the following examples. It is easier to focus on the numerical examples than on the proof in general.

Example 6.20 Show that, if $a \geq 0$, $a^{\frac{3}{2}} \cdot a^{\frac{4}{5}} = a^{\frac{3}{2}+\frac{4}{5}}$.

Solution. $a^{\frac{3}{2}} \cdot a^{\frac{4}{5}} = a^{\frac{15}{10}} \cdot a^{\frac{8}{10}} = \left(\sqrt[10]{a}\right)^{15} \cdot \left(\sqrt[10]{a}\right)^{8} = \left(\sqrt[10]{a}\right)^{15+8} = \left(\sqrt[10]{a}\right)^{23} = a^{\frac{23}{10}} = a^{\frac{3}{2}+\frac{4}{5}}$

Example 6.21 Show that if $a \geq 0$, $\left(a^{\frac{3}{2}}\right)^{\frac{4}{5}} = a^{\frac{3}{2} \cdot \frac{4}{5}}$.

Solution. This is a bit more subtle. Since fractional exponents are defined in terms of radicals, we will convert our expressions to radicals. We will need the fact that $\sqrt[5]{\sqrt[2]{a}} = \sqrt[10]{a}$. Why is this true? Well, if we compute $\left(\sqrt[10]{a}\right)^{5}$, we get $a^{\frac{5}{10}} = a^{\frac{1}{2}}$. This last sentence says that $\sqrt[10]{a}$ multiplied by itself 5 times is $a^{\frac{1}{2}}$, so $\sqrt[10]{a}$ must be $\sqrt[5]{a^{\frac{1}{2}}}$ or, put another way, $\sqrt[10]{a} = \sqrt[5]{\sqrt[2]{a}}$.

Now we proceed to the main part of this example. Using the definition of a fractional exponent we have $\left(a^{\frac{3}{2}}\right)^{\frac{4}{5}} = \sqrt[5]{\left(a^{\frac{3}{2}}\right)^{4}} = \left(\sqrt[5]{\left(a^{\frac{3}{2}}\right)}\right)^{4} = \left(\sqrt[5]{\sqrt[2]{(a^3)}}\right)^{4} = \left(\sqrt[10]{a^3}\right)^{4} = \left(\sqrt[10]{(a^{(3)})^4}\right) = \left(\sqrt[10]{a^{(3)(4)}}\right) = \sqrt[10]{a^{(12)}} = a^{\frac{12}{10}} = a^{\frac{3}{2} \cdot \frac{4}{5}}$.

As you can see, proving the rules for exponents in general is not trivial, and verifying all the cases the way we have done can be even more tedious. So once again, having given you the flavor for how these proofs are done, we simply state the theorem and outline some of the proofs.

Theorem 6.22

(1) If $a \geq 0$, then $\sqrt[q]{\sqrt[p]{a}} = \sqrt[qp]{a}$.
(2) Rules $(E1)$–$(E5)$ hold for positive fractional exponents.

Proof. (1) The proof is similar to the first part of Example 6.21. So we leave it to you.

(2) Proof of Rule $(E1)$: We will show that $a^{\frac{m}{p}} \cdot a^{\frac{n}{q}} = a^{\frac{m}{p}+\frac{n}{q}}$. Now $a^{\frac{m}{p}} \cdot a^{\frac{n}{q}} = a^{\frac{mq}{pq}} \cdot a^{\frac{np}{pq}}$
$= \sqrt[pq]{a^{mq}} \cdot \sqrt[pq]{a^{np}} = \left(\sqrt[pq]{a}\right)^{mq} \cdot \left(\sqrt[pq]{a}\right)^{np} = \left(\sqrt[pq]{a}\right)^{mq+np} = a^{\frac{mq+np}{pq}} = a^{\frac{m}{p}+\frac{n}{q}}$

(Proof of Rule $E3$) We will show that $\left(a^{\frac{m}{p}}\right)^{\frac{n}{q}} = a^{\frac{mn}{pq}}$.

We have $\left(a^{\frac{m}{p}}\right)^{\frac{n}{q}} = \sqrt[q]{\left(a^{\frac{m}{p}}\right)^{n}} = \left(\sqrt[q]{a^{\frac{m}{p}}}\right)^{n} = \left(\sqrt[q]{\sqrt[p]{(a^m)}}\right)^{n} = \left(\sqrt[pq]{a^m}\right)^{n} = \left(\sqrt[pq]{a^{(m)(n)}}\right) = a^{\frac{mn}{pq}} = a^{\frac{m}{p} \cdot \frac{n}{q}}$.

Rules $(E2)$, $(E4)$, and $(E5)$ are left for the Student Learning Opportunities. ∎

> **Corollary 6.23** *Rules (E1)–(E5) hold for all fractional exponents.*

Proof. Again, we just change everything to positive exponents and work from there. The proofs can be tedious and we just accept the theorem. ∎

6.8.3 Irrational Exponents

Having explored the meanings and rules that apply for integer and fractional exponents, it is only natural to wonder about the meaning of a^p when p is irrational. There are some serious issues with this, and to get into all of them would be beyond the scope of the book. But we can, at least, lay the groundwork. In Section 6.5, Theorem 6.5 we proved that every irrational number p is the limit of a sequence of rational numbers, r_n. So one way of defining a^p is to define it to be $\lim_{n\to\infty} a^{r_n}$. We are working with rational exponents, r_n, when defining a^p. We have already seen that we need to require that a to be ≥ 0 when dealing with arbitrary rational exponents, and, in fact, we avoid many technical issues if $a > 0$. Thus, we will require that a be positive when defining a^p where p is irrational. When we talk about the function $f(x) = a^x$, we also assume that $a > 0$.

One issue with this definition is that the irrational number, p, can be the limit of many sequences, and we have to show that we get the same answer for a^p regardless of which sequence we take.

We do that now.

> **Theorem 6.24** *Suppose that $a > 0$. If $\lim_{n\to\infty} r_n = p$ and $\lim_{n\to\infty} s_n = p$, then $\lim_{n\to\infty} a^{r_n} = \lim_{n\to\infty} a^{s_n}$. Thus, the definition of $a^p = \lim_{n\to\infty} a^{r_n}$ is independent of which sequence we take approaching p.*

Proof. Let's examine $\lim_{n\to\infty} \dfrac{a^{r_n}}{a^{s_n}}$. Since r_n and s_n both approach p, their difference $r_n - s_n$ approaches 0. Thus, $\lim_{n\to\infty} \dfrac{a^{r_n}}{a^{s_n}} = \lim_{n\to\infty} a^{r_n - s_n} = a^0 = 1$. Since from calculus $\lim_{n\to\infty} \dfrac{a^{r_n}}{a^{s_n}} = \dfrac{\lim_{n\to\infty} a^{r_n}}{\lim_{n\to\infty} a^{s_n}}$, and we showed this limit is 1 in the last sentence, it follows that the numerator and denominator of this last fraction are the same. That is, $\lim_{n\to\infty} a^{r_n} = \lim_{n\to\infty} a^{s_n}$ ∎

So now we know how to compute with irrational exponents. Thus, if we wanted to define $3^{\sqrt{2}}$ power, we notice that $\sqrt{2} = 1.4142\ldots$, where the dots indicate it goes on forever, so we can compute $3^1, 3^{1.4}, 3^{1.41}, 3^{1.414}, \ldots$ and the limit of these is the meaning of $3^{\sqrt{2}}$. (A calculator might just compute $\sqrt{2}$ to say 4 places to get 1.4142 and then compute the value of $3^{1.4142}$ as the value of $3^{\sqrt{2}}$. Actually most calculators use more than 4 places.)

With our definition of a^p, we can now show that rules (E1)–(E5) hold. We just need the limit laws from calculus.

> **Theorem 6.25** *Rules (E1)–(E5) hold even if the exponents are irrational.*

Proof. (Rule (E1)) We will prove rule (E1) and leave the rest for you as they are very similar. We wish to show only the special case that $a^p \cdot a^q = a^{p+q}$ when p and q are irrational.

Pick a sequence of rational numbers r_n converging to p and a sequence of rational numbers s_n converging to q. Then

$$\begin{aligned} a^p \cdot a^q &= \lim_{n\to\infty} a^{r_n} \cdot \lim_{n\to\infty} a^{s_n} \\ &= \lim_{n\to\infty} (a^{r_n} \cdot a^{s_n}) \quad \text{[The limit of the product is the product of the limits.]} \\ &= \lim_{n\to\infty} a^{r_n+s_n} \quad \text{[Since we have established rule (E1) for rational exponents.]} \\ &= a^{\lim_{n\to\infty}(r_n+s_n)} \quad \text{[By a limit law from calculus.]} \\ &= a^{p+q} \quad \text{[Since } r_n + s_n \text{ is a sequence converging to } p+q.] \end{aligned}$$

∎

Notice how calculus, specifically the notion of limit, was needed to prove this result from *algebra*. This is just another indication of the power of calculus. Not only was it a fundamental tool in the sciences that allows us to make major discoveries about planetary motion and physical systems in general, but it, allows us to prove relationships that were previously accepted without question.

One of the kinds of equations that secondary school students are often asked to solve are those that have variable expressions for the exponents. Here is a typical problem of the simplest type.

Example 6.26 *Solve the equation for x : $4^{2x} = 8^{3x+1}$.*

Solution. The approach here is to represent both 4 and 8 in exponential form with the same base. Since both 4 and 8 are powers of 2, our equation can be rewritten as

$$(2^2)^{2x} = (2^3)^{3x+1}.$$

Using the rules for exponents this can be simplified to

$$2^{4x} = 2^{9x+3}.$$

Since the bases are the same, the exponents must be the same also. (There is more to this statement than meets the eye. It has to do with the fact that exponential functions are 1 − 1.) (See Chapter 9 for more of a discussion on 1–1 functions.) Thus, $4x = 9x + 3$ and solving for x, we get that $x = -\frac{3}{5}$.

A slightly harder equation is

Example 6.27 *Solve for x : $\dfrac{9^{x^2} \cdot 27^{3x}}{3^5} = 1.$*

Solution. We observe that, since $9 = 3^2$, $27 = 3^3$, our equation can be written as

$$\frac{(3^2)^{x^2} \cdot (3^3)^{3x}}{3^5} = 1$$

which by the laws of exponents can be simplified to $3^{2x^2} \cdot 3^{9x} \cdot 3^{-5} = 1$. This, in turn can be simplified to $3^{2x^2+9x-5} = 3^0$ and hence $2x^2 + 9x - 5 = 0$. Factoring, we get $(2x - 1)(x + 5) = 0$ so $x = \frac{1}{2}$ and $x = -5$.

Student Learning Opportunities

1 (C) Your students claim that the solutions to the following equations are the same. Are they correct? Explain.

$$x = \sqrt{36}$$
$$x^2 = 36$$

2 How many real solutions are there for x that satisfy the equation

$$2^{6x+3} \cdot 4^{3x+6} = 8^{4x+5}?$$

3 If $\dfrac{4^x}{2^{x+y}} = 8$ and $\dfrac{9^{x+y}}{3^{5y}} = 243$, find x and y.

4 Prove part (c) of Theorem 6.17.

5 Prove part (a) of Theorem 6.18.

6 Prove part (b) of Theorem 6.18.

7 Prove part (c) of Theorem 6.18.

8 Prove that $\dfrac{a^p}{a^q} = a^{p-q}$ when p and q are irrational.

9 (C) How would you explain to a student why the definition of $a^{\frac{1}{4}}$ is $\sqrt[4]{x}$?

10 Show that every real number has only one real fifth root.

11 (C) How would you help your students understand that each of the following are true?

(a) $a^{\frac{9}{10}} \cdot a^{\frac{1}{10}} = a$

(b) $a^{\frac{2}{3}} \cdot a^{\frac{3}{4}} = a^{\frac{17}{12}}$

(c) $\left(a^{\frac{2}{3}}\right)^4 = a^{\frac{8}{3}}$

(d) $\left(\dfrac{125}{a^6}\right)^{\frac{1}{3}} = \dfrac{5}{a^3}$

(e) $\left(\dfrac{x^{1/3}}{x^{-1/3}}\right)^{1/2} = x^{\frac{1}{3}}$

12 (C) A student wants to know if the expression $\left(\sqrt{2}^{\sqrt{2}}\right)^{\sqrt{2}}$ is rational or irrational. What would you say? How would you explain your answer?

13 Solve the following equations for x:

(a) $2^{2x} \cdot 2^{4x} \cdot 2^{6x} = 8$

(b) $(3^x)^{x-1} = 9$

(c) $\dfrac{2^{x^2}}{2^x} = 64$

(d) $\sqrt{\dfrac{9^{x+3}}{27^{x+1}}} = 81$

(d) $\sqrt[3]{\dfrac{16^{x+1}}{(8^{3x})^2}} = 4$

(e) $4^{2x} - 10(4^x) + 16 = 0$

14 If $x = t^{\left(\frac{1}{t-1}\right)}$ and $y = t^{\left(\frac{t}{t-1}\right)}$ where $t > 0$ and $t \neq 1$, then show that $x^y = y^x$. [Hint: Observe that $y = x^t$. Now compute $\dfrac{y}{x}$.]

15 Show that $\sqrt{2}$ is a solution of $x^{x^2} = 2$. Find a solution of $x^{x^3} = 3$ and then try to make a general statement about the solution of $x^{x^n} = n$.

6.9 Working with Inequalities

LAUNCH

Comment on the following solution procedure: A class is given the system of equations, $x + y < 3$, $x - y < 7$ and is told to solve for x and y. Student A adds the equations to get, $2x < 10$, hence $x < 5$. Student B immediately jumps in and says, "But, if we take the point $x = 4$, $y = 0$, it doesn't work. So $x < 5$ can't be the solution." In response to this, student A says, "Hold your horses, I am not done." Student A then proceeds to subtract the two inequalities to get $2y < -4$ so that $y < -2$. He now looks at B and says, "The answer is $x < 5$ and $y < -2$. Your example, $x = 4$ and $y = 0$ doesn't fit these conditions." B looks A squarely in the eye and says, "Well then take $x = 4$ and $y = -3$. That satisfies your conditions, but doesn't work in the original inequalities!" Resolve this issue. Who is right?

If you are like most, you found the above scenario a bit confusing. When working with inequalities, care must be taken, as there are quite a few subtleties that must be attended to. We hope this next section will clarify these issues.

Now that we have essentially constructed the real numbers, we will turn to the issue of inequalities. Before we begin, we must define what $a < b$ means. Although you have an intuitive sense of what this means, in mathematics, intuition is not enough when trying to determine for sure which statements are true and which statements aren't. Proof is what is needed.

So let us begin. We define $a < b$ to mean that we can find a positive number p such that $a + p = b$. Thus, $3 < 4$ since we can find a positive number, namely 1, such that $3 + 1 = 4$. Similarly, $-4 < -1$ since we can find a positive number, namely 3, such that $-4 + 3 = -1$. We define $a > b$ to mean $b < a$. (So, we define it in terms of what we know. We essentially are saying that any "greater than" inequality is a "less than" inequality read in reverse.) Using these simple definitions, we can now prove without too much difficulty, some important relationships. This again emphasizes the importance of having good definitions to make proofs easier.

> **Theorem 6.28** *If $a < b$ and $c < d$, then*
>
> (a) $a + c < b + d$.
> (b) *If k is a positive number, then $ka < kb$.*
> (c) *If k is a negative number, then $ka > kb$.*

Part (a) states that we can add inequalities if they have the same sense (both "less than" or both "greater than.") Thus, since $3 < 4$ and $5 < 6$, $3 + 4 < 5 + 6$. Part (b) says that multiplying an inequality by a positive number does not change the sense of the inequality, and part(c) says that multiplying an inequality by a negative number reverses the sense of the inequality. So, since $3 < 4$, $5(3) < 5(4)$, but $(-5)(3) > (-5)(4)$.

Proof. (a) Since $a < b$, there is a positive p such that

$$a + p = b. \tag{6.25}$$

Similarly, since $c < d$, there is a positive number q such that

$$c + q = d. \tag{6.26}$$

We now add equations (6.25) and (6.26) to get

$$a + c + (p + q) = b + d. \tag{6.27}$$

(Notice that adding equations is really a special case of adding the same quantity to both sides of an equation. In this case we are adding $c + q$ to the left of equation (6.25) and adding d to the right of equation (6.25) which by equation (6.26) are the same quantity.)

Now, since p and q are positive, so is $p + q$. So equation (6.27) shows that we have found a positive number, $p + q$, such that, when added to $a + c$, gives us $b + d$. So, by the definition of "less than," $a + c < b + d$.

Proof. (b) Again, we begin with equation (6.25) and multiply both sides by k to get

$$ka + kp = kb. \tag{6.28}$$

Since both k and p are positive, kp is positive, and equation (6.28) shows that we have found a positive number, kp, which when added to ka gives kb. So $ka < kb$.

Proof. (c) We begin with equation (6.25) and multiply both sides by k, where k is a negative number, to get equation (6.28) and then realize that, since k is negative and p is positive, kp is negative. So $-kp$ is positive. We add the positive quantity $-kp$ to both side of equation (6.28) to get

$$ka = kb + (-kp). \tag{6.29}$$

Since we have added a positive quantity, $-kp$, to kb to get ka, it follows by definition of "less than" that $kb < ka$, which written in reverse tells us $ka > kb$. Thus, we needed to reverse our original inequality, $a < b$, when we multiplied by a negative number. ∎

Here are some typical secondary school problems:

Example 6.29 *Solve the inequality* $2x - 3 < 4x + 1$.

Solution. We subtract $4x$ from both sides and then add 3 to both sides to get $-2x < 4$. We divide by -2 to get $x > -2$. Notice the reversal of the inequality.

Example 6.30 *Find all values of x which make* $\dfrac{x-9}{x^2-9} > 0$.

Solution. This problem is more complex than the first, but we can simplify the problem if we use our prior knowledge of functions. Call the left hand side of the given inequality $f(x)$. Thus, our problem now becomes, "Find all values of x which make $f(x) > 0$." Another way of stating this is "Find the values of x where the graph of $f(x)$ is above the x-axis."

Suppose you have the graph of a function and you want to determine where the graph is above the x-axis, (that is, where $f(x) > 0$) and where the function is below the x-axis (that is, where the function is negative). The only way the graph of a function can go from positive to negative is to (a) pass through the x-axis or to (b) jump from above the x-axis to below the x-axis or vice versa. That is, (a) the function must take on the value 0, or (b) the function must have a discontinuity. Thus, to solve an inequality of the form $f(x) > 0$, we need to only look at the places where it crosses the x-axis, that is, where $f(x) = 0$ and where it is discontinuous. If we mark these points where $f(x) = 0$ or where $f(x)$ is discontinuous on a number line, this will divide the number line into subintervals. The sign of $f(x)$ cannot change sign in any such subinterval, though it can change sign from one subinterval to the next. Since $f(x)$ cannot change sign in any subinterval, we need only test the sign of $f(x)$ for one number in each subinterval and that will determine the sign of $f(x)$ in that subinterval.

That is the background. Now, using this approach let us solve the inequality above. We let $f(x) = \dfrac{x-9}{x^2-9}$. $f(x)$ will be 0 when the numerator is 0. That is, when $x - 9 = 0$ or when $x = 9$. $f(x)$ will be discontinuous when the denominator is 0. That is, when $x^2 - 9 = 0$, which is when $x = \pm 3$. We mark off the numbers ± 3 and 9 on the number line and then check the sign of $f(x)$ in the subintervals. On the interval from $-\infty$ to -3, we get that $f(x)$ is negative. (We need only compute $f(x)$ at one number in the interval. We can, if we like, compute $f(-4)$ and we will see that we also get a negative number.) Similarly, from $x = -3$ to 3 we see that $f(x)$ is positive (compute $f(1)$, for example) and after $x = 3$ but before $x = 9$, $f(x) < 0$ (e.g. compute $f(4)$). After $x = 9$, the function is positive again (e.g. compute $f(10)$). So, the solution to our problem is "$\dfrac{x-9}{x^2-9} > 0$, when $-3 < x < 3$ or when $x > 9$".

Now let us see what happens when we graph the function using the software used to write this book (Figure 6.7).

256 Building the Real Number System

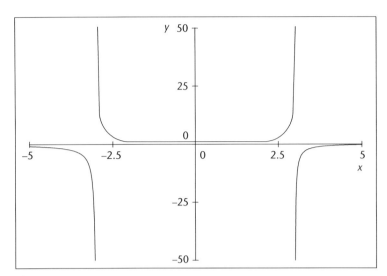

Figure 6.7

Notice that it is not clear from the graph that the function has discontinuities at $x = -3$ and at $x = 3$. Had we not done the algebraic analysis, we could never be sure about what happened at these points. Note also that our picture does not tell us what happens for $x > 9$. Is the graph above the x-axis? Even if we redraw the picture in an interval containing 9, we still can't see fully what is happening as Figure 6.8 shows:

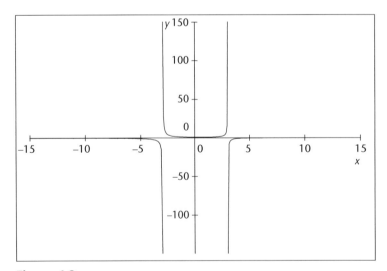

Figure 6.8

That is because the values of $f(x)$ after 9 are small. Of course, we can zoom in at 9 and get a better idea of what is happening there. But, without the algebraic analysis, we wouldn't even know that we should examine the function at $x = 9$. Furthermore, what happens at $x = 500$? Will this picture tell us? Maybe the graph crosses the x-axis at several other times and we just don't know it. The algebraic analysis tells us there are no other crossings.

Here is the picture of the graph of $f(x)$ for x between 8 and 100, just to give credence to the fact that $f(x)$ is indeed positive after $x = 9$ (Figure 6.9). Notice the small numbers on the y-axis.

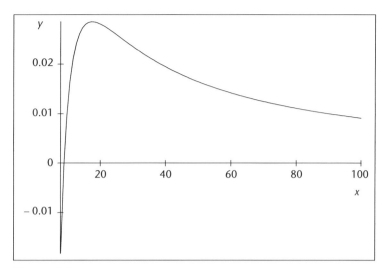

Figure 6.9

We hope that this discussion has made it clear to you why, despite the power of machine graphing technologies, we still need to be able to do the algebraic analysis!

Example 6.31 *Solve for x:* $\dfrac{1}{x} < 3$.

Solution. A natural, but incorrect way to approach this problem that most students use is to multiply both sides of the inequality by x to get $1 < 3x$ and then divide by 3 to get $x > \frac{1}{3}$. Since, if x is negative, the original inequality holds, through this approach we have lost infinitely many solutions. What has not been considered is the fact that x could be negative. When x is negative and you multiply both sides by x, you *reverse* the inequality.

Therefore, this problem really has two cases, Case 1 or Case 2.

Case 1. $x > 0$. You multiply both sides by x as we did above and you get $x > \frac{1}{3}$.

Case 2. $x < 0$. Now you multiply both sides by x and you get $x < \frac{1}{3}$. That is, the only negative x's that work are those less than $\frac{1}{3}$. However, all negative x's are less than $\frac{1}{3}$. So all negative x's work in Case 2.

Our final solution requires the joining of the two cases. Since Case 1 or Case 2 can hold, our answer is $x > \frac{1}{3}$ or $x < 0$.

If we were to graph this on a number line our graph would look as follows (Figure 6.10):

Figure 6.10

An easier approach, that doesn't involve cases, would be to rewrite our original inequality as $\dfrac{1}{x} - 3 < 0$, and then combine the fractions on the right to get $\dfrac{1-3x}{x} < 0$. If we call $f(x) = \dfrac{1-3x}{x}$, then $f(x) = 0$ when $x = 1/3$, and $f(x)$ is discontinuous when $x = 0$. We mark them both off on a number line and then check the sign of $f(x)$ in the various subintervals to again get the picture above.

Let us briefly discuss inequalities with absolute value. We know that $|N| = 3$ happens when $N = \pm 3$. When will $|N|$ be < 3? When $-3 < N < 3$. Of course, $|N|$ will be > 3 when $N > 3$ or when $N < -3$. These principles are used to solve absolute value inequalities, and were used in certain parts of calculus, for example, in finding the interval of convergence for power series.

Example 6.32 *Solve $|1 - 3x| > 3$.*

Solution. We can think of $1 - 3x$ as N. Our inequality becomes $|N| > 3$ which means that $N > 3$ or $N < -3$. Using the value of N, this yields the inequalities, $1 - 3x > 3$ or $1 - 3x < -3$. We subtract one from both sides of each inequality, then divide by -3 and make sure we remember to flip the inequality. We get as our solution that $x < -\frac{2}{3}$ or $x > \frac{4}{3}$.

Student Learning Opportunities

1 Show by example that, if $a < b$ and $c < d$, it does NOT follow that $a - c < b - d$.

2 Prove that, if $a < b$ and $b < c$, then $a < c$.

3 **(C)** A student is convinced that if $a < b$, then $a^2 < b^2$. Is the student correct? How would you convince the student of the correct answer to this question?

4 Prove using Theorem 6.28 that, if $0 < a < b$, then $a^2 < b^2$.

5 **(C)** A student wants to know whether it is true that, if $a < b$, then $\frac{1}{a} > \frac{1}{b}$. Respond to the student by giving some examples and then give a proof that, if $a, b > 0$, it is true. Is it true if a and b are < 0?

6 **(C)** One of your students has solved the following inequality as given below and has recognized that, when he picks a point in the solution set, $x = 3.5$, it doesn't work. What happened? How can you help your student realize where the error is and solve it correctly?

$$\frac{3x - 9}{x - 4} > 0$$

$3x - 9 > 0$ (Multiplying both sides by $x - 4$.)

$3x > 9$

$x > 3$

7 Solve each of the following inequalities.

(a) $\dfrac{4x + 16}{x - 1} \leq 0$

(b) $\dfrac{2x + 1}{x - 3} > 2$

(c) $\dfrac{5}{x - 3} < 4$

8 (C) One of your students has solved the following absolute value inequality as given below and has concluded that there is no solution since, $x > 5$ and $x < -1$, and there are no such numbers that satisfy both inequalities. Is the student right? If not, how can you correct the student's work?

$$|4 - 2x| < 6$$

$$-6 < 4 - 2x < 6$$

$$-10 < -2x < 2$$

$$5 < x < -1$$

9 Solve each of the following inequalities involving absolute values.

(a) $|4 - 3x| < 6$

(b) $|8 + 2x| \geq 7$

(c) $\left|\dfrac{3x - 8}{x - 1}\right| > 0$

(d) $|x - 3| < 0$.

10 (C) You asked your students to resolve the issue we presented in the launch question and to decide which of the two students, A or B, was right. Most felt that B was right. If that is true, what did A do that was wrong?

6.10 Logarithms

LAUNCH

Your friend Tilly the Trickster asked you to help her with her homework and compute $x = \log_3(-27)$. Is there a solution? Why or why not?

Students of mathematics typically find the topic of logarithms quite confusing. It involves learning new notation, new language, and many new rules. Beyond that, there are quite a few restrictions that you must be aware of, as you can see exemplified in the launch question. This section should serve as a good review of the basics of logarithms, the related rules, and interesting applications.

Before the age of calculators, logarithms were used in the sciences to simplify some of the difficult computations that were a regular part of scientific work. Since the age of calculators, logarithms are no longer used for this purpose. However, there still is a very important use for logarithms and that is to solve equations like $2^{5x} = 3$ where the variable occurs in the exponent. This is especially true when the right and left hand sides of the equation cannot be expressed in terms of a common base.

The word **logarithm** is synonymous with exponent. Let us explain. In secondary school we say that the logarithm of N to the base a is x and write

$$\log_a N = x \quad (a > 0, a \neq 1) \tag{6.30}$$

if and only if

$$a^x = N. \tag{6.31}$$

Thus, $\log_2 8 = 3$ since $2^3 = 8$, and $\log_4 \frac{1}{4} = -1$ since $4^{-1} = \frac{1}{4}$. If we look at equation (6.30), we see that x is the logarithm. If we look at (6.31), we see x is the exponent. Thus, the logarithm and exponent are really the same.

It takes some time to get used to the switching between equations (6.30) and (6.31) but, once you have it, it is easy. A way of thinking of the logarithm in words is, "the logarithm of N to the base a is the *exponent* to which we must raise a to get N." Thus, since the exponent to which we must raise 2, to get 8 is 3, the logarithm of 8 to the base 2 is 3.

One thing you should take strong note of is that we cannot take the logarithm of a negative number. For, if we were asked to compute $x = \log_2(-3)$, we would be asking for a real number x such that $2^x = -3$. But, 2 raised to any real power is positive.

The two most important logarithms are the common logarithm, which is the logarithm to the base 10, and the natural logarithm, which is the logarithm to the base e. The common logarithm is abbreviated log, while the natural logarithm is abbreviated ln. On your calculator, you will see both buttons.

Let us practice a bit with some typical secondary school problems.

Example 6.33 (a) *Solve the equation* $\log_4(3x + 2) = 1$. (b) *Solve* $10^{4x} = 7$.

Solution. (a) This is in the form of equation (6.30). We put the equation in exponential form, and the resulting equation is $3x + 2 = 4^1$. Thus, $x = 2/3$ and if we check it, we see it works.

(b) We write the equation as $\log_{10} 7 = 4x$ and therefore x is $\frac{\log_{10} 7}{4}$. Now on your calculator, you see a button labeled "log". That button represents the $\log_{10}$. You press the log button followed by 7. Then divide the answer by 4 and you find x, which in this case is approximately 0.2112.

Natural logarithms are particularly useful in equations involving e. For example, if we had to solve $e^x = 5$, we could write it in logarithmic form and get that $x = \log_e 5$ or just ln 5.

People at first cannot understand why the natural logarithm, which to many seems unnatural, plays such a big role in mathematics. It is quite remarkable that it does. In fact, it occurs in many equations describing behaviors of natural processes like radioactive decay, bacterial growth, population growth, electrical circuitry, and so on, which is probably the reason for the word "natural" in the expression "natural logarithm."

Anything that we can do with the natural logarithm, we can do with the common logarithm, but the natural logarithm offers us a simplicity that is preferable. There are many reasons for this, not the least of which is that the natural logarithm function has a much simpler derivative than the common logarithm ($\log_{10}$) function. Since the derivative measures a rate of change, which is an important concept in applications, the natural logarithm, having the simpler derivative, is often preferable.

The natural logarithm function and the related function e^x just seem to occur everywhere in applications. They are wonderful functions that do a great deal for us. Let us stop for a bit and give some real applications of these.

Newton's Law of Cooling states that, if a body with initial temperature T_0 is put in a room with surrounding temperature S_0, then t hours after it is placed in the room, its temperature will be given by the formula

$$T = S_0 + (T_0 - S_0)e^{kt} \qquad (6.32)$$

where k is a constant. We said "t hours" but t can represent any unit of time.

Newton's law is derived using calculus and is based on observations that physicists have made. The law has been verified over and over again experimentally. It is an excellent model of reality. Thus, when a cup of hot coffee is brought into a colder room, its temperature starts to decrease according to Newton's Law of Cooling until it gets to room temperature. The same thing happens when a person dies. His body temperature decreases as time passes according to the above law. This law is used in determining the time of death of a person whose body is found. The body's temperature is taken at two different times and that determines the constant k in equation (6.32) for this body. The approximate time of death is then readily obtained as illustrated in the problem below.

> **Example 6.34** *George Smith arrives at work in the morning to find his boss I. M. Meany, draped across his desk and very dead. George calls the police who arrive and measure the body's temperature at 8 am. to be $76\,°F$. At 9 am. they repeat the measurement and find the body's temperature is $73\,°F$. They observe that the thermostat in the room is at 70 degrees. They also see a note on the desk that says, "Fire that jerk, Smith." Naturally, Smith is the prime suspect and needs an alibi. For which times must he have a good alibi?*

Solution. Smith needs to find an alibi for a time period surrounding the time of death, say between 1 hour before and 1 hour after. So, we need to determine the time of death. Newton's Law is valid starting at any time we wish to start thinking about the cooling process. Thus. we can let 8 am represent $t = 0$. Therefore, T_0, the body's initial temperature at this time, is $76\,°F$, while S_0, the surrounding room temperature is $S_0 = 70\,°F$ (the temperature the thermostat was set at). By Newton's Law of Cooling, the body's temperature at any time t after 8 am (as well as before) is given by

$$T = 70 + (76 - 70)e^{kt} \quad \text{or just} \quad T = 70 + 6e^{kt}. \qquad (6.33)$$

We know that at 9 am ($t = 1$), the body's temperature is $73\,°F$. Using this information in equation (6.33) we get

$$73 = 70 + 6e^{k(1)}. \qquad (6.34)$$

We subtract 70 from both sides of the equation, divide by 6, to get the equation

$$\frac{1}{2} = e^k$$

and then, writing this in logarithmic form, we get $k = \ln(1/2) = -0.6931$. We substitute this value of k in equation (6.33) to get the body's temperature at any time t. Thus,

$$T = 70 + 6e^{-0.6931t}. \tag{6.35}$$

We are now ready to finish. At the time of death, the body's temperature was normal body temperature, or $98.6\,°F$. We use this in equation (6.35) to get

$$98.6 = 70 + 6e^{-0.6931t}. \tag{6.36}$$

We subtract 70 from both sides of equation (6.36) and divide by 6, to get

$$4.766666666 = e^{-0.6931t}.$$

Writing this in logarithmic form we get $\ln(4.766666666) = -0.6931t$ and dividing by -0.6391 we get $t = -2.2$. That is, Mr. Meany died about 2.2 hours before time $t = 0$ which was 8 am. So Mr. Meany died a bit before 6 am. But, at 6 am. "the jerk," Smith, was home having breakfast with his wife and kids. Furthermore, his mother-in-law, his father-in-law, and his neighbor were all eating with him. So Smith was probably safe. He had many witnesses and the perfect alibi.

The last problem may have seemed a bit facetious, but in fact is very real and is used by coroners on a daily basis. (They actually take the temperature at two different times and then use a formula derived from Newton's Law of Cooling to determine approximate time of death.)

Here is another real example from archaeology: Again, this is real.

> **Example 6.35** *In the 1300s, a shroud known as the shroud of Turin, was found and it was claimed to be the original burial shroud of Jesus Christ. The images on this shroud were so compelling that people had no doubt of its authenticity and considered it sacred. Then, in 1389, the bishop of Troyes, Pierre d' Arcis, wrote a memo to the pope claiming the shroud was a forgery, "cunningly painted" by one of his colleagues.*
>
> *In 1988, the shroud was subjected to carbon dating. Carbon dating is based on the fact that all things have a certain amount of radioactive carbon 14 in them, and that it decays according to the formula,*
>
> $$N = N_0 e^{kt}$$
>
> *where N is the amount of carbon 14 currently in the shroud, and N_0 is the initial amount of radioactive carbon 14. t is the time elapsed since we begin the measurement of the decay. To date the shroud, we need to take $t = 0$ to be the time it was painted.*
>
> *For carbon 14, it is known that the constant k is -0.000121 when t is measured in years. Thus, the amount of carbon 14 is*
>
> $$N = N_0 e^{-0.000121t}. \tag{6.37}$$
>
> *Now, if the shroud were real, the age of the cloth should have been about 1960 years old at the time of the dating. Scientists, using well known methods in the science community, ascertained that 92.3% of the original amount of carbon 14 remained. (a) Based on this, was the Shroud in fact a fake? (b) Approximately how old was the shroud?*

Solution. It is the real nature of this problem that makes it so interesting. If, in fact, the shroud were 1960 years old, then using this information in equation (6.37), we get that the amount of carbon 14 in 1988 should have been measured by

$$N = N_0 e^{-0.000121(1960)} \approx 0.789 N_0$$

which tells us that 78.9% of the initial amount, N_0, of carbon 14 would remain. Since 92.3% remained, we know that the shroud cannot be real.

(b) Given that 92.3% of N_0, the original carbon 14 remained, we can use this in equation (6.37) and we get

$$0.923 N_0 = N_0 e^{-0.000121t}.$$

We divide by N_0 and get

$$0.923 = e^{-0.000121t}$$

and then solve the usual way by writing this as a ln statement giving us $\ln 0.923 = -0.00121t$ and hence $t = \dfrac{\ln 0.923}{-0.000121} = 662.20$. So the shroud was about 662 years old, placing it in the 1300s and corroborating the bishop's story.

Isn't this the neatest application?

6.10.1 Rules for Logarithms

There are four basic rules for logarithms. All require that M and N be positive (Why?)

Rule (L1): $\log_c MN = \log_c M + \log_c N$

Rule (L2): $\log_c \dfrac{M}{N} = \log_c M - \log_c N$

Rule (L3): $\log_c M^p = p \log_c M$

Rule (L4): $\log_b a = \dfrac{\log_c a}{\log_c b}$

What does the first one mean? Recall we said that the word "logarithm" meant exponent. If you think of the word logarithm as exponent, Rule (L1) is saying that, when you multiply numbers with the same base, you add the exponents. Similarly, the second statement is saying that, when you divide numbers expressed with a common base, you subtract the exponents. Thus, these strange looking statements are telling us what we already know, but in a different format.

To give you a feel for the rules before providing the proofs, we will illustrate them with some numerical examples. Let M be 100 and N be 1000. Now $\log M = 2$ (the exponent to which 10 must be raised to get M is 2) and $\log N = 3$ (the exponent to which 10 must be raised to get N is 3). Also $MN = 10^5$, $\log MN = 5$ (the exponent to which we must raise 10 to, to get MN is 5). So we see that it is true that $\log MN = \log M + \log N$.

Using the same numbers as in the previous paragraph, we have $\dfrac{M}{N} = \dfrac{1}{10}$ and $\log \dfrac{M}{N} = -1$, which we see is the same as $\log M - \log N$.

The third rule tells us that exponents can be pulled out of logarithms. Thus, $\log 2^3$ is the same as $3 \log 2$, which you can check on the calculator by computing $\log 8$ and $3 \log 2$.

The fourth rule gives us a mechanism by which we can find the logarithm of any number a, to any base, b, by dividing the logarithm of a by the logarithm of b. The base that we are using for the particular conversion is irrelevant. Thus, if we use the common log, we have that $\log_3 2 = \dfrac{\log 2}{\log 3}$ and this, in turn, is equal to $\dfrac{\ln 2}{\ln 3}$ (by taking the base of the logarithm to be e).

We now give the proofs of these rules. The proofs amount to nothing more than switching between equations (6.30) and (6.31).

Proof of Rule (L1). Call $\log_c M = x$ and $\log_c N = y$. Then, from the definition of logarithm,

$$c^x = M \quad \text{and} \quad c^y = N.$$

If we multiply these two equations, we get

$$c^x c^y = MN$$

which reduces to

$$c^{x+y} = MN.$$

If we write this last statement in logarithmic form, we get

$$\log_c MN = x + y.$$

But $x = \log_c M$, and $y = \log_c N$. If we substitute these expressions in the above equation, we get

$$\log_c MN = \log_c M + \log_c N.$$

Of course, you can convince yourself and your students of this rule by doing a few numerical examples. For example, using the calculator, you can easily verify that $\log 6 = \log 2 + \log 3$. Other examples will show you the same.

Proof of Rule (L2). We leave this for you. The proof is very similar to the proof of (L1) and is instructive to do.

Proof of Rule (L3). Call $\log_c M = x$. Then, by definition of logarithm,

$$c^x = M.$$

If we raise both sides to the p power and use the laws for exponents, we get

$$c^{px} = M^p.$$

If we write this in logarithmic form, we get

$$\log_c M^p = px.$$

But, since $x = \log_c M$, this last statement becomes $\log_c M^p = p \log_c M$, which is what we wanted to prove.

Proof of Rule (L4). Call $\log_b a = x$. Then

$b^x = a$.

Take the logarithm of both side of this equation to the base c to get.

$\log_c b^x = \log_c a$.

Now use rule (L3) to pull out the exponent, x, and we get

$x \log_c b = \log_c a$.

Hence

$$x = \frac{\log_c a}{\log_c b}.$$

But $x = \log_b a$. Thus,

$$\log_b a = \frac{\log_c a}{\log_c b}.$$

Let us give some examples to practice these rules. These are typical secondary school problems.

Example 6.36 *Solve* $\log_2 x - \log_2(x-1) = 3$.

Solution. Using rule (L2) we have that

$\log_2 x - \log_2(x-1) = 3$ implies that

$\log_2 \frac{x}{(x-1)} = 3$ which implies that

$2^3 = \frac{x}{x-1}$ and upon multiplying both sides by $x - 1$ we have

$8x - 8 = x$. Hence,

$x = \frac{8}{7}$.

We can check that $x = \frac{8}{7}$ works.

Student Learning Opportunities

1 Without using a calculator, compute each of the following logarithms. Afterwards, check your answers with the calculator.

(a) $\log_4 16$

(b) $\log_6 \sqrt[3]{6}$

(c) $\log_4 \frac{1}{16}$

(d) $\log_{1/2} 8$

(e) $(\log_4 3)(\log_3 4)$

(f) $(\log_3 7)(\log_7 9)$

2 (C) A student wants to know why we require that $a \neq 1$ in in the definition of logarithm. How would you respond?

3 Change each of the following statements to an equivalent statement in logarithmic form:

(a) $4^3 = 64$

(b) $3^{-2} = \dfrac{1}{9}$

(c) $10^3 = 1000$

(d) $e^{4x} = 5$

4 Prove rule (L2) for logarithms.

5 (C) If a student asked why you can't take the logarithm of a negative number and get a real number, what would you say?

6 Prove that, if $a < 0$ and m is even, then $\log a^m = m \log |a|$.

7 (C) A student makes the following series of statements. "Given $\log x^4 = 4 \log 3$. It follows that $4 \log x = 4 \log 3$, hence $x = 3$." Since we know that there is another solution, $x = -3$, where did it go? Where is the error in the student's solution and how would you solve it correctly?

8 (C) Students often say that $\dfrac{\log_c a}{\log_c b} = \log_c a - \log_c b$. They are of course, thinking of rule (L2). But rule (L2) deals with a single logarithm of a fraction, not with the quotient of logarithms. How would you convince a student that the misconception we pointed out is, in fact, a misconception?

9 If $\log_b a = \log_a b$ where $a \neq b$, $ab > 0$ and neither a nor b are 1, then what is the value of ab?

10 Solve for x:

(a) $\log_3(x^2 - 7) = 2$

(b) $\log_6(\log_5 x) = -2$

(c) $\log_3(x + 2) + \log_3(5) = 4$

(d) $\log_4(2x + 1) - \log_4(x - 2) = 1$

(e) $\log_2 6 = \log_2(x^2 + 8) - \log_2 x$

(f) $6^{x+1} - 6^{-x} = 5$

(g) $x^x = \pi$

11 Suppose that we have a function $f(x)$ such that $f(ab) = f(a) + f(b)$ for all rational numbers a and b.

(a) Show that $f(1) = 0$.

(b) Show that $f(\dfrac{1}{a}) = -f(a)$.

(c) Show that $f(\frac{a}{b}) = f(a) - f(b)$.

(d) Show that $f(a^n) = nf(a)$ for every positive integer n.

12 Prove each of the following:

(a) $\dfrac{1}{\log_a ab} + \dfrac{1}{\log_b ab} = 1$

(b) $a^{\log b} = b^{\log a}$ where log means $\log_{10}$

13 When a beam of light enters an ocean vertically, its intensity decreases according to the formula $I = I_0 e^{-0.0101d}$ where I_0 is its initial intensity and d is the depth in centimeters the light has penetrated. How far below the ocean's surface will the intensity of a beam of light be reduced to 2% of its initial intensity?

14 A painting supposedly done by Rembrandt in 1640 was found in the 1960s and was dated using carbon dating, and found to contain 99.7% of its original carbon 14. How old (approximately) was the painting in 1960?

15 The energy, E, released by an earthquake is measured in units called joules. The intensity of all earthquakes are measured according to a standard called E_0, which is $10^{4.4}$ joules of energy. The measure of an earthquake's strength is measured by the Richter scale, and the formula that measures the Richter score for an earthquake is

$$R = \frac{2}{3} \log_{10}\left(\frac{E}{E_0}\right).$$

The great San Francisco earthquake of 1906 measured $R = 8.25$ on the Richter scale. How many joules of energy were released and approximately how many times as much energy as E_0 was released?

6.11 Solving Equations

LAUNCH

You ask Maria, one of your students, to solve the equation $(x+1)(x+3)(x+5) = (x+1)(x+3)$. The student divides both sides of the equation by $(x+1)(x+3)$ to get $(x+5) = 1$. Solving this equation, she gets $x = -4$. Has she solved the equation correctly?

Another student, Matt, is asked to solve the equation $\sqrt{x} = -7$ and squares both sides to get $x = 49$. Is Matt correct when he asserts that this is the answer?

Both Maria and Matt are wrong. When solving equations, there are a few issues that you must watch out for so that you don't get faulty, inadequate, or misleading results. That is the focus of this section.

6.11.1 Some Issues

Having discussed the development of the real number system and the solutions of equations within it, we now turn to problems that can occur in the solution process.

When asked to solve an equation like $4x - 1 = 2x + 3$, the process is simple. We subtract $2x$ from both sides, add one to both sides, and get $2x = 4$. We then divide by 2 to get $x = 2$. We check it and it works, and so we are done.

In general, when solving equations, we can add the same quantity to both sides, or subtract the same quantity from both sides, or multiply or divide both sides by the same quantity. We can also do other things when solving equations: square both sides, cube both sides, take the square root or cube root of both sides, take the logarithm of both sides, take the sine of both sides, and so on. We do lots of different things when solving equations. But, if we are not cautious when using these processes, many strange things can happen. For example, we can get answers that don't work. We can lose answers that do work. We can miss answers that are in front of our eyes, and so on. Let us begin by illustrating exactly what we mean.

We begin with several examples. The following illustrate some of the more common errors teachers see.

Example 6.37 *Jason solves the equation $x^2 = 3x$ by dividing both sides by x to get $x = 3$. He has lost the solution $x = 0$. What did he do wrong?*

Example 6.38 *Chan has the equation $\sqrt{x} = -5$ and tries to solve it by squaring both sides. He gets $x = 25$. Yet, when he checks the solution, he realizes it doesn't work, since the square root of 25, is positive. He concludes something is wrong.*

Example 6.39 *Juan solves the equation $x^2 = 9$ by taking the logarithm of both sides. He gets $\log x^2 = \log 9$, and then rewrites this as $\log x^2 = \log 3^2$. He remembers that, with logarithms, you can pull the exponent out, so he gets*

$$2 \log x = 2 \log 3. \tag{6.38}$$

He divides by 2 to get $\log x = \log 3$, and then concludes that $x = 3$. Yet, he missed the solution $x = -3$. Where did it go?

Example 6.40 *Indira solves the equation $(x + 4)(x - 3) = 8$ by setting $x + 4 = 8$ and $x - 3 = 1$, thereby getting the solution $x = 4$ from both equations. She checks her answer by substituting $x = 4$ into the original equation and finds that it works. She concludes that this quadratic equation has only one solution, $x = 4$. But, if we check $x = -5$, it also works. She lost a solution. What went wrong?*

Examples like these show us that we need to exercise a great deal of care when solving equations. Let us examine the solution process more carefully.

6.11.2 Logic Behind Solving Equations

What is it that we are really doing when we solve an equation, and why do we sometimes lose solutions or find solutions that don't work?

To understand this better, we need to examine the logic behind solving an equation. Examining a specific example will help us to illustrate this. Suppose we wish to solve the equation

$$\sqrt{x} = 2x - 1. \tag{6.39}$$

Recall that $\sqrt{x}$ means the positive square root of x and therefore $\sqrt{9}$ means 3, not ± 3.
We begin by squaring both sides of equation (6.39) to get

$$x = 4x^2 - 4x + 1. \tag{6.40}$$

This process makes use of a fundamental fact regarding equations which states that: If $a = b$ then $a^2 = b^2$ (or in words, if two quantities are equal then so are their squares). Next, we bring all the terms over to one side, to get

$$4x^2 - 5x + 1 = 0. \tag{6.41}$$

Here, we are using the fact that we can add and/or subtract the same quantity from both sides of an equation and get a valid equation. In particular, when we bring all the terms over to one side, we are subtracting the quantity, $4x^2 - 4x + 1$ from both sides of the equation, to get equation (6.41).

Finally, we factor equation (6.41) to get $(x - 1)(4x - 1) = 0$. We set each factor equal to zero and get

$$x = 1, \quad \text{and} \quad x = \frac{1}{4}. \tag{6.42}$$

(Here we are using a fact that we proved earlier that, if the product of two numbers is zero, then one or the other or both must be 0.)

This all seems pretty straight forward. Every novice in algebra believes that, by using the above approach, he or she has solved equation (6.39). But, if we actually check our answers from equation (6.42) in equation (6.39), only the solution $x = 1$ works. How could this be?

Let us take a closer look at what we have really done when solving this equation, and then in general when solving any equation.

What we are really saying when we go from equation (6.39) to equation (6.40) is that IF equation (6.39) is true for a particular value of x, THEN by squaring, so is equation (6.40) true for that value of x. That is, any solution of equation (6.39) is a solution of equation (6.40). In terms of sets, we are saying that *the solution set of equation* (6.39) *is a subset of the solution set of equation* (6.40). We are NOT saying that the solution sets of equations (6.39) and (6.40) are the same.

Let us continue. When we go from equation (6.40) to equation (6.41) by subtracting $4x^2 - 4x + 1$ from both sides, (a perfectly legitimate operation), we are again saying that IF equation (6.40) is true for a specific x, THEN so is equation (6.41). That is, every solution of equation (6.40) is a solution of equation (6.41), or put another way, the solution set of equation (6.40) is a subset of the solution set of equation (6.41).

Since the solution set of equation (6.39) is a subset of the solution set of equation (6.40), and the solution set of equation (6.40) is a subset of the solution set of equation (6.41), we have that the solution set of equation (6.39) is a subset of the solution set of equation (6.41).

Following the same reasoning as before, when we go from equation (6.41) to equation (6.42), we are saying that the solution set of equation (6.41) is a subset of the solution set of (6.42). Since the solution set of equation (6.39) is a subset of the solution set of equation (6.40) and that, in turn, is a subset of the solution set of equation (6.41) which in turn is a subset of equation (6.42), we have that the solution set of equation (6.39) is a subset of the solution set of equation (6.42). Since the solution set of equation (6.39) is only a subset of the solutions of equation (6.42), there may be solutions of equation (6.42) that don't work in equation (6.39). Thus, we must check the answers we got to see that they work in equation (6.39). Only $x = 1$ does.

To recap, when we solve an equation, if we perform legal steps, we are finding a set *containing* the solutions of this equation. This set will, in general, be larger than the set containing the solutions. The answers must be checked in the original equation to see that they work. Those that don't, are called **extraneous solutions**. In the previous example, the solution $x = \frac{1}{4}$ was extraneous.

Notice the word "legal" in the first sentence of the last paragraph. What are legal operations? That is, what are operations that we can perform on an equation that will guarantee that we generate an equation whose solutions contain the original set?

Well, we have already seen some.

(L1) If two quantities are equal, we can add to or subtract the same quantity from each of them, and the results will still be equal. (That is, if $a = b$, then $a + c = b + c$ and $a - c = b - c$.)

(L2) If two quantities are equal, we can multiply or divide each of them by the same quantity and the results will still be equal provided that when you divide, you don't divide by zero. (That is, if $a = b$, then $ac = bc$ and, if $c \neq 0$, then $\frac{a}{c} = \frac{b}{c}$.)

(L3) We can raise both sides of an equation to a positive integer power and we will get an equality. (That is, if $a = b$, then $a^n = b^n$ for positive integers n.)

As simple as these rules seem, we still need to exercise care when using them.

Example 6.41 *Debbie solves the equation*

$$\frac{x-1}{x+1} = \frac{x-1}{x+3} \tag{6.43}$$

by dividing both sides by $x - 1$ and gets the equation

$$\frac{1}{x+1} = \frac{1}{x+3} \tag{6.44}$$

from which she concludes by cross multiplying that

$$x + 1 = x + 3.$$

She then subtracts x from both sides of this last equation and gets the contradiction that $1 = 3$. She says, "This is impossible," and from this she concludes there is no solution to the original equation. Yet, the original equation has the solution $x = 1$. Where did it go?

Solution. To see what is wrong here, we need only recall that there are restrictions on division. Specifically, we cannot divide by zero. But Debbie divided by $x - 1$, which can be zero. And that happens when $x = 1$. Thus, when $x = 1$, she performed an illegal operation. This means that the resulting equation may not contain the solutions of our original equation. Indeed, that is the case here. Whenever you divide an equation by an expression that may be zero, you need to check the values of x that make the divisor 0 in the original equation. Thus, Debbie needed to check if $x = 1$ satisfied the original equation. If she checked, she would have seen it did, and would not have lost it.

Very often students make this kind of mistake and this, in fact, is the error with Example 6.37 above. On the other hand, if we have the equation $x(x^2 + 1) = 9(x^2 + 1)$, and we are interested in real solutions, we can divide both sides by $x^2 + 1$ without fear, since for real numbers x, $x^2 + 1$ is never 0. We will not lose real solutions, in the process. (But we will lose the complex solutions, $x = \pm i$.) Our point is, don't divide both sides of an equation by a variable quantity in an effort to simplify it unless you are sure that the variable quantity is not zero. If the expression you are dividing by can be zero, then you need to check the values of x that make it zero in the original equation to see if they work.

So, to summarize, here is the fix to Debbie's solution: When she divides equation (6.43) by $x - 1$ to get equation (6.44), she has to put herself on the alert that $x - 1$ can be 0 when $x = 1$. Thus, her division is illegal when $x = 1$. She needs to check if this value $x = 1$ works in the original equation. This way, if it does, she will recover the lost solution.

Example 6.42 *Solve the equation $x(x - 3)(x - 4) = (x - 3)(x - 4)$.*

Solution. The tendency is to divide both sides by $(x - 3)(x - 4)$ to get $x = 1$. But now we are wiser. We know that $(x - 3)(x - 4)$ can be zero when $x = 3$ and when $x = 4$, and both of these values need to be checked in the original equation. Since both work in the original equation, our final solution is that $x = 1$, $x = 3$, and $x = 4$. All of them work.

Other examples of illegal operations are given in Examples, 6.39, and 6.40. Let us go back to Example 6.39 where Juan solved $x^2 = 9$ by taking the logarithm of both sides. Juan correctly concluded that

$$\log x^2 = \log 3^2. \tag{6.45}$$

His next step, however, that $2 \log x = 2 \log 3$, was incorrect. One cannot take the logarithm of a negative number. Thus, while $\log x^2$ is defined for all positive and negative x, $\log x$ is only defined for positive x. Thus, the statement $\log x^2 = 2 \log x$ is not correct when $x < 0$. The correct statement is $\log x^2 = 2 \log |x|$, regardless of what x is. Now, if we use this to replace $\log x^2$ in equation (6.45), we have

$2 \log |x| = 2 \log 3$

from which it follows that $\log |x| = \log 3$. From here, we have that $|x| = 3$ and therefore $x = \pm 3$ and we have not lost any solutions.

Another place where we may risk losing solutions is by using "identities" that are not really identities. Actually, we just did that when we said that $\log x^2 = 2 \log x$. Here is a more sophisticated illustration that involves trigonometry.

Example 6.43 *Solve the equation:*

$$\tan(x + 45) = 2 \cot x - 1. \tag{6.46}$$

Solution. If we recall the rules from secondary school trigonometry that $\tan(x+y) = \frac{\tan x + \tan y}{1 - \tan x \tan y}$, and that $\cot x = \frac{1}{\tan x}$, we can substitute these into equation (6.46) to obtain

$$\frac{\tan x + \tan 45}{1 - \tan x \tan 45} = \frac{2}{\tan x} - 1 \tag{6.47}$$

and since $\tan 45° = 1$, equation (6.47) becomes.

$$\frac{\tan x + 1}{1 - \tan x} = \frac{2}{\tan x} - 1. \tag{6.48}$$

We now multiply both sides of equation (6.48) by $\tan x(1 - \tan x)$ to get

$$\tan^2 x + \tan x = 2 - 2\tan x - (\tan x)(1 - \tan x)$$

which, upon multiplying out, and combining terms, and subtracting $\tan^2 x$ from both sides simplifies to

$$\tan x = 2 - 3\tan x.$$

Adding $3 \tan x$ to both sides and dividing by 4, we get $\tan x = \frac{1}{2}$, and therefore $x = (\tan^{-1}(1/2)) + k(180°)$ where $k = 0, \pm 1, \pm 2$, and so on.

Now, all these solutions work as you can verify with your calculator. But we are missing infinitely many solutions of our equation, namely all integer multiples of 90°. (Substitute, 90°, 180°, and 270° in equation (6.46) and use your calculator to convince yourself these work.) Where did these solutions go?

We know that we need to look for moves that might be illegal. The first place to look for a false identity is in equation (6.48). On the left side we have a denominator of $1 - \tan x$. What if $\tan x = 1$? Then the left side of equation (6.48) is not defined. What we are really saying is that the relationship from secondary school, $\tan(x+y) = \frac{\tan x + \tan y}{1 - \tan x \tan y}$ is not always valid. It is not valid when the denominator is 0.

Similarly, the "identity" we used, that $\cot x = \frac{1}{\tan x}$ is not valid when $\tan x = 0$. We must examine where the denominators of equation (6.48) are zero and check them separately to see if they work in equation (6.46). When $1 - \tan x = 0$, $x = 45° + k(90°)$. When k is odd, the left side of equation (6.46) will then involve taking the cotangent of an even multiple of 90 degrees, and that would mean $\cot x$ would not be defined, causing the right side of equation (6.46) to be meaningless. So none of these solutions work. When k is even, the left side of equation (6.46) involves an odd multiple of 90°, making the tangent undefined. So none of these values work either. The bottom line is that none of the solutions of $1 - \tan x = 0$ are solutions of our original problem.

But there was the other identity we used: $\cot x = \frac{1}{\tan x}$, and the right side is not defined when $\tan x = 0$ or is undefined. That happens when x is an integer multiple of 90°. Every one of these does work since, the left side of equation (6.46) evaluates to -1 and so does the right side. (Again, you can check with your calculator.) So we have recovered our infinitely many lost solutions.

6.11.3 Equivalent Equations

Do we *always* have to check our solutions to see if they work? The answer is, "No." If we can reverse the steps that we used to go from an equation (*A*) to an equation (*B*), by going from (*B*) back to (*A*), then the solutions to equations (A) and (B) will be the same. Why, you ask? The answer is simple. When we go from (A) to (B) using legitimate operations, the solutions of (A) are a subset of the solutions of (B). When we go from (B) to (A) using legitimate operations, the solutions of (B) are a subset of the solutions of (A). Since the solutions of (A) are a subset of (B) and vice versa, these equations have the same solution set.

When two equations have the same solution set, we say they are **equivalent equations**. We have the following.

> **Theorem 6.44** *If the steps used in solving an equation are reversible, then the solution of the final equation and the solution of the original equation are the same. That is, the equations are equivalent. We need not check our answers, though we still should.*

Let us apply this to the solution of the equation

$$2x + 1 = 3. \tag{6.49}$$

We subtract 1 from both sides, to get

$$2x = 2. \tag{6.50}$$

By rule (L1) from the previous subsection, this is legal. Thus, by our previous work, the solution set of equation (6.49) is a subset of the solution set of equation (6.50).

Now, starting with equation (6.50), rule (L1) says we can add 1 to both sides of equation (6.50) to get equation (6.49). That is, we can reverse our steps to go from equation (6.50) back to equation (6.49). Thus, the solution set of equation (6.50) is a subset of the solution set of equation (6.49).

Since the solution set of equation (6.49) is a subset of the solution set of equation (6.50), and since the solution set of equation (6.50) is a subset of the solution set of equation (6.49), the solution sets of equation (6.49) and equation (6.50) are the same, and equation (6.49) and equation (6.50) are equivalent.

Let us continue. Starting with equation (6.50) we can divide both sides by 2 (a nonzero number) to get

$$x = 1. \tag{6.51}$$

Thus, the solution set of equation (6.50) is a subset of the solution set of equation (6.51).

Since we can multiply equation (6.51) by 2 to get back to equation (6.50), that is, we can reverse the steps, it follows that the solution set of equation (6.51) is a subset of the solution set of equation (6.50) and thus equation (6.50) and equation (6.51) are equivalent. Since equation (6.49) and equation (6.50) are equivalent, equation (6.50) and equation (6.51) are equivalent, equation (6.49) and equation (6.51) are equivalent, and the solutions of equation (6.49) and equation (6.51) are the same. That is, the solution of equation (6.49) is $x = 1$. We don't have to check that it works.

Since the steps used in solving equations such as $3x + 1 = 5x - 3$ are all reversible, when solving first degree equations of this form, we don't really have to check our answers. The only reason we ask students to always check their answers is that they may have made a mistake in their computations. Also, we want to train them to check answers in other cases where answers really do need to be checked. Besides, it is never a bad idea for any of us to check our answers. We are all capable of making errors.

Our discussion leads to the following.

> **Theorem 6.45** *If we add or subtract the same quantity to or from both sides of an equation, we get an equivalent equation. If we multiply or divide an equation by the same NONZERO quantity, we get an equivalent equation. (Thus, for these cases we don't have to check our solutions.)*

We have seen that adding or subtracting the same quantity to, or from, both sides of an equation results in an equivalent equation as does dividing or multiplying both sides of an equation by a nonzero quantity. On the other hand, squaring both sides of an equation can yield non-equivalent equations, as we have seen in our first example of this section. So we must check our answers as Chan did in Example 6.38. Only, he concluded, mistakenly, that something was wrong. Nothing was wrong. His equation, $\sqrt{x} = -5$ had no solution.

Consider the next example which is similar.

> **Example 6.46** Solve $\dfrac{x}{x-3} = \dfrac{3}{x-3}$.

Solution. We multiply both sides by $x - 3$, and we immediately get $x = 3$. Must we check our answer? Sure! We multiplied by $x - 3$ and that step is not reversible, as $x - 3$ can be 0. We must check. Our solution doesn't work since the left and right sides of the equation both involve divisions by 0 when $x = 3$. Thus, this equation has no solution.

Before leaving this topic, we wish to focus more clearly on what happens when we apply a function to both sides of an equation, as we did in Example 6.42. There, we took the logarithm of both sides. The problem with applying a function to both sides of an equation is that functions often have restricted domains. So, if we start with a statement like $A = B$ and then apply the function f to both sides to get $f(A) = f(B)$, we will not run into a problem if A and B are unquestionably in the domain of f. But if they aren't, we may lose solutions. For example, the equation $(x - 1)(x - 2) = (x - 1)$ has two solutions, $x = 1$ and $x = 3$. Yet, if we apply the function $f(x) = \dfrac{1}{x}$ to both sides of the equation (that is, we take the reciprocal), we get $\dfrac{1}{(x-1)(x-2)} = \dfrac{1}{x-1}$ and this does not have two solutions, since $x = 1$ does not work in this latter equation. Since we failed to account for the fact that the function $f(x) = \dfrac{1}{x}$ has a restricted domain we lost a solution. However, if the function $f(x)$ that we apply to both sides has domain all x, or if we are sure that A and B are both in the domain of $f(x)$, then we need not worry about loss of solutions when we apply $f(x)$ to both sides since in both of these cases, if $A = B$, it does follow that $f(A) = f(B)$.

Thus, when we square both sides of an equation, we are applying the function $f(x) = x^2$ to both sides of the equation. This function has domain all x. So we will not lose any solutions (though we might gain something, which is why we have to check our answers). When we solve a linear equation, we add or subtract constant quantities from both sides, or multiply both sides by a constant, or divide both sides by a nonzero constant. That is, we apply the functions $f(x) = x \pm c$, $f(x) = cx$ or $f(x) = \dfrac{x}{c}$ where $c \neq 0$, to both sides of an equation. The functions have domain all x. So we don't lose any solutions when applying these functions. In fact, neither do we gain solutions because all of these functions are reversible. But, if we take the logarithm of both sides of an equation, we may lose a solution because the domain of the logarithm is restricted. If we take the reciprocal of both sides of an equation, we may lose something because the reciprocal function has a restricted domain. Our point is, if we apply a function with a restricted domain to both sides of an equation, we may lose solutions.

Why, you may ask, don't we seem to worry about this in secondary school? Well, consider a typical equation that students are asked to solve by taking the logarithm of both sides: $2^x = 5$ to get $x \log 2 = \log 5$. Despite the fact that the logarithm has a restricted domain, 2^x and 5 are always positive. So they are in the domain of the logarithm function, and we lose nothing by taking the logarithm of both sides.

Student Learning Opportunities

1 Are there any values of x for which $\dfrac{x}{x+1} + \dfrac{1}{x+1} \neq 1$? If so, what are they?

2 How many real solutions are there to the equation $\dfrac{2x^2 - 19x}{x^2 - 5x} = x - 3$?

3 Are there any complex solutions to the equation $\dfrac{2x^2 - 19x}{x^2 - 5x} = x - 3$? If so, what are they?

4 (C) What are equivalent equations? What are some of the operations that lead to equivalent equations?

5 (C) One of your students solves the following equation and can't figure out why he came up with a solution ($x = 1$) that does not work. How do you help your student understand why this happened? His work follows:

$\sqrt{5-x} = x - 3$

$5 - x = (x - 3)^2$ (Square both sides of the equation.)

$5 - x = x^2 - 6x + 9$

$x^2 - 5x + 4 = 0$

$(x - 4)(x - 1) = 0$

$x = 4, \quad x = 1$

6 Solve each of the following equations:
(a) $1 + \sqrt{2x + 5} = -x$
(b) $\sqrt{5x - 1} + \sqrt{x - 1} = 2$

7 One of your students solves the following equation and can't figure out why she only got two solutions rather than three like everyone else. She lost the solution $x = 3$. How do you explain to her what happened?

$x(x - 3) = x(x - 3)(x - 1)$

$\dfrac{x(x - 3)}{x - 3} = \dfrac{x(x - 3)(x - 1)}{x - 3}$

$$x = x(x-1)$$
$$x = x^2 - x$$
$$x^2 - 2x = 0$$
$$x(x-2) = 0$$
$$x = 0, x = 2$$

8 Solve each of the following equations:
   (a) $x^2(2x+1) - (2x-1)(2x+1) = 0$
   (b) $\dfrac{4x-1}{3x+2} = \dfrac{4x-1}{5x-6}$

9 (C) After being cautioned about all the pitfalls involved in solving equations, one of your students concludes that, if $\dfrac{a}{c} = \dfrac{b}{c}$ where $c \neq 0$, then $a = b$. Is the student correct? How can the student justify the answer?

10 (C) Your students are now very cautious when performing algebraic manipulations when solving equations. They are given the equation $\dfrac{c}{a} = \dfrac{c}{b}$ where neither of a and b is zero. They conclude that a must equal b since, if two fractions are equal and their numerators are equal, it must be that their denominators are equal. Is this correct? Justify your answer.

11 Solve for x:
   (a) $2 \log(x-3) = \log 4$
   (b) $\log(x-3)^2 = \log 4$

12 (C) How can you prove to a student that, if we begin with a quadratic equation, $ax^2 + bx + c = 0$, the solutions we get by the quadratic formula always work?

13 (C) Give an example that would demonstrate to a student that, if we begin with a linear equation and then manipulate it into a quadratic equation, say by squaring, the solutions of the quadratic might not work in the original equation.

14 Solve for x: $\dfrac{8}{x^2-x} + \dfrac{8}{x} = \dfrac{6}{2x-2}$.

15 Solve the equation: $\dfrac{x-1}{2} + x - 2 = \dfrac{2}{x-1} + \dfrac{1}{x-2}$ in any manner you wish.

16 (C) One of your students has become intrigued by the strange things that can happen when solving equations and comes to you with the following question. He began with the equation $x = 1$, which he knew had only one solution. Then he cubed both sides of the equation to get $x^3 = 1$ which he knew should have 3 solutions (by the Fundamental Theorem of Algebra). Since he was aware that he could then reverse the procedure by taking the cube root of both sides, he thought these equations should be equivalent. But, they are not, since one equation has one solution and the other three. He knew something was wrong, but couldn't figure out what it was. How do you help your student realize what went amiss here?

17 What is the error in Example 6.40?

6.12 Part 2: Review of Geometric Series: Preparation for Decimal Representation

LAUNCH

Using a pencil, ruler, and an $8\frac{1}{2}$ inch by 11 inch piece of lined loose leaf paper, do the following activity: Starting at the edge of your paper, along one of the horizontal lines, draw a line segment that is 4 inches long. Then, add on a line segment that is 1/2 as long as your first line segment. Next, add on a line segment that is 1/2 as long as your previous line segment. Continue in this way. Will you ever get to the other edge of the paper? Why or why not?

Did you realize that the launch activity led to an example of a geometric series? You undoubtedly studied these in your precalculus courses. You will be surprised to learn that, in order to discuss decimals in any kind of meaningful way, we need to use geometric series. So we now review them.

Recall that a geometric series is a series (infinite sum) of the form $a + ar + ar^2 + ar^3 + \ldots$. This abstract expression simply says that a is the first term and each term of the series is the previous term multiplied by some fixed number r. Thus, the second term, ar, is the first term, a, multiplied by r. The third term, ar^2, is the second term, ar, multiplied by r, and so on. For example, $1 + \frac{1}{2} + \frac{1}{4} + \frac{1}{8} + \ldots$ is a geometric series where $a = 1$ and each term is the previous term multiplied by $r = \frac{1}{2}$. Notice that this series can be written as: $1 + 1 \cdot \frac{1}{2} + 1 \cdot \left(\frac{1}{2}\right)^2 + 1 \cdot \left(\frac{1}{2}\right)^3 + \ldots$, which is of the form $a + ar + ar^2 + ar^3 + \ldots$.

The sum of a series is defined in terms of a limit. That is, we add one term, two terms, three terms, four terms, and so on, each time adding one more term. What we get is a sequence of numbers called the sequence of **partial sums**. If the limit of this sequence of partial sums is a finite number, this finite number is called the sum of the series.

Let us see what happens to the series $1 + \frac{1}{2} + \frac{1}{4} + \frac{1}{8} + \ldots$. If we let s_1, s_2, s_3, and so on represent respectively, the sum of the first term, the sum of the first two terms, the sum of the first three terms, and so on of the series above, we get the following sequence of partial sums:

$s_1 = 1$

$s_2 = 1 + \frac{1}{2} = 1\frac{1}{2}$

$s_3 = 1 + \frac{1}{2} + \frac{1}{4} = 1\frac{3}{4}$

$s_4 = 1 + \frac{1}{2} + \frac{1}{4} + \frac{1}{8} = 1\frac{7}{8}$

and so on

and it can be shown that the pattern you see above continues. Thus, $s_5 = 1\frac{15}{16}$, $s_6 = 1\frac{31}{32}$, and so on. So it appears that these partial sums approach 2. They do. Because of this, we say that the sum of this series $1 + \frac{1}{2} + \frac{1}{4} + \frac{1}{8} + \ldots$ is 2 or that the series $1 + \frac{1}{2} + \frac{1}{4} + \frac{1}{8} + \ldots$ converges to 2. In summary, the sum of an infinite series is defined to be $\lim_{n\to\infty} s_n$, where s_n is obtained by adding the first n terms of the series.

We recall from our study of series in calculus that some series have finite sums and others don't. Those that have finite sums are said to **converge** and those that don't are said to **diverge**.

A series $a_1 + a_2 + a_3 + \ldots$ is abbreviated as $\sum_{i=1}^{\infty} a_i$ or when the index is clear, $\sum_1^{\infty} a_i$. Furthermore, the letter we use for the index makes no difference. Thus, $\sum_{i=1}^{\infty} a_i$ means exactly the same thing as $\sum_{n=1}^{\infty} a_n$.

Here is the main theorem about geometric series.

Theorem 6.47

(a) The sum of the first n terms of a geometric series is given by $s_n = \dfrac{a - ar^n}{1 - r}$.

(b) A geometric series $a + ar + ar^2 + ar^3 + \ldots$ converges when $|r| < 1$ and its sum is $\frac{a}{1-r}$.

(c) A geometric series diverges if $|r| > 1$.

Proof. (a) The sum of the first n terms of this geometric series is denoted by s_n. By definition,

$$s_n = a + ar + ar^2 + \ldots + ar^{n-2} + ar^{n-1}. \text{ (Count the number of terms! This is a sum of } n \text{ terms.)}$$

(6.52)

Multiply both sides of this equation by r to get

$$s_n r = ar + ar^2 + ar^3 + \ldots + ar^{n-1} + ar^n. \tag{6.53}$$

Subtract equation (6.53) from equation (6.52) and notice that many of the terms combine to give us 0, leaving us with:

$$s_n - s_n r = a - ar^n. \tag{6.54}$$

Rewrite equation (6.54) as $s_n(1 - r) = a - ar^n$ and divide this equation by $1 - r$ to get

$$s_n = \frac{a - ar^n}{1 - r}. \tag{6.55}$$

Proof. (b) Now, if $|r| < 1$, then $-1 < r < 1$ and this implies that r^n gets closer and closer to zero as n gets larger and larger. It follows that $ar^n \to 0$ as n gets large, and thus the fraction on the right of equation (6.55) approaches $\frac{a}{1-r}$. In terms of calculus, $\lim_{n\to\infty} s_n = \frac{a}{1-r}$. But $\lim_{n\to\infty} s_n$ is the sum of the series. (That is the definition of the sum of the series!) Thus, the sum of the series is $\frac{a}{1-r}$.

Proof. (c) If $|r| > 1$, then ar^n does not approach a finite number as n approaches infinity. Thus, $\lim_{n\to\infty} s_n$ in equation (6.55) does not exist. That is, the geometric series diverges.

We can now answer a question from Chapter 1. There we gave the series $1 + 2 + 4 + 8 + \ldots$ and said that the sum was $\frac{a}{1-r}$. Since $a = 1$ and $r = 2$, this sum becomes -1. We asked how this could be. The answer is, this can't be. The series does not converge, since $|r| > 1$. We can't use the formula $\frac{a}{1-r}$, as this is only valid for convergent series. ∎

To clarify this, let's give a few numerical examples.

Example 6.48 *Find the sum of the first n terms of the series,* $1 + \frac{1}{2} + \frac{1}{4} + \frac{1}{8} + \ldots$.

Solution. According to part (a) of Theorem 6.47, the sum of the first n terms is $s_n = \frac{a-ar^n}{1-r} = \frac{1-1(1/2)^n}{1-(1/2)} = \frac{1-1(1/2)^n}{(1/2)} = 2(1-(1/2)^n) = 2 - 2(1/2)^n$. Thus, if we sum 50 terms, our sum will be $s_{50} = 2 - 2(1/2)^{50}$ and if we sum 100 terms our sum will be $s_{100} = 2 - 2(1/2)^{100}$ and both of these are extremely close to 2. In fact, so close, that if you tried to compute this on a calculator, the calculator would say that both s_{50} and s_{100} *are* 2! But, of course, we know this is not true. However, the more terms we take, the closer and closer the sums will get to 2. This is just a reflection of what Theorem 6.47 says, namely, that the series should converge to $\frac{a}{1-r} = \frac{1}{(1-(1/2))} = 2$.

Example 6.49

(a) *What is the sum of the series,* $4 - \frac{4}{3} + \frac{4}{9} - \frac{4}{27} + \ldots$?
(b) *What is the sum of the series* $1 - \frac{4}{3} + \frac{16}{9} - \ldots$?

Solution. (a) This is a geometric series with $a = 4$ and $r = -\frac{1}{3}$. Since $|r| < 1$, the sum of the series is $\frac{a}{1-r} = \frac{4}{1-(-1/3)} = 3$.

(b) This is also a geometric series with $r = \frac{-4}{3}$. Since $|r| > 1$, this series diverges.

The following is a very useful result as we shall see.

Theorem 6.50 *(Comparison Test) Suppose that* $\sum_1^\infty a_i$ *and* $\sum_1^\infty b_i$ *are two series consisting of nonnegative terms and suppose that we know that* $\sum_1^\infty b_i$ *converges. Then, if* $a_i \leq b_i$ *for all i,* $\sum_1^\infty a_i$ *also converges.*

What this is saying is that, if the larger of two nonnegative series has a finite sum (converges), then so does the smaller one. This makes perfect sense.

Let us illustrate this.

Question: Does the series, $\sum_1^\infty \frac{1}{2^i+1} = \frac{1}{3} + \frac{1}{5} + \frac{1}{9} + \ldots$ converge?

Answer: Yes, the sum of the series $\sum_1^\infty \frac{1}{2^i+1} \leq \sum_1^\infty \frac{1}{2^i}$, since $\frac{1}{2^i+1} \leq \frac{1}{2^i}$ for each i. Since the series $\sum_1^\infty \frac{1}{2^i}$ is a geometric series with $|r| < 1$, it converges and hence, by the comparison test, so does the original series.

Student Learning Opportunities

1 (C) Your students are given the series $3 - 6 + 12 - 24 + \ldots$. They try to find the sum by calculating $\frac{a}{1-r}$, where in this case $a = 3$ and $r = -2$. They figure out that the sum of the series will be 1. Seeing that the negative numbers in the series are always twice as large as the positive numbers, they realize that their solution is impossible! How do you help them realize what has gone wrong?

2 (C) You gave your students the launch question in this section and your students engaged in a heated debate about whether the line segments they were drawing would ever reach the other side of the page. How would you help them realize that what they were dealing with was a converging geometric series? How would you help them represent this series and resolve their argument by determining the sum?

3 Find the sum of each of the following series. If the series has no sum, explain why.

(a) $1 - \dfrac{1}{2} + \dfrac{1}{4} - \dfrac{1}{8} + \ldots$

(b) $2 - 4 + 8 - 16 + \ldots$

(c) $2 + \dfrac{2}{\pi} + \dfrac{4}{\pi^2} + \dfrac{8}{\pi^3} + \ldots$

(d) $4 - 4\sqrt{2} + 8 - 8\sqrt{2} + 16 - \ldots$

(e) $\dfrac{5}{3^3} - \dfrac{5}{3^5} + \dfrac{5}{3^7} - \ldots$

4 Show that the series $\sum_{1}^{\infty} \dfrac{1}{n!}$ converges. [Hint: Start by showing that $n! \geq 2^{n-1}$ and hence $\dfrac{1}{n!} \leq \dfrac{1}{2^{n-1}}$.]

5 (C) You ask your students to begin with a square that has sides of length 6 inches. Have them inscribe a circle in that square. Next, inscribe a square in that circle and then inscribe a circle in that latter square. Have them continue in this manner until the figures get too small to continue. Now, ask your students to figure out what the sum of the areas of the circles would be if they could continue their drawings forever. What is the answer?

6 The sum of a geometric series is 15. The sum of the squares of the terms of the series is 45. Show that the first term of the series is 5.

7 Does the series $\dfrac{3}{5^3 + 1} + \dfrac{3}{5^5 + 1} + \dfrac{3}{5^7 + 1} + \ldots$ converge or diverge? How do you know?

8 Show that the Harmonic series $\sum_{i=1}^{\infty} \dfrac{1}{i} = \dfrac{1}{1} + \dfrac{1}{2} + \dfrac{1}{3} + \ldots$ diverges. [Hint: Show that $s_1 > \dfrac{1}{2}$, $s_2 > 1$, $s_4 > 1\dfrac{1}{2}$, $s_8 > 2$. and so on.]

6.13 Decimal Expansion

LAUNCH

1. Can every rational number be represented as a decimal? Explain your answer.
2. Give the decimal expansion of $\dfrac{3}{8}$.
3. Can every decimal number be represented as a fraction? Explain your answer.
4. Give the fractional equivalent of the decimal (a) $0.666666\ldots$ (b) $0.64646464\ldots$.

Despite the fact that you studied decimal and fractional notations in elementary school, the launch questions have probably given you some reason to doubt the depth of your understanding of these important concepts. It is for this reason, that this next section focuses on the meanings of these concepts from a higher level. You will be surprised to see how complex it can get.

The advent of decimal notation was a great moment in the history of mathematics. With decimal notation, computations that had up to that time been very cumbersome, suddenly became easy. In this section we look at some of the issues that arose in the development of decimal representation of numbers.

When we see a number like 325, we know that this is an abbreviation for 3 hundreds+ 2 tens+ 5 ones. Notice that this can be represented as $3 \cdot 10^2 + 2 \cdot 10^1 + 5 \cdot 10^0$. Since this expresses a number as powers of 10, the tendency might be to continue the pattern and also express numbers with negative powers of 10 also. The decimal point would be the demarcation line where the exponents go from nonnegative to negative. In fact, this is exactly what is done. Thus, 325.46 would mean $3 \cdot 10^2 + 2 \cdot 10^1 + 5 \cdot 10^0 + 4 \cdot 10^{-1} + 6 \cdot 10^{-2}$.

The first issue that comes up is what does an infinite decimal like 0.123412341234... mean? The answer is that this representation is an abbreviation for the *infinite series* $1 \cdot 10^{-1} + 2 \cdot 10^{-2} + 3 \cdot 10^{-3} + 4 \cdot 10^{-4} + \ldots$ or in more familiar terms, $\frac{1}{10} + \frac{2}{100} + \frac{3}{1000} + \frac{4}{10000} + \ldots$. Thus, decimals are *infinite series*.

At first glance, this may not seem like a problem. But it really is, for not all infinite series have finite sums, and it might be that some decimals that we construct, or those that we use in real life really make no sense since the sums they represent might be infinite. This would be a serious problem! So let us put that to rest right away. The following says this never happens.

Theorem 6.51 *Any decimal number $.d_1 d_2 d_3 \ldots$ represents a series that has a finite sum.*

Proof. The important thing is to realize that each digit, d_i in the decimal representation is ≤ 9. Thus, if we call $N = .d_1 d_2 d_3 \ldots$, then

$$N = \frac{d_1}{10} + \frac{d_2}{100} + \frac{d_3}{1000} + \ldots \leq \frac{9}{10} + \frac{9}{100} + \frac{9}{1000} + \ldots.$$

The series on the right is a geometric series where $|r| < 1$ (in fact, $r = \frac{1}{10}$). So it has a finite sum (it converges!) and hence by the comparison test (Theorem 6.50), so does the series on the left. That is, the series that the decimal represents, also has a finite sum. ∎

A common problem in secondary school mathematics is to find the fractional equivalent of a decimal. The following illustrates the procedure.

Example 6.52

(a) *Find the fractional equivalent of the decimal* 0.323232 ... .
(b) *Find the fractional equivalent of* 0.034212121 ... .

Solution. (a) Let

$$N = 0.323232 \ldots . \tag{6.56}$$

282 Building the Real Number System

We multiply by 100 and we get

$$100N = 32.3232\ldots. \tag{6.57}$$

Subtract equation (6.56) from equation (6.57) to get $99N = 32$ and hence, $N = \dfrac{32}{99}$.

(b) This is a bit more difficult. Rewrite $0.034212121\ldots$ as $0.034 + 0.000212121\ldots$. Now, the first part is the fraction $\dfrac{34}{1000}$ and to find the other part, we let

$$N = 0.000212121\ldots. \tag{6.58}$$

We multiply N by 100 to get

$$100N = 0.0212121\ldots. \tag{6.59}$$

We then subtract equation (6.58) from equation (6.59). Observing that the right sides of equations (6.58) and (6.59) are the same after the fifth digit, we get:

$$99N = 0.021 \text{ or } \dfrac{21}{1000},$$

and dividing by 99 we get $N = \dfrac{21}{99\,000}$. Adding this to $\dfrac{34}{1000}$ we get that $0.034212121 = \dfrac{3387}{99\,000}$ and you can easily check on a calculator that this works.

Now that we know that every decimal converges (that is, represents a finite number), we ask another question. How do we know that every number has a decimal representation? We know how to find the decimal expansion of a rational number from elementary school work. We use long division. But how do we know that that procedure really works and gives us the correct decimal expansion for *all* rational numbers? Furthermore, what if we want to find the decimal expansion of an irrational number like $\sqrt{2}$? Then what? In that case we certainly cannot use long division.

We will now address these important issues. We will need to use the concept of the greatest integer $\leq x$. Suppose x is any number. Then the largest or greatest integer $\leq x$ is exactly what it says—the largest integer $\leq x$ and is denoted by $[x]$. By definition, $[x]$ is an integer $\leq x$. Since each number x lies between two consecutive integers

$$[x] \leq x < [x] + 1. \tag{6.60}$$

Let us illustrate. $[3.2] = 3$. Clearly, $3 \leq 3.2 < 3 + 1$. As another example, $[-3.1] = -4$. Clearly, $-4 \leq -3.1 < -4 + 1$. Finally, $[6] = 6$ and we have $6 \leq 6 < 6 + 1$.

Here is the first theorem in the development of decimal representation which is a fundamental result.

Theorem 6.53 *Every real number N, where $0 \leq N < 1$ can be written as a decimal.*

Proof. Since $0 \leq N < 1$, $0 \leq 10N < 10$. Let $a_1 = [10N]$. Since a_1 is the largest number less than $10N$ and $10N$ is nonnegative and less than 10, $0 \leq a_1 < 10$.

Building the Real Number System   283

Letting $x = 10N$ in (6.60) and using the fact that $a_1 = [10N]$ we get that

$$a_1 \leq 10N < a_1 + 1. \tag{6.61}$$

Subtract a_1 from both sides of (6.61) and we get that

$$0 \leq 10N - a_1 < 1. \tag{6.62}$$

Since $10N - a_1$ is between 0 and 1, 10 times $10N - a_1 = 100N - 10a_1$ is between 0 and 10 (excluding 10), so if we let $a_2 = [100N - 10a_1]$, then $0 \leq a_2 < 10$ and by (6.60) with $100N - 10a_1$ taking the place of x we get

$$a_2 \leq 100N - 10a_1 < a_2 + 1. \tag{6.63}$$

Subtracting a_2 from both sides of (6.63) we get that

$$0 \leq 100N - 10a_1 - a_2 < 1. \tag{6.64}$$

Since $100N - 10a_1 - a_2$ is between 0 and 1, 10 times $100N - 10a_1 - a_2 = 1000N - 100a_1 - 10a_2$ is between 0 and 10. So we let $a_3 = [1000N - 100a_1 - 10a_2]$, and so on. After n steps, we have the following generalization of (6.64)

$$0 \leq 10^n N - 10^{n-1} a_1 - 10^{n-2} a_2 - \ldots - a_n < 1. \tag{6.65}$$

If we divide both sides of (6.65) by 10^n, we get

$$0 \leq N - \frac{a_1}{10} - \frac{a_2}{100} - \frac{a_3}{1000} - \ldots - \frac{a_n}{10^n} < \frac{1}{10^n} \tag{6.66}$$

or, in decimal form,

$$0 \leq N - .a_1 a_2 \ldots a_n < \frac{1}{10^n}. \tag{6.67}$$

Now, since $\frac{1}{10^n}$ approaches 0 as n goes to infinity, (6.67) is saying that the difference between N and the decimals we are generating gets smaller and smaller as n gets larger and larger. Thus, the decimals $0.a_1 a_2 \ldots a_n$ we are generating, are getting closer and closer to N as n gets large. But the finite decimals $0.a_1 a_2 \ldots a_n$ we are generating are partial sums of the series which the infinite decimal $0.a_1 a_2 \ldots a_n \ldots$ represents. Since the partial sums $0.a_1 a_2 \ldots a_n$ are getting closer and closer to N, the sum of the series represented by the infinite decimal $0.a_1 a_2 \ldots a_n \ldots$ must equal N. Thus, every number N can be written as a decimal. ∎

We see that we really needed the concept of limits to get into a full discussion of decimals.

Now that we know that every N between 0 and 1 can be expressed as a decimal we turn to the process of how to find the digits in the decimal representation of N. This will be simple once we observe that, if you are given a number like $N = 32.425$, the largest integer less than or equal to N is 32, the number before the decimal point. We will need this shortly in a proof.

284 Building the Real Number System

Now we know that every number N between 0 and 1 can be written as $N = 0.d_1d_2d_3\ldots$. Thus,

$$10N = d_1.d_2d_3\ldots \qquad (6.68)$$

and we see that d_1 is the largest integer less than or equal to this number $10N$. Thus, d_1, the first digit in the decimal representation of N is equal to $[10N_1]$. Subtracting d_1 from both sides of 6.68 we get

$$10N_1 - d_1 = 0.d_2d_3d_4\ldots \qquad (6.69)$$

To find the second digit, d_2, in the decimal representation of N, we multiply equation (6.69) by 10 to get

$$100N_1 - 10d_1 = d_2.d_3d_4\ldots \qquad (6.70)$$

And now we see that $d_2 = [100N_1 - 10d_1]$. Subtracting d_2 from equation (6.70) we get

$$100N_1 - 10d_1 - d_2 = 0.d_3d_4\ldots \qquad (6.71)$$

and multiplying both sides of 6.71 by 10 we get

$$1000N_1 - 100d_1 - 10d_2 = d_3.d_4\ldots \qquad (6.72)$$

and now we see how to get d_3, namely $d_3 = [1000N_1 - 100d_1 - 10d_2]$. We subtract d_3 from equation (6.72) and multiply the result by 10 to get $10000N_1 - 1000d_1 - 100d_2 - 10d_3 = d_4.d_5d_6\ldots$ and we see that $d_4 = [10000N_1 - 1000d_1 - 100d_2 - 10d_3]$, and so on.

This seems to be a complicated procedure, but in fact it can be described as follows. To generate the digits in the decimal expansion of a number, N, we multiply N by 10, to get a number n, take the greatest integer, g, less than or equal to n. The number g is the next digit in our decimal expansion of N. Next, compute $n - g$, and start over again with $n - g$ in place of N and continue in a similar manner.

To clarify this, we will need an example which is key to the rest of this section. Study it carefully.

Example 6.54 *Find the decimal expansion of $\frac{1}{7}$.*

First, we multiply $N = \frac{1}{7}$ by 10, to get $\frac{10}{7}$. Since $g = [\frac{10}{7}]$ is 1, 1 is the first digit in the decimal expansion of $\frac{1}{7}$. Subtract $g = 1$ from $\frac{10}{7}$ to get $\frac{3}{7}$. We have completed the first step.

Now we start over with N being replaced by $\frac{3}{7}$. We do the same as before. We multiply $\frac{3}{7}$ by 10 to get $\frac{30}{7}$. Now compute $g = [30/7] = 4$. Thus, 4 is our second digit in the decimal expansion of $\frac{1}{7}$. Subtract this 4 from $\frac{30}{7}$ to get a result of $\frac{2}{7}$. We have now completed the second step.

Start again with $\frac{2}{7}$ in place of N. Multiply this by 10 to get $\frac{20}{7}$. Since $g = [\frac{20}{7}] = 2$, this is the third digit in the expansion of $\frac{1}{7}$. Subtract 2 from $\frac{20}{7}$ to get a result of $\frac{6}{7}$.

Start again with $\frac{6}{7}$ in place of N. Multiply the result $\frac{6}{7}$ by 10 to get $\frac{60}{7}$. Since $g = \left[\frac{60}{7}\right]$ is 8, the next digit in our decimal expansion of $\frac{1}{7}$ is 8. Subtract 8 from $\frac{60}{7}$ to get $\frac{4}{7}$. and so on.

At this point you may be thinking that all this is unnecessarily complicated in comparison to the very simple method that you learned in elementary school. At that time you used the following method of long division to find the decimal expansion of a number, in this case $\frac{1}{7}$. Here are the steps you learned in elementary school to find the decimal expansion of $\frac{1}{7}$ (Figure 6.11)

$$
\begin{array}{r}
0.142857 \\
7\overline{\smash{)}1.00000} \\
\underline{7} \\
30 \\
\underline{28} \\
20 \\
\underline{14} \\
60 \\
\underline{56} \\
40 \\
\underline{35} \\
50 \\
\underline{49} \\
10
\end{array}
$$

(6.73)

Figure 6.11

Do you notice anything similar about the two methods? Yes, long division is *exactly* what we were doing in Example 6.54. Let us see how.

Look at the first step. In the above Example 6.54, we began with $N = \frac{1}{7}$ and multiplied by 10 to get $\frac{10}{7}$. Then we computed $g = \left[\frac{10}{7}\right]$. But to compute g, we need to divide 7 into 10. Take the integer part of the quotient which is 1 to get our first digit in the decimal representation of $\frac{1}{7}$. Now look at the long division above. Our long division begins as

$$
\begin{array}{ll}
& \text{Line 0} \\
7\overline{\smash{)}1.00000} & \text{Line 1.}
\end{array}
$$

We begin by asking how many times 7 goes into 10. This is exactly what we did in Example 6.54. It goes in 1 time. Our division problem now looks like:

$$
\begin{array}{ll}
\phantom{7\overline{)}}0.1 & \text{Line 0} \\
7\overline{\smash{)}1.00000} & \text{Line 1} \\
\phantom{7\overline{)}1.00}\underline{7} & \text{Line 2} \\
\phantom{7\overline{)}1.00}3 & \text{Line 3.}
\end{array}
$$

We "bring down a zero" (which essentially means we multiply the remainder 3 by 10). So our division problem now looks like:

$$
\begin{array}{ll}
\phantom{7\overline{)}}0.1 & \text{Line 0} \\
7\overline{\smash{)}1.00000} & \text{Line 1} \\
\phantom{7\overline{)}1.0}\underline{7} & \text{Line 2} \\
\phantom{7\overline{)}1.0}30 & \text{Line 3.}
\end{array}
$$

Our next step in our long division is to compute how many times 7 goes into 30 to get the next digit in the decimal representation of $\frac{1}{7}$. The number of times 7 goes into 30 is, of course, $\left[\frac{30}{7}\right]$. Now look back at Example 6.54. We were doing the same computation.

Our next division looks like:

```
      0.14        Line 0
   7|1.00000      Line 1
     7            Line 2
     30           Line 3
     28           Line 4
      2           Line 5.
```

Now we bring down a zero. Again, we are multiplying our remainder by 10. Our division looks like:

```
      0.14        Line 0
   7|1.00000      Line 1
     7            Line 2
     30           Line 3
     28           Line 4
     20           Line 5.
```

Then we compute the number of times 7 goes into 20 to get the next digit in our decimal representation. That is, we compute $\left[\frac{20}{7}\right]$. We again are following our Example 6.54.

Since 20 divided by 7 gives a quotient of 2 with a remainder of 6, the quotient, 2, is the next digit in the decimal expansion of $\frac{1}{7}$ and this goes on line 0. Our division looks as follows:

```
      0.142       Line 0
   7|1.00000      Line 1
     7            Line 2
     30           Line 3
     28           Line 4
     20           Line 5
     14           Line 6
      6.
```

We bring down a zero and our computation is to determine how many times 7 goes into 60 to get our next digit in the decimal representation of $\frac{1}{7}$. That is, we computed $\left[\frac{60}{7}\right]$ just as we did in example 6.54, and so on.

Thus, we see that the long division process is *exactly* what we were doing above in Example 6.54, and in some sense, we have justified the process of long division. Of course, the long division process works only for rational numbers, while the proof of Theorem 6.53 shows us how to find, at least theoretically, the decimal expansion of *any* real number. Thus, Theorem 6.53 is somewhat sharper. We also notice that each time we bring down a zero in the long division problem, that is analogous to the multiplying by 10 that we did in Example 6.54. Thus, we see in a very strong manner, the parallel nature of long division and the procedure we used in Example 6.54.

The steps for finding a decimal representation of a rational number are so much easier to process if we just use long division. So, now that we know the long division process is precisely what is being done in the proof of Theorem 6.53, we can use long division when we need to get our decimal representations of rational numbers.

One key thing to realize in long division of 1 by 7 is that every time we subtract a line from the previous line, we are getting a remainder when the previous number is divided by 7. Thus, when we begin, and we ask how many times does 7 go into 10, the answer is 1 and the remainder is 3. That remainder, 3, is what was on line 3 before you brought down the zero. Similarly, in the second step you asked "How many times does 7 go into 30?" The answer is 4 times, and the remainder is 2. That remainder of 2 went on line 5 before we brought down the zero. Why are we bringing this up? Well, there are only a finite number of remainders when you divide a number by 7 and they are 0, 1, 2, 3, 4, 5, 6. Thus, eventually one of the remainders we got earlier will have to show up again (you can see that the remainder of 1 in the last line of equation (6.73) is the same as in the first line), and all subsequent steps in the long division process will repeat. What this means is our decimal expansion for the fraction will repeat. Thus, when we divide two whole numbers, *we will get a repeating decimal for the decimal expansion*. We have essentially shown how to prove:

Theorem 6.55 *All rational numbers are repeating decimals. Furthermore, the number of digits in the repeating part is no more than the divisor.*

We said above that the *maximum* number of digits in the repeating part of a decimal is the size of the divisor. We need not even approach that maximum. For example $23/666 = 0.0345345345\ldots$. The repeating part is "345", which only has three digits, though theoretically, it could have as many as 666 since there are 666 remainders that one can get when one uses long division.

The size of the repeating part of a decimal is called its **period** Thus, the period of $23/666$ is 3 while the period of $2/3$ is 1 since $2/3 = 0.6666\ldots$.

We now know that every rational number can be expressed as a repeating decimal. Furthermore, it works in reverse. Every repeating decimal can be expressed as a rational number. We show this exactly as we did in Example 6.52. Thus, we have the following theorem.

Theorem 6.56 *Rational numbers are precisely those numbers that can be expressed with repeating decimals.*

A corollary of this is

Corollary 6.57 *Irrational numbers are precisely those numbers that don't have repeating decimals in their decimal expansions.*

So, if we write the number $N = 0.101001000100001\ldots$ where each time we insert one more zero to the previous string of zeros before putting in a one, from the way it is designed, this clearly cannot repeat. So this number is irrational.

Student Learning Opportunities

1. **(C)** One of your students vehemently claims that any number that can be written as a fraction is rational. How would you respond? Is she correct? Why or why not?

2. **(C)** Your students claim that the number $0.21222324252627282921 0211212\ldots$ is rational since there is a pattern that keeps repeating. What is the pattern they are referring to? Are they correct in claiming the number is rational? If so, why? If not, why not?

3. Write each of the following as a rational number. (The bar over a set of digits means the digits are repeated indefinitely.)

    (a) $0.345345345\ldots = \overline{0.345}$

    (b) $0.9898989898\ldots = \overline{0.98}$

    (c) $1.06545454\ldots = 1.06\overline{54}$

4. Which of the following are rational? Explain.

    (a) $0.12131415161718\ldots$

    (b) $0.395678\overline{9}$

    (c) $\dfrac{\sqrt{2}}{3}$

5. Use the procedure of Theorem 6.54 to find the decimal expansion of each of the following fractions. Verify your answers by using long division.

    (a) $\dfrac{1}{8}$

    (b) $\dfrac{3}{7}$

    (c) $\dfrac{5}{11}$

6. **(C)** It is stated in algebra that, if we take a number N and multiply it by say 10^3, we move the decimal place three units to the right, and if we multiply by 10^{-3} we move the decimal point 3 units to the left. Similar statements hold if you multiply a number by 10^p and 10^{-p} where p is positive. Thus, we can write 0.0023 as 23×10^{-4} and 231000 as 231×10^3. How would you use this to explain to students the rule that they learn in elementary school that, if we multiply 2.1×3.02, we need only multiply 21 by 302, and then move the decimal place in our answer three places to the left (where the number of places we move the decimal point is the sum of the number of places after the decimal point in each of the numbers you multiply together)?

6.14 Decimal Periodicity

LAUNCH

1. What are the similarities and differences between and among the following decimals?
   (a) 0.428
   (b) 0.382382 ...
   (c) 0.00467878787878 ...
   (d) 0.20200200020000 ...

2. Express each of the decimals represented in question #1 as a rational number in fraction form. Can it be done for each case? Why or why not?

3. When a rational number is written as a repeating decimal, can one predict how many places after the decimal point the repetition will begin?

Having read the previous section and trying to respond to the launch questions, you are probably convinced that the mathematics behind decimals can become quite complex. We hope that, by reading this section, some of the questions about the decimal representation of numbers will be resolved.

So, let us now continue our investigation of the decimal representation of numbers. A decimal is called **terminating**, if it ends in all zeros. Thus, $N = 0.3750000 \ldots = 0.375$ is a terminating decimal.

We can ask the following question: "What kinds of rational numbers can be represented by terminating decimals?" Well, suppose we start with a terminating decimal, say $N = 0.3750000 \ldots = 0.375$. Then it is clear in this latter form that the decimal can be written as $\frac{375}{1000}$. That is, the fraction has a denominator that is a power of 10. Conversely, if a fraction has a denominator that can be turned into a power of 10, then the fraction can be written as a terminating decimal. For example, $\frac{7}{8}$ can be built up to a fraction whose denominator is a power of 10. We just multiply the numerator and denominator by 125 to get $\frac{7}{8} = \frac{875}{1000} = 0.875$. Thus, we have that: the decimal representation of a rational number will terminate in 0's, if and only if the denominator can be built up to a power of 10.

What allows us to build the denominator of a fraction up to a power of 10? Well, if the denominator can be turned into a power of 10, then its only factors are 2 and 5. Thus, it appears that, if the denominator of a fraction has only factors of 2 and/or 5, then it can be built up to a power of 10. Let us illustrate. Suppose we have the fraction $\frac{7}{2 \cdot 5^3}$. If we multiply the numerator and denominator by 2^2, we get the following equivalent fraction: $\frac{28}{1000}$ and this is clearly .028. Similarly, if we have the fraction $\frac{3}{2^3}$, it can be converted to a denominator that is a power of 10 by multiplying numerator and denominator by 5^3 to get $\frac{3}{2^3} = \frac{3 \cdot 5^3}{2^3 5^3} = \frac{375}{1000}$ and this is 0.375. These examples illustrate the following result.

> **Theorem 6.58** *A rational number can be written as a terminating decimal if and only if the denominator can be expressed as a power of 10. This is true if and only if the only factors of the denominator are 2 and/or 5.*

As we have seen, a rational number can be expressed as a repeating decimal. The repeat can start right in the beginning of the decimal expansion or after a lag. For example, in $0.123123\ldots = 0.\overline{123}$, the repeating part starts right away, whereas in $0.021343434\ldots = 0.021\overline{34}$, the repeating digits start repeating at the 4th digit.

When a fraction is expressed as a decimal, and the repeat starts right away, the decimal is called a **simple periodic decimal**, while when there is a lag before the decimal repeats, the decimal is called a **delayed periodic decimal**.

In Example 6.52 we saw that the simple periodic decimal $0.3434\ldots$ could be written as $34/99$. We showed this by calling the decimal N, multiplying by 100, and then subtracting the former from the larger to get $99N = 34$, so $N = 34/99$. In an entirely similar manner, if we had the simple periodic decimal $N = 0.321321\ldots$ we could multiply N by 1000 and subtract N to get $999N = 321$ so $N = 321/999$. Notice that, in both of the given examples, the denominator consisted of all 9's and the numerator consisted of the repeating part.

We can reverse the process. If we know that $N = \dfrac{43}{99}$, we can immediately write the decimal expansion of N. That would be $0.4343\ldots$ and if $N = 32/999$, we could write the decimal expansion as $0.032032\ldots$. We simply write the numerator with the same number of digits as there are 9's on the denominator. So 32 becomes 032. Similarly $7/9999 = (0007)/9999 = 0.00070007\ldots$. This yields the following:

> **Theorem 6.59** *Any simple periodic decimal can be written as a fraction whose denominator consists of a number with all 9's. Furthermore, if we have a fraction less than 1 whose denominator can be turned into one that consists of all 9's, then the decimal representation of that fraction is simple periodic.*

Thus, $1/3$ is simple periodic since it can be written as $3/9$. $1/13$ is simple periodic since it can be written as $76923/999999$. Needless to say, this is not the best test for determining if a fraction has a simple periodic expansion, since we have to know that the denominator can be built up to a fraction with the denominator consisting of all 9's.

There is a much simpler criterion for determining if a fraction has a simple periodic expansion, of which we will only give half a proof. The other half is more sophisticated and is taken up in the Student Learning Opportunities. We do need the previous theorem for this proof.

> **Theorem 6.60**
> (a) *If a rational number $\dfrac{m}{n}$ **in lowest terms** has a simple periodic decimal expansion, then the denominator, n, has no factors of 2 and no factors of 5.*
> (b) *if a rational number $\dfrac{m}{n}$ is in lowest terms and n has no factors of 2 or 5, then $\dfrac{m}{n}$ has a decimal expansion that is simple periodic.*

Proof. We prove (a) saving the other half for the Student Learning Opportunities. Now we know that, if $\frac{m}{n}$ is simple periodic, then $\frac{m}{n}$ can be written as a fraction $\frac{a}{999\ldots 9}$ where the denominator consists of all 9's. Cross multiplying we get that

$$(999\ldots 9)m = an. \tag{6.74}$$

Equation (6.74) says that $(999\ldots 9)\,m$ is a multiple of n. Now, if n did have a factor of 2 or 5, then so would $(999\ldots 9)m$ have a factor of 2 or 5, and since this factor cannot come from $999\ldots 9$ (since it is not even nor divisible by 5) it must be m that has this factor. That means that m and n both have that common factor of 2 or 5, contradicting the fact that the fraction is in lowest terms. Thus, our supposition that n has a factor of either 2 or 5 led to a contradiction of the fact that the fraction was in lowest terms, and thus n cannot have a factor of 2 or 5. ∎

Theorem 6.60 tells us that a rational number in lowest terms has a simple periodic expansion, if and only if the denominator has no factors of 2 or 5. To illustrate the theorem, 1/13 is a simple periodic fraction since it has no factor of 2 and no factor of 5 in the denominator and is in lowest terms. In fact, $1/p$, where p is any prime greater than 2, is always a simple periodic fraction. Similarly, since 37 has no factors of 2 or 5, we can expect the fraction $\frac{3}{37}$ to be simple periodic. Indeed it is. $\frac{3}{37} = 0.081081081\ldots = 0.\overline{081}$.

The above theorem seems to be implying that, if the denominator of a fraction has no factors of 2 or 5, it can be expressed as a fraction with a denominator consisting of all 9's. Yes! And this is not in the least bit obvious! A corollary of this is:

> **Corollary 6.61** *A rational number in lowest terms has a delayed periodic expansion if and only if the denominator has a factor of either 2 or 5 and at least one other prime factor in the denominator.*

We can also give an idea of how big the delay is before the decimal expansion repeats. We illustrate with a numerical example which generalizes. Suppose we have the fraction $\frac{73}{75}$. We write this as $\frac{73}{75} = \frac{73}{5^2 \cdot 3} = \frac{73 \cdot 2^2}{5^2 \cdot 3 \cdot 2^2} = \frac{73 \cdot 2^2}{5^2 \cdot 2^2 \cdot 3} = \frac{73 \cdot 4}{100 \cdot 3} = \frac{292}{100 \cdot 3} = \frac{1}{100} \cdot \frac{292}{3} = \frac{1}{100} \cdot \left(97\frac{1}{3}\right) = \frac{1}{100} \cdot (97.3333\ldots) = 0.97333\ldots$ and we have a delay of 2 resulting from the factor of $\frac{1}{100}$ in front. Notice what makes this work. Once we have built up the denominator so that it has the smallest power of 10 possible as a factor, we can factor this out of the denominator and the remaining fraction will have no factors of 2 or 5 in the denominator. *So the remaining fraction must have a decimal expansion that is simple periodic.* Multiplying by $\frac{1}{100}$ just moves the decimal two places to the left and thus, there are two places before the repetition. Similarly, when the denominator of a fraction that is in lowest terms has been factored, and we get $2^1 \cdot 5^3$, then we can build the denominator up to have a power of 10^3 (by multiplying numerator and denominator by 2^2) and thus, there would be a delay of 3 before the repetition. Here is a numerical example. Consider $\frac{97}{1750} = \frac{97}{2 \cdot 5^3 \cdot 7} = \frac{97 \cdot 2^2}{2^3 \cdot 5^3 \cdot 7} = \frac{1}{1000} \cdot \left(\frac{388}{7}\right) = \frac{1}{1000}\left(55\frac{3}{7}\right) = \frac{1}{1000}\left(55.\overline{428571}\right) = 0.055\overline{428571}$. These two examples illustrate the result:

Theorem 6.62 *If the denominator of a rational number in lowest terms is factored completely, and the factor 2 occurs to the power r and the power 5 occurs to the power s, then the delay before the fraction repeats is the maximum of r and s. (Here r or s can be 0, that is, there may be no factor of 2 or 5, but they both can't be 0 since there must be a factor of one of either 2 or 5.)*

So, in the fraction $\frac{97}{1750} = \frac{97}{2 \cdot 5^3 \cdot 7}$, $r = 1$ and $s = 3$. So the delay is the maximum of r and s, which is 3. This is exactly what we found prior to the theorem. Similarly, in the fraction $\frac{73}{75} = \frac{73}{5^2 \cdot 3}$, $r = 0$ and $s = 2$, so the delay is the maximum of r and s which is 2. This is also what we found earlier.

Student Learning Opportunities

1 (C) You impress your students by asking them to give you any fraction that has only 9's in the denominator and immediately giving them the decimal expansion (i.e. 48/99 = 0.484848..., 164/99999 = 0.001640016400164). After you do a few more like these, and they seem to be noticing a pattern, you ask them to find the decimal expansion of some other fractions with only 9's in the denominator. Although they are excited that they can now do what you did, they want to know why this works. What proof will you give them?

2 (C) Your students realize that, if you have a fraction whose denominator is a power of 10, then it is easy to write it in decimal form. They want to know if all fractions can be converted into that form and if so, how? What is your response and how do you justify it?

3 Without using a calculator, determine how many digits there are before the following fractions repeat. Then check your answers with a calculator.

(a) $\frac{1}{12}$

(b) $\frac{7}{18}$

(c) $\frac{17}{75}$

4 Without a calculator, determine which of the following will terminate. Explain your work. Check your answer with a calculator.

(a) $\frac{3}{20}$

(b) $\frac{5}{35}$

(c) $\frac{7}{250}$

(d) $\frac{5}{21}$

Building the Real Number System 293

5 If $\dfrac{1}{p}$ where p is a prime can be written as $.\overline{ab}$ where a and b are different, what are the only values of p?

6 True or False: If $\dfrac{1}{b}$ is simple periodic and has period p, then so does $\dfrac{a}{b}$ if a and b have no common factor. Justify your answer.

7 Here is an outline of the proof of part (b) of Theorem 6.60. Suppose that $\dfrac{m}{n}$ is between 0 and 1, delays before repeating, and that we know the delay is 2 before repeating. (The same argument works if the delay is 3, 4, or any number.) Furthermore, assume the period is 3. Then our fraction is $0.\mathbf{ab}cdecdecde\ldots$. Then, from the previous section, the first c in the decimal expansion of $\dfrac{m}{n}$ is the remainder when $10b$ is divided by n, and the second c is the remainder when $10e$ is divided by n. This means that $10b$ and $10e$ leave the same remainder when divided by 10, and thus $10b - 10e = 10(b - e)$ must be divisible by n (Why?) Since n has no factors of 2 or 5, it can't be 10 that is divisible by n. So it must be that $b - e$ is divisible by n. But b and e are both less than n and nonnegative. The only way n can divide $b - e$ then, is if $b = e$. Thus, our decimal is really $0.\mathbf{ae}cdecdecde\ldots$. Finish the proof.

6.15 Decimals: Uniqueness of Representation

LAUNCH

One of your secondary school students asks, "What number (other than 1) is closest to 1?". Another student answers, $0.999\ldots$. Is that second student right?

At first glance, this launch question might seem quite easy. But, we're sure that, as you thought about it, the answer was not obvious at all. Hopefully, after reading this section, you should have better insight into what the answer is.

Most students know that every real number has a decimal representation, and believe that there is only one way to represent a number as a decimal. After all, if you type 3/4 into a calculator and press enter, you only get back one answer, $0.750000\ldots$.

The next example shows that this is a common misconception. There actually is more than one decimal representation of *some* numbers.

Example 6.63 *Show that* $0.7499999\ldots = 0.75$.

Solution. Let

$$N = 0.7499999\ldots \tag{6.75}$$

Multiply both sides of equation (6.75) by 10 to get

$$10N = 7.499999\ldots \tag{6.76}$$

Subtract equation (6.75) from equation (6.76) to get $9N = 6.75$ and divide by 9 to get $N = 0.75$. Thus, we have two ways of representing 0.75, one of them being $0.749999\ldots$ and the other being 0.75!

As it turns out, the only decimals that can be written in this non unique fashion are those that terminate with all 0's or all 9's. (For example, 0.75 which is $0.7500000000\ldots$ or $0.74999\ldots$.) All other real numbers have only one representation as a decimal. That is the content of the next theorem whose proof is somewhat sophisticated. One needs to be comfortable with summation notation to understand it. You may wish to read it a few times to get a better understanding. We will only give the proof for numbers x such that $0 \leq x < 1$, since the integer part of a number has no effect on the decimal part of the representation of the number.

Theorem 6.64 *(a) Every nonnegative real number x between 0 and 1 is represented by a unique infinite decimal, except those numbers whose decimal representations terminate in an infinite number of zeros or an infinite number of 9's. These and only these decimals can also be represented in two ways.*

Proof. We already know that each real number can be written as a decimal. What we will prove is that **if** there are two decimal representations of a number, x, then that number x **must be** representable in the form $x = 0.b_1b_2\ldots b_k0000000\ldots$ and that the alternate way of expressing that number is $0.b_1b_2\ldots(b_k - 1)99999999\ldots$. That will imply all other numbers have a unique representation as a decimal and this will also prove our theorem.

So suppose that x had two representations:

$$x = 0.a_1a_2\ldots \tag{6.77}$$

and

$$x = 0.b_1b_2\ldots. \tag{6.78}$$

Let a_k be the **first** digit at which the two representations of x differ. Then $a_k \neq b_k$, but

$$\begin{aligned}a_1 &= b_1, \\ a_2 &= b_2 \\ &\ldots \\ &\ldots \\ a_{k-1} &= b_{k-1}.\end{aligned} \tag{6.79}$$

Let us suppose that $a_k < b_k$. (There is a similar proof if $b_k < a_k$.) Since a_k and b_k are different integers, they differ by at least one. That is, $b_k - a_k \geq 1$, or put another way,

$$a_k + 1 \leq b_k. \tag{6.80}$$

Our plan is to produce a list of expressions and show that they are all between x and x and so they must all be the same, namely, x. Now, we have

$$x = \sum_{1}^{\infty} \frac{a_n}{10^n} \quad \text{(by (equation 6.77))} \tag{6.81}$$

$$= \left(\sum_{1}^{k-1} \frac{a_n}{10^n}\right) + \frac{a_k}{10^k} + \left(\sum_{k+1}^{\infty} \frac{a_n}{10^n}\right) \quad \text{(Just breaking the sum up.)} \tag{6.82}$$

$$\leq \left(\sum_{1}^{k-1} \frac{a_n}{10^n}\right) + \frac{a_k}{10^k} + \left(\sum_{k+1}^{\infty} \frac{9}{10^n}\right) \quad \text{(Since each } a_n \leq 9.\text{)} \tag{6.83}$$

$$= \left(\sum_{1}^{k-1} \frac{a_n}{10^n}\right) + \frac{a_k}{10^k} + \left(\frac{\frac{9}{10^{k+1}}}{1 - \frac{1}{10}}\right) \quad \text{(Sum of geometric series.)} \tag{6.84}$$

$$= \left(\sum_{1}^{k-1} \frac{a_n}{10^n}\right) + \frac{a_k}{10^k} + \left(\frac{1}{10^k}\right) \quad \text{(Simplification of the last term.)} \tag{6.85}$$

$$= \left(\sum_{1}^{k-1} \frac{b_n}{10^n}\right) + \frac{a_k}{10^k} + \frac{1}{10^k} \quad \text{(By the equations in (6.79))} \tag{6.86}$$

$$= \left(\sum_{1}^{k-1} \frac{b_n}{10^n}\right) + \frac{a_k + 1}{10^k} \quad \text{(Combining the last two terms.)} \tag{6.87}$$

$$\leq \left(\sum_{1}^{k-1} \frac{b_n}{10^n}\right) + \frac{b_k}{10^k} \quad \text{(Since } a_k + 1 \leq b_k \text{ by equation (6.80).)} \tag{6.88}$$

$$\leq \left(\sum_{1}^{k-1} \frac{b_n}{10^n}\right) + \frac{b_k}{10^k} + \sum_{k+1}^{\infty} \frac{b_n}{10^n} \quad \text{(We are just adding more.)} \tag{6.89}$$

$$= \sum_{1}^{\infty} \frac{b_n}{10^n} \quad \text{(The three sums combine into this one sum.)} \tag{6.90}$$

$$= .b_1 b_2 b_3 \ldots \quad \text{(The series represents this decimal.)} \tag{6.91}$$

$$= x \quad \text{(By equation (6.78)).}$$

We have a list of expressions sandwiched between x and x, so all lines in the above display must be equal. In particular, the contents of line equation (6.87) = the contents of line equation (6.88) or

$$\left(\sum_{1}^{k-1} \frac{b_n}{10^n}\right) + \frac{a_k + 1}{10^k} = \left(\sum_{1}^{k-1} \frac{b_n}{10^n}\right) + \frac{b_k}{10^k}. \tag{6.92}$$

Subtracting $\sum_{1}^{k-1} \frac{b_n}{10^n}$ from both sides of equation (6.92), we get

$$\frac{a_k + 1}{10^k} = \frac{b_k}{10^k} \tag{6.93}$$

and multiplying equation (6.93) by 10^k we get that $a_k + 1 = b_k$ or that

$$a_k = b_k - 1. \tag{6.94}$$

Also, from equations (6.88) and (6.89) we have

$$\left(\sum_{1}^{k-1} \frac{b_n}{10^n}\right) + \frac{b_k}{10^k} = \left(\sum_{1}^{k-1} \frac{b_n}{10^n}\right) + \frac{b_k}{10^k} + \sum_{k+1}^{\infty} \frac{b_n}{10^n}. \tag{6.95}$$

Subtracting $\left(\sum_{1}^{k-1} \frac{b_n}{10^n}\right) + \frac{b_k}{10^k}$ from both sides of equation (6.95), we get that

$$\sum_{k+1}^{\infty} \frac{b_n}{10^n} = 0. \tag{6.96}$$

Equation (6.96) tells us we have a sum of nonnegative numbers whose sum is zero and the only way this can happen is if all the b's in (6.96) are 0. That is, if $b_{k+1} = b_{k+2} = b_{k+3}$ and so on are all equal to zero.

Stop. Saying that $b_{k+1} = b_{k+2} = b_{k+3}$ and so on are all equal to zero is saying that the representation we gave for x in equation (6.78) terminates in all zeros.

Thus, **we have now shown that the representation of x given by equation (6.78) must terminate in all zeros and, in fact, looks like**

$$x = 0.b_1 b_2 \ldots b_k 0000000 \ldots. \tag{6.97}$$

Now, from equations (6.82) and (6.83) we have

$$\left(\sum_{1}^{k-1} \frac{a_n}{10^n}\right) + \frac{a_k}{10^k} + \left(\sum_{k+1}^{\infty} \frac{a_n}{10^n}\right) = \left(\sum_{1}^{k-1} \frac{a_n}{10^n}\right) + \frac{a_k}{10^k} + \left(\sum_{k+1}^{\infty} \frac{9}{10^n}\right).$$

Subtracting $\left(\sum_{1}^{k-1} \frac{a_n}{10^n}\right) + \frac{a_k}{10^k}$ from both sides, we get $\sum_{k+1}^{\infty} \frac{a_n}{10^n} = \sum_{k+1}^{\infty} \frac{9}{10^n}$, and subtracting the sum on the left hand side, we get $\sum_{k+1}^{\infty} \frac{9-a_n}{10^n} = 0$. Now, since $a_n \leq 9$, we again have a sum of nonnegative numbers whose sum is zero. Thus, $9 - a_n = 0$, for all $n \geq k+1$. That is, $a_n = 9$ for all $n \geq k+1$.

Stop again! We have shown that the tail end of the a_n of the representation for x given in equation (6.77) consists of all nines. That is,

$$x = 0.a_1 a_2 \ldots a_k 99999 \ldots. \tag{6.98}$$

To summarize, equations (6.97) and (6.98) show that any number that has two decimal expansions must have either all 0's at the end or all 9's at the end.

To finish the proof, we compare equations (6.97) and (6.98) to get

$$x = b_1 b_2 \ldots b_k 000000000 \ldots = 0.a_1 a_2 \ldots a_k 99999 \ldots. \tag{6.99}$$

From equation (6.94) $a_k = b_k - 1$ and from display (6.79) we have that $a_1 = b_1, a_2 = b_2, \ldots a_k = b_k$. When we substitute these in equation (6.99), it becomes

$$x = b_1 b_2 \ldots b_k 000000000 \ldots = 0.b_1 b_2 \ldots (b_k - 1) 9999 \ldots \tag{6.100}$$

and we are done. ∎

Thus, the fraction $0.658000000\ldots$ is the same, for example, as $0.6579999999\ldots$.

Student Learning Opportunities

1 Represent each of the following decimals in a different decimal form. If it is not possible, say so. Explain your answers. (Again, the bar over a set of digits means that that set of digits keeps repeating indefinitely.)

(a) 0.345
(b) 0.492$\overline{9}$
(c) 0.123$\overline{4}$
(d) 0.$\overline{123}$
(e) 10100100010000 ...

2 Show that, if one has an infinite sequence of numbers d_1, d_2, d_3, and so on written in decimal form, where none of them end in all 0's or all 9's, one can always find a decimal different from each of these.

3 (C) Your students simply cannot accept the fact that 0.9999999... = 1. How would you convince them that this is true? [Hint: One way that might work is to start with the statement that $\frac{1}{3}$ = 0.33333 ... which most students accept and then multiply both sides by 3. Find other ways.]

4 (C) One of your clever students has written a decimal formed by writing all the natural numbers in order after the decimal point.

0.12345678910111213141516171819202 1 ...

She wants to know if it is rational or irrational. Most students in the class say it is rational. Why do you think they are saying that and how do you prove the number is irrational? [Hint: If it were rational, there would be a part that keeps repeating. Suppose that part has, for argument's sake 1000 digits. What this means is that, after a certain point, those thousand digits would repeat over and over. But one of the numbers that occurs in this decimal we constructed is 11111 ... 1 consisting of 1001 1's. What this means is that the repeating part consists of all 1's. Said another way, the "tail" of this decimal consists of all 1's. Finish it.]

6.16 Countable and Uncountable Sets

LAUNCH

1 How many rational numbers are there? How many irrational numbers are there? Are there more rational numbers or irrational numbers?
2 How many real numbers are there? Are there more real numbers than natural numbers? Explain.

You can judge by the launch questions, that we are getting you to think about some very interesting concepts of "countability." When you were asked, "How many natural numbers are there?" did you answer "Infinitely many?" When you were asked how many real numbers there are, did you again answer, "Infinitely many?" When you were asked whether there were more real numbers than natural numbers, did you and your classmates answer differently? Did one person say, "Sure. It is obvious that there are more real numbers than natural numbers?" Did another person say, "Of course not. Both sets are infinite and the word infinite means neither set can be counted. So they both have the same size–infinite!"

This may not strike you as a big issue, but to mathematicians in the 19th century, this was a major problem. To many mathematicians, infinity was infinity. Period! To say that one infinite set has more elements than another infinite set made no sense. In fact, many mathematicians refused to even think much about infinite sets because infinite sets had their own pathological difficulties as the following examples illustrate.

Example 6.65 *Hotel Infinity boasted of the largest number of rooms in the world. They had infinitely many rooms numbered 1, 2, 3, and so on. One day a guest showed up and said "I would like a room." The receptionist said, "I am sorry sir, all our rooms are filled." The prospective guest thought for a moment and then said, "Well, then, move the guest in room number 1 to room number 2, and the guest in room number 2 to room number 3, and so on. This way room 1 will be empty, and I can take it."*

INDEED!

Example 6.66 *(Hotel Infinity continued). The hotel is full. Infinitely many guests arrive. Each is wearing a T-shirt. The T-shirts are numbered 1, 3, 5, . . . , and so on. (It is a rather odd set of guests.) They all want rooms. The receptionist who has learned from the previous example says, "No problem! This is hotel infinity! "So he moves the guest in room 1 to room 2, and the guest in room 2 to room 4, and the guest in room 3 to room 6, and so on. This leaves all the rooms numbered, 1, 3, 5, and so on open. He tells each guest, "Go to the room number on your T-shirt. "And all the guests go to their rooms with a smile.*

These facetious examples show some of the problems involved in studying infinite sets. George Cantor (1845–1918) was a mathematician who wasn't quite ready to accept that all infinite sets were the same. In his own mind he wanted proof that all sets with infinitely many elements had the same size because he didn't believe it! So he set himself to the task of deciding how to compare sizes of infinite sets. He formalized what it means for two infinite sets to be "the same size." His definition made perfect sense. He was guided by what happens with finite sets.

Two finite sets are the same size if their elements can be matched up. Thus, the sets $\{a, b, c, d, e\}$ and the set $\{1, 2, 3, 4, 5\}$ would be the same size because we can match them up in the following way:

$$\begin{array}{ccccc} a & b & c & d & e \\ \updownarrow & \updownarrow & \updownarrow & \updownarrow & \updownarrow \\ 1 & 2 & 3 & 4 & 5. \end{array}$$

He felt the same definition for infinite sets was logical. He simply said that two infinite sets A and B were the same size (or had the same **cardinality**) if their elements could be matched up. With this definition, he could say that the set of natural numbers was the same size as the set of even numbers, because they could be matched up, as we see below:

$$1 \quad 2 \quad 3 \quad 4 \quad \ldots \quad n \quad \ldots$$
$$\updownarrow \quad \updownarrow \quad \updownarrow \quad \updownarrow \quad \updownarrow \quad \updownarrow \quad \ldots \qquad\qquad (6.101)$$
$$2 \quad 4 \quad 6 \quad 8 \quad \ldots \quad 2n \quad \ldots .$$

That is, with each natural number he matched its double.

Cantor's goal was to determine once and for all if all infinite sets could be matched up with one another. If they could be, then all infinite sets had the same size or cardinality. If they couldn't be, then there were different size infinities.

If you are one of those who thinks that all infinite sets have the same size, you are in for a BIG shock, for Cantor showed they don't. This discovery wreaked havoc in the mathematics community much like the discovery of irrational numbers wreaked havoc on the Greeks. Mathematicians at first refused to believe this. But the reasoning was sound.

Now it is accepted that different infinite sets can have different sizes, and that is a standard part of a pure mathematics major's education. Let us begin this fascinating journey concerning the size of infinite sets.

Essentially, Cantor began with the definition of countable. We say that an infinite set is **countable** if it can be matched in a one-to-one fashion with the natural numbers. (Some books use the word countably infinite. We won't do that.) Saying that a set can be matched in a one-to-one manner with the natural numbers means that each natural number is matched with only one element of the set and each element of the set is matched with only one natural number. Thus, the set of natural numbers is countable (since it can be matched up with itself: 1 is matched with 1, 2 with 2, and so on). The even integers are also countable. We saw this in equation (6.101), where we presented the matching.

There is a very convenient way of describing a countable set. Any set whose elements can be listed in a row is countable. Thus, the set of numbers 1, 2, 3, ... is countable since we have listed the elements in a row. The set of numbers 2, 4, 6, ... is countable since its elements have been listed in a row. Why is a set whose elements can be listed in a row countable? Well, if we match the first number in the listing with 1, match the second number in the listing with 2, and so on, we have our matching with the natural numbers. Thus, the set is countable.

Our first theorem tells us that there are as many rational numbers as there are natural numbers. Does this surprise you?

Theorem 6.67 *The set of rational numbers with positive numerators and denominators is countable.*

Proof. What we need to show is that we can list the elements of this set in a row.

Now, let us begin by writing all rational numbers with denominators of 1 in a row. In the next row let us put all rational numbers with denominators of 2. In the third row we put all rational numbers with denominators of 3, and so on. So what we have looks like:

$$\frac{1}{1} \quad \frac{2}{1} \quad \frac{3}{1} \quad \frac{4}{1} \quad \ldots$$

$$\frac{1}{2} \quad \frac{2}{2} \quad \frac{3}{2} \quad \frac{4}{2} \quad \cdots$$

$$\frac{1}{3} \quad \frac{2}{3} \quad \frac{3}{3} \quad \frac{4}{3} \quad \cdots$$

$$\frac{1}{4} \quad \frac{2}{4} \quad \frac{3}{4} \quad \frac{4}{4} \quad \cdots$$

and we have succeeded in listing all the rational numbers in an infinite number of rows. The issue now is, how can we "string them out" in one row?

Cantor solved this ingeniously by doing what is called a **diagonal argument**. He lists the elements by "following the arrows" in the picture below. Essentially, he is moving along the diagonals.

$$\frac{1}{1} \to \frac{2}{1} \quad \frac{3}{1} \to \frac{4}{1}$$
$$\frac{1}{2} \quad \frac{2}{2} \quad \cdots \frac{3}{2}$$
$$\frac{1}{3} \quad \frac{2}{3} \quad \frac{3}{3}$$
$$\frac{1}{4} \to$$

So here is our listing: $\frac{1}{1}, \frac{2}{1}, \frac{1}{2}, \frac{1}{3}, \frac{2}{2}, \frac{3}{1}, \frac{4}{1}, \frac{3}{2}, \frac{2}{3}, \frac{1}{4}$, and so on. Only some numbers are repeated. So as we list them, we make sure that we never list a number already listed. Clearly, every rational number is in this listing. For example, if we ask whether 99/100 in the listing, we can answer "Yes, we encountered $\frac{99}{100}$ in the 100th row and 99th column and, since we traversed all the diagonals, it is picked up in our listing." The same can be said for any other rational number. Thus, all rational numbers have been listed in a row, and the set of rational numbers is countable. ∎

Neat!

You may be thinking, "What is the big deal? We can do this diagonal argument for any infinite set. So all infinite sets are countable!" If you think that, then the following will surprise you.

Theorem 6.68 *The set of real numbers between 0 and 1 is NOT countable.*

What this means is that, even though this set is infinite, its elements cannot be listed, in a horizontal row. No matter who tries, no one will succeed! Thus, this infinity represents a different size infinity! It is mind boggling! Here is the proof.

Proof. Suppose we can list the real numbers in a row. We will show this leads to a contradiction. Let us assume we can list the real numbers $r_1, r_2.r_3$, and so on. Now if we can list them in a row, we can list them in a column and vice versa. For convenience of exposition, it is easier to list them in a column.

Since each real number can be written as a decimal, we would like to write, say, $r_1 = 0.d_1d_2d_3\ldots$ where d_1 represents the first digit of the decimal expansion of r_1, d_2 the second digit of r_1, and so on. However, we have infinitely many r's and infinitely many digits in their decimal expansions, so we will be forced to use subscripts as well as superscripts. The subscript will tell us which digit we are looking at while the superscript will tell us which real number we are focusing on. Thus, if we want to write the decimal expansion of the first real number, it will look like $r_1 = 0.d_1^1 d_2^1 d_3^1 \ldots$. Here the superscript 1 tells us we are writing the decimal expansion of the first real number. The digit d_{25}^1 stands for the 25th digit in the decimal expansion of the first number. If we wanted to write the second real number, it could be written as $r_2 = 0.d_1^2 d_2^2 d_3^2 \ldots$. Again, the superscript 2 tells us we are writing the decimal expansion of the second real number. The digit d_{10}^2 would represent the 10th digit of the 2nd real number, r_2, in the list, and so on.

So here is our supposed listing of the real numbers. We have bolded the diagonal entries for a reason. You will see why shortly.

$$r_1 = 0.\mathbf{d_1^1} d_2^1 d_3^1 \ldots$$
$$r_2 = 0.d_1^2 \mathbf{d_2^2} d_3^2 \ldots$$
$$r_3 = 0.d_1^3 d_2^3 \mathbf{d_3^3} \ldots$$

(6.102)

and so on.

We will show that this list cannot possibly contain all real numbers. So numbers are missing.

Consider the following number r where $r = 0.e_1 e_2 e_3 \ldots$ and each e_i is either 1 or 2. Specifically, we look at d_1^1 in listing (6.102). If it is 1, we choose e_1 to be 2. If d_1^1 is any other digit, we take e_1 to be 1. Thus, the first digit of r differs from the first digit of r_1 in listing (6.102).

Now we look at d_2^2. If it is 1, we choose e_2, the second digit in our newly constructed number, r, to be 2. If it is any other digit, we take e_2 to be 1, and so on. Thus, r differs from the second digit of r_2 in listing (6.102). We continue in this manner choosing e_3 to be 1 or 2 but different from d_3^3 and e_4 to be 1 or 2 but different from d_4^4 and so on. The number r we constructed therefore differs from each element in the listing (6.102) by one digit, and thus must be different from every number in the listing (6.102). Thus, any listing of real numbers cannot pick up all real numbers. It follows from this that the set of real numbers cannot be listed and therefore is not countable. ∎

Is this not fascinating? What this is telling us is that the set of real numbers between 0 and 1 is a different infinity from the set of rational numbers between 0 and 1. The rational numbers between 0 and 1 can be listed in a horizontal row. The real numbers can't be. Since the set of rational numbers between 0 and 1 is a subset of the set of real numbers between 0 and 1, the only conclusion that we can make given that they are different infinities is that there are more real numbers than rational numbers. That is, the set of real numbers constitutes a larger infinite set. It is not countable.

Of course, if the number of real numbers in the interval (0, 1) is more than the number of rational numbers in (0, 1), how does the cardinality of *all* real numbers compare to the cardinality of real numbers in (0, 1). Again, our gut tells us, "Well, it has to be more." Once again, we are wrong.

The cardinality of the set of all real numbers is the *same* as the cardinality of the real numbers in (0, 1). That is, there are as many real numbers as there are real numbers in (0, 1). This seems incredible. But if we can match the set real numbers in (0, 1) with the set of all the real numbers, then this result will follow. We can give a simple picture that illustrates this. Imagine we have the real number line and that above it we have the interval from (0, 1) only we bend the interval (0, 1) into a semicircle as shown in Figure 6.12 below.

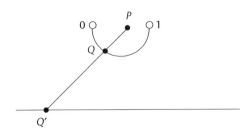

Figure 6.12

Above that, put a point P. Now, draw a line from P to any point Q on the curve representing the interval (0, 1). If we extend this line to the real number line below, it hits it at a point Q'. We match Q with Q'. This is our matching between the real numbers in the interval (0, 1) and all real numbers. Thus, these sets have the same size. We notice that the closer we are to the endpoint of the interval (0, 1) the further out we go on the real number line. (For those who don't find the picture satisfying, here is a function that sets up a one-to-one correspondence between (0, 1) and the real line: $f(x) = \tan\left(\frac{\pi}{2} \cdot (2x - 1)\right)$. As x varies from 0 to 1, $\frac{\pi}{2} \cdot (2x - 1)$ varies from $-\frac{\pi}{2}$ to $\frac{\pi}{2}$ and hence $\tan x$ varies from $-\infty$ to ∞. And, since the tan function is increasing on this interval, this function is one-to-one. So different x's yield different y's.) We state that as a theorem.

Theorem 6.69 *The set of all real numbers has the same cardinality as the set of real numbers between 0 and 1.*

Cantor took this further and proved something even more astounding. He proved that the number of points inside a square has the same cardinality as the interval (0, 1). (See Figure 6.13.)

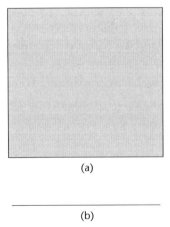

Figure 6.13 The number of points in the square in (a) is the same as the number of points on the line in (b).

(We won't prove this.) When he discovered this he remarked, "I see it, but I don't believe it!"

What happens when we unite two countable sets? Is the resulting set still countable? The answer is "Yes," and it is easy to show: Suppose that we have two countable sets, S and T and that the elements of S are $s_1, s_2, s_3 \ldots$ and those of T are $t_1, t_2, t_3, \ldots$. Then, if we can list the elements of the two sets in a row, the union of those sets is countable. Here is the listing.

$s_1, t_1, s_2, t_2, s_3, t_3 \ldots$.

Thus, we have:

Theorem 6.70 *The union of two countable sets is countable.*

Corollary 6.71 *The union of any finite number of countable sets is countable.*

The proof of the corollary is left as a Student Learning Opportunity.
The next theorem is yet another surprise.

Theorem 6.72 *The irrational numbers are not countable.*

Proof. We know the set of rationals is countable. If the set of irrationals was also countable, then their union, the set of real numbers, would be countable. But, we saw in Theorem (6.68) that the set of real numbers is uncountable. So, this contradiction tells us that the set of irrational numbers cannot be countable. ∎

Countable sets represent the smallest infinity. The set of rational numbers is countable and its size represents the smallest size that any infinite set can have. The next size infinity, as far as we know, is the size of the real numbers. So we already have two sizes of infinity. What is surprising is that it doesn't stop here. Given any size infinity, one can always find a larger one. Thus, there are infinitely many different sizes of infinity! We refer the reader to the many articles on the Internet on this topic.

6.16.1 Algebraic and Transcendental Numbers Revisited

In Chapter 3, we discussed the concept of algebraic and transcendental numbers. Every rational number, like 2/3, is the solution of a polynomial equation with integral coefficients. In this case, 2/3 is the solution of $3x - 2 = 0$ and any other rational number, p/q is the solution of $qx - p = 0$. Certain irrational numbers are also solutions of equations with integral coefficients. For example, $\sqrt{2}$ is the solution of $x^2 - 2 = 0$, and $\sqrt[3]{3}$ is a solution of $x^3 - 3 = 0$. We recall that any number that is a solution of a polynomial equation with integral coefficients is called algebraic. Since every rational number satisfies a polynomial equation with integral coefficients, every rational number is algebraic. Any number which is not algebraic, is called transcendental.

Any transcendental number must be irrational since, if it were rational, it would be algebraic as pointed out in the last paragraph. We described in Chapter 3 how difficult it was to find transcendental numbers, and how it took many years for someone to construct a transcendental number (the mathematician Louisville). Then, after great effort, it was proved that both π and e were transcendental. So we had 3 transcendental numbers. It seemed that transcendental numbers were pretty scarce. Then Cantor once again shocked us.

304 Building the Real Number System

> **Theorem 6.73** *The number of transcendental numbers is uncountable.*

Thus, not only are transcendental numbers not rare, they are abundant! They are even more abundant than rational numbers. Again, this is mind boggling. We outline the proof of this in the Student Learning Opportunities. Since this goes well beyond the secondary school curriculum, we simply stated the theorem as a matter of historical interest.

Student Learning Opportunities

1 Show that the set of multiples of 3 is countable.

2 How could you convince your students that there are as many cubes of natural numbers as there are natural numbers?

3 One of your students asks how you can make a one-to-one correspondence between the counting numbers and the whole numbers. How do you do this?

4 (C) Your students are really intrigued by the mysteries of cardinality. They ask you if it is possible to give a geometric proof that any open interval has the same cardinality as (0, 1). Show them how to do it. [Hint: Draw the two intervals parallel, with the larger one on the bottom. Connect the endpoints of the intervals to get a triangle. What is your matching of the intervals?]

5 (C) Your students want to know if there are more rational numbers than irrational numbers or are they equal in number. How do you respond and help your students understand your explanation?

6 Show that if the sets, $S_1, S_2, \ldots, S_n$ are each countable, then so is their union. (In words: The union of a finite number of countable sets is countable.)

7 Show, using the diagonal argument, that if $S_1, S_2, \ldots, S_n \ldots$ is a countable collection of countable sets, so is their union.

8 (C) Your students ask if it is possible to show that the set of *all* rational numbers is countable. How do you reply and prove your point?.

9 Suppose we have a polynomial of degree *n with integral coefficients*. We define the height of the polynomial to be the degree added to all the absolute values of the coefficients. Thus, the height of the polynomial $x^2 - 3x + 4$ is $2 + |1| + |-3| + |4|$ or 10.
Let P_k denote the set of polynomials of height k. Thus, P_1 is the set of polynomials whose degree added to the absolute value of the coefficients is 1. (There are only two such polynomials here, 1 and -1.) P_2 is the set of polynomials whose height is 2. (This contains polynomials like x and $-x$.)

    (a) Explain why each set P_k has only a finite number of elements.

    (b) Suppose that we find all the roots of all polynomials in P_k for each k. Show that we get a finite set.

(c) Show that, if we unite all the roots of all the polynomials in all the P_k's, we get all the algebraic numbers.

(d) Show all the algebraic numbers are countable.

10 **(C)** One of your students makes the suggestion that all irrational numbers are transcendental. Is the student correct? Explain.

11 Show that each of the following numbers is algebraic by finding a polynomial with integral coefficients that it satisfies.

(a) $\sqrt[4]{4}$
(b) $1 + \sqrt{3}$
(c) $\sqrt{2} + \sqrt{3}$
(d) $\sqrt{2} + \sqrt[3]{3}$

CHAPTER 7

BUILDING THE COMPLEX NUMBERS

7.1 Introduction

When secondary school students are first introduced to imaginary numbers, they find the whole topic somewhat baffling. Although they usually don't have too much trouble with the manipulation of complex numbers, they are usually bewildered by their meaning. They are often told that i was invented to solve the equation $x^2 + 1 = 0$, and so are left with the impression that imaginary numbers are just something that mathematicians dreamed up to help continue on in their abstract excursions. What students are rarely told is that, like them, for quite a long time many mathematicians were also reluctant to accept imaginary numbers. But, it wasn't too long after their discovery that the use of imaginary numbers enabled mathematicians to find the answers to many difficult questions left unanswered by the use of real numbers alone. And most surprising, it turned out that imaginary numbers have many practical applications! In fact, because of their extensive uses in engineering, many people studying to be engineers are *required* to take courses involving complex numbers. In this chapter we begin by talking about some of the interesting issues with complex numbers, and then address many of the topics included in precalculus courses, and finally connect them to higher level concepts in mathematics. By the end of the chapter, we will expose you to some connections and applications you most likely have not seen before, that will hopefully leave you with a sense of the power, as well as the mysteries of complex numbers.

7.2 The Basics

LAUNCH

Find two real numbers whose sum is 10 and whose product is 40.

You will probably be very surprised to know that this launch problem is quite famous, as it was first posed by Cardan, in his well known algebra book *Ars Magna* (The Great Art), published in 1545 (Dunham, 1994, p. 288). Let us examine the trouble he encountered when he used an algebraic approach. He called the numbers x and y, then obtained the equations $x + y = 10$ and

$xy = 40$. Since $y = 10 - x$, he substituted it in the second equation to get $x(10 - x) = 40$, which yielded the quadratic equation $x^2 - 10x + 40 = 0$. The quadratic formula yielded the solutions for x: $x = \frac{10 \pm \sqrt{-60}}{2}$. When he noticed the $\sqrt{-60}$, he realized that he hit a stumbling block. Since $\sqrt{-60}$ was a meaningless quantity in his day, he was led to believe that there is no solution to this particular problem. Nevertheless, Cardan pressed on. He chose $x = \frac{10 + \sqrt{-60}}{2}$ and solved for y to get $y = 10 - x = 10 - \frac{10 + \sqrt{-60}}{2}$. He treated these seemingly meaningless objects as if they were numbers, and computed y: $y = \frac{20}{2} - \frac{10 + \sqrt{-60}}{2} = \frac{10 - \sqrt{-60}}{2}$. Now, he added $x + y$ as one would do with real numbers, to get $x + y = \frac{10 + \sqrt{-60}}{2} + \frac{10 - \sqrt{-60}}{2} = \frac{20}{2} = 10$ which checked. Then he multiplied the fractions the way we usually do and assumed that $\sqrt{-60} \cdot \sqrt{-60} = -60$, even though these expressions had no meaning to him. He got: $xy = \frac{10 + \sqrt{-60}}{2} \cdot \frac{10 - \sqrt{-60}}{2} = \frac{100 + 10\sqrt{-60} - 10\sqrt{-60} - (-60)}{4} = \frac{160}{4} = 40$, and this also checked. The point is that, even though he had no idea about what the square roots of negative numbers meant, he multiplied and added these expressions *as if* they were real numbers and got correct answers. This was convincing enough to others so perhaps we should take a closer look at these expressions.

Mathematicians began examining expressions involving square roots of negative numbers. They began with the symbol $\sqrt{-1}$. Consistent with the computations we did in the previous paragraph, they declared that, whatever this symbol meant, it would have to satisfy $\sqrt{-1} \cdot \sqrt{-1} = -1$. To simplify the representation of $\sqrt{-1}$, they called this new symbol "i" for imaginary number. After all, it was not a real number in the usual sense. By definition, $i^2 = \sqrt{-1} \cdot \sqrt{-1} = -1$. Of course, it was desired that $\sqrt{-4} \cdot \sqrt{-4}$ also be -4. So one had to figure out how to define $\sqrt{-4}$. A logical guess was to define $\sqrt{-4}$ to be $2i$ since $2i \cdot 2i$ (assuming that we can rearrange expressions when multiplied) would give us $4i^2$ which is -4. This follows since i^2 was defined to be -1. Continuing in this manner, the square root of any negative number was defined in an analogous way. So, if N is a positive number (so that $-N$ is negative), we define

$$\sqrt{-N} = \sqrt{N} \cdot i.$$

Thus, $\sqrt{-3} = \sqrt{3} \cdot i$ by definition! Similarly $\sqrt{-16} = \sqrt{16} \cdot i = 4i$ by definition.

We have one last step. When solving the equation $x^2 - 10x + 40 = 0$ earlier, we got as solutions $\frac{10 + \sqrt{-60}}{2}$ and $\frac{10 - \sqrt{-60}}{2}$. Thus, it seemed that we had to be able to make sense of an expression like $10 + \sqrt{-60}$, which is some combination of real and imaginary numbers. Since our previous definition of $\sqrt{-60}$ was $\sqrt{60} \cdot i$, this number $10 + \sqrt{-60}$ can be written as $10 + \sqrt{60} \cdot i$. This leads to our final definition: A **complex number** is a *symbol* of the form $a + b\sqrt{-1}$, or just $a + bi$, where a and b are real numbers and i is an abbreviation for $\sqrt{-1}$. a is called the real part of the complex number, and b the imaginary part. Notice that, at this point, these are just *symbols*. They have no meaning. We have to decide now how to operate on these symbols. That is, we have to give them some semblance of structure.

7.2.1 Operating on the Complex Numbers

Before operating fully with complex numbers we must first decide what it means for two complex numbers to be the same. We will say that **two complex numbers are the same** if the real and imaginary parts are equal. Thus, if we say that $(x - 2) + (y + 1)i = 4 - 3i$, where $4 - 3i$ means $4 + (-3)i$, then $x - 2$ must be 4, and $y + 1$ must be -3.

We now define **addition and subtraction of complex numbers**. We define addition of two complex numbers in a natural way that mimics what happens in algebra if we treat i as a variable.

That is, we add complex numbers by adding the real and imaginary parts of them. In symbols we define $(a + bi) + (c + di)$ to mean $(a + c) + (b + d)i$ and, of course, subtraction is defined as you would expect: $(a + bi) - (c + di) = (a - c) + (b - d)i$. Thus, from our definitions of addition and subtraction, we have that $(2 + 3i) + (4 + 6i) = (2 + 4) + (3 + 6)i = 6 + 9i$ and that $(2 + 3i) - (4 + 6i) = (2 - 4) + (3 - 6)i = -2 - 3i$.

We will define multiplication of complex numbers the way we multiply binomials in algebra. If in algebra we had to multiply $(a + bi)(c + di)$, we would get $ac + adi + bci + bdi^2$. If we do this for complex numbers, remembering that $i^2 = -1$, this last expression simplifies to just $(a + bi)(c + di) = ac + adi + bci - bd = (ac - bd) + (bc + ad)i$. We emphasize this *definition* of **multiplication of complex numbers**:

DEFINITION: $(a + bi)(c + di) = (ac - bd) + (bc + ad)i.$ (7.1)

Thus, $(2 + 3i)(4 - 5i) = (2(4) - (3)(-5)) + (3(4) + 2(-5))i = 23 + 2i$.

Since we created the rules for working with imaginary numbers, we can't just blindly work with them as we would expect them to behave. For example, are addition and multiplication of complex numbers commutative, and associative? Is there a distributive law? One can take several specific examples and verify that the commutative, associative, and distributive laws hold for those examples, but again, this doesn't mean it holds for all complex numbers. Perhaps we were just lucky enough to pick the numbers for which it held. Thus, we need a proof, which will be forthcoming.

Before turning to division, we talk about the **conjugate** of a complex number. With each complex number $z = a + bi$, there is a number $\bar{z}$ called its conjugate. By definition, $\bar{z} = a - bi$. Thus, if $z = 3 + 2i$, then $\bar{z} = 3 - 2i$. One of the nice things about conjugates that we will use is that, if a number is multiplied by its conjugate, the result is a real number. Thus, if $z = 3 + 2i$, $z\bar{z} = (3 + 2i)(3 - 2i) = 9 - 4i^2 = 9 - 4(-1) = 13$. We can now move on to division of complex numbers.

The same way we divide ordinary numbers, we wish to divide imaginary numbers. But then what complex number should $\frac{(2+3i)}{(4-5i)}$ be? Since we want to express this in the form $a + bi$ where a and b are real numbers, we need to get rid of the complex denominator. One way to do this is to multiply the denominator of the fraction by $4 + 5i$. Since for real numbers we are allowed to multiply the numerator and denominator of a fraction by the same quantity, the part of us that likes consistency says we should be able to do this for complex numbers too. So it seems that $\frac{(2+3i)}{(4-5i)} = \frac{(2+3i)}{(4-5i)} \cdot \frac{(4+5i)}{(4+5i)} = \frac{8+22i+15i^2}{16-25i^2} = \frac{-7+22i}{41} = \frac{-7}{41} + \frac{22}{41}i$. But the symbol $\frac{(2+3i)}{(4-5i)}$ has no meaning! So what gives us the right to multiply the numerator and denominator of a meaningless expression by the conjugate and proceed as we did? The way to get out of this hole is to define, for example, $\frac{(2+3i)}{(4-5i)}$ to be the complex number obtained by multiplying the numerator and denominator by the number $4 + 5i$, the *conjugate* of the denominator, and then split the fraction into real and imaginary as if it made sense. We do this for all symbols of this type. Thus, we *define* the **quotient of two complex numbers**, $\frac{(a+bi)}{(c+di)}$ by

DEFINITION: $\dfrac{(a + bi)}{(c + di)} = \dfrac{ac + bd}{c^2 + d^2} + \dfrac{bc - ad}{c^2 + d^2}i.$ (7.2)

(This is the result of multiplying the numerator and denominator of $\frac{(a+bi)}{(c+di)}$ by $c - di$, the conjugate of $c + di$, and splitting up the fraction into real and imaginary parts, as you should check.)

In particular, if $a = 1$ and $b = 0$, we get

$$\frac{1}{c+di} = \frac{c}{c^2+d^2} + \frac{-d}{c^2+d^2}i. \tag{7.3}$$

Note that we still have not established the validity of multiplying numerators and denominators of fractions with complex numbers by the same complex number, nor if it is valid to multiply complex fractions the same way we do real fractions, namely by multiplying numerators and denominators. We have also not established if, when dividing two fractions involving complex numbers, it is valid to invert and multiply. As you can see, at this point we really know very little about what we can and cannot do with complex numbers. But don't despair, as we will soon know a great deal more. We will soon show all of these properties. First, however, we establish some basic laws. We will follow the convention that complex numbers are denoted by the letter z. The following theorem may not surprise you, but given the unusual definition of multiplication of complex numbers, this requires some work to verify.

Theorem 7.1 *For any complex numbers z_1, z_2, and z_3*

(a) $z_1 + z_2 = z_2 + z_1$ *(Commutative Law of Addition)*

(b) $z_1 \cdot z_2 = z_2 \cdot z_1$ *(Commutative Law of Multiplication)*

(c) $(z_1 + z_2) + z_3 = z_1 + (z_2 + z_3)$ *(Associative Law of Addition)*

(d) $(z_1 \cdot z_2) \cdot z_3 = z_1 \cdot (z_2 \cdot z_3)$ *(Associative Law of Multiplication)*

(e) $z_1 \cdot (z_2 + z_3) = z_1 \cdot z_2 + z_1 \cdot z_3$ *(Distributive Law)*

(f) $z_1 \cdot 1 = z_1$ *where $1 = 1 + 0i$. (Law of Multiplicative Identity)*

(g) *For each complex number z, $z + 0 = z$, where 0 means $0 + 0i$. (Zero Law)*

(h) *For each complex number $z \neq 0$, there is a complex number z^{-1} such that $z \cdot z^{-1} = 1$. (Existence of Multiplicative Inverse)*

Proof. All of these results can be proved by brute force writing $z_1 = a + bi$, $z_2 = c + di$, and $z_3 = e + fi$ and then just verifying that the left hand sides and right hand sides of each of the expressions in (a)-(h) match. For example, to prove (a),

$z_1 + z_2 = (a + bi) + (c + di)$

$\qquad = (a + c) + (b + d)i$ by the definition of addition of complex numbers

$\qquad = (c + a) + (d + b)i$ because *addition for the real numbers, a, b, c, d is commutative*

$\qquad = z_2 + z_1$ by the definition of $z_2 + z_1$.

∎ Notice that, to prove commutativity for the complex numbers, we needed to use the commutativity of the real numbers. You will find that, to prove associative laws and distributive laws for complex numbers, you will have to use the corresponding rules for real numbers. We leave the proofs of most of (b)–(e) for the Student Learning Opportunities.

To prove (f), we use the *definition* of multiplication $z_1 \cdot 1 = (a + bi)(1 + 0i) = (a \cdot 1 - b \cdot 0) + (b \cdot 1 + a \cdot 0)i = a + bi$ since $a \cdot 1 = a$ for real numbers, and since $b \cdot 0 = 0$ for real numbers.

To prove (h) we take as a candidate for z^{-1} the complex number $\frac{a}{a^2+b^2} + \frac{-b}{a^2+b^2}i$. We are motivated to do this by equation (7.3). Now, using the definition of multiplication for complex numbers, we can show that $z \cdot z^{-1} = (a+bi)(\frac{a}{a^2+b^2} + \frac{-b}{a^2+b^2}i) = 1$. You will do this in the Student Learning Opportunities. It is a good algebraic exercise.

The next theorem is more sophisticated. It tells us that we can operate with fractions whose numerators and denominators are complex numbers the same way we can do with fractions whose numerators and denominators are real numbers. That is, we multiply fractions whose numerators and denominators are complex numbers by multiplying numerators and denominators. This is not at all obvious since fractions involving complex numbers are defined in a rather unusual way. Verifying this by using the definitions alone is quite tedious. We will, however, give elegant proofs.

Before giving the proof of the next theorem, we should comment that the associative and commutative laws can be used to show (although it is a bit tedious) that, if a group of complex numbers are being multiplied together, the order in which they are multiplied does not matter, nor do we need parentheses, although we can insert them if we wish, anywhere that we wish. Thus, if we have an expression like $z(w^{-1}c)(d^{-1}z^{-1})$, we can simply write this as $zz^{-1}w^{-1}cd^{-1}$ without any parentheses or as $zz^{-1}(w^{-1}c)d^{-1}$ if we wish. The result is the same. We will use this fact a few times in the next proof.

Theorem 7.2 *If $z, z_1, z_2, z_3,$ and z_4 are complex numbers, then*

(a) $\dfrac{z_1}{z_2} = z_1 \cdot z_2^{-1}$

(b) $\dfrac{z_1 + z_2}{z_3} = \dfrac{z_1}{z_3} + \dfrac{z_2}{z_3}$

(c) $\dfrac{1}{z} = z^{-1}$

(d) $\dfrac{z_1}{z_2} = z_1 \cdot \dfrac{1}{z_2}$ *(This can also be stated as $\dfrac{z_1}{z_2} = z_1 z_2^{-1}$.)*

(e) *There is only one complex number w such that $zw = 1$. (Uniqueness of multiplicative inverse.)*

(f) $\dfrac{1}{z_1 z_2} = \dfrac{1}{z_1} \cdot \dfrac{1}{z_2}$ *(This can also be stated as $(z_1 z_2)^{-1} = (z_1^{-1} z_2^{-1})$.)*

(g) $\dfrac{z_1 z_2}{z_3 z_4} = \dfrac{z_1}{z_3} \cdot \dfrac{z_2}{z_4}$

(h) *If $z_1 \cdot z_2 = 0$, then either $z_1 = 0$ or $z_2 = 0$.*

Proof. The tendency is to try to prove these relationships the way we did the last theorem, namely by writing z_1 as $a+bi$ and z_2 as $c+di$. If we did that, the proofs would be very tedious and not particularly interesting. But, we now have the previous theorem at our disposal, and having done some of the ground work, we can now provide elegant proofs of these results. The only part that requires a bit of work is part (a).

Proof. (a) In the proof of part (h) of the previous theorem we showed that, if $z_2 = c+di$, then $z_2^{-1} = \frac{c}{c^2+d^2} + \frac{-d}{c^2+d^2}i$. Now, letting $z_1 = a+bi$, we have

$$z_1 \cdot z_2^{-1} = (a+bi) \cdot \left(\frac{c}{c^2+d^2} + \frac{-d}{c^2+d^2}i\right) = \frac{ac+bd}{c^2+d^2} + \frac{bc-ad}{c^2+d^2}i \tag{7.4}$$

(using the definition of multiplication of complex numbers). Furthermore, by definition of division, (see equation (7.2))

$$\frac{z_1}{z_2} = \frac{a+bi}{c+di} = \frac{ac+bd}{c^2+d^2} + \frac{bc-ad}{c^2+d^2}i. \tag{7.5}$$

Comparing the expressions we have for $\frac{z_1}{z_2}$ and $z_1 \cdot z_2^{-1}$ in equations (7.4) and (7.5), we see that they are the same, so $\frac{z_1}{z_2} = z_1 \cdot z_2^{-1}$. ∎

Proof. (b) By part (a)

$$\frac{z_1 + z_2}{z_3} = (z_1 + z_2)z_3^{-1}$$

$$= z_1 z_3^{-1} + z_2 z_3^{-1} \quad \text{(By the Distributive Law.)}$$

$$= \frac{z_1}{z_3} + \frac{z_2}{z_3} \quad \text{(Again, by part (a).).}$$

and we are done.

Proof. (c) Use part (a). We leave it for you to do in the Student Learning Opportunities.

Proof. (d) By part (a) $\frac{z_1}{z_2} = z_1 \cdot z_2^{-1} = z_1 \cdot \frac{1}{z_2}$ (by part (c)).

Proof. (e) To show that this is the only number that multiplies z to give you 1, suppose w_1 and w_2 are any two complex numbers that multiply z to give you 1. So $zw_1 = zw_2 = 1$. We use this in the following string of inequalities.

$$w_1$$
$$= w_1(1)$$
$$= w_1(zw_2)$$
$$= (w_1 z)w_2$$
$$= (1)w_2$$
$$= w_2.$$

So $w_1 = w_2$. We have shown that any two complex numbers, w_1 and w_2 that multiply z to give 1 must be the same, so there is only one multiplicative inverse of a given (nonzero) complex number, z.

Proof. (f)

$$(z_1 z_2)\left(\frac{1}{z_1}\right)\left(\frac{1}{z_2}\right)$$

$$= \left(z_1 \cdot \frac{1}{z_1}\right)\left(z_2 \cdot \frac{1}{z_2}\right)$$

$$= (z_1 \cdot z_1^{-1})(z_2 \cdot z_2^{-1}) \quad \text{(By part (a).)}$$

$$= 1 \cdot 1 \quad \text{(Theorem 7.1 part (h).)}$$

$$= 1.$$

Since z_1z_2, when multiplied by $\left(\frac{1}{z_1}\right)\left(\frac{1}{z_2}\right)$, gives us 1 and since z_1z_2, when multiplied by $(z_1z_2)^{-1}$, also gives us 1, it must be that $(z_1z_2)^{-1} = (\frac{1}{z_1})(\frac{1}{z_2})$ by the uniqueness of the inverse shown in part (e).

Proof. (g)

$$\frac{z_1z_2}{z_3z_4}$$

$$= (z_1z_2)(z_3z_4)^{-1}$$

$$= (z_1z_2)(z_3^{-1}z_4^{-1}) \quad \text{(By part (f).)}$$

$$= (z_1z_3^{-1})(z_2z_4^{-1})$$

$$= \frac{z_1}{z_3} \cdot \frac{z_2}{z_4} \quad \text{(By part(d).).}$$

Proof. (h) If $z_1 \cdot z_2 = 0$ and $z_1 \neq 0$, we may multiply both sides of this equation on the left by z_1^{-1} to get $z_1^{-1} \cdot z_1 \cdot z_2 = z_1^{-1} \cdot 0$, which, simplifies to $1 \cdot z_2 = 0$ or just to $z_2 = 0$. We have shown that, if $z_1 \neq 0$, then z_2 has to be zero. You can similarly show that, if z_2 is not zero, then z_1 has to be zero. Thus, when we multiply two numbers and get zero, one or the other must be zero.

It is very encouraging to see that, in many ways, complex numbers behave like real numbers and share important properties like commutative, associate, and distributive laws. This allows us to operate with them as freely as we do with real numbers.

Note: It is customary to consider the real numbers as a subset of the complex numbers. Every real number a can be thought of as $a + 0i$, and thus is complex. This makes sense since $0 \cdot i = 0$, by Student Learning Opportunity 11. Since we are considering the real numbers as a subset of the complex numbers, we are thereby considering the complex numbers to be an *extension* of the real numbers.

Part (c) of the above Theorem 7.2 is telling us that the multiplicative inverse of a complex number, is the reciprocal. The same is true for each real number, since the real numbers are a subset of the complex numbers. Part (d) is telling us that, when you divide two complex numbers, you multiply the numerator by the inverse of the denominator. In particular, if the complex numbers in the numerator and denominator are rational numbers, it is saying that, to divide two fractions, you invert and multiply, since the inverse of a rational number is its reciprocal. (You will show the same thing is true for complex numbers in the Student Learning Opportunities.) This corroborates what we have seen in the previous chapter.

Statement (h) is important when solving equations. Essentially, it says that, if we can factor an expression into a product which is equal to zero, then one or the other factors in that product will be zero. We use this all the time in algebra when solving quadratic equations and equations of higher degree.

Student Learning Opportunities

1 True or False: Every real number is complex.

2 (C) You ask your students to simplify $\sqrt{-16} \cdot \sqrt{-25}$. Two students volunteer to put their work up for the class to see. Sahil writes: $\sqrt{-16} \cdot \sqrt{-25} = \sqrt{400} = 20$. Julio writes: $\sqrt{-16} \cdot \sqrt{-25} = 4i \cdot 5i = 20i^2 = -20$. Who is correct? How do you help clarify the issue?

3 **(C)** One of your students is pursuing the concept of imaginary numbers and asks if $3 + 4i$ is greater than or less than $5 + 2i$. How do you respond in a way that will satisfy your student's curiosity?

4 **(C)** In studying imaginary numbers, your students have learned about the powers of i.(i.e., $i^2 = -1, i^3 = i, i^4 = (i^2)^2 = 1, i^5 = i$, and so on.) They want to know how that will help them find other values such as i^{17}, i^{32}, i^{43}, and i^{14}. What shortcuts do you help them discover to find these values and others that they will encounter?

5 Simplify each of the following as much as possible:
   (a) $3(4 - i) + 3(2 - i)$
   (b) i^{294}
   (c) $3i(i - 1) - 2i(i - 3)$

6 If $x - y + (x + y)i = 4 - 5i$ find x and y assuming x and y are real numbers.

7 Write each of the following in $a + bi$ form:
   (a) $\dfrac{1 - i}{1 + i}$
   (b) $\dfrac{2 - 3i}{4 - 5i}$

8 **(C)** Your student Lisa is asked to find complex numbers x and y other than $x = 4$ and $y = 3$ that make $3x + 5yi = 12 + 15i$. She claims there are none. Is she right? Explain.

9 Prove parts (b) and (c) of Theorem 7.1.

10 Prove parts (d) and (e) of Theorem 7.1. This is a bit tedious.

11 Verify the statement made in part (h) of Theorem 7.1 that $(a + bi)(\frac{a}{a^2+b^2} + \frac{-b}{a^2+b^2}i) = 1$.

12 Prove part (c) of Theorem 7.2.

13 Show that $(2 + i)^3 = 2 + 11i$.

14 In Chapter 3 we solved the cubic equation $x^3 - 15x = 4$ whose solutions we noted were all real and got, using Ferro's formula, that $x = \sqrt[3]{\frac{4+\sqrt{-484}}{2}} + \sqrt[3]{\frac{4-\sqrt{-484}}{2}}$. Now $\sqrt[3]{\frac{4+\sqrt{-484}}{2}} + \sqrt[3]{\frac{4-\sqrt{-484}}{2}} = \sqrt[3]{\frac{4+22i}{2}} + \sqrt[3]{\frac{4-22i}{2}} = \sqrt[3]{2 + 11i} + \sqrt[3]{2 - 11i}$. Show, using the previous problem, that one value of this is 4. (Later, we will establish that nonzero complex numbers have 3 cube roots.)

15 Using only the commutative and associative laws, show that $(z_1 z_2)(z_3 z_4) = (z_1 z_3)(z_2 z_4)$. You must be careful with your parentheses in this proof.

16 **(C)** One of your curious students asks if $0 \cdot z = 0$ if z is a complex number? Is it? How can you prove it? [Hint: Write 0 as $0 + 0i$ and then use the definition of multiplication of complex numbers.]

17 Show, using only the definitions and theorems that we have given that
   (a) $\dfrac{z}{z} = 1$ if z is not zero.
   (b) $\left(\dfrac{z_1}{z_2}\right)^{-1} = \dfrac{z_2}{z_1}$

(c) $\dfrac{\dfrac{z_1}{z_2}}{\dfrac{z_3}{z_4}} = \dfrac{z_1}{z_2} \cdot \dfrac{z_4}{z_3}$

(d) $(z^{-1})^{-1} = z$. This can be written as, $\dfrac{1}{\frac{1}{z}} = z$.

18 Show that the following "cancellation law" holds for complex numbers: $\dfrac{z_1 z}{z_2 z} = \dfrac{z_1}{z_2}$. (If you are thinking, 'Oh, just cancel" then you are missing the point. These are not real numbers. These are a new invention. We don't know if the rules for real numbers hold for complex numbers and the point of this problem is to show that they do!)

19 Show that we add fractions involving complex numbers the same way we add real fractions, namely, $\dfrac{z_1}{z_2} + \dfrac{z_3}{z_4} = \dfrac{z_1 z_4 + z_2 z_3}{z_2 z_4}$.

20 Find the roots of the quadratic equation $x^2 - 2x + 2 = 0$ and check that one of them works by actually substituting it into the equation and showing that you get 0.

7.3 Picturing Complex Numbers and Connections to Transformation Geometry

LAUNCH

1 You have seen how the real number line has been used to represent the real numbers. What are some of the uses of this number line? (List at least three.)
2 How might you create another line that represents all of the purely imaginary numbers (e.g., $2i$, $3i$, $\tfrac{1}{4}i$)? Draw it.
3 You have used two real number lines set at right angles (the coordinate plane) to help you plot points such as (3, 4). Now, in a similar manner, arrange the real number line and imaginary number line that you have created to represent the real and imaginary components of the complex number, $3 + 4i$, as a point.
4 What might be some uses of representing complex numbers with a picture, such as the one you drew in the previous question?

We hope that these launch questions got you thinking more deeply about the value of visual representations of abstract concepts.

We have all heard the expression that "a picture is worth a thousand words." So, you probably realize how helpful it would be if we could create a visual representation of complex numbers similar to the way that a number line is used to represent real numbers. Actually, it took quite a few years for mathematicians to come up with such a, picture, though once you see it, you wonder why it wasn't found sooner. In this section we will describe in detail, how complex numbers can be represented pictorially, and the many uses of representing them in this way.

Every complex number $a + bi$ has a real part a, and an imaginary part with coefficient b. Thus, we can think of a complex number as an ordered pair (a, b) where a is the **real part**, and b is the **imaginary part**. And, of course, that gives us the idea of how to create a picture of complex numbers! We draw what is known as the **complex plane**. This consists of the plane, divided into quadrants by two axes, a real axis, and an imaginary axis. We plot points just as we do in the real plane only now when we plot a pair of numbers, the first coordinate represents the real part and the second coordinate, the imaginary part of a complex number. Thus, when we plot a point $(3, 4)$ (as you did before) the way we do with real numbers, we are really plotting the complex number $z = 3 + 4i$. When we plot $(-2, 5)$, we are plotting $z = -2 + 5i$. When we plot $(4, 0)$, we are plotting the number $4 + 0i$ which we take to be the real number 4. Below, in Figure 7.1, we see a picture of the complex plane and the number $z = 3 + 4i$.

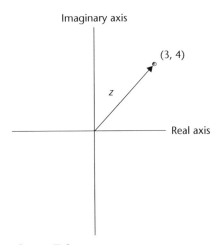

Figure 7.1

Notice that we have drawn an arrow from the origin. The above picture is called an **Argand diagram**. Many books represent complex numbers that way, as arrows, since when we add complex numbers we add them as we do vectors, and vectors are denoted by arrows. The arrow itself has no particular importance. As long as we realize that the point $(3, 4)$ represents the complex number $3 + 4i$, the arrow is not needed.

When we draw the complex plane, it has in it a real axis and an imaginary axis. The real axis is a subset of the complex plane and every real number, a, on the real axis, has coordinates $(a, 0)$ which we know means it can be written as $a = a + 0 \cdot i$. This is consistent with what we have pointed out earlier, namely, that every real number a is considered as the complex number $a + 0i$. Thus, the set of real numbers is a subset of the set of complex numbers. Put another way, we may consider the set of complex numbers to be an extension of the set of real numbers.

Recall that, for every complex number, $z = a + bi$, there is associated another complex number $a - bi$, which is called the conjugate of z. The conjugate of z is denoted by $\bar{z}$ and we can immediately represent $\bar{z}$ with a picture. (See Figure 7.2 below.)

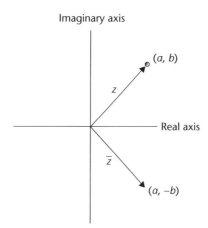

Figure 7.2

Notice that $\bar{z}$ is the reflection of z about the x-axis. (Reflections and other transformations are discussed in more detail in Chapter 10. In this chapter we assume you know the basics of these concepts. If not, see Section 2 of Chapter 10.) Notice this interesting link between complex numbers and transformation geometry. In fact, if we take a complex number $z = a + bi$ and multiply it by i, a very fascinating thing happens: We get $iz = ai + bi^2 = ai - b = -b + ai$. If we plot both the complex number z and iz on the same set of axes, we get (Figure 7.3)

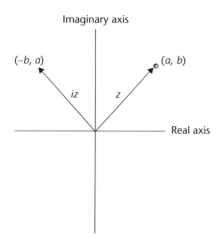

Figure 7.3

The slope of the arrow representing the complex number z is $\dfrac{b-0}{a-0}$ or just $\dfrac{b}{a}$, while the slope of the arrow representing the complex number iz is $\dfrac{a-0}{-b-0}$ or $-\dfrac{a}{b}$. Thus, the slopes of z and iz are negative reciprocals of each other. What this means is that the arrows representing z and iz are perpendicular to one another. Put another way, *when we multiply a complex number, z, by i, we rotate the arrow representing the complex number by 90 degrees counterclockwise.* Thus, we have a geometric representation of multiplication by i, which represents another situation where geometry and complex numbers are connected. First, $\bar{z}$ reflected a complex number z about the x-axis, and then multiplying z by i resulted in a rotation! In a similar manner, multiplying a complex number by $-i$ rotates the complex number 90° clockwise.

318 Building the Complex Numbers

What if we multiplied a complex number $z = a + bi = (a, b)$ by a real number k, which we at first take to be positive? Then, considering the real number k as $k + 0i$ and performing our multiplication, we get $kz = (k + 0i)(a + bi) = (k - 0b)a + (a \cdot 0 + kb)i = ka + kbi = (ka, kb)$. In the figure below (Figure 7.4) we show z and kz.

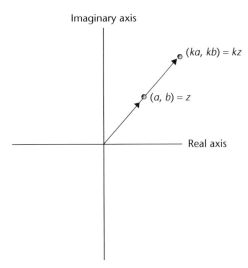

Figure 7.4

What do we notice? We see that kz stretches the arrow representing z by k. That is, *multiplying a complex number by a constant performs a dilation on it!* Wow! Yet another connection! (If k is negative, then kz stretches z in the opposite direction by a factor of $|k|$).

We have hardly begun operating with complex numbers and have already found connections to such transformations as dilations, rotations, and reflections. It is only natural to wonder if arithmetic operations with complex numbers can result in translations. Well, the answer is "Yes!" Suppose we start with a complex number $z = a + bi$ and then add to it the complex number $w = c + di$. We get a new complex number, $u = (a + c) + (b + d)i$. What this does is translate the point z a horizontal distance of c and a vertical distance of d as we see in Figure 7.5 below. In that figure we see z and w and the result of adding them. It is as if the arrow w has been moved so that its tail is at the tip of z. Indeed, this is the "vector interpretation" that you probably recognize for addition of vectors.

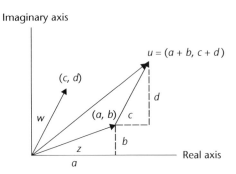

Figure 7.5

Thus, we also get translations also with complex arithmetic! If nothing else, this shows the power of complex numbers.

7.3.1 An Interesting Problem

There is a very interesting problem that one finds in an old book *One Two Three...Infinity* by George Gamow originally published in 1957 and republished in 1988. It is a delightful application of the rotational property of multiplying by *i*. It is about a youngster who, when rummaging through his grandfather's papers, finds directions for locating a treasure. He is to go to a specified island and find the only gallow on the island. From there, he is to walk to the only oak tree on the island, counting his steps. When he reaches the tree, he must turn right and walk the same number of steps and put a stake in the ground where he lands. He then returns to the gallows and walks to the only pine tree nearby, counting the steps again. When he reaches the tree, he should turn left and walk the same number of steps and again place a stake where he lands. The treasure will be midway between the stakes! Here is how this was presented in Gamow's book, where the following was written on the piece of paper that the grandson found.

"Sail to (a certain omitted latitude) north latitude and (a certain omitted longitude) west longitude where thou wilt find a deserted island. There thou wilt find a large meadow, not pent, on the north shore of the island where standeth a lonely oak and a lonely pine. There thoust wilt also see an old gallows on which we once were wont to hang traitors. Start thou from the gallows and walk to the oak counting thy steps. At the oak thou must turn *right* by a right angle and take the same number of steps. Put here a spike in the ground. Now must thou return to the gallows and walk to the pine counting thy steps. At the pine thou must turn *left* by a right angle and see that thou takest the same number of steps and put another stake in the ground. Dig halfway between the stakes: the treasure is there."

George Gamow was a bit of a comic and tells his readers that he left out the latitude and longitude lest some of us drop his book and run out to find the treasure! He also acknowledges that oak and pine trees do not grow on deserted islands, but that he has changed the names of the real trees so that we cannot know what island he is talking about! Of course, for the grandson, the numerical values of the latitude and longitude were written in as well as the real kinds of trees he was referring to. We will stick with oak and pine trees.

The story continues with the young grandson traveling to the island where he does see the trees described, but no gallows. They have disintegrated over the years from the bad weather and no clue as to where the gallows were remains. The young man, saddened that he will not find the treasure, leaves the island.

Before reading on, try to decide if you can figure out a way of finding the treasure. What are the issues you are faced with?

What the grandson didn't know, is that he could have found the treasure. And remarkably, he could have found it by using complex numbers! Here is how it is done.

We begin by placing a set of axes on the map. This will allow us to represent locations and distances. The real axis will join the oak and pine trees from the problem, as well as the origin which will be placed midway between the trees. The imaginary axis will be drawn through the origin. So one of the trees, say the oak tree is at the point $(-d, 0)$ while the other tree, the pine, is at the point $(d, 0)$ since the origin is drawn midway between them. (See Figure 7.6 below.)

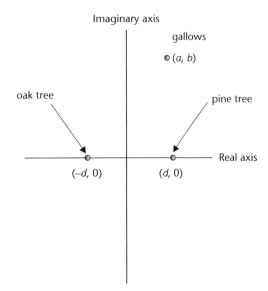

Figure 7.6

Now, since one of the main difficulties is that we don't know where the gallows are, we suppose that the gallows are at some point (a, b) relative to the origin. Now, imagine sliding the axes to the left so that the origin is at the oak tree. Then everything will be d units further away, horizontally, from the original origin than it was before. So, the new coordinates of the gallows are $(a + d, b)$ and the new coordinates of the pine tree are $(2d, 0)$. (See Figure 7.7 below where the axes have been translated.)

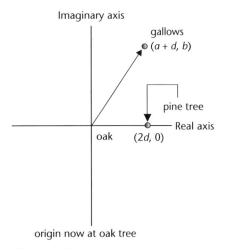

Figure 7.7

To find the location of the first spike, according to the directions, we walk from the gallows to the new origin, then turn right by a **right angle** and walk the same distance, at which point we place a spike which we denote by S_1. Here is our picture where the walk is indicated by a dashed arrow (Figure 7.8).

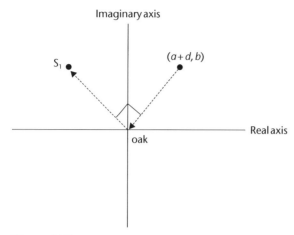

Figure 7.8

What this means is that, to locate the first spike, we can rotate the arrow representing the complex number $(a+d, b)$ 90 degrees counterclockwise, as we see from the above picture. This is accomplished by multiplying the complex number $(a+d, b)$, which represents $a+d+bi$ by i to get $-b+(a+d)i$ or just $(-b, a+d)$. To find the coordinates of this point relative to our original origin, we must remember that, when we first moved the origin to the oak tree, we had to add d to the x coordinate of every point, so now when we move back to our original origin, we have to subtract d from the x coordinate of every point to gets its coordinates relative to the original origin. Thus, the coordinates of the first spike relative to our original origin are

$$S_1 : (-b - d, a + d). \tag{7.6}$$

To find the coordinates of the first spike, we moved the origin to the oak tree. To find the location of the second spike, we will do something similar– we will move the origin to the pine tree. That means that the coordinates of the gallows relative to this new origin are $(a - d, b)$, since each point to the right of the origin is d units closer to the origin and the coordinates of the oak tree are $(-2d, 0)$. (See Figure 7.9 below.)

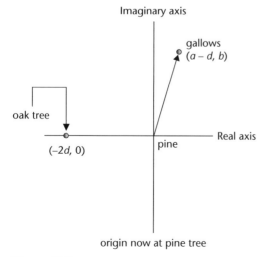

Figure 7.9

322 Building the Complex Numbers

To find the location of the second spike, we walk to the origin from the point representing the location of the gallows and go *left* by a right angle. Here is our picture where we again denote the walk by a dashed arrow (Figure 7.10).

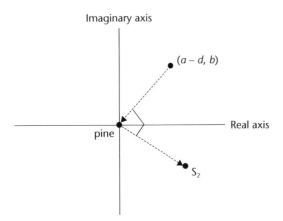

Figure 7.10

This is equivalent to rotating the arrow representing the complex number $(a - d, b)$ clockwise $90°$, as the above picture shows us. To accomplish that, we multiply the complex number $a - d + bi$ by $-i$ to get the complex number $b - (a - d)i$ or just $(b, -(a - d))$.

Since we originally subtracted d when we moved the origin to the pine tree, to find the coordinates of this second spike relative to the *original* origin, we have to add d to each x coordinate. (We are translating the origin back to its original position.) Thus, the location of the second spike relative to our original origin is

$$S_2 : (b + d, -(a - d)). \tag{7.7}$$

Now, the treasure is halfway between the first and second spikes. To find this point, we just use the midpoint formula from secondary school. We add the x and y coordinates of the spike locations given by (7.6) and (7.7) and divide by 2 to get the location of the treasure, which is $\left(\frac{(-b - d) + b + d}{2}, \frac{a + d - (a - d)}{2}\right) = (0, d)$. Notice that the a's and b's summed to zero when we used the midpoint formula. That is, we didn't need to know the coordinates (a, b) of the gallows! Our treasure is at the location $(0, d)$ or just the complex number di. But we know what d is. It is half the distance between the trees! So we know exactly where to find the treasure. We move from our origin, which we took to be midway between the trees, and move up along our imaginary axis a distance of d units. That is where we find the treasure.

Isn't this a nice problem?

Are you still there? Hmm, we bet you too are out looking for the island!

We hope you have enjoyed how we were able to use complex numbers to solve this fictitious problem. But we want you to know that complex numbers are extremely powerful in the solution of critical real-life areas such as: control theory, signal analysis, electromagnetism, quantum mechanics, solving differential equations, fluid dynamics, aerodynamics, and fractals—and the list is still growing.

7.3.2 The Magnitude of a Complex Number

Another useful concept is the **magnitude** of a complex number z, which is denoted by $|z|$. This is the distance the complex number is from the origin in the Argand diagram, and is often represented by the letter r. From the Pythagorean Theorem (look at the Figure 7.11 below)

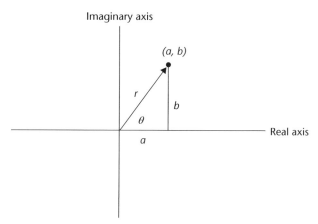

Figure 7.11

we find that $r = |z| = \sqrt{a^2 + b^2}$. Thus, $|3 + 4i| = \sqrt{3^2 + 4^2} = 5$.

There are some results about conjugates and magnitudes that we should point out, which we will use in this chapter.

Theorem 7.3 *If z_1 and z_2 are complex numbers, then*

(a) $\overline{z_1 + z_2} = \overline{z_1} + \overline{z_2}$
(b) $\overline{z_1 \cdot z_2} = \overline{z_1} \cdot \overline{z_2}$
(c) $\overline{\overline{z_1}} = z_1$
(d) $z_1 \cdot \overline{z_1} = |z|^2$
(e) $\overline{z_1}$ *is real if and only if z_1 is real.*

Proof. Most of the proofs are simple and are left to the Student Learning Opportunities. For example, to show that (a) is true, write $z_1 = a + bi$ and $z_2 = c + di$ then $z_1 + z_2 = (a + c) + (b + d)i$, $\overline{z_1 + z_2} = (a + c) - (b + d)i$, $\overline{z_1} = a - bi$, $\overline{z_2} = c - di$. Now, when we add $\overline{z_1} + \overline{z_2}$ we get $(a - bi) + (c - di) = (a + c) - (b + d)i$, which is nothing more than $\overline{z_1 + z_2}$.

To get a better sense of what the rules are saying, you can just take some specific examples of complex numbers and check the rules for those numbers. ■

Parts (a) and (b) can be generalized to any number of complex numbers. The generalizations are:

$$\overline{z_1 + z_2 + \ldots + z_n} = \overline{z_1} + \overline{z_2} + \ldots + \overline{z_n}$$

and

$$\overline{z_1 \cdot z_2 \cdot \ldots \cdot z_n} = \overline{z_1} \cdot \overline{z_2} \cdot \ldots \cdot \overline{z_n}$$

These are proved by induction. (See Chapter 8 for a review of induction.)

Student Learning Opportunities

1 (C) Your students want to know what the similarities and differences are between the coordinate plane and the complex plane. What are some of the things you can say?

2 (C) Your students are eager about the lesson for the day. You have promised to show them how some operations with imaginary numbers have geometric interpretations. What are some of the relationships you can point out and how would you help the students discover them?

3 (C) One of your curious students is intrigued by the fact that, when a complex number is multiplied by its conjugate, the resulting number seems to always be real and nonnegative. (e.g., $(2 + 7i)(2 - 7i) = 4 - 49i^2 = 4 + 49 = 53$). Your student wants to know if this is always the case. How do you help the student figure out that this is always the case?

4 Suppose we rotate the arrow representing the complex number $-3 + 2i$, 90° counterclockwise. At what complex number is the tip of the rotated arrow?

5 Suppose we wish to rotate the arrow representing a complex number, z, 180° counterclockwise. What must we multiply z by?

6 Suppose we wish to rotate a complex number 270° counterclockwise. What must we multiply by?

7 Show that $\overline{(2 + i)^2} = 3 - 4i$.

8 Show that, for every complex number, z, $\overline{\overline{z}} = z$.

9 Show that, if $\overline{z} = z$ then z is real and conversely, if z is real, then $z = \overline{z}$.

10 Show that, for all complex numbers z, $z\overline{z} = |z|^2$.

11 Show that for all complex numbers z_1 and z_2, $\overline{z_1 z_2} = \overline{z_1} \cdot \overline{z_2}$.

12 Using Theorem 7.2 part (c), show that the inverse of every complex number is a multiple of its conjugate.

13 Show that, for every complex number z, $z + \overline{z}$ is real, and $z - \overline{z}$ is purely imaginary.

14 Using the result of the previous problem, show that $|z_1 z_2| = |z_1| \cdot |z_2|$. [Hint: $|z_1 z_2|^2 = z_1 z_2 \cdot \overline{z_1 z_2} = ($ you fill in the details$) (|z_1| \cdot |z_2|)^2$. Proceed from there.]

15 Show that, for every complex number z, $\left|\dfrac{1}{z}\right| = \dfrac{1}{|z|}$.

7.4 The Polar Form of Complex Numbers and De Moivre's Theorem

LAUNCH

1. Solve the equation $x^2 = 4$. How many solutions are there? How many square roots of 4 are there?
2. Can you find a square root of a purely imaginary number? If so, what is a square root of $4i$? How many square roots of $4i$ are there?
3. Can you find a square root of a complex number? If so, what is a square root of $1 + i$? How many square roots of $1 + i$ are there?

If you are totally stymied by the launch, have no fear, as the next section is here. It will describe the fascinating connection between complex numbers and trigonometry that will then put you in a position to answer the launch questions. We hope you enjoy learning about these fascinating connections between different areas of mathematics and how powerful they are.

One of the nice connections between trigonometry and complex numbers is the fact that every complex number can be expressed in trigonometric form. This alone is quite amazing and surprisingly this representation has many applications. We will now develop some of the beautiful resulting theorems that are part of the precalculus syllabus in many secondary schools.

If the line joining the origin to the complex number (a, b) makes an angle θ with the x-axis (see Figure 7.12 below), then from trigonometry,

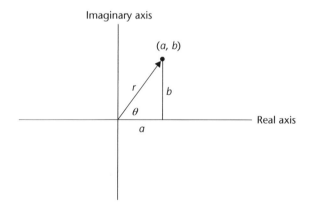
Figure 7.12

we have

$$\cos\theta = \frac{\text{adjacent}}{\text{hypotenuse}} = \frac{a}{r}$$

and

$$\sin\theta = \frac{\text{opposite}}{\text{hypotenuse}} = \frac{b}{r}$$

where $r = |z|$. It follows from these that

$$a = r \cos \theta \text{ and } b = r \sin \theta. \tag{7.8}$$

Thus, the complex number $z = a + bi$ can be written as

$$z = r \cos \theta + (r \sin \theta)i. \tag{7.9}$$

This is called the **polar form** of a complex number and is often abbreviated as $z = r$ cis θ where "cis" is a way to remember, "cosine plus **i** times sine." θ is called the **polar** angle of z. Thus,

$$r \text{ cis } \theta = r \cos \theta + (r \sin \theta)i. \tag{7.10}$$

Example 7.4 *What is the polar form of the complex number $-1 + i$?*

Solution. The coordinates of this complex number are $(-1, 1)$. That places the point in the second quadrant and, when we drop a perpendicular to the x-axis from that point, we get an isosceles right triangle as shown in Figure 7.13 below.

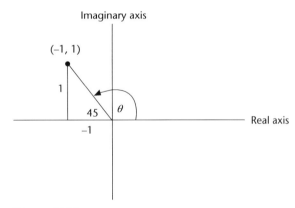

Figure 7.13

Each acute angle in that triangle is 45 degrees, which makes our polar angle θ 135 degrees. Furthermore, by the Pythagorean Theorem, the magnitude of z is $r = \sqrt{2}$. Thus, $z = \sqrt{2}$ cis 135.

Example 7.5 *The polar form of a complex number is 3 cis 240. What is the number in $a + bi$ form?*

Solution. From equation (7.8) we immediately get that $a = 3 \cos 240° = -3/2$ and $b = 3 \sin 240° = -3(\sqrt{3}/2)$. Thus, the exact representation of this complex number 3 cis 240 is $-3/2 + (-3\sqrt{3}/2)i$.

The first question one asks is why bother with polar form? The answer is that, with polar form, we can do some very difficult computations very easily. The following theorem tells us how to quickly multiply two complex numbers in polar form. It simply says that we multiply their magnitudes and add their polar angles. What could be simpler?

> **Theorem 7.6** *If $z_1 = r_1 \text{cis } \theta_1$ and $z_2 = r_2 \text{cis } \theta_2$, then $z_1 z_2 = r_1 r_2 \text{cis}(\theta_1 + \theta_2)$.*

Thus, if we want to multiply the two complex numbers whose polar forms are 3 cis (20) and 2 cis (30), we *immediately* get 6 cis (50). The simplicity of multiplying complex numbers this way surely has to be appreciated.

Proof. The proof involves one of those nice connections between the various parts of secondary school mathematics and is one of our favorite theorems.

Since $z_1 = r_1 \text{ cis } \theta_1$ and $z_2 = r_2 \text{ cis } \theta_2$, we have, when we write these in expanded form,

$$z_1 = r_1 \cos \theta_1 + (r_1 \sin \theta_1) i$$
$$z_2 = r_2 \cos \theta_2 + (r_2 \sin \theta_2) i.$$

When we multiply these two expressions together, using the definition of multiplication of complex numbers (see equation (7.1)), we get

$$z_1 z_2 = [r_1 r_2 \cos \theta_1 \cos \theta_2 - r_1 r_2 \sin \theta_1 \sin \theta_2] + [r_1 r_2 \sin \theta_1 \cos \theta_2 + r_1 r_2 \cos \theta_1 \sin \theta_2] i$$

which, upon factoring out $r_1 r_2$, gives us

$$z_1 z_2 = r_1 r_2 [\cos \theta_1 \cos \theta_2 - \sin \theta_1 \sin \theta_2] + r_1 r_2 [\sin \theta_1 \cos \theta_2 + \cos \theta_1 \sin \theta_2] i. \tag{7.11}$$

But, we know that

$$\cos(\theta_1 + \theta_2) = \cos \theta_1 \cos \theta_2 - \sin \theta_1 \sin \theta_2 \tag{7.12}$$

and that

$$\sin(\theta_1 + \theta_2) = \sin \theta_1 \cos \theta_2 + \cos \theta_1 \sin \theta_2 \tag{7.13}$$

(See Chapter 12 for a review of this.) Thus, we can reduce equation (7.11) to

$$z_1 z_2 = r_1 r_2 \cos(\theta_1 + \theta_2) + r_1 r_2 (\sin(\theta_1 + \theta_2)) i$$
$$= r_1 r_2 [\cos(\theta_1 + \theta_2) + (\sin(\theta_1 + \theta_2)) i]$$
$$= r_1 r_2 \text{cis}(\theta_1 + \theta_2).$$

∎

Let us apply this theorem. Suppose that $z_1 = 2 \text{ cis } (30°)$ and that $z_2 = 3 \text{ cis } (45°)$. Then, by the theorem $z_1 z_2 = 2 \cdot 3 \text{ cis } (30 + 45) = 6 \cos 75$. We may interpret this product in two ways. First, we can view it as taking the complex number $z_1 = 2 \text{ cis } 30$, dilating it by a factor of 3, the magnitude of z_2, and then rotating it 45°, the angle that z_2 makes with the positive x-axis. A second way to interpret the product is to start with $z_2 = 3 \cos 45°$, dilate it by the magnitude, 2, of z_1, and rotate it 30°, the polar angle z_1 makes with the positive x-axis. Either approach tells us that, when we multiply two complex numbers, we are dilating one by the magnitude of the other as well as rotating it by the polar angle the other makes with the positive x-axis.

This generalizes to any two complex numbers. That is, when we multiply z_1 by z_2, we can think of z_1, whose polar angle is θ_1, dilated by the magnitude of z_2, and then rotated by θ_2, the polar angle that z_2 makes with the positive x-axis. Alternatively, we may begin with z_2, whose polar angle is θ_2, and imagine it being dilated by the magnitude of z_1, and rotated by θ_1, the polar angle from z_1. Thus, multiplication of complex numbers can always be thought of as "dilate one by the magnitude of the other and rotate it by the polar angle of the other." (See Figure 7.14 below.)

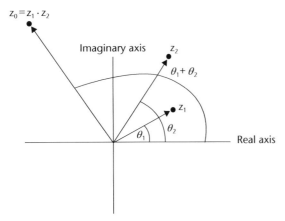

Figure 7.14

Of course, the word "dilate" also includes shrinkage if one of the numbers r_1 or r_2 is less than 1. When $r_1 = 1$, then multiplying by r_1 cis θ_1 simply rotates the other complex number by θ_1. That result is worth stating as a theorem:

Theorem 7.7 *If we take a complex number z and multiply it by 1 cis $\theta_1 = \cos \theta_1 + i \sin \theta_1$, where θ_1 is measured in degrees, this rotates z by θ_1 degrees.*

From Theorem 7.6, we can immediately deduce that, if $z = r$ cis θ, then $z^2 = z \cdot z = r$ cis $\theta \cdot r$ cis $\theta = r^2$ cis $(\theta + \theta) = r^2$ cis (2θ). Similarly, $z^3 = r^3$ cis 3θ, and so on. This gives us the following powerful theorem, which tells us how to raise complex numbers to powers *immediately*. This theorem is due to the French mathematician De Moivre (1667–1754).

Theorem 7.8 *(DeMoivre) If $z = r$ cis θ, then $z^n = r^n$ cis $n\theta$ where n is any integer.*

Proof. The computations preceding the theorem essentially give us the proof. A more formal proof for n, a natural number would use induction (Chapter 8). Then we would only need to consider the case of negative integer exponents, which we do now.

We first show that $z^{-1} = r^{-1}$ cis $(-\theta)$. We already saw in Theorem 7.2 part (e) that every complex number z has only one inverse. If we can show that z multiplied by r^{-1}cis θ gives us 1, then r^{-1}cis θ MUST be the inverse of z because of the uniqueness of the multiplicative inverse. But, this is straightforward.

$$z \cdot (r^{-1} \text{ cis } (-\theta))$$
$$= r \text{ cis } \theta \cdot (r^{-1}\text{cis } (-\theta))$$

$$= r \cdot r^{-1} \text{ cis } (\theta + (-\theta))$$

$$= 1 \text{ cis } 0$$

$$= 1[\cos 0 + (\sin 0)i]$$

$$= 1.$$

Now we proceed as before: $(z^{-1})^2 = (r^{-1} \text{ cis } (-\theta))^2 = r^{-1} \text{ cis } (-\theta) \cdot r^{-1} \text{ cis } (-\theta) = r^{-2} \text{ cis } (-2\theta)$. Similarly, $(z^{-1})^3 = r^{-3} \text{ cis } (-3\theta)$, and so on. Thus, $z^n = r^n \text{ cis } n\theta$ for all integers n. ∎

Corollary 7.9 *If $z = r \text{ cis } \theta$ is a nonzero complex number, then $z^{-1} = r^{-1} \text{cis}(-\theta)$.*

Proof. Take $n = -1$ in the theorem. ∎

We also observe something else useful:

Observation: If $z = r \text{ cis}(\theta)$, then $\bar{z} = r \text{ cis } (-\theta)$.

Example 7.10 *Compute*

(a) $z = (1 + i)^5$ *in polar form*
(b) z^{-1} *in polar form*
(c) $\bar{z}$ *in polar form.*

Solution. (a) The polar form of $1 + i$ is $\sqrt{2} \text{ cis } 45$. (Verify!). So $z = (1 + i)^5 = (\sqrt{2})^5 \text{cis } (5 \cdot 45) = 4\sqrt{2} \text{ cis } 225 = -4 - 4i$. Here we are using the well known fact that $\sin 45° = \cos 45° = \frac{1}{\sqrt{2}}$, which you can verify from the triangle you drew to get the polar form of $1 + i$ and that $\cos 225 = \sin 225 = -\frac{1}{\sqrt{2}}$. Of course, you can just use a calculator to verify this answer.

(b) and (c) Since z in polar form is $4\sqrt{2} \text{ cis } 225$, $\bar{z} = 4\sqrt{2} \text{ cis}(-225)$ and $z^{-1} = \frac{1}{4\sqrt{2}} \text{cis } (-225)$. Thus, we see a close relationship between the conjugate and the inverse. This relationship always holds and you will verify it in the Student Learning Opportunities.

The same way it is easy to multiply and raise complex numbers to powers using De Moivre's Theorem, it is easy to divide complex numbers.

Theorem 7.11 *If $z_1 = r_1 \text{ cis } \theta_1$ and $z_2 = r_2 \text{ cis } \theta_2$, then $\frac{z_1}{z_2} = \frac{r_1}{r_2} \text{ cis } (\theta_1 - \theta_2)$.*

Thus, if $z_1 = 16 \text{ cis } 40$ and $z_2 = 4 \text{ cis } 10$, then $\frac{z_1}{z_2} = 4 \text{ cis}(40 - 10) = 4 \text{ cis}(30)$. It is that simple!

Proof. From Theorem 7.2, part (a), $\frac{z_1}{z_2} = z_1 z_2^{-1}$. And by De Moivre's theorem $z_2^{-1} = r_2^{-1} \cos(-\theta_1)$. Thus, $z_1 z_2^{-1} = r_1 \text{ cis } (\theta_1) \cdot \frac{1}{r_2} \text{ cis } (-\theta_2) = \frac{r_1}{r_2} \text{ cis } (\theta_1 - \theta_2)$. ∎

7.4.1 Roots of Complex Numbers

We now get back to what we started in our launch question and ask "Can we take square roots and cube roots of complex numbers?"

The answer is "Yes," and this is what was needed to simplify the strange answers to the cubic equation we studied earlier in Chapter 3 Section 7 that Cardan's formulas for solving cubic equations provided us with.

Let us see how this works. We will define a **square root of a complex number** z to mean a complex number, w, such that, $w^2 = z$. Stated another way, a square root of z is a root of the polynomial $p(w) = w^2 - z$. Since any second degree polynomial has 2 roots counting multiplicity, every complex number has two square roots. Similarly, we say that a complex number w with the property that $w^3 = z$ is a **cube root** of z. Again, an alternate way of saying this is that w is a root of the polynomial $p(w) = w^3 - z$. Since every third degree polynomial has 3 roots counting multiplicity, there are 3 cube roots of any complex number. We will see that "counting multiplicity" plays no part here. There are two different square roots of every complex number and three different cube roots, and four different fourth roots of any complex number, and so on. Since every real number is considered a complex number, every real number also has 3 cube roots, although two of them are imaginary. We will now determine how to find roots of complex numbers.

Suppose we wanted to find the cube roots of $1 + i$. That is, we wanted to solve $w^3 = 1 + i$. We write both w and $1 + i$ in polar form. w will be r cis θ and $1 + i = \sqrt{2}$ cis 45. Thus, $w^3 = 1 + i$ becomes $(r \,(\text{cis } \theta))^3 = \sqrt{2}$ cis 45, and by De Moivre's theorem, this can be written as

$$r^3 \text{cis }(3\theta) = \sqrt{2} \text{ cis } 45$$

This equation gives us what we need to find a solution for r and θ. Since $r^3 = \sqrt{2} = 2^{\frac{1}{2}}$, we have that $r = 2^{\frac{1}{6}}$ or $\sqrt[6]{2}$. Since $3\theta = 45$, θ can be taken to be 15. So, one solution of this equation is $w = \sqrt[6]{2}$cis 15, and checking, using De Moivre's theorem, we see it works, since $w^3 = \left(\sqrt[6]{2} \text{ cis } 15\right)^3 = \sqrt{2}$ cis 45.

But when we did our analysis above, we said that $3\theta = 45$. That is not quite correct. Every time we add 360° to an angle, we get essentially the same angle. Therefore, a correct statement is not that $3\theta = 45°$ but that $3\theta = 45° +$ (any multiple of 360°). Dividing by 3, we get that $\theta = 15°+$any multiple of 120°. Thus, $\theta = 15°$, $15° + 120°$, $15° + 2(120°)$, and so on. This yields what appears to be many solutions of the equation, $w^3 = 1 + i$, namely, r cis 15°, r cis 135°, r cis 255°, and so on, where $r = \sqrt[6]{2}$. But, in fact, the answers we get by repeatedly adding multiples of 120° start to repeat, once we add 3 multiples of 120°. Thus, r cis 375°, the result of adding three multiples of 120°, is the same as r cis 15°. r cis 495°, the result of adding 4 multiples of 120°, is the same as r cis 135°. Thus, we only get three cube roots, as we should have since the Fundamental Theorem of Algebra told us we can't get more than 3 roots to the equation $w^3 = 1 + i$.

Let us do another example.

Example 7.12 *Solve the equation $w^4 = 1 - \sqrt{3}i$.*

Solution. $1 - \sqrt{3}i$ is in the 4th quadrant and the polar form of this complex number is 2 cis (300°), as we ask you to verify. Writing $w = r$ cis θ, our original equation becomes $(r \text{ cis } \theta)^4 = 2$ cis (300°) or r^4 cis $4\theta = 2$ cis (300°), from which it follow that $r^4 = 2$ and $4\theta = 300°$ plus any multiple of 360°. Therefore, $r = \sqrt[4]{2}$ and $\theta = 75°+$ any multiple of 90°. Thus, our solutions are $w = \sqrt[4]{2}$ cis 75°, $w = \sqrt[4]{2}$ cis 165°, $w = \sqrt[4]{2}$ cis 255°, and $w = \sqrt[4]{2}$ cis 345°. Adding more multiples of 90° just makes our roots repeat. The method we used to solve this equation can now be stated as a theorem.

Theorem 7.13 *If $w^n = r$ cis θ, then $w = \sqrt[n]{r}$ (cis $(\frac{\theta}{n} + \frac{360k}{n})$) where k takes on the values 0, 1, 2, ... n − 1.*

In addition to being able to find roots of complex numbers, secondary school students find it fascinating that we can get a picture of the nth roots of unity, and these provide us with a regular polygon. The next example illustrates this.

Example 7.14 *Find the solutions of $w^5 = 1$ and graph them. That is, find all five of the 5th roots of unity and plot them.*

Solution. The polar form of 1 is easy to see by inspection. It is 1 cis 0. Thus, the other roots by the above theorem are $\sqrt[5]{1}$ cis $\left(\frac{0}{5} + \frac{360k}{5}\right)$ or just 1 cis $72k$ where $k = 0, 1, 2, 3, 4$. Thus, our roots are $w_1 = 1$ cis 0, $w_2 = 1$ cis 72, $w_3 = 1$ cis 144, $w_4 = 1$ cis 216 and $w_5 = 1$ cis 288. All of these roots are a distance 1 from the origin, since $r = 1$ for each of them, and they form angles that differ from each other by 72 degrees. That is, when they are plotted, they give the vertices of a regular pentagon, which we have drawn in Figure 7.15 below.

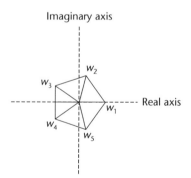

Figure 7.15

Now that we know how to work with complex numbers, we can derive some rather nice trigonometric formulas. Here is an example to show you how this works. We will use this example later in Chapter 14 to solve a problem that has baffled mathematicians for thousands of years.

Example 7.15 *Show that $\cos 3\theta = \cos^3 \theta - 3\cos\theta \sin^2 \theta$, and that $\sin 3\theta = 3\cos^2 \theta \sin\theta - \sin^3 \theta$.*

Solution. Let $z = r(\cos\theta + i\sin\theta)$. We compute z^3 in two ways, first, by De Moivre's theorem to get $z^3 = r^3(\cos 3\theta + i\sin 3\theta)$, which is the same as

$$z^3 = r^3 \cos 3\theta + ir^3 \sin 3\theta. \tag{7.14}$$

Next, we compute z^3 by writing z as $r\cos\theta + ir\sin\theta$, multiplying it by itself three times and using the formula that $(a+b)^3 = a^3 + 3a^2b + 3ab^2 + b^3$. Letting $a = r\cos\theta$ and $b = r\sin\theta$, we get (leaving the details to you) that

$$z^3 = (r^3\cos^3\theta - 3r^3\cos\theta\sin^2\theta) + i(3r^3\cos^2\theta\sin\theta - r^3\sin^3\theta). \tag{7.15}$$

Since the complex number z^3 is the same in both equations (7.14) and (7.15), the real and imaginary parts of z^3 are the same. That is

$$r^3\cos 3\theta = r^3\cos^3\theta - 3r^3\cos\theta\sin^2\theta \tag{7.16}$$

and

$$r^3\sin 3\theta = 3r^3\cos^2\theta\sin\theta - r^3\sin^3\theta. \tag{7.17}$$

Dividing equations (7.16) and (7.17) by r^3 we get, respectively:

$$\cos 3\theta = \cos^3\theta - 3\cos\theta\sin^2\theta$$

and

$$\sin 3\theta = 3\cos^2\theta\sin\theta - \sin^3\theta.$$

We can get other similar trigonometric identities if we wish.

We are not suggesting this is the way to find trigonometric identities. It is just another nicety about complex numbers. It easily allows us to find trigonometric identities if we wish. If we have a computer algebra system available that does algebraic manipulations for us, like cubing or raising to the fifth power, then we can exploit that to get many trigonometric relationships.

Despite the fact that De Moivre was a first rate mathematician, he was not able to secure a regular teaching job. As a result, he made a meager living at tutoring—something that upset him his whole life. As he became older, he became more lethargic and according to Eli Maor in his book, *Trigonometric Delights*, (1998) "He declared on a certain day that he would need 20 more minutes of sleep each day. On the 73rd day—when the additional time accumulated to 24 hours, he died; the official cause was recorded as 'somnolence' (sleepiness)."

Student Learning Opportunities

1 (C) One of your students multiplies the following as indicated below:

(7 cis 60)(2 cis 30) = 14 cis 1800.

Is your student correct? Why or why not? How would you help your student understand the rules for multiplication of numbers in polar form and the geometric interpretation of this product?

2 Suppose we wish to rotate the arrow representing the complex number $4 + 5i$, 50° counterclockwise, and then stretch the result by a factor of 3. By which complex number must we multiply $4 + 5i$ to achieve this?

3 Multiply:
   (a) (3 cis 20)(2 cis 30)
   (b) (4 cis 70)(2 cis −30)
   (c) (5 cis 40)(3 cis 70)

4 Divide:
   (a) $\dfrac{6 \text{ cis } 20}{2 \text{ cis } 10}$
   (b) $\dfrac{12 \text{ cis } 90}{2 \text{ cis } (-30)}$

5 Convert the following complex numbers to polar form:
   (a) −8
   (b) 2 − 2i
   (c) $\sqrt{3} + i$
   (d) 1 + 3i
   (e) 2 + 11i

6 Write each of the following complex numbers in the form $a + bi$.
   (a) 3 cis 120
   (b) 5 cis 330
   (c) 7 cis 180

7 Find the cube roots of unity (the number 1) and plot them on an Argand diagram. What kind of triangle do we get when we connect the three roots of unity?

8 Plot the six 6th roots of unity.

9 (C) You give your students the expression below to simplify.

$$\left(\frac{\sqrt{2}}{2} + \frac{\sqrt{2}}{2}i\right)^3$$

You ask one half of your students to convert it to polar form and then simplify, and the other half of your students to multiply $\frac{\sqrt{2}}{2} + \frac{\sqrt{2}}{2}i$ by itself three times without converting it to polar form. Two students (one from each group) put their work on the board. Assuming they did it correctly, show the work that was put on the board and compare the answers. If you ask your students which method they would prefer to use if they were given a similar problem again, what do you expect they would say, and what reasons do you think they would give?

10 Use the polar form of a complex number to simplify each of the following:
   (a) $\left(\dfrac{1}{2} + \dfrac{\sqrt{3}}{2}i\right)^4$
   (b) $\dfrac{(1+i)^6}{(1-i)^5}$

11 Your students are asked to solve the following equation for w: $w^3 = 4 + 4i$. They do it as follows: First they write both w and $4 + 4i$ in polar form. w is r cis θ and $4 + 4i = \sqrt{32}$ cis 45.

Substituting into the original equation they get $(r \text{ cis } \theta)^3 = \sqrt{32} \text{ cis } 45$ which can be written as $r^3 \text{ cis } 3\theta = \sqrt{32} \text{ cis } 45$. They then claim that $r^3 = \sqrt{32} = 32^{1/2}$, which means that $r = (32)^{1/6}$. Since $3\theta = 45$, $\theta = 15$. They therefore claim that their solution is $w = \sqrt[6]{32} \text{ cis } 15$. What are your comments on your students' solution? Is it correct? Have they found all of the solutions? Give a detailed explanation and complete solution.

12 Solve the following equations for *all* values of w:

(a) $w^3 = 8$
(b) $w^4 = -16$
(c) $w^3 = 2 - 2i$
(d) $w^4 = -\sqrt{3} - i$

13 Find all eight 8th roots of 1. How many of them are complex?

14 Prove that $\cos 2\theta = \cos^2 \theta - \sin^2 \theta$ and that $\sin 2\theta = 2 \sin \theta \cos \theta$ using De Moivre's result. Then do it using equations (7.12) and (7.13).

15 (Tricky) Prove that $z \cdot \bar{z} =$ is a sum of squares. Then show that the product of a sum of squares of real numbers can be written as a sum of squares in two different ways. That is, show that $(a^2 + b^2)(c^2 + d^2) = m^2 + n^2$ for two separate pairs of numbers for m and n. [Hint: Write $a^2 + b^2$ as $(a + bi)(a - bi)$ and do a similar thing for the second factor. Then rearrange the factors to show that you are multiplying two complex numbers, z and $\bar{z}$.]

7.5 A Closer Look at the Geometry of Complex Numbers

LAUNCH

Complete the blanks in the following sentences by describing the geometric transformation that occurs. Choose from among the following: rotation, translation, dilation, reflection.

1. Conjugating a complex number z performs a _____ of z about the x-axis.
2. Multiplying a complex number z by 5 _____.
3. Adding a complex number, say z_0 to another complex number z _____.
4. Multiplying a complex number z by cis θ _____.

If you were able to correctly complete the above sentences you have a pretty clear idea of how complex number arithmetic can be interpreted geometrically. We will discuss these things in more detail in this section.

For clarification and easier reference, we will state all of these relationships which are answers to the launch as a theorem.

Theorem 7.16 (a) *The function $f(z) = z + z_0$ where $z_0 = a + bi$ translates each point z in the plane a horizontal distance of a and a vertical distance of b.* (b) *The function $f(z) = (\text{cis }\theta)\,z$ takes every point in the plane and rotates it about the origin an angle of θ degrees counterclockwise if θ is positive and θ degrees clockwise if θ is negative.* (c) *The function $f(z) = \bar{z}$ reflects each point z about the x-axis.* (d) *The function $f(z) = kz$ takes each complex number and moves it k times its original distance from the origin if k is positive. That is, it performs a dilation.*

To fully appreciate the power of complex number arithmetic to represent geometric transformations, let us give an example. Suppose that you were interested in taking the point representing a complex number, z, and performing the following operations on it: (1) rotate it 30 degrees, (2) reflect the image about the x-axis, and (3) translate the result horizontally 5 and vertically 3. We can accomplish these transformations as follows: Multiplying z by 1 cis 30 will rotate z 30° counterclockwise by part (b) of the previous theorem. Taking the conjugate of the result will reflect the result about the x-axis by part (c) of the previous theorem. Adding the complex number $5 + 3i$ to the result will accomplish (3) by part (a) of the previous theorem. Putting it all together, the function $f(z) = \overline{(\text{cis }30)z} + 5 + 3i$ will accomplish our goal.

Similarly, if we wanted to translate every point z by z_0, then take the result and dilate it by a factor of 5 and finally reflect the result about the x-axis, the function $g(z) = \overline{5(z + z_0)}$ will do the job. Let us examine some more complicated operations.

Example 7.17 *Using complex numbers, explain how to reflect a point z about a line l which passes through the origin.*

Solution. This is really nice. Suppose that the line l makes an angle θ degrees with the positive x-axis, as we see in Figure 7.16(a) below. If we multiply every point in the plane by cis $(-\theta)$, the entire figure gets rotated θ degrees clockwise and we get Figure 7.16(b) where the line l becomes the positive x-axis and z becomes z'. We now reflect z' about the x-axis, (which is the new l) by taking the conjugate of z' and we get the point z''. (See Figure 7.16 (b).) Finally, we rotate the picture back θ degrees and z'' becomes z'''. (See Figure 7.16(c) below.) This last point, z''', is the reflection of z over l. The net effect of all this is that we have reflected the point z about the line l.

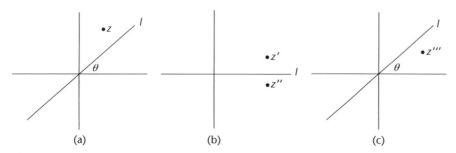

(a)   (b)   (c)

Figure 7.16

Now, let us describe this using complex arithmetic. Rotating z, θ degrees clockwise is accomplished by multiplying z by 1 cis $(-\theta)$. Reflecting about the x-axis means conjugating the result to get $\overline{\text{cis}(-\theta)\,z}$. Rotating the result θ degrees conterclockwise means multiplying by cis θ. Thus, the function $f(z)$ that takes any point z and reflects it about a line l is given by $f(z) = (\text{cis }\theta)\,\overline{\text{cis }(-\theta)\,z}$.

Example 7.18 *Suppose we have two lines, one line, l, making an angle of 15 degrees with the positive x-axis, and the other line m, making an angle of 45 degree with the positive x-axis, and that both lines pass through the origin. Describe how, with complex number arithmetic, we can take a point z, and reflect this about about l first and then m next.*

Solution. We essentially do what we did in the last example. First, we rotate the picture 15 degrees clockwise. z becomes z′. (See part (a) in Figure 7.17 below.) Then reflect z′ about the x-axis (the new position of line l) to get z″. (See Figure 7.17(b) below.) Of course, m having been moved in the process now makes an angle of 30 degrees with the positive x-axis. We then rotate the resulting picture 30 degrees clockwise to bring m to the positive x-axis. During this rotation z″ becomes z‴. (See Figure 7.17(c) below.) We then reflect z‴ about the x-axis (the new position of m) to get w. We then rotate the whole picture back 45° counterclockwise and w becomes w′. (See Figure 7.17(d) below.) The image of z after two reflections, one about l and the other about m in that order, becomes w′.

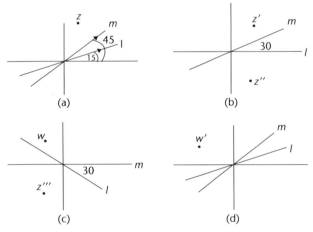

Figure 7.17

Let us take these steps one by one. Rotating a complex number clockwise 15° is achieved by multiplying it by cis (−15). Reflecting the result about the x-axis is achieved by conjugation. Rotating the result clockwise 30 degrees is achieved by multiplying by cis (−30). Reflecting the result about the x-axis is achieved again by conjugation. Finally, rotating counterclockwise by 45 degrees is achieved by multiplying by cis (45). In short, the function that accomplishes all this is $f(z) = \text{cis } 45 \left(\overline{\text{cis }(-30) \left(\overline{\text{cis }(-15)z} \right)} \right)$. Now, in this form, everything looks complicated. But according to Theorem 7.3 since the product of the conjugates is the conjugate of the product we have

$$f(z) = \text{cis } 45 \left(\overline{\text{cis }(-30) \left(\overline{\text{cis }(-15)z} \right)} \right)$$

$$= \text{cis } 45 \, \overline{\text{cis }(-30)} \overline{\left(\overline{\text{cis }(-15)z} \right)} \quad \text{(Theorem 7.3 part (c).)}$$

$$= \text{cis } 45 \, \text{cis }(30) \, \text{cis }(-15)z \quad \text{(Theorem 7.3 part (b) and the observation on page 329.)}$$

$$= (\text{cis } 60) z \quad \text{(Theorem 7.6. We add the angles when we multiply.)}$$

Of course, this representation and operation is *much* simpler. In fact, his identical argument can be generalized to the following result, which we ask you to prove in the Student Learning Opportunities. The proof simply mimics the above example replacing 45 by θ_1 and 15 by θ_2.

Theorem 7.19 *Suppose that l and m are two lines which pass through the origin and that the angle from l to m is θ. Then if we reflect a point z about l and then about m, the net effect is a rotation of z by an angle of 2θ. The function which performs this operation is $f(z) = (cis\, 2\theta)\, z$.*

What this theorem says is that performing two successive reflections about two intersecting lines is equivalent to a rotation of twice the angle from *l* to *m*. (θ is positive if, to get from *l* to *m*, we must travel counterclockwise. Otherwise, it is negative.)

Although we stated this theorem for lines passing through the origin, it is true if the lines intersect at some point other than the origin. The proof really amounts to translating the picture so that the point of intersection lies at the origin, performing the reflections, which rotates the point by an angle of 2θ and then translating the picture back so that the origin is where it was originally. The net effect is that z has been rotated by an angle of 2θ.

There is a similar theorem for reflecting about parallel lines. We will ask you to try the proof in the Student Learning Opportunities.

Theorem 7.20 *Suppose that l and m are two lines parallel to each other and initially parallel to the x-axis. Then, if we reflect a point first over l and then over m, the net effect is a translation of z in a direction perpendicular to l and m and a distance twice the distance from l to m.*

We have already established that the magnitude of a complex number z represents the distance z is from the origin. One can easily find the distance between any two complex numbers z_1 and z_2 by using nothing more than the formula for the distance between two points. If we have two complex numbers $z_1 = a + bi$ and $z_2 = c + di$, then the distance between z_1 and z_2 is given by $\sqrt{(c-a)^2 + (d-b)^2}$. See Figure 7.18 below.

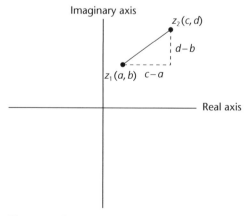

Figure 7.18

Of course, the derivation of this formula is the same as that in the real plane using the Pythagorean Theorem. In the picture above, both z_1 and z_2 are in the first quadrant. Since the complex number $z_2 - z_1 = (c - a) + (d - b)i$ has magnitude $\sqrt{(c-a)^2 + (d-b)^2}$, we see that the distance between z_1 and z_2 is $|z_2 - z_1|$. Of course $\sqrt{(c-a)^2 + (d-b)^2} = \sqrt{(a-c)^2 + (b-d)^2}$ and this latter square root is $|z_1 - z_2|$. So $|z_2 - z_1| = |z_1 - z_2|$. We can now verify an intuitively obvious result.

> **Theorem 7.21** *Rotations, translations, and reflections preserve distance. That is, if z_1 and z_2 are any two complex numbers, then the distance between z_1 and z_2 stays the same under any of the above transformations.*

Proof. The proof is not difficult. We show just one part, that rotations about the origin preserve distance. We first observe that $|\operatorname{cis} \theta| = 1$. This follows since $\operatorname{cis} \theta = \cos \theta + i \sin \theta$ and $|\operatorname{cis} \theta| = \sqrt{\cos^2 \theta + \sin^2 \theta} = 1$.

Before the rotation, the distance between z_1 and z_2 is $|z_1 - z_2|$. After a rotation by an angle of θ, the points z_1 and z_2 become $(\operatorname{cis} \theta) z_1$ and $(\operatorname{cis} \theta) z_2$. The distance between these new points is $|(\operatorname{cis} \theta) z_1 - (\operatorname{cis} \theta) z_2| = |(\operatorname{cis} \theta)(z_1 - z_2)| = |(\operatorname{cis} \theta)| |(z_1 - z_2)| = 1 \cdot |(z_1 - z_2)| = |(z_1 - z_2)|$. Thus, the distance between z_1 and z_2 before a rotation, namely $|z_1 - z_2|$, is the same as the distance between them after a rotation. Thus, rotations preserve distance. ∎

We will have much more to say about rotations and translations and reflections in Chapter 10. For now we get into some applications of complex numbers.

Student Learning Opportunities

1 Show that, if z_1 and z_2 are complex numbers, then the distance between them stays the same when both are translated by the same amount, z_0.

2 Show that, if z_1 and z_2 are complex numbers, then the distance between them stays the same when both are reflected about the real axis. Then show that, when we reflect the two points about any line, the distance between them is preserved.

3 Show, using complex arithmetic, that if we translate a point z by z_0 and then translate the result by w_0, the result is the same as if we translated z by $z_0 + w_0$.

4 Show, using complex arithmetic and the ideas of this section, that if we take a point, z, and reflect it about a line l and then another line m both passing through the origin, then the net effect is a rotation of z by an angle of 2θ where θ is the angle going from l to m.

5 (C) You give your class the following question: Find two lines, l and m passing through the origin such that, if we reflect any point P about l and then about m, the net effect is a rotation of 60° counterclockwise. They respond by drawing two lines that pass through the origin and are at an angle of 120 degrees with each other. Are they correct? If they are not correct, what is a correct answer?

6 (C) Your students have just completed reflecting a point *P* across lines *l* and *m* that intersect at the origin and that are at an angle of 45 degrees with each other. First, they reflect *P* across the line *l* and then they reflect it across the line *m*. One of your curious students asks if the result would be the same if he had taken the point *P* and first reflected it across the line *m* and then across the line *l*. How do you respond? Is it the same? If not, what would the new result be? How would you help your student understand this?

7 Find the function that takes a complex number *z* and reflects it about a line *l* parallel to the *x*-axis. [Hint: Translate the line so that it becomes the *x*-axis, and then do the reflection and translate back again.]

8 Show, using complex arithmetic and the ideas of this section, that if we take a point, *z*, and reflect it about a line *l* and then another line *m* both parallel to the *x*-axis, that the net effect is a *translation* of *z* by a distance of 2*d* in a direction perpendicular to the lines, where *d* is the distance between the lines. (A similar proof shows this result is true if the lines are parallel, but not necessarily to the *x*-axis. We just have to rotate them so that they are.) This result, together with Theorem 7.19, tells us that the performance of two reflections, about lines, is either a reflection or a translation.

9 (C) You ask your students to do the following: Find two parallel lines *l* and *m* such that, when we reflect any point about *l* and then take the result and reflect about *m*, we get a translation of that point 10 units in a positive vertical direction. Your students have each found different pairs of lines and are quite confused about this. How could this have happened? What are some possible pairs of lines they could have found?

10 Suppose we have 4 lines, *l*, *m*, *l'*, and *m'* through the origin where the angle between *l*, and *m* is the same as the angle between *l'* and *m'*. Show that, if we reflect a point *P*, first about *l* and then about *m*, we get the same result as we would get if we reflect that point first about *l'* and then about *m'*.

11 Suppose we have 4 lines, *l*, *m*, *l'* and *m'* all parallel to one another. Suppose that the distance between *l* and *m* is the same as the distance between *l'* and *m'*. Show that if we reflect a point *P*, first about *l* and then about *m*, we get the same result that we would get if we reflect that point first about *l'* and then about *m'*.

12 A glide reflection of a point about a line *l* is a reflection in *l* of *P* followed by a translation parallel to *l*.
   (a) Write the function that reflects a point *z* about the *x*-axis and then translates it parallel to the *x*-axis a distance 5 in the positive direction.
   (b) Write the function that reflects a point *z* about a line that makes a 45° angle with the positive *x*-axis, and then translates the result by $1 + i$.
   (c) Write the function that reflects a point *z* about a line, *l*, that makes an angle of θ degrees with the positive *x*-axis, and then translates the result by a complex number z_0.

13 (C) A student asks if a glide reflection (see the previous problem) is the result of three reflections? Give an explanation that would help your student understand the answer to this question.

7.6 Some Connections to Roots of Polynomials

Find all cube roots of 1. Explain your work and justify your answer.

When posed with the launch question, you might have been wondering why we specified that you should find multiple cube roots of 1. Well, if you remember, in the previous section we saw that every complex number has 3 cube roots and 4 fourth roots, and so on. Since we are now considering real numbers as a subset of the complex numbers, it follows that every real number has 3 cube roots, and 4 fourth roots, which can be complex numbers. Consider the next example which shows how to complete the launch problem and find the 3 cube roots of 1.

Example 7.22 *Find all cube roots of 1.*

Solution. We can write 1 = 1 cis 0. So, by De Moivre's theorem, the roots are given by $\sqrt[3]{1}$ cis $(\frac{0}{3} + \frac{360k}{3})$ = cis $120k$ where k = 0, 1, and 2. The roots are thus 1 cis 0 = 1, 1 cis 120, which we call ω, and 1 cis 240, which from De Moivre's theorem is ω^2. Thus, the three cube roots of 1 are 1, ω, and ω^2. If we evaluate ω = cis 120, we get $\frac{-1}{2} + \frac{\sqrt{3}}{2}i$, and if we evaluate ω^2 = cis 240 we get $\frac{-1}{2} - \frac{\sqrt{3}}{2}i$. Thus, the three cube roots of 1 are $\frac{-1}{2} \pm \frac{\sqrt{3}}{2}i$ and 1. Since ω is a cube root of 1, ω^3 = 1. As we shall soon see, this fact will be very useful.

The number ω plays a part in cube roots of all complex numbers. If we know a cube root of a number z, then we can automatically find the other two cube roots of z. Specifically, if one cube root of z is p, then the others are ωp and $\omega^2 p$. Why? Well, saying that p is a cube root of z means that

$$p^3 = z. \tag{7.18}$$

Now, if we compute the cube of ωp, we get $\omega^3 p^3$. Since ω^3 = 1, as we observed in the last example, this simplifies to p^3, which by equation (7.18) gives us z. Similarly, if we cube $\omega^2 p$, we get $\omega^6 p^3 = 1 \cdot p^3 = z$. Thus, each of p, ωp and $\omega^2 p$ gives us z when cubed, so each is a cube root of z. You will show in the Student Learning Opportunities that p, ωp, and $\omega^2 p$ are all different, and since any complex number has only 3 roots, p, ωp and $\omega^2 p$ are the three cube roots of z. Thus, we can write the 3 cube roots of any number very simply once we know one of them.

We are now ready to tie up a loose end from Chapter 3. We said in that chapter that complex numbers were really developed as a result of trying to understand the roots of cubic equations. Let us show how this all fits together with an example.

Example 7.23 *Suppose that we have the cubic equation, $x^3 - x = 0$. (a) Use factoring to find all solutions. (b) Use the formula for solving cubic equations given in Chapter 3 Section 7 to find the solutions.*

Solution. (a) We factor $x^3 - x$ into $x(x-1)(x+1)$ to get $x(x-1)(x+1) = 0$ and then set each factor equal to zero to get $x = 0$, $x = 1$, and $x = -1$.

(b) Now let us use the formula from Chapter 3 Section 7. That formula was for solving equations of the form $x^3 + px = q$, and the solution was given by $x = \sqrt[3]{\frac{q+\sqrt{q^2+\frac{4p^3}{27}}}{2}} + \sqrt[3]{\frac{q-\sqrt{q^2+\frac{4p^3}{27}}}{2}}$. In this case, $p = -1$ and $q = 0$ and our solution becomes:

$$\sqrt[3]{\frac{\sqrt{\frac{-4}{27}}}{2}} + \sqrt[3]{\frac{-\sqrt{\frac{-4}{27}}}{2}}. \tag{7.19}$$

However, we know that $\frac{\sqrt{\frac{-4}{27}}}{2}$ is a complex number and has 3 cube roots. Thus, there are 3 values for the first cube root that occurs in (7.19) and the same is true for the second cube root. Thus, there are a total of 9 possible values for (7.19), which we will detail in question #7 in the Student Learning Opportunities. Three of them are 0, two are 1, and two are -1. Thus, we see how the solutions we obtained in (a) can be explained by the strange and complicated expression in (7.19)

One of the facts that students are often presented with in precalculus courses is that, if we have a polynomial equation

$$p(z) = a_n z^n + a_{n-1} z^{n-1} + \ldots a_1 x + a_0 = 0 \tag{7.20}$$

and if the coefficients are *real numbers*, then the roots of this polynomial occur in conjugate pairs. We state that as a theorem.

Theorem 7.24 *If $a + bi$ is a root of the polynomial, $p(z) = a_n \cdot z^n + a_{n-1} \cdot z^{n-1} + \ldots a_1 \cdot z + a_0$, then so is $a - bi$.*

Proof. This is easy to prove, for if $p(z) = 0$, then $\overline{p(z)} = 0$ by part (e) of Theorem 7.3, since 0 is a real number. But

$$\overline{p(z)} = \overline{a_n z^n} + \overline{a_{n-1} z^{n-1}} + \ldots \overline{a_1 z} + \overline{a_0}$$

$$= \overline{a_n} \cdot \overline{z}^n + \overline{a_{n-1}} \cdot \overline{z}^{n-1} + \ldots \overline{a_1} \cdot \overline{z} + \overline{a_0}$$

$$= a_n \cdot \overline{z}^n + a_{n-1} \cdot \overline{z}^{n-1} + \ldots a_1 \cdot \overline{z} + a_0 \quad \text{(Since all the } a_n\text{s are real)}$$

$$= p(\overline{z}) \quad \text{(Definition of } p(\overline{z}))$$

$$= 0 \quad \text{(Since we started with } \overline{p(z)} \text{ which is 0.)}$$

In summary, the above sequence of equations tells us that $p(\overline{z}) = 0$ which means that $\overline{z}$ is a root of the equation $p(z) = 0$. That is, complex solutions come in conjugate pairs. ∎

Here is a typical question found in precalculus texts.

Example 7.25 *Write an equation that has as its roots $3 + 2i$ and $4 + i$.*

Solution. Since the roots come in conjugate pairs, if $3 + 2i$ is a root, so is $3 - 2i$. If $4 + i$ is a root, so is $4 - i$. From Chapter 3, if r is a root of a polynomial, then $z - r$ is a factor. Thus, an equation which works in factored form is

$$(z - (3 + 2i))(z - (3 - 2i))(z - (4 + i))(z - (4 - i)) = 0$$

The first two factors multiply together to give $z^2 - 6z + 13$, while the second two factors multiply together to give $z^2 - 8z + 17$. Thus, our equation becomes

$$(z^2 - 6z + 13)(z^2 - 8z + 17) = 0$$

or just $z^4 - 14z^3 + 78z^2 - 206z + 221 = 0$.

Another theorem that one sees in precalculus courses is that, if $a + \sqrt{b}$ is a root of an equation, then so is $a - \sqrt{b}$, assuming that a and b are rational numbers and $\sqrt{b}$ is irrational. The number $a + \sqrt{b}$ and $a - \sqrt{b}$ are also known as **conjugate surds**. The proof of this is identical to the proof of Theorem 7.24 if we make a few replacements. If, when $x = a + \sqrt{b}$, we denote by $\overline{x}$ the number $a - \sqrt{b}$, then the following is easy to verify: $(\overline{x})^2 = \overline{(x^2)}$ and by induction that $\overline{(x^n)} = (\overline{x})^n$. Also, if k is a rational number, then $\overline{(kx)} = k\overline{x}$. (We need k to be rational since, if k is any real number, like $\sqrt{5}$ then, when we multiply by k, we won't get an expression of the form $a + \sqrt{b}$ where a is rational.) What we are saying is that, if in the proof of Theorem 7.24, we replace z by x and $\overline{z}$ by $\overline{x}$, and use the observations we just described, the identical proof of Theorem (7.24) yields:

Theorem 7.26 *If we have a polynomial with rational coefficients, and if $a + \sqrt{b}$ is a root of the polynomial, then so is $a - \sqrt{b}$.*

Here is a simple example.

Example 7.27 *Find a polynomial with real integer coefficients which has as its roots, i and $1 + \sqrt{2}$.*

Solution. The complex roots and roots of the form $a + \sqrt{b}$ come in conjugate pairs. Thus, the roots of the polynomial will be $z = i$, $z = -i$, $z = 1 + \sqrt{2}$, and $z = 1 - \sqrt{2}$. Our polynomial will be

$$(z - i)(z + i)(z - (1 + \sqrt{2}))(z - (1 - \sqrt{2})) = 0.$$

The product of the first two factors is $z^2 + 1$ and the product of the second two factors is $z^2 - 2z - 1$. Thus, our polynomial is $(z^2 + 1)(z^2 - 2z - 1) = 0$, which simplifies to

$$z^4 - 2z^3 - 2z - 1 = 0.$$

Student Learning Opportunities

1 Show that, if $x = a + \sqrt{b}$, where a and b are nonnegative rational numbers and $\sqrt{b}$, is irrational, then x^2 can be written in the form $c + \sqrt{d}$, where c and d are rational and $\sqrt{d}$ is irrational.

2 Prove that no two of the complex numbers $p, \omega p$, and $\omega^2 p$ are the same. Thus, if p is one cube root of a complex number, the other two are ωp and $\omega^2 p$.

3 Write a polynomial with real coefficients, two of whose roots are $1 + i$ and $4 - \sqrt{2}$.

4 Write a polynomial with real coefficients, two of whose roots are $2i$ and $\sqrt{3}$.

5 Write a polynomial $p(x)$ whose roots are $1 + \sqrt{2}$ and $1 - \sqrt{3}$ and such that $p(2) = 5$.

6 (C) You asked your students to find the roots of the equation $z^2 - 2iz - 1 = 0$. They were able to determine rather quickly that $z = i$ is a root, which they verified by substitution. Using Theorem 7.24, they immediately figured out that $z = -i$ should also be a root. But, much to their shock and disappointment, when they substituted $-i$ into the equation, it did not work. They asked you how this could have happened? What is the answer?

7 Using the polar form of $\sqrt{\frac{-4}{27}}$, show that $\sqrt[3]{\frac{\sqrt{\frac{-4}{27}}}{2}} = \frac{1}{\sqrt{3}}\operatorname{cis}(30 + 120k)$ for $k = 0, 1, 2$, and that $\sqrt[3]{\frac{\sqrt{\frac{-4}{27}}}{2}} = \frac{1}{\sqrt{3}}\operatorname{cis}(90 + 120k)$ for $k = 0, 1, 2$. Show that three of the values of $\sqrt[3]{\frac{\sqrt{\frac{-4}{27}}}{2}} + \sqrt[3]{\frac{\sqrt{\frac{-4}{27}}}{2}}$ are 0, 1, and -1.

8 (C) You have presented Theorem 7.24 to your students. In order to determine if your students really understood it, you gave them 1 minute to answer the following question: "Does there exist a 4th degree polynomial with integer coefficients that has as its root, $2i$, $3i$, $4i$, and $5i$? If so, find it. If not justify your answer." How could you expect your students to answer this question so quickly? Explain.

9 Show that $x^4 + 5x^2 + 1 = 0$ has no real roots.

7.7 Euler's Amazing Identity and the Irrationality of e

LAUNCH

1 Give the definition of π.
2 Give the definition of e.
3 Give the definition of i.
4 Are the above numbers related in any way? Explain.

Most likely, in response to the launch question, you did not see any relationships among π, e, and i. You will be very surprised to see that in fact, they are related in a very well known formula that you will learn about in this section. Let us begin by examining the value of e^x more closely.

When you studied calculus, you studied series and discovered the interesting fact that the functions $\sin x$, $\cos x$, and e^x could be represented as series. Specifically,

$$\sin x = x - \frac{x^3}{3!} + \frac{x^5}{5!} - \frac{x^7}{7!} + \ldots \tag{7.21}$$

$$\cos x = 1 - \frac{x^2}{2!} + \frac{x^4}{4!} - \frac{x^6}{6!} + \ldots \tag{7.22}$$

$$e^x = 1 + x + \frac{x^2}{2!} + \frac{x^3}{3!} + \frac{x^4}{4!} \ldots \tag{7.23}$$

In calculus, these series are restricted to real values of x. Euler wondered what would happen if we were to replace the variable x in equation (7.23) by a complex number. Specifically, replace x by ix where x is a real number. This yields

$$e^{ix} = 1 + ix + \frac{(ix)^2}{2!} + \frac{(ix)^3}{3!} + \frac{(ix)^4}{4!} + \frac{(ix)^5}{5!} + \ldots . \tag{7.24}$$

He then expanded the terms of equation (7.24) and obtained

$$e^{ix} = 1 + ix + \frac{i^2 x^2}{2!} + \frac{i^3 x^3}{3!} + \frac{i^4 x^4}{4!} + \frac{i^5 x^5}{5!} \ldots \tag{7.25}$$

and then made use of the facts that $i^2 = -1$, $i^3 = -i$, $i^4 = 1$, $i^5 = i$, and so on, and found that equation (7.25) simplified to

$$e^{ix} = 1 + ix - \frac{x^2}{2!} - \frac{ix^3}{3!} + \frac{x^4}{4!} + \frac{ix^5}{5!} + \ldots .$$

He then grouped all the odd powers of x together and all the even powers of x together and got

$$e^{ix} = (1 - \frac{x^2}{2!} + \frac{x^4}{4!} - \ldots) + i(x - \frac{x^3}{3!} + \frac{x^5}{5!} \ldots) \tag{7.26}$$

and observed that the series in parentheses in equation (7.26) match those in equations (7.21) and (7.22). His conclusion was that

$$e^{ix} = \cos x + i \sin x \tag{7.27}$$

where x is measured in *radians*. This is known as **Euler's Identity**. He substituted $x = \pi$ in equation (7.27) and got as a result that

$$e^{i\pi} = \cos \pi + i \sin \pi$$

$$= -1 + 0$$

$$= -1. \tag{7.28}$$

He thus discovered an incredible relationship among three of the most important numbers in mathematics, e, i, and π namely, that $e^{i\pi} = -1$.

You might be thinking that this is pretty amazing. Well, let's check into this a bit further.

Euler used a formula that was true for real numbers and in it he substituted a complex number. How could such an illegal move lead to a correct result? Furthermore, what does $e^{i\pi}$

mean? Specifically, how could equation (7.28) be correct if we don't even know what it might mean to have e raised to an imaginary power? All these issues were eventually resolved when mathematicians defined e^z when z is complex, by the series

$$e^z = 1 + z + \frac{z^2}{2!} + \frac{z^3}{3!} + \ldots$$

That is, they defined e^z as an extension of the definition of e^x when x is real. There are many issues with this definition. First, what does it mean for an infinite series of complex numbers to have a sum? One might suspect that limits will play a part, but what does it mean for a function of a complex variable to have a limit? Limits use the idea of closeness and what is an appropriate definition of complex numbers being close? Many more issues come up, and all of them can be resolved. The definition of e^z given above is fine. The series will have a sum which is a complex number regardless of what z is. And it can be proven that the series may be split up as we did in equation (7.26). The details are involved and we have the mathematician Cauchy to thank for figuring much of this out. He was a major player in the development of complex numbers and their calculus. Even more interesting was the fact that complex numbers turned out to have many real-life applications.

What equation (7.27) is telling us is that cis θ and $e^{i\theta}$ are one and the same, namely, $\cos\theta + i\sin\theta$. Thus, we may describe the function which rotates all complex numbers θ degrees about the origin as $f(z) = e^{i\theta}z$ instead of $f(z) = (\text{cis } \theta)z$ as we saw earlier in the chapter. Furthermore, by equation (7.27) any complex number z, by De Moivre's theorem, can now be written as $z = re^{i\theta}$ instead of r cis θ. Now, we know that r cis θ describes a point r units from the origin, making an angle of θ degrees with the positive x-axis. Since $re^{i\theta}$ and r cis θ are the same, if you are given a complex number $re^{i\theta}$ where $r \geq 0$, you immediately know that the complex number is at a distance r units from the origin, and that the arrow representing the complex number makes an angle θ with the positive real axis. Thus, you can immediately plot $z = 4\,e^{i\pi}$. It is 4 units from the origin and the arrow representing it makes an angle of π with the positive real axis. In Figure 7.19 below we show the locations of several points written in the form $re^{i\theta}$.

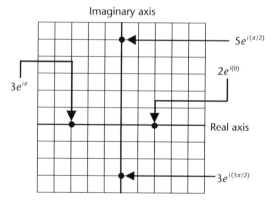

Figure 7.19

Using equation (7.23) we can give a proof of a fact that you have probably taken for granted.

Theorem 7.28 *e is irrational.*

346 Building the Complex Numbers

Proof. Use the series, equation (7.23) and let $x = 1$. Assume that e is rational. Then by $e = \dfrac{p}{q}$ and by equation (7.23), we have

$$\frac{p}{q} = 1 + 1 + \frac{1}{2!} + \frac{1}{3!} + \frac{1}{4!} + \ldots. \tag{7.29}$$

Multiply both sides of equation (7.29) by $q!$. Then equation (7.29) becomes

$$\underbrace{\frac{p}{q}q!}_{I_1} = \underbrace{q! + q! + \frac{q!}{2!} + \frac{q!}{3!} + \frac{q!}{4!} + \ldots + \frac{q!}{q!}}_{I_2} + \frac{q!}{(q+1)!} + \frac{q!}{(q+2)!} + \ldots \tag{7.30}$$

since the q divides $q!$, the left side of equation (7.30) is an integer, I_1. Since each of the numbers $2!, 3!, 4!, \ldots q!$ divide $q!$, the first $q + 1$ terms on the right will also be integers. Thus, we have an integer I_2 on the right side of equation (7.30) consisting of the first $q + 1$ terms (which we have indicated), followed by the terms $\dfrac{1}{q+1} + \dfrac{1}{(q+1)(q+2)} + \dfrac{1}{(q+1)(q+2)(q+3)} + \ldots$. Moving I_2 to the left side, we get another positive integer, $I = I_1 - I_2$ on the left side. Our equation now reads:

$$I_1 - I_2 = I = \frac{1}{q+1} + \frac{1}{(q+1)(q+2)} + \frac{1}{(q+1)(q+2)(q+3)} + \ldots.$$

$$< \frac{1}{q+1} + \frac{1}{(q+1)^2} + \frac{1}{(q+1)^3} + \ldots. \text{ (Why?)}$$

$$= \frac{\frac{1}{q+1}}{1 - \frac{1}{q+1}} \quad \text{(The sum of a geometric series.)}$$

$$= \frac{1}{q} \quad \text{(Simplifying)}$$

$$< 1$$

But this string of equations and inequalities tells us that $I < 1$. That is a contradiction since I is a positive integer. Our contradiction arose from assuming that e was rational. Thus, e is irrational. ∎

Student Learning Opportunities

1 (C) Your students have just been exposed to one of the most incredible formulas in mathematics, Euler's Identity, which states that $e^{j\pi} = -1$. Although they are all very excited by it, some of your very brightest students seem somewhat skeptical. They are concerned about several issues with this formula. What are the issues they might raise? How would you respond to their concerns?

2 (C) After being exposed to Euler's formula, your students are asked to plot the complex number $z = 7e^{j(\frac{\pi}{2})}$. They convert it to 7 cis $(\frac{\pi}{2})$, figure out what that is, and then plot it. You amaze them, however, by plotting the point immediately, without doing the work they did. Your students want to know how you can locate this, and other points like it, so quickly. They also want to know the mathematical reasoning behind your "trick." How do you explain it?

3 Evaluate

(a) $e^{2\pi i}$

(b) $4e^{3\pi i}$

(c) $2e^{-\pi i}$

(d) $-2e^{-\pi i}$

(e) $5e^{2\pi i}$

(f) $3e^{-\pi i}$

(g) $7e^{(\pi/2)i}$

4 Show, using Euler's identity, that $e^{i(\theta_1+\theta_2)} = e^{i\theta_1} \cdot e^{i\theta_2}$. (You can't just argue that, when you multiply numbers with the same base, you add the exponents since that is a rule for real number exponents! Numbers with imaginary exponents are a new creation. You need to use either definitions or theorems to prove this for complex numbers.)

5 Without doing any computations at all, write each of the following in the form $re^{i\theta}$, where $r \geq 0$

(a) $2i$

(b) $-3i$

(c) -4

6 Explain why in the proof of theorem 7.28, l is a *positive* integer.

7 Using the facts that $\sin(-\theta) = -\sin\theta$ and that $\cos(-\theta) = \cos\theta$

(a) Show that $\dfrac{e^{i\theta} + e^{-i\theta}}{2} = \sin\theta$.

(b) Show that $\dfrac{e^{i\theta} - e^{-i\theta}}{2i} = \cos\theta$

7.8 Fractal Images

LAUNCH

1 If you were asked to measure the exact length of your desk, what would you use?
2 If you were asked to measure the exact circumference of the round clock in your room, how would you do it?
3 If you were asked to trace your open hand on a piece of paper and then measure the perimeter from the base of your thumb to the base of your pinky, how would you do it?
4 If you were asked to measure the perimeter of a maple leaf, how would you do it?
5 In which of the above cases would you be able to get the most accurate measurement? Why?
6 In which of the above cases would you get the least accurate measurement? Why?

Before thinking about the launch question, you might have thought that all lengths are measurable. We hope that you are now beginning to wonder if that is really the case. This section will introduce you to a most fascinating branch of mathematics that will have you thinking about geometric figures in a very new way. We expect that you will be very intrigued.

Euclidean Geometry is used to help us study figures in our environment whose boundaries are smooth. So, for example, the sides of a polygon are straight lines which can be considered smooth. A circle has no bumps in it, so it too can be considered smooth. But, in many parts of nature, we see shapes that are not smooth, but rather jagged. And the closer we look at some of these, the more jagged they may appear. For example, consider a coastline. If you wanted to measure the length, you might use a scaled drawing of the coastline, to get an estimate. Or you might actually take a ruler and start measuring the physical coastline that way. But the coastline is not straight, and a ruler would miss some of the nooks and crannies. You might think that a device that measures smaller distances, say a centimeter at a time, would work better, but we would run into the same problem. There are tiny nooks and crannies that even this device cannot measure. Our point is that, no matter how small a device we use to measure the coastline, we will never be able to measure it accurately, because no matter how many times we zoom in on the coastline, we find that it always has more nooks and crannies. That is, the coastline is really not smooth and has a completely different character from, say a straight line or a circle. One might say that the coastline has "**infinite complexity**," meaning that, no matter how many times you zoom in, you see jaggedness. Figures that have this infinite complexity are called **fractals**. Thus, if we want to study such real world phenomena, we have to develop a new kind of geometry. Only in the last 100 years, and in particular in the last 30 or so years, have we begun to make progress in this area, and surprisingly, imaginary numbers play a big role in it.

Suppose we take an imaginary number, say $z = 3 + i$. Its magnitude, or distance from the origin is $\sqrt{3^2 + 1^2}$ or $\sqrt{10}$. Suppose we square z to get z^2. We know that its magniture is $|z|^2 = 10$. Now suppose we repeat this operation on z^2. That is, we square it again. We get z^4 whose magnitude, or distance from the origin, is 100. Thus, this is even further from the origin than z^2 is. The more times we square, the further from the origin we get. Thus, the magnitudes of the resulting complex numbers, go off to infinity.

A sequence of such points whose magnitudes get larger and larger is said to diverge to infinity. So the sequence of complex numbers we generated starting with $z = 3 + i$ and successively squaring diverges to infinity. In contrast to this, if we take $z = \frac{1}{2} + \frac{1}{2}i$, then $|z| = \sqrt{\frac{1}{2}}$. So the magnitude of $z^2 = \frac{1}{2}$ and that of z^4, $\frac{1}{4}$, and so on, and now we see that the magnitudes of the numbers are getting smaller and smaller. What this means is that the points, z, z^2, z^4, and so on are getting closer and closer to the origin.

Finally, if we take a complex number, z, whose magnitude is 1, then $|z|^2 = 1$, and in a similar manner $|z^4| = 1$, and so on. That is, all points generated remain at a distance 1 from the origin.

In summary, points that start out on the circle whose center is at the origin, and whose radius is 1, stay on the circle as we repeatedly square, those inside get close to the origin, and those outside the circle initially have magnitudes that get larger and larger and diverge to ∞. This may not seem important, but watch what happens if we vary the process a bit. Pick a complex number c and form the function, $f(z) = z^2 + c$. This function takes a complex number z, squares it, and then adds a fixed complex number, c, to the result. Now, suppose that we start with a complex number, z_0

and repeatedly apply the function $f(z)$ to it. That is, compute $z_1 = f(z_0)$ then compute $z_2 = f(z_1)$, and so on where we keep taking the result we get and substituting it back into the expression. One of two things will happen. Either the sequence of complex numbers we generate will go to ∞ or they won't. We are going to draw a picture to illustrate what happens. If the sequence of numbers generated by z_0 doesn't go to infinity, we will place a black dot in the complex plane at z_0 to indicate this. Otherwise, we will not do anything. The picture we get is called the $c-$ **Julia set**, and the pictures we get are rather remarkable. Each c value has it's own Julia set. For example, below is the Julia we get when $c = -0.0519 + 0.688i$. We notice that, though it is elaborate, it comes in one piece. We say that this set is connected (Figure 7.20).

Figure 7.20

You might be thinking where are the real and imaginary axes? We have left them out so as not to detract from the picture. Here is the picture with the axes put in (Figure 7.21):

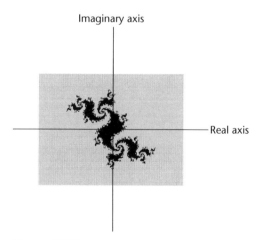

Figure 7.21

Below is the Julia set for $c = -0.577 + 0.478i$. Here the Julia set comes in many pieces. (Picture from http://fractals.iut.u-bordeaux1.fr/jpl/jpl1a.html) (Figure 7.22).

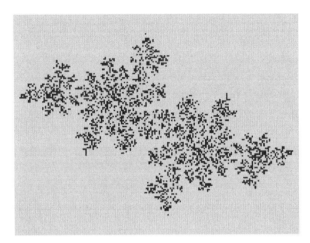

Figure 7.22

What is interesting in these two pictures, which is similar to the coastline example, is that the boundaries of these Julia sets are jagged. In fact, if we zoomed in repeatedly, we would see the same thing we alluded to with the coastline example, namely, continued jaggedness at all magnifications, and hence, infinite complexity. Thus, the boundaries of these sets are fractals. Scientists and mathematicians are hoping that the geometry associated with these kinds of figures and the repetitive procedures used to generate these pictures might help us in understanding fractals.

One other related picture is called the **Mandelbrot set.** In this set, a point representing a complex number is only blackened if its Julia set is connected. Otherwise it isn't. The Mandelbrot set looks like Figure 7.23 below. (A more illustrious and color version of this figure may be found at: http://commons.wikimedia.org/wiki/File:Mandelbrot_set_2500px.png.)

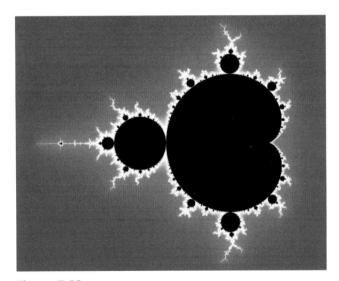

Figure 7.23

Thus, every complex number inside the Mandelbrot set will give a connected Julia

It is the complex numbers on the boundary of the Mandelbrot set that form the most interesting Julia sets. You can actually see which Julia set corresponds to each complex number in the Mandelbrot set (as well as the complex number's coordinates) at:

http://math.bu.edu/DYSYS/applets/JuliaIteration.html. It is certainly worth a visit. For some other very pretty pictures of fractals, see http://www.wirefree.net.au/~lawrence/.

Other than beautiful pictures that fractals often provide, what can fractals be used for? We quote from http://library.thinkquest.org/3288/usesfrac.html:

"Fractals provide a simple solution to capture the enormous detail and irregularity of clouds and landscapes. Fractal geometry is an efficient way to draw realistic natural objects on a computer screen. Landscape designers start with basic shapes and iterate them over and over. Science-fiction films design imaginary landscapes likewise for backdrops. (This was true for the Star Wars series where landscapes were made by computer using iteration.) In the 1980s Benoit Mandelbrot working with some metallurgists concluded that a metal surface's fractal dimension may be a useful measurement of a metal's strength. This can be used to distinguish or characterize metals. The evolution of different ecosystems have been described and predicted using fractals. For example, Herald Hasting used fractals to model ecosystem dynamics at Okefenokee Swamp. Fractals along with ecosystem studies are essential in determining the spread of acid rain and other pollutants. Other uses of fractals are describing astronomy, meteorology, economics, ecology, and in the study of galaxy clusters."

A promising new area of technology related to fractals is with antennas. Specifically, a new and powerful type of antenna known as a fractal antenna has been used by the military to do some very sophisticated transmitting. These are tiny antennas with an exceedingly powerful ability to transmit.

7.8.1 Other Ways to Generate Fractal Images

In Chapter 3 we used Newton's method to find roots of polynomials. What is surprising is that this same method can be used to find roots of polynomials with complex coefficients. Just as we did in Chapter 3, we start with a polynomial of degree n with complex coefficients and choose a complex number z_0 to start with, and then generate the following sequence of points.

$$z_1 = z_0 - \frac{f(z_0)}{f'(z_0)}$$

$$z_2 = z_1 - \frac{f(z_1)}{f'(z_1)}$$

$$z_3 = z_2 - \frac{f(z_2)}{f'(z_2)}$$

. . . .

We call this sequence of numbers the Newton sequence for z_0. What will happen (most of the time) is that this Newton sequence of numbers will converge to a root of the polynomial. If r_1 is a root of the polynomial, then the set of all z_0 whose Newton sequence converges to r_1 is called the r_1 **attractor** set or the **Newton Basin** for r_1. Since the polynomials have n roots counting multiplicity, there will be at most n different attractor sets for a polynomial. We can color the attractor sets with different colors. It turns out that the pictures we get are quite involved and the boundaries form fractals. One can view some of these beautiful fractals at: http://aleph0.clarku.edu/~djoyce/newton/newton.html.

Below we have a black and white picture to illustrate this. These are the Newton Basins for the 6 roots of the polynomial $(z^2 - (1+3i)^2)(z^2 - (5+i)^2)(z^2 - (3-2i)^2)$ whose roots are obviously $\pm(1+3i)$, $\pm(5+i)$, and $\pm(3-2i)$. The six roots are located where the white dots are (Figure 7.24).

Figure 7.24

For a picture in color see http://aleph0.clarku.edu/~djoyce/newton/examples.html. We hope you appreciate how surprising it is that complex numbers relate to the study of fractals.

Student Learning Opportunities

1 **(C)** You ask your students to take out a piece of paper from their loose leaf book and measure its length. Then you ask them to rip their paper in half vertically, and thereby create a jagged edge along the length. You now ask them to tell you whether the new length is longer or shorter than the original length (which was approximately 11 inches). You then request that they measure this new length as accurately as they can using a magnifying glass if they wish. After doing this activity, your students are completely intrigued and want to know what this has to do with their current unit on complex numbers. How do you respond?

2 Visit the website http://facstaff.unca.edu/mcmcclur/java/Julia/ and generate the Julia sets for each of the following points by running your mouse over to the selected point. Describe the similarities and differences you notice in the pictures.

(a) $c = 0.270 + 0.008i$
(b) $c = 0.322 + 0.606i$
(c) $c = -0.63 - 0.467i$

7.9 Logarithms of Complex Numbers and Complex Powers

**LAUNCH**

State whether the following values are real or imaginary. (a) i^2 (b) i^3 (c) i^4 (d) i^5 (e) i^i
Support your answers with explanations.

Surely, you were quite confident with your responses to the first four examples in the launch. You are probably also quite sure about your answer to part (e), although you are not positive. After you have read this section, you will be ready to find out the answer, which will most likely be a shocker for you.

Strictly speaking, the topic of logarithms of complex numbers and complex powers is not part of the secondary school curriculum. Nevertheless, we would like to go into greater depth with complex numbers and show you how mathematicians have extended the concepts taught in secondary school. The ideas you will see in this section are not strictly abstract. They are used by engineers on a daily basis and are the basis of many powerful (though sophisticated) real-life applications.

Recall from Section 6, Euler's amazing identity that for any real θ in radians,

$$e^{i\theta} = \cos\theta + i\sin\theta. \tag{7.31}$$

We would like to extend the definition of e raised to any complex power z. One of the rules for exponents for real numbers is that, if we multiply numbers with the same base, we add the exponents. This very useful rule, which we mentioned in Section 6 holds. So, if $z = x + iy$, it must follow that

$$e^{x+iy} = e^x \cdot e^{iy}. \tag{7.32}$$

Using Euler's identity, equation (7.31) this can be written as

$$e^x \cdot e^{iy} = e^x(\cos y + i\sin y) \tag{7.33}$$

or just

$$e^z = r \operatorname{cis} y \tag{7.34}$$

where $r = e^x$ and it is assumed that y is measured in radians. Thus, e raised to a complex power, z, is defined to be the specific complex number defined by equation (7.33) or equivalently, equation (7.34)

This is a new idea, so let us examine some examples.

Example 7.29 *Evaluate* (a) e^{2+3i} (b) $e^{i(\pi/2)}$ (c) $4e^{-1+i}$

Solution. (a) $e^{2+3i} = e^2 \cdot e^{3i} = e^2(\cos 3 + i\sin 3)$ by equation (7.33). For (b), we notice that the real part of the exponent is 0, thus $e^{i\pi/2} = e^0(\cos(\pi/2) + i\sin(\pi/2)) = 1(0 + i(1)) = i$. For part (c), we have $4e^{-1+i} = 4e^{-1}(\cos 1 + i\sin 1)$.

We are now ready to motivate the definition of the logarithm of a complex number. Suppose we want to solve the following equation for z:

$$e^z = 1 + i. \tag{7.35}$$

Writing $z = x + iy$, the left side of equation (7.35) is $e^x \operatorname{cis} y$ while the right side is $\sqrt{2} \operatorname{cis}(\pi/4)$. Thus,

$$e^x \operatorname{cis} y = \sqrt{2} \operatorname{cis}(\pi/4)$$

which implies that

$$e^x = \sqrt{2} \tag{7.36}$$

and that

$$\operatorname{cis} y = \operatorname{cis}(\pi/4). \tag{7.37}$$

But equation (7.36) is an equation with a real exponent, and from this we get $x = \ln \sqrt{2}$. From equation (7.37) we see that $y = \pi/4$ is a solution of equation (7.37). This is called the **principal solution**. But, there are infinitely many solutions of equation (7.37) and each differs from the next by a multiple of 2π. That is, $y = \pi/4 + 2k\pi$ where $k = 0, \pm 1, \pm 2$, and so on. Thus, our equation (7.37) has *infinitely* many solutions. For now, we are only interested in the principal solution. That is,

$$z = x + iy = \ln \sqrt{2} + i(\pi/4).$$

Let us redo this whole process for the general case. If we want to solve the equation $e^z = w$ where w is a complex number, we let $z = x + iy$ and $w = re^{i\theta}$. Then $e^z = w$ becomes $e^x \operatorname{cis} y = r \operatorname{cis} \theta$. This implies that $e^x = r$ and that $\operatorname{cis} y = \operatorname{cis} \theta$. It follows from this that one solution is $x = \ln r$ and that $y = \theta$, and that this is, in fact, the principal solution. Thus, we are led to the following:

> The principal solution of $e^z = w$, where $w = re^{i\theta}$ is $z = \ln r + i(\theta)$, where $0 \leq \theta < 2\pi$. (7.38)

The quantity $\ln r + i\theta$ is called the **principal logarithm** of w and is denoted by Log w. (Notice the capital "L.") Let's give some examples.

Example 7.30 *Find Log w, where $w = -1 + \sqrt{3}i$*

Solution. The polar form of $-1 + i\sqrt{3}$ is $2 \operatorname{cis}(2\pi/3)$. Thus, $r = 2$, and $\theta = 2\pi/3$. Since Log$w = \ln r + i\theta$, it follows that $\operatorname{Log}\left(-1 + i\sqrt{3}\right) = \ln 2 + i(2\pi/3)$.

Example 7.31 *Solve for the principal value of z: $e^z = 1 - i$.*

Solution. $z = \operatorname{Log}(1 - i)$. Since $1 - i = \sqrt{2} \operatorname{cis}(-\pi/4)$, $\operatorname{Log}(1 - i) = \ln \sqrt{2} + i(-\pi/4)$.

If we are working with only real numbers, then we cannot take the logarithm of a negative number. But, if we work with complex numbers, such a thing is possible. We can see how this works in the next example.

Example 7.32 *Find Log (-1).*

Solution. Since $-1 = 1 \operatorname{cis} \pi$, $\operatorname{Log}(-1) = \ln 1 + i\pi = \pi i$. We can check our answer. If $\operatorname{Log}(-1) = \pi i$, then by definition of the principal value of the logarithm, $e^{\pi i}$ should be -1. It is by Euler's amazing identity!

Before we finish with complex numbers, we take this one step further. Recall that, for real numbers, if

$$a^u = b \tag{7.39}$$

where a and b are positive numbers, we are allowed to take the natural logarithm of both sides of the equation to get $u \ln a = \ln b$. If we write this last equation in exponential form, we get that

$$e^{u \ln a} = b. \tag{7.40}$$

Comparing equation (7.39) with equation (7.40), we see that

$$a^u = e^{u \ln a}. \tag{7.41}$$

Now, suppose that we wish to define z^w where z and w are complex numbers. Taking the lead from equation (7.41), we can define the **principal value** of z^w by

$$z^w = e^{w \operatorname{Log} z}. \tag{7.42}$$

Since we now know what it means to raise e to a complex power, we know how to evaluate z^w for any two complex numbers where $z \neq 0$. Let us give some examples.

Example 7.33 *Find the principal value of* $(-2)^i$.

Solution. Since $-2 = 2 \operatorname{cis} \pi$, we have by equation (7.42) that

$(-2)^i$

$= e^{i \operatorname{Log}(-2)}$ (By equation (7.42))

$= e^{i(\ln 2 + i\pi)}$ (By equation (7.38), definition of Log.)

$= e^{i \ln 2 + \pi i^2}$ (Multiplying)

$= e^{-\pi + i \ln 2}$ (Simplifying and rearranging.)

$= e^{-\pi} e^{i \ln 2}$

$= e^{-\pi}(\cos(\ln 2) + i \sin \ln(2))$ (Euler's identity (7.31).)

Notice that our answer is complex, as we expect it would be. But this might not be the case. We now leave you with quite a surprise.

Example 7.34 *Show that the principal value of* i^i *is a real number.*

Are you shocked? Most people are. Here is the solution:

Solution.

$i^i = e^{i \operatorname{Log} i}$ (by equation (7.42)

$e^{i(\ln 1 + i(\pi/2))}$ (by equation (7.38)

$= e^{i(i(\pi/2))}$ (Since $\ln 1 = 0$)

$= e^{-\pi/2}$.

This, of course, is real. And that fact, is unreal!

Student Learning Opportunities

1. Solve for the principal value of z
   (a) $e^z = i$
   (b) $e^{2z} = 1 + i$
   (c) $e^{4z} = -1$

2. What is the value of each of the following expressions?
   (a) Log $(3i)$
   (b) Log $(-2i)$
   (c) Log $(1 + \sqrt{3}i)$
   (d) Log $(-\sqrt{3} - 1i)$

3. (C) A student asks, why z can't be zero in the definition of z^w. How do you respond?

4. (C) You ask your students to simplify the expression Log $(-1-i)^2$ and the first thing they do is to write it as 2 Log $(-1-i)$. Evaluate each of the expressions, Log $(-1-i)^2$ and 2 Log $(-1-i)$ and decide if they are correct. If they are not correct, explain why.

5. Show that Log $(-i)(i) \neq$ Log $(-i) +$ Log$i$. What can you say about the applicabilty of the Laws of Logarithms that you learn in high school to complex numbers?

6. (C) You ask the students in your enrichment class if $(z^c)^d$ is the same as z^{cd} when c and d are complex, and everyone says, "Yes." Are they correct? If not, how can you help them understand why not? [Hint: Let $z = -i$, $c = -i$, and $d = i$.]

7. Find the principal value of $(1-i)^{4i}$.

8. Show that the principal value of $(-1)^{\frac{1}{\pi}} = e^i$.

9. Accepting the fact that $e^{z_1} \cdot e^{z_2} = e^{z_1 + z_2}$, show that, if z, c, and d are complex numbers, then $z^c z^d = z^{c+d}$.

CHAPTER 8

INDUCTION, RECURSION, AND FRACTAL DIMENSION

8.1 Introduction

Despite the fact that most people encounter recursive relationships in the course of their lives, it is a topic that secondary school students don't get to study in a formal way, unless they take a course in precalculus, which, as we well know, most students never do. Many practical problems can be solved by using recursive relationships, making recursion a very important topic in applied mathematics. Its applications are numerous, ranging from the mundane, like finance, to the obscure, like fractals. Closely related is the topic of mathematical induction, which is a powerful tool both in mathematics, and also in the study of recursive relationships. In this chapter we take a very close look at both of these topics in hopes that you will appreciate their value and be able to incorporate them in some of the updated curricula now being used in many secondary schools.

8.2 Recursive Relations

 LAUNCH

A standard roll of paper towels consists of a cardboard tube with outer diameter 4 cm. A typical roll contains 100 sheets of paper each 25 cm long, so the total length of paper is 2500 cm. Imagine the paper being wound onto the cardboard tube. After each complete winding, the total diameter of the roll increases by an amount $2t$, where t cm is the thickness of the paper. Let S_n cm denote the total length of paper wrapped around the tube when it is wrapped around n times, so that $S_0 = 0$.

1. When the paper is wrapped around the roll the first time ($n = 1$), what will the outer diameter, D_1, of the roll be?
2. What length of paper, C_1, is needed to wrap around the roll the first time?
3. When the paper is wrapped around the roll the second time ($n = 2$), what will the outer diameter, D_2 of the roll be?
4. What length of paper, C_2, is needed to wrap around the roll for the second time?
5. How much paper, S_2 (in centimeters) will you have wrapped altogether, after this second wrapping?

358 Induction, Recursion, and Fractal Dimension

6 After three wrappings, how much paper, S_3, will you have wrapped altogether?
7 After n wrappings, how much paper, S_n, will you have wrapped altogether?
8 How does the amount of paper after n wrappings, S_n, compare with the amount of paper after $n-1$ wrappings? Write a relationship that expresses it.

We hope that when doing the launch problem, you realized that to answer some of the questions, you were using the same operations repeatedly. Did you also notice that the final relationship you expressed involved both S_n and S_{n-1}? Without knowing it, you have been examining a recurrence relation, which we will investigate further in this section.

What do you think of when you hear the term "recursion?" Probably, the word repetition comes to mind. The fact is that recursion is a very sophisticated type of repetition. What do we mean by this? In this section we will answer this question by investigating many different applications of recursion as well as the related topic of induction.

We begin by examining a case of recursion that you probably encounter in your daily life. That is, consider what happens when you put money into a bank account and you receive interest. How does your money grow? Here is an example that addresses this situation.

Example 8.1 *You just got a gift of 126 dollars which you put into an account paying annual interest at the rate of 6% per year, hoping that it will grow into a thousand dollars in 10 years. Assuming that you don't withdraw any of the money, will this be possible?*

Solution. Let A_n be the amount in the account at the end of n years, and A_0 the initial amount put into the account. So $A_0 = 126$. At the end of year 1, we have in the account our initial amount A_0, plus the interest, $0.06A_0$, for a total amount of $1.06A_0$. Let us call this amount, A_1, to remind us that it is the amount that is in the account after 1 year, or equivalently, at the beginning of the second year. So,

$A_1 = A_0 + 0.06A_0$ that is,

$A_1 = 1.06A_0$.

In a similar manner, at the end of year 2, we have our initial amount from the beginning of the year, A_1 plus the interest $0.06A_1$ for a total of $1.06A_1$. That is,

$A_2 = A_1 + 0.06A_1$

$A_2 = 1.06A_1$.

If we let A_n be the amount in the account at the end of the nth year, we can compute it as follows: It is the amount at the beginning of the year, which is A_{n-1}, plus the interest, $0.06A_{n-1}$ for a total of $1.06A_{n-1}$. Thus the amount in the account at the end of n years is given by the equations:

$A_n = 1.06A_{n-1}$ \hfill (8.1)

$A_0 = 126$. \hfill (8.2)

Induction, Recursion, and Fractal Dimension 359

The equation (8.1) is what is called a recursive relation. A **recursive relation** is a rule which tells us how to compute new values from old values. Equation (8.2) is called an initial condition.

Using the recursive relation and the initial condition, we can now generate a sequence of numbers, which tells us how much money is in the account after *n* years. Here is our table:

n	0	1	2	3	4	5	etc.
$A_n =$	126	133.56	141.57	150.06	159.07	168.61	

In each case, we simply multiplied the previous value A_{n-1} by 1.06 to get A_n. Thus, $A_3 = 1.06 A_2$ and $A_4 = 1.06 A_3$, and so on. Finish the table and answer the question.

Example 8.2 *When you take any medication, say in pill form, written on the bottle are directions for how often you should take the pills. This is because scientists know that, with time, a certain portion of the medication is excreted from the body. Another pill will have to be taken at a certain time interval in order to maintain the effective level of medication in the bloodstream. Suppose that one begins an initial dosage of 80 milligrams of a medication, and that at the end of each 24 hour period 40% of what is in the body has been excreted. As a result, a new 80 milligram dose must be taken after each 24-hour period. Write a recursive relation that expresses the amount of medication in the bloodstream after n 24-hour periods right after each new dosage is taken.*

Solution. Let A_0 represent the initial dose. Thus $A_0 = 80$. After one 24-hour period, 40% of the medication is excreted, thus 60% of the medication stays. So, the amount in the system right before we take the new dose is $0.60 A_0$. We then take an 80 milligram dose and the amount of medication in our system, represented by A_1 is given by

$$A_1 = 0.60 A_0 + 80.$$

This analysis holds for the other 24-hour periods.

In general, the amount of medication after *n* days (right after the new dose of medication is taken) is 60% of the amount in the bloodstream after the last dose +80. In symbols,

$$A_n = 0.60 A_{n-1} + 80.$$

This is our recursive relation which we will now use to generate some values.

$A_1 = 0.60 A_0 + 80 = 0.60(80) + 80 = 128$; $A_2 = 0.60 A_1 + 80 = 0.60(128) + 80 = 156.8$, and so on. Below is a table with the first 6 values of A_n:

n	0	1	2	3	4	5	etc.
$A_n =$	80	128	156.8	174.08	184.45	190.67	

360 Induction, Recursion, and Fractal Dimension

In the previous two examples, A_n was defined in terms of A_{n-1}. Now we give you an example of a recursive relation that defines A_n in terms of both A_{n-1} and A_{n-2}. It leads to the famous Fibonacci sequence.

Example 8.3 *We start with the two initial conditions $F_1 = 1$ and $F_2 = 1$. We define the nth Fibonacci number F_n where $n > 2$ to be the sum of the previous two Fibonacci numbers. Thus, $F_n = F_{n-1} + F_{n-2}$ when $n > 2$. Generate a table that gives the first 7 Fibonacci numbers.*

Solution. Our recursive relation, $F_n = F_{n-1} + F_{n-2}$, yields the following table

F_1	F_2	F_3	F_4	F_5	F_6	F_7	etc.
1	1	2	3	5	8	13	

where it is clear that each Fibonacci number is the sum of the previous two. Notice that, in the first two examples, the A_n was defined in terms of only one previous value, namely, A_{n-1}. In this example, A_n is defined in terms of two previous values A_{n-1} and A_{n-2}. In general, in a recursive relation, it is possible to define A_n in terms of several of the previous values of A.

In each of the previous three examples, our recursive relations generated sequences of numbers. However, recursive relations can be defined for other objects as well. Here is an example which we will talk more about when we discuss fractals.

Example 8.4 *Begin with an equilateral triangle. Call that T_1. Connect the midpoints of the triangle cutting the triangle into 4 equilateral triangles, and then remove the "middle one," which we have shaded, leaving us with three equilateral triangles. Call the resulting figure T_2. Now do the same thing on each of the remaining 3 unshaded triangles in T_2, namely, join the midpoints of the sides, forming 4 equilateral triangles, and then remove the middle one in each triangle so divided. Thus, each equilateral triangle will have been divided into three other equilateral triangles. Call this figure T_3. On T_3 repeat the procedure and call the result T_4. Describe the figure $\dot{T}_4$.*

Solution. Our first triangle, T_1 is shown in Figure 8.1(a) below.

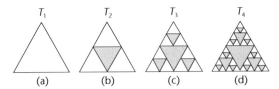

Figure 8.1

We divide this into 4 equilateral triangles by connecting the midpoints and then removing the middle one. In Figure 8.1(b) we see the resulting figure, T_2, where the gray area represents the triangle removed. Now, on each of the remaining (white) triangles in T_2, we join the midpoints of the sides and split each such triangle into 4 triangles and then remove the middle one from each triangle. Figure 8.1(c) shows the resulting Figure T_3 with the gray representing the triangles removed. Now, on each white triangle in triangle T_3, we perform the same process. We split it into 4 equilateral triangles and remove the middle one. We get the Figure T_4 in Figure 8.1(d).

If we do this forever, the final figure we get is what is called the **Sierpinski triangle**. It looks a lot like Figure (d), only there are many more gray areas.

There are some interesting properties of the Sierpinski triangle. First, it is a fractal in the sense of Chapter 7. That is, it has what is known as infinite complexity. Let's look at its area and perimeter. We work with a triangle all of whose sides are equal to 1.

In the first step we have an equilateral triangle with sides 1, whose perimeter is 3 and whose area we know is $\frac{\sqrt{3}}{4}$. In the second step we have 3 triangles similar to the original, each of whose sides is 1/2 the original. Thus, the perimeter of each of these smaller triangles is 1/2 that of the previous triangle. But there are 3 of them. Thus, the perimeter of this second figure is 3/2 that of the first or $\left(\frac{3}{2}\right)(3)$. At each iteration, the perimeter of the new figure is 3/2 that of the prior figure. So, after the second iteration, the perimeter is $\left(\frac{3}{2}\right)^2 (3)$. After the third iteration, the perimeter is $\left(\frac{3}{2}\right)^3 (3)$. After n iterations the perimeter is $\left(\frac{3}{2}\right)^n (3)$ whose limit is ∞.

What about the sum of the areas of the triangles making up each successive figure? Do they also go to infinity? The answer may surprise you. The area of the first triangle is $\frac{\sqrt{3}}{4}$. At the second iteration we have 3 similar triangles, each of whose sides is 1/2 the side of the original triangle. Hence, the area of each of the smaller similar triangles is 1/4 that of the original triangle. But there are 3 times as many smaller triangles. Thus, the area of the figure at the second iteration is $3 \cdot \frac{1}{4} \cdot \frac{\sqrt{3}}{4}$ or just $\frac{3}{4} \frac{\sqrt{3}}{4}$. After the next iteration, we have three times as many triangles each with area 1/4 that of the previous. Thus the sum of the areas of the triangles in the next figure is $\frac{3}{4} \cdot \left(\frac{3}{4} \cdot \frac{\sqrt{3}}{4}\right)$. Continuing in this manner, we can show (by induction) that, at the nth step of this iterative process, the sum of the areas of the triangles is $\left(\frac{3}{4}\right)^{n-1} \cdot \frac{\sqrt{3}}{4}$. Now as n gets large, $\left(\frac{3}{4}\right)^{n-1}$ gets small and approaches zero, so $\left(\frac{3}{4}\right)^{n-1} \cdot \frac{\sqrt{3}}{4}$, the sum of the areas of the triangles approaches zero. Since the Sierpinski triangle is the figure created after infinitely many iterations, we have established that the area of the Sierpinski triangle is zero, while its perimeter is infinite! Not only that, the Sierpinski triangle still contains many points even though its area is zero. The vertex of each unshaded triangle at each step is still there! How strange! This kind of property is not unusual for fractals, as you will see later in this chapter (see section on fractals). But for now we move onto solving recursive relations.

8.2.1 Solving Recursive Relations

As you have seen, recursive relations have many important applications. Our goal in this section is to learn how to solve these recursive relations when they involve only numbers. Specifically, we will examine procedures for finding the nth term of a recursive relationship in terms of n alone. Finding such a formula is known as **solving the recursive (or recurrence) relation**. Using a formula is much more efficient then generating tables to find our values as we did in the previous section. Let us now see how we can find these powerful formulas.

The simplest type of recursive relation which you might already be familiar with is the **arithmetic sequence:**

$$a_n = a_{n-1} + d \quad \text{where } a_1 \text{ is given,} \tag{8.3}$$

and the number d is the common difference. Let us give an example.

Example 8.5 Find the first 5 terms of the sequence generated by the recursive relation $a_n = a_{n-1} + 5$ given that $a_1 = 3$.

Solution. Here d is 5. Thus, according to our recursive relation, $a_2 = a_1 + 5 = 3 + 5 = 8$. Similarly, $a_3 = a_2 + 5 = 8 + 5 = 13$, and so on. This generates the table

a_1	a_2	a_3	a_4	a_5	a_6	a_7	etc.
3	8	13	18	23	28	33	

where it is clear that each entry is 5 more than the previous entry. (These tables are easily generated by spreadsheets.)

Solving a recurrence relation like equation (8.3) is easy. We rewrite it as

$$a_n - a_{n-1} = d. \tag{8.4}$$

This yields the following system of equations obtained by substituting $n = 1$, $n = 2$, and so on in equation (8.4):

$a_2 - a_1 = d$

$a_3 - a_2 = d$

$a_4 - a_3 = d$

$\ldots$

$a_n - a_{n-1} = d.$

We have $n - 1$ equations here (as we can see, since the first equation involves a subtraction of a_1, the second of a_2, and the last, a_{n-1}). If we add them up, we are left with $a_n - a_1 = (n-1)d$ (since almost every term is matched with its opposite), which can be rewritten as

$a_n = a_1 + (n-1)d.$

In words: The nth term of an arithmetic sequence is the first term, plus $(n-1)$ times the common difference, d, which represents the solution of equation (8.3). We state this as a theorem.

Theorem 8.6 *The solution of the recurrence relation given by $a_n = a_{n-1} + d$ where a_1 is fixed, is given by $a_n = a_1 + (n-1)d$.*

Thus, if we wanted to find the 23rd term in the above sequence from Example 8.5, we would get

$a_n = a_1 + (n-1)d,$ which means that

$a_{23} = 3 + (22)5 = 113.$

Another simple type of recursion is one you are also probably familiar with, the **geometric sequence**, which looks like

$$a_n = k a_{n-1}$$

where a_1 is known and k is a constant. Here, each term generated is a constant k, times the previous term. We call k the **constant multiplier**. We saw a case like this in Example 8.1.

We can solve this recursion by rewriting it as

$$\frac{a_n}{a_{n-1}} = k \quad \text{for} \quad n > 2. \tag{8.5}$$

This yields the following equations obtained by substituting $n = 2, 3, 4$ and so on into equation (8.5):

$$\frac{a_2}{a_1} = k$$

$$\frac{a_3}{a_2} = k$$

$$\frac{a_4}{a_3} = k$$

$$\ldots$$

$$\frac{a_n}{a_{n-1}} = k.$$

Again, we see there are only $n - 1$ equations. If we multiply them together, we are left with: $\frac{a_n}{a_1} = k^{n-1}$, (verify!) and this can be written as

$$a_n = a_1 k^{n-1}.$$

We state this as a theorem:

Theorem 8.7 *The solution of the recurrence relation $a_n = k a_{n-1}$ where a_1, the first term, is known, is $a_n = a_1 k^{n-1}$. Stated another way, the nth term of a geometric sequence is $a_1 k^{n-1}$ where k is the constant multiplier.*

Let us see how this works in an example.

Example 8.8 *Given the geometric sequence, $2, \frac{2}{3}, \frac{2}{9}, \ldots$, find the 50th term.*

Solution. The first term is 2, and the constant multiplier is $k = \frac{1}{3}$. (That is, $a_1 = 2$ and $a_n = \frac{1}{3} a_{n-1}$ is the recursive relation defining this geometric sequence.) By Theorem (8.7) the 50th term is, $a_{50} = 2(\frac{1}{3})^{49}$.

In Example (8.1) we wanted to find out the amount in an account after 10 years, given our initial investment of 126 dollars. Thus, we were given a_1. We were also told that we were earning interest at the rate of 6%, so the value of our investment after each year was 1.06 times that of the previous year. That is, k was 1.06. Thus, by Theorem (8.7), after 10 years we would have

$$a_{10} = a_1 k^9 = (126)(1.06)^9 = \$212.87.$$

You may have noticed that, in some examples, we have called the initial condition a_0 rather than a_1. The decision to use a_0 or a_1 is arbitrary and is usually based on what seems more natural in a given problem.

The third type of recurrence relation we will now examine is a combination of the first two. It is the relation: $a_n = ka_{n-1} + b$ where k and b are constants and a_1 is known. This yields the following equations.

$$a_2 = ka_1 + b \tag{8.6}$$

$$a_3 = ka_2 + b \tag{8.7}$$

$$a_4 = ka_3 + b \tag{8.8}$$

...

$$a_n = ka_{n-1} + b.$$

Substituting equation (8.6) into equation (8.7), we get

$$a_3 = ka_2 + b$$
$$= k(ka_1 + b) + b$$
$$= k^2 a_1 + bk + b$$
$$= k^2 a_1 + b(k + 1). \tag{8.9}$$

(We are writing a_3 in this form to enable us to find a pattern.) Substituting equation (8.9) into equation (8.8) we get

$$a_4 = ka_3 + b$$
$$= k(k^2 a_1 + bk + b) + b$$
$$= k^3 a_1 + bk^2 + bk + b$$
$$= k^3 a_1 + b(k^2 + k + 1)$$

and so on. Our final result is that

$$a_n = k^{n-1} a_1 + b(k^{n-2} + k^{n-1} + \ldots + 1) \tag{8.10}$$

(which we can prove by induction if we wish. See the section on induction later in this chapter for a review.) The quantity in parentheses in equation (8.10) is a (finite) geometric series and we found in Chapter 6, Theorem 6.47 that the sum of that series is $\dfrac{1 - k^{n-1}}{1 - k}$. Thus, our final result is

$$a_n = k^{n-1} a_1 + b\left(\dfrac{1 - k^{n-1}}{1 - k}\right). \tag{8.11}$$

Sometimes in a recursion we call the first term a_0 as we did in Example 8.1. If we do that here, our recursion formula becomes

$$a_n = k^n a_0 + b\left(\frac{1-k^n}{1-k}\right). \tag{8.12}$$

We can see how this is applied in Example 8.2, where we had a problem which involved the administration of medication. Our recursion for that problem was $a_n = 0.60 a_{n-1} + 80$. Thus, after n days, using the value of $b = 80$ and $k = 0.60$, in formula (8.12), the amount of medication in the person's system is

$$a_n = (0.60)^n (80) + 80\left(\frac{1-(0.60)^n}{0.40}\right). \tag{8.13}$$

Let us generate some of these amounts using formula (8.13): We get the following table:

a_1	a_2	a_3	a_4	a_5	a_6	a_7	etc.
128	156.8	174	184.4	190.6	194.4	196.2	

As you can see, the number of milligrams of medication in the body is getting larger and larger as the days go on. Now, suppose that a safe dose is under 225 milligrams. Can we be assured that this patient will not go above the safe dosage? To answer the question, take the limit of equation (8.13) as $n \to \infty$. Using the fact that $(0.60)^n \to 0$ as $n \to \infty$, we see that equation (8.13) yields

$$\lim_{n\to\infty} a_n = \lim_{n\to\infty} (0.60)^n (80) + 80\left(\frac{1-(0.60)^n}{0.40}\right) = 0 + 80\left(\frac{1}{0.40}\right) = 200.$$

Thus, this patient is safe from overdose with the current dosage. Of course, if a safe dosage were 180 milligrams, then not only would our patient be at risk, but the doctor who prescribed the medication would be open to a lawsuit! So, a careful analysis of such a problem as this is critical!

Another way to solve certain recursive relations is by a method shown below in Example (8.9). Here, we simply use the recursive relationship given repeatedly until we get to our answer.

Example 8.9 *Solve the recursive relation*

$$a_n = 3n a_{n-1} \tag{8.14}$$

where $a_1 = 1$.

Solution. We begin with the recursive relationship $a_n = 3n a_{n-1}$. Now, since this recursive relation holds for all $n > 1$, we have, replacing n by $n-1$ in equation (8.14) that

$$a_{n-1} = 3(n-1) a_{n-2}. \tag{8.15}$$

Similarly, replacing n by $n-2$ in equation (8.14), we get

$$a_{n-2} = 3(n-2) a_{n-3}, \tag{8.16}$$

and so on. Thus, using this string of equations (8.14)–(8.16) repeatedly, we have

$$a_n = 3n \cdot a_{n-1}$$
$$= 3n \cdot 3(n-1)a_{n-2}$$
$$= 3n \cdot 3(n-1) \cdot 3(n-2)a_{n-3}$$
$$= \ldots$$
$$= 3n \cdot 3(n-1) \cdot 3(n-2) \ldots \cdot 3(2)a_1$$
$$= 3n \cdot 3(n-1) \cdot 3(n-2) \ldots \cdot 3(2) \cdot 1$$
$$= 3^{n-1}[n \cdot (n-1) \cdot (n-2) \ldots \cdot 2] \cdot 1 \quad \text{(since all have factors of 3 except the last)}$$
$$= 3^{n-1}n!.$$

Although we do not wish to explore recursion to the depths you might see it in a Discrete Math course, we do wish to talk about one particular kind of recursion which, surprisingly, is solved by finding roots of polynomial equations. This kind of recursion, known as a **linear recurrence relation**, relates to material we studied in Chapter 3 and looks as follows:

$$a_n = k_1 a_{n-1} + k_2 a_{n-2} + \ldots k_p a_{n-p} \tag{8.17}$$

where a_n is defined solely in terms of p of the previous terms and the k's are constants. We saw an example of this type of relation with the Fibonacci sequence, where each term was defined in terms of the previous two terms. Here are examples of some other linear recurrence relations:

$$a_n = 3a_{n-1} + 4a_{n-2} \tag{8.18}$$
$$a_n = 5a_{n-1} + 6a_{n-2} + 2a_{n-3}. \tag{8.19}$$

In equation (8.18) we see that a_n is defined in terms of the two previous terms, while in equation (8.19) we see that a_n was defined in terms of the three previous terms.

With every linear recurrence relation, there is associated a polynomial equation called the **characteristic equation**. If a_n is defined in terms of the two previous terms, then our equation is of second degree. If a_n is defined in terms of the 3 previous terms, it is of third degree. If a_n is defined in terms of the k previous terms, it is of kth degree.

Here is how a characteristic polynomial is formed. If a_n is defined in terms of the k previous terms, replace a_n by x^k in equation (8.17). Then replace a_{n-1} by x^{k-1} and a_{n-2} by x^{k-2}, and so on. So, in equation (8.18) since a_n was defined in terms of the previous two terms, we replace a_n by x^2 and a_{n-1} by x, and so on. Thus, the characteristic equation for equation (8.18) is $x^2 = 3x + 4$ and that of equation (8.19) is $x^3 = 5x^2 + 6x + 2$, since it is defined in terms of the previous 3 terms.

> **Theorem 8.10** *If the characteristic equation associated with a recurrence relation has degree k and this equation has k distinct roots $r_1, r_2, \ldots, r_k$ then the solution of the recurrence relation (8.17) is of the form $a_n = c_1(r_1)^n + c_2(r_2)^n + \ldots + c_k(r_k)^n$, where the c's are constants.*

Induction, Recursion, and Fractal Dimension   367

Thus, the roots of the characteristic equation gives us the solution to our recursion. We will guide you through the proof of this in the Student Learning Opportunities, but for now, let's give some examples.

Example 8.11 *Solve the recurrence relation $a_n = 5a_{n-1} - 6a_{n-2}$, given that $a_1 = 4$ and $a_2 = 7$.*

Solution. Since a_n is defined in terms of the two previous terms, our characteristic equation is of second degree. We begin by replacing a_n in the given recursion relation by x^2. Our characteristic equation becomes, $x^2 = 5x - 6$. If we solve this quadratic, we get the solutions $r_1 = 2$ and $r_2 = 3$. Thus by the theorem, the solution of our recursion is of the form

$$a_n = c_1(2)^n + c_2(3)^n. \tag{8.20}$$

Using the fact that $a_1 = 4$ we get, substituting in equation (8.20) that

$$a_1 = c_1(2)^1 + c_2(3)^1 \quad \text{or just} \quad 4 = 2c_1 + 3c_2. \tag{8.21}$$

Using the fact that $a_2 = 7$ in equation (8.20), we get that

$$a_2 = c_1(2)^2 + c_2(3)^2 \quad \text{or just} \quad 7 = 4c_1 + 9c_2. \tag{8.22}$$

Solving the system of equations represented by equations (8.21) and (8.22) simultaneously, we get $c_1 = \dfrac{5}{2}$ and $c_2 = \dfrac{-1}{3}$. (Verify!) Thus, the solution of our recurrence relation is

$$a_n = \frac{5}{2}(2)^n + \frac{-1}{3}(3)^n. \tag{8.23}$$

You can check that $a_1 = 4$ and that $a_2 = 7$ when substituting $n = 1$ and $n = 2$ in equation (8.23).
Let us solve one more recurrence relation.

Example 8.12 *Find the solution to the recurrence relation $a_n = 9a_{n-1}$, given that $a_1 = 5$.*

Solution. Since a_n is given in terms of one of the previous terms, the characteristic equation must be of first degree. The characteristic equation is $x = 9x^0$, whose only root is $r_1 = 9$. Thus, the solution to our recurrence relation is $a_n = c_1(9)^n$ and, since $a_1 = 5$, we have, by substituting $n = 1$ into this equation, that $5 = c_1(9)^1$. Hence $c_1 = \dfrac{5}{9}$ and the solution of our recurrence relation is $a_n = \dfrac{5}{9}(9)^n$, which is what we would have expected anyway from Theorem 8.7.

Here is another example whose final result is quite surprising. Earlier, we mentioned the Fibonacci sequence, 1, 1, 2, 3, 5, 8, 13,... which was generated by using the recursion $F_n = F_{n-1} + F_{n-2}$ where $F_1 = 1$ and $F_2 = 1$. According to the work with recurrence relations, we should be able to find a solution for the recursion. We set up our characteristic equation, which is $x^2 = x + 1$, and then solve for x by the quadratic formula. We get $x = \dfrac{1 \pm \sqrt{5}}{2}$. Thus, the solution to our recurrence relation is:

$$F_n = c_1 \left(\frac{1+\sqrt{5}}{2}\right)^n + c_2 \left(\frac{1-\sqrt{5}}{2}\right)^n.$$

Setting $n = 1$ we get

$$1 = c_1 \left(\frac{1+\sqrt{5}}{2}\right) + c_2 \left(\frac{1-\sqrt{5}}{2}\right) \qquad (8.24)$$

and setting $n = 2$ we get

$$1 = c_1 \left(\frac{1+\sqrt{5}}{2}\right)^2 + c_2 \left(\frac{1-\sqrt{5}}{2}\right)^2$$

which simplifies to

$$1 = c_1 \left(\frac{6+2\sqrt{5}}{4}\right) + c_2 \left(\frac{6-2\sqrt{5}}{4}\right). \qquad (8.25)$$

Solving equations (8.24) and (8.25) simultaneously, say using Cramer's rule (see Appendix 1), we get

$$c_1 = \frac{\begin{vmatrix} 1 & \dfrac{1-\sqrt{5}}{2} \\ 1 & \dfrac{6-2\sqrt{5}}{4} \end{vmatrix}}{\begin{vmatrix} \left(\dfrac{1+\sqrt{5}}{2}\right) & \dfrac{1-\sqrt{5}}{2} \\ \left(\dfrac{6+2\sqrt{5}}{4}\right) & \dfrac{6-2\sqrt{5}}{4} \end{vmatrix}} = \frac{1}{\sqrt{5}}$$

and

$$c_2 = \frac{\begin{vmatrix} \left(\dfrac{1+\sqrt{5}}{2}\right) & 1 \\ \left(\dfrac{6+2\sqrt{5}}{4}\right) & 1 \end{vmatrix}}{\begin{vmatrix} \left(\dfrac{1+\sqrt{5}}{2}\right) & \dfrac{1-\sqrt{5}}{2} \\ \left(\dfrac{6+2\sqrt{5}}{4}\right) & \dfrac{6-2\sqrt{5}}{4} \end{vmatrix}} = \frac{-1}{\sqrt{5}}.$$

Thus, the solution to our recursion is

$$F_n = \frac{1}{\sqrt{5}} \left(\frac{1+\sqrt{5}}{2}\right)^n - \frac{1}{\sqrt{5}} \left(\frac{1-\sqrt{5}}{2}\right)^n. \qquad (8.26)$$

Now equation (8.26) seems unbelievable, since all the terms of the Fibonacci sequence are integers and in equation (8.26) square roots occur throughout our solution. But, this really does work. It can also be proved by strong induction, though it is tricky (see the next section for a proof). However, if you want to convince yourself that this works, just enter, equation (8.26) in your calculator and see what it gives you for different values of n. We worked out the first five terms using a computer algebra system, (the program used to write this book) and indeed it gave, $F_1 = 1$, $F_2 = 1$, $F_3 = 2$, $F_4 = 3$, $F_5 = 5$.

There is one last recursion that we would like to talk about that relates to a famous puzzle known as the **Tower of Hanoi puzzle**. This puzzle is actually sold commercially and is attributed to the French mathematician Edouard Lucas, who discovered it in about 1883. Every computer science student is usually exposed to this Hanoi problem, since its solution involves a recursion which is a fundamental technique in computer programming. Also, recursion is used extensively in sorting algorithms and database programs.

> **Example 8.13** *(Tower of Hanoi) In this problem we have a set of n disks all of different sizes, and we have three pegs. All of the disks are on the first peg, and they are in order of size with the largest disk on the bottom. The goal is to move all the disks from the first peg to the second peg, moving only one disk at a time. There is only one catch. You can never put a larger one on top of a smaller one. The question is, what is the minimum number of moves needed to achieve this?*

Solution. Here is the picture with 7 disks (Figure 8.2).

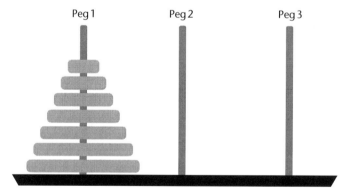

Figure 8.2

One can play with this puzzle a bit before reading what follows. Using the problem-solving approach of starting with a simpler problem, try working it for 3 disks, 4 disks, and so on. The following website gives an interactive program, where you can play with as many disks as you want: http://www.mazeworks.com/hanoi/index.htm. It is worth visiting.

Getting back to the problem at hand, we let the minimum number of moves needed to solve the problem when we have n disks be a_n. We will try to set up a recursive relationship, which shows how we can determine the minimum number of moves with $n+1$ disks, when we know how to do it with n disks.

First we realize that, with $n+1$ disks on a peg, since the largest one is on the bottom, it will be the last to be moved when the disks are taken off the first peg. We refer to the largest disk as L. We move all the disks to the third peg except for L. (Just think of this as the problem of moving the n disks above L.) This requires a_n moves. Then we will move L to the second peg. That requires one move. Then all of the n disks on the third peg will get moved back to the second peg. This requires another a_n moves. Thus, the total number of moves, a_{n+1} needed to move the $n+1$ disks to the second peg is given by

$$a_{n+1} = a_n + 1 + a_n$$

or just

$$a_{n+1} = 2a_n + 1.$$

Although we already know how to solve this, we will do it in the next section with induction, where the solution is quite easy once we guess at the formula for a_n.

We have talked about the number of moves required to solve this puzzle, but we have not discussed what the actual moves are. It is easy to find these moves. First solve the problem for 1 disk. There the movement is simple. We move the one disk to peg 2 and we are done. The recursive analysis given in the previous paragraph tells you the moves for $n = 2$ disks. You move the 1 disk above the largest disk to peg 3, then move the largest disk to peg 2, and then move the disk on peg 3 to peg 2. This involves 3 moves.

Now, suppose we have 3 disks. We move the top two disks to peg 3, but we do it in a manner similar to the way we did it in the previous paragraph where we dealt with the question of moving two disks to peg 2. Then move the last disk to peg 2. Now we move the two disks on peg 3 to peg 2. This can be done by moving the smallest disk on peg 3 to peg 1, and then the remaining disk to peg 2 and then the disk on peg 1 to peg 2. You should draw yourself a picture to see what is going on and then try to figure out how to do it for 4 disks. Using those moves, we can generate the moves for 5 disks, and so on.

Although there is much more we could say about recurrence relations, our goal has been to just give a quick refresher to those of you who have seen it, and give a primer to those who haven't.

Student Learning Opportunities

1 Generate the first five terms of the sequence of numbers defined by each of the following recursive relations. Is the resulting sequence an arithmetic sequence, geometric sequence, or neither? Justify your answer.

(a) $a_n = 5a_{n-1}$, where $a_1 = 2$
(b) $a_n = 2a_{n-1} + a_{n-2}$, where $a_1 = 0$ and $a_2 = 1$
(c) $a_n = a_{n-1} + 6$, where $a_1 = 2$

2 (C)

(a) The seats in your local theater are arranged so that there are 50 seats in the first row, 52 seats in the second row, 54 seats in the third row, and so on, for a total of 100 rows. You want your students to create a formula that will tell you how many seats, R_n, are in row n. Jason answers by giving the recursion formula: $R_n = R_{n-1} + 2$. Is his answer complete? If not, what is missing?

(b) Johnny answers the same question in (a) as follows: $R_n = 48 + 2n$, where $n = 1, 2, \ldots, 100$. Is he right?

3 Write a recursive relation for each of the following scenarios:

(a) You invest 1000 dollars in a vehicle that pays interest at the end of each month at the rate of $\frac{1}{2}$% of whatever you had at the end of the previous month. Assuming no money is withdrawn, write a recursive relationship that gives the amount of money in this investment after t months. Then write a recursive relationship with this same scenario, but where 50 dollars is withdrawn at the end of each month. Find how much will be in the account under each scenario after 6 months.

(b) Bacteria are growing at the rate of 3% per hour. Write a recursive relationship that gives the number of bacteria after t hours, if the initial bacterial population is 450 bacteria.

(c) A patient is advised to take a 50 milligram dose of a drug every morning at the same time. If 40% of the drug is excreted each day, how much of the drug will be in his system before he takes his fifth 50 milligram dose? What is the recursive relation that gives the amount of drug in the patient's system after n additional 50 milligram doses have been taken? What happens as n goes to infinity?

(d) Jack begins a walking routine. The first day he walks a half a mile, then each day thereafter, walks an additional $\frac{1}{4}$ mile. Write the recursive relation that describes the number of miles he walks on the nth day.

(e) A pond is stocked with 3000 trout. Each year 500 new trout are removed, but in general they grow at a rate of 30% per year counting deaths. Write a recursive relation that describes the number of trout in the pond after n years.

4 Solve each of the following recurrence relations:

(a) $a_n = 4a_{n-1}$, where $a_1 = 3$
(b) $a_n = a_{n-1} + 4$, where $a_1 = 5$
(c) $a_n = na_{n-1}$, where $a_1 = 1$
(d) $a_{n+1} = 3a_n - 2a_{n-1}$ when $n > 2$, but $a_1 = 5$ and $a_2 = 7$
(e) $a_n = 4na_{n-1}$, where $a_1 = 2$
(f) $a_n = (2n-1)a_{n-1}$. No initial condition is given.

5 Answer the various parts of the launch question. Write the recursive relationship that expresses S_n in terms of S_{n-1} and solve the recursive relationship.

6 Here is an outline of the proof of Theorem 8.10 for the special case of the recurrence relation

$$a_n = aa_{n-1} + ba_{n-2} \qquad (8.27)$$

where a and b are constants.

We guess a solution of equation (8.27) of the form, $a_n = x^n$. Then substituting into equation (8.27), we get $x^n = ax^{n-1} + bx^{n-2}$. Setting this equation equal to 0, we get $x^n - ax^{n-1} - bx^{n-2} = 0$ and dividing by x^{n-2} we get that

$$x^2 - ax - b = 0. \qquad (8.28)$$

This, of course, is the characteristic equation. Thus, IF our guess is correct, the characteristic equation must hold.

Reversing the steps, if r is a root of the characteristic equation (8.28) then $r^2 - ar - b = 0$. Multiplying this by r^{n-2} we get $r^n - ar^{n-1} - br^{n-2} = 0$ and then, if we solve for r^n, we get $r^n = ar^{n-1} + br^{n-2}$. This tells us that $a_n = r^n$ satisfies equation (8.27). In short, *if r is a root of* *the characteristic equation, then r^n is a solution of the recursion.* Thus, if r and s are the distinct roots of the characteristic equation, then

$$r^n = ar^{n-1} + br^{n-2} \tag{8.29}$$

and

$$s^n = as^{n-1} + bs^{n-2}. \tag{8.30}$$

We want to show that $a_n = c_1 r^n + c_2 s^n$ satisfies the recursion, equation (8.27). We compute $aa_{n-1} + ba_{n-2}$ and we must show that this is equal to a_n. But $aa_{n-1} + ba_{n-2} = a(c_1 r^{n-1} + c_2 s^{n-1}) + b(c_1 r^{n-2} + c_2 s^{n-2}) = c_1(ar^{n-1} + br^{n-2}) + c_2(as^{n-1} + bs^{n-2}) = c_1 r^n + c_2 s^n$ by equations (8.29) and (8.30). And this last expression, is a_n.

Try to generalize this to the case when a_n is defined in terms of the 3 previous terms.

7 (C) You have had your students investigate the Tower of Hanoi Problem and they are comfortable with how to solve it for 1, 2, and 3 disks. However, when trying to explain how to solve it for $n = 4$ disks, they get very confused. How can you help them organize their procedures and come up with an accurate written account that will demonstrate the recursive nature of the solution?

8.3 Induction

LAUNCH

Imagine an extremely long line of dominoes, each standing vertically, with even spacing between each one. Your little niece comes over and pushes the first domino over. She is excited to see that the first domino knocks into the second domino and in turn, the second domino falls forward. You might think that, since the dominoes are all spaced evenly apart, then the second domino will fall and knock the third over, and so on down the line. Are you correct? Do ALL of the dominoes end up falling? If you say, "Yes," prove it.

If in the launch problem, you realized that all of the dominoes would end up falling over, then you already have good intuition about the type of reasoning involved in proof by induction, the focus of this section. You may be wondering why we need yet another type of proof, especially since we have seen how well suited direct proofs have been in constructing certain theorems in mathematics relating to natural numbers. However, there are theorems involving natural numbers that are extremely difficult and complex to prove using a direct method. When this happens, we often try to use the method of mathematical induction.

For example, we might like to prove that, for all natural numbers, n,

$$1 + 2 + 3 + \ldots + n = \frac{n(n+1)}{2}$$

or perhaps that Pick's Theorem (Chapter 5 Section 7 Theorem 5.34) is true for any polygon with n sides. While it is easy to give a direct proof of the first statement, as we did in Chapter 1, the second statement about Pick's Theorem would be harder to do without induction.

The idea behind mathematical induction is very intuitive and is illustrated by the dominoes launch. Here is yet another perspective on this. Suppose that we told you that we saw a stairway, and that we climbed the first step. Suppose we also told you that whenever we climbed a step, we climbed the next step. How many of the steps on this stairway did we climb? The answer seems obvious: we climbed all the steps!

Why? Well, we told you that we climbed the first step. We also told you that whenever we climbed a step, we climbed the next step. Thus, we climbed the second step. By the same reasoning, since we climbed the second step, we had to climb the third step, because we told you that, whenever we climbed a step, we climbed the next step. Having already established that we climbed the third step, we must have climbed the fourth step, for we told you that, whenever we climbed a step we climbed the next step, and so on. Clearly, then, we climbed all the steps. This, essentially, is the principle of mathematical induction. To state it more formally:

Principle of Mathematical Induction

1 Suppose that we have a statement about natural numbers. And suppose that it is true for the natural number $n = 1$.
2 Suppose that, whenever the statement is true for the natural number $n = k$, it is also true for the next natural number $n = k + 1$.

Conclusion: The statement must be true for all natural numbers.

Using the stair example above, it appears to make perfect sense. Saying that the statement is true for $n = 1$ is analogous to saying that you climbed the first step. Statement (2) says that, whenever you climb a step, you climb the next step. Of course, the conclusion that the statement is true for all natural numbers is essentially saying that we have climbed all the steps.

It is usually easy to verify that (1) holds in a problem, though not always.

The way step (2) is implemented is that we usually have to *develop a method* that will show us how to get from any step to the next step. This can be very routine, to very difficult, depending on the problem.

Let us demonstrate a proof by induction using a relationship for which we have already provided a direct proof in Chapter 1. Secondary school students often see this example as their first exposure to proof by induction.

Theorem 8.14 *Using mathematical induction, show that if n is natural number, then*

$$1 + 2 + 3 + \ldots + n = \frac{n(n+1)}{2}. \tag{8.31}$$

Proof. First we must show that equation (8.31) is true when $n = 1$. When $n = 1$, the left side of the equation consists only of the number 1, while the right side is the expression $\frac{1(1+1)}{2}$. Clearly

$$1 = \frac{1(1+1)}{2}.$$

So equation (8.31) is true when $n = 1$.

We now suppose that equation (8.31) is true for some value of n, say $n = k$, and then show that it is true for the next n, namely $n = k + 1$.

Saying that the statement is true for $n = k$ means that, we are supposing that, for some k,

$$1 + 2 + 3 + \ldots + k = \frac{k(k+1)}{2}. \tag{8.32}$$

What we have to show is that equation (8.31) is true for $n = k + 1$. That is, we have to show that

$$1 + 2 + 3 + \ldots + (k+1) = \frac{(k+1)(k+1+1)}{2}. \tag{8.33}$$

Let us work on the left side of equation (8.33).

$$1 + 2 + 3 + \ldots + (k+1)$$
$$= \underbrace{1 + 2 + 3 + \ldots + k} + (k+1)$$
$$= \frac{k(k+1)}{2} + (k+1), \quad \text{since we are assuming equation (8.32) is true.}$$
$$= \frac{k(k+1)}{2} + \frac{2(k+1)}{2} \quad \text{(Common denominator)}$$
$$= \frac{k^2 + k + 2k + 2}{2} \quad \text{(Simplifying)}$$
$$= \frac{k^2 + 3k + 2}{2}$$
$$= \frac{(k+1)(k+2)}{2}. \tag{8.34}$$

Clearly, equation (8.34) is equal to the right side of equation (8.33) and we have shown what we set out to prove. Having shown that equation (8.31) is true for $n = 1$, and having shown that, when equation (8.31) is true for $n = k$, equation (8.31) is true for $n = k + 1$, it follows by the principle of mathematical induction that equation (8.31) is true for all natural numbers. ∎

As we pointed out in the introduction to this section, one must admit that the direct proof from Chapter 1, is certainly, well, more direct!

It is a general misconception that induction is only used to prove relationships involving the sum of integers or squares of integers and so forth. Nothing could be further from the truth, as you will see by the examples that follow.

Theorem 8.15 *Prove using Mathematical Induction, that for all natural numbers n,*

$$n < 2^n. \tag{8.35}$$

Proof. (a) First we show that the relationship $n < 2^n$ is true when $n = 1$. That is,

$$1 < 2^1,$$

which is clearly true.

(b) Next we suppose that inequality (8.35) is true for some natural number k. That is, we assume that

$$k < 2^k \tag{8.36}$$

for some k, and then we try to show that inequality (8.35) is also true for the next natural number, $k + 1$. That is, we try to show that

$$k + 1 < 2^{k+1}. \tag{8.37}$$

How do we do this? Let us work on the left side of inequality (8.36), which we are assuming is true. Adding 1 to both sides yields:

$$\begin{aligned} k + 1 &< 2^k + 1 \\ &< 2^k + 2 \quad \text{(Obviously!)} \\ &\leq 2^k + 2^k \quad \text{(Since } 2 \leq 2^k \text{ when } k \geq 1\text{)} \\ &= 2(2^k) \\ &= 2^{k+1}. \end{aligned}$$

This string of inequalities shows that $k + 1 < 2^{k+1}$, which is inequality (8.37) and this is what we wanted to show. Having shown that inequality (8.35) is true for $n = 1$, and having shown that, whenever inequality (8.35) is true for $n = k$, it is true for $n = k + 1$, it follows that, by mathematical induction, inequality (8.35) is true for all natural numbers n.

You might feel that the statement in this theorem was so simple that there was no need to prove it. Well, yes, the statement does seem obvious. But, with proof, we can be sure. We hope you appreciate the simplicity of this next inductive proof of a non-obvious result.

Theorem 8.16 *Prove that, for all natural numbers n,*

$$4^n + 15n - 1 \quad \text{is divisible by } 9. \tag{8.38}$$

Proof. Although this can be done using modular arithmetic, we will do it by induction, as it is a bit more direct. First, let us show that equation (8.38) is true for $n = 1$. When n is 1, our statement becomes,

$$4 + 15 - 1 \quad \text{is divisible by } 9.$$

This is clearly true since $4 + 15 - 1 = 18$.

Now suppose that equation (8.38) is true for some natural number k. This means that

$$4^k + 15k - 1 \quad \text{is divisible by } 9 \tag{8.39}$$

which can be rewritten as

$$4^k + 15k - 1 = 9m \quad \text{for some integer } m. \tag{8.40}$$

We must show that equation (8.39) is true for $n = k + 1$. That is, we must show that

$$4^{k+1} + 15(k + 1) - 1 \quad \text{is divisible by 9.} \tag{8.41}$$

Now, let us examine what we are trying to prove. We must show that $4^{k+1} + 15(k + 1) - 1$ is divisible by 9. But, we know that

$$4^{k+1} + 15(k + 1) - 1$$
$$= 4 \cdot 4^k + 15k + 14.$$

Now we have to remember that we are trying to use what we know, namely 8.40 in our proof. So we need to manipulate the terms of this last expression so that it is clearly a multiple of 9. Here is how we do that:

$$4 \cdot 4^k + 15k + 14$$
$$= 4(4^k + 15k - 1) - 45k + 18$$
$$= 4(4^k + 15k - 1) - 9(5k - 2)$$
$$= 4(9m) - 9(5k - 2) \quad \text{(Using equation (8.40))}.$$

Since each term in this last expression is divisible by 9, the entire expression is divisible by 9. So we have proved what we set out to prove and, by the principle of mathematical induction, $4^n + 15n - 1$ is divisible by 9 for all natural numbers n. ∎

This last proof is just one example of how induction can be used to prove non-obvious statements. In this next section we prove relationships that are often addressed in the secondary school curricula, but whose proofs are hardly obvious, and would be tedious or impossible to do otherwise.

8.3.1 Taking Induction to a Higher Level

We recall that the Fibonacci numbers are the numbers in the sequence 1, 1, 2, 3, 5, 8, 13, 18, and so on. If we denote the nth Fibonacci number by F_n, then we can describe the terms of this sequence as follows: $F_1 = 1$, $F_2 = 1$, and for $n \geq 2$, $F_n = F_{n-1} + F_{n-2}$. That is,

$$\boxed{\text{any number in the sequence is the sum of the preceding two.}} \tag{8.42}$$

Finding a formula for the Fibonacci sequence has been done in Section 8.2 using theorems related to solving recursive relations. The formula of equation (8.26) for generating the Fibonacci sequence is hardly obvious, but in fact can be proved by srong induction. (See the next section for a proof.)

Interesting relationships hidden in the Fibonacci sequence are numerous. In fact, it has *so* many interesting properties, that there is a journal called *The Fibonacci Quarterly*, devoted exclusively to the sequence and its applications. Did we say applications? Yes! There are numerous real-life applications of the Fibonacci sequence! To talk about them now would divert us from our main purpose, so we continue with induction, but urge you to pursue the fascinating aspects of the Fibonacci sequence. But, we will now introduce you to some amazing relationships that lay within the Fibonacci sequence, and how they can be proven by induction.

Induction, Recursion, and Fractal Dimension

Let us start with the sum of the squares of the first n Fibonacci numbers. Beginning with the Fibonacci sequence $1, 1, 2, 3, 5, 8, 13, \ldots$, we notice that

$$1^2 = 1 \cdot 1 \tag{8.43}$$

$$1^2 + 1^2 = 1 \cdot 2 \tag{8.44}$$

$$1^2 + 1^2 + 2^2 = 2 \cdot 3 \tag{8.45}$$

$$1^2 + 1^2 + 2^2 + 3^2 = 3 \cdot 5 \tag{8.46}$$

$$1^2 + 1^2 + 2^2 + 3^2 + 5^2 = 5 \cdot 8. \tag{8.47}$$

We notice that the right hand side of equation (8.43) is the product of the first two Fibonacci numbers. The right hand side of equation (8.44) is the product of the second and third Fibonacci number. The right hand side of equation (8.45) is the product of the third and fourth Fibonacci number. The right hand sides of the next two equations are the product of the fourth and fifth and fifth and sixth Fibonacci numbers, respectively. This seems to indicate that, if we sum the squares of the first n Fibonacci numbers, we get the product of the nth Fibonacci number and the $n+1$st Fibonacci number. Quite surprising! Don't you agree? If we use the formula of equation (8.26) obtained in the previous section for the nth Fibonacci number and try to prove this, we are in for a major headache. If we prefer not to use this complicated formula, is there a way we can show this pattern always holds? The answer is induction.

Theorem 8.17 *Using induction, show that the sum of the squares of the first n Fibonacci numbers is $F_n F_{n+1}$.*

Proof. When $n = 1$, this statement becomes simply the statement:

$$1^2 = F_1 \cdot F_2.$$

Since $F_1 = F_2 = 1$, this is clearly true.

Now suppose that the statement in the theorem is true for some natural number k. This means that

$$1^2 + 1^2 + 2^2 + \ldots + F_k^2 = F_k \cdot F_{k+1}. \tag{8.48}$$

We need to show that the statement of the theorem is true for $n = k + 1$. That is,

$$1^2 + 1^2 + 2^2 + \ldots + F_{k+1}^2 = F_{k+1} \cdot F_{k+2}. \tag{8.49}$$

We begin by working on the left side of equation (8.49).

$$1^2 + 1^2 + 2^2 + \ldots + F_{k+1}^2$$

$$= \underbrace{1^2 + 1^2 + 2^2 + \ldots F_k^2}_{} + F_{k+1}^2$$

$$= F_k \cdot F_{k+1} + F_{k+1}^2 \quad \text{(Using equation (8.48))}$$

$$= F_{k+1}(F_k + F_{k+1}) \quad \text{(Factoring)}$$

$$= F_{k+1}(F_{k+2}) \quad \text{(Since } F_{k+2} = F_k + F_{k+1} \text{ by (8.42))}.$$

We have shown that the statement of the theorem is true for $n = 1$ and we have shown that, whenever the statement is true for some natural number k, it is true for the next natural number. Hence, the statement of the theorem is true for all natural numbers n by the principle of mathematical induction. ■

Isn't it remarkable how simple this induction proof was in proving a result that was hardly obvious? This is what makes induction so powerful!

Let us now see how induction can be used with the Tower of Hanoi problem from the previous section.

Example 8.18 *In the Tower of Hanoi Problem (Example 8.13) we obtained the following recursion formula for the minimum number of moves needed to move n disks from peg 1 to peg 3 : $a_n = 2a_{n-1} + 1$ where a_1 is obviously 1. Generate some values of a_n and then guess at a formula for a_n. Finally, prove your guess by induction.*

Solution. We have, $a_2 = 2a_1 + 1 = 3$, $a_3 = 2a_2 + 1 = 2(3) + 1 = 7$. Similarly, we can show that $a_4 = 15$, $a_5 = 31$, and so on. It appears that

$$a_n = 2^n - 1. \tag{8.50}$$

Let us prove equation (8.50) by induction.

When $n = 1$, equation (8.50) becomes

$$a_1 = 2^1 - 1 = 1$$

which we already know is true since, if we only have one disk, it only takes one move to get it from peg 1 to peg 2.

Now suppose that equation (8.50) is true for $n = k$. Then a_k, the minimum number of moves needed to move k disks from peg 1 to peg 3, is $2^k - 1$, or

$$a_k = 2^k - 1. \tag{8.51}$$

We need to show that the minimum number of moves to move $k + 1$ disks is $2^{k+1} - 1$. That is, we must show that equation (8.50) is true for $n = k + 1$ or, put another way, that $a_{k+1} = 2^{k+1} - 1$. It only takes a few steps to show this. We will start with what we already know about a_{k+1}.

$a_{k+1} = 2a_k + 1$ (By definition of the recurrence relation)

$\qquad = 2(2^k - 1) + 1$ (By equation (8.51))

$\qquad = 2^{k+1} - 2 + 1$

$\qquad = 2^{k+1} - 1.$

How simple! So by mathematical induction, this relationship holds for all n.

8.3.2 Other Forms of Induction

There are two variations of Mathematical Induction. The first deals with the case where we are trying to prove a relationship only for all natural numbers greater than, or equal to, some number N (not for *all* natural numbers). For example, $2n+1$ is only less than 2^n when $n \geq 3$. To prove the statement that $2n+1 < 2^n$ when $n \geq 3$, by induction, is exactly like the procedure for doing basic induction. The only difference is that, instead of showing the statement is true when $n = 1$ (since it isn't), we begin by showing that it is true when $n = 3$. Then we show that, if the statement is true for $n = k$, then it is true for $n = k + 1$. We will show how this works in the next example.

Example 8.19 *Show that $2n + 1 < 2^n$ when $n \geq 3$.*

Solution. The statement is true when $n = 3$, since it says that $2(3) + 1 < 2^3$.
Now suppose that the statement is true for $n = k$. That is,

$$2k + 1 < 2^k. \tag{8.52}$$

We need to show that it is true for $n = k + 1$. That is, we need to show that $2(k + 1) + 1 < 2^{k+1}$ or, put another way, that $2k + 3 < 2^{k+1}$ when $k \geq 3$. Here are the steps.

$$2k + 3 = \underbrace{2k + 1}_{} + 2$$
$$< 2^k + 2 \quad \text{(By inequality 8.52)}$$
$$< 2^k + 2^k \quad \text{(Since } k \geq 3 \text{ implies } 2 < 2^k\text{)}$$
$$= 2 \cdot 2^k$$
$$= 2^{k+1}.$$

This string of equations and inequalities shows that $2k + 3 < 2^{k+1}$ and since this is what we wanted to show, we are done. You can get practice with doing proofs like this in the Student Learning Opportunities.

Now we turn to another form of induction that is used and this is called the strong form of induction.

Strong Principle of Mathematical Induction

1. Suppose that we have a statement about natural numbers. Also, assume that this statement is true for the natural number $n = 1$.
2. Suppose that, whenever the statement is true for *all* natural numbers *less than* $k + 1$, then it is also true for $k + 1$.

Conclusion: The statement is true for all natural numbers.

Thus, part (2) of this principle says that, if on the basis of the statement being true for *all* natural numbers before $k + 1$, we can show the statement is true for $k + 1$, the statement is true for all natural numbers. (The analogy with climbing steps is as follows: When we climb all the previous steps, we climb this step.) Since in this type of proof we assume the statement is true for all $n < k + 1$, we can then use as many of those statements as we wish to prove it is true for $n = k + 1$ and therefore for all natural numbers. Here are two examples.

Example 8.20 *Prove that every natural number can be expressed as an integral power of 2 or a sum of distinct integral powers of 2. (This is essentially base 2 representation of the number.)*

Solution. This is certainly true for the natural number 1, since 1 can be written as 2^0. Now suppose that the result is true for all numbers less than $k + 1$. We must now show it is true for $N = k + 1$. Let 2^m be the largest power of 2 less than or equal to N. Then

$$N = 2^m + b \qquad (8.53)$$

where $0 \leq b < 2^m \leq N$. If $b = 0$, then N is a power of 2 and we are done. If $b \neq 0$, then $b < N$. But, since we are assuming that all numbers less than N can be written as a sum of powers of 2, b can be written that way. It follows from this, using equation (8.53) that N can also be written as a sum of powers of 2. In summary, N is either an integral power of 2 or the sum of integral powers of 2. Since we have shown that the statement we are trying to prove was true for $n = 1$ and was true for $N = k + 1$ when it was true for all numbers less than N, it follows from the strong principle of mathematical induction that the statement we are trying to prove is true for all natural numbers n.

We didn't need induction for this. We already saw how to convert to base 2 in Chapter 2. But suppose we didn't. This would show that such a representation is possible.

Example 8.21 *Prove the Fundamental Theorem of Arithmetic, that every positive integer ≥ 2 is either prime or can be factored into primes.*

Solution. (Recall that a prime number must be greater than 1, which is why we are only considering numbers ≥ 2.) First look at 2. It is prime. Now, suppose that the every number less than $k + 1$ (but ≥ 2) is prime or can be factored into primes. We have to show that $k + 1$ is either prime, or can be factored into primes. We have two cases. Either $k + 1$ is prime, or it isn't. If it is prime, then we are done. If it is not, then it is composite and can be written as a product $a \cdot b$, where a and b are less than $k + 1$ and greater than or equal to 2. But a and b each being less than $k + 1$ and greater than or equal to 2 can be factored into primes. Thus $k + 1$ being the product of a and b can be written as a product of primes, namely those contained in a and b. We have shown that $k + 1$ is either prime or a product of primes, and so by the strong principle of mathematical induction, every positive integer, $n \geq 2$ is either prime or a product of primes.

Notice that we needed strong induction to assure us that both a and b were prime or factorable into primes. We needed the factorability of all numbers less than $k + 1$, so that we could apply the induction to both a and b. Regular induction wouldn't work so easily.

Example 8.22 *We showed earlier that the formula for the nth Fibonacci number is given by*

$$F_n = \frac{1}{\sqrt{5}} \left(\frac{1 + \sqrt{5}}{2} \right)^n - \frac{1}{\sqrt{5}} \left(\frac{1 - \sqrt{5}}{2} \right)^n.$$

Prove this using strong induction.

Solution. The induction is a bit tricky. Let $\varphi_1 = \frac{1+\sqrt{5}}{2}$ and $\varphi_2 = \frac{1-\sqrt{5}}{2}$. These are the roots of $x^2 - x - 1 = 0$, as we can verify using the quadratic formula. So, in particular,

$$\varphi_1^2 - \varphi_1 - 1 = 0 \quad \text{and}$$
$$\varphi_2^2 - \varphi_2 - 1 = 0.$$

From these it follows that

$$\varphi_1^2 = \varphi_1 + 1 \text{ and} \tag{8.54}$$
$$\varphi_2^2 = \varphi_2 + 1. \tag{8.55}$$

Using our new notation, we are trying to prove that

$$F_n = \frac{1}{\sqrt{5}}\varphi_1^n - \frac{1}{\sqrt{5}}\varphi_2^n. \tag{8.56}$$

We do a strong induction on this. The case $n = 1$ is left for you to check.

We assume that the formula of equation (8.56) is true for all natural numbers prior to $n = k + 1$ and show that it is true for $n = k + 1$. Now consider F_{k+1} which we know is $F_k + F_{k-1}$. But by our strong induction hypothesis, both F_k and F_{k+1} are given by equation (8.56). Substituting into the expression for F_{k+1} we get

$$F_{k+1} = F_k + F_{k-1}$$
$$= \frac{1}{\sqrt{5}}\varphi_1^k - \frac{1}{\sqrt{5}}\varphi_2^k + \frac{1}{\sqrt{5}}\varphi_1^{k-1} - \frac{1}{\sqrt{5}}\varphi_2^{k-1}$$
$$= \frac{1}{\sqrt{5}}\varphi_1^k + \frac{1}{\sqrt{5}}\varphi_1^{k-1} - \left(\frac{1}{\sqrt{5}}\varphi_2^k + \frac{1}{\sqrt{5}}\varphi_2^{k-1}\right)$$
$$= \frac{1}{\sqrt{5}}\varphi_1^{k-1}(\varphi_1 + 1) - \frac{1}{\sqrt{5}}\varphi_2^{k-1}(\varphi_2 + 1)$$
$$= \frac{1}{\sqrt{5}}\varphi_1^{k-1}(\varphi_1^2) - \frac{1}{\sqrt{5}}\varphi_2^{k-1}(\varphi_2^2) \quad \text{(By equations (8.54) and (8.55))}$$
$$= \frac{1}{\sqrt{5}}\varphi_1^{k+1} - \frac{1}{\sqrt{5}}\varphi_2^{k+1}. \quad \text{(Simplifying)}$$

Since we have shown that the formula of equation (8.56) for F_{k+1} follows from the same formula for the previous F_k and F_{k-1}, we have shown that the formula is true for all F_n.

Isn't this a neat proof?

Some trigonometric identities would be very difficult to prove without induction. In Chapter 3, Example 3.15, we proved that, for any rational value of θ, between 0 and 90°, $\cos\theta$ is irrational except for $\theta = 60°$. To do this, we used a result that we had not yet proven, but which we will soon prove using the following lemma.

Lemma 8.23 *Show that* $\cos(n + 1)\theta = 2\cos n\theta \cos\theta - \cos(n - 1)\theta$ *for any positive integer n.*

Proof. Consider the standard trigonometric relationships

$\cos(A + B) = \cos A \cos B - \sin A \sin B$ and
$\cos(A - B) = \cos A \cos B + \sin A \sin B$.

Upon adding these equations, we get

$$\cos(A+B) + \cos(A-B) = 2\cos A \cos B$$

from which it follows that

$$\cos(A+B) = 2\cos A \cos B - \cos(A-B).$$

Now, if we let $A = n\theta$ and $B = \theta$ we have

$$\cos(n+1)\theta = 2\cos n\theta \cos\theta - \cos(n-1)\theta \tag{8.57}$$

which is what we were trying to prove. ∎

What follows is a trigonometric proof of the interesting result that any trigonometric function of the form $2\cos n\theta$ can be represented as a polynomial in $2\cos\theta$ with lead coefficient 1. (This is what we used to show that $\cos\theta$ is irrational for any rational θ strictly between 0 and 90° except for $\theta = 60°$.) This proof is easiest with strong induction.

Example 8.24 *Show that, for any rational value of θ, there exists a polynomial with integer coefficients and lead coefficient 1 such that*

$$2\cos n\theta = 1x^n + a_{n-1}x^{n-1} + \ldots + a_0 \tag{8.58}$$

where $x = 2\cos\theta$. In words: we can represent $2\cos n\theta$ as a polynomial of degree n in x with lead coefficient 1 where $x = 2\cos\theta$.

Solution. When we substitute $n = 1$ in (8.58) we get $2\cos\theta = x$, which we know is true. This also says that $2\cos\theta$ is a polynomial in x of degree 1 with lead coefficient 1.

Now we use strong induction to complete the proof. Suppose that equation (8.58) is true for all $n \leq k$. We will show that it is true for $n = k + 1$. So, let us consider $2\cos(k+1)\theta$. By our lemma, this is the same as $2 \cdot [2\cos k\theta \cos\theta - \cos(k-1)\theta]$ or just

$$= 2 \cdot [2\cos\theta \cos k\theta - \cos(k-1)\theta] \tag{8.59}$$

$$= 2\cos\theta \cdot 2\cos k\theta - 2\cos(k-1)\theta. \tag{8.60}$$

Now by the strong induction assumption, $2\cos k\theta$ is a polynomial of degree k in x with lead coefficient 1. When this is multiplied by $2\cos\theta$, which is x, we get a polynomial of degree $k + 1$ in x with lead coefficient 1. So the first term in equation (8.60) is a polynomial of degree $k + 1$ in x with lead coefficient 1. Also, by the strong induction, the second term in equation (8.60), $2\cos(k-1)\theta$, can be written as a polynomial of degree $k - 1$ in x with lead coefficient 1. So, putting these two polynomials together, we get that equation (8.60) is a polynomial of degree $k + 1$ with lead coefficient 1, and we are done.

As a corollary of this example we have:

Corollary 8.25 *Suppose that θ is rational. Then $x = 2\cos\theta$ satisfies an equation with lead coefficient 1 with integer coefficients.*

Proof. Suppose $\theta = a/b$. Choose n to be $360b$. Then $n\theta = 360a$ and $2\cos n\theta = 2$, since the cosine of any multiple of 360 is 1. Equation (8.58) becomes:

$$2 = 1x^n + a_{n-1}x^{n-1} + \ldots\ldots + a_0$$

or,

$$1x^n + a_{n-1}x^{n-1} + \ldots\ldots + a_0 - 2 = 0 \tag{8.61}$$

which is satisifed by $x = 2\cos\theta$. In short, we have shown that $x = 2\cos\theta$ satisfies equation (8.61), which is an equation with integral coefficients and lead coefficient 1. ∎

As we mentioned, this corollary was used in Chapter 3 Example 3.15 to prove that the only rational value of θ in degrees, where $0 < \theta < 90$, that makes $\cos\theta$ rational is $\theta = 60$ degrees. You might like to go back to that example now and review it in light of this theorem.

In Chapter 5 Section 11 we stated that Pick's Theorem (Theorem 5.34) was true for polygons, but we only gave a proof that was valid for triangles. We can now finish what we started and give a proof that Pick's Theorem is true for all convex polygons.

> **Example 8.26** *Finish the proof of the following version of Pick's Theorem started in Chapter 5: For any convex polygon with n sides, where $n \geq 3$, with vertices at lattice points, the area is given by Area $= I + \dfrac{B}{2} - 1$, where I is the number of interior points, and B is the number of boundary points of the polygon.*

Solution. We have already proven the theorem for triangles in Chapter 5. Thus, we have proven the theorem for the case when $n = 3$. Now assume that Pick's Theorem is true for a convex polygon with k sides. We would like to show that it is true for a convex polygon, P_{k+1}, with $k + 1$ sides. We do this by strong induction. That is, we assume that Pick's Theorem is valid for all convex polygons with less than $k + 1$ sides and show that it is true for all convex polygon with $k + 1$ sides.

In the picture below we show part of a polygon with $k + 1$ sides. We draw a diagonal from a vertex A to vertex C which divides the polygon into a triangle and another polygon, P_k with k sides. (See Figure 8.3 below)

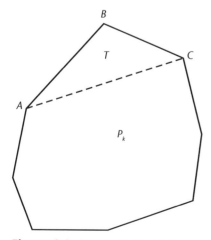

Figure 8.3 P_{K+1} is divided into a polygon P_k with k sides, and a triangle T.

We want to show Pick's Theorem for a convex polygon with $k+1$ sides. The proof is not much different than the proof we gave for triangles. First, observe that all boundary point on the diagonal AC except for A and C are interior points of the polygon P. Thus, if the number of lattice points on diagonal AC is B_d, then the number, I_{k+1}, of interior points on a polygon with $k+1$ sides is:

$$I_{k+1} = I_T + I_k + B_d - 2 \tag{8.62}$$

where I_T is the number of interior points of the triangle and I_k is the number of interior points of the polygon with k sides.

Now we wish to represent the number of boundary points on a polygon with $k+1$ sides in terms of the polygons into which it is broken. When we add the boundary points of the triangle and the polygon with k sides, we are including B_d twice. And they are not boundary points of the polygon with $k+1$ sides. Thus, we have to subtract twice B_d from our count, and then add back the vertices A and C, which means we have to add back 2. Thus

$$B_{k+1} = B_T + B_k - 2B_d + 2. \tag{8.63}$$

Here B_T and B_k are the number of boundary points of the triangle and k sided polygon, respectively. If we denote the area of the triangle by A_T and the areas of the polygons P_k and P_{k+1} by A_k and A_{k+1}, respectively, we have, since we are assuming Pick's theorum holds for A_T and A_k,

$$\begin{aligned}
A_{k+1} &= A_T + A_k \\
&= \left(I_T + \frac{B_T}{2} - 1\right) + \left(I_k + \frac{B_k}{2} - 1\right) \\
&= I_T + I_k + \frac{B_T + B_k}{2} - 2 && \text{(Rearranging terms)} \\
&= I_T + I_k + B_d - 2 + \frac{B_T + B_k}{2} - B_d && \text{(We just added and subtracted } B_d\text{. This just changes the} \\
& && \text{way the expression looks, not its value.)} \\
&= I_{k+1} + \frac{B_T + B_k}{2} - B_d && \text{(Substituting equation (8.62) in the previous line.)} \\
&= I_{k+1} + \frac{B_T + B_k - 2B_d}{2} && \text{(Rewriting the previous line.)} \\
&= I_{k+1} + \frac{B_{k+1} - 2}{2} && \text{(By equation (8.63))} \\
&= I_{k+1} + \frac{B_{k+1}}{2} - 1.
\end{aligned}$$

We are done. We have shown Pick's Theorem is true for convex polygons with $k+1$ sides given that it is true for convex polygons with k sides. So, by the principle of strong math induction, Pick's Theorem is true for convex polygons. Pick's Theorem happens to be true for polygons that are not convex as well.

Student Learning Opportunities

1. Show that three consecutive Fibonacci numbers cannot be the sides of a triangle.

2. Show that, for any natural number n, $n! + (n+1)! = n!(n+2)$. Then show that $n! + (n+1)! + (n+2)! = n!(n+2)^2$. Is it true that $n! + (n+1)! + (n+2)! + (n+3)! = n!(n+2)^3$? Can you prove it?

3. Give a proof by induction that, for any positive integer n, $(ab)^n = a^n b^n$ where a and b real numbers.

4. (C) Your students have completed the proof by induction that, for any positive integer n, $(ab)^n = a^n b^n$ where a and b real numbers. They are now curious to know if they can use a similar proof to show that $\left(\dfrac{a}{b}\right)^n = \dfrac{a^n}{b^n}$ and $b \neq 0$. How do you respond? Can it be done? If so, how? If not, why not?

5. Prove, using induction, that if $z_1, z_2, z_3, \ldots z_n$ are complex numbers, then $\overline{z_1 + z_2 + \ldots z_n} = \overline{z_1} + \overline{z_2} + \ldots \overline{z_n}$ and $\overline{z_1 \cdot z_2 \cdot \ldots \cdot z_n} = \overline{z_1} \cdot \overline{z_2} \cdot \ldots \cdot \overline{z_n}$ where $\overline{z}$ represents the conjugate of z.

6. Give induction proofs for each of the following where n is a natural number.

    (a) $1^2 + 2^2 + 3^2 + \ldots n^2 = \dfrac{n(n+1)(2n+1)}{6}$

    (b) $1^3 + 2^3 + 3^3 + \ldots + n^3 = \left(\dfrac{n(n+1)}{2}\right)^2$

    (c) $n < 3^n$ when $n \geq 1$

    (d) $(1)(2) + (2)(3) + (3)(4) + \ldots + (n)(n+1) = \dfrac{(n)(n+1)(n+2)}{3}$

    (e) $1(1!) + 2(2!) + \ldots + n(n!) = (n+1)! - 1$

    (f) $\dfrac{1}{(1)(3)} + \dfrac{1}{(3)(5)} + \dfrac{1}{(5)(7)} + \ldots + \dfrac{1}{(2n-1)(2n+1)} = \dfrac{n}{2n+1}$

    (g) $\dfrac{1}{n+1} + \dfrac{1}{n+2} + \ldots + \dfrac{1}{2n} > \dfrac{13}{24}$ when $n = 2, 3, 4, \ldots$

    (h) $n^2 < 2^n$ when $n \geq 5$

7. (C) One of your inquisitive students asks if it is legal to use mathematical induction when trying to prove a statement that you are claiming is true for all negative integers. Is it? If not, why not?

8. Suppose that n is a positive integer.

    (a) Show, by induction that $x^n - y^n$ is divisible by $x - y$ if $y \neq x$. [Hint: $x^{k+1} - y^{k+1} = x^k(x - y) + y(x^k - y^k)$.]

    (b) Using part (a), show using a direct proof that $3^n - 1$ is divisible by 2 and that $2^{3n} - 1$ is divisible by 7.

    (c) Give a direct proof that $6(7)^n - 2(3)^n$ is divisible by 4. [Hint: Rewrite it as $2(7)^n - 2(3)^n + 4(7^n)$.]

386 Induction, Recursion, and Fractal Dimension

9 Show that, if the product of 2 or more complex numbers is zero, then one of them must be 0.

10 In calculus it is shown that, if $f(x)$ and $g(x)$ are two functions, then $\lim_{x \to a}(f(x) + g(x)) = \lim_{x \to a} f(x) + \lim_{x \to a} g(x)$. That is, the limit of the sum is the sum of the limits for two functions. Show that the limit of the sum of n functions is the sum of the limits of the n functions where n is a positive integer ≥ 2.

11 In calculus, one studies the product rule for finding the derivative of a product. That rule is, $(f(x)g(x))' = f'(x)g(x) + f(x)g'(x)$. Show that there is a product rule for 3 functions, and that it is $(f(x)g(x)h(x))' = f'(x)g(x)h(x) + f(x)g'(x)h(x) + f(x)g(x)h'(x)$. Generalize the product rule to n functions and prove it by induction.

12 Prove De Moivre's theorem from Chapter 7: If $z = r\,\text{cis}\,\theta$, then for any positive integer, $z^n = r^n \text{cis}\,n\theta$.

13 Use induction to show that, if $\left(x + \dfrac{1}{x}\right) = 1$, then $\left(x^n + \dfrac{1}{x^n}\right)$ is an integer for any natural number n. [Hint: Multiply $\left(x^k + \dfrac{1}{x^k}\right)\left(x + \dfrac{1}{x}\right)$ and then subtract $\left(x^{k-1} + \dfrac{1}{x^{k-1}}\right)$. What do you get? Use strong induction.]

14 Show that the sum of the interior angles of a convex polygon is $180(n-2)$ where n is the number of sides.

15 Prove the following facts about the Fibonacci sequence. Probably the easiest approach would be to use a direct proof that uses the property that each Fibonacci number is the sum of the previous two. Of course, if you prefer, try proving it by induction.
  (a) $F_{n+2} + F_n + F_{n-2} = 4F_n$ where $n \geq 3$
  (b) $F_{n+3} + F_n = 2F_{n+2}$
  (c) $F_{n+4} + F_n = 3F_{n+2}$

16 Let $Q = \begin{pmatrix} 1 & 1 \\ 1 & 0 \end{pmatrix}$. Show that $Q^n = \begin{pmatrix} F_{n+1} & F_n \\ F_n & F_{n-1} \end{pmatrix}$ when $n = 2, 3, 4, \ldots$

17 Show that the value of each term of the following recursive sequence is < 2: $a_1 = \sqrt{2}$, $a_2 = \sqrt{2 + \sqrt{2}}$, $a_3 = \sqrt{2 + \sqrt{2 + \sqrt{2}}}$, and so on, where $a_{n+1} = \sqrt{2 + a_n}$ for $n > 1$.

18 (C) A student comes to you saying that he has found the following proof that all horses are the same color. He is completely baffled. Help him find the error in the proof: "Let P be the statement that every horse in a set of n horses has the same color. If $n = 1$, then we have only one horse, so of course, it has the same color as itself. *Suppose now that every horse in any set of k horses has the same color.* We need to show that every horse in a set of $k+1$ horses has the same color. So take such a set, S, of $k+1$ horses and number the horses $h_1, h_2 \ldots h_{k+1}$. Consider the subset, S_1 of S consisting of the horses $h_1, h_2 \ldots h_k$. Since this has k elements, all the horses in this set have the same color, say brown by our supposition in italics above. We need only show that the horse h_{k+1} has this same color as the other horses. Now consider the subset S_2 of S consisting of the horses $h_2, h_3, \ldots, h_{k+1}$. This set also has only k horses. So by assumption, these also must have the same color. But the horse numbered h_2 was in both subsets S_1 and S_2. (See Figure 8.4 below.)

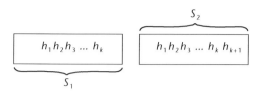

Figure 8.4

By virtue of it being in set S_1, it is brown. Since all horses in the second set S_2 have the same color, by assumption, and h_2 is in that set, all horses in the second set are brown also. In particular, since h_{k+1} is in that second set, it must also be brown."

We have shown that all sets consisting of $k+1$ horses have the same color, assuming that all sets with k horses have the same color. Therefore, all sets of horses, regardless of the size of the set, have the same color.

19 (C) One of your students tells you that someone from the math team has just proven to him that all positive integers are small. He relays the following "proof" by induction: Step 1: The number 1 is small. Step 2: If k is small, then so is $k+1$. Step 3: Since the result is true for $n = 1$, and it is true for $k+1$ when it is true for k, it is true for all $n \geq 1$. How do you respond? What is wrong with the proof?

20 What postages can be made using only 3 cents stamps and 5 cents stamps? Prove it.

8.4 Fractals Revisited and Fractal Dimension

**LAUNCH**

1. Fill in the missing numbers in Pascal's Triangle (Figure 8.5).
2. Color in the odd and even numbers using two different colors.
3. Do you notice a pattern? What does it look like?

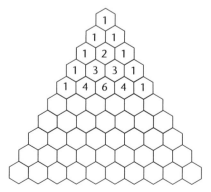

Figure 8.5

If you have completed the launch problem, you are probably totally mystified by the relationship that appears to exist between Pascal's Triangle and the Sierpinski triangle. We agree that this is quite amazing and mysterious. As you read through the remainder of this section, you will find still more mind boggling relationships involving fractals. Enjoy!

Earlier in this chapter, we saw how we can use a recursive routine to generate a fractal, specifically the Sierpinski triangle. We found that the perimeter of the Sierpinski triangle was infinite, yet its area was zero. We now show another example of a fractal generated by using a recursive relation.

Suppose we begin with a line segment as shown below, and for convenience, assume that its length is 1. Our recursive routine has the following instructions: We divide each line in the figure into 3 equal parts, remove the middle section, and replace it by two line segments equal to 1/3, the length of the segment removed. The result is shown in Figure 8.6(b) below.

We now perform the same recursive procedure on each line segment in the resulting figure. We remove the middle third of each line segment remaining in Figure 8.6(b) and replace it by two segments of the same length. After this second iteration, we have the figure shown in Figure 8.6(c).

Now we repeat this once again. From each part, we remove the middle third and replace each middle third by two segments of the same length to get Figure 8.6(d).

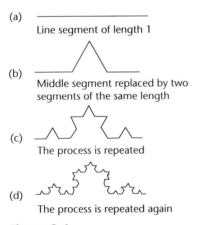

Figure 8.6

We continue repeating the process and after infinitely many iterations we generate a figure called the **Koch curve**, which is also a fractal. Of course, we can't draw it, since the Koch curve is the result of performing this recursion infinitely many times. But the figure in (d) does give us a good sense of what it might look like.

There is something very interesting about the length of the Koch curve. When we formed it, we began with one segment of length 1. In Figure (b) we had 4 segments of length 1/3 to give us a total length of 4/3. In the Figure (c) we had 16 segments of length 1/9 to give us a length of 16/9 and in Figure (d) we had 64 segments of length 1/27 to give us a length of 64/27. Each successive figure has a length equal to 4/3 the length of the previous segment. Thus, the length of the nth figure is $(4/3)^{n-1}$. As n gets large, this goes to ∞. Thus, the perimeter of the Koch curve, the result of iterating over and over is ∞. Yet, the Koch curve can be enclosed in a rectangle with finite area and so it behaves just like the Sierpinski triangle.

One of the applications of fractals is the study of chaotic behavior. You may be wondering how it is even possible to study something like chaos. The following game will illustrate this in a rather remarkable way.

8.4.1 The Chaos Game

The chaos game is played in the following way. Begin with an equilateral triangle and label its vertices A, B, and C. Now take a spinner which has been divided into 3 equal sectors with letters A, B, and C on them. Pick an arbitrary point P in the triangle (although any point in the plane will do). Now spin the spinner, and if it lands on A, place a dot halfway between your starting point P and vertex A. If B comes up, place a dot halfway between P and vertex B. If C comes up, then place a dot 1/2 way between P and the third vertex. Call the new dot P_1 and iterate the procedure, namely, spin again. If A comes up, place a dot halfway between P_1 and vertex A. If B comes up, place a dot halfway between P_1 and vertex B. If a C comes up, place a dot halfway between P_1 and vertex C. Call this new dot P_2. Iterate this procedure on P_2. That is, spin the spinner. If A comes up, place a dot halfway between point P_2 and vertex A, if B comes up place a dot halfway between P_2 and vertex B. If a C comes up, then place a dot 1/2 way between P_2 and vertex C. Call the new point we get, P_3. Repeat this procedure on P_3 to get P_4, and so on. The dots generated depend on the spin of the spinner. The outcome of the spin is random. We would expect to find dots all over the triangle scattered randomly. What happens is a major surprise. As you place more and more dots, the picture you get looks like the following (Figure 8.7):

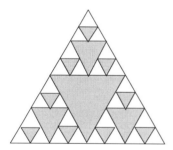

Figure 8.7

Yes! It is the Sierpinski triangle! Are you shocked? Most people are! Here we see what appears to be a chaotic procedure, leading to a figure that seems to have regularity!

There are many websites on the Internet where you can play this game and have the computer draw the dots one at a time. Here is one: http://serendip.brynmawr.edu/complexity/sierpinski.html

It is interesting to note that you need not start with a triangle to generate a fractal with this random process. You can start with any polygon and play the same game. Furthermore, you don't have to plot a point half way between the previous point and the vertex. You can plot a point 1/3 of the way, or 1/6 of the way. The figures you get are always interesting. Here is the picture generated when you start with a pentagon and play the game, each time placing the next dot 1/3 of the way to the vertex that comes up when a spinner with 5 equal sectors is spun (Figure 8.8).

Figure 8.8

If we began with the same pentagon and took each successive point 3/8 of the way from the previous point to the next vertex chosen at random, we would get the following diagram (Figure 8.9):

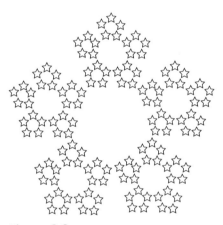

Figure 8.9

Does this remind you of anything? Doesn't it look like a snowflake?

8.4.2 Fractal Dimension

In geometry you have studied the concept of dimension. You probably recall that a line is one dimensional, a square is two dimensional, and a cube is three dimensional. Is it possible for a figure to have fractional dimension? What would something like that mean? We will soon find out as we investigate more properties of fractals.

When forming fractals like the Koch curve or the Sierpinski triangle using iteration, we always do the iterative process by breaking the figure into smaller figures which are similar to the whole. This process accounts for the self similarity that exists in many fractals. Let us look at this process more closely.

Suppose we have a square, and we break it into 4 congruent squares as shown below (Figure 8.10):

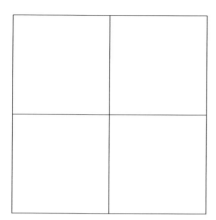

Figure 8.10

Then the side of each of the congruent squares is 1/2 the side of the original square. Put another way, the original square is the side of the smaller square magnified by 2. This number 2 is called the **magnification factor**. If we broke the original square into 9 congruent squares as in the figure below (Figure 8.11),

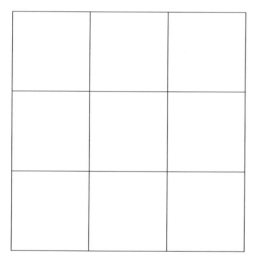

Figure 8.11

the magnification factor is 3.

When the magnification factor is 2, we have broken the original square into 4 squares. When the magnification factor is 3, we have broken the square into 9 squares. If the magnification factor is m, the number of parts into which we have broken the original square into is m^2. (Thus, when $m = 2$ we get 2^2 smaller copies of the original. When $m = 3$ we get 3^2 smaller copies of the original square.) The exponent 2 in m^2 is significant. It reflects the fact that the square is two dimensional.

Now let's extend this to a 3 dimensional object. Suppose that we have a cube and that we divide it into smaller cubes by cutting it by planes. If the planes cut the sides in half, that is, if the magnification factor is 2, we get 8 cubes as shown in the figure below (Figure 8.12).

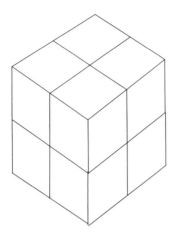

Figure 8.12

If we divide each side into thirds as shown in Figure 8.13 below,

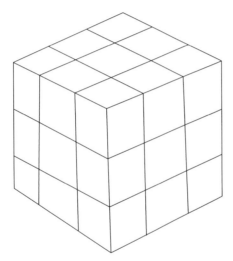

Figure 8.13

that is, if the magnification factor is 3, we get 27 smaller cubes. In general, if the magnification factor is m, we get m^3 smaller cubes.

In general, if the magnification factor is m and the object is D-dimensional, you will get $N = m^D$ copies of the original. Thus the dimension, D, of an object satisfies

$$m^D = N$$

where m is the magnification factor, and N is the number of smaller copies of the original that we get. If we solve for D, we get

$$D = \frac{\log N}{\log m}. \tag{8.64}$$

It is this relationship that leads to the a definition of fractal dimension. The **fractal dimension** of an object formed by iteration using similar objects is given by $D = \frac{\log N}{\log m}$, where N is the number of copies of the original that the iteration performs, and m is the magnification factor. Let us now see how we can use this relation to find the dimension of the Koch curve defined earlier. We realize that, at each iteration, we break the curve into 4 times as many parts we started with, so $N = 4$. But the original object was three times the size of the iterates since, when forming the Koch curve, we broke the curve up into parts of size 1/3 of the previous size. Thus, the magnification factor is 3. Using the formula of equation (8.64) above, we see that the dimension of the Koch curve is

$$\frac{\log 4}{\log 3} \approx 1.2619.$$

And so, we have found a figure that has a fractional dimension. So what does it mean? Actually, fractal dimension was developed to determine the complexity of the fractal. The higher the number, the more complex the object. Some people have suggested that you can envision fractional dimension by thinking of a piece of crumpled paper.

There are many different definitions of dimension (the Renyi definition, the box counting definition, the correlation definition, etc.) in use today to describe the complexity of a fractal; however, the above definition is the one most related to the secondary school curriculum.

Student Learning Opportunities

1 (C) In response to your students' desire to see real-world applications of the mathematics they learn, you ask them to find examples of irregular shapes in the natural world that they think can be modeled by fractals. They are at a total loss of what to look for. What are some suggestions you can give them for real-life objects they can bring in? What are the properties of these objects that would classify them as resembling fractals?

2 Determine the dimension of the Sierpinski triangle.

3 Begin with the line segment below, and replace the middle third by an open square as shown below in Figure 8.14.

Figure 8.14

Now repeat the process on each segment in the resulting figure.

(a) What figure do you get in the next iteration?

(b) If this process is continued, what will the dimension of the resulting fractal be?

(c) Guess at what you think the final fractal would look like.

4 Begin with a square. Replace each side of the square with the following figure (Figure 8.15):

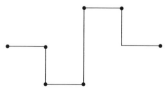

Figure 8.15

(a) What will the resulting figure look like after one iteration?

(b) If we continue iterating the figures in this way, what is the dimension of the resulting fractal?

5 Below are some fractals. Find their dimensions (Figures 8.16, 8.17, 8.18).

(a)

Figure 8.16

[Hint: This was generated from an initial figure by replacing each piece by 3 congruent pieces, each half the size.]

(b)

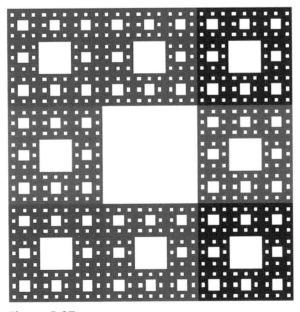

Figure 8.17

[Hint: This figure is known as the Sierpinski Carpet. This is formed by starting with a square. The square is cut into 9 congruent subsquares in a 3-by-3 grid, and the central subsquare is removed. The same procedure is then applied recursively to the remaining 8 subsquares.]

(c)

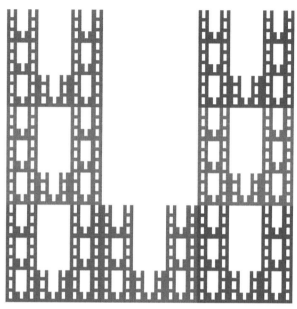

Figure 8.18

[Hint: Begin with a square and divide it into nine congruent subsquares. Remove the middle square and the one above it. Now repeat the process on each subsquare.]

CHAPTER 9

FUNCTIONS AND MODELING

9.1 Introduction

The study of functions and modeling is now a cornerstone in the secondary school curriculum. Students begin studying the concept of function in elementary school and continue throughout their secondary school careers. With the advent of technology, students now can easily model real-world situations using hand-held graphing calculators. Despite what most students think, the notion of functions took many years to develop. Now it is the language of all science. In this chapter we study a bit of the evolution of functions, how they are used to model practical problems, how practical problems are solved by studying data and fitting them to curves known as regression curves or best fit curves, and conclude with a careful discussion of inverse functions that will prepare us for our higher level study of transformations in the next chapter.

9.2 Functions

 LAUNCH

Examine the five different scenarios below:

1. If you enter a person's name (*x*), in a data base, there is a footprint, (*y*), on file from when they were born.
2. When you use a vending machine, you push a certain letter, (*x*), and your candy of choice, (*y*), comes out.
3. You are planning a trip and on the computer you enter an amount of money, (*x*), you can spend on airfare. The computer gives you the many different places, (*y*), you could fly to, using that amount of money.
4. At the end of the semester, students' math grades are posted in a table, which lists each student's ID number, (*x*), in the left column and the student's grade, (*y*), in the right column.
5. You give your friend any number, (*x*), he squares it and then adds 2, and tells you the result, (*y*).

Describe the similarities and differences of the five scenarios above.
In which do you think that *y* is a function of *x*? Why?

After doing the launch problem, you are probably getting some ideas about functions. You might have even noticed that the scenario described in the third example, was quite different from the other four, and as you will soon discover, that in fact, y, was not a function of x. We hope you will appreciate reading this section and finding out how functions and non-functions are part of our daily lives.

9.2.1 The Historical Notion of Function

When you send a package through the post office, the postage you pay depends on the weight. When you open a faucet, the amount of water that flows out per second depends on how much you open the valve. When you drop a ball, the distance it falls before it hits the ground depends on the amount of time that elapses since it has been dropped. Note that in each of these cases, we have two quantities where one depends on the other.

Historically, when we had two variables, y and x, and one of them, say y, depended on x, then we would say that y was a function of x. We called x the independent variable and y the dependent variable.

Thus, using this historical notion of function, the postage that one pays on a package is a function of the weight of the package, since the postage depends on the weight. The weight is the independent variable, and the postage is the dependent variable. The amount of water that flows out of a faucet, say every second, is a function of how much the valve is opened. The independent variable is the amount the valve was opened. The dependent variable is the amount of water flowing out per second.

This historical notion is satisfactory for most applications, and is how most students understand the notion of function. However, in today's world, the notion of function has been refined.

9.2.2 Functions Today

Returning to the postage scenario, suppose when you went to the post office and asked, "How much is it to mail this package?" the postal clerk answered, "That will be $3 and $4." You would look puzzled wouldn't you? The postage should be either $3 or $4, not both. It is because of the desire to avoid situations like this that the definition of function has changed over the years. Today, for y to be a function of x, with each x under consideration, one must *associate* one and only one y. Thus, if you had the equation $y = x^2$, y is a function of x, since with each x we associate one and only one y, namely x^2. On the other hand, if we have the equation $y = \pm\sqrt{x}$, then y is not a function of x since, if $x = 9$, there will be *two* values of y, namely $y = 3$ and $y = -3$. Having two y values for the same x value is analogous to paying two different postages on the same package. In applications, this just can't exist. Thus, today, in order for y to be a function of x, with each x under consideration, we must associate one and only one y.

The word "associate" does *not* imply any kind of dependence or *cause and effect relationship between y and x*, as was the case in earlier years. *We still call x* the **independent variable** and y the **dependent variable** when y is a function of x. The key point in the modern definition of function is that, when we say that y is a function of x, the y associated with each x is unique. Let us illustrate by giving some examples of different types of functions.

> **Example 9.1** *We are going to roll a die 3 times and record the outcome. We will let x be the roll number, and y be the outcome of the roll. Construct such a function.*

Solution. Suppose that, on the first three rolls, we got the numbers, 3, 5, and 2, respectively. Then, with $x = 1$, we would associate $y = 3$. With $x = 2$, we would associate $y = 5$, and with $x = 3$ we would associate $y = 2$. With each x, we are associating one y, and so y is a function of x. However, in this case, there is no dependence of y on x. The result, y, that the die falls on is random. In no way does it depend on the roll number, x, and it is certainly not caused by the roll number x. We still call x the independent variable and y the dependent variable.

If someone else tossed the die, one might get a different function, since the y values most likely would be different. However, since we are associating with each x one and only one y, the new y would still be a function of x, although a different function.

> **Example 9.2** *It is known that the rate at which certain birds chirp is directly related to the temperature outside. Suppose that the number of chirps per minute, C is given in terms of the temperature, T, and that the function relating the two is*
>
> $$C = 5T - 32 \tag{9.1}$$
>
> *where T is measured in Fahrenheit and T is restricted to between* 30 *and* 150 *degrees. Suppose also that the values of C are restricted to* 118 *to* 718 *chirps per minute. A student says, "C is clearly a function of T since the number of chirps depends on the temperature." However, the student continues, "the temperature is not a function of C, since increasing the number of chirps does not cause the temperature to increase." Is this student correct?*

Solution. While what the student said might have been true using the historical definition of function, it is not true using the modern definition of function. For each T, there is associated one and only one value of C computed by equation (9.1). So C is a function of T. Furthermore, once a value of C is chosen (between 118 and 718), only one value of T is associated with it from equation (9.1). So T *is* a function of C. The temperature is not caused by the number of chirps per minute, but by today's definition, T *is* still a function of C.

In this example, which variable is the independent one and which is the dependent one? It depends on which we are considering to be a function of the other. If we are considering C as a function of T and studying that function, then T is called the independent variable. If we are considering T to be a function of C, and studying that relationship, then C is called the independent variable. Thus, the words dependent and independent depend on which function you are working with. This may sound confusing, but it is routinely done with functions and their inverses. (See Section 9.7.)

When relating the weight of an object to its postage, we can consider the postage as a function of weight. Since the post office restricts weight, we realize that the weight has limits. They will not mail a letter weighing 100 pounds. The restricted values placed on the independent variable constitute what is called the **domain** of the function. In the example where a ball is dropped and the distance the object fell was considered a function of the time, t, elapsed since it was dropped, the independent variable t must be ≥ 0. This would be the domain of our function.

9.2.3 Functions – The More General Notion

Most functions that one studies in secondary school are those such as $y = 2x + 1$, where x and y are numbers. That is, numbers are associated with numbers. This gives secondary school students the impression that functions only deal with numbers. This is far too limited an interpretation of functions. In reality, functions are used on a day-to-day basis in many contexts and the functions used do not necessarily associate numbers with numbers. The functions may associate objects with objects. This leads to a more general definition of function, which we refer to as the modern, but not completely formal (from a mathematician's point of view), definition. Increasingly, this definition is seen more and more in secondary school texts today.

If A and B are any sets, then a function, f, from A to B, is a rule which associates with each element x of the set A one and only one element, y, of the set B. The element y associated with x is called the image of x under f and is denoted by f(x). A is called the domain of the function, and the set of images of the domain values is called the range.

Notice the wording: We have a function from A to B, and the y "associated" with x is denoted by $f(x)$. It does not say that there is a dependence of y on x (though we reiterate, in many applications there is). Here are some examples of the modern definition of function.

Example 9.3 *When we assign a set of workers to tasks, the assignment is a function from the set, A, of workers to the set B of jobs. The domain is the set of workers being assigned. The range is the set of jobs to which they are assigned. If we call this function f, and if John is assigned the job of manning the telescope, then f(John) = manning the telescope.*

Example 9.4 *Suppose that, with each point on the earth's surface, there is associated the single ordered pair of numbers (latitude, longitude), of that point. This is a function from the set A of physical points on the earth's surface, to the set B of latitude–longitude pairs. The physical points on the earth's surface constitute the domain, and the image of any point on the earth's surface is its latitude–longitude pair. The set of all latitude–longitude pairs is the range. If we denote this function by f, then f(New York City) = (40.70519 – 74.01136) since this pair gives New York City's latitude and longitude.*

Example 9.5 *Computers can only deal with data that have been converted to zeros and ones. When we use a word processor and type the letter "A," the computer translates this into the following string of 0's and 1's called the ASCII value for the letter "A": 01000001. (ASCII stands for American Standard Code for Information Interchange.) If you type in the symbol "<," the computer converts this to 00111100, the ASCII value for "<." The ASCII value for each symbol we type consists of 8 digits, which are 0 or 1. Consider the rule from the set of symbols to the set of 8 digit numbers consisting of 0's and 1's. With each symbol, we associate its ASCII value. Is the rule which associates with each symbol, its ASCII value, a function of the symbol?*

Solution. Since with each symbol we associate one and only one ASCII value, this is a function from the set of symbols to the set of ASCII values. If we denote this function by the letter f, then $f(A) = 01000001$ and $f(<) = 00111100$. The domain of this function is the set of symbols that have ASCII values, and the range is the set of all the ASCII values we obtain.

Example 9.6 *In Chapter 10 we will be discussing the notions of rotation, reflections, and translations. These are also functions in which figures are given and then transformed in some way to give us new figures. Thus, with each figure, say a triangle, we might associate a dilated triangle, or similar triangle, or rotated triangle, or reflected triangle. In these functions we can take the domain to be the set of all "figures" where a "figure" means any set of points. Each "figure" is then associated with one and only one transformed "figure." The set of transformed figures is the range.*

Example 9.7 *Consider the equation $y = \dfrac{1}{x-1}$. Is this y a function of x? If so, discuss the domain and range.*

Solution. For each real number x other than 1, we can compute the value of y and we get only one y value. Thus y is a function of x. The values that x can take on are restricted. x cannot be 1. Thus, the domain of this function is the set of $x \neq 1$. If we graph the function (see Figure 9.1 below), we will see that the range, which is the set of y values taken on by the function, is all $y \neq 0$.

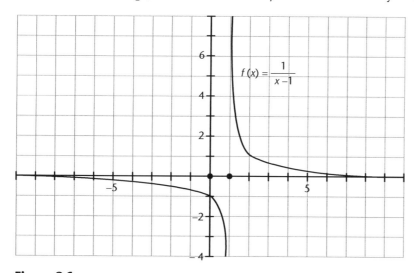

Figure 9.1

Since y is a function of x, we can write $y = f(x) = \dfrac{1}{x-1}$. Since, when $x = 3$, $y = \dfrac{1}{2}$, we would write $f(3) = \dfrac{1}{2}$.

9.2.4 Ways of Representing Functions

The functions one studies in secondary school can be represented in many different ways. The purpose of this section is to highlight some of these representations that you may not be familiar with and that clearly illustrate the concept of function.

Example 9.8 *Consider the function whose domain is the set of letters a, b, c, and d. With "a" we associate the number 1, with "b", 1, "c", 2, and with "d", 3. This is our complete function. Describe 4 ways to denote this function.*

Solution. If we call this function f, then we can describe f as follows:

$f(a) = 1$

$f(b) = 1$

$f(c) = 2$

$f(d) = 3$.

Here we have completely described the function by listing which element of the range is associated with each element in the domain. This is known as the **listing method**. Here, the domain consists of four elements, a, b, c, and d and the range is the set of numbers, 1, 2, and 3. Obviously, if the domain is infinite, this listing method is not feasible.

Another way of describing functions is with **ordered pairs** where the first element is the element of the domain, and the second element is the associated element of the range. Thus, we can completely describe the function by listing the ordered pairs, $\{(a, 1), (b, 1), (c, 2), (d, 3)\}$. Again, if the domain is finite, and not too large, this method of representing a function is feasible.

A third way to represent a function is to give a **table** of values

Domain element	Range element
a	1
b	1
c	2
d	3

A fourth way used to represent a function is with a **picture** akin to the following (Figure 9.2):

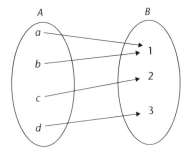

Figure 9.2

Each arrow goes from the domain element to its image. So, we can clearly see the association between domain and range values.

Example 9.9 *Discuss 3 other ways of representing functions.*

Solution. A fifth way to describe a function, and probably the most useful way, is by the action it performs or by the **rule** used to compute the dependent variable. Thus, if we have a rule that subtracts 2 from any number we start with, we might like to explicitly state the rule: $f(x) = x - 2$, which tells us that, with each x, we associate the number $x - 2$.

A sixth way of thinking of a function (and this is a very common way), is to envision it as a **machine** where the x is the input, and the y is the output. This "machine" somehow transforms x into y. This is especially useful in certain geometric and computer science applications.

For example, when we type a document using some kind of word processor, say Microsoft Word, and want to convert it, say to PDF format so that it can be read by anyone on any computer, the document needs to be converted. The conversion process is a mechanical process. The machine takes your original document as the input and outputs a document which is a PDF document. Here is a picture (Figure 9.3).

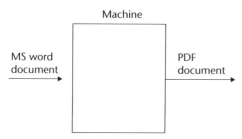

Figure 9.3

We can see this is a function of the Word document, since for each Word document that is input, one and only one PDF document emerges. Again, in geometric applications, we might like to describe the mechanical process of stretching an object. Thus, when we input a certain figure, our machine, (stretch function) stretches it and gives us back the stretched figure (our output). Figure 9.4 is a picture illustrating it.

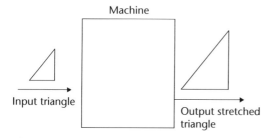

Figure 9.4

Finally, a seventh way of representing a function which is pervasive in the secondary school curriculum, is to **graph** the function. Of course, this representation only makes sense when the x and y values are numbers. From the graph, many different conclusions can be made. For example, we may notice that, as the x values increase, so do the y values (the graph is increasing), or each successive increase in x by 1 causes y to increase by greater and greater amounts. (The curve is concave up.) Since all prospective secondary school teachers have had a course in calculus where graphs and their properties were discussed in great detail, we won't say more about it now. However, for the most part, we will use this graphing approach. But, before we continue with other issues related to functions, let us start applying the ideas developed thus far.

Student Learning Opportunities

1. Which of the following are functions of x?

   (a) The area of a circle with radius x.
   (b) The value of y if $y^4 = 2x^3$.
   (c) The temperature in Chicago on January 1st, 2007, x hours after midnight where x is between 0 and 12. [Hint: Which part of Chicago are we talking about? Is the temperature the same everywhere in Chicago?]
   (d) The price of an x ounce container of cottage cheese in Los Angeles on February 1st 2006.
   (e) The tuition a student ends up paying if he or she goes to college x in a specific year.
   (f) The birth date of a person x.
   (g) The profit you make from selling x wazoos, where each wazoo sells for 10 cents and costs 5 cents to make, including all costs of production.

2. Consider a post office in a small town, and suppose, for simplicity, that it only accepts letters up to 8 ounces in weight and charges the following rates.

Weight, W (ounces)	Postage, P (cents)
$0 < W < 2$	39
$2 \leq W < 4$	43
$4 \leq W < 6$	47
$6 \leq W < 8$	55

   Since with each weight we associate one postage, the postage P is a function of weight. If we use the letter f to denote this function, then $P = f(W)$.

   (a) Compute $f(1.5)$, $f(2)$, and $f(5)$.
   (b) Compute $f(9)$.
   (c) What is the domain?
   (d) What is the range?

3. In the introduction to Chapter 3 we gave the following example: A manufacturer has just received a large order for metal boxes that must hold 50 cubic inches. He plans to make these boxes out of rectangular pieces of metal 8 inches by 10 inches, by cutting out squares from the corners and folding up the sides. There we found that, if x was the side of the square cut out, then the volume of the resulting box was given by $V(x) = x(8 - 2x)(10 - 2x)$, where x was length of the side of the square cut out. For each x we get a box with a specific volume. Thus, the volume is a function of the side x of the square cut out.

   (a) Compute $V(2)$. Interpret your answer.
   (b) What is the domain of the function?
   (c) At which x does the maximum volume occur? Use your graphing calculator or calculus to answer this.
   (d) Estimate the range of the function.

4. With the numbers $x = 1$, 2, and 3, associate the letters a, b, and c, respectively. (These letters represent our y values which we arbitrarily assign.) Is y a function of x? Explain.

5 What are the domain and range of the following functions of x? (Assume that all variables take on real values.) A graphing calculator might help.

(a) $y = \sqrt{x + 3}$

(b) $y = \dfrac{1}{x + 4}$

(c) $f(x) = \sqrt{9 - x^2}$

(d) $g(x) = \log_{10}(4x - 8)$

(e) $h(x) = \dfrac{\sqrt{x}}{(x - 3)}$

(f) $k(x) = \dfrac{1}{(x - 2)^2}$

(g) $l(x) = e^x$

(h) $y = \dfrac{x}{x - 1}$

6 (C) Your students are confused about whether the following two statements are true or false. How would you respond to help them understand whether in fact they are true or false?

(a) For y to be a function of x, y must be dependent on x in the sense that x causes y to happen.

(b) If many x's give the same y, then y cannot be a function of x.

7 Show that the number $y = 1$ is not in the range of $y = \frac{x-1}{x+1}$.

8 Below, we see a table of values which represents a "split function" with independent variable, x. If we call this function f, compute $f(1)$, $f(2)$, and $f(5)$.

x	y
$0 < x < 2$	7
$2 \leq x < 4$	4
$4 \leq x < 6$	-11

9 (C) You spoke with your students about what it means for y to be a function of x. Now they are curious about what it means for x to be a function of y and they want to know if in the equation $y^2 = x$, they can say that x is a function of y, even though y is not a function of x. How do you respond to their questions and clarify their areas of confusion?

10 (C) A student is given the equation $y = x + 1$, and asks whether y is a function of x or if x is a function of y. How can you use a graph to help your student understand the answer to this question?

11 (C) A student is given the equation $x^2 + y^2 = 9$, and asks whether y a function of x or if x is a function of y. How can you use a graph to help your student understand the answer to this question?

12 (C) A student is given the equation $y = x^4$, and asks whether y a function of x or if x is a function of y. How can you use a graph to help your student understand the answer to this question?

13 In secondary school the word "relation" is used to describe a set of ordered pairs. In (a) and (b) examine the following ordered pairs of numbers, where the first coordinate is the *x* value and the second coordinate is the associated *y* value. In each case, explain why *y* is not a function of *x*.

   (a) {(1, 2), (1, 3), (1, 4)}
   (b) {(−2, 1), (3, 2), (4, 3), (3, 1)}
   (c) Using the 4th method of representing functions described in this section, draw the picture that represents each of the relations from (a) and (b).

14 What must be true about a relation for *y* to be a function of *x*, assuming that the first coordinate of the ordered pair is *x* and the second coordinate is *y*?
   In the following relations, is *y* a function of *x*? Why?

   (a) {(1, 2), (2, 3), (3, 4)}
   (b) {(−2, 1), (3, 2), (4, 3), (5, 1)}

15 Note that, for a relation to be a function, there can only be one arrow emanating from each element in the domain. Draw the relation from 14(a) using the picture method and show that this is the case for this relation. Show that the relation in 13(b) doesn't satisfy this condition.

16 If we call the function *f* in 14(a), what is $f(1) + f(3)$? (It is not unusual to see functions written as a set of ordered pairs in books. For example, $f = \{(1, 2), (2, 3), (3, 4)\}$.)

9.3 Modeling with Functions

LAUNCH

Read each of the following scenarios and decide which type of algebraic function, $y = f(x)$ would best fit each one. Choose from among quadratic, linear, power, or polynomial functions.

1 You board an airplane in New York heading west. Your distance, *y*, from the Atlantic Ocean, in miles, increases at a constant rate.
2 You are popping popcorn in the microwave oven and are counting the number, *y*, of pops per minute, which changes over the course of the 4 minutes you let it pop.
3 It is a dark night and you are driving to the country on a dark lonely road. You see a car coming towards you from the distance. The intensity, *y*, of the approaching headlights increases.

In secondary school some of the more important graphs you studied were those of linear functions, quadratic functions, exponential functions, polynomial functions, and root functions. Did you ever wonder about their importance? We hope that this launch question helped you realize that these functions can serve as models for real-life situations we encounter on a daily basis. In this section we will briefly review these functions and discuss how they are used to model other real-life situations.

9.3.1 Some Types of Models

We begin this section by discussing linear functions, which you probably first encountered in middle school, and which are used to model a myriad of situations.

Linear functions are those that are of the form $y = f(x) = mx + b$. The graph of a linear function is a straight line, where m is the slope and b is the y intercept.

> **Example 9.10** *(Linear Function Applied to Springs) Suppose that we have a spring attached to some object which is stationary, and suppose that we pull the spring out a distance x units beyond its natural length. Now, if it is a very stiff spring, then one would have to pull it with a lot of force, while if it is not so stiff, a lesser force is needed. Physicists have determined that the force needed to pull the spring x units beyond its natural length is given by $F = kx$, where k is a constant that depends on the spring's strength. This is known as Hooke's Law. Since this law is of the form $y = mx + b$ where F takes the place of y and k takes the place of m (and $b = 0$), it falls under the category of a linear function, of which the graph is a straight line.*
>
> *For a given spring, we have the following data.*
>
> | x | 1 | 2 | 3 | 4 | 5 |
> |---|---|---|---|---|---|
> | F | 2 | 4 | 6 | 8 | 10 |
>
> *Find the function which fits the curve to the data.*

Solution. We know that a linear function fits this data and the function is of the form $F = kx$. We need only find k by substituting any pair of corresponding x and F values in this equation and we see that k must equal 2. That is, $F = 2x$.

Quadratic functions are those that are of the form $y = f(x) = ax^2 + bx + c$ where $a \neq 0$. Saying that $a \neq 0$ means that there has to be an x^2 term. The graphs of these functions are parabolas. If $a > 0$, the graph opens up. If $a < 0$, the graph opens down. See Figure 9.5 below.

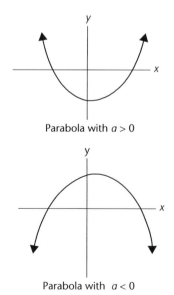

Parabola with $a > 0$

Parabola with $a < 0$

Figure 9.5

Example 9.11 *(Quadratic Function Applied to Firefighting)* *It can be shown that, if a hose is held at an angle of θ degrees with the horizontal, and water is coming out of the hose at a constant velocity of v, in feet per second, then the height, h, of the water above ground measured in feet, at a distance x feet from the nozzle is given by*

$$h = -16(1+m^2)\frac{x^2}{v^2} + mx + h_0$$

where $m = \tan\theta$ is the slope of the nozzle and h_0 is the height of the nozzle above ground. Assume that the nozzle is held at an angle of $45°$ and that $v = 30$ feet per second and that the hose is at an initial height of 4 feet above the ground. Draw the graph of this function and estimate the distance from the nozzle the water is at its highest point. Also, estimate how high the water goes and how far away from the nozzle the water hits the ground.

Solution. Here is our picture (Figure 9.6).

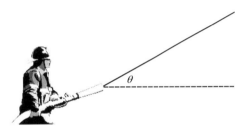

Figure 9.6

Since $m = \tan 45 = \dfrac{\sqrt{2}}{2}$, our function for the height of the water above ground at a distance x from the hose is

$$h = \frac{-24x^2}{900} + \frac{\sqrt{2}}{2}x + 4 \qquad (9.2)$$

obtained by substituting $\dfrac{\sqrt{2}}{2}$ for m in equation (9.2). The graph of this function is given by Figure 9.7.

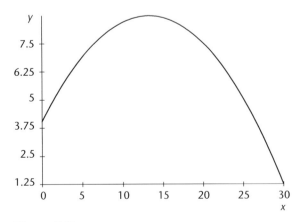

Figure 9.7

Here the y-axis is the "h" axis. As we can see, the path that the water follows is along a parabolic arc, something that we know from experience. The picture indicates that the water hits the ground at approximately 30 feet from the firefighter, and the maximum height occurs at roughly 13 feet. Of course, if you need more exact answers, you can get them from analyzing the algebraic function. Since the turning point of the parabola $y = ax^2 + bx + c$ occurs at $x = -\frac{b}{2a}$, the maximum height will occur exactly at $x = \dfrac{-\frac{\sqrt{2}}{2}}{2(-24/900)} = 13.258$ and will be ≈ 8.6875 feet, obtained by substituting 13.258 into equation (9.2). To find out exactly where the water hits the ground, set $h = 0$ and solve for x, using either the quadratic formula or a calculator, to get $x \approx 31.308$ feet.

An **exponential function** is a function that can be put in the form $y = f(x) = ca^x$ where $a > 0$. Thus $y = 2(3)^x$ is an exponential function, and so is $y = -4 \cdot 2^{-3x}$ since the latter can be written as $y = -4 \cdot (2^{-3})^x$ or $-4 \cdot (1/8)^x$. The graph of an exponential function looks like one of the graphs below. We notice that the graph either rises very quickly or rapidly decreases to 0, which is usually what we look for when trying to fit a curve with an exponential function (Figure 9.8). Later, we will discuss this further.

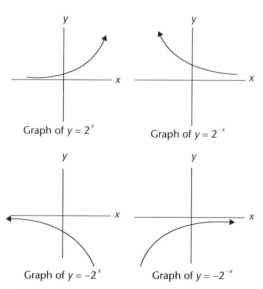

Figure 9.8

Example 9.12 (*Exponential Function Applied to Fossil dating*) *If N_0 grams of a radioactive element are present in an object now, then t years from now the number, N, of grams left in the object is given by*

$$N = N_0 e^{kt}$$

where k is a constant that depends on the radioactive element. (In fact, if it takes T years for half of the radioactive element to decay, then $k = -\ln 2/T$.) This formula is used to date fossils and old paintings, and so on. (See also Chapter 6 Example 6.35 for a very interesting example of this.) It is known that, when something living dies, the radioactive carbon $C14$ in its body (something all living things have), begins to decay. It is also known that the half life of carbon 14 is 5730 years old. (a) Suppose that an object containing 100 grams of radioactive carbon 14 dies. What percentage of radioactive carbon will remain after 500 years? (b) A fossil is found and, using certain well known techniques in the sciences, it is determined that 20% of the original $C14$ it contained at the time of death remains. How old is the fossil?

Solution. (a) We use the formula $N = N_0 e^{kt}$ and make use of the fact that $k = \dfrac{-\ln 2}{5730} \approx -0.000121$. Since we are given that N_0, the initial amount of $C14$ is 100, we have $N = 100 e^{-.00121 t}$. After 500 years, the number of grams of $C14$ is

$$N = 100 e^{-0.00121(500)} \approx 54.607 \text{ grams.}$$

(b) We measure everything from the time of death of the fossil, since that is when the carbon begins to decay. We are given that, currently, the amount of $C14$ left is 20% of the original amount, or $0.20 N_0$. Substituting this into the formula we get

$$0.20 N_0 = N_0 e^{-0.00121 t}.$$

Dividing this equation by N_0 we get that $0.20 = e^{-0.00121 t}$, and solving for t we get that $t \approx 13,300$ years old, which tells us the age of the fossil.

A **polynomial function** is a function of the form $f(x) = a_0 + a_1 x + a_2 x^2 + \ldots + a_n x^n$, where n is a whole number. The highest power that occurs is called the **degree** of the polynomial. Thus $f(x) = x^3 - 4x^2 + 5$ has degree 3 since the highest power of x that occurs is 3. A quadratic function is a special case of a polynomial function and has degree 2. In general, the graphs of polynomials generally speaking have "wiggles" in them. (But having wiggles doesn't make a graph a polynomial. For example $f(x) = x + \sin x$ or $g(x) = \cos 2x$ have wiggles in them and they are not polynomials.) Below, we see the graphs of several different polynomials (Figures 9.9, 9.10, 9.11).

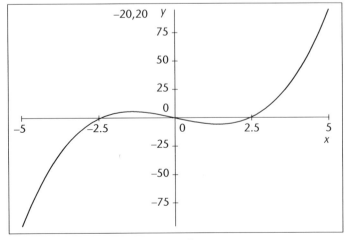

Figure 9.9 The graph of $f(x) = x^3 - 6x$.

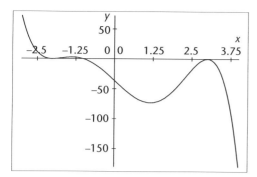

Figure 9.10 The graph of $f(-x) = -(x+2)^2(x+1)(x-3)^2$.

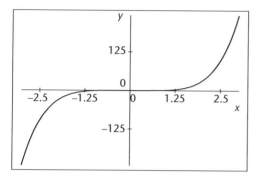

Figure 9.11 The graph of $f(x) = x^5$.

Finally, a function is a **power function** if it is of the form $f(x) = x^p$, where p is a constant. Below we see the graphs of several power functions drawn on the same set of axes (Figure 9.12).

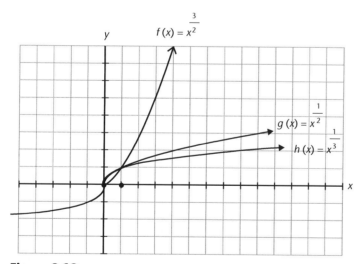

Figure 9.12

We will give some applications of polynomial functions and power functions later in the chapter.

The examples we have presented in our short review are actual practical applications, and represent what are known as mathematical models of reality. A **mathematical model** is simply a mathematical representation of what we observe. A good mathematical model is one that explains what we observe and has predictive value in the sense that we can use it to tell what will happen

412 Functions and Modeling

under new sets of circumstances. The models we gave above were fairly accurate and have good predictive value as has been verified experimentally over and over.

We are not always that lucky. Many times in doing scientific research, it is difficult to find models which connect or explain the data we observe and we have to accept what appears to be our best model, though it may not be perfectly accurate. Still, these models can sometimes be useful to make predictions. So how do we go about finding models? We address this in the next section.

9.3.2 Which Model Should We Use?

When trying to model real-world phenomena, we often begin by plotting the data. This plot is called a **scatter plot**. We examine the graph and then use our knowledge of functions and their graphs to try to decide which model might be best.

Let us illustrate with some examples from real life.

> **Example 9.13** *When the Magellan spacecraft was sent to the planet Venus in 1991, it sent back its measurements of the temperature of the planet at various altitudes as it descended. The following table gives the approximate data sent back:*
>
> | Altitude (km) | Temperature (K) |
> |---|---|
> | 60 | 250 |
> | 55 | 300 |
> | 50 | 340 |
> | 40 | 410 |
> | 35 | 460 |
>
> *At an altitude of 35 kilometers, it stopped transmitting data. One goal of that mission was to find out the temperature at the surface of Venus. Find it.*

Solution. If we plot the data points to get a scatter plot, we get the following picture (Figure 9.13).

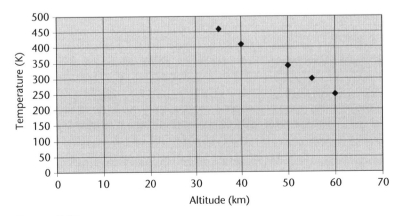

Figure 9.13

We can see that the data are almost perfectly linear. (Actually it had transmitted a lot more data up to that point, all consistent with the linear picture near the surface of the planet.) Therefore, we might take a leap of faith, and try to fit the data accurately with the equation of a line. Since we can see that not all the data lie on one straight line, we find a line that best fits the data, known as the **regression line**. When we are fitting a curve to data, the curve of best fit is known as the **regression curve**.

Most graphing calculators have the capability of making scatter plots and of finding regression curves that fit the data. Of course, we have to tell it which model to use to fit the data: linear, exponential, quadratic, and so on. The manual that comes with your calculator will tell you how to do this.

Our goal in this problem is to predict the temperature at the surface of Venus. So we ask the computer to draw a scatter plot and then fit the data with a line, since our data seem to lie very close to a line. The computer (specifically Microsoft Excel), gives us the following (Figure 9.14):

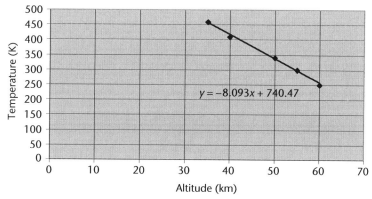

Figure 9.14

Our regression line for this data is $T = (-8.093)h + (740.47)$, where h is the height above the surface of Venus, and T is the temperature at that height. To find the temperature at the surface of Venus, we set $h = 0$, as $h = 0$ represents the surface of Venus. After setting h to zero and solving for T we get $T = 740.47$ Kelvin. Thus, we estimate that the temperature on the surface of Venus is 740.47 K.

The most recent information from NASA puts the temperature at 740 K (hot enough to melt lead). Our model led to some very accurate conclusions.

Example 9.14 *A person with diabetes needs insulin to help process the glucose in his body. Insulin breaks down very quickly in the body, and so the goal is to give medication via some delivery system which releases the insulin slowly. To get an idea of how quickly the insulin breaks down in a patient's body, 20 units of insulin is injected into the body of a patient, and every 10 minutes blood is drawn to see the level of insulin. Here are the data for a particular patient that represents what happens in nearly all patients, though the rate of breakdown differs from patient to patient.*

414 Functions and Modeling

Time, t (elapsed in minutes)	Number of Units of Insulin
0	20
10	9.5
20	3.6
30	1.3
40	0.2
50	0.1
60	0.03

Try to fit a mathematical model which describes the level of insulin in this person's system over time.

Solution. The rapid decrease to 0 seems to indicate that an exponential function probably is the right model here and drawing the scatter plot below seems to verify this. So, we ask the calculator to do an exponential regression. We get the following equation and show the data plotted together with the regression equation (Figure 9.15).

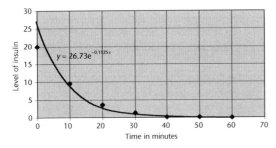

Figure 9.15

The fit is excellent. It is models like these that allow doctors and pharmaceutical companies to decide on dosages.

Example 9.15 *Farmers need rain for their crops to grow. But too much rain can result in fungus growth and subsequent loss of crops. In the following table we see how the number of inches of rain affected the corn crop in a certain region of the country over a period of several years.*

Year	Rainfall (in)	Corn yield (bushels)
1998	18.1	4325
1999	20.3	5167
2000	13.4	3462
2001	22.6	4856
2002	26.2	4126
2003	35.8	2678
2004	28.5	4576
2005	30.2	3698

Draw a scatter plot of the data, and fit the data with a curve.

Solution. Here is a scatter plot of the data (Figure 9.16):

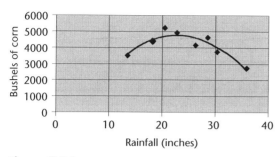

Figure 9.16

No function we have discussed so far seems to fit it extremely well. But it certainly looks like the quadratic function might offer us a better fit than any other function we are aware of. We do a quadratic regression and find that the quadratic function that fits the data best is $y = -13.053x^2 + 594.8x - 2037$.

So, if one had to estimate how many bushels of corn this part of the country would produce, assuming they planted the same amount each year and they had 30 inches of rain, we need only substitute 30 in the above formula for y to get $y = -13.053(900) + 594.8(30) - 2037 \approx 4059$ bushels. But this is only an estimate. Unlike the earlier models we showed, this function does not appear to fit the data that well, even though it is called the quadratic of best fit. So what does "best fit" really mean? We will address this after we give one last example. The following is an application from astronomy and exemplifies the fitting of data by a power function.

Example 9.16 *All planets revolve about the sun. The time it takes to complete one revolution around the sun is called its period which we represent by T. The planet's average distance from the sun is denoted by D and is measured in astronomical units where one astronomical unit is the average distance the earth is from the sun. One of Kepler's laws is that for all planets.*

$$T^2 = D^3. \tag{9.3}$$

Here is actual astronomical data collected for the various planets

Planet	Mean distance from sun (AU)	Period (years)
Mercury	0.387	0.241
Venus	0.723	0.615
Earth	1	1
Mars	1.523	1.881
Jupiter	5.203	11.861
Saturn	9.541	29.457
Uranus	19.190	84.008
Neptune	30.086	164.704
Pluto	39.507	248.350

Thus, the average distance Mars is from the sun is about 1.5 times that of the earth from the sun, and it takes about 1.8 years for Mars to go around the sun. Show that the data are consistent with Kepler's Law by fitting these data to a power function.

Solution. Here is our scatter plot, together with the power function that fits it best (Figure 9.17).

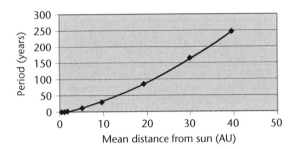

Figure 9.17

Our power function that best fits the data is $T = 1.0004 D^{1.496}$ (where y is T and x is D) which, if we round, gives us $T = 1.0 D^{1.5}$. Kepler's Law says that $T^2 = D^3$. If we solve for T, we get $T = D^{3/2}$ which is exactly what our regression is telling us from the data.

Thus, if we didn't know Kepler's Law, we could get it by the method of finding the curve of best fit. This really is a nice application of regression, don't you think?

Scientists use regression to figure out laws of nature and how certain things relate to each other. The graphs are invaluable in this respect and when one of the many available models works and explains what is going on in some natural phenomenon, it is always something to rejoice about.

Student Learning Opportunities

1 (C) On their exam you ask your students the following question: "What is the purpose of finding a model to best fit data?" One of your students, Maxine, writes: "The purpose of fitting data to a curve is so that you can give an exact statement about what will happen in a future situation like it for which you do not have the data." Comment on Maxine's response. Is she correct? Why or why not?

2 (C) One of your more curious students asks: "What do I do if, after making a scatter plot from the data, I can't envision any function that will fit it? Does this mean that there is no relationship at all between the two variables?" How do you respond?

3 It is well known that if tires are kept at the right pressure, then their life is extended. If the pressure is too low or too high, the life of the tire is reduced. A famous manufacturer tested its top of the line tire and observed the following about its life, for tires that were kept at the following pressures.

Pressure in tire (lb/in^2)	Tire life (Miles)
26	45000
28	48000
30	51000
32	58000
34	54000
36	51000
38	46000

Fit the data to a curve. What kind of curve seems to be a good fit? Using the model you found, estimate the number of miles a driver will get if he drives his car with his tires kept at a pressure of 24 pounds per square inch.

4 In 1885 Sir Francis Galton did a study of the average height of offspring compared to the average height of the parents. Here are the data for seven different families.

Average of parents' height	Average of children's height in adulthood
64.5	66.2
66.5	67.2
67.5	69.2
68.5	67.2
68.5	69.2
69.5	71.2
70.5	70.2

If you had to guess which type of regression equation would best model these data, without even graphing it, what would you guess? Key the data in and see what the calculator gives you for the regression line.

5 If we have a small hole in the bottom of a can, we observe that the water flows out faster when the can is fuller than when it is less full. The physical law that describes this is known as Toricelli's law. Here are the data for a specific can filled with water in which a small hole was made in the bottom.

Height of the water (ft)	Velocity of fluid leaking out (ft/s)
8	22.5
7	21.2
6	19.6
5	18
4	16
3	14
2	11.2
1	8

Draw a scatter plot for the data. Try fitting it with a linear, quadratic, exponential, and power function. Which of these seems to be the best fit? Why?

6 Psychologists often study word retention. The test they use is to have people memorize a group of words and then see how many they remember after different periods of time. Suppose we have 50 people in our experiment, and they memorize 100 words they have not seen before and their meanings. They are then asked, after various time intervals, how much they remember and the average results for the 50 people who are tallied. Here are the results of one such experiment. Try to fit the data by a power function.

Time (h)	Average number of words remembered
1	80.2
2	45.3
3	33.4
4	22.6
5	18.5

If the trend continues, what is the average number of words this group remembers in 24 hours?

7 Try fitting the astronomical data in Example 9.16 by an exponential curve. How does his compare to the power function?

8 (C) When you begin teaching, a good class project is to have the students do measurements on their femur length and their height. The femur is the bone from your hip to your knee. For now, you can do this for at least 5 people you know. Does there appear to be a relationship between these? Does it appear to be linear? Support your response by drawing a scatter plot and trying to fit the data with a function (Figure 9.18).

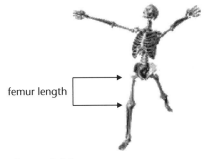

Figure 9.18

9 The percentage of households with a computer is given in the following table.

Year	Percentage of households with computers
1984	8.2
1989	15
1993	22.8

Year	Percentage of households with computers
1997	36.6
1998	42.1
2000	51.1

Does this appear to be linear growth, exponential growth, or neither? Using your model, estimate the percentage of households with computers today. Can this trend continue indefinitely? Explain.

10 Get some information from the Internet on how computing power (number of operations per second) has changed over the last 30 years. What type of model best fits your data? If the trend continues, how powerful should the average computer be in ten years from now?

11 Biologists have observed that the chirping rate of a certain kind of cricket is related to the temperature. The data are given in the table below.

Temperature (°F)	Chirping rate (chirps/minute)
45	25
50	36
55	49
60	70
65	86
70	95
75	103
80	110

Find the curve that best fits the data. If this trend continues, how many chirps should we expect to hear at 90 degrees Fahrenheit?

12 It has been said that health care costs in the United States have been going up exponentially in the last 30 years or so. Examine this claim by getting data from the last 30 years and trying to fit an exponential curve to the data. What do your eyes tell you about the fit?

13 The number of households in the US measured every 5 years in March is given in the following table. What kind of model do you feel best fits this data? Find the equation of your model.

1960	52,610
1965	57,251
1970	62,874
1975	71,120
1980	80,776
1985	86,789
1990	93,347
1995	98,990
2000	104,705

14 Go to the website http://www.bts.gov/publications/transportation_statistics_annual_report/2000/chapter6/key_air_pollutants_fig1.html and examine how lead emissions in the US decreased from 1970–1998. What kind of model do you feel best fits this data? Find the equation of your model.

9.4 What Does Best Fit Mean?

 **LAUNCH**

Although in the previous section we have used technology to find the curves of best fit, as pointed out before, we have not yet determined what best fit means. We would like to know what your perception is about the meaning of "best fit." So, we ask you to do the following:

1 In the scatter plot below, draw a line that you believe fits the data the best (Figure 9.19).

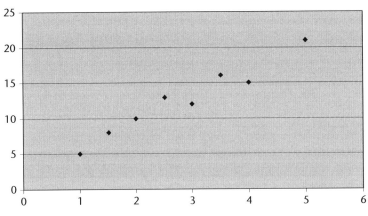
Figure 9.19

2 Describe what you were thinking when you drew this line.
3 Compare the line you drew with that of your neighbor. Were they the same or different? Explain.
4 Write a description of what you think is the meaning of the line of best fit.

We are guessing that, in the launch, some of you, when asked to draw the line that best fits these data, tried to draw a line that went through as many of the points as possible. You might have drawn a line similar to Figure 9.20 below.

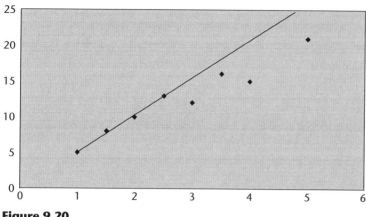
Figure 9.20

Others of you might have thought that a line of best fit is one that separates the data points equally into points that are above the line and below the line. So you might have drawn the following for the line of best fit (Figure 9.21):

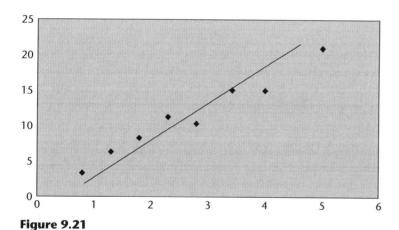
Figure 9.21

Both are reasonable answers. Others of you might even have different impressions of what a line of best fit is. But in reality, there is a specific definition of line of best fit that mathematicians have created.

In this section we will define line of best fit and give some indication of what it is that the calculator does when it finds this line. In the next section we talk about fitting curves to data which don't seem to lie along a line.

9.4.1 What is Behind Finding the Line of Best Fit?

The **line of best fit (the regression line)** is defined to be that line with the property that the sum of the squares of the vertical distances of the y coordinates of the data points from the line is a minimum. See the picture below where we have drawn four data points and the distances these data points are from a line. These distances are denoted by d_1, d_2, d_3, and d_4 (Figure 9.22).

422 Functions and Modeling

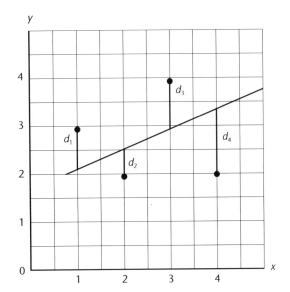

Figure 9.22

The line which minimizes $d_1^2 + d_2^2 + d_3^2 + d_4^2$ is what we call the line of best fit or the regression line.

In a similar manner, when we refer to a quadratic or other function that best fits the data, we mean exactly the same thing. It is the quadratic or other function, respectively, that minimizes the sum of the squares of the vertical distances of the y coordinates of the data points to the curve.

Before we can answer the question posed in the title of this section, we need to recall some concepts from calculus. Recall that, in your first course in calculus, when you wanted to maximize or minimize a polynomial function, you used first derivatives to find what are known as critical points of the polynomial. What you did was set the first derivative = 0. You then tested these critical points to see if they gave maxima and minima. When you took multivariable calculus, you learned a similar technique for finding maxima and minima of a polynomial $f(x, y)$ in two variables. You would take the partial derivative of $f(x, y)$ with respect to x, and the partial derivative with respect to y, and set them both equal to zero, and then solve the resulting equations simultaneously to get critical points. Recall that the partial derivative of $f(x, y)$ with respect to x (denoted by $\frac{\partial f}{\partial x}$) is the derivative of $f(x, y)$ with respect to x, assuming that y is constant, and the partial derivative of $f(x, y)$ with respect to y (denoted by $\frac{\partial f}{\partial y}$) is the derivative of $f(x, y)$ with respect to y, assuming that x is constant. So if $f(x, y) = x^3 y^4$ then

$$\frac{\partial f}{\partial x} = 3x^2 y^4 \quad \text{and}$$

$$\frac{\partial f}{\partial y} = 4x^3 y^3.$$

Now let us use this material to find the line of best fit in the next example.

Example 9.17 *Suppose we have the same 4 data points we just used in Figure 9.22: (1, 3), (2, 2), (3, 4), and (4, 2). We are interested in the line of best fit for these points. Derive the equation of the line of best fit.*

Solution. Suppose that we have a proposed line of best fit $f(x) = ax + b$. The difference between the y value, 3, of the first data point and the height, $f(1)$ of the proposed line of best fit, is $3 - f(1)$. In a similar manner the difference between the y coordinates of the rest of the data points and the function values are, respectively, $2 - f(2)$, $4 - f(3)$, and $2 - f(4)$. These differences need to be squared, and summed, and the resulting quantity must be minimized. Thus, we want to minimize

$$S = (3 - f(1))^2 + (2 - f(2))^2 + (4 - f(3))^2 + (2 - f(4))^2. \tag{9.4}$$

Now, $f(1) = a + b$, $f(2) = 2a + b$, $f(3) = 3a + b$, and $f(4) = 4a + b$. Substituting these values into equation (9.4) we get

$$S = (3 - a - b)^2 + (2 - 2a - b)^2 + (4 - 3a - b)^2 + (2 - 4a - b)^2$$

which is the quantity we want to minimize. The variables in S are a and b and those are the variables that must be determined.

We take the partial derivative of S with respect to a and set it equal to 0, and do the same for the partial derivative of S with respect to b. We get

$$\frac{\partial S}{\partial a} = 2(3 - a - b)(-1) + 2(2 - 2a - b)(-2) + 2(4 - 3a - b)(-3) + 2(2 - 4a - b)(-4) = 0$$

$$\frac{\partial S}{\partial b} = 2(3 - a - b)(-1) + 2(2 - 2a - b)(-1) + 2(4 - 3a - b)(-1) + 2(2 - 4a - b)(-1) = 0.$$

These two equations simplify to

$$60a + 20b - 54 = 0$$

$$20a + 8b - 22 = 0.$$

When we solve these equations simultaneously, we get that the solution is $a = -\frac{1}{10}$, $b = 3$. Thus, our regression line is $f(x) = -\frac{1}{10}x + 3$. Here is the picture Microsoft Excel gave us (Figure 9.23).

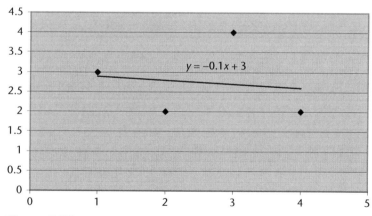

Figure 9.23

We can also get the regression line on the calculator. Here are the steps for the TI. series First, we enter the data. To do this, we press $\boxed{\text{STAT}}\boxed{1}$, which will bring you to the list menu. Then you enter the data into the lists L_1 and L_2. You go to the home screen afterwards and press $\boxed{\text{STAT}}$ again and go to the "calc" menu. Then press the linear regression and this will give you the linear regression line. The calculator gives us exactly the same regression line.

We derived the line of best fit for a specific numerical example, but we can do this for an arbitrary set of data points $(x_1, y_1), (x_2, y_2), \ldots, (x_n, y_n)$. Doing a similar calculation to that above (but leaving out the messy details), we find that the regression line for these data is $f(x) = ax + b$ where

$$a = \frac{n\sum_{i=1}^{n} x_i y_i - \left(\sum_{i=1}^{n} x_i\right)\left(\sum_{i=1}^{n} y_i\right)}{n\sum_{i=1}^{n} x_i^2 - \left(\sum_{i=1}^{n} x_i\right)^2}. \tag{9.5}$$

$$b = \frac{\sum_{i=1}^{n} y_i - a\sum_{i=1}^{n} x_i}{n}. \tag{9.6}$$

This is how the computer or calculator quickly computes the slope, a, of the regression line and the y intercept, b, of the regression line. Here, $\sum_{i=1}^{n} x_i y_i$ is obtained by multiplying the coordinates of each data point and summing the results, $\left(\sum_{i=1}^{n} x_i\right)$ is the sum of all the x coordinates of the data points, $\left(\sum_{i=1}^{n} y_i\right)$ is the sum of all the y coordinates of the data points, $\sum_{i=1}^{n} x_i^2$ is the sum of the squares of all the x coordinates of the data points, and $\left(\sum_{i=1}^{n} x_i\right)^2$ is the square of the sum of all the x coordinates of the data points. One never wants to do this by hand. But just for the sake of illustration, we organize the computations for the example we did previously.

We have 4 data points, so $n = 4$

x_i	y_i	$x_i y_i$	x_i^2
1	3	3	1
2	2	4	4
3	4	12	9
4	2	8	16
$\left(\sum_{i=1}^{n} x_i\right) = 10$	$\left(\sum_{i=1}^{n} y_i\right) = 11$	$\sum_{i=1}^{n} x_i y_i = 27$	$\sum_{i=1}^{n} x_i^2 = 30$

Now, using the numbers in this table and using the formula of equation (9.5), we have

$$a = \frac{n\sum_{i=1}^{n} x_i y_i - \left(\sum_{i=1}^{n} x_i\right)\left(\sum_{i=1}^{n} y_i\right)}{n\sum_{i=1}^{n} x_i^2 - \left(\sum_{i=1}^{n} x_i\right)^2}$$

$$= \frac{4(27) - 10(11)}{4(30) - 10^2}$$

$$= \frac{-2}{20}$$

$$= \frac{-1}{10}.$$

In the Student Learning Opportunities you will find b by using the numbers in the table and the formula of equation (9.6).

9.4.2 How Well Does a Function Fit the Data?

While we have a method of finding the line of best fit, when we actually draw the line of best fit, we may not be so happy with how the line fits the data. Mathematicians have come up with a numerical measure of how well a line fits the data. The numerical measure is known as the **linear correlation coefficient** denoted by the letter r. If $|r|$ is close to 1, the fit is supposed to be good and we say that the independent and dependent variables are highly **correlated**. If $|r|$ is small, the fit is bad. When doing linear regression, most calculators will automatically give you an r value. What is important to understand is that r measures how well the data fit a *linear* model, and not to make conclusions about data that really don't fit linearly. Thus, it is possible that a data set has an r value of 0, which means no *linear* function will fit it well. However, a quadratic function might fit it very well. You might like to plot some points from the curve $y = x^2$ and then ask the calculator to do a linear regression. You will get a correlation coefficient close to zero. Yet, y and x are very much related via the function $y = x^2$.

High correlation between two variables does not mean that one causes the other. All it indicates is that the two variables seem somehow to be linked. For example, data show that ice cream consumption and drownings are highly correlated. You might conclude that, if you eat more ice cream, you will drown. The high correlation comes from the fact that people eat more ice cream in the summer, and people go to the beach more in the summer. So, you would expect there to be more swimming accidents in the summer.

Below we see 4 different data sets. They all have the same linear regression line, the same coefficient of correlation, $r = 0.81$, which would indicate a good fit. Yet they all fit the data in very different ways. In fact, the last picture doesn't seem like a good fit at all! Thus, the correlation coefficient, r, is not enough to make conclusions. Visual inspection of the data is necessary (Figure 9.24).

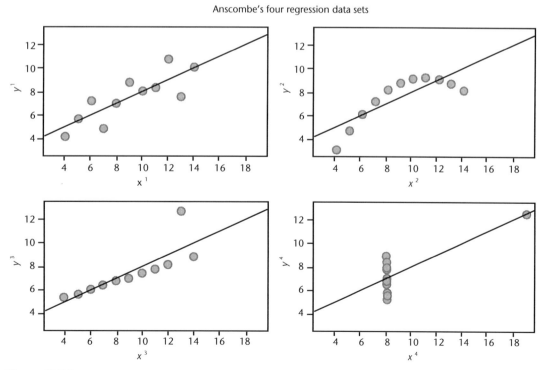

Figure 9.24

Student Learning Opportunities

1. Using the following data points, find the equation of the line of best fit. First use formulas of equations (9.5) and (9.6). Then find the line of best fit by keying in the data points and having the calculator find the regression line. How do your results compare?

   (a) {(2, 3), (4, 5), (5, 6)}

   (b) {(2, 3), (3, 9), (4, 2), (7, 6)}

   (c) {(−1, 3), (3, 4), (−4, 2), (7, −6), (8, 10)}

2. (C) One of your students has calculated and drawn a line of best fit and notices that it does not cross through ANY of the data points. He comes to you, certain that he has done something wrong. How do you respond?

3. (C) To help your students understand how the least-square regression line works, you have told them to experiment with the applet found at the NCTM website: http://standards.nctm.org/document/eexamples/chap7/7.4/. One of your students comes to you all excited about a discovery he has made. When he makes the data points co-linear and creates a line parallel to these points and moves it around, he can get the same sum of the squares in two different places. He wants to know why this happens. What do you say?

4. Entomologists have discovered that the number of chirps per minute a certain type of cricket chirps depends on the outside temperature. But the cricket will not chirp if the temperature is below 38° Fahrenheit. The following data were collected:

Number of chirps per minute	0	4	9	19	58
Outside temperature (Fahrenheit)	38	39	40	42	50

   (a) Draw a scatter plot of the data, letting x be the outside temperature.
   (b) Does there appear to be a linear relationship between the number of chirps per minute and the outside temperature at temperatures of 38°F or above? Explain.
   (c) Find the line of best fit that fits this data.
   (d) Predict the number of chirps if the outside temperature is 60°F.
   (e) If you hear 80 chirps per minute, could you estimate the outside temperature? Explain.
   (f) Discuss why any linear model of this data cannot be valid for all temperatures beyond 38°F.

5. It has been said that asbestos exposure causes lung cancer. This theory has been tested. The following are data collected by a group of scientists that measure the percentage of mice that contracted lung cancer. Draw a scatter plot of the data and then find the line of best fit. Do you feel a line is a good model for this data? Explain.

Asbestos exposure (fibers per mL)	50	250	500	750	1200	1500
Percent that developed lung cancer	1.2	5	8	9	25	35

6 The following table gives the life expectancy of people in the US from 1920 to 2000.

Year	1920	1940	1960	1980	2000
Life expectancy in years	54.1	62.9	69.7	73.7	76.9

(a) Draw a scatter plot of the data.
(b) The data are approximately linear. Find the equation of the line of best fit.
(c) If the trend continues, what should the yearly life expectancy be in the year 2010?

7 The number of metric tonnes of coal in the US from 1983 to 2003 is given in the following table.

Year	1983	1988	1993	1998	2003
Metric tonnes of coal produced	782.1	950.3	945.4	1117.5	1071.8

(a) Find the line of best fit for the data.
(b) Predict the approximate production for the year 2008, based on this model, assuming the trend continues.

9.5 Finding Exponential and Power Functions That Fit Curves

LAUNCH

As you may be aware, biological populations can grow exponentially if not restrained by predators or lack of food or space. Below are some data regarding an outbreak of the gypsy moth, which devastated forests in Massachusetts in the US. Instead of counting the number of moths, the number of acres defoliated by the moths was counted. These data were supplied by Chuck Schwalbe, US Dept of Agriculture.

Year	1978	1979	1980	1981
Acres	63,042	226,260	907,075	2,826,095

1 Plot the number of acres defoliated, y against the year x. Does the pattern of growth appear to be exponential?
2 Use a graphing calculator to find the exponential model that best fits these data.
3 Use this model to predict the number of acres defoliated in 1982. The actual number for 1982 was 1,383,265. Give a possible reason why the predicted value and the actual value could be so different.

In doing this launch problem, you probably encountered several difficulties. First, the numbers given were outside the range of values that the calculator can handle for exponential regression. So, hopefully, you represented the data in a different way that enabled you to work with smaller numbers. One way to do this is shown in the table below.

Year	1	2	3	4
Acres (1000s)	63.042	226.260	907.075	2,826.095

After changing the representation of the data, you were probably able to calculate that the predicted value for the following year, 1982, was 10,723,000. But, this was nowhere near the actual value, which was approximately 1,383,000. Were you able to figure out some reasons for why there was such a discrepancy between the predicted value and the actual value? Would you ever have imagined that there was a viral infection in the gypsy moths which drastically reduced their numbers?

We hope that you are now quite aware that you must be very careful when extrapolating from data. This, along with how the calculator calculates regression curves will be discussed in this section.

9.5.1 How Calculators Find Exponential and Power Regressions

Most people don't realize that the method of using partial derivatives that we gave earlier for when we found the line of best fit does not work well when trying to find power or exponential functions of best fit. The equations we get by setting the partial derivatives equal to zero are unwieldy and difficult to solve. In this section we will show how, by "linearizing" the data, the calculator can produce an exponential curve or power curve that fits the data well, when the data are amenable to those fits.

Suppose that we want to fit a set of data with an exponential function like $y = ab^x$. Since the partial derivative approach that we took to find a line of best fit will not work well in this case, we look for an alternate approach. We can take the ln of both sides of $y = ab^x$ and get

$$\ln y = \ln a + x \ln b. \tag{9.7}$$

(See Chapter 6 for a review of properties of logarithms.) Now, $\ln a$ and $\ln b$ are constants which we can call c and d, respectively. If we let $\ln y = u$, then equation (9.7) becomes,

$$u = c + dx.$$

This is the equation of a line in the $x - u$ plane. (That is the plane where x stays as it was and y is replaced by u, which is $\ln y$.) What we are saying is that, if we start with the data points (x_1, y_1), (x_2, y_2), and so on, and they are well fit by an exponential function, then the points $(x_1, \ln y_1)$, $(x_2, \ln y_2)$, and so on, will be well fit by a line. The converse is also true. That is, if we plot the points $(x_1, \ln y_1)$, $(x_2, \ln y_2)$, and so on, and these appear to fit a line fairly well, an exponential function is a good fit for our original data. Thus, we can find the line of best fit for the data $(x_1, \ln y_1)$, $(x_2, \ln y_2)$, and so on, and then convert back to $y = ab^x$. This is a technique that scientists have used in the past to determine if one should try using a line of best fit, or possibly an exponential function.

Let us illustrate this in Example 9.14 where we studied glucose breakdown. We repeat that table for convenience.

x = Time (elapsed in minutes)	y = Number of units of insulin
0	20
10	9.5
20	3.6
30	1.3
40	0.2
50	0.1
60	0.03

Now, to see if an exponential fits these data well, we plot the points $(x, \ln y)$, which are given in the following table.

Time, x, (elapsed in minutes)	Ln of number of the units of insulin
0	$\ln 20 = 2.9957$
10	$\ln 9.5 = 2.2513$
20	$\ln 3.6 = 1.2809$
30	$\ln 1.3 = 0.26236$
40	$\ln 0.2 = -1.6094$
50	$\ln 0.1 = -2.3026$
60	$\ln 0.03 = -3.5066$

The graph we get is (Figure 9.25):

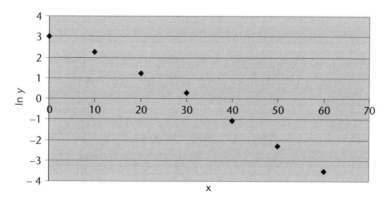

Figure 9.25

There is no question that, since this graph is almost linear, an exponential function should fit the data well. Of course, our eyes seemed to tell us that from the original set of data. But this corroboration is the icing on the cake!

Now, how does the calculator find the best fit exponential that it generates? It uses the line of best fit and fits the data in the above table with a regression line, which turns out to be, $u = c + dx = 3.2858 - 0.1125x$. Now, since $c = \ln a$ and $d = \ln b$, we immediately find that $a = e^c = e^{3.2858} = 26.73$

and $b = e^d = e^{-0.1125}$. Thus, the exponential function which fits our data well is $y = ab^x = 26.73 * (e^{-0.1125})^x$, which is what the computer generated earlier.

How do we know if a power function, $y = ax^b$, will be a good fit to data and how does a calculator find it? Again, we can analyze this by taking the logarithm of both sides of the equation to get:

$\ln y = \ln a + b \ln x.$

Calling $\ln y = v$, $\ln a = c$, and $\ln x = u$, our equation becomes

$$v = c + bu \qquad (9.8)$$

which is a linear equation in the $u - v$ plane. But the $u - v$ plane is the $\ln x - \ln y$ plane. So what we are saying is, if a function can be well fit by a power function, then the data points ($\ln x_i$, $\ln y_i$) very closely approximate a line, and conversely. So, to see if a power function is a good fit, instead of graphing our original points (x_i, y_i), we plot the points ($\ln x_i$, $\ln y_i$). If that graph looks pretty linear, you can be quite certain that the power function is a good fit for the original data.

Let us illustrate this using the data from Example 9.16, one of Kepler's Laws of Planetary Motion. There we had the data

Planet	x = Mean distance from sun (AU)	y = Period (years)
Mercury	0.387	0.241
Venus	0.723	0.615
Earth	1	1
Mars	1.523	1.881
Jupiter	5.203	11.861
Saturn	9.541	29.457
Uranus	19.190	84.008
Neptune	30.086	164.704
Pluto	39.507	248.350

We plotted these data points earlier, and from the graph, it was not clear if, within the given interval, a quadratic, power, exponential, or cubic might fit the data best in that interval. But, if instead of looking at points (x, y), we look at the points $(\ln x, \ln y)$, we get the table

Planet	$\ln x = \ln$(Mean distance from sun) (AU)	$\ln y = \ln$ Period (years)
Mercury	$\ln 0.387 = 3 - 0.94933$	$\ln 0.241 = 3 - 1.4230$
Venus	$\ln 0.723 = 3 - 0.32435$	$\ln 0.615 = 3 - 0.48613$
Earth	$\ln 1 = 30.0$	$\ln 1 = 30.0$
Mars	$\ln 1.523 = 30.42068$	$\ln 1.881 = 30.6318$
Jupiter	$\ln 5.203 = 31.6492$	$\ln 11.861 = 32.4733$
Saturn	$\ln 9.541 = 32.2556$	$\ln 29.457 = 33.3829$
Uranus	$\ln 19.190 = 32.9544$	$\ln 84.008 = 34.4309$
Neptune	$\ln 30.086 = 33.4041$	$\ln 164.704 = 35.1041$
Pluto	$\ln 39.507 = 33.6765$	$\ln 248.350 = 35.5148$

The graph we get from plotting the points in this table is (Figure 9.26).

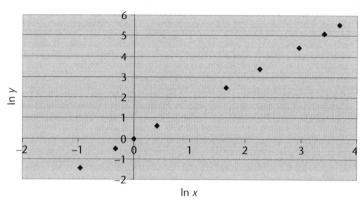

Figure 9.26

It is almost perfectly linear! So, we are now thinking that our fit with a power function should be good. We fit the data in this table with a line and we find that our regression equation is $v = c + bu = 0.0004 + 1.4996u$. Since $a = e^c = e^{0.0004} = 1.0004$ and $b = 1.4996$, we get as our regression power function for the original data, $y = ax^b = 1.0004x^{1.4996}$, which is what we got earlier when we first did this problem.

Whatever model we use, we always have to answer the question, how good is our model? As we have pointed out, one measure of goodness of fit that is used for linear models is the correlation coefficient r. When fitting nonlinear curves to data, there is another measure that is used to decide on the goodness of fit. It is the R^2 value. This is known as the **coefficient of determination**, which intuitively measures the percentage variation of the data explained by the model being used. Thus, if R^2 is 0.93, this means that 93% of the data variation is explained by the model. Most calculators will give us this value. (For linear functions we will get a value for r^2, but this coincides with the value of R^2.) The general rule is that whichever model gives us the highest value of R^2 is the one we use as the best fitting model. But, don't be misled by this. Draw the data and the curve and make sure that what you are seeing makes sense. (Think of Ascombe's data set!) There is no measure of goodness of fit which is foolproof.

9.5.2 Things to Watch Out for in Curve Fitting

The ability to fit data to curves is an important consideration in applied mathematics. But there is a lot to be cautious about. Curves are fitted to data you already have. Many times people assume that, if a certain curve fits data already collected, then it will fit all data collected in the future. That is, the curve that you use to fit the data has some predictive value. There are times when this is true and many times it is not true.

People who work in the sciences are always looking to discover the laws of nature, and when they fit curves to data, they really hope that the ones that fit the data are curves that work all the time. If they do, then they have discovered a law. One instance where this happens is in Hooke's Law for the Spring Constant, which we discussed in Example 9.10. Another example of this was Kepler's Law of Planetary Motion from Example 9.16. Hooke's Law was discovered by trying different springs and observing that the data seemed to always be fit by a linear function. This was then used to predict what would happen with other springs, and the predictions were borne out. Thus, this model had good predictive value.

Newton's Law of Gravitation says that if we have two bodies with masses m_1 and m_2, then there is a force acting on the bodies which either attracts them or repels them. This is given by $F = \frac{km_1 m_2}{r^2}$, where k is a constant and r is the distance between the centers of gravity of the objects. (A center of gravity is a point where all the mass of the object seems to be concentrated. That bodies have such a center of gravity was also discovered by experimentation.) This law has great practical value and is the basis of many useful physical results. It is this law that allows us to send satellites into space and was even used to predict the existence of the planet Neptune. Though it has great use, in certain applications it does not predict as well as it should, and, in fact, was superseded by Einstein's Theory of Relativity which, in many instances, was a better model. We just want to make the point that, while a function may appear to fit data, we cannot always assume that all data will fit this function. Similarly, if we fit a curve to data that are gathered over time, and we try to use the model to predict what will happen too far in the future, since the trend may not continue, the model may lose its effectiveness. *Thus, all predictions made with a model over time make an assumption (right or wrong) that the trend continues.*

Consider the example of the growth of bacteria in a Petri dish. At first, the growth appears to be exponential. But, as time elapses, the food supply diminishes, the bacteria begin to die and the reproduction rate changes. Since there is a finite limit to how much bacteria can survive in a Petri dish, the exponential model that we use initially is not a good model in the long run.

Student Learning Opportunities

1 Use the graphing calculator to find an exponential model that fits the following data well:

x	2	2.5	3	3.5	4.5	5
y	12.1	23.2	43.5	79.6	150.2	524.6

2 Use the graphing calculator to find a power function which fits the following data well:

x	2	2.5	3	3.5	4
y	38.8	76.1	133.2	220.66	338.6

3 In questions 1 and 2 graph the data $(x, \ln y)$ and $(\ln x, \ln y)$. Do these seem to predict which model fits each curve best? Explain.

4 Graph the following data and, by eye, decide whether you should use a linear, quadratic, exponential, cubic polynomial, or power function to model the data. If your calculator computes R^2 values, use this to find which curve fits the data best. How did your estimates compare to what the calculator gave you?

(a)

r	2	2.5	3	3.5	4
s	12.1	15.2	17.6	20.1	23.6

(b)

p	2	2.5	3	3.5	4
q	75	238	595	20.1	23.6

(c)

e	2	2.5	3	3.5	4
f	60	134	305	700	1545

5 In the previous example part (c), draw the $(x, \ln y)$ plots and $(\ln x, \ln y)$ plot. Does this give you any further information about which curve to use to fit the data? Explain.

6 Try fitting a linear, quadratic, exponential, and power function to the following data. Which fits best in your opinion if you look at the graph? Does the R^2 value support your opinion?

x	2	2.5	3	3.5	4
y	9.7	12.6	19.2	35	42

7 (C) Your students ask you if the coefficient of determination is the only way to measure goodness of fit for data. How do you respond? [Hint: Look up measures of goodness of fit on the Internet.]

8 Begin with a power function and generate some data points. Then "doctor" them somewhat. Use your new points and fit them with a power function. Does the calculator give you a function close to the one you started with?

9.6 Fitting Data Exactly With Polynomials

LAUNCH

1 Examine the following table which lists the values of the function $f(x) = 3x + 2$ for values of $x = 1$, $x = 2$, $x = 3$, and so on. Complete the last row.

x	1	2	3	4	5
$y = f(x) = 3x + 2$	5	8	11	14	17
1st differences: (difference between a y value and the previous y value)	8 − 5 = 3	11 − 8			

2 The entries in the last row are called first differences. What do you notice about the values of these 1st differences?
3 Examine the following table which lists the values of the function $g(x) = x^2 + x + 1$ for values of $x = 1$, $x = 2$, $x = 3$, and so on. Complete the last two rows of the table.

x	1	2	3	4	5
$g(x) = x^2 + x + 1$	3	7	13	21	31
1st differences		7 − 3 = 4	13 − 7 = 6		
2nd differences (differences of successive entries in the previous row)		6 − 4 = 2			

4 What do you notice about the values of these 2nd differences?
5 Examine the following table, which lists the values of the function $h(x) = x^3$ for values of $x = 1$, $x = 2$, $x = 3$, and so on. Complete the last three rows.

x	1	2	3	4	5
$h(x) = x^3$	1	8	27	64	125
1st differences		7	19		
2nd differences		12	18		
3rd differences (difference between successive entries from the previous row)		6			

6 What do you notice about the values of these 3rd differences?
7 Given the function $y = x^6$, if we make a similar table, what can you say about the values of the sixth differences?
8 Given the function, $y = x^n$, (where n is a positive integer), if we make a similar table, what can you say about its nth differences?

Thus far, we have examined curve fitting with linear, quadratic, exponential, and power functions. We now turn to polynomial models and interesting approaches to determine when polynomials fit data exactly. We begin by making some observations about linear, quadratic, and cubic polynomials.

You probably noticed from the launch question, that for the linear function, the difference in the successive y values was constant, that is, 3. We can put the last 3 on the third row in parentheses because, although it is the correct difference, the next number, 20, in the second row, which you most likely used to compute this difference was not shown.

Now, as you discovered, for the quadratic function, $g(x) = x^2 + x + 1$, the difference between the successive y values was not constant, but the differences on the third row, which we called the second differences were constant. Next, we gave the table for the cubic function $h(x) = x^3$ and when you calculated the first differences, the second differences, and then the third differences, which are the differences of the second differences, you noticed that the third differences were constant.

To summarize, using the given functions, you noticed that, for the linear function the first differences were constant. For the quadratic function, the second differences were constant. For the cubic function, the third differences were constant. This probably led you to conjecture that if we have an nth degree polynomial then the nth differences are constant. This is true and you might want to try proving it by induction.

The question is, "Is the reverse true?" That is, if we have a function $f(x)$ whose nth differences are constant, does this imply that the function is a polynomial of degree n? Well, the answer to this is "No." (For example, the function $f(x) = x + \sin 2\pi x$ fits the data points $(1, 1), (2, 2), (3, 3)$ and so on exactly and $f(x+1) - f(x) = 1$ for all x. Clearly, $f(x)$ is not linear.) But if the function is defined only on the natural numbers or even the integers, the answer to this is "Yes" and the proof is not so difficult, but is a bit tedious.

Theorem 9.18 *Suppose that we have a function $f(n)$ defined only on the natural numbers such that the first differences are constant. Then $f(n)$ must be a linear function. That is, $f(n) = an + b$.*

Proof. We are assuming that the first differences are constant. That is, we are assuming that

$f(2) - f(1) = a$

$f(3) - f(2) = a$

$f(4) - f(3) = a$

...

$f(n) - f(n-1) = a.$

If we add this string of n identities, we are left with

$f(n) - f(1) = an.$

Adding $f(1)$ to both sides we get that

$f(n) = an + f(1).$

Calling $f(1) = b$, we have

$f(n) = an + b.$

which is the result we wanted. ∎

Theorem 9.19 *Suppose we have function $f(n)$, defined only on the natural numbers such that the second differences are constant. Then $f(n)$ must be a quadratic function. That is, $f(n) = an^2 + bn + c$.*

Proof. By the previous theorem, if the 2nd differences are constant, then the first differences can be fit by a linear function. Let us call the first differences $g(n)$. So $g(n) = f(n) - f(n-1)$ for $n \geq 2$.

Since $g(n)$ can be fit by a linear function, $g(n) = an + b$. We have, using the definition of $g(n)$ and the fact that $g(n) = an + b$,

$$g(2) = f(2) - f(1) = a(2) + b$$
$$g(3) = f(3) - f(2) = a(3) + b$$
$$\ldots$$
$$\ldots$$
$$g(n) = f(n) - f(n-1) = a(n) + b.$$

Adding the middle portions of these n equations we have

$$f(n) - f(1) = a(2 + 3 + \ldots n) + \underbrace{(b + b + \ldots b)}_{n-1 \text{ times}} \tag{9.9}$$

$$= a\left[\frac{n(n+1)}{2} - 1\right] + b(n-1). \tag{9.10}$$

(Here we are using the result from Chapter 1 that the sum of the first n integers is $\frac{n(n+1)}{2}$. Since the sum in the first parentheses of equation (9.9) is the sum of of the first n natural numbers except for 1, we need to subtract 1 from the answer which is why you see the -1 in equation (9.10).) Adding $f(1)$ to both sides of equation (9.10) and simplifying, we have that $f(n) = \frac{an^2 + an}{2} - a + bn - b + f(1)$, which is clearly quadratic. ∎

In the Student Learning Opportunities we will have you give an analogous proof for the third differences. You can probably guess how the general proof concerning constant nth differences goes: It is an induction proof. If the nth differences are constant, then the $(n-1)$st differences are fit exactly by a linear function by Theorem 9.18, and the $(n-2)$nd differences are fit exactly by a quadratic function by Theorem 9.19, and so on. Writing it out is a bit tedious, but hopefully these previous two proofs will give you the idea about how to proceed. The general proof also makes use of the fact (which also can be proved by induction) that the sum of the nth powers of the integers from 1 to k is given by a polynomial of degree $k + 1$. Let us state our final result.

Theorem 9.20 *Suppose we have a function $f(n)$ defined only on the natural numbers such that the nth differences are constant. Then $f(n)$ must be an nth degree polynomial.*

Let us give two examples.

Example 9.21 *A polynomial $p(x)$ passes through the points (1, 3), (2, 5), (3, 9), and (4, 15). Find the polynomial.*

Solution. Since the x values of the points are successive, 1, 2, 3, and 4, we have a chance at using the difference method. The first differences in the y values are 2, 4, and 6. The second differences are 2 and 2. Since these are the same, we try to fit the data with a polynomial of second degree.

Now you can either do this by hand, or enter these points into the calculator and do a quadratic regression. The calculator comes out with $p(x) = x^2 - x + 3$ and $R^2 = 1$. The fit is perfect (that is what $R^2 = 1$ tells us) and we can check by verifying that all the points lie on this curve.

> **Example 9.22** Let us return to a problem from Chapter 1: Start with a circle and pick two points on the circumference as shown in figure (a). Draw the chord connecting them. It divides the circle into two parts. Now put 3 points on the circumference and connect each pair. What is the maximum number of regions the circle is divided into?

Solution. The picture in (b) (Figure 9.27) below shows 4 parts. When we do the same thing for 4 points on the circle we get 8 regions, and we pointed out that, if you did the same thing for 5 points, you would get 16 regions and for 6 points 31 regions.

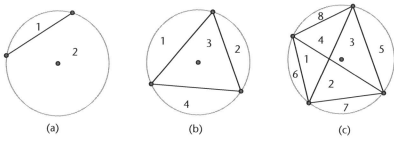

Figure 9.27

The first couple of examples seemed to indicate that, if we had n points on the circle, then the number of regions would be 2^{n-1}. But this formula did not work for the case of $n = 6$. Thus, the number of regions is not 2^{n-1}. So we continue to draw pictures and count the number of regions. We draw a big circle and draw the lines and with a group of friends, number the regions thus, counting them. We find, after a lot of work, the results in the following table:

n	2	3	4	5	6	7	8
Number of regions	2	4	8	16	31	57	99
First differences	2	4	8	15	26	42	
Second differences	2	4	7	11	16		
Third differences	2	3	4	5			

and we begin to gather hope, since it is clear that all the fourth differences are 1. Thus we feel that a reasonable guess would be that a 4th degree polynomial fits this data. We make a list of our coordinates, (n, number of regions) in a table and then do a quartic (4th degree polynomial) regression. We get a 4th degree polynomial with decimal coefficients. We then change the coefficients to fractions and, for the number R, regions we get:

$$R = \frac{1}{24}n^4 - \frac{1}{4}n^3 + \frac{23}{24}n^2 - \frac{3n}{4} + 1.$$

If we have the patience, we can now draw one more picture with $n = 9$ points, we find that the number of regions is 163. We substitute this into the expression for R and find that $R(9) = 163$.

We may think that the function we have found continues to predict the correct number of regions. But in reality, we have not proven anything. What we have demonstrated is the way a research mathematician may go about solving a problem. He or she draws pictures, gathers data, tries to look for a pattern, tries to exploit the pattern to get some kind of formula, and then tries to prove that it is true. Now, we have to prove that the polynomial that we have found works. The proof is more than we want to get into, so we refer the interested reader to http://en.wikipedia.org/wiki/Dividing_a_circle_into_areas, for the proof.

Student Learning Opportunities

1 (C) One of your students asks you if it is possible to have real-life situations in which the data are exactly fit by a polynomial. The answer is "Yes." Give at least two examples of real-life situations which are modeled perfectly by a polynomial.

2 For each of the following data, a polynomial fits the data exactly. Find the polynomial.

(a)
x	1	2	3	4	5	6
y	−1	2	17	50	107	194

(b)
x	1	2	3	4	5	6
y	2	−1	−4	−7	−10	−13

(c)
x	1	2	3	4	5	6
y	2.5	14	31.5	55	84.5	120

3 Make a table showing n and the sum of the first n natural numbers where n goes from, 1 to 6. Then fit it with a quadratic function. Did you get the same answer we got in Theorem 1.1 of Chapter 1?

4 Make a table showing n and the sum of the squares of the first n natural numbers. Let n go from 1 to 8. Then fit the data with a cubic equation. How does your answer compare with the following formula usually given in texts?

$$\sum_{k=1}^{n} k^2 = \frac{n(n+1)(2n+1)}{6}$$

5 Make a table showing n and the sum of the cubes of the first n natural numbers where n goes from 1 to 6. Then fit it with a quartic (4th degree) polynomial. Is your 4th degree polynomial $\left[\frac{n(n+1)}{2}\right]^2$?

9.7 1–1 Functions

LAUNCH

Examine the five different scenarios below:

1. In the United States, when someone is born the person, x, is issued a social security number, y.
2. Given any specific year, x, you can find out how many cell phones, y, were sold in the United States in that year.
3. Some day doctors may be able to enter a person's name, x, in a data base, and complete information about their DNA, y, will appear.
4. When you look up a person's name, x, in a telephone directory, a phone number, y appears.
5. You give your friend a number, x, and he raises it to the 6th power to get a number y.

Describe the similarities and differences of the above scenarios.

In which, if any of the scenarios above, do you think that y is a function of x? In those that are, which are 1–1 functions? Why?

After doing the launch problem, you are probably getting some ideas about the concept of a 1–1 function. You might have even noticed that the scenarios described in the first and third examples, were somewhat different from the other three, and as you will soon discover, were in fact, 1–1 functions. This section will focus on the interesting features of 1–1 functions and their importance in our daily lives.

9.7.1 The Rudiments

Earlier in the chapter we reviewed the notion of function so that we could examine modeling with functions. Now we wish to address a related concept, that of a one-to-one function which we will need to use in the next chapter. This discussion will accomplish the following: (a) review the basic ideas of functions, (b) address some of the issues that arise at the secondary school level, (c) extend the basic ideas to a higher level, and (d) prepare the reader for work that will follow.

Recall that, if $y = f(x)$ is a function of x, then we say that f is **one-to-one**, and write f is $1 - 1$, if different values of the independent variables give rise to different values of the dependent variables (that is, if different x's yield different y's). Thus, the function which associates the postage that one must pay on a package with its weight is *not* 1–1, since it is not true that different weights give rise to different postages. For example, you might pay the same postage on a letter that weighed 1.4 ounces as you would on a letter weighing 1.6 ounces.

The function that associates with each point on the earth's surface, its latitude and longitude pair, *is* 1–1 since different points on the earth's surface necessarily have different latitude and longitude pairs.

The function from Exercise 3, of section 9.2, which gives the volume of a box is *not* 1–1, since different size squares can give the same volume, as we see from graphing $V(x)$. (Try it!)

The function that associates with each symbol its ASCII equivalent (Example 9.5), is 1–1 since different symbols have different ASCII representations. (Think of the havoc that would be wreaked if this were not true.)

Perhaps the simplest way to describe a 1–1 function is to compare it to the definition of function. If y is a function of x, then for each x (in the domain), there is associated one and only one y in the range. The same y can come from many xs. In a 1–1 function, each y (in the range of the original function) can come from one and only one x (in the domain).

Example 9.23 *In secondary school, when using the ordered pair approach to functions, one sees questions like (a) "Given the set of ordered pairs, $\{(1, 2), (2, 3), (3, 5)\}$ where the first coordinate is x and the second is y, is y a function of x?" If it is a function, is it a 1–1 function?" (b)Ditto for the set of ordered pairs $\{(1, 2), (2, 2), (3, 5)\}$. (c) Ditto for the set of ordered pairs $\{(1, 1), (1, 2), (1, 3)\}$.*

Solution. (a) Well, y is a function of x here, since with each x coordinate there is only one y coordinate. Furthermore, it is a 1–1 function since different x's give different y's. (b) y is also a function of x since, with each x, we associate one y. However, it is not 1–1 since different x's can give you the same y ($x = 1$ and $x = 2$ gave you the same y value, $y = 2$.) (c) This is not a function, since with $x = 1$, there are 3 values of y : 1, 2 and 3. So it makes no sense to ask if this is 1–1.

While we are on the topic, let us point out something that one often sees in secondary school books. The set of ordered pairs $\{(1, 1), (1, 2), (1, 3)\}$ is given and the question posed is, "Is this a function." The question is not well phrased, since it is not clear whether y (the second coordinate) is a function of x (in which case the answer is, "No"), or is x a function of y (in which case the answer is, "Yes"). Draw a picture to convince yourself. As a mathematics teacher, this is something you should be attentive to.

In secondary school we teach that, if you are given a graph and you want to know if y is a function of x, you draw vertical lines. If each vertical line that hits the graph touches it only once, then y is a function of x. This is simply another way of saying that, for each x, there is only one y and this test is known as **the vertical line test**.

To see if the graph of a function is 1–1, in addition to performing the vertical line test, we perform the **horizontal line test** as well. If each horizontal line which hits the graph touches it only once, then the function is 1–1. This is another way of showing that, for each fixed y value, there is only one x value that gives rise to it and so the function is 1–1. Below (Figure 9.28), we have the graph of a function, and we have drawn the horizontal line $y = 5$. It hits the curve only once, where $x = 9$. It touches the graph at no other point. Thus only $x = 9$ will yield $y = 5$. No matter which horizontal line we draw, the result is the same, namely, there is only one x for that y. So each y came from only one x. (Equivalently, different x's give different y's.)

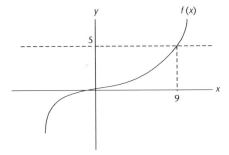

Figure 9.28

Contrast this to the following graph (Figure 9.29), where we have drawn the line $y = 5$. Now, there are two values of x, which give us $y = 5$, and they are -3 and 3. Thus, it is not true that each y came from only one x. (Equivalently, it is not true that different x's give different y's.)

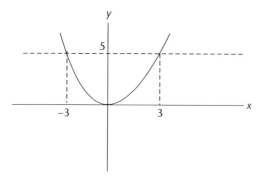

Figure 9.29

One thing we observe is that, if the graph of a function is strictly increasing (on the rise as we move from left to right), then the function is 1–1 since it will pass the horizontal line test. The same is true for a function which is strictly decreasing as we move from left to right. Below we see several graphs (Figure 9.30). We indicate which are functions of x and which are 1–1.

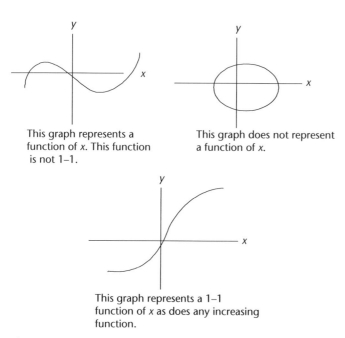

Figure 9.30

The vertical and horizontal line tests only work for *numerical* valued functions. They don't help us for functions that are not numerical valued. There we have to be more creative about deciding if a function is 1–1 and we will do that in a later section. Before we do that, however, we should mention why it is important to know if a function is 1–1.

9.7.2 Why Are 1–1 Functions Important?

When y is a function of x, once you specify x, you know y. One of the main reasons for studying 1–1 functions is that you can go in reverse. That is, if you specify a y in the range, you can immediately tell which x value it originated from. To see this most easily, refer to the ASCII conversion in Example 9.5. Below is a more detailed table showing the ASCII equivalent of the first seven capital letters of the alphabet:

Symbol	ASCII equivalent
A	01000001
B	01000010
C	1000011
D	01000100
E	01000101
F	00100011
G	11000110

If we give you a symbol, you can tell us its ASCII equivalent. This was our original function, which we called $f(x)$. Here, x stood for the symbol, and $f(x)$, its ASCII value. Now notice that we can go in reverse. If we give you the ASCII equivalent of a symbol, you can always tell us what the symbol is.

A function that goes in reverse such as this is called the **inverse function** (of the original function). If we call this function g, then $g(01000001) = A$ and $g(01000010) = B$, and so on.

As another example, in Chapter 2 we spoke about the process of encryption whereby your credit card number is encrypted so that it can't be deciphered. The inverse function is the decryption function, the "machine" that brings your encrypted number back to its un-encrypted form. We are sure you are in agreement about the importance of the decryption function.

Our point is that, whatever is accomplished by a function is undone by its inverse function, leaving the original argument unaltered. This is usually stated abstractly in secondary school as $f^{-1}(f(x)) = x$.

9.7.3 Inverse Functions in More Depth

Earlier, we talked about functions having inverses. In order to have an inverse, the function must be 1–1. That is, different inputs, yield different outputs. Another way of saying this is by using the contrapositive. That is, if the outputs are the same, then the inputs must be the same. In function notation, a function $y = f(x)$ is 1–1 if, when

$$f(b) = f(a), \text{ it follows that } b = a. \tag{9.11}$$

Let us practice using this definition of $1-1$ on some functions.

Example 9.24 *Consider the function $f(x) = 3x + 1$. Show that it is 1–1 using equation (9.11).*

Solution. Suppose that $f(b) = f(a)$. Then $3b + 1 = 3a + 1$. Subtracting 1 from both sides, we get that $3b = 3a$ and dividing both sides by 3 we get that $b = a$. Of course, if we looked at the graph of $f(x)$ we would know it is 1–1, since it passes the horizontal line test.

Example 9.25 *Show $f(x) = x^4$ is not 1–1.*

Solution. We try to see if condition equation (9.11) holds. So, suppose that $f(b) = f(a)$. So $b^4 = a^4$. Does it follow that $b = a$? No! Take $b = -1$ and $a = 1$. With these values, it follows that $f(b) = f(a)$ but of course, b is not equal to a. Once again, looking at the graph will tell us this function is not 1–1 since it doesn't pass the horizontal line test.

We now extend the application of function to points (x, y) in the Cartesian plane.

Example 9.26 *Consider the function $f(x, y) = (5x, 6y)$ from points in the xy plane to points in the xy plane. (a) Compute $f(2, 3)$ and $f(-1, 2)$. (b) Show this function is 1–1.*

Solution. $f(2, 3) = (5 \cdot 2, 6 \cdot 3) = (10, 18)$. Similarly, $f(-1, 2) = (5 \cdot -1, 6 \cdot 2) = (-5, 12)$. (b) To show that this function is 1–1, we can't refer to any graph, because there is no graph. We must use the more abstract definition of 1–1, namely if $f(b) = f(a)$ then $b = a$. Suppose then that $f(b) = f(a)$. It is understood now that b and a are ordered pairs of points. Suppose that $b = (r, s)$ and $a = (t, u)$. Our goal is to show that $b = a$, or that the ordered pairs, (r, s) and (t, u) are the same. Now, from $f(b) = f(a)$ and the definition of $f(b)$ and $f(a)$ we get that $(5r, 6s) = (5t, 6u)$. Since two ordered pairs are equal if their components are the same, $5r = 5t$ and $6s = 6u$. Dividing these two equations by 5 and 6, respectively, we get that $r = t$ and $s = u$. Thus, the pair $(r, s) = (t, u)$ which means that $b = a$ since $b = (r, s)$ and $a = (t, u)$. Since equation (9.11) holds, the function is 1–1.

We now give an example of a 1–1 function which uses matrices. This will prepare you for the next chapter on transformations. You will need to review matrix multiplication from the Appendix and the notion of the inverse of a matrix.

Example 9.27 *Suppose that $M = \begin{pmatrix} 2 & 3 \\ -1 & 2 \end{pmatrix}$. Let S be the set of all 2×1 matrices. We will define a function f, from S to S. More specifically, if $X = \begin{pmatrix} x \\ y \end{pmatrix}$ is any 2×1 matrix, we define $f(X)$ to be the 2×1 matrix obtained by multiplying the matrices M and X.*

(a) Compute $f\begin{pmatrix} 2 \\ 1 \end{pmatrix}$ and $f\begin{pmatrix} -3 \\ 2 \end{pmatrix}$.
(b) Show that the function f is 1–1.

Solution. (a) By the definition of the function, $f\begin{pmatrix} 2 \\ 1 \end{pmatrix} = \begin{pmatrix} 2 & 3 \\ -1 & 2 \end{pmatrix}\begin{pmatrix} 2 \\ 1 \end{pmatrix} = \begin{pmatrix} 7 \\ 0 \end{pmatrix}$ and $f\begin{pmatrix} -3 \\ 2 \end{pmatrix} = \begin{pmatrix} 2 & 3 \\ -1 & 2 \end{pmatrix}\begin{pmatrix} -3 \\ 2 \end{pmatrix} = \begin{pmatrix} 0 \\ 7 \end{pmatrix}$.

(b) To see that f is 1 – 1, suppose that $f(b) = f(a)$. We want to show that $b = a$. Now, a and b are 2×1 matrices. Suppose that $a = \begin{pmatrix} r \\ s \end{pmatrix}$ and $b = \begin{pmatrix} t \\ u \end{pmatrix}$. Now, starting with $f(b) = f(a)$ and using the definition of f, we have that if $f(b) = f(a)$ then $f\begin{pmatrix} t \\ u \end{pmatrix} = f\begin{pmatrix} r \\ s \end{pmatrix}$. This means that

$$M\begin{pmatrix} t \\ u \end{pmatrix} = M\begin{pmatrix} r \\ s \end{pmatrix} \tag{9.12}$$

where $M = \begin{pmatrix} 2 & 3 \\ -1 & 2 \end{pmatrix}$. Now we are baffled. How do we go from here? Well, since the determinant of $\begin{pmatrix} 2 & 3 \\ -1 & 2 \end{pmatrix}$ is not zero (see Appendix for a review of determinants and the main facts associated with them), the matrix $M = \begin{pmatrix} 2 & 3 \\ -1 & 2 \end{pmatrix}$ has an inverse, which we denote by M^{-1}. Multiplying both sides of equation (9.12) by the inverse we get

$$M^{-1}M\begin{pmatrix} t \\ u \end{pmatrix} = M^{-1}M\begin{pmatrix} r \\ s \end{pmatrix}$$

which reduces to

$$I\begin{pmatrix} t \\ u \end{pmatrix} = I\begin{pmatrix} r \\ s \end{pmatrix} \tag{9.13}$$

where I is the identity matrix, $\begin{pmatrix} 1 & 0 \\ 0 & 1 \end{pmatrix}$. Since I times any matrix is the matrix, equation (9.10) becomes

$$\begin{pmatrix} t \\ u \end{pmatrix} = \begin{pmatrix} r \\ s \end{pmatrix}. \tag{9.14}$$

But saying that $\begin{pmatrix} t \\ u \end{pmatrix} = \begin{pmatrix} r \\ s \end{pmatrix}$ is the same as saying that $b = a$ since $b = \begin{pmatrix} t \\ u \end{pmatrix}$ and $a = \begin{pmatrix} r \\ s \end{pmatrix}$. In short, we have shown the $f(b) = f(a)$ implies $b = a$, so f is 1–1.

9.7.4 Finding the Inverse Function

For each function $y = f(x)$ which is 1–1, for each x there is only one y (definition of function) and, for each y, there is only one x that it came from. The rule which associates the x with the given y is called the **inverse function**. So how does one find the inverse function? Well, one surefire way is to solve for x in terms of y. Then, if you know y you automatically get x. Let us illustrate.

Example 9.28 *Consider the function $y = f(x) = x^3$. Find the inverse function.*

Solution. We solve for x in terms of y to get $x = \sqrt[3]{y}$. This is our inverse function. Now, all you have to do is specify y and, you get x immediately. Thus, if $y = 8$, then the x it came from in the original function is $x = \sqrt[3]{8}$ or 2.

> **Example 9.29** *The relationship between Fahrenheit and Centigrade temperature is given by $F = \frac{9}{5}C + 32$. Find the inverse function.*

Solution. We solve for C in terms of F to get

$$C = \frac{5}{9}(F - 32). \tag{9.15}$$

This is our inverse function. Thus if the Fahrenheit temperature is 104 degrees, we need only substitute this into equation (9.15) to get $C = \frac{5}{9}(104 - 32) = 40$, the centigrade temperature.

> **Example 9.30** *In Example 9.27 we spoke of the function $Y = f(X)$, where $f(X) = MX$ where $M = \begin{pmatrix} 2 & 3 \\ -1 & 2 \end{pmatrix}$ and $X = \begin{pmatrix} x \\ y \end{pmatrix}$ is an arbitrary 2×1 matrix. Find the inverse function.*

Solution. We have

$$Y = MX. \tag{9.16}$$

We need to solve for X in terms of Y. We multiply both sides of (9.16) by M^{-1} and we get that the inverse function is

$$X = M^{-1}Y.$$

Recalling that $M = \begin{pmatrix} 2 & 3 \\ -1 & 2 \end{pmatrix}$, we can compute M^{-1} by either hand or calculator to get $M^{-1} = \begin{pmatrix} \frac{2}{7} & -\frac{3}{7} \\ \frac{1}{7} & \frac{2}{7} \end{pmatrix}$. So the inverse function is

$$X = M^{-1}Y = \begin{pmatrix} \frac{2}{7} & -\frac{3}{7} \\ \frac{1}{7} & \frac{2}{7} \end{pmatrix} Y. \tag{9.17}$$

Thus if we wanted to find the 2×1 matrix, X, that yields the 2×1 matrix $Y = \begin{pmatrix} 7 \\ 0 \end{pmatrix}$ under the function f, we need only substitute this value of Y into equation (9.17) to get

$$X = M^{-1}Y = \begin{pmatrix} \frac{2}{7} & -\frac{3}{7} \\ \frac{1}{7} & \frac{2}{7} \end{pmatrix} \begin{pmatrix} 7 \\ 0 \end{pmatrix} = \begin{pmatrix} 2 \\ 1 \end{pmatrix}.$$

This is consistent with what we found in Example 9.27.

We have illustrated in the last few examples how to find the inverse of a $1 - 1$ function $y = f(x)$. We solve for x in terms of y. Thus, y becomes the independent variable and x the dependent variable. Since there is a reversal of roles, *it also follows that the domain of the inverse function is the range of the original function and the range of the inverse function is the domain of the original function.*

If $y = f(x)$ has an inverse, we can denote the inverse function by $x = f^{-1}(y)$ if we wish. This notation emphasizes that (a) the inverse function is a function of y and (b) that y is now the independent variable. It is probably easier though to denote the inverse of $y = f(x)$ by $x = g(y)$ where it is understood that we are solving for x in terms of y.

9.7.5 Graphing the Inverse Function

In the previous paragraph we stated that, if $y = f(x)$, the inverse function, g, is a function of y. You might wonder what the graph of the inverse function looks like in relation to the graph of $f(x)$. When we graph $y = f(x)$ where x and y are real, we always take the horizontal axis to be the axis representing the independent variable, and the vertical axis representing the dependent variable. Thus, when graphing the original function $y = f(x)$, the horizontal axis is the x-axis. But, when graphing the inverse function, the horizontal axis should be the $y-$ axis since, in the inverse function, the y variable is now the independent variable. This point is never made in secondary school texts. So, let us take the time to elaborate on it now. In Figure 9.31 below, you see the graph of $y = x^3$ with the horizontal axis labeled "x."

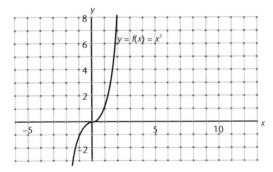

Figure 9.31

In Figure 9.32 you see the graph of the inverse function $x = \sqrt[3]{y}$ or $y^{1/3}$ with the horizontal axis labeled "y."

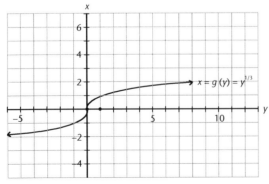

Figure 9.32

If we take the ordered pair approach to functions, then the original function consists of ordered pairs, (x, y) where x is the independent variable and y is the dependent variable. When dealing with the inverse function, the y value is the independent variable and the x value is the dependent variable. Thus, when graphing the inverse function, we are really graphing the points (y, x). Now one can verify that the points $(2, 3)$ and $(3, 2)$ are reflections of each other about the line $y = x$ as are the points $(4, 3)$ and $(3, 4)$. In general the points (x, y) and (y, x) are reflections of each other about the line $y = x$. *Thus, the graph of the inverse function is the reflection of the graph of the original function about the line $y = x$.*

Let us return to the graphs of $y = x^3$ and the inverse function $x = y^{(1/3)}$. Yes, the graphs of these are reflections of each other about the line $y = x$ *as long as the axes are labeled correctly. That is, the horizontal* axis is x for the original function and y for the inverse function. In the graph below you see the graphs of both functions together with $y = x$. Notice how we have labeled the axes in the picture below (Figure 9.33).

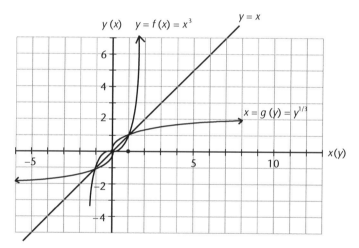

Figure 9.33

For the original function, we have the x- and y-axes as the horizontal and vertical axes, respectively. For the inverse function we have the y- and x-axes as the horizontal and vertical axes, respectively, except that these letters are in parentheses.

Thus, when we teach in secondary school that the graphs of the original function and the inverse are reflections of each other about the line $y = x$ (which looks the same whether the horizontal axis is the x-axis or y-axis) we mean IF the axes are labeled correctly. Some people may feel that this is too complex. So they simply label the horizontal axis the x-axis, and then when it is time to graph the inverse function, they simply switch the variables as is shown in the example below.

Example 9.31 *Find the inverse of $y = x^3$ and then graph the function and its inverse on the same set of axes.*

Solution given in most secondary school texts: Step 1: Switch the variables to get $x = y^3$. Step 2: Solve for y to get $y = x^{(1/3)}$. Now graph both on the same set of axes as shown below and we immediately see that they are reflections of each other about the line $y = x$ (Figure 9.34).

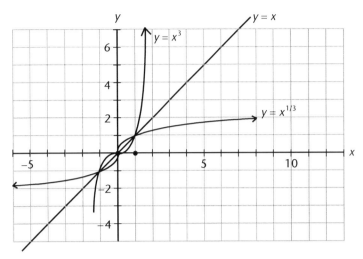

Figure 9.34

Most people are happy doing it this way, as it is mechanical and makes the point. But think about what we just did. We switched the variables! Does that make any physical sense at all? When you are given the relationship between Fahrenheit and Centigrade, $F = \frac{9}{5}C + 32$, and you want to find the inverse function, $(C = \frac{5}{9}(F - 32))$, does it make sense to switch the variables and call centigrade Fahrenheit and Fahrenheit Centigrade? They are completely different things! If you did, then instead of $C = \frac{5}{9}(F - 32)$, which is a correct relationship, you would get $F = \frac{5}{9}(C - 32)$ which is false! Similarly, if you are dealing with a problem involving pressure (P) and volume (V) of a gas in a container, does it make sense to switch the variables and use volume as pressure and pressure as volume? Of course not! In practical problems, you *never* switch the variables when finding the inverse function. Never! So this is why we feel it is unfortunate that this procedure of switching the variables is embedded in textbooks. The way around it is to simply tell your students that, when you are graphing the inverse function, the horizontal axis stands for what was originally y.

The fact that the graph of the inverse function is the reflection of the original function about the line $y = x$ has important consequences that are rarely addressed in secondary school. Specifically: (1) If the original function is continuous, then so is its inverse. (2) When we plot the points of the original function, we are plotting points (x, y), and when we plot the inverse function, we are plotting points (y, x). If we pick any two points (x_1, y_1) and (x_2, y_2) on the graph of the original function and connect them with a line l_1, and then pick their corresponding reflected points (y_1, x_1) (y_2, x_2) on the inverse function, and connect them with a line l_2, then the slopes of l_1 and l_2 are reciprocals. From this, using the notion of limit from calculus, it follows that the slopes of any tangent line to a point on the original function, and the corresponding tangent line (at the reflected point) on the inverse function, are reciprocals. Since the derivative of the function is the slope of the tangent line, this gives us an intuitive way of proving the important calculus result that the derivative of the inverse function is equal to the reciprocal of the derivative of the original function. This can be used to prove many theorems in calculus regarding derivatives. It is this type of depth of understanding that secondary school teachers need to have to be able to demonstrate the links between algebraic concepts and the calculus to their students, especially when they teach calculus.

Student Learning Opportunities

1 Show that any line graph that is not vertical or horizontal represents a 1–1 function.

2 (C) A student asks you why quadratic functions are never 1–1. What is your explanation?

3 (C) One of your students understands how to use the horizontal and vertical line tests to determine if a function is 1–1, but she just doesn't understand why it works. How can you help her understand this?

4 (C) One of your students claims that, unless a function is 1–1 it does not have an inverse. Is she correct? Why or why not?

5 Determine if the following are functions of x, and if they are determine if they are 1–1. Explain your answer.

(a) The distance an object falls after x seconds have elapsed, if it is dropped from a height of 100 feet and hasn't yet hit the floor

(b) The relationship $y = x^3 - 4x^2 + 6x$

(c) $\{(1, 2), (1, 3), (1, 4)\}$ where the first coordinate is x

(d) $\{(2, 1), (3, 1), (4, 1)\}$ where the first coordinate is x

(e)

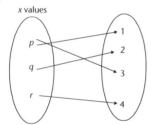

Figure 9.35

(f)

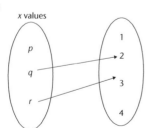

Figure 9.36

(g)

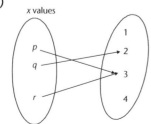

Figure 9.37

6 Show that the points $P = (2, 3)$ and $Q = (3, 2)$ are reflections (that is, mirror images) of each other about the line $y = x$. That is, show that the line $y = x$ is the perpendicular bisector of the line joining P and Q.

7 Find the inverse of each of the following functions. Don't switch the variables.
 (a) $y = 4x + 1$
 (b) $r = 1 + e^{3s}$
 (c) $p = q^5 + 3$
 (d) $t = \log(1 - 3x)$

8 Write the inverse of each of the following matrix transformations, and then find $f^{-1}(Y)$ where $Y = \begin{pmatrix} 2 \\ 1 \end{pmatrix}$. If an inverse function does not exist, say so.
 (a) $Y = \begin{pmatrix} 1 & 5 \\ 3 & 9 \end{pmatrix} \begin{pmatrix} x \\ y \end{pmatrix}$
 (b) $Y = \begin{pmatrix} 4 & 3 \\ 3 & 2 \end{pmatrix} \begin{pmatrix} x \\ y \end{pmatrix}$
 (c) $Y = \begin{pmatrix} 4 & 8 \\ 1 & 2 \end{pmatrix} \begin{pmatrix} x \\ y \end{pmatrix}$

9 Graph each of the following functions as well as their inverses and show, using a graphing calculator, that they are reflections of each other about the line $y = x$. (On some graphing calculators, to graph both on the same set of axes, you will need to switch the variables. So for this application, graphing, it is reasonable to switch the variables.)
 (a) $y = 2x + 1$
 (b) $y = \log_2 x$
 (c) $y = e^x$
 (d) $y = x^3 - 1$

CHAPTER 10

GEOMETRIC TRANSFORMATIONS

10.1 Introduction

In the previous chapter we gave a very general definition of function, along with several examples that were not numerical valued. Let us briefly review some of those examples. (For more details you should read Sections 9.2 and 9.3 of Chapter 9.) Suppose that, with each point on the earth's surface, there is associated the ordered pair of numbers (latitude, longitude). This is a function from the points on the earth's surface to ordered pairs of numbers. As another example, when we assign a set of workers to tasks, the assignment is a function from the set of workers to the set of tasks assigned. In this function each worker is associated with the task he or she gets. Thus, this function associates objects with objects.

In general, if A and B are any sets, then a function from A to B is a rule which associates with each element of the set A, one and only one element of the set B. Thus, in the first example we mentioned, A is the set of points on the earth's surface, while B is the set of ordered pairs of numbers consisting of latitude and longitude. With each element of A (that is, with each point on the earth's surface), we associate one and only one element of B (one and only one latitude and longitude pair). In the second example, A is the set of workers being assigned, and B is the set of tasks to which they are assigned. With each element of A (with each worker), one and only one element of B (one and only one task) is assigned.

The only real exposure secondary school students get to this more general concept of function is when they study transformations within the coordinate plane. That is, they study translations, rotations, reflections, and dilations. Typically, they are given a point such as (1, 2), and asked to find its image under a specified translation, rotation, reflection, or dilation. Transformations of this type are exceedingly useful in real-life applications. When you play a game on a computer, and you see an object moving, its motion is really a combination of several different kinds of transformations consisting of translations, rotations, reflections, and dilations. Thus, computer animation is one application of how one uses transformations to give the appearance of motion. Actually, what the computer does to accomplish this is a combination of a large number of matrix multiplications and additions.

Another application would be the design of robots which, for example, are used in industry and medicine to perform tasks that humans would find tedious and even impossible. Specifically, robots are used in building cars and a variety of manufacturing processes as well as in microscopic surgery. Every robotic movement can be described by some transformation or composition of transformations that is represented by matrices and programmed into the computer. In this chapter we describe how motion can be represented in these ways.

10.2 Transformations: The Secondary School Level

**LAUNCH**

Below you will see a section of the well known M.C. Escher's Symmetry Drawing E45. As you view this drawing, you will probably see examples of many different geometric transformations. On tracing paper, trace the central angel on the picture and keep your tracing paper where it is. Now, do the following:

1. See if you can hold your pencil at some point on the traced angel and turn the paper so the new position of the traced angel exactly overlaps another angel. How many degrees do you think you turned the paper? Where else could you have done this on the drawing? Explain. What type of transformation is this?
2. Holding the tracing paper on top of the original picture so that your traced angel exactly overlaps an angel, fold the tracing paper so that your tracing of the angel overlaps exactly with one of the original angels. Can you find another angel in the picture where you can do a similar process? Explain. What type of transformations are these?
3. Assuming that the picture continues as shown to the right and to the left, is it possible to slide an angel across the drawing so that it overlaps with another angel? Is there more than one way to do this? Explain. What type of transformation is this?
4. In the smaller picture of Escher's Symmetry E45 drawing, what has changed? What has remained the same? What type of transformation is used to go from one picture to the other (Figures 10.1, 10.2)?

Figure 10.1 M.C. Escher's "Symmetry Drawing E45". Copyright 2009 The M.C. Escher Company-Holland. All rights reserved.

Geometric Transformations   453

Figure 10.2 M.C. Escher's "Symmetry Drawing E45". Copyright 2009 The M.C. Escher Company-Holland. All rights reserved.

After having explored the many transformations that exist in Escher's drawing, you probably have a pretty good idea of the types of transformations that will be discussed in this section. As you will see, transformations are not only found in art, but also in all areas of our lives.

10.2.1 Basic Ideas

The most basic transformation is the **translation**. In this transformation, we essentially slide an object to another position. For example, below we see the picture of an octagon. If we imagine it made out of metal, so that it is rigid, and then move the whole thing in the direction of the arrow, we have a picture of the octagon after a translation (Figure 10.3).

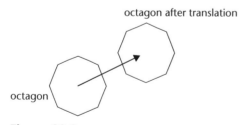

Figure 10.3

The arrow shows us the direction in which we moved the object, as well as how far we moved it. When we ride in an elevator, the elevator translates us from our starting floor to our final floor.

The next transformation included in the secondary school curriculum is **reflection in a line**. Here we are given a line l which acts as a mirror. We simply take the "mirror image" of the object with respect to our mirror l and that is called the reflection of the object with respect to l or about l. Thus, in the figure below, we have reflected the letter E about the line l (Figure 10.4).

454 Geometric Transformations

Figure 10.4

Notice how its mirror image is reversed.
Here is a reflection that teachers might find quite meaningful (Figure 10.5):

Figure 10.5 Copyright © 2009 Scott Kim, scottkim.com. All rights reserved.

Technically speaking, when we reflect a point P about a line l, we drop a perpendicular from P to l and then extend it the same distance past l to P'. P' is the reflection of P about l. Figure 10.6 below shows this. The length OP must be the same as the length of OP'.

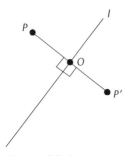

Figure 10.6

Next, there is **rotation** of a figure about a point O. The easiest way to think of rotating a point P about a point O, say 60 degrees counterclockwise, is to imagine that O is the center of a circle with radius OP. Draw the circle, and using OP as one side of a central angle, draw a central angle of $60°$, where the other side of the central angle is counterclockwise from P. The point P' where the other side of the angle intersects the circle is the rotation of P about O at an angle of $60°$. (See Figure 10.7 below.)

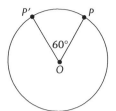

Figure 10.7

If the direction along the circle from P to P' is counterclockwise, we say that we have rotated P 60° counterclockwise about O or that we have just performed a rotation of 60°. When the rotation is clockwise, we either say the rotation is a clockwise rotation of P about O or that P has been rotated −60°. In general, when we rotate a figure, we rotate every point on that figure the same number of degrees. Below we see what happens when we rotate the letter F counterclockwise an angle of 60 degrees. Here, every single point on the letter "F" in Figure 10.8 is rotated 60 degrees.

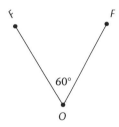

Figure 10.8

Finally, there is the **dilation**. In graphics programs this is known as a "stretch" or shrink." Here we stretch or shrink the figure evenly in all directions, and what we get is a figure similar to the original figure. That is, it has the same shape, but not the same size. When you take a picture of an object, the image you get is a dilation of the object. In the smaller picture in Figure 10.9 below you see a picture of a face and the larger picture is a dilation of the face.

Figure 10.9

Technically speaking, every dilation has a center of dilation. Thus, when we dilate a point P by a factor of k > 0, where the center of dilation is O, we draw OP, and continue the line to OP' so that the length of OP' = k times the length of OP. (See Figure 10.10 below.)

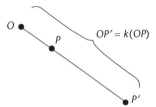

Figure 10.10

These are the basic four transformations studied in secondary school. There are more transformations that are used in practice, and we will discuss some of them later when we get to the higher level. For now, we will describe how mathematicians represent these transformations.

Student Learning Opportunities

1 (C) Your students make the following comments. In each case state whether what they have said is correct.

   (a) Under a translation, no points remain where they are.
   (b) If we reflect an object touching a line *l* about *l*, all points of the object on *l* are fixed.
   (c) If we rotate a figure about a point P, we get a congruent figure.
   (d) If we dilate a figure using a point P as the center of dilation, we get a similar figure.
   (e) Suppose we have a triangle ABC and we reflect it about a line *l* to get a new triangle, A'B'C' where A', B', and C' are the images of A, B, and C, respectively. If we have to travel clockwise to go from A to B to C in a triangle, then we have to travel clockwise to go from the image points A' to B' to C'.

2 In the following diagrams, the image of a figure is drawn. In each case determine if the image is correct. If it isn't, draw the correct picture. Check your work with tracing paper if possible.

   (a) Figure 10.11 (Translation in the direction of the arrow. The lower triangle is the original and the upper triangle is the image.)

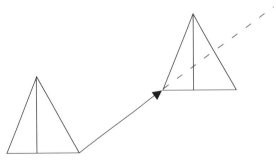

Figure 10.11

(b) Figure 10.12 (Reflection in a line. The word on the left is the original figure and the word on the right is the image.)

Figure 10.12

(c) Figure 10.13 (Rotation of 90 degrees counterclockwise about P. The figure on the top is the original and the figure on the bottom is the image.)

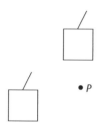

Figure 10.13

(d) Figure 10.14 (Dilation of the letter A on the lower left by a factor of $\frac{1}{2}$ where the center of dilation is P.)

Figure 10.14

3 (a) If we reflect the point (a, b) about the x-axis, what are the coordinates of the image point?
(b) If we reflect the point (a, b) about the y-axis, what are the coordinates of the image point?
(c) If we reflect the point (a, b) about the line $y = x$, what are the coordinates of the image point?

10.3 Bringing in the Main Tool – Functions

LAUNCH

To create the many transformations that exist in his drawing of Angels and Devils, Escher must have used very specific measurements. He must have known a great deal about the mathematics of transformations in order to do that. Let us now examine a very simplified version of his drawing and determine the specific mathematical representations for each of the transformations that exist in this diagram (Figure 10.15).

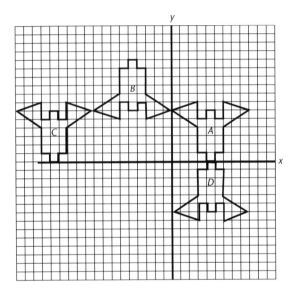

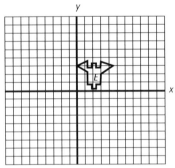

Figure 10.15

In order to give a frame of reference for units of measurement, we have superimposed a grid on the picture and its smaller image. Use this grid to help you respond to the following questions:

1. How many degrees, and about which point, must you rotate angel A to get to angel B?
2. Give the equations of two lines of reflection that you notice in the graph.
3. Describe the magnitude and direction of the translation from angel A to angel C.
4. Describe how much larger angel *A* is than angel *E* in the second graph.

After having completed this launch, you probably have a pretty good idea of how to express transformations in a mathematical way. As you can well imagine, such mathematical representations of transformations are essential for mathematicians. Actually, the simplest way for a mathematician to discuss the transformations described above is not by what they do to the whole figure, but by what they do to each point in the figure. As you will soon see, this allows them to apply mathematics to find efficient ways of rotating and translating figures in practice. We will initially only deal with two dimensional concepts, though everything we do can be extended to the real world of three dimensions. Afterwards, we will discuss some of the extensions of this material to three dimensions. The xy- plane will be denoted by R^2, the customary notation. R^2 is simply an abbreviation for the set of ordered pairs, (x, y), where x and y are **Real**, hence the R. Our main concern will be with functions from R^2 to R^2, that is, rules that associate points in the plane with points in the plane. We will call these rules, **transformations**. We use the word transformation, since we will be transforming figures into new figures by transforming all the points in the figure. If under a transformation, a point (x, y) is transformed into another point (z, w), then (z, w) is called the **image** of (x, y) under the transformation. This is consistent with our terminology for functions in general.

The first transformation we discuss is translation. Since it seems clear that moving an entire figure in a certain direction is the same as moving each point in the figure in that direction, we need only define what we mean by the "same direction" and describe what "direction" means. One way of doing that is to imagine an arrow pointing in the direction we want to move. That arrow is called a vector, and every vector has two components, a horizontal one, h, and a vertical one, k. The vector, or equivalently, the horizontal and vertical components of the vector (how far the object moves horizontally and vertically) describe the direction representing our translation. In Figure 10.16 we see a vector with horizontal component h and vertical component k.

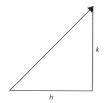

Figure 10.16

Suppose we want to move each point (x, y) in the plane in the direction of this arrow. That means that we move the x coordinate of each point the same distance, h, in the horizontal direction, and the y coordinate the same distance, k, in the vertical direction. That is (x, y) is mapped into, or transformed into the point $(x + h, y + k)$. Or, equivalently, $(x + h, y + k)$ is the image of (x, y) under this transformation. Figure 10.17 below shows this.

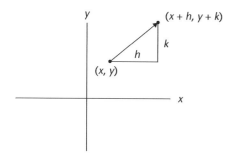

Figure 10.17

In symbols this transformation is often denoted by $T_{(h, k)}$ and we denote the image of any point (x, y) under the transformation as $T_{(h, k)}(x, y)$. Thus, $T_{(h, k)}(x, y) = (x + h, y + k)$. The subscript (h, k) tells us the x coordinate is increased by h and the y coordinate is increased by k. (Of course when h or k is negative, the respective coordinate "increase" really amounts to a decrease.) A translation is a function from R^2 to R^2, as are all the other transformations we discuss in the main body of this section. A typical secondary school question is:

Example 10.1

(a) *Find the image of the point (1, 4) under the translation* $T_{(-3, 2)}$.
(b) *Find the image of the triangle whose vertices are* $A = (-1, 2)$, $B = (4, 7)$, *and* $C = (0, 6)$ *under the translation* $T_{(-3, 2)}$.

Solution. (a) $T_{(-3, 2)}(1, 4) = (1 + -3, 4 + 2) = (-2, 6)$.

(b) At first glance, the solution seems simple. We just find the image of each of the three points separately and then connect them. Although that can be done easily, there are some issues that we have to address. One is: How do we know that the image of the triangle is still a triangle? Our intuition says it must be. But does the mathematics correctly capture this? (See Student Learning Opportunity 10 in this regard.) For now, we just accept that translations map triangles to triangles. We will return to that later in Section 10.5.2. Thus, the image of our triangle has vertices $A' = T_{(-3, 2)}(-1, 2)$, $B' = T_{(-3, 2)}(4, 7)$, and $C' = T_{(-3, 2)}(0, 6)$. And this yields $A' = (-4, 4)$, $B' = (1, 9)$, and $C' = (-3, 8)$.

Below in Figure 10.18 we see the original triangle together with its image.

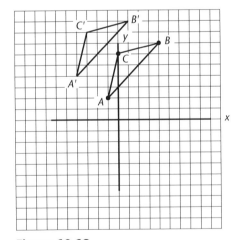

Figure 10.18

As we can see, the triangle has been translated parallel to itself, and every point of the triangle has been moved 3 to the left and 2 up.

The next type of transformation we discuss is the reflection. Initially, we will deal only with the simplest types of reflections, those in the x-axis, the y-axis and the lines $y = x$, and $y = -x$ but will go further in the next section.

Geometric Transformations

We use the letter r to represent reflection. Later on when we do rotations, we will use capital R to denote rotation. Thus, reflection will be denoted by a lower case r and rotation by R. The reflection of a point in the x-axis, which we denote by r_x, is easy to describe. (See Figure 10.19 below.)

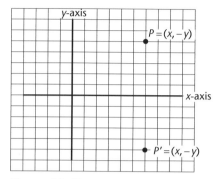

Figure 10.19

We see immediately that, if we reflect a point using the x-axis as a mirror, the image of a point $P = (x, y)$ is the point $P' = (x, -y)$. Thus,

$$r_x(x, y) = (x, -y). \tag{10.1}$$

If we reflect the point (x, y) about the y-axis, and denote that transformation by r_y, we can write this as $r_y(x, y) = (-x, y)$. (Draw a picture and see for yourself.) Thus, the reflection about the x-axis of the point $(1, 2)$ is $(1, -2)$ and the reflection of the point $(1, 2)$ about the y-axis is $(-1, 2)$.

The reflections about the line $y = x$, and $y = -x$ require a bit more thinking. We will obtain formulas for these later from a single result.

The next transformation we describe is the dilation function, with the center of dilation at the origin. The dilation function takes a point and multiplies each coordinate by a real number k. We denote the dilation transformation by D_k. Thus, $D_k(x, y) = (kx, ky)$. This will stretch or shrink the figure by a factor of $|k|$. If $k > 1$ or $k < -1$, the figure is enlarged. If $-1 < k < 1$, the figure is shrunk. Of course, if $k = 1$, then the figure is left unchanged. So the image of $(1, 2)$ under a dilation with a factor of 5 (that is, $D_5(1, 2)$), will be $(5, 10)$ which is 5 times further from the origin than it was before. If this is applied to every point in a figure, every point will be 5 times further away from the origin and the figure will be stretched. If k were -5, each point would still be 5 times as far from the origin as it was before, only in the opposite direction. So the figure would be stretched and flipped about the origin. If k were $1/5$, then each point of the figure would be $1/5$ as far from the origin as it was before, and the figure would be shrunk.

Finally, for now, we discuss rotation about a point O. Initially, O will be the origin. Here each point, P, is rotated about O a certain number of degrees. To find the image of a point P under such a rotation requires the use of trigonometry. Suppose that the starting point P has coordinates (x, y) and that it is a distance r from the origin. Suppose further that the line segment OP makes an angle of α with the positive x-axis. Then using the right triangle in the Figure 10.20 below

462 Geometric Transformations

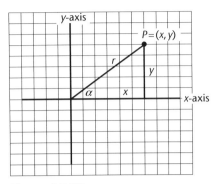

Figure 10.20

we see that

$$\cos \alpha = \frac{\text{adjacent}}{\text{hypotenuse}} = \frac{x}{r} \tag{10.2}$$

and

$$\sin \alpha = \frac{\text{opposite}}{\text{hypotenuse}} = \frac{y}{r} \tag{10.3}$$

and by cross multiplying we get

$$x = r \cos \alpha \tag{10.4}$$

and

$$y = r \sin \alpha. \tag{10.5}$$

We now rotate OP by an angle of θ to get a new point, x', y'. (See Figure 10.21 below where θ is the angle between OP and OP'.)

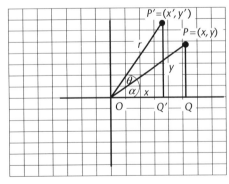

Figure 10.21

Then using triangle $OP'Q'$ whose sides are x' and y', and the figure above, the new point, P' has coordinates (x', y') where

$$x' = r \cos(\alpha + \theta) \tag{10.6}$$

and

$$y' = r \sin(\alpha + \theta). \tag{10.7}$$

But we know from trigonometry that

$$\cos(\alpha + \theta) = \cos\alpha \cos\theta - \sin\alpha \sin\theta \tag{10.8}$$

and that

$$\sin(\alpha + \theta) = \sin\alpha \cos\theta + \cos\alpha \sin\theta \tag{10.9}$$

and when equations (10.8) and (10.9) are substituted into equations (10.6) and (10.7), we get

$$x' = r(\cos\alpha \cos\theta - \sin\alpha \sin\theta) = r\cos\alpha \cos\theta - r\sin\alpha \sin\theta \tag{10.10}$$

and

$$y' = r(\sin\alpha \cos\theta + \cos\alpha \sin\theta) = r\sin\alpha \cos\theta + r\cos\alpha \sin\theta. \tag{10.11}$$

Using the facts that $x = r\cos\alpha$, and $y = r\sin\alpha$ from equations (10.4) and (10.5), and substituting these in equations (10.10) and (10.11), equations (10.10) and (10.11), respectively can be written as

$$x' = r\cos\alpha \cos\theta - r\sin\theta \sin\alpha = x\cos\theta - y\sin\theta$$

and

$$y' = r\sin\alpha \cos\theta + r\cos\alpha \sin\theta = y\cos\theta + x\sin\theta = x\sin\theta + y\cos\theta$$

In summary,

Theorem 10.2 *When we rotate a point (x, y) by an angle of θ, the new coordinates of the point are x' and y' where*

$$x' = x\cos\theta - y\sin\theta \tag{10.12}$$

and

$$y' = x\sin\theta + y\cos\theta. \tag{10.13}$$

Let us illustrate this. Suppose we want to rotate the point $P = (x, y) = (1, 2)$, 90° counterclockwise. Using equations (10.12) and (10.13), we get $x' = 1\cos 90° - 2\sin 90°$ and $y' = 1\sin 90° + 2\cos 90°$. This gives us (since $\sin 90° = 1$ and $\cos 90° = 0$), that $P' = (x', y') = (-2, 1)$. (See Figure 10.22 below.)

464 Geometric Transformations

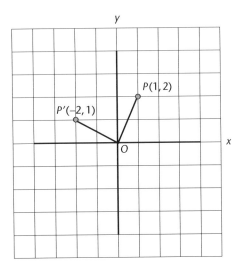

Figure 10.22

We can check that this is correct. After all, if θ is 90°, then the lines OP and OP' must be perpendicular. That means that the slopes of OP and OP' should be negative reciprocals of one another. But the slope of OP where O is the origin, is $2/1$ and the slope of OP' is $1/-2$ and these are negative reciprocals!

As another example, if we took the same point $P = (x, y) = (1, 2)$ and rotated it by an angle of 30 degrees, then the point $P' = (x', y')$ would be $(1 \cos 30° - 2 \sin 30°, 1 \sin 30° + 2 \cos 30°) = (\sqrt{3}/2 - 1, 1/2 + \sqrt{3})$ since $\sin 30° = 1/2$ and $\cos 30° = \sqrt{3}/2$.

We are now ready to examine transformations from a somewhat higher level.

Student Learning Opportunities

1 Find the image of each of the points under the translation $T_{(-1,\ 2)}$. Then find the reflection of the original points about the x-axis and y-axis.

(a) (2, 4)

(b) (−3, 5)

(c) (7, 6)

2 Find the image of the point (−2, 3) after a rotation of 30 degrees about the origin.

3 Find the image of the point (5, 9) under a rotation of −60 degrees about the origin.

4 Find the image of the point (2, 5) after a rotation of 90 degrees about the origin. Then rotate the image 30°. Show that the result is the same as rotating the original point, (2, 5) 120° about the origin. Use the rotation formulas from this section to do this problem.

5 Find the image of a triangle whose vertices are $A(2, 4)$, $B(4, 2)$, and $C(6, 8)$ under a dilation with a factor of 1/2. Draw the original triangle ABC and its dilated image, triangle $A'B'C'$ on a coordinate grid. Compare the size and shape of triangle ABC and its image, triangle $A'B'C'$.

6 If $T_{(h, k)}(4, 2) = (7, -5)$, what point is (h, k)?

7 (C) One of your students asks, "If you are given two transformations, does it matter which one you do first?" How do you reply? Does it matter? Support your answer.

8 (C) A student has a suspicion that, if we reflect a point (x, y) about the x-axis and then take the result and reflect it about the y-axis, that we have rotated the point a total of 180 degrees about the origin. How can you use an algebraic argument to help the student understand why this really is the case?

9 The image of a point (x, y) under a rotation about the origin by an angle of 90 degrees yields the point $(2, 3)$. What point is (x, y)?

10 Consider the transformation T such that $T(x, y) = (0, 0)$ for all (x, y). What is the image of any figure under this transformation?

11 (C) You give your students triangle ABC where to travel from A to B to C we must go counterclockwise. You ask them to reflect this triangle about a line l, which doesn't touch the triangle and ask them to determine if the image triangle, $A'B'C'$ also has the property that to go from A' to B' to C' we must travel counterclockwise. Answer the question. Then make a statement about the preservation of orientation of the vertices of a triangle after a reflection.

12 (C) How would you help your students understand that translations preserve distance? Specifically, show using the distance formula in the Cartesian plane, that if A and B are two points, and we translate each of them using $T_{(h, k)}$, then the image points A' and B' are the same distance from each other as A and B are.

13 (C) How would you help your students understand that reflections preserve distance? Specifically, show, using congruent triangles, that if A and B are two points not on a line l, and A' and B' are their reflections about l, then the distance from A to B is the same as the distance from A' to B'.

14 (C) How would you help your students understand that reflections preserve angles between lines? Specifically, show that, if triangle ABC does not touch or cross a line l, and we reflect it about l, the image triangle $A'B'C'$ that we get is congruent to ABC and hence the angles ABC and $A'B'C'$ are congruent.

15 (C) How would you help your students understand that rotations preserve length? Specifically, using congruent triangles, show that, if AB is rotated about a point P, then the image points A' and B' are the same distance from each other as A and B are.

16 Derive the formula which rotates a point P about a point (a, b) where (a, b) is not the origin. [Hint: Translate both points simultaneously so that (a, b) is at the origin, then do the rotation, and then translate both points simultaneously so that the rotation point, (a, b), is back where it started.] Using the same idea, tell how to dilate a figure with center at a point other than the origin.

10.4 The Matrix Approach – a Higher Level

LAUNCH

1 Perform the following matrix operations:

(a) $\begin{pmatrix} 1 & 0 \\ 0 & -1 \end{pmatrix} \begin{pmatrix} x \\ y \end{pmatrix} = \begin{pmatrix} \\ \end{pmatrix}$

(b) $\begin{pmatrix} -1 & 0 \\ 0 & 1 \end{pmatrix} \begin{pmatrix} x \\ y \end{pmatrix} = \begin{pmatrix} \\ \end{pmatrix}$

(c) $\begin{pmatrix} k & 0 \\ 0 & k \end{pmatrix} \begin{pmatrix} x \\ y \end{pmatrix} = \begin{pmatrix} \\ \end{pmatrix}$

(d) $\begin{pmatrix} \cos\theta & -\sin\theta \\ \sin\theta & \cos\theta \end{pmatrix} \begin{pmatrix} x \\ y \end{pmatrix} = \begin{pmatrix} \\ \end{pmatrix}$

2 If we consider $\begin{pmatrix} x \\ y \end{pmatrix}$ as the ordered pair (x, y), what transformations are being performed on the point (x, y) in examples *a*, *b*, *c*, and *d*?

After having done the launch questions, you might be wondering why we have asked you to represent transformations with matrices. You will learn more about how to use matrices and learn about their value when discussing transformations as you read on.

10.4.1 Reflections, Rotations, and Dilations

In this section we expand our study of transformations by examining them from a matrix point of view. Because of the many advantages of using the matrix approach, the latest secondary school texts have been using matrices as well.

What are these advantages you might ask? For one, the matrix approach makes computer implementation of animation easy (since this requires many transformations and consequently many matrix multiplications, which computers can do effortlessly). Second, and perhaps most important, the matrix approach in this section generalizes to three dimensions and thus gives us the ability to apply this material to many real-life situations that use 3 dimensional motion. For example, in practical applications with robotics, we use transformations and the matrices related to them to get the robots to move the way we wish. Third, there is a certain beauty to the approach, as it allows us to connect different areas of mathematics such as, geometry, functions, matrix manipulations, trigonometry, and algebra.

In Chapter 7 we used complex numbers to discuss certain transformations and that is quite a beautiful approach to this subject area. However, complex numbers are a 2 dimensional idea, and won't accomplish our need to work in three dimensions. Matrices, on the other hand, will. Finally, the matrix approach gives us information that we just cannot get otherwise. Thus, matrices not only give us a different perspective on rotations, reflections, and translations, but have genuine advantages over other approaches. And, by the way, we hope you will appreciate how so many branches of mathematics (geometry, functions, matrix operations, trigonometry, and algebra) work together to accomplish the most complex transformations.

Normally, points in the plane are written as (horizontal) ordered pairs (x, y), but since we will need matrix operations, and to multiply matrices, the matrices have to be the right sizes, we will often need to write our ordered pairs vertically as $\begin{pmatrix} x \\ y \end{pmatrix}$. Initially, we will be dealing only with 2×2 matrices, and we recall that, when we multiply a 2×2 matrix $\begin{pmatrix} a & b \\ c & d \end{pmatrix}$ by a 2×1 matrix $\begin{pmatrix} x \\ y \end{pmatrix}$, we get a 2×1 matrix which is

$$\begin{pmatrix} ax + by \\ cx + dy \end{pmatrix}.$$

Thus,

$$\begin{pmatrix} 1 & 2 \\ 3 & 4 \end{pmatrix} \cdot \begin{pmatrix} 5 \\ 6 \end{pmatrix} = \begin{pmatrix} 1 \cdot 5 + 2 \cdot 6 \\ 3 \cdot 5 + 4 \cdot 6 \end{pmatrix} = \begin{pmatrix} 17 \\ 39 \end{pmatrix}.$$

(See Appendix for a review of matrix multiplication.)

Let us go back to the reflection transformation, r_x which reflects every point about the x-axis. That transformation is given by $r_x(x, y) = (x, -y)$. Using our convention that we will write ordered pairs vertically, this can be written as

$$r_x \begin{pmatrix} x \\ y \end{pmatrix} = \begin{pmatrix} x \\ -y \end{pmatrix}.$$

The way to write this using matrix multiplication is:

$$r_x \begin{pmatrix} x \\ y \end{pmatrix} = \begin{pmatrix} x \\ -y \end{pmatrix} = \begin{pmatrix} 1 & 0 \\ 0 & -1 \end{pmatrix} \cdot \begin{pmatrix} x \\ y \end{pmatrix} \tag{10.14}$$

as you can verify by multiplying the two matrices on the right.

Similarly, r_y, the reflection about the y-axis is given by:

$$r_y \begin{pmatrix} x \\ y \end{pmatrix} = \begin{pmatrix} -x \\ y \end{pmatrix} = \begin{pmatrix} -1 & 0 \\ 0 & 1 \end{pmatrix} \cdot \begin{pmatrix} x \\ y \end{pmatrix}. \tag{10.15}$$

The dilation of a point by a factor of k can be written as

$$D_k \begin{pmatrix} x \\ y \end{pmatrix} = \begin{pmatrix} kx \\ ky \end{pmatrix} = \begin{pmatrix} k & 0 \\ 0 & k \end{pmatrix} \cdot \begin{pmatrix} x \\ y \end{pmatrix}. \tag{10.16}$$

Finally, the rotation transformation R_θ (notice the capital R) that rotates every point about the origin an angle θ (where θ is always measured from the positive x-axis and θ is positive when we rotate counterclockwise and negative when we rotate clockwise) can be written as

$$R_\theta \begin{pmatrix} x \\ y \end{pmatrix} = \begin{pmatrix} x \cos \theta - y \sin \theta \\ x \sin \theta + y \cos \theta \end{pmatrix} = \begin{pmatrix} \cos \theta & -\sin \theta \\ \sin \theta & \cos \theta \end{pmatrix} \cdot \begin{pmatrix} x \\ y \end{pmatrix} \tag{10.17}$$

which follows from equations (10.12) and (10.13).

In each of the transformations we have given above, the image of a point after the transformation can be obtained by multiplying the coordinates of the point by a matrix. This matrix is called the **matrix of the transformation**. Thus, the matrix of the rotation transformation from equation (10.17) is

$$\begin{pmatrix} \cos\theta & -\sin\theta \\ \sin\theta & \cos\theta \end{pmatrix} \tag{10.18}$$

and the matrix of the reflection about the y-axis transformation from equation (10.15) is

$$\begin{pmatrix} -1 & 0 \\ 0 & 1 \end{pmatrix} \tag{10.19}$$

while the matrix of the transformation that dilates a point by a factor of k by equation (10.16) is

$$\begin{pmatrix} k & 0 \\ 0 & k \end{pmatrix}. \tag{10.20}$$

Let us illustrate some of these ideas with one example.

Example 10.3 *We wish to rotate the three points with coordinates $A = (1, 2)$, $B = (-1, 3)$, and $C = (-2, -4)$, $90°$ counterclockwise about the origin. (a) Find the image of each point after this counterclockwise rotation, and then (b) describe a way of rotating all three points $90°$ at once.*

Solution. The matrix that we need to perform the rotation by equation (10.17) is

$$R_{90°} = \begin{pmatrix} \cos 90° & -\sin 90° \\ \sin 90° & \cos 90° \end{pmatrix} = \begin{pmatrix} 0 & -1 \\ 1 & 0 \end{pmatrix}. \tag{10.21}$$

Thus, the images A', B', and C' of the points $A = (1, 2)$, $B = (-1, 3)$, and $C = (-2, -4)$, respectively, are $A' = (-2, 1)$, $B' = (-3, -1)$, and $C' = (4, -2)$ obtained as follows:

$$A' = R_{90}\begin{pmatrix} 1 \\ 2 \end{pmatrix} = \begin{pmatrix} 0 & -1 \\ 1 & 0 \end{pmatrix} \cdot \begin{pmatrix} 1 \\ 2 \end{pmatrix} = \begin{pmatrix} -2 \\ 1 \end{pmatrix},$$

$$B' = R_{90}\begin{pmatrix} -1 \\ 3 \end{pmatrix} = \begin{pmatrix} 0 & -1 \\ 1 & 0 \end{pmatrix} \cdot \begin{pmatrix} -1 \\ 3 \end{pmatrix} = \begin{pmatrix} -3 \\ -1 \end{pmatrix}, \quad \text{and}$$

$$C' = R_{90}\begin{pmatrix} -2 \\ -4 \end{pmatrix} = \begin{pmatrix} 0 & -1 \\ 1 & 0 \end{pmatrix} \cdot \begin{pmatrix} -2 \\ -4 \end{pmatrix} = \begin{pmatrix} 4 \\ -2 \end{pmatrix}$$

where we have written our points vertically to do the matrix multiplications. A quick way of finding the images A', B', and C' is to put the three points A, B, and C we want rotated in a matrix whose columns are A, B, and C, and then multiply that matrix by the rotation matrix. Thus, all our work becomes much shorter if we simply multiply

$$\begin{pmatrix} 0 & -1 \\ 1 & 0 \end{pmatrix} \cdot \begin{pmatrix} 1 & -1 & -2 \\ 2 & 3 & -4 \end{pmatrix}$$

to get

$$\begin{pmatrix} -2 & -3 & 4 \\ 1 & -1 & -2 \end{pmatrix}. \tag{10.22}$$

Notice that the columns of this new matrix in (10.22) are A', B', and C' but written vertically.

This same method works for all transformations. That is, we can find the image of *all* points in a figure under a transformation by letting the coordinates of the points of the figure be the columns of a matrix, and then multiplying that matrix by the matrix of the transformation. This is the way computers do it.

You might be thinking, "But a figure has infinitely many points on it. How does a computer multiply by an infinite matrix?" The answer is that, in a computer's memory, figures are represented by a finite number of points, though that finite number may be large. Each point on the computer screen takes up a tiny amount of space called a pixel, and each pixel has a set of coordinates. Each point on a curve takes up one or more of these pixels. Thus, as far as the computer is concerned, the curve itself consists of only a finite number of pixels.

10.4.2 Compositions of Transformations

As you can imagine, complex computer animations, as well as other real-life applications, require the use of multiple transformations, which when put together form what is called a composition of transformations. Recall that in secondary school we studied the **composition**, $(f \circ g)(x)$, of two functions often abbreviated $f \circ g$. This was defined as $f(g(x))$. That is, first we apply g to x to get $g(x)$, (the "inner" function) and then we apply f to the result, to get $f(g(x))$. So, for example, if $f(x)$ is the function x^2 and $g(x)$ is the function $2x - 1$, then $(f \circ g)(3) = f(g(3))$ is obtained by first computing $g(3) = 2(3) - 1 = 5$, and then computing $f(g(3)) = f(5) = 5^2 = 25$.

The transformation of compositions works in exactly the same way. That is, we can compose two transformations, R and S to get a new transformation $R \circ S$. We interpret the composition the same way, namely, we do the inner one, S, first, and then follow it by the outer one, R. So if we wanted to say, dilate a point $(2, 3)$ by a factor of 5, and then rotate it $90°$, we would be interested in first performing $D_5(2, 3)$ to get $(10, 15)$ and then following it with $R_{90}(10, 15)$, which by equation (10.21) is

$$\begin{pmatrix} 0 & -1 \\ 1 & 0 \end{pmatrix} \begin{pmatrix} 10 \\ 15 \end{pmatrix} = \begin{pmatrix} -15 \\ 10 \end{pmatrix}.$$

Our first theorem discusses the results of composing two translations. Intuitively, if we translate an object twice, we have still, ultimately, translated it. To illustrate, if we wanted to compute $T_{(5,\ 6)} \circ T_{(-2,\ 4)})$ on a point (x, y), the first function would translate the point by the vector $(-2, 4)$ and then the second one would take the result and translate it by the vector $(5, 6)$. So

$(T_{(5,\ 6)} \circ T_{(-2,\ 4)})(x, y)$

$= (T_{(5,\ 6)}(T_{(-2,\ 4)}(x, y))$

$= T_{(5,\ 6)}(x - 2, y + 4)$

$= (x - 2 + 5, y + 4 + 6)$

$= (x + 3, y + 10)$

470 Geometric Transformations

and we observe that the result here is $T_{(3,\ 10)}(x,\ y)$. That is, the net result of composing these two translations, is that we get yet another translation, $T_{(3,\ 10)}$. We can state this more generally as a theorem.

Theorem 10.4 *The composition of two translations is a translation. More specifically* $(T_{(a,\ b)} \circ T_{(c,\ d)}) = T_{(a+c,\ b+d)}$.

It is also geometrically obvious that, when you rotate a point by an angle θ_1 about the origin, and then rotate the result by an angle θ_2 about the origin, that we are really rotating the initial point through an angle of $\theta_1 + \theta_2$ about the origin. We state this as a theorem also.

Theorem 10.5 $R_{\theta_2} \circ R_{\theta_1} = R_{\theta_1 + \theta_2}$.

If our matrix representation is going to be useful at all, then we should be able to derive this theorem using matrices. Let us see how the matrix approach accomplishes this.

Rotating a point (x, y) by θ_1 first, and then by θ_2 next, means performing the composition $(R_{\theta_2} \circ R_{\theta_1})$ on (x, y). From the definition of composition,

$$(R_{\theta_2} \circ R_{\theta_1}) \begin{pmatrix} x \\ y \end{pmatrix} = R_{\theta_2} \left(R_{\theta_1} \begin{pmatrix} x \\ y \end{pmatrix} \right).$$

Using equation (10.17), this can be written as

$$R_{\theta_2} \left(\begin{pmatrix} \cos\theta_1 & -\sin\theta_1 \\ \sin\theta_1 & \cos\theta_1 \end{pmatrix} \cdot \begin{pmatrix} x \\ y \end{pmatrix} \right).$$

Using equation (10.17) again, this is the same as

$$\begin{pmatrix} \cos\theta_2 & -\sin\theta_2 \\ \sin\theta_2 & \cos\theta_2 \end{pmatrix} \cdot \left(\begin{pmatrix} \cos\theta_1 & -\sin\theta_1 \\ \sin\theta_1 & \cos\theta_1 \end{pmatrix} \cdot \begin{pmatrix} x \\ y \end{pmatrix} \right)$$

which, by associativity of matrix multiplication (see Appendix), gives us

$$\left(\begin{pmatrix} \cos\theta_2 & -\sin\theta_2 \\ \sin\theta_2 & \cos\theta_2 \end{pmatrix} \cdot \begin{pmatrix} \cos\theta_1 & -\sin\theta_1 \\ \sin\theta_1 & \cos\theta_1 \end{pmatrix} \right) \cdot \begin{pmatrix} x \\ y \end{pmatrix}. \tag{10.23}$$

Multiplying these matrices we get

$$\begin{pmatrix} \cos\theta_2\cos\theta_1 - \sin\theta_2\sin\theta_1 & -\cos\theta_2\sin\theta_1 - \sin\theta_2\cos\theta_1 \\ \sin\theta_2\cos\theta_1 + \cos\theta_2\sin\theta_1 & -\sin\theta_2\sin\theta_1 + \cos\theta_2\cos\theta_1 \end{pmatrix} \cdot \begin{pmatrix} x \\ y \end{pmatrix}$$

which from (10.8) and (10.9) gives us

$$\begin{pmatrix} \cos(\theta_2 + \theta_1) & -\sin(\theta_2 + \theta_1) \\ \sin(\theta_2 + \theta_1) & \cos(\theta_2 + \theta_1) \end{pmatrix}$$

which, of course, is the same as

$$\begin{pmatrix} \cos(\theta_1 + \theta_2) & -\sin(\theta_1 + \theta_2) \\ \sin(\theta_1 + \theta_2) & \cos(\theta_1 + \theta_2) \end{pmatrix}.$$

And this is precisely the matrix

$$(R_{\theta_1 + \theta_2}) \begin{pmatrix} x \\ y \end{pmatrix}.$$

Thus, we see that $(R_{\theta_2} \circ R_{\theta_1}) \begin{pmatrix} x \\ y \end{pmatrix} = (R_{\theta_1 + \theta_2}) \begin{pmatrix} x \\ y \end{pmatrix}$ for all points $\begin{pmatrix} x \\ y \end{pmatrix}$. Hence, the two transformations $(R_{\theta_2} \circ R_{\theta_1})$ and $(R_{\theta_1 + \theta_2})$ are the same. Thus, the composition of two rotations is a rotation.

Wait a minute! We almost missed something! Look at what (10.23) is saying. It is claiming that *when you compose two transformations that can be written in terms of matrices, the matrix of the composition is obtained by multiplying their matrices!!!* How nice! And, if we compose many transformations, we can get the net result instantly by just multiplying the matrices and applying the product to the point! This is important enough to state as a theorem.

Theorem 10.6 *If $T_1, T_2, \ldots, T_n$ are transformations with matrices $M_1, M_2, \ldots, M_n$, then when we compute $(T_n \circ \ldots \circ T_2 \circ T_1) \begin{pmatrix} x \\ y \end{pmatrix}$, we get the same as $(M_n \ldots M_2 M_1) \begin{pmatrix} x \\ y \end{pmatrix}$.*

What this means is that, if you perform T_1 first, then T_2, and so on to T_n, then you multiply the matrices M_1, M_2, and so on to M_n, but in the reverse order.

Let us give a numerical example.

Example 10.7 (a) *We wish to take the point (2, 3) and do the following: (1) First rotate it 30° counterclockwise, then (2) reflect the result about the y-axis. Finally, (3) dilate this new result by a factor of 5. Do this using matrices. (b) Then find a single matrix that accomplishes (1) – (3) for any point (x, y).*

Solution. (a) The matrix, M_1, associated with our first transformation, a rotation by equation 30° is, by equation (10.18),

$$M_1 = \begin{pmatrix} \cos 30° & -\sin 30° \\ \sin 30° & \cos 30° \end{pmatrix}.$$

Then the image of (2, 3) under this transformation is

$$M_1 \begin{pmatrix} 2 \\ 3 \end{pmatrix}$$

$$= \begin{pmatrix} \cos 30° & -\sin 30° \\ \sin 30° & \cos 30° \end{pmatrix} \begin{pmatrix} 2 \\ 3 \end{pmatrix}$$

$$= \begin{pmatrix} \sqrt{3} - \frac{3}{2} \\ 1 + \frac{3}{2}\sqrt{3} \end{pmatrix}.$$

(We are using the well known results that $\cos 30° = \frac{\sqrt{3}}{2}$ and $\sin 30° = \frac{1}{2}$.) The matrix, M_2 associated with our reflection about the y-axis is, by equation (10.19),

$$M_2 = \begin{pmatrix} -1 & 0 \\ 0 & 1 \end{pmatrix}.$$

The image of the point we just got, $\begin{pmatrix} \sqrt{3} - \frac{3}{2} \\ 1 + \frac{3}{2}\sqrt{3} \end{pmatrix}$, under this transformation is

$$M_2 \begin{pmatrix} \sqrt{3} - \frac{3}{2} \\ 1 + \frac{3}{2}\sqrt{3} \end{pmatrix}$$

$$= \begin{pmatrix} -1 & 0 \\ 0 & 1 \end{pmatrix} \begin{pmatrix} \sqrt{3} - \frac{3}{2} \\ 1 + \frac{3}{2}\sqrt{3} \end{pmatrix}$$

$$= \begin{pmatrix} -\sqrt{3} + \frac{3}{2} \\ 1 + \frac{3}{2}\sqrt{3} \end{pmatrix}.$$

Finally, the matrix M_3, dilating a point by a factor of 5 by (10.20) is

$$M_3 = \begin{pmatrix} 5 & 0 \\ 0 & 5 \end{pmatrix}$$

and the image of the most recent point we got, $\begin{pmatrix} -\sqrt{3} + \frac{3}{2} \\ 1 + \frac{3}{2}\sqrt{3} \end{pmatrix}$, under this transformation is

$$M_3 \begin{pmatrix} -\sqrt{3} + \frac{3}{2} \\ 1 + \frac{3}{2}\sqrt{3} \end{pmatrix}$$

$$= \begin{pmatrix} 5 & 0 \\ 0 & 5 \end{pmatrix} \begin{pmatrix} -\sqrt{3} + \frac{3}{2} \\ 1 + \frac{3}{2}\sqrt{3} \end{pmatrix}$$

$$= \begin{pmatrix} -5\sqrt{3} + \frac{15}{2} \\ 5 + \frac{15}{2}\sqrt{3} \end{pmatrix}$$

$$\approx \begin{pmatrix} -1.1603 \\ 17.99 \end{pmatrix}.$$

Solution. (b) Since in the statement of the example we are doing (1), rotating by 30° counterclockwise, followed by (2) reflecting the result about the y-axis, followed by (3) dilating the result by a factor of 5, we can accomplish this all in one step by with the single matrix M obtained by multiplying the matrices M_1, M_2, and M_3 of the respective transformations, in reverse order.

That is, the single matrix that accomplishes all three transformations described in (1)–(3) is $M = M_3 M_2 M_1$ or just

$$\begin{pmatrix} 5 & 0 \\ 0 & 5 \end{pmatrix} \begin{pmatrix} -1 & 0 \\ 0 & 1 \end{pmatrix} \begin{pmatrix} \cos 30° & -\sin 30° \\ \sin 30° & \cos 30° \end{pmatrix}$$

which simplifies to

$$M = \begin{pmatrix} -\frac{5}{2}\sqrt{3} & \frac{5}{2} \\ \frac{5}{2} & \frac{5}{2}\sqrt{3} \end{pmatrix}.$$

This matrix will accomplish (1)–(3) for *any* point. Let us verify that in the case of the point (2, 3). We have:

$$M \begin{pmatrix} 2 \\ 3 \end{pmatrix}$$

$$= \begin{pmatrix} -\frac{5}{2}\sqrt{3} & \frac{5}{2} \\ \frac{5}{2} & \frac{5}{2}\sqrt{3} \end{pmatrix} \begin{pmatrix} 2 \\ 3 \end{pmatrix}$$

$$= \begin{pmatrix} -5\sqrt{3} + \frac{15}{2} \\ 5 + \frac{15}{2}\sqrt{3} \end{pmatrix}$$

$$\approx \begin{pmatrix} -1.1603 \\ 17.99 \end{pmatrix}$$

which is exactly the same result we got earlier. We can, if we wish, express M in decimal form. In that case,

$$M = \begin{pmatrix} -\frac{5}{2}\sqrt{3} & \frac{5}{2} \\ \frac{5}{2} & \frac{5}{2}\sqrt{3} \end{pmatrix} \approx \begin{pmatrix} -4.3301 & 2.5 \\ 2.5 & 4.3301 \end{pmatrix}$$

and if you multiply $\begin{pmatrix} 2 \\ 3 \end{pmatrix}$ by this, we get

$$\begin{pmatrix} -4.3301 & 2.5 \\ 2.5 & 4.3301 \end{pmatrix} \begin{pmatrix} 2 \\ 3 \end{pmatrix}$$

$$\approx \begin{pmatrix} -1.1602 \\ 17.99 \end{pmatrix}.$$

You might feel it is easier to do this example without the matrix approach and you may be right. But then you will be missing the point, namely that this matrix. M, will transform *any* point in the same way that we transformed (2, 3). This is important in computer animation when we want to transform a figure that may have thousands of points in it at once. Rather than multiplying each point by 3 matrices, we multiply each point by 1 matrix which does the work of 3, and this gives us efficiency.

10.4.3 Reflecting about Arbitrary Lines

We already have formulas for the transformations that reflect points about the *x*- and *y*-axes, respectively. (See equations (10.14) and (10.15).) But, what do we do if we need to reflect points about other lines? In this section we discuss how to find a formula which reflects a point about *any* line $\mathcal{L}$, which passes through the origin. We begin by assuming that the line $\mathcal{L}$ makes an angle θ with the positive *x*-axis as shown in Figure 10.23 below.

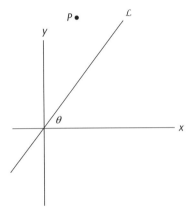

Figure 10.23

The approach here is quite nice. We use the problem-solving technique of reducing this to a simpler problem that we already know how to do. Specifically, rather than reflecting our point P around $\mathcal{L}$, we first rotate the plane through an angle of $-\theta$ bringing $\mathcal{L}$ to where the *x*-axis was and moving P to a new point, P'. (See figure below.) Since we can now consider $\mathcal{L}$ as the *x*-axis, we are able to reflect P' about the new position of $\mathcal{L}$, which we have a formula for. This will give us a point P'' as shown in Figure 10.24 below.

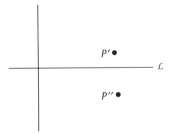

Figure 10.24

Of course, the coordinates of P'' are not the coordinates of the image of P when P is reflected around the original position of line $\mathcal{L}$, since everything has been rotated by an angle of θ. So, to find the coordinates of P when rotated about the original position of line $\mathcal{L}$ we rotate back by an angle of θ. The point, Q, that P'' is transformed into when the plane is rotated back are the coordinates of P when P is reflected about the line $\mathcal{L}$. (See Figure 10.25.)

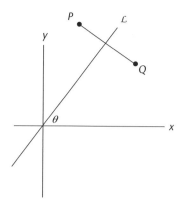

Figure 10.25

Summarizing, to find the image, Q, of the point P when P is reflected about the line $\mathcal{L}$ which makes an angle θ with the x-axis, we perform 3 transformations. First, we rotate the plane $-\theta$ about the origin. That puts $\mathcal{L}$ where the x-axis is. Then we reflect about the x-axis. Finally, we rotate the plane back. That is, $Q = (R_\theta \circ r_x \circ R_{-\theta})P$. This seems pretty complicated. Let's see how we can use matrices to reduce this to one transformation.

$$Q = (R_\theta \circ r_x \circ R_{-\theta})P$$

$$= (R_\theta \circ r_x \circ R_{-\theta})\begin{pmatrix} x \\ y \end{pmatrix} \quad \left(\text{Since } P = \begin{pmatrix} x \\ y \end{pmatrix}\right)$$

$$= (R_\theta \circ r_x)\begin{pmatrix} \cos-\theta & -\sin-\theta \\ \sin-\theta & \cos-\theta \end{pmatrix} \cdot \begin{pmatrix} x \\ y \end{pmatrix} \quad \text{(Using equation (10.17).)}$$

$$= (R_\theta \circ r_x)\begin{pmatrix} \cos\theta & \sin\theta \\ -\sin\theta & \cos\theta \end{pmatrix} \cdot \begin{pmatrix} x \\ y \end{pmatrix} \quad (\text{Since } \cos(-\theta) = \cos\theta \text{ and since } \sin(-\theta) = -\sin(\theta).).$$

Multiplying the matrices on the right we get that this equals

$$(R_\theta \circ r_x)\begin{pmatrix} x\cos\theta + y\sin\theta \\ -x\sin\theta + y\cos\theta \end{pmatrix}.$$

Now by equation (10.1) this is the same as

$$R_\theta \begin{pmatrix} x\cos\theta + y\sin\theta \\ x\sin\theta - y\cos\theta \end{pmatrix}$$

and by equation (10.17) this can be written as

$$\begin{pmatrix} \cos\theta & -\sin\theta \\ \sin\theta & \cos\theta \end{pmatrix} \cdot \begin{pmatrix} x\cos\theta + y\sin\theta \\ x\sin\theta - y\cos\theta \end{pmatrix}. \tag{10.24}$$

Since $\begin{pmatrix} x\cos\theta + y\sin\theta \\ x\sin\theta - y\cos\theta \end{pmatrix}$ is a 2×1 matrix, we can multiply these matrices in (10.24) and get

$$\begin{pmatrix} \cos\theta(x\cos\theta + y\sin\theta) - \sin\theta(x\sin\theta - y\cos\theta) \\ \sin\theta(x\cos\theta + y\sin\theta) + \cos\theta(x\sin\theta - y\cos\theta) \end{pmatrix}.$$

Simplifying we get

$$\begin{pmatrix} x\cos^2\theta + y\sin\theta\cos\theta - x\sin^2\theta + y\sin\theta\cos\theta \\ x\sin\theta\cos\theta + y\sin^2\theta + x\sin\theta\cos\theta - y\cos^2\theta \end{pmatrix}.$$

Finally, we group terms and factor to get

$$\begin{pmatrix} x(\cos^2\theta - \sin^2\theta) + 2y\sin\theta\cos\theta \\ 2x\sin\theta\cos\theta + y(\sin^2\theta - \cos^2\theta) \end{pmatrix}.$$

Now, using the well known identities $\cos 2\theta = \cos^2\theta - \sin^2\theta$ and $\sin 2\theta = 2\sin\theta\cos\theta$, this last matrix simplifies to

$$\begin{pmatrix} x\cos 2\theta + y\sin 2\theta \\ x\sin 2\theta + y(-\cos 2\theta) \end{pmatrix}.$$

Moving the negative sign on the second row, this equals

$$\begin{pmatrix} x\cos 2\theta + y\sin 2\theta \\ x\sin 2\theta - y(\cos 2\theta) \end{pmatrix}.$$

This in turn equals

$$\begin{pmatrix} \cos 2\theta & \sin 2\theta \\ \sin 2\theta & -\cos 2\theta \end{pmatrix} \cdot \begin{pmatrix} x \\ y \end{pmatrix}$$

and, since $P = (x, y)$, this becomes

$$\begin{pmatrix} \cos 2\theta & \sin 2\theta \\ \sin 2\theta & -\cos 2\theta \end{pmatrix} P$$

and we are finally done.

We state the above result as a theorem.

Theorem 10.8 *Suppose we call the transformation that reflects points about the line $\mathcal{L}$ (where $\mathcal{L}$ passes through the origin and makes an angle of θ with the positive x-axis), $r_\mathcal{L}$. Then for any point P, with coordinates (x, y)*

$$r_\mathcal{L}(P) = (x\cos 2\theta + y\sin 2\theta, x\sin 2\theta - y\cos 2\theta) \tag{10.25}$$

or in matrix form,

$$r_\mathcal{L}\begin{pmatrix} x \\ y \end{pmatrix} = \begin{pmatrix} \cos 2\theta & \sin 2\theta \\ \sin 2\theta & -\cos 2\theta \end{pmatrix} \cdot \begin{pmatrix} x \\ y \end{pmatrix}. \tag{10.26}$$

Thus, if we want to reflect the point (1, 2) about the line making an angle of $\theta = 30$ degrees with the positive *x*-axis, our image point would be

$$r_\mathcal{L}((1, 2)) = (1\cos 60 + 2\sin 60, 1\sin 60 - 2\cos 60)$$
$$= (1/2 + \sqrt{3}, \sqrt{3}/2 - 1) \approx (2.2321, -0.13397)$$

or, in matrix form

$$\begin{pmatrix} \cos 60° & \sin 60° \\ \sin 60° & -\cos 60° \end{pmatrix} \cdot \begin{pmatrix} 1 \\ 2 \end{pmatrix} \approx \begin{pmatrix} 2.2321 \\ -0.13397 \end{pmatrix}.$$

There is a nice corollary of this theorem.

Corollary 10.9 *If we reflect a point $P = (a, b)$ about the line $y = x$, we get the point (b, a). If we reflect the point (a, b) about the line $y = -x$, we get $(-b, -a)$.*

We will outline a more direct proof of this in the Student Learning Opportunities.

Proof. (a) We observe for the first part that the line $y = x$ makes an angle of $\theta = 45°$ with the positive *x*-axis, so by Theorem 10.8, when we reflect a point about this line, the image of the point will be

$$r_\mathcal{L}(P) = (a\cos 90° + b\sin 90°, a\sin 90° - b\cos 90°)$$
$$= (b, a) \quad (\text{since } \sin 90° = 1 \text{ and } \cos 90° = 0).$$

(b) For the second part, we note that the angle that the line $y = -x$ makes with the positive *x*-axis is 135°. Using that, in formula of equation (10.25) we get that

$$r_\mathcal{L}(P) = (a\cos 270° + b\sin 270°, a\sin 270° - b\cos 270°)$$
$$= (-b, -a) \quad (\text{since } \sin 270° = -1 \text{ and } \cos 270° = 0).$$

∎

You may have observed that the matrix of the transformation $r_\mathcal{L}$, given in equation (10.26) and the matrix of the rotation given in equation (10.18) seem to be similar and this leads us to wonder if reflections and rotations, though different transformations, are related. The following theorem tells us they are.

Theorem 10.10 *Suppose that $\mathcal{L}_1$ and $\mathcal{L}_2$ are two lines going through the origin. Suppose also that the angle between $\mathcal{L}_1$ and $\mathcal{L}_2$ is θ degrees. Then if P is any point in the xy-plane, and if we reflect P first about $\mathcal{L}_1$ and then about $\mathcal{L}_2$, the net result is that P will be rotated an angle of 2θ degrees about the origin in the direction from $\mathcal{L}_1$ to $\mathcal{L}_2$. In more colloquial terms, the composition of two reflections about two lines intersecting at the origin is a rotation equal to twice the angle between the lines.*

Thus, if we have two lines $\mathcal{L}_1$ and $\mathcal{L}_2$ passing through the origin and the angle between $\mathcal{L}_1$ and $\mathcal{L}_2$ is 30° (measured counterclockwise), then if we reflect a point, *P*, first about $\mathcal{L}_1$ and then about

$\mathcal{L}_2$, the net effect is that we have *rotated* P 60° (counterclockwise). But if we rotate about $\mathcal{L}_2$ first and then $\mathcal{L}_1$, we have rotated P by an angle of 60° clockwise or just $-60°$. Here is the proof:

Proof. Our goal here is to show that $r_{\mathcal{L}_2} \circ r_{\mathcal{L}_1} = R_{2\theta}$, where θ is the angle between $\mathcal{L}_1$ and $\mathcal{L}_2$. We do this by showing that these functions, $r_{\mathcal{L}_2} \circ r_{\mathcal{L}_1}$ and $R_{2\theta}$ give the same image to each point P. So suppose that $P = (x, y)$. Suppose also that $\mathcal{L}_1$ makes an angle of θ_1 with the positive x-axis and that $\mathcal{L}_2$ makes an angle of θ_2 with the x-axis as shown below in Figure 10.26.

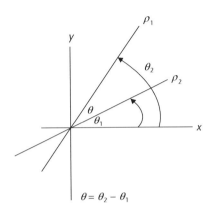

Figure 10.26

Then $\theta = \theta_2 - \theta_1$. Now, $(r_{\mathcal{L}_2} \circ r_{\mathcal{L}_1})(P)$ equals $(r_{\mathcal{L}_2} \circ r_{\mathcal{L}_1}) \begin{pmatrix} x \\ y \end{pmatrix}$ since $P = (x, y)$. Using the fact that, when we compose transformations, we multiply their matrices in reverse order, we see that this is the same as

$$\begin{pmatrix} \cos 2\theta_2 & \sin 2\theta_2 \\ \sin 2\theta_2 & -\cos 2\theta_2 \end{pmatrix} \begin{pmatrix} \cos 2\theta_1 & \sin 2\theta_1 \\ \sin 2\theta_1 & -\cos 2\theta_1 \end{pmatrix} \cdot \begin{pmatrix} x \\ y \end{pmatrix}.$$

Multiplying the matrices yields

$$\begin{pmatrix} \cos 2\theta_2 \cos 2\theta_1 + \sin 2\theta_2 \sin 2\theta_1 & \cos 2\theta_2 \sin 2\theta_1 - \sin 2\theta_2 \cos 2\theta_1 \\ \sin 2\theta_2 \cos 2\theta_1 - \cos 2\theta_2 \sin 2\theta_1 & \sin 2\theta_2 \sin 2\theta_1 + \cos 2\theta_2 \cos 2\theta_1 \end{pmatrix} \cdot \begin{pmatrix} x \\ y \end{pmatrix}.$$

And using, equations (10.8) and (10.9), this is

$$\begin{pmatrix} \cos(2\theta_2 - 2\theta_1) & -\sin(2\theta_2 - 2\theta_1) \\ \sin(2\theta_2 - 2\theta_1) & \cos(2\theta_2 - 2\theta_1) \end{pmatrix} \cdot \begin{pmatrix} x \\ y \end{pmatrix}.$$

But since $\theta = \theta_2 - \theta_1$, this can be rewritten as

$$\begin{pmatrix} \cos 2\theta & -\sin 2\theta \\ \sin 2\theta & \cos 2\theta \end{pmatrix} \cdot \begin{pmatrix} x \\ y \end{pmatrix}.$$

And this is

$$R_{2\theta} \begin{pmatrix} x \\ y \end{pmatrix}.$$

We have shown that $r_{\mathcal{L}_2} \circ r_{\mathcal{L}_1} = R_{2\theta}$ for every point (x, y) and thus, these functions are the same. ∎

From this theorem we can deduce the interesting result that every rotation can be thought of as two successive reflections. Thus, to rotate a point 90 degrees about the origin, we need only reflect twice about two lines that intersect at the origin and make an angle of 45 degrees. You may now be thinking that there are many pairs of these intersecting lines that go through the origin. So does it matter which pair we use? The answer is "No." Any two lines through the origin will do. The result is always the same! This tells us that the reflection is a more basic transformation than the rotation since rotations can be obtained from them.

There is nothing special about requiring that $\mathcal{L}_1$ and $\mathcal{L}_2$ intersect at the origin. Theorem 10.10 is true, regardless of where they intersect. To prove this, we translate the lines so that their intersection is at the origin, then reflect twice about these lines giving us a rotation. We then translate the plane back to where the intersection originally was. We state this as a corollary.

Corollary 10.11 *The reflection of a point P about two lines $\mathcal{L}_1$ and $\mathcal{L}_2$ that intersect at a point, is a rotation of twice the angle between $\mathcal{L}_1$ and $\mathcal{L}_2$ where the intersection points of the two lines is the center of rotation.*

Earlier in this chapter we pointed out that, if you perform several transformations in a row, the net effect is to multiply their matrices. Here is another example of this using our latest theorem.

Example 10.12 *Suppose we want to rotate a point 30° about the origin and then reflect the result about the line y = 2x and then take the result and scale it by a factor of 5. Find a single matrix that will perform all these operations at once.*

Solution. Before we begin this, we recall from secondary school that, if a line makes an angle of θ degrees with the x-axis, then m, the slope of the line, is given by $\tan\theta$. We will ask you to verify this in the Student Learning Opportunities. Since the line $y = 2x$ has slope 2, the angle θ it makes with the x-axis is $\tan^{-1}(2)$. Now to our problem.

If we call these transformations T_1, T_2, and T_3, respectively, we get the result by performing $T_3 \circ T_2 \circ T_1$. But, the matrix of T_1 is

$$\begin{pmatrix} \cos 30 & -\sin 30 \\ \sin 30 & \cos 30 \end{pmatrix}$$

and the matrix of T_2 is

$$\begin{pmatrix} \cos 2\theta & \sin 2\theta \\ \sin 2\theta & -\cos 2\theta \end{pmatrix}$$

where θ is $\tan^{-1} 2$. The matrix for T_3 is

$$\begin{pmatrix} 5 & 0 \\ 0 & 5 \end{pmatrix}.$$

Thus, the matrix for $T_3 \circ T_2 \circ T_1$ is

$$\begin{pmatrix} 5 & 0 \\ 0 & 5 \end{pmatrix} \begin{pmatrix} \cos 2(\tan^{-1} 2) & \sin 2(\tan^{-1} 2) \\ \sin 2(\tan^{-1} 2) & -\cos 2(\tan^{-1} 2) \end{pmatrix} \begin{pmatrix} \cos 30° & -\sin 30° \\ \sin 30° & \cos 30° \end{pmatrix}$$

which turns out to be (approximately)

$$\begin{pmatrix} -0.59808 & 4.9641 \\ 4.9641 & 0.59808 \end{pmatrix}.$$

You can check this on your calculator. Thus, the image of (1, 2) under $T_3 \circ T_2 \circ T_1$ is $(T_3 \circ T_2 \circ T_1)(1, 2) = \begin{pmatrix} -0.59808 & 4.9641 \\ 4.9641 & 0.59808 \end{pmatrix} \begin{pmatrix} 1 \\ 2 \end{pmatrix} = \begin{pmatrix} 9.3301 \\ 6.1603 \end{pmatrix}.$

Student Learning Opportunities

1. **(C)** How would you convince your students that you can use matrices to show that, under a rotation of 90° clockwise, a point (x, y) goes to $(y, -x)$?

2. **(C)** How would you convince your students that you can use matrices to show that, under a counterclockwise rotation of 90°, a point (x, y) goes to $(-y, x)$?

3. If we rotate (x, y) 180° counterclockwise, what point do we get? First find the point without using matrices and then check using matrices.

4. **(C)** One of your students, Gina, asks you, "I thought the definition of slope was 'rise over run.' How come now slope is being given as $\tan \theta$ when the line makes an angle θ with the positive x-axis?" How do you help Gina see the relationship between these two ways of expressing slope? (We used this fact in Example 10.12.)

5. **(C)** Your students can easily see that the line $y = x$ makes an angle of 45° with the positive x-axis and that the line $y = -x$ makes an angle of 135° with the positive x-axis, but they are curious about how to prove it. How do you do it? (We used this fact in the proof of Corollary 10.9.)

6. Find the image of a point (x, y) if it is reflected about the line $y = -x$.

7. We showed that the reflection of a point $P = (a, b)$ about the line $y = x$ is the point $P' = (b, a)$. Using slopes, verify that PP' is indeed perpendicular to $y = x$ and that the distance from P to the line $y = x$ is the same as the distance from P' to the line $y = x$. (You can find the midpoint of PP' and show that that midpoint lies on $y = x$. How will that prove it?)

8. **(C)** Your students can easily see that, if they rotate a point counterclockwise 90 degrees twice, it is the same as rotating it once 180 degrees. But, they can't figure out how to prove it by using multiplication of matrices. How do you do it?

9. Show that the determinant of a rotation matrix is 1. (See Appendix for a review of determinants.)

10 Show that the determinant of the reflection matrix is -1. (See Appendix for a review of determinants.)

11 Find a single matrix that first reflects a point P about the line $y = x$ and then about the line $y = -x$. What rotation of P is equivalent to these two reflections?

12 Find the reflection of the point $(-1, 3)$ about the line $y = 3x$.

13 Find two lines l and m such that a reflection of any point about l, followed by a reflection about m, is a rotation of 60 degrees about their point of intersection.

14 Using Figure 10.27 below, show, using congruent triangles, that reflecting point P about line l and then line m is the same as rotating P an angle of 2θ where θ is the angle from l to m.

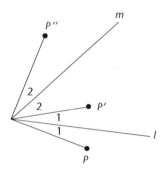

Figure 10.27

(The geometric proof using the picture seems simpler, but is not complete. It assumes that the image of P is between the two lines, which it doesn't have to be. There are many other cases to consider. The matrix approach, deals with all cases at once which illustrates one advantage of using matrices.)

15 (C) Your students understand that the transformation that takes a point (x, y) and transforms it to a point (kx, ky) is a dilation transformation. Now they ask you what would happen to a figure if it were transformed in such a way that each point (x, y) on the figure became (kx, my) where k does not equal m. Before answering this, find the images of specific figures first with say $k = 2$ and $m = 3$ and then respond to their question. (Assume k and m are positive.)

16 Confirm, using the formulas of equations (10.12) and (10.13), that the distance from the origin O to a point $P = (x, y)$ doesn't change when P is rotated θ degrees.

17 Show that, if l and m are two lines intersecting at P, and l' and m' are two other lines intersecting at P, and the angle from l to m is the same as the angle from l' to m', then reflecting about l and then m is the same as reflecting about l' and then m'. [Hint: You may assume that P is at the origin by just translating the whole plane.]

18 (C) Your student is sure that, if you translate a point and then rotate the resulting point about the origin θ degrees, the result is the same as rotating the point about the origin θ degrees and then translating it. How would you convince your student that this is not necessarily the case? (This is just a manifestation of the fact that matrix multiplication is not commutative. That matrix multiplication is not commutative bothers people at first, but it is precisely this fact that allows us to make valid conclusions when applying transformations.)

19 Using Figure 10.28 below, give a purely geometric proof of the fact that, if l and m are two parallel lines, and if we reflect P about l and m, in that order, the net effect is that we have translated P a distance of $2d$ in the direction from l to m, where d is the distance between l and m.

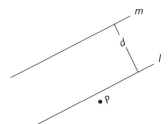

Figure 10.28

10.5 Matrix Transformations

 **LAUNCH**

Now that we have become familiar with some matrices that reflect, rotate, and dilate figures, let us examine how other matrices transform figures.

1. Plot the points $A(0, 0)$, $B(0, 4)$, and $C(3, 0)$ to form a right triangle.
2. Represent these points as 2×1 matrices.
3. Let the matrix, $M = \begin{pmatrix} 1 & 2 \\ -2 & 3 \end{pmatrix}$. Define a mystery transformation from the xy-plane to the xy-plane as follows: To get the image points of A, B, and C, multiply the matrix you got in number 2 by the matrix, M. Write out the image points, A', B', and C' of A, B, and C, respectively.
4. Plot the image points. Do they form a triangle? Do they form a right triangle? Explain what happened.

After having done the launch question, you might be wondering why the image of your right triangle was so different from your original figure. In this section we will try to clear up any confusion you may have by discussing different matrix transformations and their effects on geometric figures.

10.5.1 The Basics

We have shown that the basic transformations discussed in secondary school: rotation, reflection, and dilation can be described by matrices in the sense that, for each transformation, T, there is a matrix M such that $T\begin{pmatrix} x \\ y \end{pmatrix} = M \cdot \begin{pmatrix} x \\ y \end{pmatrix}$. (We will see in a later section that a similar property holds for translations, but requires some perspectives we haven't yet discussed.) Recall that M was called the matrix of the transformation. For example, we found that the matrix of the transformation that rotates a point θ degrees about the origin is $\begin{pmatrix} \cos\theta & -\sin\theta \\ \sin\theta & \cos\theta \end{pmatrix}$, while the matrix of the transformation that reflects a point about the y-axis is $\begin{pmatrix} -1 & 0 \\ 0 & 1 \end{pmatrix}$. (See equations (10.18) and (10.19), respectively.)

In this section we go in reverse. That is, we start with a matrix, any matrix, and *define* a transformation from the xy-plane to the xy-plane using this matrix. We then study how figures transform under general matrix transformations. You will see some surprising results.

So, suppose that M is any two by two matrix. We *define* the following transformation from R^2 to R^2:

$$T_M \begin{pmatrix} x \\ y \end{pmatrix} = M \cdot \begin{pmatrix} x \\ y \end{pmatrix} \quad \text{for any point} \quad \begin{pmatrix} x \\ y \end{pmatrix} \text{ in } R^2.$$

Any such function is called a **matrix transformation** (and, of course, M is the matrix of that transformation). Thus, if $M = \begin{pmatrix} 3 & 2 \\ 4 & -3 \end{pmatrix}$ our matrix transformation is defined by

$$T_M \begin{pmatrix} x \\ y \end{pmatrix} = \begin{pmatrix} 3 & 2 \\ 4 & -3 \end{pmatrix} \cdot \begin{pmatrix} x \\ y \end{pmatrix} \tag{10.27}$$

for any point (x, y). If we want to find the image of the point $\begin{pmatrix} 2 \\ -1 \end{pmatrix}$ under this transformation, we get

$$T_M \begin{pmatrix} 2 \\ -1 \end{pmatrix} = \begin{pmatrix} 3 & 2 \\ 4 & -3 \end{pmatrix} \cdot \begin{pmatrix} 2 \\ -1 \end{pmatrix}$$

$$= \begin{pmatrix} 4 \\ 11 \end{pmatrix}.$$

In general, rotations, translations, and reflections are matrix transformations that maintain the shapes of figures. That is, under these transformations, squares map into squares and right triangles map into right triangles and so on. This is intuitively clear by the nature of these transformations. They are essentially rigid. That is, they don't bend lines, they keep the lengths of lines the same and the measures of angles the same. We will indicate how to prove some of these facts in the Student Learning Opportunities. The general matrix transformation, T_M, however, can distort the shape of a figure, as the following example shows:

> **Example 10.13** *Suppose that we have the matrix transformation defined in equation (10.27). Find the image of the square with vertices $A = (0, 0)$, $B = (1, 0)$, $C = (1, 1)$, and $D = (0, 1)$.*

Solution. The images of these points under this transformation are

$$A' = T_M(A) : \begin{pmatrix} 3 & 2 \\ 4 & -3 \end{pmatrix} \begin{pmatrix} 0 \\ 0 \end{pmatrix} = \begin{pmatrix} 0 \\ 0 \end{pmatrix}$$

$$B' = T_M(B) : \begin{pmatrix} 3 & 2 \\ 4 & -3 \end{pmatrix} \begin{pmatrix} 1 \\ 0 \end{pmatrix} = \begin{pmatrix} 3 \\ 4 \end{pmatrix}$$

$$C' = T_M(C) : \begin{pmatrix} 3 & 2 \\ 4 & -3 \end{pmatrix} \begin{pmatrix} 1 \\ 1 \end{pmatrix} = \begin{pmatrix} 5 \\ 1 \end{pmatrix}$$

$$D' = T_M(D) : \begin{pmatrix} 3 & 2 \\ 4 & -3 \end{pmatrix} \begin{pmatrix} 0 \\ 1 \end{pmatrix} = \begin{pmatrix} 2 \\ -3 \end{pmatrix}.$$

We can get all the image points at once by multiplying

$$\begin{pmatrix} 3 & 2 \\ 4 & -3 \end{pmatrix} \cdot \begin{pmatrix} 0 & 1 & 1 & 0 \\ 0 & 0 & 1 & 1 \end{pmatrix}$$

where the columns of the second matrix are the points A, B, C, and D as we have already indicated. (See Example 10.3.)

Now in secondary school students are taught that, to get the image of a figure $ABCD$ under a transformation, you compute the images of the points A, B, C, and D under this transformation and connect them. That this is legitimate is not clear, and in fact, needs proof. But for now, we accept it and use it.

In Figure 10.29 below we show the image of the original square $ABCD$, which is $A'B'C'D'$. Notice it is not a square, but indeed appears to be a parallelogram. (In fact, it is a parallelogram!)

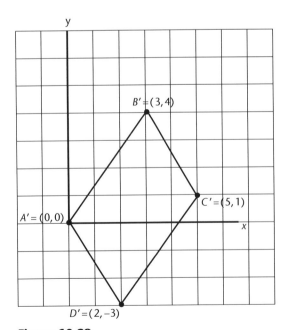

Figure 10.29

Our next example describes a transformation that has even more dramatic effects.

Example 10.14 Let M be the matrix $\begin{pmatrix} 1 & 2 \\ 2 & 4 \end{pmatrix}$ and T_M be the matrix transformation defined by $T_M \begin{pmatrix} x \\ y \end{pmatrix} = M \cdot \begin{pmatrix} x \\ y \end{pmatrix}$ for any point $\begin{pmatrix} x \\ y \end{pmatrix}$ in R^2. Find the image of the square with vertices $A = (0, 0)$, $B = (1, 0)$, $C = (1, 1)$, and $D = (0, 1)$ under this transformation.

This is the same square we had in the previous example. The image points, $T_M(A)$, $T_M(B)$, $T_M(C)$, and $T_M(D)$ are, respectively, $A' = (0, 0)$, $B' = (1, 2)$, $C' = (3, 6)$, and $D' = (2, 4)$. (Verify!) When we graph these points and connect them, we find that our image is no longer even a quadrilateral. It is part of a line segment! Here is the picture (Figure 10.30).

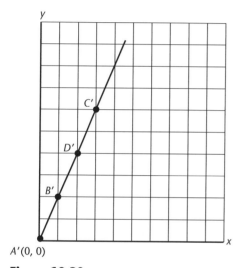

Figure 10.30

(To verify that the points are collinear, you can find the slope between any two of the image points and you will see they are all the same, so they all lie on a line.)

The last two examples show that distortions and downright destruction of figures can occur under matrix transformations. Thus, we cannot blindly assume that, if A and B are points, then under a matrix transformation, T, the image of the line AB is the line $T(A)T(B)$.

These types of results serve as motivation to investigate these transformations in greater detail so we can understand their specific effects on geometric figures. We will do that in the next section.

10.5.2 Matrix Transformations in More Detail – A Technical Point

We know that under rotations, translations, and reflections, the images of line segments get mapped to line segments of the same length, and that angles between lines remain unchanged. But in the previous section we saw that under the more general matrix transformation, the images of figures can get distorted, and so it is not clear that lines really do map into lines under general matrix transformations, or even that polygons map into polygons. See especially Example 10.14 in this connection. Our plan is to show that, under matrix transformations, the image of a line is always a line or a point. You will never get a curve.

To do this, we must formulate a way of describing points on a line parametrically. Suppose that $A = (x_0, y_0)$ and $B = (x_1, y_1)$ are any two points in the plane, and that, instead of considering them as points, we consider them as 1×2 matrices. Then, if we form a new matrix $P = \lambda A + (1 - \lambda)B$, we get, using the rules for working with matrices that

$$P = \lambda A + (1 - \lambda)B$$
$$= \lambda(x_0, y_0) + (1 - \lambda)(x_1, y_1)$$
$$= (\lambda x_0, \lambda y_0) + ((1 - \lambda)x_1, (1 - \lambda)y_1)$$
$$= (\lambda x_0 + (1 - \lambda)x_1, \lambda y_0 + (1 - \lambda)y_1).$$

It follows from this that, if $P = (x, y)$, then $(x, y) = (\lambda x_0 + (1 - \lambda)x_1, \lambda y_0 + (1 - \lambda)y_1)$ and therefore

$$x = (1 - \lambda)x_0 + \lambda x_1 \quad \text{and} \tag{10.28}$$

$$y = (1 - \lambda)y_0 + \lambda y_1 \tag{10.29}$$

since two matrices are equal if their components are equal. Conversely, if equations (10.28) and (10.29) hold, then $P = (x, y) = \lambda A + (1 - \lambda)B$.

So what? We begin with a lemma which uses this.

Lemma 10.15 *If $P = (x, y)$ is any point on the line segment AB where $A = (x_0, y_0)$ and $B = (x_1, y_1)$, then P can be written as $P = \lambda A + (1 - \lambda)B$ for some $0 \leq \lambda \leq 1$. Conversely, if P is any point of the form $\lambda A + (1 - \lambda)B$ for some $0 \leq \lambda \leq 1$, then P is on the line segment joining A to B.*

Proof. We will give the proof only for the picture where the line segment has a positive slope, which means that $x_1 > x_0$. But it is true in general. In the Student Learning Opportunities, you can give the proof for the case when the line has a negative slope. Refer to Figure 10.31 below.

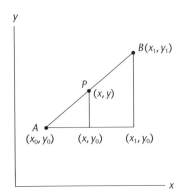

Figure 10.31

We will first show that any point P on the line joining A to B is of the form $P = (x, y)$, where x and y satisfy equations (10.28) and (10.29).

Since $P = (x, y)$ is between A and B, $x_0 \leq x \leq x_1$. We may subtract x_0 from each part of this inequality to get $0 \leq x - x_0 \leq x_1 - x_0$. Since $x_1 - x_0 > 0$, we may divide this inequality by $x_1 - x_0$ to get

$$0 \leq \frac{x - x_0}{x_1 - x_0} \leq 1. \tag{10.30}$$

Call

$$\frac{x - x_0}{x_1 - x_0} = \lambda. \tag{10.31}$$

From (10.30), λ is a number between 0 and 1. (λ will be zero when $x = x_0$, that is, when P is the point A, and λ will be one when $x = x_1$, that is, when P is the point B.) Multiplying both sides of equation (10.31) by $x_1 - x_0$ we get

$$x - x_0 = \lambda(x_1 - x_0)$$

and adding x_0 to both sides of this equation we get that

$$x = x_0 + \lambda(x_1 - x_0).$$

When we distribute the λ, we get

$$x = x_0 + \lambda x_1 - \lambda x_0$$

and by rearranging and factoring, this can be rewritten as

$$x = (1 - \lambda)x_0 + \lambda x_1.$$

In a similar manner we show that

$$y = (1 - \lambda)y_0 + \lambda y_1.$$

These last two equations show that x and y satisfy equations (10.28) and (10.29). In summary, we have shown that any point $P = (x, y)$ on the line segment joining A to B can be written in the form of equations (10.28) and (10.29). Thus, by what preceded the lemma, $P = \lambda A + (1 - \lambda)B$ for some λ where $0 \leq \lambda \leq 1$.

Now we go in reverse. Suppose that $P = (x, y) = \lambda A + (1 - \lambda)B$ where $0 \leq \lambda \leq 1$. Then by what preceded the lemma, $x = (1 - \lambda)x_0 + \lambda x_1$ and $y = (1 - \lambda)y_0 + \lambda y_1$ where $0 \leq \lambda \leq 1$. Let us work on the first of these two equations. Distributing we get

$$x = x_0 - \lambda x_0 + \lambda x_1. \quad 0 \leq \lambda \leq 1.$$

Factoring out λ, we see that this can be written as

$$x = x_0 + \lambda(x_1 - x_0). \quad 0 \leq \lambda \leq 1. \tag{10.32}$$

Subtracting x_0 from both sides of equation (10.32) and dividing by $(x_1 - x_0)$ we get

$$\frac{x - x_0}{x_1 - x_0} = \lambda \quad \text{where} \quad 0 \leq \lambda \leq 1.$$

That is,

$$0 \leq \frac{x - x_0}{x_1 - x_0} \leq 1. \tag{10.33}$$

488 Geometric Transformations

Multiplying by $x_1 - x_0$, which is positive since $x_1 > x_0$, we get that

$$0 \leq x - x_0 \leq x_1 - x_0. \tag{10.34}$$

Adding x_0 to both sides we get

$$x_0 \leq x \leq x_1.$$

Thus, the x coordinate of P is between x_0 and x_1. Similarly, from $\frac{y - y_0}{y_1 - y_0} = \lambda$ where $0 \leq \lambda \leq 1$, we can show that $y_0 \leq y \leq y_1$. Thus, the y coordinate of P is between y_0 and y_1. Since the x and y coordinates of P are between those of A and B, respectively, P is between A and B. ∎

Although we considered A and B as 1×2 matrices, the same holds if we write them as columns. That is, all points $P = \begin{pmatrix} x \\ y \end{pmatrix}$ on the line joining $A = \begin{pmatrix} x_0 \\ y_0 \end{pmatrix}$ and $B = \begin{pmatrix} x_1 \\ y_1 \end{pmatrix}$ can be written as $P = (1 - \lambda) \begin{pmatrix} x_0 \\ y_0 \end{pmatrix} + \lambda \begin{pmatrix} x_1 \\ y_1 \end{pmatrix}$ or, more simply, $(1 - \lambda)A + \lambda B$ for $0 \leq \lambda \leq 1$, and any point that can be written that way necessarily is a point between A and B.

> **Theorem 10.16** *Suppose that T is the matrix transformation defined by $T(P) = MP$ for any point $P = (x, y)$. Then, if P is between A and B, then $T(P)$ is between $T(A)$ and $T(B)$.*

Proof. If P is on the line segment between A and B, then P can be written as $P = (1 - \lambda)A + \lambda B$ where $0 \leq \lambda \leq 1$ from the previous lemma. Thus, $T(P) = MP = M((1 - \lambda)A + \lambda B)$. Since matrix multiplication is distributive (see Appendix), this last equation can be written as

$$T(P) = M((1 - \lambda)A) + M(\lambda B). \tag{10.35}$$

Since we can factor out scalars in matrix multiplication, this last equation becomes

$$T(P) = (1 - \lambda)M(A) + \lambda M(B) \tag{10.36}$$

and, since $T(A) = MA$ and $T(B) = MB$, by definition of the transformation, this last equation becomes

$$T(P) = (1 - \lambda)T(A) + \lambda T(B). \tag{10.37}$$

What equation (10.37) says, using the last lemma, is that $T(P)$ is between $T(A)$ and $T(B)$ which is what we were trying to prove. ∎

Using Lemma 10.15 and this theorem we immediately have Theorem 10.17.

> **Theorem 10.17** *If T is a matrix transformation, say $T\begin{pmatrix} x \\ y \end{pmatrix} = M\begin{pmatrix} x \\ y \end{pmatrix}$, then the image of the line segment AB is the line segment $T(A)T(B)$, and when $T(A) = T(B)$, the image of the line segment AB is just a point.*

Proof. Since every point P, between A and B is mapped into a point $T(P)$ between $T(A)$ and $T(B)$, by the previous theorem we see that the image of the line segment AB is *contained in* the line segment $T(A)T(B)$. To show that the image of the line segment AB is equal to $T(A)T(B)$, we need to show that *every* point in $T(A)T(B)$ is the image of some point between A and B. This way nothing will be missed.

To do this, we pick any point Q between $T(A)$ and $T(B)$. By Lemma 10.15

$$Q = (1-\lambda)T(A) + \lambda TB \qquad (10.38)$$

where $0 \le \lambda \le 1$. But, since $T(A) = MA$ and $T(B) = MB$, this can be rewritten as

$$Q = (1-\lambda)MA + \lambda MB.$$

Using properties of matrix multiplication, this can be rewritten as

$$Q = M((1-\lambda)A + \lambda B)$$

which is the same as saying

$$Q = T((1-\lambda)A + \lambda B). \qquad (10.39)$$

If we let $P = (1-\lambda)A + \lambda B$, then P is between A and B by the previous lemma, and equation (10.39) says that $T(P) = Q$. That is, every point Q between $T(A)$ and $T(B)$ is the image of some point P between A and B. So everything between $T(A)$ and $T(B)$ is picked up when we take the image of a line segment under a matrix transformation and the image of any line segment AB is the "line" segment $T(A)T(B)$. ∎

You may ask why we put the word "line" in quotes above. That answer is that $T(A)$ and $T(B)$ might be the same point. So, in this case, $T(A)T(B)$ is a point.

So, if $T(A) = T(B)$, our figure can collapse. How can we avoid this situation? By requiring that T be $1-1$ (that is, T has an inverse). For then it cannot happen that $T(A) = T(B)$ unless $A = B$. (See Chapter 9 Section 7.) Thus we have:

Corollary 10.18 *If T is an invertible matrix transformation, the image of line segments are line segments, and the image of different line segments are different line segments. Thus, the image of polygons are polygons.*

This theorem justifies what we do in secondary school and reduces the work immensely when transforming polygons. We just take the images of the vertices and we are done. This is *much* easier that taking the image of the entire figure. Computer graphics programs that transform polygons with known vertices take this approach when transforming a figure using one of the standard (invertible) transformations.

One can extend this theorem, though it is by no means easy to prove the next corollary.

Corollary 10.19 *Under an invertible matrix transformation, to find the image of **any** enclosed figure, polygon or not, we need only find the image of the boundary of the figure. We need not consider the interior points.*

490 Geometric Transformations

The saavy student might have made a connection here with material we studied in the previous chapter. There we discussed inverse functions. If $Y = MX$ is a matrix transformation where M is invertible, then solving for X in terms of Y, which gives the inverse transformation, we get $X = M^{-1}Y$. Thus the inverse of the matrix, and the inverse of the matrix transformation are very closely connected. We will develop this more in the Student Learning Opportunities.

Student Learning Opportunities

1 (C) Your students make the following claims. In each case, state whether or not their statement is true or false, then justify your answer.

(a) Under a general matrix transformation, the image of a line will always be a line.

(b) Under any general matrix transformations, the image of a square will always be some type of quadrilateral.

(c) Under matrices that reflect, translate, rotate, and dilate figures, to find the image of any figure, you need only find the image of the boundary of the figure, and not consider the interior points.

2 Find the image of the points $\begin{pmatrix} 2 \\ 1 \end{pmatrix}, \begin{pmatrix} 3 \\ -2 \end{pmatrix}, \begin{pmatrix} 4 \\ 5 \end{pmatrix}$, and $\begin{pmatrix} -2 \\ 3 \end{pmatrix}$ under the matrix transformation given by $T\begin{pmatrix} x \\ y \end{pmatrix} = \begin{pmatrix} 4 & 2 \\ -1 & 3 \end{pmatrix}\begin{pmatrix} x \\ y \end{pmatrix}$. What effect did the transformation have on your final figure?

3 Find the image of the points $\begin{pmatrix} 2 \\ 1 \end{pmatrix}, \begin{pmatrix} 3 \\ -2 \end{pmatrix}, \begin{pmatrix} 4 \\ 5 \end{pmatrix}$, and $\begin{pmatrix} -2 \\ 3 \end{pmatrix}$ under the matrix transformation given by $T\begin{pmatrix} x \\ y \end{pmatrix} = \begin{pmatrix} 4 & 2 \\ 6 & 3 \end{pmatrix}\begin{pmatrix} x \\ y \end{pmatrix}$. What is similar and what is different about your answers in Questions 1 and 2?

4 Using the transformation in Question 1, it is found that the image of a point $\begin{pmatrix} x \\ y \end{pmatrix}$ is $\begin{pmatrix} 4 \\ 2 \end{pmatrix}$. What are the coordinates of the original point $\begin{pmatrix} x \\ y \end{pmatrix}$?

5 Using the transformation from Question 2, is the point $\begin{pmatrix} 2 \\ 3 \end{pmatrix}$ the image of any point $\begin{pmatrix} x \\ y \end{pmatrix}$? How do you know? Answer the same question for $\begin{pmatrix} 0 \\ 2 \end{pmatrix}$.

6 Find the image of the polygon with vertices $\begin{pmatrix} 4 \\ 5 \end{pmatrix}, \begin{pmatrix} 6 \\ -3 \end{pmatrix}, \begin{pmatrix} 1 \\ 2 \end{pmatrix}$, and $\begin{pmatrix} 4 \\ 1 \end{pmatrix}$ under the matrix transformation given by $T\begin{pmatrix} x \\ y \end{pmatrix} = \begin{pmatrix} 5 & 3 \\ 2 & 4 \end{pmatrix}\begin{pmatrix} x \\ y \end{pmatrix}$. Also show that the transformation is 1–1 and find its inverse.

7 (C) One of your students, Krysten, claims that, if two line segments are parallel to each other, then after a translation, they will also be parallel to each other. Krysten is correct, but wants to know if one can give a proof of this without using matrices. Make up a proof of this using the slope of the line segment before and after the translation.

8 **(C)** One of your students, Michael, claims that, if two line segments are perpendicular to each other, then after a translation, they will also be perpendicular to each other. He asks your assitance in proving this. Make up a proof of this using slopes before you guide him with discovering the proof himself.

9 Give a direct proof, without using matrices, that
  (a) if two line segments are parallel to each other, then after a rotation about the same point, they will also be parallel to each other.
  (b) if two line segments are perpendicular to one another, they will remain perpendicular after a rotation about a point P or a reflection in a line l.
  (c) if two line segments AB and AC meet at an angle of θ degrees, then under a translation or rotation or reflection, the angle between the images of these line segments will be the same. Is this still true for an arbitrary matrix transformation? Support your answer with an example or a proof.

10 Show that, under a rotation or translation or reflection, rectangles map to rectangles.

11 **(C)** Your students ask you to explain in words, what the inverse of the transformation that rotates a figure θ degrees clockwise is. What answer do you give? If they asked you how you could show it using matrices, how would you do it?

12 Show that an invertible matrix transformation takes parallel line segments into parallel line segments. Thus, parallelograms map into parallelograms.

13 If the x coordinate of each point in the plane is multiplied by a constant k, while the y coordinate is unchanged, we have an example of what is called an expansion or contraction in the x direction by a factor of k. If $k > 1$, we have an expansion. If $0 < k < 1$ we have a compression. In transformation symbolism, this transformation is defined by the rule that $T(x, y) = (kx, y)$. Write this transformation in matrix form. Find the image of the square with vertices (0, 0) (1, 0), (1, 1), (0, 1) under this transformation when $k = 5$. Also, write the transformation which is the inverse transformation.

14 If the y coordinate of each point in the plane is multiplied by a constant k, while the x coordinate is unchanged, we have an example of what is called an expansion or contraction in the y direction by a factor of k. If $k > 1$, we have an expansion. If $0 < k < 1$, we have a contraction. In transformation symbolism, this transformation is defined by the rule that $T(x, y) = (x, ky)$. Write this transformation in matrix form. Find the image of the square with vertices (0, 0), (1, 0), (1, 1), (0, 1) under this transformation when $k = 1/2$. Find the inverse of this matrix and show that the inverse is the matrix that expands the figure in the y direction by a factor of 2. Did you expect this? Explain.

15 **(C)** How would you convince your students that, if we perform an expansion or contraction in the x direction by a factor of k, followed by an expansion or contraction in the direction of y by the same factor of k, we get a dilation by a factor of k?

16 The transformation which increases the x coordinate of a point (x, y) by a multiple of y is called a shear in the x direction by a factor of k. That is, $T(xy) = (x + ky, y)$.
  (a) Find the matrix of this transformation.
  (b) Find the image of the square with vertices (0, 0), (1, 0), (1, 1), (0, 1) under this transformation when $k = 3$.

(c) Show that points on the x-axis are not moved under this transformation.
(d) Find a matrix that first shears a figure in the x direction by a factor of 2 and then reflects the result about the line $y = x$.
(e) Find a matrix that first reflects a figure about the line $y = x$ and then shears the result in the x direction by a factor of 2.
(f) Suppose that we first shear a figure in the x direction by a factor of 2 and then reflect about the line $y = x$. Now start over. Take the same figure and reflect it about the line $y = x$ and then shear the result in the x direction by a factor of 2. Are the results the same?

17 Find the image of the point $(-3, 4)$ under the transformation T that first rotates the point by 90 degrees, then reflects the result about the y-axis, and finally shears in the x direction by a factor of 3.

18 Prove Theorem 10.15 when the line joining (x_0, y_0) to (x_1, y_1) has a negative slope.

10.6 Transforming Areas

LAUNCH

1 Plot the points $A(1, 1)$, $B(4, 1)$, and $C(1, 5)$ on a coordinate grid.
2 Calculate the area of triangle ABC.
3 Calculate $\frac{1}{2} \left| \det \begin{pmatrix} 1 & 1 & 1 \\ 4 & 1 & 1 \\ 1 & 5 & 1 \end{pmatrix} \right|$. (This matrix was obtained from the points, $A(1, 1)$, $B(4, 1)$, and $C(1, 5)$ by putting a 1 at the end of each row. A review of determinants can be found in the Appendix.)
4 What do you notice about the values you found in answer to questions #2 and #3 above? Do you think that this is a coincidence?

Now that you have completed the launch question, you most likely have two questions in your mind. First, you may be wondering why the value of 1/2 of the absolute value of the determinant is equal to the area of the triangle. Second, you may be wondering why on earth we would want to use the determinant to find the area of a figure. Third, you may want to know if this procedure for finding the area of a triangle always works. The answers to all of these questions will be given in this section.

We have already seen that, under transformations, figures can get distorted. It is a natural question to ask how areas are changed under matrix transformations. We begin that study by first noticing how determinants can be used to find areas of triangles. If you need a review of 3×3 determinants, see the Appendix. This area of study exemplifies a nice connection between linear algebra and areas of triangles.

Theorem 10.20 *If a triangle in the plane has coordinates (x_1, y_1), (x_2, y_2), and (x_3, y_3), then its area is given by $\frac{1}{2}\left|\det\begin{pmatrix} x_1 & y_1 & 1 \\ x_2 & y_2 & 1 \\ x_3 & y_3 & 1 \end{pmatrix}\right|$. (That is, the area is 1/2 of the absolute value of the determinant.)*

Proof. We give only a partial proof that would be sufficient for the brighter secondary school student. (The student can then be asked to complete the proof say, when the labeling is different or the triangle is in a different quadrant or position.) We refer to Figure 10.32.

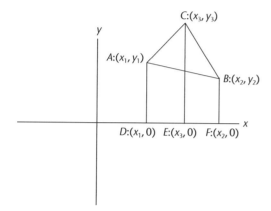

Figure 10.32

We draw AD, CE, and BF. We know the area of a trapezoid is 1/2 the height times the sum of the bases. The height of trapezoid $DACE$ is $DE = x_3 - x_1$, while the bases, DA and CE have lengths y_1 and y_3, respectively. So the area of trapezoid $DACE$ is $\frac{1}{2}(x_3 - x_1)(y_1 + y_3)$. In a similar manner the height of trapezoid $ECBF$ is $EF = x_2 - x_3$ while the bases, EC and FB, are y_3 and y_2. Thus, the area of trapezoid $ECBF$ is $\frac{1}{2}(x_2 - x_3)(y_3 + y_2)$. We leave it to you to show that the area of trapezoid $DABF$ is $\frac{1}{2}(x_2 - x_1)(y_1 + y_2)$. Now we know that the area of the triangle ABC = the area of trapezoid $DACE$ + the area of trapezoid $ECBF$ − the area of trapezoid $DABF$. Using what we just established, we have that

the area of triangle $ABC = \frac{1}{2}(x_3 - x_1)(y_1 + y_3) + \frac{1}{2}(x_2 - x_3)(y_3 + y_2) - \frac{1}{2}(x_2 - x_1)(y_1 + y_2)$

which, upon simplification, yields

the area of triangle $ABC = \frac{1}{2}x_1y_2 - \frac{1}{2}x_2y_1 - \frac{1}{2}x_1y_3 + \frac{1}{2}y_1x_3 + \frac{1}{2}x_2y_3 - \frac{1}{2}x_3y_2.$

Now, if we compute $\frac{1}{2}\det\begin{pmatrix} x_1 & y_1 & 1 \\ x_2 & y_2 & 1 \\ x_3 & y_3 & 1 \end{pmatrix}$, we get $\frac{1}{2}x_1y_2 - \frac{1}{2}x_2y_1 - \frac{1}{2}x_1y_3 + \frac{1}{2}y_1x_3 + \frac{1}{2}x_2y_3 - \frac{1}{2}x_3y_2$ and clearly these are the same.

We used the picture and labeling given above where everything is in the first quadrant. If the triangle had been labeled differently, then the rows of the matrix might have been switched and this might change the sign of the determinant, which is why we need the absolute value. ∎

494 Geometric Transformations

Here is a numerical example.

Example 10.21 *Find the area of the triangle whose vertices are at (1, 2), (3, 7), and (5, 9).*

Solution. According to the theorem, the area of the triangle is obtained by computing

$$\frac{1}{2} \left| \det \begin{pmatrix} 1 & 2 & 1 \\ 3 & 7 & 1 \\ 5 & 9 & 1 \end{pmatrix} \right|.$$

This can be done by hand or computer, and we get that the area is 3.

Theorem 10.22 *Suppose T is a matrix transformation defined by*

$$T \begin{pmatrix} x \\ y \end{pmatrix} = \begin{pmatrix} e & f \\ g & h \end{pmatrix} \begin{pmatrix} x \\ y \end{pmatrix} \tag{10.40}$$

and that ABC is any right triangle. Then the area of the triangle with vertices T(A), T(B), and T(C) is equal to the area of triangle ABC · |det(M)| where $M = \begin{pmatrix} e & f \\ g & h \end{pmatrix}$.

Proof. We give the proof for the specific right triangle shown below in Figure 10.33 where one vertex is at the origin since it easier to follow and leave the more general case to the interested reader.

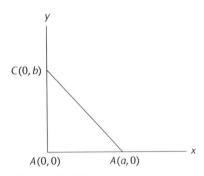

Figure 10.33

Let the vertices be $A(0, 0)$, $B(a, 0)$, and $C(0, b)$. Using equation (10.40), we have that $T(A) = \begin{pmatrix} 0 \\ 0 \end{pmatrix}$, $T(B) = \begin{pmatrix} ae \\ ag \end{pmatrix}$ and $T(C) = \begin{pmatrix} bf + ae \\ ag + bh \end{pmatrix}$ (Check it!). We observe also that

$$\det M = \det \begin{pmatrix} e & f \\ g & h \end{pmatrix} = he - fg. \tag{10.41}$$

From the previous theorem, the area of the image triangle is

$$\left| \frac{1}{2} \det \begin{pmatrix} 0 & 0 & 1 \\ ae & ag & 1 \\ bf + ae & ag + bh & 1 \end{pmatrix} \right|$$

$$= \frac{1}{2} |abhe - abfg|$$

$$= \frac{1}{2} |ab| \cdot |he - fg|$$

$$= \frac{1}{2} ab \cdot |he - fg| \quad \text{since} \quad a > 0 \quad \text{and} \quad b > 0$$

$$= \text{Area of triangle } ABC \cdot |\det(M)| \quad \text{(By equation (10.41))}.$$

As a note, we observe that, if the matrix M is not invertible, the determinant of the matrix is 0, which is a well known fact from linear algebra. In that case the image of our triangle collapses into a line segment, which has area 0. (See Example 10.14 in this regard.) So actually, the theorem holds even in that case too. ∎

Corollary 10.23 *The area of a square under an invertible transformation is multiplied by* $|\det M|$.

Proof: A square can be broken into two congruent right triangles, each of whose areas is multiplied by $|\det M|$ when transformed. So, the area of the square is also multiplied by $|\det M|$ when transformed.

Corollary 10.24 *Under an invertible matrix transformation, with matrix M, the area of any figure gets multiplied by* $|\det M|$.

Proof. A proof for the general figure requires a careful limit argument and really is quite sophisticated, so we won't give it. But here is the idea.

Suppose we have any closed figure, F. That figure's area can be approximated *to any desired degree of accuracy* by inscribing squares. (To convince yourself of this, as well as your students, imagine the figure placed on a very fine grid consisting of tiny squares. The sum of the areas of the squares is close to the area of the figure. The finer the grid, the closer the sum of the areas is to the area of F. See Figure 10.34 below.)

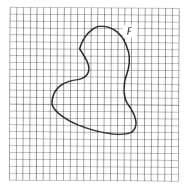

Figure 10.34

When we transform the figure, to get $T(F)$, the squares contained in F get transformed into parallelograms (see Student Learning Opportunity 12 from the previous section), the sum of whose areas approximate, to a high degree of accuracy, the area of $T(F)$. Since these inscribed parallelograms have areas equal to the areas of the squares they came from multiplied by $|\det M|$, the area of $T(F)$ is highly approximated by the area of F multiplied by $|\det M|$. Of course, the finer the grid, the better the approximation, which leads to the fact that the area of $T(F) = |\det M| \cdot$ area of T. ∎

Let us see what this leads to. Suppose we begin with a circle whose radius is 1. We know its area is π. (In fact, some people take this to be the definition of π.)

Let us place the circle so that its center is at the origin. Now we perform the following matrix transformation $T\begin{pmatrix} x \\ y \end{pmatrix} = \begin{pmatrix} a & 0 \\ 0 & b \end{pmatrix} \begin{pmatrix} x \\ y \end{pmatrix} = \begin{pmatrix} ax \\ ay \end{pmatrix}$. What this does is stretch the x coordinate by a factor of a and the y coordinate by a factor of b. Let us assume that a and b are positive and that $a > b > 0$. What does this transformation do to the circle? Well, since it stretches it in the x direction by a factor of a and in the y direction by a factor of b, it transforms the circle into an ellipse with semimajor axis of length a and semiminor axis of length b. See Figure 10.35 below where we show the original circle of radius 1 with center at the origin and the ellipse that results from this stretch.

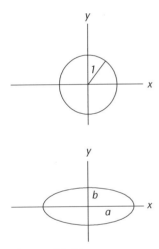

Figure 10.35

Thus, according to our theorem, the area of the resulting ellipse is $\left|\det \begin{pmatrix} a & 0 \\ 0 & b \end{pmatrix}\right| \cdot$ area of the original unit circle $= ab \cdot \pi$. And thus we have proved:

Theorem 10.25 *The area of an ellipse with semimajor axis a and semiminor axis b is given by πab.*

This is a result that is seldom pointed out in books.

Corollary 10.26 *The area of a circle with radius r is πr^2.*

Proof. Take $a = b = r$ in the theorem. ∎

Student Learning Opportunities

1. Using a determinant, find the area of a triangle whose vertices are (1, 2), (2, 7), and (3, 9).

2. Using a determinant, find the area of a triangle whose vertices are (−3, 1), (2, 4), and (−4, 5).

3. Using one or more determinants, find the area of a quadrilateral whose vertices are (4, 2), (6, 7), (3, 7), and (−1, 3).

4. (C) You ask your students to use the formula in this section to compute the area of the triangle whose vertices are (1, 2), (2, 4), and (3, 6). When they get their result that the area is 0, they are totally confused and ask you, "How could this be?" How do you answer them?

5. The area of a polygon is 5.
   (a) After transforming this polygon using the matrix transformation: $T\begin{pmatrix} x \\ y \end{pmatrix} = \begin{pmatrix} 4 & 2 \\ -3 & 1 \end{pmatrix}\begin{pmatrix} x \\ y \end{pmatrix}$, we get a new polygon. What is the area of this new polygon?
   (b) What is the area of the resulting polygon if it is transformed using the matrix transformation: $T\begin{pmatrix} x \\ y \end{pmatrix} = \begin{pmatrix} -8 & 6 \\ 3 & 2 \end{pmatrix}\begin{pmatrix} x \\ y \end{pmatrix}$?

6. (C) There is a famous rule that says that the determinant of a product of two matrices is the product of the determinants. Your students want to know what this means in terms of compositions of transformations and areas when the determinants are positive. How do you respond?

7. The determinant of the inverse of a matrix is the reciprocal of the determinant of the matrix. Explain how you can deduce this from the study of transformations and areas when the determinants are positive.

10.7 Connections to Fractals

LAUNCH

In the following questions we let $M = \begin{pmatrix} 1 & 0 & 8 \\ 0 & 1 & 9 \\ 0 & 0 & 1 \end{pmatrix}$, which we will refer to as our mystery matrix.

1. Multiply the mystery matrix M by $\begin{pmatrix} 4 \\ 5 \\ 1 \end{pmatrix}$. What do you get?

2. Multiply the mystery matrix M by $\begin{pmatrix} 10 \\ 10 \\ 1 \end{pmatrix}$. What do you get?

3 Multiply the mystery matrix M by $\begin{pmatrix} a \\ b \\ 1 \end{pmatrix}$. What do you get?

4 If each of the matrices of the form $\begin{pmatrix} x \\ y \\ 1 \end{pmatrix}$ represents a point $\begin{pmatrix} x \\ y \end{pmatrix}$, then what is the mystery matrix doing to each point?

Having completed the launch problem, you are probably convinced that you have found a way to represent translations using matrix multiplications. But, you are also probably curious as to why we would want to go to such lengths to represent translations this way, when using the function notation was so simple. You are also probably wondering what any of this has to do with fractals, which is the title of this section. Read on, and you will find out.

10.7.1 Translations

As you have most likely discovered from the launch question, it is possible to represent translations by using matrices. As you have seen, to do this however, is a bit subtle. We need to introduce a new coordinate, 1 to each point. Thus, each point $\begin{pmatrix} x \\ y \end{pmatrix}$ must now be written as $\begin{pmatrix} x \\ y \\ 1 \end{pmatrix}$ and each point in the analysis of the form $\begin{pmatrix} x \\ y \\ 1 \end{pmatrix}$, really refers to the point $\begin{pmatrix} x \\ y \end{pmatrix}$. The coordinates $\begin{pmatrix} x \\ y \\ 1 \end{pmatrix}$ are called **homogeneous coordinates** of $\begin{pmatrix} x \\ y \end{pmatrix}$. Thus, homogeneous coordinates of $\begin{pmatrix} 2 \\ 3 \end{pmatrix}$ are $\begin{pmatrix} 2 \\ 3 \\ 1 \end{pmatrix}$.

Now, if we want to translate a point $\begin{pmatrix} x \\ y \end{pmatrix}$ by $\begin{pmatrix} h \\ k \end{pmatrix}$, to get the point $\begin{pmatrix} x+h \\ y+k \end{pmatrix}$, we form the matrix.

$$M = \begin{pmatrix} 1 & 0 & h \\ 0 & 1 & k \\ 0 & 0 & 1 \end{pmatrix}$$

which we call the **translation matrix**. Now, if we multiply the translation matrix M by $\begin{pmatrix} x \\ y \\ 1 \end{pmatrix}$, we get

$$M \cdot \begin{pmatrix} x \\ y \\ 1 \end{pmatrix} = \begin{pmatrix} 1 & 0 & h \\ 0 & 1 & k \\ 0 & 0 & 1 \end{pmatrix} \cdot \begin{pmatrix} x \\ y \\ 1 \end{pmatrix} = \begin{pmatrix} x+h \\ y+k \\ 1 \end{pmatrix}$$

which are homogeneous coordinates for the translated point $\begin{pmatrix} x+h \\ y+k \end{pmatrix}$. Thus the translation matrix will accomplish our goal.

Since translating points is so easy without matrices, why bother using this complicated method? Again, the answer is that animations are easily implemented on the computer doing this kind of matrix multiplication. Let us redo Example 10.1 from this point of view.

Example 10.27 *Find the image of the triangle whose vertices are $A = (-1, 2)$, $B = (4, 7)$, and $C = (0, 6)$ under the translation $T_{(-3, 2)}$.*

Solution. We put the homogeneous coordinates of the points as the columns in a matrix. Our matrix is

$$\begin{pmatrix} -1 & 4 & 0 \\ 2 & 7 & 6 \\ 1 & 1 & 1 \end{pmatrix}.$$

We now multiply this by the appropriate translation matrix

$$\begin{pmatrix} 1 & 0 & -3 \\ 0 & 1 & 2 \\ 0 & 0 & 1 \end{pmatrix}.$$

Our result is

$$\begin{pmatrix} 1 & 0 & -3 \\ 0 & 1 & 2 \\ 0 & 0 & 1 \end{pmatrix} \begin{pmatrix} -1 & 4 & 0 \\ 2 & 7 & 6 \\ 1 & 1 & 1 \end{pmatrix} = \begin{pmatrix} -4 & 1 & -3 \\ 4 & 9 & 8 \\ 1 & 1 & 1 \end{pmatrix}.$$

Now we just drop the 1's in the final row and we get that the triangle that is the image has vertices $(-4, 4)$, $(1, 9)$, and $(3, 8)$.

We talked about how to use 2×2 matrices to perform rotations and reflections and mentioned that, when it comes to translations, we need to use the homogenous coordinates $\begin{pmatrix} x \\ y \\ 1 \end{pmatrix}$ of the point $\begin{pmatrix} x \\ y \end{pmatrix}$ to accomplish what we needed. But suppose we want to combine translations, rotations, reflections, and so on? Then what? Some of the matrices have two columns and some have three, and this is going to cause a problem when multiplying the matrices. The way computers get around this is to always use the homogeneous coordinates when doing animation. Then to rotate a point $\begin{pmatrix} x \\ y \end{pmatrix}$ by an angle of θ, instead of multiplying the 2×2 matrix $\begin{pmatrix} \cos\theta & -\sin\theta \\ \sin\theta & \cos\theta \end{pmatrix}$ by $\begin{pmatrix} x \\ y \end{pmatrix}$ to get

$$\begin{pmatrix} x\cos\theta - y\sin\theta \\ y\cos\theta + x\sin\theta \end{pmatrix} \tag{10.42}$$

they instead use the 3×3 matrix

$$\begin{pmatrix} \cos\theta & -\sin\theta & 0 \\ \sin\theta & \cos\theta & 0 \\ 0 & 0 & 1 \end{pmatrix}$$

500   Geometric Transformations

and multiply the homogeneous coordinates of $\begin{pmatrix} x \\ y \end{pmatrix}$, namely $\begin{pmatrix} x \\ y \\ 1 \end{pmatrix}$ to get

$$\begin{pmatrix} \cos\theta & -\sin\theta & 0 \\ \sin\theta & \cos\theta & 0 \\ 0 & 0 & 1 \end{pmatrix} \begin{pmatrix} x \\ y \\ 1 \end{pmatrix} = \begin{pmatrix} x\cos\theta - y\sin\theta \\ x\sin\theta + y\cos\theta \\ 1 \end{pmatrix}$$

which are the homogeneous coordinates of (10.42). Similarly, when doing a reflection, they change the reflection matrix $\begin{pmatrix} \cos 2\theta & \sin 2\theta \\ \sin 2\theta & -\cos 2\theta \end{pmatrix}$ from equation (10.26) to

$$\begin{pmatrix} \cos 2\theta & \sin 2\theta & 0 \\ \sin 2\theta & -\cos 2\theta & 0 \\ 0 & 0 & 1 \end{pmatrix}$$

and multiply the homogeneous coordinates $\begin{pmatrix} x \\ y \\ 1 \end{pmatrix}$ of $\begin{pmatrix} x \\ y \end{pmatrix}$ by this to get

$$\begin{pmatrix} \cos 2\theta & \sin 2\theta & 0 \\ \sin 2\theta & -\cos 2\theta & 0 \\ 0 & 0 & 1 \end{pmatrix} \begin{pmatrix} x \\ y \\ 1 \end{pmatrix} = \begin{pmatrix} x\cos 2\theta + y\sin 2\theta \\ x\sin 2\theta + -y\cos 2\theta \\ 1 \end{pmatrix}$$

which are the homogeneous coordinates of the correct reflected point.

Similarly, we work with a 3×3 dilation matrix, modified as above, when we wish to dilate a point having two coordinates when combining several operations one of which is translation.

Example 10.28 *Suppose that we wanted to perform a dilation on $\begin{pmatrix} x \\ y \end{pmatrix}$ by a factor of d, and then rotate the result by an angle of θ and then translate by the vector $\begin{pmatrix} h \\ k \end{pmatrix}$. Find a single matrix that accomplishes all of these transformations.*

Solution. Since there is a translation here, we are going to use homogeneous coordinates for the point $\begin{pmatrix} x \\ y \end{pmatrix}$ and modify all our matrices as we indicated above. Thus, to dilate by a factor of d, we will use the matrix

$$D = \begin{pmatrix} d & 0 & 0 \\ 0 & d & 0 \\ 0 & 0 & 1 \end{pmatrix}.$$

To rotate by an angle of θ, we will use the matrix,

$$R = \begin{pmatrix} \cos\theta & -\sin\theta & 0 \\ \sin\theta & \cos\theta & 0 \\ 0 & 0 & 1 \end{pmatrix}.$$

Finally, to translate by the vector $\begin{pmatrix} h \\ k \end{pmatrix}$, we use the matrix

$$T = \begin{pmatrix} 1 & 0 & h \\ 0 & 1 & k \\ 0 & 0 & 1 \end{pmatrix}.$$

The single matrix that will perform all three operations on $\begin{pmatrix} x \\ y \\ 1 \end{pmatrix}$ is TRD which is

$$\begin{pmatrix} 1 & 0 & h \\ 0 & 1 & k \\ 0 & 0 & 1 \end{pmatrix} \begin{pmatrix} \cos\theta & -\sin\theta & 0 \\ \sin\theta & \cos\theta & 0 \\ 0 & 0 & 1 \end{pmatrix} \begin{pmatrix} d & 0 & 0 \\ 0 & d & 0 \\ 0 & 0 & 1 \end{pmatrix}$$

or just

$$\begin{pmatrix} d\cos\theta & -d\sin\theta & h \\ d\sin\theta & d\cos\theta & k \\ 0 & 0 & 1 \end{pmatrix}.$$

We have essentially proved the following theorem:

Theorem 10.29 *To perform a dilation, with dilation factor d, a rotation by an angle of θ degrees, and translation by vector $\begin{pmatrix} h \\ k \end{pmatrix}$ in that order on a point $\begin{pmatrix} x \\ y \end{pmatrix}$, we multiply its homogenous coordinates $\begin{pmatrix} x \\ y \\ 1 \end{pmatrix}$ by the matrix*

$$\begin{pmatrix} d\cos\theta° & -d\sin\theta° & k \\ d\sin\theta° & d\cos\theta° & k \\ 0 & 0 & 1 \end{pmatrix}. \tag{10.43}$$

So, of what use is this theorem? Well, it connects rather strongly with the construction of fractals, a topic which we have discussed several times in this book. Let us see how.

Michael Barnsley, a researcher and entrepreneur, has discovered a method of generating fractals. He used ideas related to fractals in his business which is concerned with data compression. Here is how it works.

Begin with four matrix transformations, T_1, T_2, T_3, and T_4, each of whose matrices is of the form (10.43) or where there are separate dilations in the x and y direction (so the d's in the first column can be different from the d's in the second column). In all the matrices, the d's must be < 1. We are going to pick transformations *at random* to perform but with certain probabilities. That is, you might wish to pick T_1 50% of the time, T_2 30% of the time, T_3 15% of the time and T_4 5% of the time. How do we do this? Well, we can actually create a machine for picking these transformations with these percentages (a random number generator can be used to do this as well). We simply take a circle and divide it into 4 parts where the first part, A, takes up 50% of the circle, the second part, B, 30%, the third part, C, 15%, and the fourth part, D, 5% of the circle and then place a spinner at the center of the circle as shown in Figure 10.36.

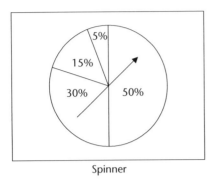

Figure 10.36

We spin it. If it lands on A, then we pick T_1, if it lands on B, we pick T_2, and so on. This will result in picking transformation 1 about 50% of the time, transformation 2 about 30% of the time, transformation 3 about 15% of the time, and transformation 4 about 5% of the time. The probabilities we decide to use for T_1 to T_4 are up to us to choose.

We begin with a point in the plane. We spin our spinner and apply to the current point whatever transformation the spinner lands on. So, if the spinner first lands on A, we take our point (x, y) and then apply T_1 to get a new point P_1. We plot it. We spin again. If the spinner lands on B, we apply T_2 to the point P_1 we previously obtained, to get a new point P_2. We plot it. We keep spinning and generating new points always obtained by applying a transformation to the most recent point and plot the points we get. After a large number of iterations of this procedure, we often will get fractals.

We emphasize that what we are doing is composing the transformations T_1 to T_4 in some *random order* determined by the spins.

To illustrate, suppose that T_1, T_2, T_3 and T_4 are matrix transformations with matrices $M_1, M_2, M_3,$ and M_4, respectively, where

$$M_1 = \begin{pmatrix} 0.8\cos 3° & -0.8\sin 3° & 0 \\ 0.8\sin 3° & 0.8\cos 3° & 3 \\ 0 & 0 & 1 \end{pmatrix}$$

$$M_2 = \begin{pmatrix} 0.3\cos 52° & -0.3\sin 52° & 0 \\ 0.3\sin 52° & 0.3\cos 52° & 2 \\ 0 & 0 & 1 \end{pmatrix}$$

$$M_3 = \begin{pmatrix} 0.8\cos -46° & -0.8\sin -46° & 0 \\ 0.8\sin -46° & 0.8\cos -46° & 3 \\ 0 & 0 & 1 \end{pmatrix}$$

$$M_4 = \begin{pmatrix} 0 & 0 & 0 \\ 0 & 0.5 & 0 \\ 0 & 0 & 1 \end{pmatrix}.$$

We notice that M_4 performs a dilation of 0.5 in the y direction, but nothing in the x direction. Thus, rectangles get shrunk down to lines. If we randomly pick these transformations T_1, T_2, T_3, and T_4 and apply them many times, starting with any point, we get a picture that looks similar to the following fern (Figure 10.37), which is a fractal.

Figure 10.37

It is quite amazing that, by using these matrices, we can generate such a realistic picture.

Student Learning Opportunities

1 (C) One of your students is very resistant to using the homogeneous coordinates to do a translation. He observes that, if you want to perform a translation $T_{(h,\ k)}$ on a figure, you need only take each point $\begin{pmatrix} x \\ y \end{pmatrix}$ on the figure and add $\begin{pmatrix} h \\ k \end{pmatrix}$, to get the point $\begin{pmatrix} x+h \\ y+k \end{pmatrix}$. How would you explain to the student why using homogeneous coordinates are beneficial in practice.

2 Write a single 3×3 matrix that performs the following three operations in order: (1) A dilation by a factor of 2. (2) A rotation by an angle of $30°$ counterclockwise. (3) A translation by the vector $\begin{pmatrix} 2 \\ 3 \end{pmatrix}$.

3 Write a single 3×3 matrix that performs the following three operations in order: (1) A dilation by a factor of $1/3$. (2) A rotation by an angle of $60°$ clockwise. (3) A translation by the vector $\begin{pmatrix} -5 \\ 1 \end{pmatrix}$.

(a) Try to find a program on the Internet that performs the Barnsley method, or write your own if you have such skills. Then pick a point at random and apply the method 10,000 times using the matrices M_1, M_2, M_3 and M_4. See if the picture you get is similar to the Fern we got or, if it is fractal-like.

(b) Experiment a bit by changing the probabilities for T_1, T_2, T_3, and T_4 in part (a) and see if you come out with different figures.

10.8 Transformations in Three dimensions

LAUNCH

You have just learned about transformations matrices in two dimensions. Since we live in a 3 dimensional world, as you can well imagine, it will be important to find similar matrices that work in 3 dimensions. Use your intuition to make conjectures about what some of these matrices will look like, by responding to the questions below.

1. Given that, in 2 dimensions, the reflection transformation, r_x which reflects every point about the x-axis is
$$r_x \begin{pmatrix} x \\ y \end{pmatrix} = \begin{pmatrix} x \\ -y \end{pmatrix} = \begin{pmatrix} 1 & 0 \\ 0 & -1 \end{pmatrix} \cdot \begin{pmatrix} x \\ y \end{pmatrix},$$ what will be a similar matrix in 3 dimensions that reflects a point (x, y, z) about the yz-plane? Fill in the blanks in the matrices below:
$$r_{yz} \begin{pmatrix} x \\ y \\ z \end{pmatrix} = \begin{pmatrix} \\ \\ \end{pmatrix} = \begin{pmatrix} \\ \\ \end{pmatrix} \cdot \begin{pmatrix} x \\ y \\ z \end{pmatrix}$$

2. Given that, in 2 dimensions the dilation transformation D_k which dilates a point by a factor of k is written as $D_k \begin{pmatrix} x \\ y \end{pmatrix} = \begin{pmatrix} kx \\ ky \end{pmatrix} = \begin{pmatrix} k & 0 \\ 0 & k \end{pmatrix} \cdot \begin{pmatrix} x \\ y \end{pmatrix}$, what will be a similar matrix in 3 dimensions that dilates a point (x, y, z) by a factor of k look like? Fill in the blanks in the matrices below:
$$D_k \begin{pmatrix} x \\ y \\ z \end{pmatrix} = \begin{pmatrix} \\ \\ \end{pmatrix} = \begin{pmatrix} \\ \\ \end{pmatrix} \cdot \begin{pmatrix} x \\ y \\ z \end{pmatrix}$$

Now that you have given some thought to what matrix transformations look like in 3 dimensions, you can read this section to see if your intuitions about them were correct and learn more about other transformations.

We have pointed out earlier that one of the advantages of the matrix approach to rotations, reflections, dilations, and translations is that we can generalize our results to 3 dimensions. What that means is that we can apply the material in this chapter to real-world problems. In this section we discuss a bit about how this works.

In 3 dimensional analysis, we have an x, y, and z-axis. You are all undoubtedly familiar with this. Three dimensional space is often denoted by R^3. We can talk about matrix transformations from R^3 to R^3 and we can ask questions that we asked in 2 dimensions in 3 dimensions. For example, we can talk about reflecting a point (x, y, z) about the xy-plane. If (x, y, z) is our original point, then under a reflection about the xy-plane, the image is $(x, y, -z)$. This can be written in matrix form.

$$T \begin{pmatrix} x \\ y \\ z \end{pmatrix} = \begin{pmatrix} x \\ y \\ -z \end{pmatrix} = \begin{pmatrix} 1 & 0 & 0 \\ 0 & 1 & 0 \\ 0 & 0 & -1 \end{pmatrix} \cdot \begin{pmatrix} x \\ y \\ z \end{pmatrix}.$$

There are similar statements that can be made for reflections about the *xz*-plane and the *yz*-plane.

When discussing rotations in 3 dimensions, we need to be very careful. Suppose we have a line through the origin, and we wish to rotate a point $P = (x, y, z)$, θ degrees counterclockwise about this line. First, what does rotate about a line in 3 dimensions mean, and what does counterclockwise mean?

When we rotate a point, P, about a line, l, we imagine that the line is the central axis of a cone and that the point $P = (x, y, z)$ is a point on that cone. The radius of the cone is the distance from the point P to the line l. (See Figure 10.38 below.)

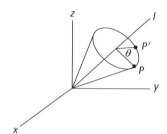

Figure 10.38

Now rotating (x, y, z) θ degrees about the line means traveling around the circular base of this cone θ degrees. The direction counterclockwise is measured looking *towards the origin*. In the picture above, you see P rotated θ degrees counterclockwise to get a new point P'.

When we rotate a point, P, say about l where l is the *y*-axis, the picture looks like the one in Figure 10.39 below.

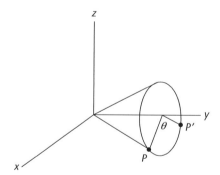

Figure 10.39

In this case the image of any point (x, y, z) is given by $(x\cos\theta + z\sin\theta, y, -x\sin\theta + z\cos\theta)$. Notice that the *y* coordinate is the same, since the circle representing the cone is parallel to the *xz*-plane, and so every point on that circle has the same *y* coordinate. The *x* and *z* coordinates were obtained in exactly the same way we derived formulas of equations (10.12) and (10.13), only now everything is taking place in the *xz*-plane. That is, the circle representing the base of the cone is projected onto the *yz*-plane and our computations are done there. In matrix form, our transformation is

$$T\begin{pmatrix} x \\ y \\ z \end{pmatrix} = \begin{pmatrix} x\cos\theta + z\sin\theta \\ y \\ -x\sin\theta + z\cos\theta \end{pmatrix} = \begin{pmatrix} \cos\theta & 0 & \sin\theta \\ 0 & 1 & 0 \\ -\sin\theta & 0 & \cos\theta \end{pmatrix} \cdot \begin{pmatrix} x \\ y \\ z \end{pmatrix}. \qquad (10.44)$$

506 Geometric Transformations

There are similar results for rotating about the *x*-axis and *z*-axis, as well as any line through the origin.

The same way we talk about how figures or areas are transformed under matrix transformations in two dimensions, we can do it for three dimensions. Many of our theorems are analogous, only the proofs are more difficult. For example we have the following:

> **Theorem 10.30** *If T is an invertible matrix transformation from R^3 to R^3 with matrix M and where S is a solid with volume V, then the image of S under this transformation has volume $|\det M| \cdot V$*

For completeness, we mention the formula for rotating a point *P* about any line *l* going through the origin. If (a, b, c) is any point on such a line and $a^2 + b^2 + c^2 = 1$, then the matrix that rotates $P = (x, y, z)$ about this line is

$$M = \begin{pmatrix} a^2(1-\cos\theta) + \cos\theta & ab(1-\cos\theta) - c\sin\theta & ac(1-\cos\theta) + b\sin\theta \\ ab(1-\cos\theta) + c\sin\theta & b^2(1-\cos\theta) + \cos\theta & bc(1-\cos\theta) - a\sin\theta \\ ac(1-\cos\theta) - b\sin\theta & bc(1-\cos\theta) + a\sin\theta & c^2(1-\cos\theta) + \cos\theta \end{pmatrix}. \quad (10.45)$$

The proof of this can be found in the book *Principles of Interactive Computer Graphics*, (Newman and Sproull, 1979). One can derive equation (10.44) as a special case of this, as well as the formulas, for rotating about the *x*-axis and the *z*-axis. We ask you to do that in the Student Learning Opportunities.

Student Learning Opportunities

1 Show that the matrix that rotates a vector about the *y*-axis is a special case of equation (10.45). [Hint: You need to find a point (a, b, c) on the *y*-axis such that $a^2 + b^2 + c^2 = 1$.]

2 Find the matrix in 3 dimensions that rotates a point P, θ degrees about the *x*-axis.

3 Suppose we have the matrix transformation that stretches a point *P* in the *x* direction by *a*, in the *y* direction by *b*, and in the *z* direction by *c*. That is, $T(x, y, z) = (ax, by, cz)$. What is the matrix of this transformation?

4 (C) You have shown your students how to derive the area of an ellipse from the area of a circle with radius 1 (see the end of the section "Transforming areas" in this chapter) and how, from this, one can get the formula for the area of a circle of radius *r*. They are curious to know if, from the volume of a sphere of radius 1, you can derive the formula for the volume of an ellipsoid in a similar manner, and from that, get the volume of a sphere of radius *r*. Assuming that the volume of a sphere of radius 1 is $\frac{4}{3}\pi$, show, using Theorem 10.30, that the volume of an ellipsoid, with semi axes of length *a*, *b* and *c* is $\frac{4}{3}\pi abc$ and from this, derive the formula for the volume of a sphere of radius *r*. [Hint: Perform a stretch on the sphere in the *x*, *y*, and *z* directions separately by factors of *a*, *b*, and *c* respectively.]

5 Find the matrix of the transformation from R^3 to R^3 that first rotates a point *P* counterclockwise about the *z*-axis by an angle θ and then reflects the result about the *xy*-plane.

6 An orthogonal projection in the xy-plane is a transformation that takes the point (x, y, z) into (x, y, 0). Write the matrix of this transformation and then find the matrix that rotates a point 90° about the y-axis and then projects the result onto the xy-plane.

10.9 Reflecting on Reflections

LAUNCH

Suppose that triangle ABC, given below in Figure 10.40, is an equilateral triangle having sides of length 6 inches.

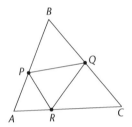

Figure 10.40

Answer the following questions.

1. What is the perimeter of triangle ABC?
2. Suppose that points P, Q, and R are midpoints of the sides, what is the perimeter of triangle PQR?
3. Suppose that P, Q, and R divide the sides as follows: $AR = 2$, $RC = 4$, $CQ = 2$, $QB = 4$, $BP = 2$, and $PA = 4$. What is the perimeter of triangle PQR? (You may need to use some of the trigonometry laws to calculate this one.)
4. Compare the perimeters you got in Questions 2 and 3. Which one has the greater perimeter?
5. If P, Q, and R can be chosen arbitrarily on the sides of the triangle ABC and we compute the perimeters of the different triangles PQR, where do you think P, Q, and R should be located to give us the minimum perimeter? Can you prove it?

We hope that this launch problem has piqued your curiosity. As you read this section, you will find out more information about this and other interesting problems.

Have you ever spent a lot of time and effort trying to solve a problem, and then someone shows you a simple and elegant solution? There are many difficult problems in mathematics, which when looked at from the "right" point of view make the solution easier. In this section we present two such problems. The first is a rather famous one and is really connected to a well known law about how light is reflected in mirrors. We give a more mundane example that secondary school students could relate to. It is quite difficult to find the solution to the second problem we present but, with the material we have in this chapter, solving the problem becomes much simpler.

Example 10.31 *Jack lives at point A, 500 feet from the shore which is straight. He works at point B which is 1000 feet from his home measured horizontally and 800 feet from the shore. (See Figure 10.41 below.) Each morning he walks to the shore before going to work, and then continues onto work. He enjoys this walk and likes to do it daily. But sometimes he is running late and has to make his trip in the shortest amount of time. He needs to determine the point C on the shoreline which will minimize his travel time, assuming that he walks at a constant rate. Find this point C and show in the diagram below, that $\theta_1 = \theta_2$ at C.*

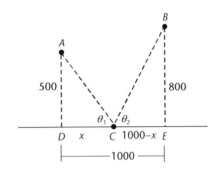

Figure 10.41

Solution. Since Jack walks at a constant rate, if he travels the shortest distance, he will travel in the shortest time. Thus, we can recast our problem as, what is the point C on the shore that will minimize $AC + CB$? You might want to think about that for a minute or so before we give the solution. Is the point C midway between D and E, or 1/3 of the way, or some other fraction of the distance from D to E?

Here is a calculus solution to this problem. We let x be the horizontal distance from A to C. Then by the Pythagorean Theorem, using the picture above, we have that $AC = \sqrt{(500)^2 + x^2}$, and $CB = \sqrt{(1000 - x)^2 + (800)^2}$ and we want to minimize

$$M = AC + CB$$
$$= \sqrt{(500)^2 + x^2} + \sqrt{(1000 - x)^2 + (800)^2}.$$

From calculus, you may remember that to solve this minimum problem we must take the derivative, and set it equal to zero and then solve for x. Here are some of the details:

$$M' = \frac{x}{\sqrt{(500)^2 + x^2}} - \frac{(1000 - x)}{\sqrt{(1000 - x)^2 + (800)^2}} = 0.$$

To solve this, we add $\frac{(1000-x)}{\sqrt{(1000-x)^2+(800)^2}}$ to both sides to get

$$\frac{x}{\sqrt{(500)^2 + x^2}} = \frac{(1000 - x)}{\sqrt{(1000 - x)^2 + (800)^2}}. \tag{10.46}$$

To show that $\theta_1 = \theta_2$, observe that $\sin\theta_1 = \frac{x}{\sqrt{(500)^2 + x^2}}$ and that $\sin\theta_2 = \frac{(1000-x)}{\sqrt{(1000-x)^2 + (800)^2}}$ and by equation (10.46) these are equal. So

$$\sin\theta_1 = \sin\theta_2$$

But, if $\sin\theta_1 = \sin\theta_2$, it follows that $\theta_1 = \theta_2$ since both angles are acute. To find x, we must solve equation (10.46). We can do that on a calculator using a solver, or we can square both sides, do a lot of algebra, and end up with a quadratic whose solution is $x \approx 384.615$.

Let us solve Jack's problem more efficiently. Look at Figure 10.42 below, where C is any point on the shoreline.

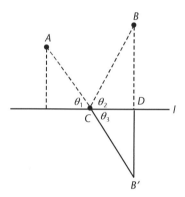

Figure 10.42

Reflect CB about l to get CB'. Now $AC + CB = AC + CB'$. This distance, $AC + CB'$ represents the length of a path from A to B'. But the shortest path from A to B' is a straight line, and thus AB' must be a straight line. It follows that

$$\theta_1 = \theta_3 \tag{10.47}$$

by vertical angles. But by the definition of a reflection $BD = B'D$ and angles BDC and $B'DC$ are right angles and, of course, $CD = CD$. Thus triangle BDC is congruent to triangle $B'DC$ and hence

$$\theta_3 = \theta_2. \tag{10.48}$$

Using equations (10.47) and (10.48), we get that $\theta_1 = \theta_2$ and we have solved Jack's problem more efficiently without the use of calculus. Of course, to find x, we realize that, if $\theta_1 = \theta_2$, then $\sin\theta_1 = \sin\theta_2$ and hence equation (10.46) is true. We then solve it as we did earlier.

Now let us imagine that the shoreline is no longer a shoreline but rather a mirror. There is a principle in physics that states that, if light bounces off of a mirror when traveling from A to B, it follows the path that takes the least amount of time. This least time principle coupled with the fact that light travels at a constant speed of 186,000 miles per second, tells us that, when light bounces off a mirror, it travels along the shortest path. What we are saying is that the light reflection problem and our problem are essentially the same. From this it follows that the angle at which the light hits the mirror, the so-called angle of incidence, equals the angle that it bounces off the mirror, the angle of reflection. This is a famous law of physics.

The next problem is known as **Fagnano's problem** which we began in the launch.

510 Geometric Transformations

Begin with acute triangle ABC and let R be any point on the base AC. Pick any points P and Q on the sides of the triangle and compute the perimeter of triangle PQR. Of all the points P and Q on the sides AB and BC, which ones give us a triangle of minimum perimeter? Next, determine the position of R that minimizes the perimeter of all possible triangles PQR.

This is not an easy problem. But here we show the elegant solution obtained by the mathematician Fejer, nearly 250 years after Fagnano and his son solved this problem.

Here is a typical triangle shown in Figure 10.43.

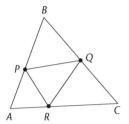

Figure 10.43 Figure is not drawn to scale.

Now, we reflect segment BR about line AB to get segment BS. (See Figure 10.44 below.) We then reflect segment BR again about line BC, to get segment BT. Since reflections preserve distance, $BS = BT$. Pick any points U and V on AB and BC, respectively, and draw segments UR and VR.

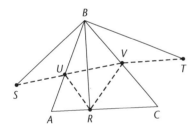

Figure 10.44

First, we observe that triangles BSU and BRU are congruent. This follows because $BS = BR$ as we pointed out, and obviously $BU = BU$. Finally, $\angle SBU = \angle UBR$ since reflections preserve angles between lines. (We ask you to verify this in the Student Learning Opportunities.) So by $SAS = SAS$, triangles SBU and RBU are congruent. It follows that the corresponding parts, SU and RU are congruent. In a similar manner using triangles RBV and TBV, which are congruent, we have that $VR = VT$. Thus, the perimeter of UVR, which is $UR + UV + VR = SU + UV + VT$. But $SU + UV + VT$ will be a minimum when SVT is a straight line. And this will happen when we join S and T by a straight line to get our points U and V. Thus, we have indicated how to construct the triangle of minimum perimeter inscribed in triangle ABC when we begin with a point R on the base AC. We reflect segment BR about lines AB and BC and connect the endpoints, S and T of these line segments to get points P and Q. Triangle PQR will have minimum perimeter.

Now we turn to the question of how to find the minimum of the perimeters of *all* triangles PQR. Below we see a typical figure that results when we connect S and T to get our points U and V that will make the perimeter of triangle UVR a minimum for a fixed position R. We need to observe that whatever triangle BST we form, the measure of $\angle SBT$ is always the same, and in fact is twice the measure of angle ABC. This follows since $\angle SBU = \angle UBR$ and $\angle RBC = \angle CBT$ as we have indicated in Figure 10.45 below.

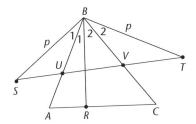

Figure 10.45

Now, by the Law of Cosines,

$$(ST)^2 = SB^2 + BT^2 - 2SB \cdot BT \cos \angle SBT$$
$$= p^2 + p^2 - 2p \cdot p \cos \angle SBT$$
$$= 2p^2 - 2p^2 \cos \angle SBT$$
$$= 2p^2(1 - \cos \angle SBT).$$

Since $(1 - \cos \angle SBT)$ is fixed, $(ST)^2$ will be a minimum when p^2 is a minimum and hence when p is a minimum. But the length p is equal to the length BR and BR will be minimum when BR is an altitude of the triangle. Thus we have shown that, when R is the foot of the altitude drawn to side AC, our ST will be minimum. But ST has length equal to the perimeter of the triangle. Thus the perimeter of the triangle will be minimum when R is where the altitude from B meets side AC.

Now this was our analysis when we chose R on side AC. Using a similar analysis, U must be the foot of the perpendicular to AB from S, and V must be the foot of the perpendicular from A to side BC. Thus our triangle of minimum perimeter is the triangle formed by the points that the three altitudes of the triangle meet the sides. This triangle is known as the **orthic** triangle.

Student Learning Opportunities

1 Change the first sentence in Example 10.14 to:

(a) Jack lives at point A, 800 feet from the shore which is straight. He works at B which is 1200 feet from his home measured horizontally and 900 feet from the shore. Find the location of point C in Jack's problem under these conditions.

(b) Jack lives at point A, d_1 feet from the shore which is straight. He works at B, which is d_2 feet from his home measured horizontally and d_3 feet from the shore. Write an equation that could be used to find the point C in Jack's problem and then solve it for x, if you can in terms of d_1, d_2, and d_3 using a computer algebra system, if you have one available.

2 In Figure 10.46 below,

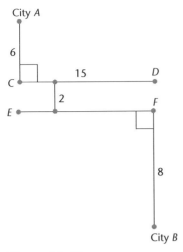

Figure 10.46

CD and EF are the north and south banks of a river with uniform width of 2 miles. City A is 6 miles north of CD; City B is 8 miles south of EF and 15 miles east of City A. We would like to travel from city A to City B by traveling to some point P on the north bank, crossing the riverbank at a right angle to some point Q on the south bank, and then traveling onto B. Find the length of the shortest such path.

3 Visit the website: http://www.cut-the-knot.org/Curriculum/Geometry/Fagnano.shtml to find four solutions of Fagnano's problem. Of all the solutions you have seen, which do you find the easiest and why?

4 Do some research on the Internet. Did Fagnano solve his problem using purely geometric methods? If not, what did he use?

CHAPTER 11

TRIGONOMETRY

11.1 Introduction

A common complaint of students is that they will never make use of the mathematics they learn in school. Trigonometry is an area of mathematics that, in fact, they will most probably use in their lifetime and whose applications are numerous. While originally defined to be the study of triangles and the relationships between their sides and angles, Wikipedia encyclopedia (which contains the most up to date input from people in different fields), states that some of the fields that make use of trigonometry are:

"…acoustics, architecture, astronomy biology, cartography, chemistry, civil engineering, computer graphics, geophysics, crystallography, economics (in particular, in analysis of financial markets), electrical engineering, electronics, land surveying and geodesy, many physical sciences, mechanical engineering, medical imaging (CAT scans and ultrasound), meteorology, music theory, number theory (and hence cryptology), oceanography, optics, pharmacology, phonetics, probability theory, psychology, seismology, statistics, and visual perception."

Also, the Fourier Transform, which is based on sine functions, and cosine functions, is one of the primary mathematical tools used in many modern devices such as: portable phones, digital cameras, digital TVs, computer image processing, the Internet, satellite communications, teleconferencing systems, and compact disc players.

We will develop all of the traditional trigonometric concepts that are part of the secondary school curriculum and will examine how trigonometric functions can be used to solve cubic equations and form Lissajous curves. We end the chapter by showing how vectors can be used to prove geometric theorems that students usually encounter in their study of Euclidean geometry.

As you can see, this is quite a long list of uses of trigonometric functions. Let's start by examining some of the most typical and impressive applications of trigonometry.

11.2 Typical Applications Using Angles and Basic Trigonometric Functions

LAUNCH

As unbelievable as it may seem, several thousand years ago Erathosthenes computed the radius of the earth! Here is how he did it, way back then. He assumed the world was spherical and that the rays of sun that hit the earth were all parallel to each other. At the summer solstice at noon, if a vertical stick was placed in the ground in Syene, Egypt shown as point B in Figure 11.1 below, there was no shadow. (Syene modern day is known as Aswan.)

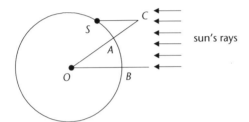

Figure 11.1

However, in the nearby town of Alexandria, shown at A, which he assumed was on the same longitude line, there was a shadow cast on a vertical pole. He measured the angle SCA made by the shadow and the stick. Since the rays of the sun were being assumed parallel, by alternate interior angles, angle AOB is the same as angle SCA which we call θ. Now, $\overset{\frown}{AB}$ being part of a circle, has length proportional to the circumference of the circle. Specifically, $\frac{\theta}{360} = \frac{AB}{2\pi r}$ where r is the radius of the earth. Since he measured θ to be $\approx 7.2°$, and the distance $\overset{\frown}{AB}$ was approximately 5000 stades, where a stade was approximately 559 feet, he substituted into the previous formula and solved for r. What did he get? Don't forget to convert to miles. (1 mile = 5280 feet.)

If you completed the launch problem correctly, you probably got the same estimate for the radius of the earth that Erathosthenes got which, although was not as good as today's estimates, was amazingly quite close. This is especially impressive, given the fact that he used little technology and his assumption that Syene and Alexandria are on the same longitude line is not true. Nevertheless, you have gotten a taste of how powerful angles are in their ability to help measure the unmeasurable. We hope you will enjoy reading other amazing applications throughout this section.

11.2.1 Engineering and Astronomy

Let us now examine some of the other unmeasurable distances that trigonometry has helped us find. We will begin with some difficult engineering problems that can be solved using elementary trigonometric relationships. Here are some interesting applications.

There are several tunnels throughout the United States that are built through mountains. For example, when the railroads were being built in the 1800s and pushing south, the Georgia Chettogetta mountains were a formidable obstacle. To connect Atlanta and Chattanooga, a tunnel had to be built through the mountains. The tunnel was considered one of the engineering marvels of the 1800s. But as far back as the 6th century BC, such tunnels had been built. The tunnel of Samos was a water tunnel 4000 feet long, excavated through a limestone mountain on the Greek island of Samos. Two separate teams dug from each end towards the middle. Of course, how did they know the direction to dig in so that they would meet in the middle? The Greek mathematician Heron devised a way to do that and we present it here. It is a very nice application of trigonometry, which we will now investigate.

Example 11.1 *The irregular shapes in Figure 11.2 below represent an aerial view of a set of mountains. Heron's idea was to pick convenient points A and B on each side of the mountain where the road could go through and find a point C (at the same level as A and B) for which the angle ACB is a right angle. Imagine a straight line joining A to B through the mountain. That line represents the tunnel we seek. If AC = 2000 feet and BC = 1500 feet, find the angles A and B that will tell each crew the direction in which to dig.*

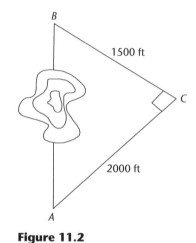

Figure 11.2

Solution. Since $AC = 2000$ feet and $BC = 1500$ feet, then $\tan A = \dfrac{1500}{2000}$ from which it follows that angle $A = \tan^{-1}(1500/2000) = 36.870$ degrees. Since angle B is the complement of angle A, it is 53.130 degrees. Thus, each crew at A and B knows the direction to dig in so that they meet somewhere inside the mountain!

This is such a nice application for a simple idea! Here is another application.

Example 11.2 *The radius of the earth was computed by Eratosthenes in the third century BC by an ingenious method, which we point out in the Student Learning Opportunities. Here is a modern way which we can use to verify his findings. We send a satellite into orbit at 600 miles above the earth. It has on-board instruments to measure the angle formed by the vertical (looking straight down to the earth) and the line of sight to the horizon. This angle is roughly $60.276°$. (See Figure 11.3 below.)*

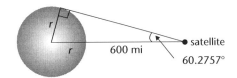

Figure 11.3

Use this to find the radius of the earth, assuming that the earth is a sphere.

Solution. Using the right triangle shown, $\sin 60.2757° = \dfrac{r}{r+600}$. But since $\sin 60.2757° = 0.86842$ we have

$$0.868\,42 = \frac{r}{r+600}.$$

Cross multiplying we get

$$0.868\,42r + 521.05 = r.$$

Subtracting $0.86842r$ from both sides we get

$$521.05 = 0.13158r$$

and dividing by 0.13158, we get $r \approx 3959.9$ miles.

Using right triangles and methods even more basic than this, it is possible to find the distance from the earth to the sun, and from the earth to the moon. Indeed, trigonometry was originally used to study such problems of distance in astronomy. We will point out more of these applications of trigonometry in the Student Learning Opportunities. Imagine how remarkable it was, in ancient times, and still is, to be able to measure the unmeasurable and discover information about important objects in our universe! It is this amazing aspect of mathematics that Richard Feynman the well known physicist must have been thinking about when he said, "If you want to understand nature, you must be conversant in the language in which nature speaks to us [mathematics]."

11.2.2 Forces Acting on a Body

Physicists are always studying problems that involve forces acting on bodies and they are often concerned with finding the net force that results from several different forces acting on the body at the same time.

Force is an example of something called a **vector**, which is anything having a magnitude and direction. For example, a force of 20 pounds acting on an object at an angle of 45 degrees with the horizontal is an example of a vector. We say that the magnitude of the force is 20 pounds, while the direction of the vector is 45 degrees with the horizontal. Often, we hear people talking about something traveling at say 30 miles per hour. This is NOT a vector, since it does not tell us the direction in which the object is traveling. This 30 miles per hour is just the speed of the object. If, on the other hand, an object was traveling at a rate of 30 miles per hour east, then this quantity, called the velocity, is a vector. The magnitude of the vector is 30 miles per hour, and the direction is east. Other examples of vectors are acceleration, magnetic force, electrical force, and tension in a cable. All this points to other applications in the sciences, specifically physics, once again emphasizing the interdisciplinary nature of mathematics.

Typically, we represent a vector by an arrow whose length is the magnitude of the vector and whose angle is the direction of the vector. For example, a force of 10 pounds acting at a 50 degree angle to the horizontal, may be pictured as in Figure 11.4 below, where the length of the arrow is 10.

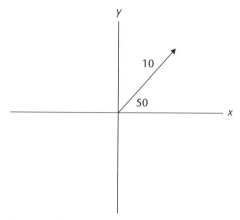

Figure 11.4

It is common to denote vectors by boldface letters. So, if we talk about the velocity of something, we might call it **v**. If we have a vector **v**, we can denote its magnitude by |**v**| as seen in most mathematics books, or, just the letter v without boldface, which is what we will do, as it is more consistent with what is seen in physics books. Thus, a letter with boldface refers to the vector, and the letter without boldface refers to its magnitude.

What physicists have discovered is that each force, **F**, is made up of two separate forces, $\mathbf{F}_x$ and $\mathbf{F}_y$ that work in directions that are perpendicular to each other. We will take the two directions to be horizontal and vertical for our purposes, though in real applications any force can be broken into forces acting in any two directions which are perpendicular.

When we break a force, **F**, into a horizontal force, $\mathbf{F}_x$ and a vertical force $\mathbf{F}_y$, we call the magnitudes of these forces, F_x, and F_y, **the horizontal and vertical components of F** and we write $\mathbf{F} = (F_x, F_y)$.

Below, in Figure 11.5, we see the representation of the force of 10 pounds making an angle of 50 degrees with the horizontal, discussed earlier, drawn together with its component forces.

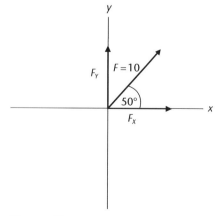

Figure 11.5

From the picture above, it is not clear how to get the component forces F_x and F_y from F. But suppose that we were not careful and, because of the notation $\mathbf{F} = (F_x, F_y)$, thought of F_x and F_y as the x coordinate and y coordinates of a point F. Then we might have drawn the following picture (Figure 11.6).

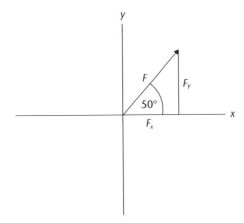

Figure 11.6

And then we might have said, "From the picture we see that $\dfrac{F_x}{F} = \cos 50$, and so $F_x = F \cos 50$. Similarly, $\dfrac{F_y}{F} = \sin 50$, and so $F_y = F \sin 50$."

What should bother you about taking the component forces, making a triangle out of them, and then using properties of a triangle to find the horizontal component of the force F? Well, F is a *force*, not a side of a triangle in the everyday sense! So what on earth are we doing here?

The remarkable thing is that what we just did happens to actually work when dealing with real life. That is, the magnitude of the original force and the horizontal and vertical components of the original force behave *as if* they were sides of a triangle. So if we wish to compute, F_x, and F_y the x and y components of a force, F, where the force makes an angle of θ with the horizontal, then the x and y components actually are given by

$$F_x = F \cos\theta \tag{11.1}$$

and

$$F_y = F \sin\theta. \tag{11.2}$$

That these formulas work is nothing short of amazing! This represents a wonderful joining of mathematics and the real world.

Now, suppose that we have several forces, $\mathbf{F}_1, \mathbf{F}_2, \mathbf{F}_3, \ldots, \mathbf{F}_n$ all acting on the same body and we want to get the net result of all these forces. Amazingly, according to experimental results, this is very easy to do if we know the components of the forces. According to this law of physics, we simply add the x components of all the forces and that will be the x component of the net force. We also add the y components and that will be the y component of our net force. It is because of this addition of components that the net force that results from forces, say $\mathbf{F}_1, \mathbf{F}_2, \mathbf{F}_3, \ldots, \mathbf{F}_n$ acting on a body is denoted by $\mathbf{F}_1 + \mathbf{F}_2 + \mathbf{F}_3 + \ldots + \mathbf{F}_n$. Thus, forces add in much the same way as matrices do. Here is an illustration with two forces.

Example 11.3 *Suppose that we have two forces* F_1 *and* F_2 *acting on a body placed at the origin, where the forces are* $F_1 = (5, 6)$ *and* $F_2 = (1, 2)$, *and where all numbers are in pounds.* (a) *Find the net force,* $F_1 + F_2$, *acting on the object.* (b) *Find the direction in which the body will move.*

Solution. (a) The net force is $F = F_1 + F_2 = (5, 6) + (1, 2) = (6, 8)$, obtained by adding all the x components of the forces and the y components of the individual forces F_1 and F_2. Thus, the net force has an x component of 6 pounds and a y component of 8 pounds.

We can graph this net force $F = (6, 8)$ as an arrow as shown below in Figure 11.7.

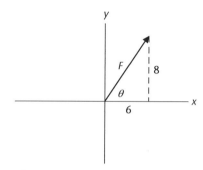

Figure 11.7

Solution. (b) The direction the body will move in, when subject to these individual forces, F_1 and F_2, is the direction of the net force (the angle θ in the above picture), as has been verified experimentally over and over. This is quite astounding, is it not? We can find θ by using the picture above. Since F, the magnitude of the force F, behaves as if it were the hypotenuse of a triangle with sides $F_x = 6$ and $F_y = 8$, the angle the net force makes with the horizontal satisfies $\tan \theta = \frac{8}{6}$. Thus, θ which can be found by pressing the $\boxed{\tan^{-1}}$ button on the calculator is $\tan^{-1}(8/6) \approx 53°$ and the object will move in a direction of 53° to the horizontal.

To compute the size of the net force, we find the length of the arrow. Using the Pythagorean Theorem, this is, $\sqrt{6^2 + 8^2} = 10$ pounds. In short, the object will move as if a single force of 10 pounds acted on it and pulled it in a direction of approximately 53° to the horizontal.

We have mentioned the remarkable fact that the horizontal and vertical components, F_x and F_y of a force, F, behave in all respects like the sides of a right triangle and that this has been repeatedly verified experimentally. This same idea carries over to all kinds of other vectors, like electrical and magnetic forces. Thus, this concept of representing vectors by arrows, and then by sides of triangles, combined with trigonometric analysis is an important and powerful tool in physics.

Now let us move forward and consider the following example.

Example 11.4 *Three forces act on a body. The first force,* F_1 *is 15 pounds, and acts in a direction of 10° with the horizontal. The second,* F_2, *is a force of 25 pounds and acts at an angle of 45 degrees with the horizontal. The third force,* F_3 *is a 30 pound force and acts at an angle of 60 degrees with the horizontal. Find the net force acting on the body and then determine the direction in which the body will move.*

How is this example different from the previous ones we have examined? You got it! As in most real-life problems, the forces are not given in component form. How do we find the components of the net force? Well, one way to do it is to break the forces into components, using $F_x = F \cos\theta$ and $F_y = F \sin\theta$ for each force F, and then add the results. Here is the solution.

Solution. We will find the x and y components of each force using the formulas, $F_x = F \cos\theta$ and $F_y = F \sin\theta$, and then add the x and y components to get the net force. Here are the results for $\mathbf{F_1}$:

$(F_1)_x = F_1 \cos\theta = 15 \cos 10° = 14.772$

$(F_1)_y = F_1 \sin\theta = 15 \sin 10° = 2.6047$.

In a similar manner we have for $\mathbf{F_2}$ that

$(F_2)_x = F_2 \cos\theta = 25 \cos 45° = 17.678$

$(F_2)_y = F_2 \sin\theta = 25 \sin 45° = 17.678$.

and for $\mathbf{F_3}$ we have

$(F_3)_x = F_3 \cos\theta = 30 \cos 60° = 15$

$(F_3)_y = F_3 \sin\theta = 30 \sin 60° = 25.981$.

Now, the x and y components of the net force, F_x and F_y are, respectively,

$F_x = (F_1)_x + (F_2)_x + (F_3)_x = 14.772 + 17.678 + 15 = 47.45$

$F_y = (F_1)_y + (F_2)_y + (F_3)_y = 2.6047 + 17.678 + 25.981 = 46.264$.

So $\mathbf{F} = (F_x, F_y) = (47.45, 46.26)$ and the magnitude of the net force, $\mathbf{F}$ in pounds is, $\sqrt{47.45^2 + 46.264^2} = 66.271$ pounds. The angle, θ, that that the net force makes with the horizontal satisfies $\tan\theta = \frac{F_y}{F_x} = \frac{46.264}{47.45} = 0.97501$. So $\theta° = \tan^{-1}(0.97501) \approx 44.275°$. Thus, our body will move at an angle of 44.275° to the horizontal, and will move as if a single net force of 66.271 pounds were pulling it in that direction.

Thus far, the formulas of equations 11.1 and 11.2 have worked beautifully and would be great if we could always apply them. But, there is still a case we haven't addressed. If we have the force, **F**, shown in figure 11.8 below,

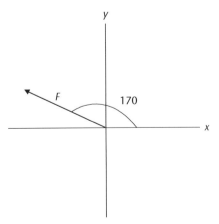

Figure 11.8

and we tried to compute the x component of it as we did above, we would have to compute $F_x = F \cos 170$. But what would cosine of 170 degrees mean? Thus far, we have only defined the cosine of an angle in a right triangle where there are no angles of 170 degrees. Furthermore, if the component forces acted as if they were sides of a triangle, with hypotenuse F, then the x component, which depends on cos 170, would have to be negative, since our force is represented by an arrow in the second quadrant. Situations like this are what motivated mathematicians and scientists who worked in tandem to start talking about trigonometric functions of angles more than 90 degrees and trying to define them in a way that made practical sense. We follow that train of thought in the next section. We will return to vectors again in the last section of this chapter after we have developed more needed concepts.

Student Learning Opportunities

1 **(C)** Solve the launch problem.

2 **(C)** Your students are fascinated to find out that basic trigonometry can be used to find the distance from the earth to the moon and are eager to find out some details. They turn to you for guidance. Use Figure 11.9 below and the following suggestion, to show how it can be done.
Suggestion: It is known that, when the moon is at its zenith at point B, it is observed to be at the horizon at point A, which is 6155 miles from B along the surface of the earth. Knowing that the radius of the earth is roughly 3960 miles, we can find the angle θ in the figure by realizing that the arc AB is $2\pi r \frac{\theta}{360}$, which is the formula for the length of the arc. Use this value of θ to find the distance from the earth to the moon.

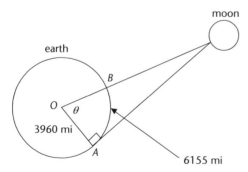

Figure 11.9

3 **(C)** After solving the last problem, your students ask if methods similar to that of Question 2 can be used to find things like the distance stars are from the earth, the distance from the earth to the sun, the maximum distance planets are from the sun, and so on. You say, "Yes" and offer them one more example, that of calculating the distance Venus is from the sun. You give them the following facts and picture: Venus' orbit around the sun is almost perfectly circular and the same is true for the earth's orbit. So we can draw two concentric circles representing their orbits with the sun as center. (See Figure 11.10 below.)

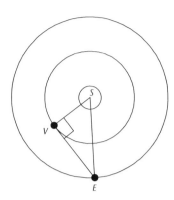

Figure 11.10

When the angle between the sun, earth, and Venus, (angle *SEV*) is maximum, the angle between the sun, Venus and earth (angle *SVE*), is 90°. The maximum angle *SEV* is about 46.3 degrees.

Now, using this information, calculate the approximate distance from Venus to the sun. Note that the distance from the Earth to the sun is approximately 93,000,000 miles.

4 Find the components of the net force and the direction of the net force acting on a body (assumed to be at the origin) if the individual forces are:

(a) **F**$_1$ = (−2, 3), **F**$_2$ = (−3, 4), and **F**$_3$ = (7, 5).
(b) **F**$_1$ = (9, 6), **F**$_2$ = (2, − 5), and **F**$_3$ = (6, 14).

5 Find the approximate net force on a body (assumed to be at the origin) if there are two forces, **F**$_1$ and **F**$_2$ acting on the body and they are given below. Here we want the size of the net force as well as the angle it makes with the positive *x*-axis.

(a) **F**$_1$ is a force of 30 pounds acting at an angle of 34 degrees with the horizontal and **F**$_2$ is a force of 50 pounds acting at an angle of 80 degrees with the *x*-axis.
(b) **F**$_1$ is a force of 40 pounds acting at an angle of 10 degrees with the positive *x*-axis and **F**$_2$ is a force of 30 pounds acting at an angle of 70 degrees with the positive *x*-axis.

6 A woman is standing on a hill 150 feet from a flagpole. She has with her a surveying tool that measures angles. She finds with this tool that, looking to the top of the flagpole, the angle is 45°, while looking towards the bottom, the angle is 24°. (See Figure 11.11 below.) What is the height of the flagpole?

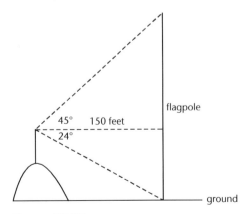

Figure 11.11

7 Let us change the previous problem. Suppose, that instead of assuming that the woman is standing 150 feet from the flagpole, assume that the flagpole is 150 feet high. Approximately how far is the surveying tool from the flagpole?

8 In Figure 11.12 we see an object being suspended by two cables. If the weight of the object is 100 pounds, find the tension in the cables. (The tension is essentially how much force each cable is exerting to do its part to keep the weight from crashing down.) [Hint: Find the horizontal components of the tension in each cable. Since they must balance each other out, they must be equal. The sum of the vertical components of the forces must be equal to the weight.]

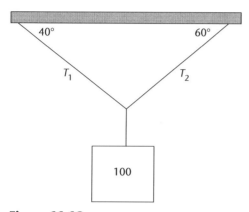

Figure 11.12

11.3 Extending Notions of Trigonometric Functions

**LAUNCH**

1 Find, without the use of a calculator, the sine, cosine, and tangent of 0°. Explain how you got your answers.
2 Find, without the use of a calculator, the sine, cosine, and tangent of 90°. Explain how you got your answers.
3 Make conjectures about the values of the sine, cosine, and tangent of 270°. Explain your reasoning.

In middle school you most likely learned the mnemonic, SOHCAHTOA, which was a way of remembering how to compute the sine, cosine, and tangent of an angle. But, as you have seen from the launch question, this mnemonic is not very helpful for special angles of 0°

and 90° or for angles that are larger than 90°. After reading this section, you will learn how to expand the definitions of the trigonometric relationships so that these cases will be easily dealt with.

11.3.1 Trigonometric Functions of Angles More than 90 Degrees

In the last section we spoke of how practical considerations led to the need to consider angles of more than 90 degrees, which we will now examine in more detail.

Suppose that we have an angle θ. As we know, every angle has two sides, an initial side and a terminal side. The angle is always measured as a rotation from the initial side to the terminal side. (See Figure 11.13 below.)

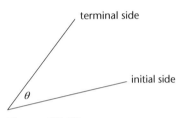

Figure 11.13

An angle is said to be positive if the direction from the initial side to the terminal side is counterclockwise, and negative if the direction from the initial side to the terminal side is clockwise. (This is just a convention. Mathematicians could have done it the opposite way.) Thus, with the labeling in the figure above, θ is a positive angle.

An angle of 360° represents a complete counterclockwise revolution. Angles can be more than 360°. An angle of 720° represents two complete revolutions, while an angle of 800° represents two complete counterclockwise revolutions and another 80°. (See Figure 11.14 below.)

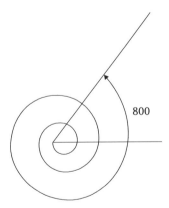

Figure 11.14

(An angle of −800° represents the same only in the clockwise direction.)

In what follows, we will assume that the initial side of any angle we discuss is along the positive x-axis. (Below in Figure 11.15 we see an angle, θ, whose initial side is the x-axis.) We pick an arbitrary point, (x, y), on the terminal side, and suppose that it is at a distance r from the origin, where r is greater than 0.

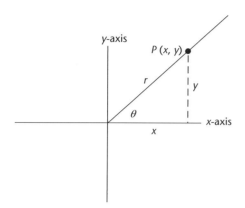

Figure 11.15

Then the horizontal leg of the triangle is x, the vertical leg is y, and the hypotenuse is r, as indicated in Figure 11.15. Accordingly, we have

$$\cos\theta = \frac{x}{r}$$

$$\sin\theta = \frac{y}{r}$$

$$\tan\theta = \frac{y}{x} \quad (11.3)$$

It is this basic definition that leads us to analogous definitions of functions more than 90 degrees.

For any angle, we define the sine, cosine, and tangent of θ as we did above in display (11.3). Namely, we pick an arbitrary point (x, y) on the terminal side of the angle at a distance r from the origin and use the definitions given in display (11.3). It is important to note that x and y in the definitions given in equation (11.3) do not always refer to sides of a triangle, rather to points on the terminal side of the angle. (This will play a part when we discuss things such as sine of 90°.) The reciprocals of the functions $\sin\theta$, $\cos\theta$, and $\tan\theta$ are known as cosecant θ, secant θ, and cotangent θ (abbreviated $\csc\theta$, $\sec\theta$, and $\cot\theta$, respectively.) These will be developed more in the Student Learning Opportunities, since most of you are familiar with these terms.

The definitions given in display (11.3) for sine, cosine, and tangent immediately imply that it is possible for the sine or cosine of an angle to be negative. Specifically, in the first quadrant, according to display (11.3), all of the functions, sine, cosine, and tangent are positive, since x, y, are positive in that quadrant and r is always positive. (Recall that the quadrants are numbered 1–4 counterclockwise, where quadrant 1 is where $x > 0$ and $y > 0$.) In the second quadrant, the sine is positive, since y and r are positive in that quadrant, while cosine and tangent are negative, since x is negative. In the third quadrant, the tangent is positive and the sine and cosine are negative. In the fourth quadrant, the cosine is positive and the sine and tangent are negative. You should make sure you see why before you proceed. Figure 11.16 illustrates these facts:

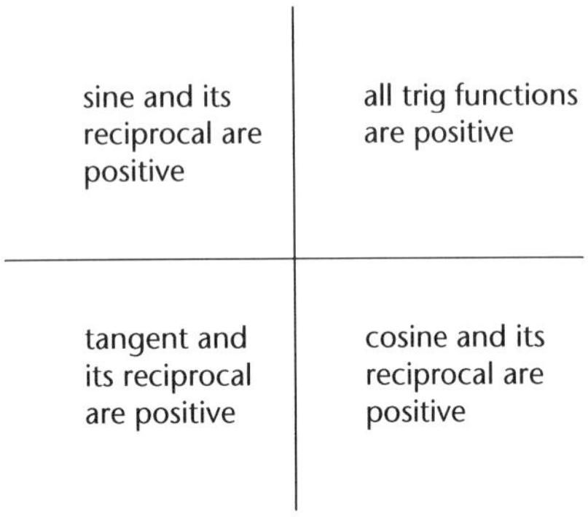

sine and its reciprocal are positive	all trig functions are positive
tangent and its reciprocal are positive	cosine and its reciprocal are positive

Figure 11.16

A way to remember this is **A-S-T-C**. All Students Talk Constantly!

Here is a typical secondary school problem.

Example 11.5 *The point $P(-8, 15)$ is on the terminal side of an angle θ. Find the sine, cosine, and tangent of θ.*

Solution. We draw the picture first as shown below in Figure 11.17.

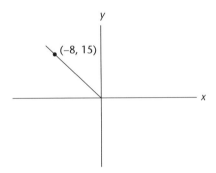

Figure 11.17

The x coordinate of P is -8, the y coordinate is 15, and the distance this point is from the origin is $\sqrt{(-8)^2 + 15^2} = 17$. Thus,

$$\cos \theta = \frac{-8}{17}$$
$$\sin \theta = \frac{15}{17}$$
$$\tan \theta = \frac{15}{-8}.$$

So, we do see that the sine is positive and the cosine and tangents are negative, as they should be in the second quadrant.

An angle whose terminal side is not in the quadrant but on one of the axes is known as a **quadrantal angle**. Thus, quadrantal angles are $0°$, $90°$, $180°$, $270°$, $360°$ and so on.

Example 11.6 *Find, without the use of a calculator, the sine, cosine, and tangent of each of the following quadrantal angles* (a) 90 (b) 180 (c) 270.

Solution. (a) The terminal side of a 90 degree angle is on the *y*-axis. Pick the point (0, 1) on the terminal side of $\theta = 90$. Here $x = 0$, $y = 1$, and $r = 1$. So

$$\sin 90 = \frac{y}{r} = \frac{1}{1} = 1$$

$$\cos 90 = \frac{x}{r} = \frac{0}{1} = 0 \quad \text{and}$$

$$\tan 90 = \frac{y}{x} = \frac{1}{0} \quad \text{which is not defined.}$$

(b) Pick the point $(-1, 0)$ on the terminal side of $\theta = 180$ (which is along the negative *x*-axis). Then $x = -1$, $y = 0$, and $r = 1$. Thus,

$$\sin 180 = \frac{y}{r} = \frac{0}{1} = 0$$

$$\cos 180 = \frac{x}{r} = \frac{-1}{1} = -1$$

$$\tan 180 = \frac{y}{x} = \frac{0}{-1} = 0.$$

(c) Pick the point $(0, -1)$ on the terminal side of $\theta = 270$. Then $r = 1$ and

$$\sin 270 = \frac{y}{r} = \frac{-1}{1} = -1$$

$$\cos 270 = \frac{x}{r} = \frac{0}{1} = 0$$

$$\tan 270 = \frac{y}{x} = \frac{-1}{0} \quad \text{which is not defined.}$$

We have, in this example, given a non-calculator solution to parts of the launch. Using Figure 11.18 below,

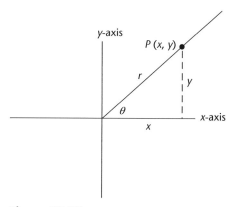

Figure 11.18

we see that

$$x^2 + y^2 = r^2.$$

Dividing this by r^2 we get

$$\frac{x^2}{r^2} + \frac{y^2}{r^2} = 1.$$

This can be rewritten as

$$\left(\frac{x}{r}\right)^2 + \left(\frac{y}{r}\right)^2 = 1.$$

Using the definitions of $\sin\theta$ and $\cos\theta$ from equation (11.3), this becomes

$$\sin^2\theta + \cos^2\theta = 1.$$

If we took a similar picture with the terminal side in another quadrant, we would have the exact same result. Thus, regardless of what θ is, $\sin^2\theta + \cos^2\theta = 1$. This you have undoubtedly seen and is called an identity. (You can see a visual representation of this identity by graphing the equation $Y = \sin^2\theta + \cos^2\theta$ on your calculator. What do you see? Is this what you expected?) We will develop the other Pythagorean identities related to the reciprocal functions in the Student Learning Opportunities.

11.3.2 Some Useful Trigonometric Relationships

Below in Figure 11.19 you see an angle of $-\theta$ and θ for θ in the first quadrant and points (x, y) and $(x, -y)$ on the terminal sides of θ and $-\theta$.

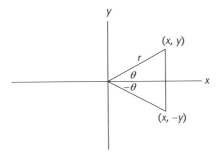

Figure 11.19

There is a similar picture for θ in the third quadrant, or any other quadrant. θ and $-\theta$ are always reflections of each other about the x-axis. It follows from this immediately that

$$\cos(-\theta) = \cos(\theta) \tag{11.4}$$

(since according to the above diagram they are both $\frac{x}{r}$) and that

$$\sin(-\theta) = -\sin(\theta) \tag{11.5}$$

since $\sin(-\theta)$ is $\frac{-y}{r}$ and $\sin\theta$ is $\frac{y}{r}$.

In Figure 11.20 below, we see both θ and $180 - \theta$. You will find a similar picture regardless of which quadrant θ is in. (You should take some numerical values of θ to convince yourself of this.)

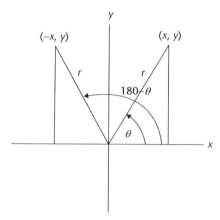

Figure 11.20

We notice right away, when we express everything in terms of x, y, and r, that

$$\sin(180 - \theta) = \sin \theta \tag{11.6}$$

(since both are $\frac{y}{r}$) and that

$$\cos(180 - \theta) = -\cos \theta \tag{11.7}$$

(since one is $\frac{x}{r}$ and the other is $\frac{-x}{r}$).

We will make good use of these facts soon, but what we want to point out now is how the proofs that we gave in Chapter 5 for the Laws of Sines and Cosines, can be modified so that they are now valid even if θ is obtuse. You may recall that the proofs we gave in Chapter 5 required the triangles to be acute. (We had you do the obtuse case in Student Learning Opportunity 4 Section 2.1 Chapter 5 asking you to accept facts equations (11.6) and (11.7)).

So, for the sake of completeness, we will now give the proof of the Law of Cosines for obtuse triangles. We begin with the triangle ABC as shown in Figure 11.21 below and draw altitude h to side AB extended.

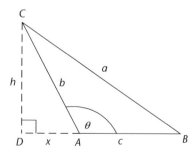

Figure 11.21

In right triangle BDC, we have

$$BC^2 = BD^2 + DC^2$$

or just

$$a^2 = (c + x)^2 + h^2.$$

This last equation expands to

$$a^2 = c^2 + 2cx + x^2 + h^2. \tag{11.8}$$

From right triangle ADC, we have $x^2 + h^2 = b^2$. Substituting for $x^2 + h^2$ in equation (11.8) we get that

$$a^2 = c^2 + 2cx + b^2. \tag{11.9}$$

From triangle ADC we have $\cos(\angle CAD) = \frac{x}{b}$, which tells us that $x = b\cos(\angle CAD)$. Substituting this value for x into equation (11.9), we get

$$a^2 = c^2 + 2cb\cos(\angle CAD) + b^2. \tag{11.10}$$

But $\angle CAD = 180 - \theta$ and $\cos(\angle CAD) = \cos(180 - \theta) = -\cos\theta$ by equation (11.7). Replacing $\cos(\angle CAD)$ by $-\cos\theta$, we get that

$$a^2 = c^2 - 2cb\cos\theta + b^2.$$

Rewriting this last equation we get

$$a^2 = b^2 + c^2 - 2cb\cos A$$

which is the usual form of the Law of Cosines. So, what does this tell us? In any type of triangle, if we know two sides and one included angle, we can find the third side. Or, if we know three sides, we can find all of the angles. Also, notice how similar this formula is to the Pythagorean Theorem. The size of a^2 increases or decreases based on the size of angle A. Notice, if angle A is $90°$, $\cos A = 0$ and we get the Pythagorean Theorem.

Since we now have meanings for sines and cosines of angles more than $90°$, it is natural to ask if the Law of Sines, presented earlier in Chapter 5, is also valid when one of the angles of the triangle is more than $90°$. The answer is, "Yes" and we give a partial proof here when angle A is obtuse. We know by equation (11.6) that $\sin(A) = \sin(180 - A)$. Thus,

$$\frac{a}{\sin(A)} = \frac{a}{\sin(180 - A)}. \tag{11.11}$$

Using Figure 11.22,

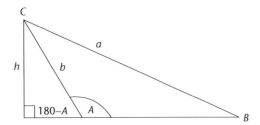

Figure 11.22

we see, that $\sin(180 - A) = \frac{h}{b}$. Thus equation (11.11) becomes

$$\frac{a}{\sin(A)} = \frac{a}{\sin(180 - A)} = \frac{a}{\frac{h}{b}} = \frac{ab}{h}. \tag{11.12}$$

Similarly, from the picture above, $\sin B = \frac{h}{a}$. Thus

$$\frac{b}{\sin B} = \frac{b}{\frac{h}{a}} = \frac{ab}{h} \tag{11.13}$$

Since by equations (11.12) and (11.13) $\frac{a}{\sin(A)}$ and $\frac{b}{\sin B}$ are both $\frac{ab}{h}$, we see that

$$\frac{a}{\sin(A)} = \frac{b}{\sin B}.$$

You should finish the proof and show that $\frac{b}{\sin B} = \frac{c}{\sin C}$ so that

$$\frac{a}{\sin(A)} = \frac{b}{\sin B} = \frac{c}{\sin C}.$$

The Law of Sines is more important than you might think. Other than the logical applications to architecture, construction and astronomy, where triangles abound, it is a key tool in surveying. It is often used to estimate distances over hilly terrain and was used in the Great Survey of India which took about 100 years to complete. It uses a method called triangulation where only one distance is measured to start with and then the region you are interested in surveying is broken into adjacent triangles. Using an instrument called a theodolite, which measures angles, all the other distances can be calculated from that one distance by using the Laws of Sines and/or Cosines repeatedly. We illustrate that with a scaled down version of this type of problem.

Example 11.7 *In Figure 11.23 below,*

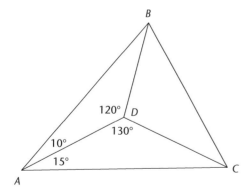

Figure 11.23

we see a schematic of a certain part of an area consisting of rocks and difficult to navigate terrain. The distance AD has been measured to be 500 feet. Using the theodolite, all other angles have been determined and are, as shown in the picture. Find all the remaining lengths.

Solution. We won't solve the whole problem. We will only give you the flow of the solution process. Since we know AD, we can use the law of sines in triangle ABD to find both AB and BD. Similarly, we can find the sides AC and DC using the Law of Sines in triangle ADC. Now that we know BD and DC, and angle BDC (which is 110°), we can find BC using the Law of Cosines. It is this kind of process that is done on a grand scale when surveying land.

The proof that we gave in Chapter 5 for $\sin(A + B)$ using Ptolemy's theorem was given for acute angles A and B. (See Example 5.20.) The proof can be modified with equations (11.6) and (11.7) to prove this theorem in the case when A or B is obtuse.

Are the usual rules for $\sin(A+B)$ and $\cos(A+B)$ true, even if the angles are more than 180 degrees? The answer is "Yes" to both questions and we turn to a proof of that now. The proof is somewhat algebraic, but the speed with which we get the result in *all* cases makes the work worthwhile. In many ways, this is an elegant proof.

Theorem 11.8 *For any angles θ_1 and θ_2, $\cos(\theta_1 + \theta_2) = \cos\theta_1 \cos\theta_2 - \sin\theta_1 \sin\theta_2$.*

Proof. The beauty of this proof is how it ties the notion of distance from Cartesian coordinates to trigonometry. The proof is somewhat surprising.

Since, when dealing with trigonometric functions of any angle, we are able to take any point on the terminal side, we take all our points at a distance r from the origin, which means they are all on a circle with radius r. Place angles θ_1, θ_2 and $-\theta_2$ as shown in Figure 11.24 below. (For clarification, θ_2 is the angle from OP to OQ, where O is the origin and $-\theta_2$ is the angle of the same size as θ_2, drawn from the positive x-axis, but drawn clockwise.)

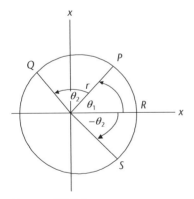

Figure 11.24

We list the coordinates of P, R, Q, and S. (These follow from equation (11.3).)

$P = (r\cos\theta_1, r\sin\theta_1)$

$R = (r, 0)$

$Q = (r\cos(\theta_1 + \theta_2), r\sin(\theta_1 + \theta_2))$

$S = (r\cos(-\theta_2), r\sin(-\theta_2)) = (r\cos(\theta_2), -r\sin(\theta_2))$.

These follow immediately from equation (11.3), since $x = r\cos\theta$ and $y = r\sin\theta$. Notice when finding the coordinates of S, we also used equations (11.4) and (11.5).

Now the arc RQ is intercepted by a central angle of $\theta_1 + \theta_2$, as is arc PS. (For arc PS, although the angle is $-\theta_2$, its *size* is θ_2. The negative sign simply tells us the direction from the initial side to the terminal side.) Thus, the chords RQ and PS have the same length from the well known result in geometry, which says that in a circle, equal arcs have equal chords. Since $RQ = PS$ we have, using the distance formula

$$\underbrace{\sqrt{(r\cos(\theta_1 + \theta_2) - r)^2 + (r\sin(\theta_1 + \theta_2) - 0)^2}}_{PQ} = \underbrace{\sqrt{(r\cos\theta_2 - r\cos\theta_1)^2 + ((-r\sin\theta_2) - r\sin\theta_1)^2}}_{RS}.$$
(11.14)

The rest is algebra.

Square both sides of equation (11.14) to get rid of the square roots and then expand what is left to get

$$r^2\cos^2(\theta_1 + \theta_2) - 2r^2\cos(\theta_1 + \theta_2) + r^2 + r^2\sin^2(\theta_1 + \theta_2)$$
$$= r^2\cos^2\theta_2 - 2r^2\cos\theta_1\cos\theta_2 + r^2\cos^2\theta_1 + r^2\sin^2\theta_2 + 2r^2\sin\theta_1\sin\theta_2 + r^2\sin^2\theta_1.$$

Divide this last equation by r^2 to get

$$\cos^2(\theta_1 + \theta_2) - 2\cos(\theta_1 + \theta_2) + 1 + \sin^2(\theta_1 + \theta_2)$$
(11.15)
$$= \cos^2\theta_2 - 2\cos\theta_1\cos\theta_2 + \cos^2\theta_1 + \sin^2\theta_2 + 2\sin\theta_1\sin\theta_2 + \sin^2\theta_1.$$

Now on the right side of equation (11.15), we have $\cos^2\theta_2 + \sin^2\theta_2$ and $\cos^2\theta_1 + \sin^2\theta_1$ both of which are 1, and on the left side we have $\cos^2(\theta_1 + \theta_2) + \sin^2(\theta_1 + \theta_2)$ which is also 1. So equation (11.15) simplifies to

$$2 - 2\cos(\theta_1 + \theta_2) = 2 - 2\cos\theta_1\cos\theta_2 + 2\sin\theta_1\sin\theta_2.$$
(11.16)

Subtracting 2 from both sides of equation (11.16) and dividing by -2, we get

$$\cos(\theta_1 + \theta_2) = \cos\theta_1\cos\theta_2 - \sin\theta_1\sin\theta_2$$

and we are done. How nice! ■

Corollary 11.9 *For any angles θ_1 and θ_2, $\cos(\theta_1 - \theta_2) = \cos\theta_1\cos\theta_2 + \sin\theta_1\sin\theta_2$.*

Proof. Since the formula $\cos(\theta_1 + \theta_2) = \cos\theta_1\cos\theta_2 - \sin\theta_1\sin\theta_2$ is true for *all* angles θ_1 and θ_2, we can replace θ_2 by $-\theta_2$ to get

$$\cos(\theta_1 - \theta_2) = \cos\theta_1\cos(-\theta_2) + \sin\theta_1\sin(-\theta_2).$$

Now we use equations (11.4) and (11.5) to immediately get that

$$\cos(\theta_1 - \theta_2) = \cos\theta_1\cos\theta_2 - \sin\theta_1\sin\theta_2.\ ■$$

Corollary 11.10 $\cos(90-\theta)° = \sin\theta°$, *regardless of what θ is.*

Proof. $\cos(90-\theta)° = \cos 90° \cos(\theta°) + \sin 90° \sin(\theta°) = 0\cos(\theta°) + 1\sin(\theta°) = \sin\theta°$. ∎

Corollary 11.11 $\sin(90-\theta)° = \cos\theta°$ *regardless of what θ is.*

Proof. From the previous corollary, which is true for all angles θ, we get, replacing θ by $90-\theta$:

$$\cos(90-(90-\theta))° = \sin(90-\theta)°$$

or just,

$$\cos(\theta) = \sin(90-\theta)$$

which is what we were trying to prove. ∎

Student Learning Opportunities

1 (C) Solve part (a) of the launch problem.

2 (C) Your students want to know why a rotation in the counterclockwise direction is called positive, rather than negative, which would make more sense to them. According to the text, what is the answer?

3 Find the approximate net force on a body (assumed to be at the origin) if there are two forces, F_1, and F_2 acting on the body and they are given below. Here we want the size of the net force as well as the angle it makes with the positive x-axis.
   (a) F_1 is a force of 30 pounds acting at an angle of 34 degrees with the positive x-axis. F_2 is a force of 50 pounds acting at an angle of 150 degrees with the positive x-axis.
   (b) F_1 is a force of 40 pounds acting at an angle of 10 degrees with the positive x-axis and F_2 is a force of 30 pounds acting at an angle of 110 degrees with the positive x-axis.

4 (C) Your students are confused about how to find the sine, cosine, and tangent of $-90°$ without the use of a calculator. What do you tell them?

5 We define the secant of θ as the reciprocal of $\cos\theta$, the cosecant of θ to be the reciprocal of $\sin\theta$, and the cotangent of θ to be the reciprocal of $\tan\theta$. The secant, cosecant, and cotangent of θ are abbreviated as $\sec\theta$, $\csc\theta$, and $\cot\theta$, respectively.
   (a) Find, without a calculator, $\sec 90°$, $\csc 90°$, $\cot 90°$.
   (b) Find, without a calculator, $\sec 180°$, $\csc 180°$, $\cot 180°$.
   (c) Find, WITH a calculator, $\sec 70°$, $\csc 70°$, $\cot 70°$.

6 Prove that, for any angle where $\cos\theta$ is not zero, $\tan\theta = \frac{\sin\theta}{\cos\theta}$.

7 Prove that, for any angle θ for which $\tan\theta$ and $\sec\theta$ are defined, $1 + \tan^2\theta = \sec^2\theta$.

8 Prove that, for any angle θ for which $\cot\theta$ and $\csc\theta$ is defined, $1 + \cot^2\theta = \csc^2\theta$.

9 **(C)**

   (a) Your students make the continual error of saying that $\sin(A + B) = \sin A + \sin B$. They are thinking "Distributive Law." How can you convince them that there is no distributive law for trigonometric functions?

   (b) How can you prove by using Corollaries 11.10 and 11.11, that $\sin(A + B) = \sin A \cos B + \cos A \sin B$? [Hint: Write $\sin(A + B) = \cos(90 - (A + B)) = \cos((90 - A) - B)$. Now use Theorem 11.8.]

   (c) Derive a formula for $\sin(A - B)$ using your answer from part (b).

10 **(C)** A very common error that students make is to say that $\sin 2A = 2 \sin A$. Show, by example, or by graphing $y = \sin 2A$ and $y = 2 \sin A$ on the same set of axes, that this is not true. Then show how you would guide your students in using the formula for $\sin(A + B)$ to help them see that $\sin 2A = 2 \sin A \cos A$. Finally, find specific angles when $\sin 2A = 2 \sin A$.

11 Using the known values of $\sin 30° = \frac{1}{2}$ and $\sin 45° = \frac{\sqrt{2}}{2}$, find the exact values of $\sin 75°$ and $\sin 15°$.

12 **(C)** Your students use the following formulas in secondary school when proving identities and also in calculus when finding certain trigonometric integrals:

$$\cos 2\theta = \cos^2 \theta - \sin^2 \theta$$
$$= 2\cos^2 \theta - 1$$
$$= 1 - 2\sin^2 \theta.$$

They need help in trying to prove them. How can you help them understand how to go about proving these formulas?

13 Using the identities from the previous exercise, show that $\sin^2 \theta = \frac{1 - \cos 2\theta}{2}$ and that $\cos^2 \theta = \frac{1 + \cos 2\theta}{2}$. (These identities are used to evaluate certain trigonometric integrals in calculus.)

14 A nice proof of the Law of Cosines can be given using coordinate geometry. In any triangle $c^2 = a^2 + b^2 - 2ab \cos C$. Prove this using the following hint: We start with a triangle and place it on a coordinate plane as shown in Figure 11.25 below

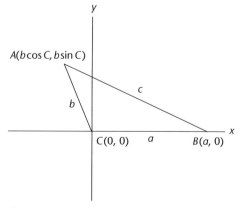

Figure 11.25

and call the coordinates of A, (x, y).

Since $\cos C = \frac{x}{r}$, $x = r \cos C$, which equals $b \cos C$ since $r = b$. Similarly, since $\sin C = \frac{y}{b}$, $y = b \sin C$. Now the distance from A to B is c, and this can be computed by the distance formula, $c = \sqrt{(b \cos C - a)^2 + (b \sin C)^2}$. Finish it.

15 A surveyor sees a building across the river. Standing at point A, he measures the angle of elevation from the ground to the top of the building to be 30 degrees. He steps back 100 feet and again measures the angle of elevation and finds it to be 15°. (See Figure 11.26.) Assuming that it makes a 90 degree angle with the floor, approximately how tall is the building?

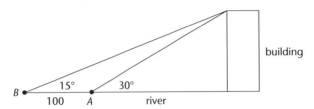

Figure 11.26

16 The method used in the previous problem can be generalized. Suppose we wish to find the height, h, of some object. We take two points A and B on the ground as shown in Figure 11.27 below and measure the distance between them.

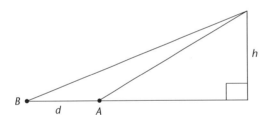

Figure 11.27

(a) Show that
$$h = \frac{d}{\cot B - \cot A}. \tag{11.17}$$

(b) Show that h can also be calculated by
$$h = \frac{d \sin A \sin B}{\sin(A - B)}. \tag{11.18}$$

(c) Setting equations (11.17) and (11.18) equal to each other, and dividing both sides by d, we get that
$$\frac{1}{\cot B - \cot A} = \frac{\sin A \sin B}{\sin(A - B)}.$$

Assuming that none of the denominators are 0, if A and B are any angles at all and not just angles related to our picture, in general, is this a true relationship? Why or why not?

17 A cruise ship starts out from point P and travels 30 miles at a bearing of 50° degrees northwest to a point Q. It then turns and travels a distance of 48 miles, at a bearing of 20° southwest to a point R as shown in Figure 11.28 below. How far is it from its original debarkation point?

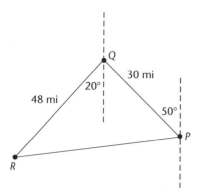

Figure 11.28

18 (C) In your calculus class your students have just learned the series for sin x and cos x. They are:

$$\sin x = x - \frac{x^3}{3!} + \frac{x^5}{5!} - \frac{x^7}{7!} + \ldots$$

and

$$\cos x = 1 - \frac{x^2}{2!} + \frac{x^4}{4!} - \frac{x^6}{6!} + \ldots .$$

One of your students notices that these relationships can be used to explain why $\cos(-x) = \cos(x)$ and why $\sin(-x) = -\sin x$ for all angles x without reference to any pictures. Explain how.

11.4 Radian Measure

11.4.1 Conversion

While degree measure seems to be the most natural way to measure angles, when calculus came on the scene, a different kind of measure was created that turned out to be more useful: radian measure. Suppose that we have a circle of radius r and we mark off a distance of r along the circumference of the circle, the central angle formed is called an angle of 1 **radian**. (See Figure 11.29 below.)

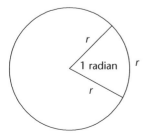

Figure 11.29

Since the circumference of a circle is $2\pi r \approx 6.28r$, we can mark off the radius of the circle 2π or approximately 6.28 times. Since the central angle formed by marking off the radius 2π times along the circumference is 360 degrees, we have that

2π radians = 360 degrees.

Dividing both sides of this equation by 2π, we get that

$$1 \text{ radian} = \frac{360}{2\pi} \text{ degrees} \approx 57.296 \text{ degrees}.$$

One advantage of radian measure is that it makes certain formulas in calculus much simpler. For example, if we take the derivative, y', of $y = \sin x$ and x is measured in radians, as is *always* done in calculus, then $y' = \cos x$. But if we were to measure x in degrees, then the derivative of $\sin x$ would not be $\cos x$, rather it would be $\frac{\pi}{180} \cos x$, and a similar statement can be made for the other trigonometric functions. This extra factor of $\frac{\pi}{180}$ would really be annoying to work with, especially when taking higher order derivatives. But in radian measure, this factor does not appear, giving radian measure a big advantage over degree measure. There is another more technical reason why radian measure is useful in the sciences. Radian measure is "dimensionless," which allows scientists to divorce it completely from angular measure and apply it to situations not relating to angles as we shall see.

Being able to convert from degree measure to radian measure and vice versa is important. Here is a review of how it is done. To convert from degrees to radians, we multiply the number of degrees by $\frac{\pi}{180}$. (Do you recognize this factor from the previous paragraph?) To convert from radians to degrees, we multiply the number of radians by $\frac{180}{\pi}$.

Thus, to convert 90° to radians, we multiply by $\frac{\pi}{180}$ to get

$$90° = 90 \times \frac{\pi}{180} \text{ radians} = \frac{\pi}{2} \text{ radians}.$$

Similarly, to convert $\frac{\pi}{4}$ radians to degrees, we multiply by $\frac{180}{\pi}$. Thus,

$$\frac{\pi}{4} \text{ radians} = \frac{\pi}{4} \times \frac{180}{\pi} \text{ degrees} = 45°.$$

The following conversions occur so often in secondary school that it is good to know them.

Number of degrees	Number of radians
0	0
30	$\frac{\pi}{6}$
45	$\frac{\pi}{4}$
60	$\frac{\pi}{3}$
90	$\frac{\pi}{2}$

We will use both degrees and radians in our discussion, depending on what seems easiest in a particular context.

11.4.2 Areas and Arc Length in Terms of Radians

How can we represent area and lengths of arcs on a circle using radian measure? We know that the area of a circle is πr^2, where r is the radius of the circle. Therefore, if we have a sector of a circle with central angle $k°$ as shown in Figure 11.30 below,

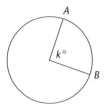

Figure 11.30

then the area of that sector is $\pi r^2 \frac{k}{360}$ and the length of the arc of the sector is $2\pi r \frac{k}{360}$. We observe that the area of the sector can be rewritten as $\pi r^2 \frac{k}{360} = \frac{1}{2} r^2 \left(\frac{\pi}{180} k\right)$, and the factor $\left(\frac{\pi}{180} k\right)$ is simply θ, the radian measure of the angle k. Substituting into the above expression, we get that the area of a sector is

$$A_{\text{sector}} = \frac{1}{2} r^2 \theta$$

where θ is the *radian* measure of an angle. In a similar manner, the length of the arc of the sector, $2\pi r \cdot \frac{k}{360} = r \cdot \frac{\pi}{180} k = r\theta$ where θ is the radian measure of the angle. Thus, if L_{sector} represents the length of the arc of a sector, we have

$$L_{\text{sector}} = r\theta. \tag{11.19}$$

Let us illustrate these formulas with some examples. The first is one that is often given in secondary school as a "fun type" of problem.

Example 11.12 *A dog located at point D is attached to a rope 25 feet long, which is attached to the corner of an enclosed rectangular plot of grass, with dimensions 10 feet by 30 feet, as shown in Figure 11.31 below.*

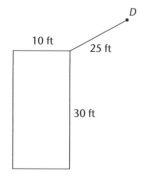

Figure 11.31

The dog cannot get inside the plot. If the rope is taut, find the area within which the dog can roam.

Solution. As the dog roams, it can trace out the area shown in Figure 11.32 below.

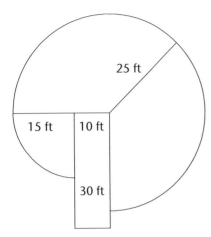

Figure 11.32

This area consists of $\frac{3}{4}$ of a circle with radius 25 feet and $\frac{1}{4}$ of a circle with radius 15 feet. The central angle associated with the $\frac{3}{4}$ part of the circle is 270° or $\frac{3\pi}{2}$ radians, and with the $\frac{1}{4}$ part of a circle is 90° or $\frac{\pi}{2}$ radians. Using the formula that the area of a sector is given by $\frac{1}{2}r^2\theta$, we find that the area the dog has to roam is given by:

$$\frac{1}{2}(25)^2\left(\frac{3\pi}{2}\right) + \frac{1}{2}(15)^2\frac{\pi}{2}$$

$$\approx 1649.3 \quad \text{square feet.}$$

Example 11.13 *Every car has a system of belts and pulleys that makes such things as the air conditioner and fan belt run. Consider Figure 11.33 below, where we have a system of two pulleys and a belt wrapped around them tightly. Assuming the angle where the belt crosses itself is $\frac{\pi}{6}$ radians, and the radii of the pulleys are 3 and 5 inches, find the approximate length of the belt.*

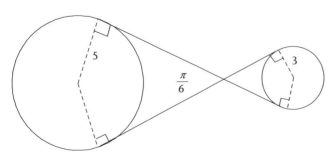

Figure 11.33

Solution. Draw a line joining the centers of the pulleys, which divides the angle $\frac{\pi}{6}$ into two equal parts of length $\frac{\pi}{12}$. See *FG* in Figure 11.34 below:

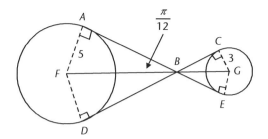

Figure 11.34

Now, using triangle ABF, we have that

$$\tan\left(\frac{\pi}{12}\right) = \frac{AF}{AB} = \frac{5}{AB}.$$

Solving for AB, we get that $AB \approx \frac{5}{\tan(\frac{\pi}{12})} \approx 18.66$ inches. In a similar manner, using triangle BEG we get that

$$\tan\left(\frac{\pi}{12}\right) = \frac{GE}{BE} = \frac{3}{BE}.$$

Thus, $BE = \frac{3}{\tan(\frac{\pi}{12})} \approx 11.196$. Since $AE = AB + BE$, we have that

$$AE \approx 29.856 \text{ inches.}$$

Similarly,

$$CD \approx 29.856 \text{ inches.}$$

To finish finding the length of the belt, we need to find the length of the (larger) arc AD and the (larger) arc EC. Now $\angle AFB = \frac{\pi}{2} - \angle ABF = \frac{\pi}{2} - \frac{\pi}{12} = \frac{5\pi}{12}$ and the same is true for angle BFD. Thus, the larger angle $\angle AFD = 2\pi - (\frac{10\pi}{12}) = \frac{14}{12}\pi$. The same is true for the larger angle CGE. So the length of the (larger) arc

$$\widehat{AD} = r\theta = 5\left(\frac{14\pi}{12}\right)$$

by equation (11.19), and the length of the larger arc

$$\widehat{EC} = 3\left(\frac{14\pi}{12}\right).$$

The length of the belt is therefore,

$$AE + (\text{larger})\widehat{EC} + CD + (\text{larger})\widehat{AD}$$
$$\approx 29.856 + 3\left(\frac{14\pi}{12}\right) + 29.856 + 5\left(\frac{14\pi}{12}\right)$$
$$\approx 89.034 \text{ inches.}$$

Another interesting application of these ideas is to mapmaking. Let us quickly review latitude and longitude. The earth, considered as a sphere, is broken into circles parallel to the equator called

542 Trigonometry

latitude lines. Similarly, all circles on the earth's surface that pass through both the north and south pole are called longitude lines or meridians. The equator is given a latitude of 0. Any longitude line may be taken as a longitude of 0. But the longitude line that passes through Greenwich, England, by international agreement, is the line taken as the one with longitude 0 and that line is called the **prime meridian**.

Each point, P, on the earth's surface is given a latitude and longitude. These are both angular measurements. The longitude measures how many degrees from the prime meridian we have to turn to get to the meridian that P is on, while the latitude is how many degrees up or down we must travel to get to the latitude line P is on. See Figure 11.35 below.

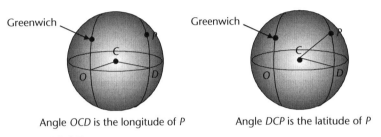

Angle *OCD* is the longitude of *P*   Angle *DCP* is the latitude of *P*

Figure 11.35

Suppose we are interested in making a map of part of the surface of the earth. There are many ways to do this, each with its advantages. One way to do this is with a method called cylindrical projection. In this method, the earth is considered to be enclosed by a cylinder (Figure 11.36).

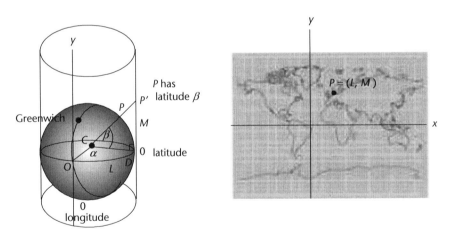

Figure 11.36

Now each point P on the part of the earth that we want to map is projected onto the cylinder in the following way. We draw a line from the center, C, of the earth to the point P on the earth's surface and then continue the line until it hits the cylinder at P' shown in Figure 11.36.

Once we do this, we unroll our cylinder as if it is a sheet of paper and we have our map. The equator we take as our x-axis, and the y-axis is the prime meridian through Greenwich England, $0°$. The question is, how do we get the coordinates (x, y) of any point P' on the earth's surface on this planar representation of the map? The answer is, we use trigonometry. If the point P has longitude α and latitude β, then the x coordinate of P' is the length of the arc L shown in the picture above when the cylinder is unwrapped. Furthermore, the y coordinate is the height M shown in the picture above. We know that $L = 2\pi R \cdot \frac{\alpha}{360}$, where R is the radius of the earth and thus, this is our x coordinate. Again, using trigonometry, we find that $M = y = R \tan \beta$. Doing this for various points on the surface of the earth, we get our map.

Of course, the question arises, how do we wrap the cylinder around the earth? After all, the earth is pretty big, and furthermore the cylinder when unrolled will be quite big. True. That is why our map is a scaled down version of the real thing!

Student Learning Opportunities

1 (C) Your students want to know how to draw an angle that measures 1 radian and approximately how many degrees 1 radian represents. What is the answer?

2 (C) Your students know that π is approximately equal to 3.14. Now they are learning that π radians $= 180°$ and this confuses them. What is the source of their confusion concerning the value of π and how can you resolve it?

3 Convert each of the following from degrees to radians:

(a) $210°$

(b) $-300°$

(c) $450°$

4 Convert each of the following from radians to degrees:

(a) $\dfrac{7\pi}{6}$

(b) $-\dfrac{3\pi}{4}$

(c) $\dfrac{2\pi}{5}$

5 A pulley system consists of two pulleys having radii R and r. There is a belt tightly placed around the system and the angle where the belt meets itself is θ radians as shown in the Figure 11.37 below.

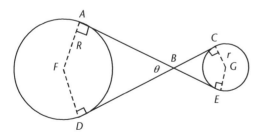

Figure 11.37

Show that the length of the belt is given by

$$L = (R + r)\left(\theta + \pi + 2\cot \dfrac{\theta}{2}\right).$$

6 A cow is attached to an L-shaped building surrounded by grass as shown. If the rope attached to the cow's neck is 60 feet long and the dimensions of the building are all in feet, how much area can the cow graze? The lengths given are the lengths of the solid lines, AB, BC, CD, and DE in Figure 11.38.

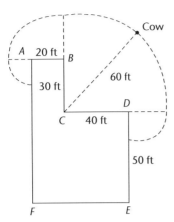

Figure 11.38

11.5 Graphing Trigonometric Curves

In the following questions below, use the graphing calculator to draw the graphs of the given functions. Then sketch these graphs on your graphing paper.

1. Graph the functions: $y = \sin x$, $y = 2 \sin x$, $y = 3 \sin x$. If you graph the equation $y = A \sin x$, how does changing A affect the graph of $y = \sin x$?
2. How are the graphs of $y = \sin 2x$, and $y = \sin 3x$ different from the graph $y = \sin x$? Predict how $y = \sin Bx$ will look.
3. Graph the functions: $y = \sin x$, $y = 3 + \sin x$, $y = -1 + \sin x$. If you graph the equation $y = D + \sin x$, how does changing D affect the graph of $y = \sin x$?
4. Graph the functions: $y = \sin x$, $y = \sin(x - \frac{\pi}{2})$, $y = \sin(x - \pi)$, and $y = \sin(x + \frac{\pi}{4})$. If you graph the equation $y = \sin(x - C)$, how does changing C affect the graph of $y = \sin x$?
5. Sketch the graph of $y = 2 + 4 \sin(x + \pi)$. Now graph it on your graphing calculator. Was your sketch accurate?
6. Describe what you have learned about the effects of changing A, B, C, and D on the graph of $y = \sin x$.

After having done the launch questions, your memory has probably been refreshed on what you learned in secondary school about the graphs of trigonometric functions. By reading this section, you will learn more details about these graphs.

11.5.1 The Graphs of Sin θ and Cos θ

Thus far, we have examined the graphs of exponential and polynomial functions. It is only natural to wonder what the graph of a trigonometric function looks like. Let us proceed to find out.

We already mentioned that, when finding the sine of an angle, θ, we are allowed to pick *any* point on the terminal side of θ. We will always get the same answer for $\sin\theta$. That follows by similar triangles as Figure 11.39 below shows.

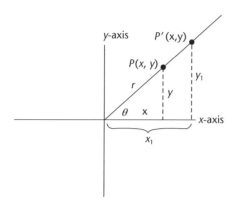

Figure 11.39

So, if we can take any point on the terminal side of an angle, why not take a point whose distance from the origin is 1? (See Figure 11.40 below.)

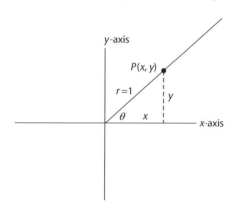

Figure 11.40

Since

$$\sin\theta = \frac{y}{1} = y$$

and

$$\cos\theta = \frac{x}{1} = x,$$

we see that, when we choose a point on the terminal side of the angle drawn in a circle whose radius is 1, the y coordinate represents $\sin\theta$ and the x coordinate is $\cos\theta$. That leads us to consider taking points on the circle whose center is at the origin and whose radius is 1. (See Figure 11.41 below.) The x coordinate of P is $\cos\theta$ and the y coordinate is $\sin\theta$.

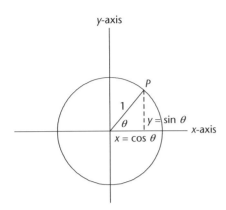

Figure 11.41

By looking at the *y* coordinate of *P* as θ varies, we get a picture of how $\sin\theta$ varies with θ and if we look at the *x* coordinate of *P*, we see how $\cos\theta$ varies.

Let us first concentrate on $\sin\theta$. We notice that, as θ goes from 0 to 90°, the *y* coordinate, which is $\sin\theta$, varies from 0 to 1. As θ varies from 90° to 180°, $y = \sin\theta$ decreases from 1 to 0. As θ varies from 180° to 270°, $y = \sin\theta$ decreases from 0 to −1 and as θ varies from 270° to 360°, $y = \sin\theta$ increases from −1 to 0. In Figure 11.42 you see a graph of how $\sin\theta$ varies as θ varies from 0 to 360°.

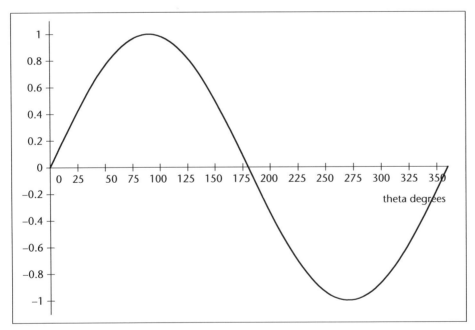

Figure 11.42 The graph of $y = \sin\theta$, $0° \leq \theta \leq 360°$.

We have traced out what is called one complete cycle of the sine curve. The **period** is the number of degrees it takes to complete one cycle. Thus, the period of the function $y = \sin\theta$ is 360 degrees or 2π radians. As you already know, the sine curve looks like a wave and repeats once the angles go beyond 360.

In the Student Learning Opportunities you will use a similar process to find the graph of the cosine curve, which we show in Figure 11.43.

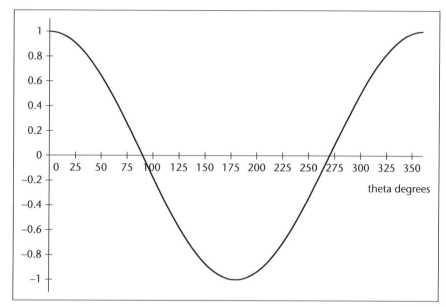

Figure 11.43 The graph of $y = \cos\theta$, $0° \leq \theta \leq 360°$.

Below in Figures 11.44 and 11.45 we see the graphs of $\sin\theta$ and $\cos\theta$ with the θ-axis marked off in radians.

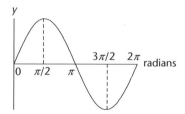

Figure 11.44 The graph of $y = \sin\theta$, $0 \leq \theta \leq 2\pi$.

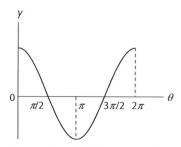

Figure 11.45 The graph of $y = \cos\theta$, $0 \leq \theta \leq 2\pi$.

Whether you choose to graph the function in degree or radians, it makes no difference.

Any function that repeats over and over is called **periodic**. Thus, $f(\theta) = \sin\theta$ and $g(\theta) = \cos\theta$ are periodic, but so is the function shown in Figure 11.46 below, where the graph repeats forever in both directions.

548 Trigonometry

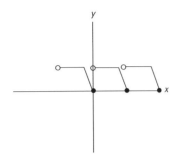

Figure 11.46

It is interesting to note that any periodic behavior, even the one illustrated above, can be studied using sines and cosines. If you are familiar with Fourier Series, you will understand what we mean. However, this is beyond the scope of this book. Sticking to secondary school material, we will focus only on the periodic functions of sine, cosine, tangent, and their reciprocals.

When we graph $y = 5 \sin \theta$, it looks just like the graph of $y = \sin \theta$, only instead of reaching a maximum of 1 and a minimum of -1, it reaches a maximum of 5 and a minimum of -5. Its graph is below and we see that its period remains 360°. The graph of $y = -5 \sin \theta$ is just a reflection of the graph of $y = 5 \sin \theta$ about the θ axis. Its graph is shown in Figure 11.47.

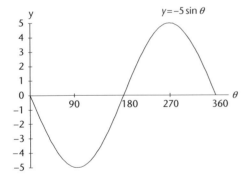

Figure 11.47

When we graph the function $y = \sin 2\theta$, it is essentially traced out twice as fast as $y = \sin \theta$. So, in 360 degrees, you would have two cycles and the period of the graph would be 180°. The frequency is the number of cycles traced out in 360 degrees. Thus $y = \sin 2\theta$ has a frequency of 2. In Figure 11.48 we see the graph of $y = \sin 2\theta$ for $0° \leq \theta \leq 360°$. Observe that there are two full sine cycles in 360 degrees.

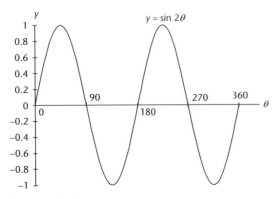

Figure 11.48

Note: In general, when we graph functions, we use x to label the horizontal axis, and y to label the vertical axis, though we have been labeling the horizontal axis as the θ-axis. From here on in, we will suppress θ and use x when it doesn't cause confusion.

In general, the graph of $y = A \sin Bx$ has a period of $\frac{360}{B}$, or $\frac{2\pi}{B}$ if we are measuring in radians. It reaches a maximum height of $|A|$ and a minimum height of $-|A|$. The graph of $y = D + (A \sin Bx)$ raises or lowers the graph of $y = A \sin Bx$ by D, that is, translates it up or down. If D is positive, the graph is translated up, while if it is negative, it is translated down. Thus, the graph of $y = 5 + 3 \sin x$ raises the graph of $y = 3 \sin x$ by 5, while $y = -3 + \sin x$ lowers the graph of $y = \sin x$ by 3. A vertical translation of the graph does not affect its period.

The graph of $y = \sin 4(x - 16°)$ is the graph of $y = \sin 4x$ translated to the right 16 degrees as one can see by making a table of values for x and y. So one of its cycle's begins at the point $(16, 0)$. The graph of $y = \sin 4(x + 16°)$ is the graph of $y = \sin 4x$ translated to the left by 16. So one of its cycles begins at $(-16, 0)$. You can check these statements by graphing all three functions on your calculator.

We summarize our observations in the theorem.

Theorem 11.14 *The graph of $y = D + A \sin B(x - C)$ takes the graph $y = A \sin Bx$ and vertically translates it by D, and horizontally translates it by C to the right if C is positive, and a distance of C to the left if C is negative. The period of this graph is $\frac{360}{B}$ degrees. If $B(x - C)$ is measured in radians, then the period is $\frac{2\pi}{B}$.*

There is a similar theorem for $y = D + A \cos B(x - C)$. C is called the **phase shift**.
Here is an illustration.

Example 11.15 *Graph the function $y = 4 + 5 \sin 3x$ for x between 0 and 360 degrees. Do it without a calculator.*

Solution. If we can graph $y = 5 \sin 3x$, then we can graph $y = 4 + 5 \sin 3x$. The latter graph will be the former graph raised by 4 units. So we turn to graphing $y = 5 \sin 3x$. Its period is $\frac{360}{B} = \frac{360}{3}$ or 120 degrees. Thus, one complete cycle will cover 120 degrees. Since we want the graph between $0°$ and $360°$, there will be three cycles. The maximum height of $y = 5 \sin 3x$ is 5 and the minimum height is -5. Translating this vertically by 4 units, our graph looks like that in Figure 11.49.

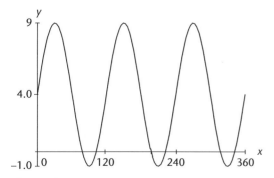

Figure 11.49

The maximum height is 9 and the minimum height is -1.

550 Trigonometry

With the power of today's graphing calculators, you may be wondering about the value of knowing how to graph these equations by hand. Our students always question this. But experience shows that dependence on the calculator is dangerous. For example, the future teacher can see this by asking a class to graph $y = x + 100$ on the calculator. Since the graph is not visible in the standard window, you will hear some students who do not know that the graph is a line and who graphed it in the standard window claim that "There is no graph." Of course, we know there is one. With trigonometric functions, graphing equations on the calculator can lead to even stranger results. Below, we show what can happen when you graph the same trigonometric function in different windows. The results are striking.

Example 11.16 *Graph $y = \sin 30x$ in each of the following windows:* (a) $[-5, 5] \times [-1, 1]$ (b) $[-0.01, 0.01] \times [-1, 1]$ (c) $[0.1, 07] \times [-1, 1]$ (d) $[-16, 16] \times [-1, 1]$ (e) $[-7, 7] \times [-1, 1]$ (f) $[-20, 20] \times [-1, 1]$. *(x is measured in radians.)*

Solution. Here are the 6 graphs (Figures 11.50–11.55).

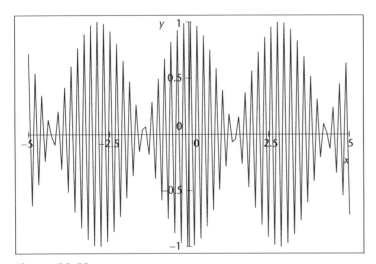

Figure 11.50

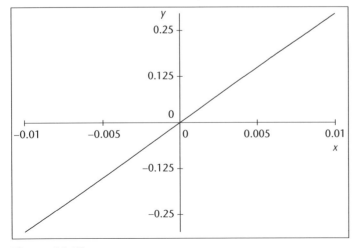

Figure 11.51

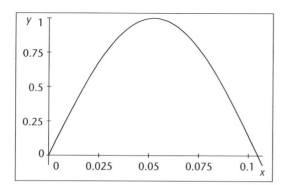

Figure 11.52

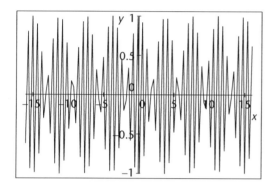

Figure 11.53

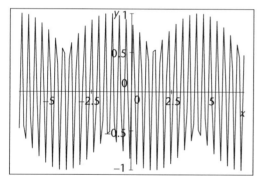

Figure 11.54

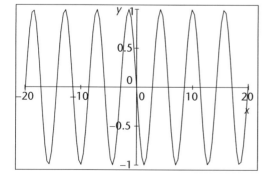

Figure 11.55

11.5.2 The Graph of y = Tan θ

In Section 11.3 we defined $\sin\theta$ to be $\frac{y}{r}$ and $\cos\theta$ to be $\frac{x}{r}$ where (x, y) is any point on the terminal side of θ, and $r > 0$ is the distance of that point to the origin. If we divide the expression for $\sin\theta$ by $\cos\theta$, assuming that $\cos\theta$ is not zero, we get that

$$\frac{\sin\theta}{\cos\theta} = \frac{\frac{y}{r}}{\frac{x}{r}} = \frac{y}{x}$$

Since $\frac{y}{x}$ is $\tan\theta$, we see that

$$\tan\theta = \frac{\sin\theta}{\cos\theta} \quad \text{when} \quad \cos\theta \neq 0.$$

Thus, from the values of $\sin\theta$ and $\cos\theta$, we can get the values of $\tan\theta$ and we can easily draw the graph of $\tan\theta$. Since $\sin 0 = 0$, and $\cos 0 = 1$, we immediately see that $\tan 0 = \frac{0}{1} = 0$. As θ goes from 0 to 90°, $\sin\theta$ increases to 1, and $\cos\theta$ decreases to 0. When the numerator of a fraction increases while the denominator decreases, the fraction increases. Thus, the quotient of them, $\tan\theta$ increases. Since $\sin\theta$ is approaching 1 and $\cos\theta$ is approaching 0, as θ approaches 90°, $\tan\theta$, their quotient, gets larger and larger and goes to ∞. Tan 90° does not exist, as you can easily show. Thus as x approaches 90°, the graph of $\tan\theta$ approaches ∞, and $x = 90°$ is a vertical asymptote. Similarly, as x approaches $-90°$, $y = \tan\theta$ approaches $-\infty$. Thus, the graph of $\tan\theta$ for $-90° < \theta < 90°$ is shown in Figure 11.56.

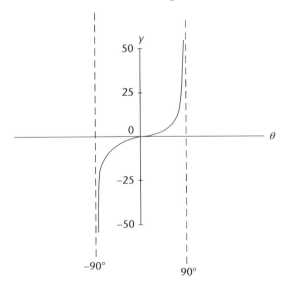

Figure 11.56

Once we pass 90° and move to 270°, we find by calculation that the graph repeats. In fact, the graph repeats every 180° and is shown below. Thus the graph has a period θ of 180° or π radians. The graph of $y = \tan\theta$ is shown in Figure 11.57.

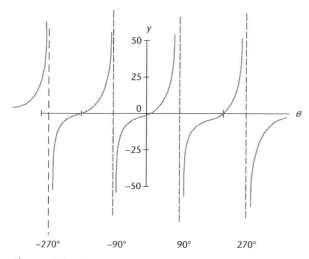

Figure 11.57

Student Learning Opportunities

1 Write the equations for each of the following graphs (Figures 11.58–11.60):

(a)

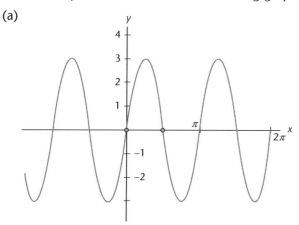

Figure 11.58

(b)

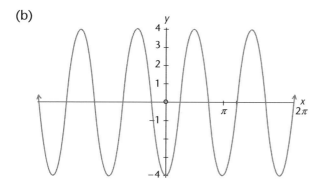

Figure 11.59

(c)

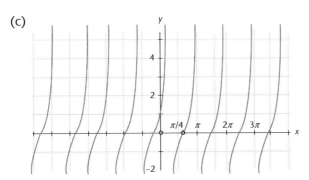

Figure 11.60

2 For each of the following graphs, find a trig function of the form $y = D + A \sin B(x - C)$ or $y = D + A \cos B(x - C)$ that fits the graph. Then check that you are correct with your calculator (Figure 11.61–11.64).

(a)

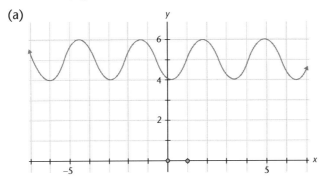

Figure 11.61

(b)

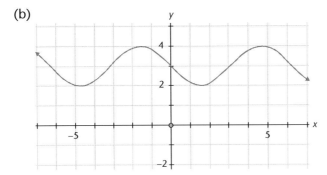

Figure 11.62

(c)

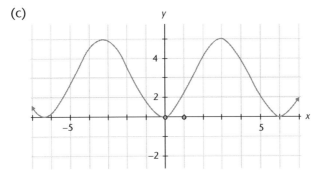

Figure 11.63

(d)
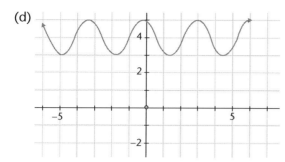

Figure 11.64

3 (C) When graphing $y = \sin(x - \frac{\pi}{2})$, many of your students make the mistake of shifting the curve $y = \sin x$, $\frac{\pi}{2}$ radians to the left. They claim that, since $-\frac{\pi}{2}$ occurs, the curve should be shifted in the negative direction. How do you help your students understand why this is incorrect?

4 Show that the graph of $y = \sin 2\pi/x$ for x in $(0, 1)$ has infinitely many zeroes in that interval. Now graph this on your calculator. Are you seeing this? If not, try zooming in a few times. What are you seeing?

11.6 Modeling with Trigonometric Functions

LAUNCH

Examine the following scenarios and decide what they have in common:

a. Imagine a bicycle wheel whose radius is one unit, with a marker attached to the rim of the rear wheel. As the wheel rotates, the height $h(t)$ of the marker above the center of the wheel is measured.
b. Due to the tidal changes, the depth of water at my favorite surfing spot varies from 5 ft to 15 ft daily depending on the time of day, and this variation of depth doesn't change from day to day. A description is given of the depth of the water as a function of time.
c. A spring is hanging from the ceiling. Young Jack comes around and pulls the spring and lets it go and it starts to move.

We hope that in examining the scenarios in the launch, you realized that each of the above situations represented cyclical motion and could be modeled by a trigonometric function.

As we have pointed out in the introduction to this chapter, trigonometric functions are widely applied. In this section we will describe how they can be used to model some very interesting situations that occur in reality, some of which affect us on a daily basis. We will now give several examples.

Example 11.17 As shown in Figure 11.65 below, a spring is hanging from the ceiling and attached to it is a block. It is in equilibrium, meaning it is not moving. The block is pulled down a distance of 5 inches and then is released.

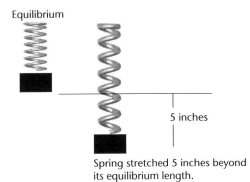

Figure 11.65

The spring starts to oscillate, and its position relative to its equilibrium point is graphed over time. Figure 11.66 shows the resulting graph for the first 3 seconds. How can we model the motion with a trigonometric function?

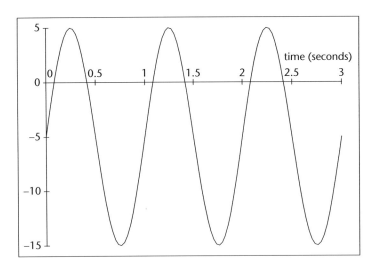

Figure 11.66

Solution. We note that the graph of the motion begins at a height of -5. This simply represents the fact that initially the block is 5 inches below its natural position, which we are taking to be at a height of 0. A distance of $+5$ would mean it is 5 inches above its natural position. For the first three seconds, the graph does appear to be a sine curve that has been translated, so its equation will be of the form $y = D + A \sin B(x - C)$. Only in this case, the x-axis no longer represents angles, rather time, which we denote by t.

It seems that, during the 3 second period, the graph undergoes three cycles. Thus, the period of the graph, the time needed to go through one cycle, is 1 second. Since the period is $\frac{2\pi}{B}$ and this is 1 second, $\frac{2\pi}{B} = 1$ and hence $B = 2\pi$. In general, to find $|A|$, take half the difference of the minimum and maximum value of the curve. This yields $|A| = \frac{1}{2}(5 - -15) = 10$. Since the curve looks like a translated sine curve and not a reflected sine curve, A should be positive. Since the graph starts 5 units below the x-axis, we have translated the sine curve down 5 inches. Thus $D = -5$. Finally, since the cycle begins at the origin, $C = 0$. Thus our graph is $y = -5 + 10 \sin 2\pi(t - 0)$, where t represents time, or just $-5 + 10 \sin 2\pi t$.

Our model seems to give us the faulty impression that the spring continues to oscillate forever, which we know does not happen. Since friction dampens the oscillations and eventually stops the spring from oscillating, a more accurate model is needed. In fact, such models exist and they typically consist of a product of two functions. One function is the sine function and the other is the function e^{-kt}, where k is positive. The correct value of k depends on something called the spring constant, which measures the stiffness of the spring. The details of all this involve differential equations, which we will not describe. Rather, we just show you the graph of the damped oscillation $y = e^{-2t}(-5 + 10 \sin 2\pi t)$, which is typical of models of these things. (See Figure 11.67.)

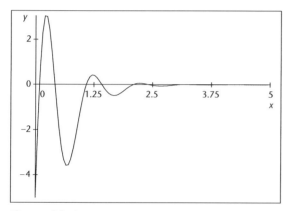

Figure 11.67

This certainly is a better model of what is happening.

Example 11.18 *The number of hours and minutes of daylight in a city located at a latitude of 40° on the 15th of the month is shown in the following table:*

January	9.6167
February	10.7
March	11.883
April	13.233
May	15.367
June	15
July	14.817
August	13.8
September	12.517
October	11.167
November	10.017
December	9.33

Model this with a trigonometric function.

Solution. To get an idea of the model that will work best we start by plotting the points. A scatter plot of the data is shown in Figure 11.68 below with month 1 being January.

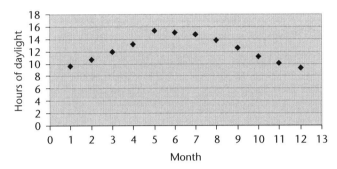

Figure 11.68

The points seem to outline half a sine curve translated in the vertical direction. Therefore, we know we are looking to find the values of A, B, C, and D in the equation $y = D + A \sin B(x - c)$. Since the graph has been translated 9.6167, $D = 9.6167$. The 11 month period from January 15th to December 15th graphed is half of the sine curve's period. So, the sine curve has a period of 22 months. Since $\frac{2\pi}{B}$ is the period, $\frac{2\pi}{B} = 22$, so $B = \frac{\pi}{11}$. Since there is a phase shift here of 1, $C = 1$, and the value of A which measures the rise of the curve is $15.367 - 9.33 = 6.037$. So our model sine curve is $y = 9.6167 + 6.037 \sin \frac{\pi}{11}(x - 1)$. You will note that, since the data points are not symmetric, our model, like most models, is only an approximation and cannot possibly be a perfect fit.

Example 11.19 *Below in Figure 11.69 we have a schematic of a Ferris wheel. It takes 1 minute to make a complete revolution. If, when the wheel begins, the point P is at the bottom, which is 6 feet of the ground, how high will it be t seconds after the wheel begins turning if the Ferris wheel has a radius of 50 feet and turns counterclockwise?*

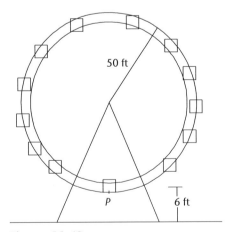

Figure 11.69

Solution. Since P starts at a height of 6 feet, reaches a maximum height of 106 feet at 30 seconds, and then returns, its motion is cyclical and we can picture it as in Figure 11.70.

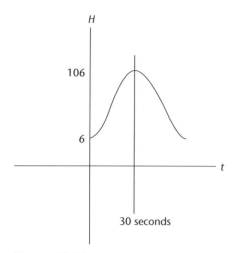

Figure 11.70

It looks like a cosine curve reflected and translated. So, we will model the height by a curve of the form $D + A\cos B(x - C)$. Since it takes 60 seconds to complete a cycle, the period is 60 seconds. Because the period is calculated by computing $\frac{2\pi}{B}$, we have that $\frac{2\pi}{B} = 60$ and $B = \frac{\pi}{30}$. The amplitude, A, which is half the difference between the maximum and minimum points is $\frac{1}{2}(106 - 6) = 50$ feet. Furthermore, since it is a reflected cosine curve, A is negative. So, our model is $y = -50\cos\left(\frac{\pi}{30}\right)t$. If we graphed this, it would begin at a height of -50. Since the Ferris wheel begins at a height of 6 and our model begins at a height of -50, we must translate our graph up 56 units. So $D = 56$. There is no phase shift here, so $C = 0$. Our model is then $H = 56 - 50\cos\frac{\pi}{30}t$, where H is the height off the ground and t is the number of seconds.

Sound is caused by vibration and is believed to be wavelike in nature. Thus, when we strike the middle "A" on the piano, we are causing a string inside the piano to vibrate with a frequency of 440 herz. (A herz is one complete cycle per second, so 440 herz means 440 cycles per second.) Every note has a different frequency for which there is a formula based on the note's proximity to middle A. That formula is $F = 2^{\frac{n}{12}} \cdot 440$ Hz where n is the number of half steps from middle A, with n being positive if the note is above middle A and negative if n is below middle A. Thus, the C above middle A (for those familiar with the piano keyboard), is 3 half steps above A and has frequency of approximately $F = 2^{3/12} \cdot 440 \approx 523$ herz. Similarly, the F below middle A is 4 half-steps below A, so its frequency is approximately $2^{-4/12} \cdot 440 \approx 349$ herz. Interestingly, history indicates that one of the first persons to explore the rich mathematical theory behind music was Pythagoras.

When we turn on the radio and tune into a station, we are really experiencing trigonometry first hand. Radio signals are sent out in the form of sine and cosine waves. So, when we turn our dial to say 1010, we are trying to access a signal, which is being broadcast at a frequency of 1010 kiloherz. The prefix "kilo" in "kiloherz" means thousand. These are very high frequency waves. The words and music from the broadcasting station are low frequency waves (compared with the broadcasting waves). These low frequency waves are "carried" on the high frequency signals, which change or "modulate" the signal. One kind of change that occurs is that the amplitude of the carrier wave changes as a result of putting this other wave coming from the broadcast studio on top of it. The resulting signal is an amplitude modulated signal, which we know as AM radio. This modulation is achieved by combining the carrier wave and the signal wave. The net result is an amplitude modified curve which is sent over the airwaves. Inside your radio is a decoder that strips off the carrier wave and leaves the signal wave. Another type of modulation is the FM signal, where we modulate the frequency of the carrier wave. FM is less prone to distortion from noise. Each method of transmission, however, has its advantages which we won't discuss here.

Another application we will briefly mention affects us on a daily basis: electricity! Did you know that the electricity in your home consists of something called alternating current whose behavior is represented by sine curves? Through the use of a transformer, electrical energy can be transmitted over long distances. Thus, as a result of this engineering feat with trigonometric functions, we have all the electrical conveniences in our homes. Isn't this impressive? After all, who would think that trigonometry had anything to do with such things as the lights in our house?

Finally, department stores place devices by their doors to check if a person is shoplifting. The device you walk through as you enter or leave a store that sounds an alarm if you are shoplifting an item works on trigonometric principles. New applications of trigonometry in technology are constantly being found which is why, in many senses, trigonometry is considered the bedrock of the technological applications we have today.

Student Learning Opportunities

1 Suppose that, in Example 11.17, the period of the spring is 30 seconds. Write the equation of motion.

2 Suppose that, in Example 11.19, P is at the top of the wheel when the Ferris wheel starts. What function will describe the height now?

3 Suppose that, in Example 11.19, P is at the 3 o'clock position when the Ferris wheel starts. What function will describe the height now?

4 Suppose that, in Example 11.19, P is at the 9 o' clock position when the Ferris wheel starts. What function will describe the height now?

5 What is the function that describes the position of P if the Ferris wheel has a height of 120 feet and when it begins P is at the lowest point 6 feet off the ground?

6 In Example 11.19, write a function that describes how far P is horizontally from the center of the Ferris wheel at time t.

11.7 Inverse Trigonometric Functions

One of the following, (a) or (b) is false. Which one, and why?

(a) $\sin^{-1}(\sin x) = x$ (b) $\sin(\sin^{-1} y) = y$.

Trigonometry 561

If you are like most people, the launch question probably has you a bit confused. As you read this section on inverse trigonometric functions, you will have a clearer picture of what is really meant by the inverse of a trigonometric function.

As we have seen from previous sections, one of the most amazing features of trigonometry is that it allows us to measure the unmeasurable. In secondary school there are two types of problems that students are exposed to when they study trigonometry that illustrate this and are shown in the following examples.

Example 11.20 *A person standing 50 feet from a tree measures the angle from the ground to the top of a tree with a surveying instrument. She finds that angle θ is 23°. Estimate the height of the tree.*

Example 11.21 *An entrance ramp on a highway is to span 60 feet and rise 3 feet. (See Figure 11.71.) What angle must the ramp make with the ground to achieve this?*

In the first example, we look at the diagram below and notice that we can solve this problem by using the tangent ratio. (See Figure 11.71 below.)

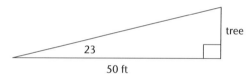

Figure 11.71

If we call the height of the tree, *x*, then

$$\tan 23° = \frac{x}{50}$$

and multiplying both sides by 50 yields $x = 50 \tan 23° = 21.224$ ft, which is easily obtained by the calculator.

In the second problem, we have the reverse situation. We draw a picture of the ramp in Figure 11.72 below:

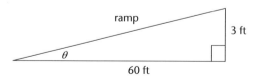

Figure 11.72

Now we have to find the angle. If we call the angle that the ramp must make with the horizontal, θ, then

$$\tan \theta = \frac{3}{60}$$

and we have to solve for θ. In secondary school, we are taught that, if we use the $\tan^{-1}$ button on the calculator, we can solve this equation. Doing so, we get that $\theta = \tan^{-1}(\frac{3}{60}) \approx 2.8°$. So we have solved our problem.

In a similar manner, when we wish to find a solution of, say an equation like $\sin\theta = 0.5$, we compute $\sin^{-1}(0.5)$ to get $30°$. What are these functions, $\tan^{-1}$ and $\sin^{-1}$, and what does the exponent -1 in each case mean?

First, we tell you what they are not. These *do not* stand for reciprocals. While it is true that in algebra, x^{-1} means $\frac{1}{x}$, or the reciprocal of x, this is not true here. Thus, don't think that $\sin^{-1}$ is the reciprocal of sin. The reciprocal of the sine function is given a specific name. It is called the cosecant. In a similar manner, the reciprocal of the tangent is called the cotangent. So once again, $\tan^{-1}$ is not the reciprocal of tangent. The -1 exponent refers to the concept of inverse function that we discussed in Chapter 9.

Let us explain. If one looks at the graph of $\sin x$, one sees that the equation $\sin x = \frac{1}{2}$ has many solutions. In fact, it has infinitely many solutions as we see in Figure 11.73 below.

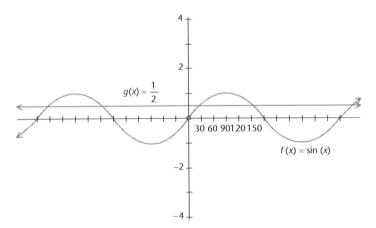

Figure 11.73

One of them is $30°$, and another is $150°$. If we add or subtract multiples of $360°$, we get all our solutions. Thus, the function $y = \sin x$ is not a 1–1 function, since different x's can give rise to the same y. However, if we restrict x to be between $-90°$ and $90°$, then the equation $\sin x = \frac{1}{2}$ has only one solution, namely $x = 30°$. See Figure 11.74 below where $\sin x$ has been restricted.

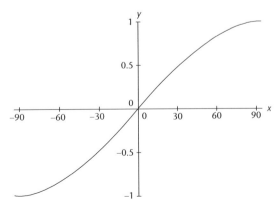

Figure 11.74

What we are saying is that, if the function $y = \sin x$ is restricted to angles between $-90°$ and $90°$, then we get a 1–1 function and this function has an inverse. The inverse function, obtained by solving for x in terms of y is denoted by $x = \sin^{-1} y$ or also as $x = \arcsin y$. These are read "x equals inverse sin of y" and "x equals arc sine y" and both mean the same thing, namely, that *x is the angle (between $-90°$ and $90°$) whose sine is y*. Thus, the equations $y = \sin x$ and $x = \sin^{-1} y$ are interchangeable when the angle x is restricted to between $-90°$ and $90°$. Similarly, when the function $y = \tan x$ is restricted to between $-90°$ and $90°$ (not including these values), it too becomes 1–1 and we can talk about its inverse, denoted by $x = \tan^{-1} y$. Again, this is read as x is the angle ($-90° < x < 90°$), whose tangent is y. This brings us to the common practice in secondary school of interchanging variables when writing an inverse function, which we mentioned in Chapter 9. As we said, in practical problems this makes no sense, as we will demonstrate in this situation. When x is an angle say in a triangle, $y = \sin x$ represents the ratio of two sides. There is a big difference between the angle x, and the ratio, y, of two sides of a triangle. So other than for the sake of graphing, say on a graphing calculator, we should not interchange variables.

When using the calculator to solve equations, we must be careful about getting *all the solutions*. The general rule is that, if $f(x)$ is periodic and has period p, to find all solutions of $f(x) = a$, we find all solutions within 1 period, and we can then generate the rest by adding multiples of p. Let us illustrate.

Example 11.22 *Solve the equation* $2 \sin x + 3 = 4$.

Solution. We solve the equation for $\sin x$ first to get $\sin x = \frac{1}{2}$. The tendency at this point is to just compute $x = \sin^{-1}(\frac{1}{2})$ on our calculator and get $x = 30°$. We might now reason, "The graph of $y = \sin x$ is periodic with period $360°$ and so, once we have a set of solutions in one period, we can generate all the other solutions by adding multiples of $360°$. So our solutions are $x = 30° + 360k°$ where $k = 0, \pm 1, \pm 2$, and so on." But we are not completely correct, since we have missed infinitely many solutions. We have not yet found *all* the solutions in one period. When we press $\sin^{-1}(1/2)$, the calculator only gives us an angle between $-90°$ and $90°$, and the region from $-90°$ to $90°$ is only half of one period. A period for the function $\sin x$ is 360 degrees. We need to find all solutions in one period before we can be sure we have all the solutions.

One way to find all the solutions in one period is to graph the functions $Y_1 = 2 \sin x + 3$ and $Y_2 = 4$ on the same set of axes and see where they intersect. One will find, assuming that the calculator is in degree mode, that between 0 and $360°$ (one period) we get two solutions, $x = 30°$ and $x = 150°$. Now we can generate all solutions and they are $x = 30° + 360k°$ where $k = 0, \pm 1, \pm 2, \ldots$ and $x = 150° + 360k°$ where $k = 0, \pm 1, \pm 2, \ldots$

A second way to find all solutions in one period is to use what is called the **reference angle**, which is the angle the calculator gives you when you press the $\sin^{-1}$ button, but without the sign. Thus, if you press $\sin^{-1} -\frac{1}{2}$ you get $-30°$ and ignoring the sign we get that the reference angle is $30°$. In the problem we are working on, however, we are computing $\sin^{-1} \frac{1}{2}$ and we get $30°$. Our answer is positive and so we don't have to change the sign.

Now we argue that, if $\sin x = \frac{1}{2}$, then since $\sin x$ is positive, x is either in quadrant 1 or 2. Using the x-axis as one side of an angle, we draw an angle of 30 degrees in quadrant 1 and another in quadrant 2 as shown in Figure 11.75 below.

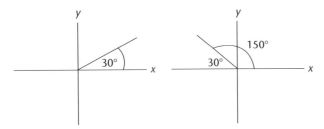

Figure 11.75

We then measure our angles starting from the positive x-axis. They are 30° and 150° and we can generate our solutions as before, $x = 30° + 360k$ where $k = 0, \pm 1, \pm 2, \ldots$. This method always works. (So does the graphing method if you have the right graph.)

One must be careful with quadrantal angles when solving equations.

Example 11.23 *Solve the equation* $\sin 2x \cos 3x - \sin 2x = 0$ *for all values of x where* $0° \leq x < 360°$.

Solution. It is a good idea to factor out $\sin 2x$. For then we get

$\sin 2x(\cos 3x - 1) = 0$.

This tells us that $\sin 2x = 0$ or $\cos x - 1 = 0$. Calling $\theta = 2x$ and referring to the graph of $\sin \theta$, we see it crosses the x-axis, when $\theta = 0°$ or $\theta = 180°, 360°$, and $540°$. Since $\theta = 2x$, $x = 0°, 90°, 180°$, and $270°$. Notice that $0 \leq x < 360$, which makes $0 \leq \theta(= 2x) < 720$. That is why we looked at angles up to, but not including, 720°

To solve the second equation, we get $\cos 3x = 1$. This yields $3x = \cos^{-1} 1 = 0°$ and again, referring to the cosine graph, we see that the graph of $\cos \theta$ is 1 only when $\theta = 3x$ is 0° or 360° or 720°. (Once again, if $0 \leq x < 360$, $0 \leq 3x < 1080$, so we need to go up to, but not include 1080°.) So solutions are $x = 0°, 120°$, and 240° for this latter equation. So the given equation has as its solutions 0°, 120°, 180°, 240°, and 360°.

Memorization of the graphs of $\sin \theta$ and $\cos \theta$ will help in solving trigonometric equations.

Student Learning Opportunities

1 (a) **(C)** You ask your students to solve the equation $4 \sin x + 3 = 5$ for all values of x. One of your students, Jacqueline, solves the equation as follows:

$4 \sin x + 3 = 5$

$4 \sin x = 2$

$\sin x = 1/2$

$x = \sin^{-1}(1/2)$

$x = 30°$. (From the calculator)

Is she correct? Why or why not?

(b) Another student, Sung, offers Jacqueline some help. He says, "You almost have the solution. The solution is $x = 30° + 360k$ where $k = 0, \pm 1, \pm 2 \ldots$." Is Sung correct? If not, what is the correct answer?

2 After seeing Example 11.21, Melissa now thinks she undertands what the $\sin^{-1}$, $\cos^{-1}$, and $\tan^{-1}$ buttons are used for. She says, "They are used only to find angles in a right triangle. So if you compute $\cos^{-1}(0.5)$, you are finding the angle in a right triangle whose cosine is 0.5 and this is 60°." When you persist and ask her to find $\cos^{-1}(-0.5)$ using the calculator, she gets 150° and now she is baffled. How could a right triangle have a 150 degree angle? What is wrong with how Melissa is thinking about inverse trigonometric functions?

3 (a) (C) One of your students, Jadin, is asked to solve the equation $y = \tan^{-1} x$ for x in terms of y. He solves for x as follows:

$$y = \tan^{-1} x$$
$$y = \frac{1}{\tan} x$$
$$\tan y = x.$$

Jadin checks the answer in the back of the book and sees he is right. How can you help Jadin understand what is wrong with his work?

(b) Another student, Marta, solves the same problem as follows:

$$y = \tan^{-1} x$$
$$y = \frac{1}{\tan x}$$
$$y \tan x = 1$$
$$\tan x = \frac{1}{y}$$
$$x = \tan^{-1}\left(\frac{1}{y}\right).$$

When Marta checks her answer in the back of the book, it doesn't check. How will you help her to see what is wrong with her work?

4 Solve each of the following equations for all values of x:
(a) $\cos 3x = \frac{1}{2}$
(b) $\sin 2x = -\frac{1}{2}$
(c) $2 \sin x - 1 = 3$
(d) $3 \cos 4x = 1$
(e) $\sin x \cos x - \sin x = 0$
(f) $2 \sin^2 x - 3 \sin x + 2 = 0$
(g) $3 \tan x - 4 = 0$ (Remember, the period of $\tan x$ is 180° degrees, not 360°.)
(h) $\csc x = 2$
(i) $3 \sec x - 1 = 5$

5 Using the relationship that $\sin^2 x + \cos^2 x = 1$, reduce each of the following equations to one involving only either $\sin x$ or $\cos x$. Then solve them. Check your answers graphically.

(a) $1 + \sin x = 2\cos^2 x$
(b) $\cos x = \sin x - 1$ [Hint: Square both sides and be sure to check your answers.]
(c) $3 \sin 3x - 1 = 0.5$

6 (C) Solve the launch question.

7 Using the damped oscillation function presented earlier after example 11.17, $y = e^{-2t}(-5 + 10 \sin 2\pi t)$, find the times that the spring passes its equilibrium point.

8 When a projectile is fired from the ground at an angle of θ degrees with the horizontal, the horizontal distance it travels is given by

$$H(\theta) = \frac{v_0^2 \sin 2\theta}{32}.$$

(a) If v_0 is the initial velocity and $v_0 = 1100$ ft/s, what values of θ to the nearest degree will make $H(\theta) = 1125$ ft?
(b) What value of θ will make $H(\theta)$ a maximum?

9 When a beam of light is shined in to the water at an angle of $20°$ to the horizontal, it can be shown using Snell's Law for refraction of light, that the beam bends towards the perpendicular to the water's surface at an angle of θ degrees where θ is a solution to the equation:

$$\frac{\sin \theta}{\sin 70°} = 0.752\,487\,319\,305\,7.$$

Find θ to the nearest degree.

10 Suppose that a rocket could be launched from the ground with a constant velocity of 400 feet per minute and that an observer is standing 100 feet from the rocket on level ground with his eye on the bottom of the rocket. He keeps his eye on the bottom as the rocket rises straight up. What angle will his eye make with the horizontal after 2 minutes if the man's eye is 6 feet off the ground?

11 Given a rectangular sheet of paper 6 inches by 9 inches. Suppose that someone grabs one corner of the 6 inch side and carries it to the other side and presses down forming a fold as shown in Figure 11.76 below.

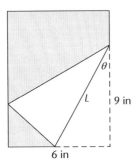

Figure 11.76

Find L in terms of θ.

12 Show by using right triangles, that if a is acute, $\sin^{-1} a + \cos^{-1} a = \frac{\pi}{2}$. Is this still true if a is not acute? How do you know?

11.8 Trigonometric Identities

**LAUNCH**

In the following three equations, graph both the right side of the equation and the left side of the equation on the same set of axes. So, for example, if the equation is $1 + \tan^2 \theta = \sec^2 \theta$, then graph the two equations: $y = 1 + \tan^2 \theta$ and $y = \sec^2 \theta$. In each of the cases below, determine if the graphs are the same or different. If they are the same, state what that means.

(a) $\dfrac{\sec x + \csc x}{\tan x + \cot x} = \sin x + \cos x$.

(b) $\dfrac{1 - \cos x}{\sin x} = \dfrac{\cos x}{1 + \cos x}$

(c) $\cos(3x) = 4\cos^3 x - 3\cos x$

Now that you have graphed the equations in the launch question, you are probably trying to recall what it might mean when both sides of the original equation have the same graph. You might remember other relationships where this has happened, such as:

$$\sin^2 \theta + \cos^2 \theta = 1$$
$$\cos(\theta) = \cos(-\theta)$$
$$\sin(-\theta) = -\sin(\theta).$$

We begin this section by asking you how you would refer to the relationships listed below. Do you now recall how they are referred to? Are they formulas or equations? Actually, the above fall into a category of things known as identities. Identities may be thought of as different ways to represent the same expressions. We are somewhat familiar with this idea from arithmetic. For example, $\frac{1}{2}$, $\frac{100}{200}$, 50% and 0.5 all represent the same quantity. However, in different situations, one representation makes more sense or is more useful than another. The same is true for these trigonometric identities.

More formally, a trigonometric identity is an equation involving trigonometric functions of an angle, say θ, which is valid for all values of θ for which the functions on both sides of the equation are defined. Here is a list of identities that we have discussed so far either in the text or in the Student Learning Opportunities. They constitute the basic identities given whenever one studies trigonometry

$$\sin^2 \theta + \cos^2 \theta = 1$$
$$\cos(\theta) = \cos(-\theta)$$
$$\sin(-\theta) = -\sin(\theta)$$
$$\sin(180 - \theta)° = \sin \theta°$$
$$\cos(180 - \theta)° = -\cos(\theta°)$$

$$\cos(\theta°) = \sin(90-\theta)°$$

$$\tan\theta = \frac{\sin\theta}{\cos\theta} \text{ when } \cos\theta \neq 0$$

$$\cos(\theta_1 \pm \theta_2) = \cos\theta_1\cos\theta_2 \mp \sin\theta_1\sin\theta_2$$

$$\sin(\theta_1 \pm \theta_2) = \sin\theta_1\cos\theta_2 \pm \cos\theta_1\sin\theta_2$$

$$\sin 2\theta = 2\sin\theta\cos\theta \tag{11.20}$$

$$\cos 2\theta = 1 - 2\sin^2\theta = 2\cos^2\theta - 1. \tag{11.21}$$

Based on these identities, it is possible to generate other useful identities such as those listed below.

$$1 + \tan^2\theta = \sec^2\theta$$

$$! + \cot^2\theta = \csc^2\theta$$

$$\cot\theta = \frac{\cos\theta}{\sin\theta} \quad \text{when} \quad \sin\theta \neq 0$$

$$\sin^2\theta = \frac{1 - \cos 2\theta}{2} \tag{11.22}$$

$$\cos^2\theta = \frac{1 + \cos 2\theta}{2} \tag{11.23}$$

$$\sin 2x = 2\sin x \cos x. \tag{11.24}$$

We ask you to prove these identities in the Student Learning Opportunities.

Part of the secondary school curriculum deals with trigonometric identities, but unfortunately, most secondary school students fail to appreciate their value. This might be due to the fact that they rarely get to see how extremely useful they are in all kinds of applications. For example, many real-life problems necessitate the use of calculus, which involves the evaluation of certain integrals that without the use of trigonometric identities would be too difficult to do. Identities are especially useful in Fourier series, which forms the foundation for so much of our current technology. We saw them used repeatedly in the chapter on transformations to get some interesting and practical results. We also used them earlier in the chapter in solving certain kinds of trigonometric equations used to model real-life situations and we will use them again later in this chapter.

Before calculators were so readily available, students, mathematicians, and scientists used trigonometric tables to find the values they needed. These tables were first created by the Hellenistic mathematician Hipparchus, and were later refined by Ptolemy, the mathematician we spoke about in Chapter 5. Ptolemy was aware of certain trigonometric values that could be established geometrically, and used these, together with trigonometric identities, to build a table of trigonometric functions for all angles. Specifically, he used the half angle formulas of equations (11.22) and (11.23) to cut angles into smaller and smaller pieces to find their sines and cosines, and then combined these results and used the trigonometric identities for $\sin(A \pm B)$ and $\cos(A \pm B)$ to find the trigonometric functions of many other angles. In that way, he built the trigonometric tables that were used by astronomers of his day. It is worth noting that Ptolemy had to do his work in terms of chords in a circle, since the terminology we use today for our identities was not yet available. To get a better idea of how he did this, check back to Chapter 5 to see how chords of a circle can be expressed in terms of sines and cosines. Of course, his famous theorem given in Section 4 of Chapter 5, Theorem 5.19, was the basis of much of his work.

So now let us get to our study of trigonometric identities. There are an enormous number of trigonometric identities that one can create, in fact, an infinite number. Some of the more obscure identities have strong uses in applications and can be deduced from the identities we gave above. We will now only give a few identities and then ask you to do others in the Student Learning Opportunities. We also point out some of the common mistakes that students make when proving identities.

Generally speaking, when you are asked to prove a trigonometric identity, you are given an equation of the form $A = B$ where A and B usually are trigonometric expressions and you have to show that the left side equals the right side (for all values of θ for which both sides make sense), usually by manipulating only one side of the equation.

Example 11.24 *Prove that*

$$\frac{\sec x + \csc x}{\tan x + \cot x} = \sin x + \cos x.$$

Solution. Since the right side of this equation is expressed only in terms of $\sin x$ and $\cos x$, it makes sense to express the left hand side (*LHS*) in a similar way. Using the facts that $\sec x$, $\csc x$, and $\cot x$ are the reciprocals of $\cos x$, $\sin x$, and $\tan x$, respectively, and using the fact that $\tan x = \frac{\sin x}{\cos x}$, we have

$$LHS = \frac{\sec x + \csc x}{\tan x + \cot x}$$

$$= \frac{\dfrac{1}{\cos x} + \dfrac{1}{\sin x}}{\dfrac{\sin x}{\cos x} + \dfrac{\cos x}{\sin x}}.$$

Combining the fractions in the numerator and denominator of the overall fraction we get

$$\frac{\dfrac{\sin x + \cos x}{\sin x \cos x}}{\dfrac{\sin^2 x + \cos^2 x}{\sin x \cos x}}.$$

Inverting the bottom fraction and multiplying by the top fraction we get

$$\frac{\dfrac{\sin x + \cos x}{\sin x \cos x}}{\dfrac{\sin^2 x + \cos^2 x}{\sin x \cos x}} = \frac{\sin x + \cos x}{\sin x \cos x} \cdot \frac{\sin x \cos x}{\sin^2 x + \cos^2 x} = \frac{\sin x + \cos x}{\sin^2 x + \cos^2 x}$$

and since $\sin^2 x + \cos^2 x = 1$, this last fraction reduces to

$\sin x + \cos x$

which is the right hand side of the identity we were trying to prove. So, we are done.

> **Example 11.25** *Prove that*
>
> $$\frac{2\tan x}{1 + \tan^2 x} = \sin 2x.$$

Solution. The denominator of the fraction stands out. We know that, according to one of the previous identities, $1 + \tan^2 x = \sec^2 x$ so that should be our starting point. Our left hand side (*LHS*) is

$$LHS = \frac{2\tan x}{1 + \tan^2 x}$$

$$= \frac{2\tan x}{\sec^2 x}.$$

Now since $\tan x = \frac{\sin x}{\cos x}$ and $\sec x = \frac{1}{\cos x}$, we get

$$\frac{2\frac{\sin x}{\cos x}}{\frac{1}{\cos^2 x}}$$

which, when we invert and multiply and divide common factors, yields

$2 \sin x \cos x.$

But we know that $2 \sin x \cos x = \sin 2x$ by equation (11.24). So we have shown that the left hand side and the right hand side of the identity we were trying to prove are the same, and we are done.

Note that, in both of these examples, we expressed all expressions in terms of sines and cosines. This is an effective strategy for students to use. They need to also realize that it is best to only change one side of the identity at a time. A common mistake that students make is illustrated in the following "solution" to show an identity is valid.

> **Example 11.26** *Show that*
>
> $$\frac{1 - \cos x}{\sin x} = \frac{\sin x}{1 + \cos x}.$$

Student's Mistaken Solution: Cross multiply (or equivalently multiply both sides of the equation by $\sin x (1 + \cos x)$) to get

$1 - \cos^2 x = \sin^2 x$

which we know is true since $\sin^2 x + \cos^2 x = 1$ Done!

So what is wrong? The student is using the fact that both sides of an equation can be multiplied by the same quantity. But we don't know the result is an equality! That is the point of the exercise! So, when you multiply both sides of the "equation" by $\sin x(1 + \cos x)$, you are assuming they are already equal. That is, you are assuming what you are trying to prove in the proof and that is illegal! A correct way to solve the problem follows:

Solution. Work only on the left side and to get it to look like the right side, we multiply the numerator and denominator by $1 + \cos x$

$$
\begin{aligned}
LHS &= \frac{1 - \cos x}{\sin x} \\
&= \frac{1 - \cos x}{\sin x} \cdot \frac{1 + \cos x}{1 + \cos x} \\
&= \frac{1 - \cos^2 x}{\sin x(1 + \cos x)} \\
&= \frac{\sin^2 x}{\sin x(1 + \cos x)} \\
&= \frac{\sin x}{1 + \cos x} \\
&= RHS.
\end{aligned}
$$

We are done!

Identities can often be used to explain mysterious behavior. For example, suppose that we graph $y = 2 \sin x$ and $y = 4 \cos x$ on the same set of axes as shown in Figure 11.77 below.

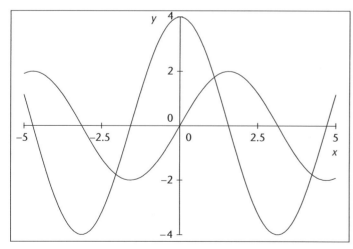

Figure 11.77

If we were to add these two functions to get $y = 2 \sin x + 4 \cos x$, what would you guess the resulting curve would look like? Might you guess at something like the graph shown in Figure 11.78 below

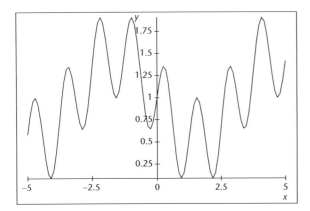

Figure 11.78

or would you expect to get something like the following graph shown in Figure 11.79?

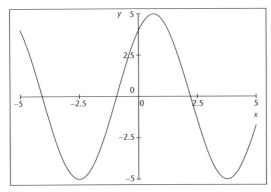

Figure 11.79

Most people would guess that it looks something like the first graph. The second graph is just too regular and looks like a sine curve. Surprisingly, it does look like the second curve. Other examples like this seem to indicate that, when we graph equations like $y = A \sin x + B \cos x$, we get a sine-like curve. But why? The following identity, stated as a theorem, explains it. It is very useful to electrical engineers.

Theorem 11.27 $A \sin x + B \cos x = \sqrt{A^2 + B^2} \sin(x + \theta)$, where θ satisfies $\cos\theta = \frac{A}{\sqrt{A^2+B^2}}$ and $\sin\theta = \frac{B}{\sqrt{A^2+B^2}}$.

Proof. We use a creative approach to this problem that is surprising. Rewrite $A \sin x + B \cos x$ as

$$\sqrt{A^2 + B^2}\left(\frac{A}{\sqrt{A^2+B^2}} \sin x + \frac{B}{\sqrt{A^2+B^2}} \cos x\right). \tag{11.25}$$

Now, if we can find θ such that $\cos\theta = \frac{A}{\sqrt{A^2+B^2}}$ and $\sin\theta = \frac{B}{\sqrt{A^2+B^2}}$, then equation (11.25) can be written as

$$\sqrt{A^2 + B^2}\,(\cos\theta \sin x + \sin\theta \cos x)$$

which by our identities is the same as

$$\sqrt{A^2 + B^2}(\sin(x + \theta)).$$

Thus the graph of $y = A \sin x + B \cos x = \sqrt{A^2 + B^2} \sin(x + \theta)$ should look like a sine curve! ∎
Let us illustrate this with an example.

Example 11.28 *Write* $-\sin x + \cos x$ *as a sine function.*

Solution. Here $A = -1$ and $B = 1$. Hence, $\sqrt{A^2 + B^2} = \sqrt{2}$ and we must find a θ such that

$$\cos \theta = \frac{A}{\sqrt{A^2 + B^2}} = \frac{-1}{\sqrt{2}} \text{ and}$$

$$\sin \theta = \frac{B}{\sqrt{A^2 + B^2}} = \frac{1}{\sqrt{2}}.$$

The fact that $\sin \theta$ is positive and $\cos \theta$ is negative tells us that our θ should be in the second quadrant, and that our reference angle is $45°$. Thus, $\theta = 135°$ will work. Our function $-\sin x + \cos x = \sqrt{2} \sin(x + 135°)$. One can check on the graphing calculator that the graphs of $y = -\sin x + \cos x$ and $y = \sqrt{2} \sin(x + 135°)$ are one and the same and look like the graph shown in Figure 11.80.

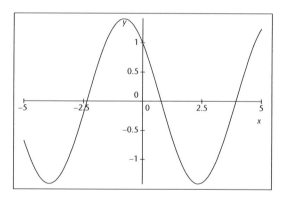

Figure 11.80

Student Learning Opportunities

1 (C) One of your students proves the equation below is an identity by doing the following work:

$$\cot x + \tan x = \sec x \cdot \csc x$$

$$\frac{\cos x}{\sin x} + \frac{\sin x}{\cos x} = \frac{1}{\cos x} \cdot \frac{1}{\sin x}$$

$$\frac{\cos^2 x + \sin^2 x}{\sin x \cos x} = \frac{1}{\cos x \sin x}$$

$$\frac{1}{\cos x \sin x} = \frac{1}{\cos x \sin x}.$$

He is asserting that it is okay when proving an identity to work on both sides of the equal sign, manipulate each side, and then show that the manipulated results are the same on each side. Is the student right?

2 Using your calculator, first determine if each of the following appears to be an identity. If it does, prove it.

(a) $\tan(90 - u)° = \cot u°$

(b) $\sin^2 a + \cos^2 a + \tan^2 a = \sec^2 a$

(c) $\sin 3x = 3 \sin x$

(d) $\sec^4 x - \tan^4 x = \sec^2 x + \tan^2 x$ [Hint: Factor the left side.]

(e) $\cos(x + 30)° + \sin(x - 60)° = 0$

(f) $\sin 2x + \cos 2x = 1$.

(g) $(\tan z + \cot z) \sin z \cos z = 1$

(h) $\dfrac{\sin x}{\sin x + \cos x} = \dfrac{\tan x}{1 + \tan x}$

(i) $\dfrac{\sin x}{1 + \cos x} = \dfrac{\tan x}{1 + \tan s}$

(j) $(\tan x + \cot x)^2 = \sec^2 x + \csc^2 x$

3 Using the formulas for $\sin(\theta_1 \pm \theta_2)$ show that:

(a) $\sin \theta_1 \cos \theta_2 = \dfrac{1}{2}[\sin(\theta_1 + \theta_2) + \sin(\theta_1 - \theta_2)]$

(b) $\cos \theta_1 \sin \theta_2 = \dfrac{1}{2}[\sin(\theta_1 + \theta_2) - \sin(\theta_1 - \theta_2)]$

(c) $\cos \theta_1 \cos \theta_2 = \dfrac{1}{2}[\cos(\theta_1 + \theta_2) + \cos(\theta_1 - \theta_2)]$

(d) $\sin \theta_1 \sin \theta_2 = \dfrac{1}{2}[\cos(\theta_1 - \theta_2) - \cos(\theta_1 + \theta_2)]$

These are called the product to sum formulas.

4 Replacing θ_1 in each of the identities from the previous question by $\frac{x+y}{2}$ and θ_2 by $\frac{x-y}{2}$, show that:

(a) $\sin x + \sin y = 2 \sin\left(\dfrac{x + y}{2}\right) \cos\left(\dfrac{x - y}{2}\right)$

(b) $\sin x - \sin y = 2 \cos\left(\dfrac{x + y}{2}\right) \sin\left(\dfrac{x - y}{2}\right)$

(c) $\cos x + \cos y = 2 \cos\left(\dfrac{x + y}{2}\right) \cos\left(\dfrac{x - y}{2}\right)$

(d) $\cos x - \cos y = -2 \sin\left(\dfrac{x + y}{2}\right) \sin\left(\dfrac{x - y}{2}\right)$

These are known as sum to product formulas.

5 Using the formulas in Student Learning Opportunity 4(b) and 4(c), show that
$$\frac{\sin 3x - \sin x}{\cos 3x + \cos x} = \tan x.$$

6 Using the appropriate formula from Question 3, write the product $\sin 2x \cos 3x$ as a sum.

7 Using the appropriate formula from Question 4, write the sum $\sin 5x + \sin 3x$ as a product.

8 Using the identities listed in this section, show that

(a) $\tan^2 x = \dfrac{1 - \cos 2x}{1 + \cos 2x}$

(b) $\cos(3x) = 4\cos^3 x - 3\cos x$ [Hint: $3x = 2x + x$.]

(c) $\dfrac{\sin 4x}{\sin x} = 4 \cos x \cos 2x$

(d) $\tan(x \pm y) = \dfrac{\tan x \pm \tan y}{1 \mp \tan x \tan y}$ and hence $\tan 2x = \dfrac{2 \tan x}{1 - \tan^2 x}$

(e) $\dfrac{\sin 10x}{\sin 9x + \sin x} = \dfrac{\cos 5x}{\cos 4x}$ [Hint: The results of Student Learning Opportunity 4 might help.]

9 Express each of the following as a function of $\sin x$.

(a) $\dfrac{1}{2} \sin x + \dfrac{\sqrt{3}}{2} \cos x$

(b) $-\sqrt{3} \sin x + \cos x$

(c) $3 \sin x + 4 \cos x$

10 Corroborate graphically the identity that we proved in Example 11.26, $\dfrac{1 - \cos x}{\sin x} = \dfrac{\sin x}{1 + \cos x}$.

11.9 Solutions of Cubic Equations Using Trigonometry

LAUNCH

Solve the cubic equation $x^3 = 15x + 4$, by using algebraic methods or by trial and error [Hint: check integer values of x between 0 and 5.]

(a) What solutions did you find?
(b) How many roots does a cubic equation have? Were you able to find them all? If not, why not?

Now that you have tried solving what appeared to be a relatively simple cubic equation and you were only able to find one real root, you are probably wondering how on earth you might find the other roots. If you are thinking about using the graphing calculator to see where the graph crosses the x-axis, you have a good idea. But, how could this problem be solved without such technology? Would you ever imagine that trigonometry would play a role in its solution? Most people wouldn't. But, you will be surprised to find out, as you read this section, that indeed, trigonometry can play a major role in solving such cubic equations as this one. Read on to find out how it is done. While you are reading this, you will be exposed to a most fascinating aspect of the history of solving cubic equations.

In Chapter 3, we discussed how one finds a solution to the general cubic equation. From the material in that chapter, it follows that the formula for a solution of the cubic equation

$$x^3 - px - q = 0 \quad \text{where } p \text{ and } q \text{ are nonnegative} \tag{11.26}$$

is

$$x = \sqrt[3]{\frac{q \pm \sqrt{q^2 - \frac{4p^3}{27}}}{2}} + \sqrt[3]{\frac{q \mp \sqrt{q^2 - \frac{4p^3}{27}}}{2}}. \tag{11.27}$$

This formula led to some strange kinds of answers to cubic equations, which involved square roots of negative numbers for cubic equations whose roots are known to be real. This kind of situation was what motivated the study of imaginary numbers, as we have pointed out in the chapter on imaginary numbers, Chapter 7, and somewhat in Chapter 3.

It is interesting to note that, after these formulas were discovered, the French mathematician Francois Viète (1540–1603) (who was a lawyer by trade and did mathematics on the side), discovered a formula that would give a solution of the cubic equation (11.26) that involved trigonometric functions. We discuss that now, since the solution process used a trigonometric identity and hence used material from this chapter. The identity he used was $\cos(3\theta) = 4\cos^3\theta - 3\cos\theta$, which can be derived as follows:

$$\cos(3\theta) = \cos(2\theta + \theta)$$
$$= \cos(2\theta)\cos\theta - \sin 2\theta \sin\theta$$
$$= (\cos^2\theta - \sin^2\theta)\cos\theta - 2\sin\theta\cos\theta\sin\theta \quad \text{(using (11.20) and (11.21))}$$
$$= \cos^3\theta - \sin^2\theta\cos\theta - 2\sin^2\theta\cos\theta$$
$$= \cos^3\theta - 3\cos\theta\sin^2\theta$$
$$= \cos^3\theta - 3(\cos\theta)(1 - \cos^2\theta)$$
$$= \cos^3\theta - 3\cos\theta + 3\cos^3\theta$$
$$= 4\cos^3\theta - 3\cos\theta.$$

Viète observed the following: From the identity $\cos(3\theta) = 4\cos^3\theta - 3\cos\theta$ we can solve for $\cos^3\theta$ to get

$$\cos^3\theta = \frac{3}{4}\cos\theta + \frac{1}{4}\cos(3\theta). \tag{11.28}$$

Now he wrote equation (11.26) as

$$x^3 = px + q \tag{11.29}$$

and, since p and q were positive, he observed that we could write $p = 3a^2$ for some a and $q = a^2 b$ for some b. (This is genius at work. To see this, just take $a = \sqrt{\frac{p}{3}}$ and $b = \frac{3q}{p}$.) Then equation (11.29) becomes

$$x^3 = 3a^2 x + a^2 b. \tag{11.30}$$

Then out of the clear blue, he said, "Let us suppose there is a solution of the form $x = 2a\cos\theta$ to equation (11.30)" (His goal was to show that his assumption is true by actually finding the θ that works. Let's bear with him.) From $x = 2a\cos\theta$, we get that $\cos\theta = \frac{x}{2a}$. Substituting this into the identity equation (11.28), we get

$$\frac{x^3}{8a^3} = \frac{3}{4}\frac{x}{2a} + \frac{1}{4}\cos(3\theta)$$

and multiplying both sides of this equation by $8a^3$ we get

$$x^3 = 3a^2 x + 2a^3 \cos(3\theta)$$

which is our equation (11.30), provided $a^2 b = 2a^3 \cos(3\theta)$. Thus $x = 2a\cos\theta$ will be a solution provided $a^2 b = 2a^3 \cos(3\theta)$ or equivalently if $b = 2a\cos(3\theta)$. Solving this for $\cos(3\theta)$ we get that

$$\cos 3\theta = \left(\frac{b}{2a}\right). \tag{11.31}$$

Using the facts that $a = \sqrt{\frac{p}{3}}$ and $p = \frac{3q}{b}$ and putting everything in equation (11.31) in terms of p's and q's we have, after doing some algebraic simplifications, that

$$\cos(3\theta) = \left(\frac{3q\sqrt{3}}{2p\sqrt{p}}\right). \tag{11.32}$$

In summary, we have found a solution to equation (11.29), namely

$$x = 2a\cos\theta = 2\sqrt{\frac{p}{3}}\cos\theta \tag{11.33}$$

where θ satisfies equation (11.32). This is a remarkable way of solving cubic equations by using trigonometry.

Let return to the launch problem.

Example 11.29 *Solve the cubic equation $x^3 = 15x + 4$ using Viète's formula, equation (11.32)*

Solution. Here $p = 15$ and $q = 4$. Substituting this into equation (11.32), we get that

$$\cos 3\theta = \left(\frac{12\sqrt{3}}{30\sqrt{15}}\right) = 0.17889.$$

Hence $3\theta = \cos^{-1} 0.17889 = 1.3909$ radians, so $\theta = 0.46363$ radians. From our experience with solving trigonometric equations, we get our other solutions by solving $3\theta = 1.3909 \pm 2\pi k$, $k = 0, 1, 2, \ldots$ or just $\theta = 0.46363 \pm 2\pi k/3$. Substituting $k = 0, 1,$ and 2, we get $\theta = 0.46363$, $\theta = 0.46363 \pm 2\pi/3 = 2.558$, and $\theta = 4.6524$. Evaluating equation (11.33) at each of these 3 values, we get $x = 2\sqrt{5}\cos 0.46363 = 4.0$, $x = 2\sqrt{5}\cos 2.558 = -3.7319$, and $x = 2\sqrt{5}\cos 4.6524 = -0.26812$. Not only did we get one solution, we got all 3 solutions of our equation without even having to deal with the complicated expression in equation (11.27)! (Recall, a cubic polynomial can only have 3 solutions.) This really is impressive.

Of course, you might have wondered what would happen in Viète's solution if the right hand side of equation (11.32) was more than 1? Well, we won't get into it, but that just means that we have some imaginary roots. Surprisingly, equation (11.32) does have solutions in that case, but they are imaginary and a discussion of that is left to a course in Complex Analysis, since it is beyond the scope of this book. There is, however, another way to get the real roots of a cubic like equation (11.26) when $\cos 3\theta > 1$. It uses the notion of hyperbolic cosine instead of cosine. The development parallels what we did in this section. Some of the details are given in the Student Learning Opportunities.

Student Learning Opportunities

1. Using the Viète formula, find one or more real roots of the following cubic equations.
   (a) $x^3 - x = 0$
   (b) $y^3 - 3y = 2$
   (c) $z^3 - 5z = 17$

2. The definition of hyperbolic sine (sinh) and hyperbolic cosine (cosh) follow:

   $$\sinh x = f(x) = \frac{e^x - e^{-x}}{2}$$

   $$\cosh x = g(x) = \frac{e^x + e^{-x}}{2}$$

   (a) Show that $\cosh^2 x - \sinh^2 x = 1$.
   (b) Show that $\cosh 2x = \cosh^2 x + \sinh^2 x$.
   (c) Show that $\sinh 2x = 2 \sinh x \cosh x$.
   (d) Show that $\cosh(2x) = 1 + 2(\sinh(x))^2$.
   (e) Show that $\cosh(x + y) = \cosh(x)\cosh(y) + \sinh(x)\sinh(y)$.
   (f) Show that $\cosh(3\theta) = 4\cosh^3 \theta + 3\cosh \theta$ and hence that $\cosh^3 \theta = \frac{3}{4}\cosh \theta - \frac{1}{4}\cosh(3\theta)$.
   (g) Mimicking what we did in this section, solve the cubic equation given in equation (11.29) in terms of $\cosh \theta$ when the right side of equation (11.31) is more than 1.

3. (C) As you have read in this chapter, Francois Viète was quite a genius. Use the Internet to find out more about him. What other areas of mathematics was he involved in? What were his other professions? What were some of the notable things he did?

11.10 Lissajous Curves

The three curves below in Figures 11.81–11.83 are called Lissajous curves. Examine them carefully and describe what their similarities and differences are.

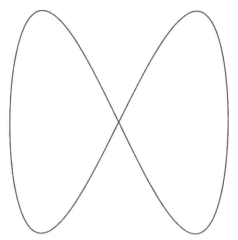
Figure 11.81

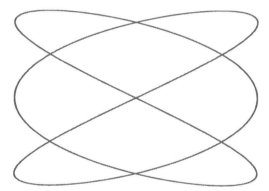
Figure 11.82

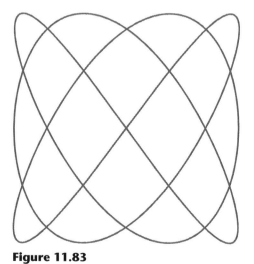
Figure 11.83

In looking at these curves, you probably have noticed that they contain curves that look similar to the graphs of trigonometric functions, perhaps on their sides. In this section you will learn more about them and how they can be used in practical settings.

We mentioned earlier that sound is believed to be wavelike. Where did this idea come from? The French physicist Jules Lissajous (1822–1880) is given credit for this discovery. He would strike tuning forks, hear sound and wanted to be able to "see" the sound. After numerous experiments, he happened upon the "right" way to "see" sound. He would place the tuning fork in front of series of carefully placed mirrors. He would then strike it, hear the sound, and shine a beam of light on it. The image went from one mirror to the next, to a screen on the wall. Suddenly, he saw images of waves dancing on the screen as the tuning fork vibrated. He actually saw the sound waves! He then started playing with images formed by striking two tuning forks positioned at right angles to one another and discovered some beautiful patterns now known as **Lissajous curves** (also known as **Bowditch curves**). For these experiments he received a Nobel prize.

The Lissajous curves can be described by using two trigonometric functions, one for x and the other for y. Since the x and y axes are perpendicular, we can simulate what he did in the laboratory. Here are typical equations for x and y. (There are several different variations on this.)

$x = A \sin(at + d)$

$y = B \sin(bt + e)$

where A, B, a, b, and d are constants. Lissajous curves have applications in physics, astronomy, and other sciences.

Students can graph these parametric equations on their graphing calculators by setting the machine to parametric mode. The pictures one gets are very sensitive to the ratio of a/b and can be very pretty when done with a good grapher program. Below are 2 such pictures which are not so elaborate.

Our first picture shown in Figure 11.84 is generated by the parametric equations $x = \sin(2t)$ and $y = \sin(3t)$. It looks a lot like the shield of the Atomic Energy Commission.

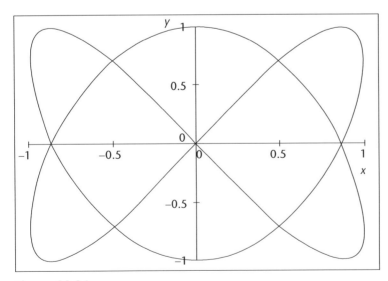

Figure 11.84

The second, shown in Figure 11.85 is the logo of television channel *ABC*, and is generated by $x = \sin(t)$, $y = \sin 3(t - \frac{\pi}{2})$.

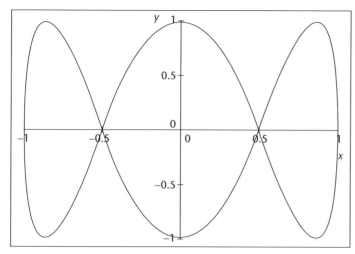

Figure 11.85

A check of websites dealing with Lissajous curves will show you some much nicer and more interesting pictures. We recommend a visit to: http://math.jccc.net:8180/webMathematica/JSP/swilson/lissajous.jsp.

Student Learning Opportunities

1. Draw each of the following Lissajous curves by going to the website we mentioned earlier or by using your graphing calculator.
   (a) $x = 2\sin(3t + \pi)$, $y = -2\sin(4t - \pi)$
   (b) $x = 4\sin(\pi t + \pi/2)$, $y = 2\sin(2t)$
   (c) $x = 5\sin(3t)$, $y = 4\sin(5t)$
   (d) $x = \sin(5t - 6)$, $y = \sqrt{2}\sin\left(t - \frac{\pi}{2}\right)$

2. Make up your own values and, using the website mentioned, find some interesting Lissajous curves. Print out at least two of the ones you like best and bring them to class. Describe how you made them.

11.11 Vectors

 LAUNCH

If you visit the NCTM website http://standards.nctm.org/document/eexamples/chap7/7.1/part2.htm, you will be able to control the motion of a plane by using two vectors, one is a red vector representing wind on the screen and the other is a blue vector which is used to direct the airplane.

582 Trigonometry

It is interesting to see how the wind affects the motion of the plane. Let us ask a few questions that you can probably answer, even if you are not able to access the Internet site.

1. If the plane is traveling due north and the wind is blowing in an easterly direction, how will it affect the direction in which the plane is traveling? How will it affect the speed at which the plane is traveling?
2. If the plane is traveling due north and the wind is blowing in a northerly direction, how will it affect the direction in which the plane is traveling? How will it affect the speed at which the plane is traveling?

Even if you weren't able to access the applet recommended in the launch, you probably have an intuitive idea of what happens to an airplane when it is affected by wind. In actuality, you have an idea of how vectors work together. You will learn more details about this in the upcoming section.

11.11.1 Basic Vector Algebra

In Section 11.2 you have seen how vectors are used in trigonometry and applied in physics. Although we defined vectors as anything that has magnitude and direction, we have not yet studied them from a purely geometric or algebraic standpoint. In this section we will do just this by expanding on Section 11.2 and reviewing the topic of vectors that most of you have probably learned in linear algebra or calculus. We will then turn to some interesting applications that are most likely new to you.

As you will recall, vectors can be represented by arrows whether they are forces or other things that have magnitude and direction. The length of the arrow, just as in forces, represents the magnitude of the vector and the angle the arrow makes with the horizontal, the direction.

The arrowhead is the tip of the vector, while the other end is the tail. Thus, if a plane is flying at 600 miles per hour northeast, we could denote this velocity as a vector which we show in Figure 11.86.

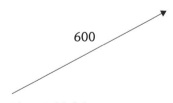

Figure 11.86

As in Section 11.2, vectors will be represented by bold letters, while their lengths will be represented without bold. Two vectors are considered equal if they have the same magnitude and direction. In two dimensions this means that arrows representing them are both parallel and congruent. So, each of the vectors, $\mathbf{V}_1$, $\mathbf{V}_2$, and $\mathbf{V}_3$ shown in Figure 11.87 below are the same because they are parallel and have the same length.

Trigonometry 583

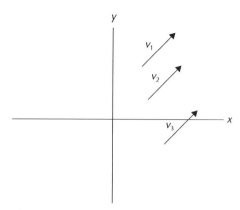

Figure 11.87

When a vector is moved parallel to itself, so that its tail is at the origin, we say the vector is in **standard form**. When a vector is in standard form, its tip will lie at some point (a, b) and a and b are called the components of the vector. If the tail of a vector is at $P(x_1, y_1)$ and the tip is at $Q(x_2, y_2)$, then the vector when put in standard form has components $(x_2 - x_1, y_2 - y_1)$, as shown in Figure 11.88 below.

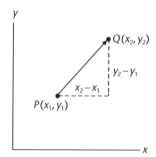

Figure 11.88

This vector is abbreviated as Q–P.

The components of a vector are analogous to the components of the forces that we described in Section 11.2, but now we are speaking of vectors that need not be forces. So, a vector representing the velocity of a car can be broken into components, just like was done for a force. The components are called velocity in the x direction and velocity in the y direction.

Let us now see how to add vectors. It can be done in two ways. Suppose, for example, we have two vectors, $\mathbf{V}_1$ and $\mathbf{V}_2$ shown in the Figure 11.89 below.

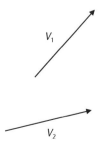

Figure 11.89

One way to add these vectors and get $V_1 + V_2$ is to move the tail of V_1 to the tip of V_2 and draw the arrow from the tail of V_2 to the tip of the moved V_1. See Figure 11.90 below.

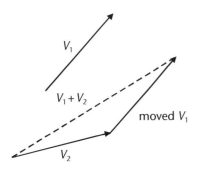

Figure 11.90

Another way to add two vectors V_1 and V_2 is to move V_1 so that its tail coincides with the tail of V_2 and then complete the parallelogram having V_1 and V_2 as sides. The diagonal of the parallelogram will be the $V_1 + V_2$ as the diagram in Figure 11.91 shows.

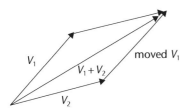

Figure 11.91

You may be wondering if addition of vectors is commutative. Well, in a similar manner to what we did earlier, we can construct $V_2 + V_1$ by moving the tail of V_2 to the tip of V_1 and drawing an arrow from the tail of V_1 to the tip of the moved V_2. We will get the same diagonal as we did earlier for the sum. So $V_1 + V_2 = V_2 + V_1$. We now explore the subtraction of vectors.

If V_1 and V_2 are vectors and we draw the vector from the tip of V_1 to the tip of V_2, what vector is that? (See Figure 11.92 below.)

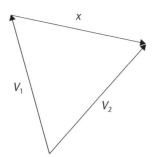

Figure 11.92

Well, if we call that vector x, then $V_1 + x = V_2$ and it seems logical to call $x = V_2 - V_1$. Indeed, that is what we call it, and we now have our definition of **subtraction of vectors**: $V_2 - V_1$ is the arrow drawn from the tip of V_1 to the tip of V_2.

Given a vector **V**, the vector −**V** is the arrow with the same length but opposite direction as shown in Figure 11.93 below.

Figure 11.93

Now we examine the concept of multiplication of vectors. If **V** is a vector and λ is a scalar, then λ**V** is the vector |λ| times as long as *V* and in the same direction if λ > 0 and in the opposite direction if λ < 0. Thus the vector 3*V* is 3 times as long as *V* as is the vector −3*V*. Only 3*V* is in the same direction as *V*, while −3*V* is in the opposite direction. See Figure 11.94 below, where 3*V* is shown.

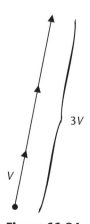

Figure 11.94

Let us practice a bit with "arrow algebra." Although we have said that the result of adding two vectors can be interpreted as the diagonal of a parallelogram, in practice, this is rarely used. Physicists simply represent vectors by arrows and then use the geometric definition of adding arrows. Namely, to add $\mathbf{V}_1$ and $\mathbf{V}_2$, we move $\mathbf{V}_1$ so that its tail is at the tip of $\mathbf{V}_2$ and then draw the arrow from the tail of $\mathbf{V}_2$ to the tip of the moved $\mathbf{V}_1$. The next example clarifies this.

Example 11.30 *Using the vectors **u**, **v**, and **w** below in Figure 11.95,*

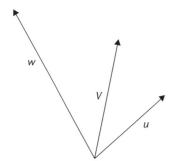

Figure 11.95

draw each of the following vectors (a) **u** + **v** + **w**, (b) **u** − **v**, (c) **v** − **u**.

Solution. To add **u**+**v**+**w**, first add **u**+**v** as shown below in Figure 11.96,

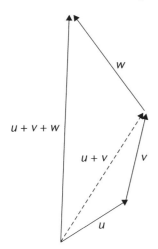

Figure 11.96

and then add **w**. We notice all we need to do to get our final result is to move **v** to the tip of **u** and then move **w** to the tip of the moved **v**. Once we draw an arrow from the tail of **u** to the tip of the moved **w**, we are done.

Solution. (b) and (c) **u** − **v** is the vector drawn from the tip of **v** to the tip of **u**, and **v** − **u** is the vector going from the tip of **u** to the tip of **v**. See Figure 11.97 below.

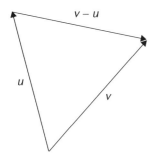

Figure 11.97

Example 11.31 *Suppose that we have any triangle ABC and turn the sides into vectors as shown in Figure 11.98 below.*

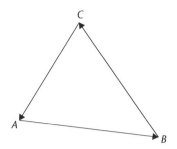

Figure 11.98

Show that the sum of the vectors is the zero vector (0, 0).

Solution. To simplify this problem, we need to situate one of the vertices at the origin. We can do this by moving the triangle parallel to itself so that A is at the origin. Since we are moving the entire figure parallel to itself, none of the vectors are changed. Using the result of the previous problem part (a), the sum of the vectors, is where the final tip lands when the vectors are put "tail to tip" as they are above. This place is (0, 0) as Figure 11.99 below shows.

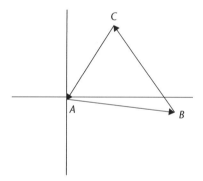

Figure 11.99

So, the sum of the vectors is (0, 0), the zero vector.

We will take this geometric interpretation of vectors to an interesting conclusion in Subsection 11.11.3.

11.11.2 Components of Vectors

We just explained how to add and subtract two vectors and multiply a vector by a scalar. For addition, we used the notation $V_1 + V_2$, the same as we used for adding forces. But, in Section 11.2, when we added forces, we added components of the vectors. Thus, we seem to have used the same notation for two different things which is confusing, unless of course, they are really the same. Does the "tail to tip" geometric approach that we used to add vectors have anything to do with adding the components? The answer is "Yes." In fact, the two approaches are equivalent, which we will now show.

588 Trigonometry

In Figure 11.100 below, we let $\mathbf{V}_1 = (a, b)$ and $\mathbf{V}_2 = (c, d)$. To add $\mathbf{V}_1$ and V_2, we move $\mathbf{V}_1$, so that its tail is at R and draw horizontal line RT and QT perpendicular to RT. Triangle OPD is congruent to triangle RQT and we see by the diagram, that the point, Q, where the tip of $\mathbf{V}_1 + \mathbf{V}_2$ lies, when it is put in standard form, is $(a + c, b + d)$. Thus, $\mathbf{V}_1 + \mathbf{V}_2 = (a + c, b + d)$. So, to add two vectors, we just add their components, exactly as we did with forces!

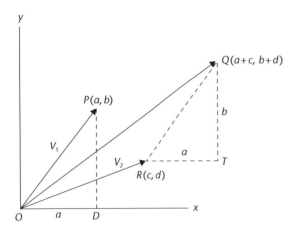

Figure 11.100

Using components it can be shown that if $\mathbf{V}_2 = (c, d)$ and $\mathbf{V}_1 = (a, b)$, then $\mathbf{V}_2 - \mathbf{V}_1 = (c - a, d - b) = \mathbf{V}_2 + (-\mathbf{V}_1)$. For practice, you can show that, if $\mathbf{V} = (a, b)$, then $-\mathbf{V} = (-a, -b)$.

One also notices a similarity between vectors and imaginary numbers. In Chapter 7 we found that imaginary numbers could be represented by arrows and that their sum is found by adding the components. Thus, imaginary numbers can also be considered vectors.

How do we find the components of a vector? We do it exactly as in Section 11.2. The magnitude of a vector $\mathbf{V}_1 = (a, b)$ is the length of the vector, which we can see from the picture in Figure 11.101 below is just $V_1 = \sqrt{a^2 + b^2}$.

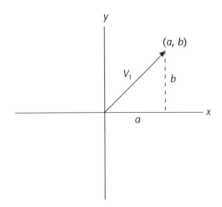

Figure 11.101

(Remember, we are using the convention that boldface represents the vector, while unbolded represents the length of the vector.)

Using trigonometry, we see that $\frac{a}{V_1} = \cos \theta$ and that $\frac{b}{V_1} = \sin \theta$ so

$a = V_1 \cos \theta$

and

$$b = V_1 \sin \theta$$

just as with forces. Furthermore,

$$\tan \theta = \frac{b}{a}.$$

We can again make the connection with imaginary numbers, where, as shown in Chapter 7, we had the same results. There, a was called the real part of the imaginary number and b the complex part.

Here is a problem similar to one we did with forces earlier, only this time it involves velocities. At first glance, you will probably think this problems looks quite different.

Example 11.32 *An airplane starts out traveling at 300 miles per hour north, but a wind traveling at a direction N 40° E is blowing at 60 miles per hour takes it in another direction. (a) What will be the resulting speed and direction of the airplane? (b) With what speed and in what direction must the plane fly so that, taking the wind into account, it flies at 300 miles per hour north?*

Solution. (a) Figure 11.102 shows us our picture.

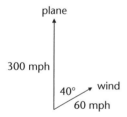

Figure 11.102

We are really looking for the vector which results from adding these two vectors. The velocity of the plane is the vector $V_1 = (0, 300)$. To find the components of the vector, V_2, the velocity of the wind, we need to use

$$(V_2)_x = V_2 \cos 50° = 60 \cos 50° = 38.567$$
$$(V_2)_y = V_2 \sin 50° = 60 \sin 50° = 45.963.$$

Note that the 50° is the angle the wind vector makes with the horizontal. So $V_2 = (38.567, 45.963)$. Adding V_1 and V_2 we get

$$V_1 + V_2 = (38.567, 345.963).$$

The magnitude of this vector is $V_1 + V_2 = \sqrt{38.567^2 + 345.963^2} = 348.11$, which is the actual speed of the plane with the help of the wind. The direction of the plane is found from

$$\tan \theta = \frac{b}{a} = \frac{345.963}{38.567} = 8.9704.$$

Thus,

$$\theta = \tan^{-1}(8.9704) = 83.64°$$

to the horizontal.

Solution. (b) Suppose the plane flies with velocity vector $\mathbf{V} = (a, b)$. Here is our picture where we want the diagonal of the parallelogram (the resulting velocity) to be $(0, 300)$. (See Figure 11.103.)

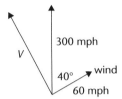

Figure 11.103

From our picture,

$$V + \text{velocity of the wind} = (0, 300).$$

Thus,

$$(a, b) + (38.567, 45.963) = (0, 300)$$

and so,

$$(a, b) = (-38.567, 254.037).$$

To find θ we solve

$$\tan \theta = \frac{b}{a} = \frac{254.037}{-38.567} = -6.5869.$$

We get $\theta = 98.634°$. So, the plane must fly at an angle of $98.634°$ with the horizontal, or N 8.634 W and with a speed of $\sqrt{(-38.567)^2 + 254.037^2} \approx 257$ miles per hour.

You may have noticed that vectors behave much like matrices with respect to addition and subtraction. In fact, 1×2 and 1×3 matrices are used to represent vectors and that is one of the reasons we define matrix arithmetic the way we do.

Since we have talked about the geometry of vectors, it is natural to ask geometric questions about vectors. For example, how can we tell if two vectors $\mathbf{V}_1 = (a, b)$ and $\mathbf{V}_2 = (c, d)$ are parallel? One simple way is to realize that, if they are parallel, then when brought into standard position, they will both emanate from the origin and will make the same angle θ with the positive x-axis. Thus, $\tan \theta$ is the same for both vectors. That is,

$$\tan \theta = \frac{b}{a} = \frac{d}{c}. \tag{11.34}$$

If we substitute $\tan \theta$ with λ, then equation (11.34) says that

$$\frac{b}{a} = \frac{d}{c} = \lambda$$

which can be rewritten as

$$\frac{c}{a} = \frac{d}{b} = \lambda.$$

From this, it follows that $c = a\lambda$ and $d = b\lambda$. Thus $(c, d) = (a\lambda, b\lambda) = \lambda(a.b)$. Remembering that $\mathbf{V}_1 = (a, b)$ and $\mathbf{V}_2 = (c, d)$, this last statement says that

$$\mathbf{V}_2 = \lambda \mathbf{V}_1.$$

Thus, we have shown that, if $\mathbf{V}_2$ is parallel to $\mathbf{V}_1$ then $\mathbf{V}_2 = \lambda \mathbf{V}_1$ for some constant λ. For practice, we are leaving the converse for you to prove. Thus, we have

> **Theorem 11.33** *Two nonzero vectors $\mathbf{V}_1$ and $\mathbf{V}_2$ are parallel, if and only if there is some constant λ such that*
>
> $$\mathbf{V}_2 = \lambda \mathbf{V}_1.$$

Another way of saying this is that two vectors are parallel if and only if one is a scalar multiple of the other. For example, the vector, $\mathbf{V}_1 = (2, 2)$ is parallel to $\mathbf{V}_2 = (8, 8)$, since $(8, 8)$ is $4\mathbf{V}_1$. But $\mathbf{V}_1$ is not parallel to $(3, 4)$, since no multiple of $\mathbf{V}_1$ will give $(3, 4)$.

Now that we know how to tell if two vectors are parallel (they have the same slope or one is a multiple of the other), it is natural to ask how we know if two vectors are perpendicular. This too is not too difficult to answer, since in 2 dimensions if two vectors, $\mathbf{V}_1 = (a, b)$ and $\mathbf{V}_2 = (c, d)$ are perpendicular, then according to what we learned in secondary school, the product of their slopes is -1. Thus, the

slope of $\mathbf{V}_1 \cdot$ slope of $\mathbf{V}_2 = -1$

or, put another way

$$\frac{b}{a} \cdot \frac{d}{c} = -1.$$

Multiplying both sides of this equation by ac we get

$$bd = -ac$$

or

$$ac + bd = 0.$$

We have shown that, if two vectors (a, b) and (c, d) are perpendicular, then $ac + bd = 0$. Just reversing the steps we can show that, if $ac + bd = 0$, then the product of the slopes of the vectors is -1 and thus, they are perpendicular. So what we have shown is that the following theorem holds.

Theorem 11.34 *Two nonzero vectors* $\mathbf{u} = (a, b)$ *and* $\mathbf{v} = (c, d)$ *are perpendicular if and only if* $ac + bd = 0$.

The quantity $ac + bd$ given in the theorem is known as the **dot product** of the vectors **u** and **v** and is abbreviated as $\mathbf{u} \cdot \mathbf{v}$ (read *u* dot *v*).

Example 11.35 *Are the following pairs of vectors perpendicular?* (a) $\mathbf{u} = (-3, 4)$ *and* $\mathbf{v} = (4, 3)$ (b) $\mathbf{u} = (1, 2)$ *and* $\mathbf{v} = (-5, 3)$

Solution. (a) Here $\mathbf{u} \cdot \mathbf{v} = (-3)(4) + (4)(3) = 0$, so **u** and **v** are perpendicular.

Solution. (b) Here $\mathbf{u} \cdot \mathbf{v} = (1)(-5) + (2)(3) \neq 0$. So these vectors are not perpendicular.

Most pre-service teachers have been exposed to dot products either in their linear algebra courses, or in their calculus courses, so we will just state some rules for dot products that we will use and that you can verify.

$$\mathbf{u} \cdot \mathbf{v} = \mathbf{v} \cdot \mathbf{u} \tag{11.35}$$

$$\mathbf{u} \cdot (\mathbf{v} + \mathbf{r}) = \mathbf{u} \cdot \mathbf{v} + \mathbf{u} \cdot \mathbf{r} \tag{11.36}$$

$$\mathbf{u} \cdot \mathbf{u} = u^2. \tag{11.37}$$

Note that the third result is immediate once you see that, if $\mathbf{u} = (a.b)$, then $u = \sqrt{a^2 + b^2}$ and $\mathbf{u} \cdot \mathbf{u} = a^2 + b^2$, so certainly $\mathbf{u} \cdot \mathbf{u} = u^2$.

11.11.3 Using Vectors to Prove Geometric Theorems

We have shown how vectors can be used in physics and navigation. We now turn to a very interesting consequence of the geometric interpretation of vectors and their role in proof. You have probably seen a proof of the fact that the diagonals of a parallelogram bisect each other in a secondary school geometry course. We prove it here using vector methods. Notice its elegance.

Example 11.36 *Prove that the diagonals of a parallelogram bisect each other.*

Solution. Before starting any geometric proof, we must represent the geometric figure using vectors. In this case we start with parallelogram *ABCD* and consider each side as a vector as shown in Figure 11.104 below. Note that it does not matter how you choose the direction of your vectors. We draw one diagonal and indicate its midpoint by *E*. The goal is to show that *E* is also the midpoint of *DB*. We will do that by showing that $\mathbf{DE} = \frac{1}{2}\mathbf{DB}$.

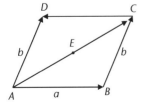

Figure 11.104

Now, diagonal **AC** = **a** + **b**, and so **AE** = $\frac{1}{2}$(**a** + **b**). Furthermore, **DE** = **AD** − **AE** = **b** − $\frac{1}{2}$(**a** + **b**) = $\frac{1}{2}$(**b** − **a**). But the diagonal **BD** = **b** − **a**. So we have shown that **DE** = $\frac{1}{2}$**BD** and we are done!

We hope you appreciated this interesting use of vectors. Let's try another geometric proof using vectors, to reinforce the approach.

Example 11.37 *Using vectors, show that an angle inscribed in a semicircle is a right angle.*

Solution. Below in Figure 11.105 you see angle *ACB* inscribed in a semicircle. Our goal is to show that angle *C* = 90°.

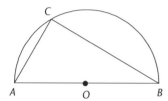

Figure 11.105

We draw *CO* and then represent the sides using vectors. (As we said, you may choose any directions for your vectors just as long as the relationships you write are consistent with your drawing.) Figure 11.106 shows our new diagram using vectors.

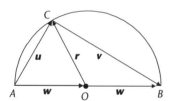

Figure 11.106

We need to show that **u** · **v** = **0**, from which it will follow that *AC* is perpendicular to *CB* and angle *C* is 90 degrees. Observe that, according to our diagram, **AO** = **w** and **OB** = **w**, since the arrows have the same length (both are radii) and point in the same direction. From our diagram we have **u** = **w** + **r** and **r** + **v** = **w**. So **v** = **w** − **r**.

Therefore

u · **v**

= (**w** + **r**) · (**w** − **r**)

= **w** · **w** − **w** · **r** + **r** · **w** − **r** · **r**   (By repeated use of equation (11.36).)

= **w** · **w** − **r** · **r**   (By equation 11.35)

= $w^2 - r^2$

= 0   (Since w^2 is the square of the radius and so is r^2.)

We have shown that **u** · **v** = **0**, hence *AC* is perpendicular to *BC* and angle *C* = 90°.

Student Learning Opportunities

1. If an airplane starts out flying southeast at 300 miles per hour, but encounters a tail wind of 50 miles per hour south acts on it, at what speed will the plane fly and in what direction?

2. A boat needs to travel at 20 knots per hour east but the current is pulling it north at 3 knots per hour. At what speed and direction must the boat travel to accomplish this?

3. Find, if possible, scalars, λ, μ such that $\lambda(2, 5) + \mu(-1, 3) = (-8, -9)$.

4. Find all values of r such that the magnitude of $r(3, 5) = 1$.

5. Determine if the following pairs of vectors (given in terms of their components) are parallel, perpendicular, or neither? Explain.

    (a) $(-2, 4)$ and $(4, -8)$
    (b) $(4, 6)$ and $(6, 9)$
    (c) $(-2, 5)$ and $(5, 2)$
    (d) $(3, 4)$ and $(5, 7)$
    (e) (a, b) and $(9a, 9b)$ where $(a, b) \neq (0, 0)$
    (f) (c, d) and $(-d, c)$ where $(c, d) \neq (0, 0)$

6. (C) One of your students, Georgia, knows how to use slopes to figure out if a triangle is a right triangle and wants to know whether it can be done using vectors, and if it can, how do the two methods compare? How do you respond? [Hint: It is a good idea to begin by asking Georgia to examine a specific case, say a triangle MAT, where M, A, and T, are the points $M = (4, 10)$, $A = (8, 2)$, and $T = (2, 4)$.]

7. Using the parallelogram shown in Figure 11.107 below, express each of the following in terms of **x** and/or **y**.

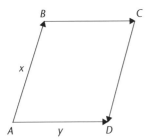

Figure 11.107

(a) **BC**
(b) **CD**
(c) **AC**
(d) **BD**
(e) **AM** where M is the intersection of the diagonals of the parallelogram.

8 Using the diagram shown in Figure 11.108 below,

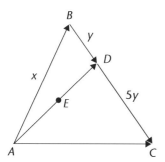

Figure 11.108

express each of the given vectors in terms of **x** and **y**, given that E is the midpoint of AD.

(a) **AD**
(b) **AE**
(c) **BE**
(d) **EC**
(e) **CE**

9 (C) Your students ask if there is a way to use vectors to prove the Pythagorean Theorem. How do you do it? [Hint: Draw your vectors so $c = a + b$ then dot c with itself.]

10 (C) You told your students that vectors can be used to simplify certain proofs. They want you to prove your point by using vectors to prove the well known theorem in geometry that says that the line joining the midpoints of two sides of a triangle is parallel to the third side and half its length. Show how you would prove this theorem by using Figure 11.109 below.

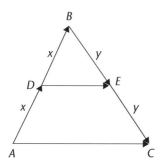

Figure 11.109

11 A well known theorem from geometry says that, if the diagonals of a quadrilateral $ABCD$ bisect each other (that is have the same midpoint), then the quadrilateral is a parallelogram. Prove this using vectors. You will need the following. If P and Q are points in the plane, then the vector joining the points P and Q is denoted by $Q - P$. From secondary school, the midpoint of the line segment joining P and Q is obtained from averaging the x coordinates and averaging the y coordinates and this can be abbreviated by $\frac{1}{2}(P + Q)$. [Hint: Begin with quadrilateral $ABCD$ shown in Figure 11.110.]

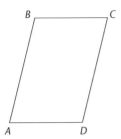

Figure 11.110

Saying the diagonals have the same midpoint is the same as saying $\frac{1}{2}(A + C) = \frac{1}{2}(B + D)$. Show that this implies that both $B - A = C - D$ and that $C - B = D - A$.]

12 In this problem we guide you through the proof that the altitudes of a triangle are concurrent, that is, meet at a point. We use the conventions stated in the previous problem, namely that $Q - P$ is the vector joining P to Q. Begin with triangle ABC, which has been put on the coordinate plane. Draw altitudes AD and CE. They, of course, meet in some point F as Figure 11.111 shows.

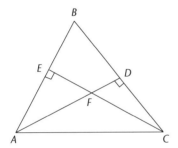

Figure 11.111

(a) $(F - A) \cdot (B - C) = 0$
(b) $(F - C) \cdot (A - B) = 0$
(c) Show that BF extended to AC is an altitude by showing that the vector **BF** is perpendicular to the vector **AC**. [Hint: Expand the expressions in (a) and (b), add the results and then factor.]

13 Using vectors, show that, if we have a parallelogram with legs of lengths a and b, and diagonals with lengths d_1 and d_2, then $2a^2 + 2b^2 = d_1^2 + d_2^2$.

14 Explain why each of the following holds.

(a) $\mathbf{u} \cdot \mathbf{v} = \mathbf{v} \cdot \mathbf{u}$
(b) $(\mathbf{u} + \mathbf{v}) \cdot (\mathbf{u} + \mathbf{v}) = \mathbf{u} \cdot \mathbf{u} + 2\mathbf{u} \cdot \mathbf{v} + \mathbf{v} \cdot \mathbf{v}$
(c) $(\mathbf{u} + \mathbf{v}) + \mathbf{w} = \mathbf{u} + (\mathbf{v} + \mathbf{w})$
(d) $\mathbf{u} \cdot (\mathbf{v} + \mathbf{r}) = \mathbf{u} \cdot \mathbf{v} + \mathbf{u} \cdot \mathbf{r}$
(e) $5(\mathbf{u} + \mathbf{v}) = 5\mathbf{u} + 5\mathbf{v}$
(f) $-\mathbf{v} = -1(\mathbf{v})$

15 Show that, if $(\mathbf{u} + \mathbf{v}) \cdot (\mathbf{u} - \mathbf{v}) = 0$, then **u** and **v** have the same magnitude.

16 In this problem we guide you through the derivation of the formula

$$\cos\theta = \frac{\mathbf{u}\cdot\mathbf{v}}{(u)(v)}$$

which gives the cosine of the smaller angle between two vectors.

Suppose that $\mathbf{V}_1 = (a, b)$ and $\mathbf{V}_2 = (c, d)$ are two vectors, and we are interested in the smaller angle, θ, between them. Figure 11.112 gives us our picture.

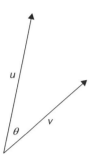

Figure 11.112

Draw the vector from the tip of u to the tip of v forming a triangle as shown in Figure 11.113 below.

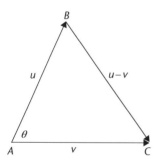

Figure 11.113

Begin by using the Law of Cosines in the triangle:

$$(u - v)^2 = u^2 + v^2 - 2uv\cos\theta$$

and recall that $u^2 = \mathbf{u}\cdot\mathbf{u}$, $v^2 = \mathbf{v}\cdot\mathbf{v}$ and $(u - v)^2 = (\mathbf{u} - \mathbf{v})\cdot(\mathbf{u} - \mathbf{v})$. Express everything in the Law of Cosines in terms of dot products, expand and simplify and then solve for $\cos\theta$.

17 Using the previous exercise, find the angle between each of the following vectors.
  (a) $\mathbf{u} = (-2, 3)$ and $\mathbf{v} = (4, 5)$
  (b) $\mathbf{u} = (-4, 3)$ and $\mathbf{v} = (3, 2)$
  (c) $\mathbf{u} = (-4, 3)$ and $\mathbf{v} = (3, 4)$

18 **(C)** Your students have accepted all they have learned about the algebra of vectors. But, they still want to know if it is the case that, if $\mathbf{u}\cdot\mathbf{v} = \mathbf{u}\cdot\mathbf{r}$, then $\mathbf{v} = \mathbf{r}$ (where $\mathbf{u}$, $\mathbf{v}$, and $\mathbf{r}$ are non-zero vectors). What is your reply? Explain.

CHAPTER 12

DATA ANALYSIS AND PROBABILITY

12.1 Introduction

Students begin studying probability and statistics early in their schooling, first very informally, and later with increasing abstraction. This area of mathematics can be most interesting to learn and to teach, as the applications pervade our daily life. For example, life insurance companies decide on their rates, based on the percentage of people that they believe will die in the coming year, the same way auto insurance companies base their rates on the proportion of their subscribers who they feel will be in accidents. How can they tell what proportion of their subscribers will have accidents? The answer is, they can't. But what they can do is estimate what should happen, based on the proportion of drivers who have had accidents in the past, because what insurance companies have found is that there is, for the most part, a certain regularity in the proportion of accidents from year to year. This is based on many years of data from millions of drivers. Thus, their data come from a very large population. Sometimes the proportions do vary. For example, when the seat belt laws were implemented, the proportions of deaths due to car accidents decreased, and that, in turn, affected their analysis.

The study of probability is, in many senses, the study of proportions. Based on these proportions, we make claims about the likelihood or chance of certain events occurring and this is why probability is often defined as the study of chance. When we say the chance of rain is 45% today, we are saying that, in the past, when conditions were similar to what they are now, rain resulted about 45% of the time. Of course, that guarantees nothing about what will happen today, but it does give us something to make decisions with. For example, should we carry an umbrella today? For some people, a 45% chance of rain is not enough to carry an umbrella. For others, it is.

In this chapter we will discuss some basic concepts and point out some interesting and thought provoking ideas and issues of both probability and statistics. We will begin by giving a quick review of the basics of probability.

12.2 Basic Ideas of Probability

LAUNCH

Maxine loves to play the lottery. She claims she has a better chance of winning if she continues to bet the same set of five lottery numbers every time. But Molly says that it is better if you randomly select any set of five numbers for the same period of time. Sophie says it really doesn't matter how you pick your numbers. Who is right? Why?

People often have very strong opinions about issues concerning probability. It is most interesting that many of these opinions are based on fundamental misconceptions about probabilistic events. We hope that in doing the launch question you noticed one of these misconceptions. This next section will address many of these issues.

12.2.1 Different Approaches to Probability

Every one has heard a statement like, "If we toss a fair coin, the probability that 'heads' shows up is 1/2." What does this mean? Does it mean that we can be assured that if we toss a coin ten times that 50% of the time heads will show up? The answer is, "Of course not." Any child knows that when we toss a coin anything can show up and in any order. For example, if we toss a coin 5 times, we may very well get 5 heads in a row. Or, we may gets head–tail–head–tail–head. Saying that the probability of getting a head is 1/2 means that, if we toss the coin over and over and over a large number of times and measure the proportion of times that a head comes up, we would expect that proportion to be close to 1/2. So, if we toss the coin say, a million times, we would expect about 500 thousand of the flips to turn up heads. Will this in fact happen? Isn't it possible that we can get a million flips all of which turn up heads? Of course! But when lots and lots of people flip a coin a million times, and we total the number of flips and the number of heads that come up, we are saying we believe it will be *about* 50%. So how do we check this assertion? Do we just toss coins day in and day out and check our assertion that way? We could, but even if we did find our ratio of heads to flips is $\frac{1}{2}$, there is no guarantee that, when others do the same, their results will be the same, or when we put all our results together that we will be close to 50%. So, in answer to our question, "How do we check this assertion?" the answer is, we don't.

When we say that the probability of getting a head is 1/2, we are using a *model which expresses our belief that many tosses will result in a proportion of heads that is roughly* 1/2. This model is known as the **frequency approach.** Now, take note that not all models are necessarily good. But this model has been around for several hundred years, and is based on what we have observed in the past, and what we believe will always happen in a large number of flips. What is different about this model is that we can never really check our beliefs, for even though in a million flips the ratio of heads to flips might be close to $\frac{1}{2}$, unlikely as it is, by the time we get to a billion flips, things might change. It is not like Hooke's law (see Chapter 9) for a spring, where we can actually test spring after spring and see that Hooke's law is true. Probability is a different kind of model. In general, if we wish to determine the probability of an event occurring during an experiment using the frequency approach, we perform the experiment many times. Each time we perform the experiment, we have performed what is called a **trial**. Each time the event we are interested in occurs, we say that we have a **success**. If n is the number of trials performed and S is the number of successes, the probability, p, of the event is defined to be

$$p = \lim_{n \to \infty} \frac{S}{n}. \tag{12.1}$$

Because of our inability to check statements like "If we toss a coin a large number of times, the proportion of times heads will come up is always close to 1/2," some people prefer the classical approach which we will discuss shortly.

Using the frequency approach, finding the probability of an event requires that we take a limit as $n \to \infty$. Obviously, we can't continue to perform an indefinite number of trials, so the practical person needs to be able to estimate the probability of an event happening. This is done by performing an experiment a fixed but large number of times, calculating the ratio of successes to the total number of trials, and using that ratio as an estimate of the true probability of the event occurring. This ratio is called the **experimental probability or empirical probability** of an event. The person then uses this number in his or her calculations. As it turns out, there are many situations where using experimental probability is the only way that a probability can be calculated. For example, insurance companies use experimental probability all the time to price their life insurance policies. If they want to determine the probability that, say a man 40 years old who smokes and drinks will die of a heart attack by the age of 60, they use the data gathered from extensive records of people who died by the age of 60 and fell into this category, to price the life insurance policy for a 40-year old man who smokes and drinks and asks to be insured. Also, only experimental probability can be used to find the probability of an irregular object landing in a certain position. As a simple case, suppose one had a cup and for some reason wanted to know the probability that it would land on its side when tossed and allowed to fall, he or she could toss the cup up and allow it to fall, say 1000 times. If 914 times the cup landed on its side, that person would say that the probability of this happening is approximately $\frac{914}{1000}$.

From an instructional perspective, asking students to compute experimental probabilities gives them a feel for the problem and an inkling into what is a reasonable result for the theoretical probability. For example, a teacher might ask a student to toss a pair of die say 50 times and count the number of times the faces on the die add up to 7 before ever computing the theoretical probability of this event. If later they do compute the theoretical probability, they can determine whether or not the result they arrived at was reasonable.

One problem with experimental probability is that, if different people perform the experiment 1000 times, the results can vary widely. So, each will have his or her own probability to use in computations. The theoretical probability discussed soon, by contrast, is a fixed number and does not vary from person to person.

Now, suppose we have a dart board, part of which is colored yellow and the rest of which is colored blue, and we asked "What is the probability that a dart thrown at random which hits the board, lands on yellow?" One might say, the probability is 1/2, the reason being that there are only two possibilities, one of which is landing on yellow. This may seem reasonable. But suppose that we have the dart board shown in Figure 12.1 below.

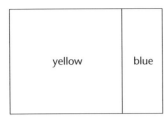

Figure 12.1

Now would you say that the chances or probability of a dart hitting the yellow region is 1/2? Look at how much greater the yellow area is than the blue area. This leads to the notion of likelihood. Landing on yellow in the above dartboard seems more likely than landing on blue provided the dart hits the dart board. Thus, just because we have two outcomes does not mean that the probability of each outcome is 1/2.

As another example, suppose that one tosses a pair of dice. The sum of the numbers that come up when the dice fall can be any number from 2 to 12 for a total of 11 possibilities. If one were to ask, "What is the probability of a sum of 7 coming up on the dice?", one might answer "1/11, the same as the probability of a sum of 2 coming up." But the probability of getting a 2 is not the same as the probability of getting a 7, since there is only one way to get a 2, namely, each die falls with 1 dot up which we denote by the ordered pair (1, 1), and there are several ways to get a sum of 7 (e.g., (1, 6), (2, 5), (3, 4), (6, 1), and so on.) Thus, the event of getting a sum of 7 is more likely to occur than the event of getting a 2.

This brings us to what is called the **classical or theoretical approach** to probability and to a definition of probability that one finds in most secondary school books known as the **classical or theoretical definition of probability**: Suppose that an experiment can result in n equally likely outcomes, and that we are interested in the probability of event, E, occurring. If E occurs in precisely m out of the n outcomes (that is, we have "**success**" m out of n times), then the probability of that event occurring is $\frac{m}{n}$. Here are some examples.

Example 12.1 *If one tosses an ordinary die, there are six possible outcomes, each (we believe) equally likely, so the probability of getting a "2" face up when the die falls is $\frac{1}{6}$, and the probability of getting a prime number (2, 3, or 5) is $\frac{3}{6}$ or $\frac{1}{2}$.*

Example 12.2 *In a regular hexagon ABCDEF, a point is chosen at random inside the hexagon. What is the probability that the point, P, chosen, is closer to the vertex A than to any other vertex?*

Solution. The hexagon is symmetric. Thus, the events that (1) P is closer to A than the other vertices, (2) P is closer to B than to any other of the vertices, (3) P is closer to C than to any of the other vertices ... (6) P is closer to F than to the other vertices, are all equally probable. Hence the probability is 1/6.

Example 12.3 *Three numbers x, y, and z are chosen at random from the interval $[0, 10]$. What is the probability that $x \leq y \leq z$?*

Solution. There are only 6 possibilities:

$$x \leq y \leq z$$

$$x \leq z \leq y$$

$$y \leq x \leq z$$

$$y \leq z \leq x$$

$$z \leq x \leq y$$

$$z \leq y \leq x.$$

There is no reason any one should occur over any other, so all possibilities are equally likely. The probability therefore is $1/6$ that $x \leq y \leq z$.

There is yet a third school of thought on probability known as the **subjective approach to probability**. There, the probability of an event is a measure of how likely we feel the result is, which, of course, is subjective. Certainly, this can vary from person to person, and often in business decisions, executives make decisions based on what they consider the subjective probability that an event will happen.

12.2.2 Issues with the Approaches to Probability

There are issues with all of the above approaches to probability. For example, with the frequency approach, suppose that we want to determine the probability of an event happening. How many times must we perform the experiment before we are sure that the ratio of successes to the number of trials gives us a number close to the probability of the event happening? Are we ever sure we are close? Do we know that the limit exists in the definition of probability according to the frequency definition, equation (12.1)? What if we performed the experiment say a million times and then repeated it another million times? Would the ratio of the number of successful outcomes to all outcomes be the same or even close to each other in each of the million trials? If not, how can we be sure our definition means anything? Suppose that it is not possible to perform an experiment many times, (for example, dropping an atomic bomb on a country), how can we determine the probability of an event related to this?

The classical definition also has problems. For one, it can only be applied to situations with a finite number of outcomes. What if there are infinitely many outcomes as will occur in later examples? Also, if we figure out the classical probability of an event to be $1/6$, will the frequency approach to the experiment also yield the same $1/6$? That is, are the approaches compatible? When we use the classical approach, we are assuming that the outcomes are equally likely. Perhaps this is not the case and we are building on a false foundation. (Indeed there are *real* examples in atomic physics where everything points to the outcomes being equally likely, but something in nature makes them not equally likely. (See Feller, (1968), pp. 40-41.))

Of course, the subjective approach to probability has even more problems. Can we conclude anything about probabilities using the subjective approach? If they are subjective, what does a probability of 50% mean in practice? What does this tell us to expect? If one person assesses the probability of an event to be 0.50 and another 0.90, is either right, and aren't the conclusions each makes on the basis of these likely to be different?

Given the difficulties with each of the three approaches to probability, mathematicians for whom precision is paramount, tried to develop a way to put probability on a solid mathematical foundation, which we will discuss in the next section.

Student Learning Opportunities

1. Suppose we have a coin with heads on both sides. What is the probability that, when we toss the coin, we get a head? A tail?

2. Four numbers x, y, z, and w are chosen at random from the interval [0, 10]. According to the classical approach, what is the probability that $x \leq y \leq z \leq w$?

3. (C) You pose the following problem to your class: You are given two boxes of lollipops, one has 7 greens and 3 reds. The other box has 70 greens and 30 reds. Each box is shaken and the lollipops are mixed up. You want to get a red lollipop, but you are only allowed to pick out one candy without looking. From which box would you choose? One of your students, Lenny claims that he would definitely pick from the second box, since it has more red lollipops in it and he therefore has a better chance of picking a red lollipop. What is your response? Is Lenny correct? Explain.

4. (C) You told your students that you were going to flip two nickels and that, if one landed on heads and the other on tails, you would bring in ice cream for them. Your students insisted that it was a bad deal, since there was only a probability of 1/3 that this would happen. They justified it by saying that either the nickels both landed on heads, both landed on tails, or each landed differently. What is your reply?

5. A pair of dice is tossed. According to the classical approach, (a) what is the probability of getting a total of 7 dots facing up? (b) what is the probability of getting a 1 on one of the dice if the sum rolled is 7?

6. If a pair of dice is rolled, what is the probability of a double? What is the probability of rolling 11 or more?

7. By listing all possibilities and using the classical approach, find the probability of getting exactly two heads when 3 coins are simultaneously tossed.

8. A number is picked at random from the integers 1 to 36 inclusive. (a) What is the probability that it is a perfect square? (b) What is the probability that it is divisible by 3? (c) What is the probability that it is divisible by 2 or 3 or both? Explain.

9. (C) Your student claims that she tossed a coin 40 times and got all heads. She asks you if this means her coin is unfair? How do you respond?

10. If you roll 2 dice, what is the most likely sum you will get? According to the classical approach to probability, what is the probability of getting that sum?

11. A pair of dice is thrown. What is the probability that at least one of them shows an even number? What is the probability that both show numbers less than 3?

12. Suppose we randomly choose two integers between 5 and 9 inclusive. What is the probability that their product will be odd? Explain.

13 In the game of craps seen in almost all casinos, a pair of dice is thrown by a player. If you make a "pass" bet, and a 7 or 11 comes up, you win whatever you bet. If 2, 3, or 12 comes up, you lose. All other outcomes result in no win or loss. Find the probability of your winning and the probability of a player's losing on this type of bet.

14 (C) Comment on the following argument made by one of your students: We defined the probability of an event to be $\frac{m}{n}$ where there are n equally likely events, and m of them represent "success" with this event. But doesn't the word "likely" mean probable? So isn't the definition of probability circular in the sense that we are defining a concept in terms of itself?

15 (C) Your students are questioning the fact that the probability of getting a head is 1/2 when a coin is tossed as is the probability of getting a tail. They point out that, since there might be a chance that the coin will land on its side, the probability of heads might not really be 1/2. How do you respond?

16 (C) Give two examples of where it might be better to use a frequency approach to estimate the probability of an event happening, and two examples of where it might be better to use the classical approach to estimate such a probability.

17 Explain what is **wrong** with each of the following:

(a) A die has 5 of its faces painted red and one side painted blue. The die is tossed. The probability of the die landing with a red face up is $\frac{1}{2}$, since it either falls with the red face up or with the blue face up. Since there are only two choices and one of them results in red, the probability of getting a red is $\frac{1}{2}$.

(b) There are 3 prisoners, A, B, and C in a jail and two are going to be released. A wants to know if he is one of those to be released. He figures if two are released, then they will be A and B, B and C, or A and C and thus, he has a 2/3 chance of being released. He is thinking of asking the warden to name one of the other persons who will be released, figuring that, if the warden says B, then he has a 50% chance of being the other one being released. This lowers his chances from 2/3. So he decides not to ask.

(c) If there is a 50% chance that something will go wrong, then by Murphy's Law, we can expect this to happen about 90% of the time. (Murphy's law says that what can go wrong will go wrong.)

12.3 The Set Theoretic Approach to Probability

In a crime at Ithaca University, it is determined that the perpetrator is a student who was wearing Cornell sweatpants and a New York sweatshirt. A student is arrested who was wearing both of these items of clothing on the evening of the crime.

The defense provides evidence that shows the probability that a randomly selected student in Ithaca is wearing Cornell sweatpants is 1/10, and the probability that a randomly selected student in Cornell is wearing a New York sweatshirt is 1/5. The prosecutor concludes that the probability that a student is wearing both Cornell sweatpants and a New York sweatshirt is (1/10)(1/5) = 1/50 = 2%, which is large enough to cause reasonable doubt for the jury.

Problem: Do you think the prosecutor's claim about the probability is true or false? In three sentences or less, justify your conclusions. Assume the 1/10 and 1/5 are accurate probabilities.

Problems that depend on probabilistic analysis, such as that shown in the launch question, are surprisingly part of our daily lives. Very often, probabilistic statements are derived from data that have implications for such things as: criminal justice, decisions we make about how to dress each day, decisions buyers make about the number and types of items to purchase for their company to sell, and costs insurance companies charge depending on certain given conditions. To derive the probabilistic statements from the data, a set theoretic approach is used, which is described in this section.

The mathematician Kolmogorov suggested the following set theoretic approach, which has gained large favor among mathematicians and is seen in secondary school courses. It is felt by mathematicians worldwide that any viable approach to probability must satisfy the axioms presented soon.

First, we need some background. When we perform an experiment, the *set* of all outcomes is known as the **sample space** for this experiment. We denote the sample space by the letter S. Thus, if we toss a die (our "experiment") and want to know what number of dots will face up when the die lands, our sample space, $S = \{1, 2, 3, 4, 5, 6\}$. If we toss a pair of dice (yet another "experiment"), say a red die and a blue die and wish to know how the dice fell, our sample space might be $S = \{(1, 1), (1, 2), (1, 3), \ldots, (1, 6), (2, 1), (2, 2), (2, 3), \ldots, (2, 6), \ldots, (6, 1), (6, 2), (6, 3), \ldots, (6, 6)\}$, which consists of 36 ordered pairs of numbers, where the first number in each ordered pair is the number that faced up on the red die and the second number is the number that faced up on the blue die. Of course, if we were not interested in what came up on the individual dies, but were interested in the total sum of dots that came up when the dice were tossed, then our sample space might be $S = \{2, 3, 4, \ldots, 11, 12\}$, since these are the only possible sums one can get. In summary, the sample space depends on what it is that we are interested in recording when we perform our experiment.

An **event**, according to this theoretical approach, is defined to be a subset of the sample space. Thus, if we tossed a die, and we were interested in the event, E, that a prime number came up, then E could be described by the subset $E = \{2, 3, 5\}$ of the sample space $S = \{1, 2, 3, 4, 5, 6\}$. The event, F that an even number came up could be described by the subset $F = \{2, 4, 6\}$ of the sample space $S = \{1, 2, 3, 4, 5, 6\}$. Every event is made up of singleton events. For example, the event that a die turns up an even number, that is, the event $F = \{2, 4, 6\}$ is made up of the three single events $\{2\}$, $\{4\}$, and $\{6\}$. If any of these singleton events occur, F has occurred. That is, if say a 2 came up on a toss, then the event that "an even number came up" occurred.

When two events, A and B, cannot occur at the same time, the events are called **mutually exclusive**. Thus, the event A, that tossing a single die results in an even number, and the event B, that the toss of the die results in an odd number, are mutually exclusive, since you can't get a toss that is both an even number and an odd number at the same time. On the other hand, if we pick a card from a deck and, if we let C be the event that the card is a picture card and D be the event that the card is from the suit of diamonds, then C and D are not mutually exclusive, since drawing a Jack of Diamonds means that we have achieved both C and D simultaneously.

It is easy to describe the words "mutually exclusive" in terms of sets. We already observed two paragraphs ago, that whenever a singleton in the event set A occurs, then event A has occurred. Thus, for two events A and B to be mutually exclusive, there cannot be any singletons that occur in both, or else they would both occur. That is, $A \cap B = \phi$. Of course, if $A \cap B \neq \phi$, then there is a common element and A and B can occur simultaneously. Thus, they would not be mutually exclusive. In short, two events, A and B are mutually exclusive if, and only if, $A \cap B = \phi$

There is more. Since any common element in both sets A and B means that A and B can occur simultaneously, it means that the set of common elements to A and B is the set of outcomes that result in both events occurring. Thus, the event that A and B occur simultaneously is described by the set $A \cap B$. The advantage of this approach is that we can now use results about sets, and we will, to show that certain probability statements are true. In a similar manner, $A \cup B$ consists of all singletons there are in A or B or both. If we pick any such singleton in $A \cup B$, then A or B or both events have occurred.

In summary, *the event that both A and B have occurred is the set $A \cap B$, and the event that A or B or both have occurred is the set $A \cup B$.*

We now present Kolmogorov's theoretical axioms for probability. You will notice that he intentionally excluded providing the definition of probability to avoid the issues discussed earlier. Thus, we can think of probability as an undefined term about which we have an intuitive idea. This is no different from the agreement that we make in geometry to leave certain words, like "point" and "line" undefined and to work with our intuitive notion of what they are.

Kolmogorov's axioms for probability: If E is an event and $P(E)$ is the probability of the event occurring, then

1. $0 \leq P(E) \leq 1$.
2. (Finite Additivity) If A and B are mutually exclusive events, that is, if $A \cap B = \phi$, then $P(A \cup B) = P(A) + P(B)$. (In words, if the events A and B are mutually exclusive, then the probability that A or B occurs is the sum of the probabilities that A occurs and that B occurs.)
3. $P(S) = 1$ where S is the sample space. (Recall the sample space is the set of *all* outcomes.)

    A note about Axiom 2: While (2) is the axiom one ordinarily sees in secondary school books, the correct version of (2) in the definition of the probability axioms is:

(2)′ (Countable Additivity.) If A_1, A_2, A_3, and so on are mutually exclusive events, that is, no two of them can occur at the same time, then $P(A_1 \cup A_2 \cup A_3 \ldots) = P(A_1) + P(A_2) + \ldots$

It is this rule which leads to some of the surprising results we get later on. This rule is also used quite a bit in the proofs of some of the major results in probability.

Axioms (1)–(3) were motivated by what seemed to be logical consequences of the frequency and classical definitions. Let us illustrate with the classical definition of probability. Since probabilities were measured as proportions of time that "success" occurred, and these proportions were between 0 and 1, the probability of an event should be between 0 and 1. Hence, Axiom (1) holds. Let us illustrate Axiom (2). Suppose we toss a die and we let A be the event when we get a number of dots facing up which is a perfect square, and B be the event when we get a number which is prime. The event A occurs happens when the die falls on either 1 or 4. Thus, "success" occurs 2 times out of our 6 possibilities. Thus, $P(A) = \frac{2}{6}$. The event B occurs when either a 2, 3, or 5 shows up. Thus, we have success with 3 outcomes out of the six possible outcomes. So $P(B) = \frac{3}{6}$. The events A and B are mutually exclusive, since they can't occur at the same time. Finally, the event that A or B occurs can happen when we have any of the following outcomes: 1, 4, 2, 3, 5. So the outcomes associated with the event A or B occurs 5 out of 6 times. Thus, $P(A \text{ or } B) = \frac{5}{6}$. We see that $P(A \text{ or } B) = P(A) + P(B)$ and it is examples like this that motivated Axiom 2.

Using the frequency approach, we can also deduce rule 2. Here is how. Suppose that A and B are any two mutually exclusive events and that we perform an experiment n times and that A occurs m_A times and B occurs m_B times. Then the event A or B occurs $m_A + m_B$ times out of the n times. Now $P(A)$ is defined by

$$P(A) = \lim_{n \to \infty} \frac{\text{number of successes}}{\text{total number of trials}} = \lim_{n \to \infty} \frac{m_A}{n} \tag{12.2}$$

and $P(B)$ is defined by

$$P(B) = \lim_{n \to \infty} \frac{\text{number of successes}}{\text{total number of trials}} = \lim_{n \to \infty} \frac{m_B}{n}. \tag{12.3}$$

Finally, $P(A \text{ or } B)$ is defined by

$$P(A \cup B) = \lim_{n \to \infty} \frac{\text{number of successes}}{\text{total number of trials}} = \lim_{n \to \infty} \frac{m_A + m_B}{n}. \tag{12.4}$$

(The number of successes is $m_A + m_B$, since the events A and B have no overlap.) But (assuming all the limits exist) this last limit is the same is

$$\lim_{n \to \infty} \left(\frac{m_A}{n} + \frac{m_B}{n} \right) \tag{12.5}$$

$$= \lim_{n \to \infty} \frac{m_A}{n} + \lim_{n \to \infty} \frac{m_B}{n} \tag{12.6}$$

$$= P(A) + P(B). \tag{12.7}$$

From equations (12.4)–(12.7) we get that

$$P(A \cup B) = P(A) + P(B).$$

Since S represents the set of *all* outcomes, Axiom (3) is saying that the probability that *some outcome* occurs in an experiment is 1. Of course, this seems like an obvious axiom to include.

12.3.1 Some Elementary Results in Probability

Suppose we toss a die and A is the event that an odd number comes up. Then the **complementary event** is the event that occurs when A does not happen (i.e., when an odd number doesn't come up). Thus, the complementary event in this case is the event A' that an even number comes up. If B is the event that a prime number comes up, then B' is the event that a prime number doesn't come up. If E is any event, then the complementary event, E' and E are mutually exclusive, that is, cannot occur at the same time, and $E \cup E' = S$ since every outcome in S, the sample space, is either in E or not in E (making it in E'). From this it follows that:

Theorem 12.4 *If E is an event and E' is the complementary event, then $P(E') = 1 - P(E)$.*

Proof. Since $E \cup E' = S$, it follows from Axiom (3) that

$$P(E \cup E') = P(S) = 1.$$

But, from Axiom (2), since E and E' are mutually exclusive, this simplifies to

$$P(E) + P(E') = 1.$$

Solving for $P(E')$ in this equation, we get $P(E') = 1 - P(E)$. ∎

Thus, if the probability of winning a contest is $\frac{3}{4}$, then the probability of not winning the contest is $1 - \frac{3}{4}$ or $\frac{1}{4}$. If the probability of getting a disease is 0.1, then the probability of not getting the disease is $1 - 0.01$ or 0.9.

If A and B are events, then we denote by $A - A \cap B$ the event that A occurs, but not both A and B occur. We need the following:

Theorem 12.5 $P(A - A \cap B) = P(A) - P(A \cap B)$.

Proof. $A = (A - A \cap B) \cup (A \cap B)$ as the Venn diagram in Figure 12.2 below shows.

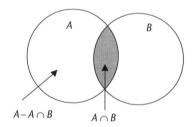

Figure 12.2

Thus,

$$P(A) = P((A - A \cap B) \cup (A \cap B))$$
$$= P(A - A \cap B) + P(A \cap B)$$

since $(A - A \cap B)$ and $(A \cap B)$ are mutually exclusive. Subtracting $P(A \cap B)$ from both sides, we get that

$$P(A) - P(A \cap B) = P(A - A \cap B)$$

which we can rewrite as $P(A - A \cap B) = P(A) - P(A \cap B)$ by just switching the entries on both sides. And this, is what we were trying to prove. ∎

There is a corollary of this that we will use in the section on geometric probability that will lead to a surprising result. That corollary in words says that, if whenever the event C occurs, the event A occurs, A's probability of occurring is greater than or equal to that of C occurring. This makes perfect sense. Here it is in symbols.

Corollary 12.6 *If C is a subset of A then $P(C) \leq P(A)$.*

Proof. If C is a subset of A, then $A - C = A - A \cap C$, since $A \cap C$ would be the same as C. (Draw a Venn diagram to convince yourself.) So by the theorem,

$P(A - C)$

$= P(A) - P(A \cap C)$

$= P(A) - P(C)$.

Since by Axiom (1), $P(A - C) \geq 0$, we have that $P(A) - P(C) \geq 0$ or $P(C) \leq P(A)$. ∎
This brings us to an important result.

Theorem 12.7 *If E and F are two (not necessarily exclusive) events, then*

$P(E \cup F) = P(E) + P(F) - P(E \cap F)$.

Proof. $E \cup F = (E - E \cap F) \cup F$. (See the Venn diagram in Figure 12.3 below.)

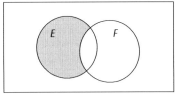

The shaded area is $E - E \cap F$

Figure 12.3

Thus,

$P(E \cup F) = P(E - E \cap F) + P(F)$

$\qquad = P(E) - P(E \cap F) + P(F)$ (by the last theorem and since

$\qquad E - E \cap F$ and F are mutually exclusive.)

$\qquad = P(E) + P(F) - P(E \cap F)$ (by rearranging the last line).

∎

Here is a typical secondary school problem that makes use of this theorem.

Example 12.8 *A card is picked from an ordinary deck of playing cards. What is the probability that the card picked is a face card or a diamond?*

Solution. There are 12 face cards and 13 diamonds, and 3 cards which are both face cards and diamonds (Jack, Queen, and King of diamonds). So, if F is the event of obtaining a face card, and D is the event of obtaining a diamond, then $P(F \text{ or } D) = P(F \cup D) = P(F) + P(D) - P(F \cap D) = \frac{12}{52} + \frac{13}{52} - \frac{3}{52} = \frac{22}{52}$.

Here is another example.

Example 12.9 *In the small town of Cherokee, there is an 80% chance that a person subscribes to the local newspaper, a 50% chance that a person will subscribe to the metropolitan newspaper from a nearby town, and a 40% chance a person subscribes to both. If a household is selected at random from this town: (a) What is the probability that it will subscribe to at least one of the two newspapers? (b) What is the probability that it will subscribe to exactly one of the two newspapers?*

Solution. (a) If L stands for the event that a person subscribes to the local paper and M is the event that a person subscribes to the metropolitan paper, then we are given that $P(L) = 0.80$, $P(M) = 0.50$, and that $P(L \cap M) = 0.40$. In part (a) we seek $P(L \cup M)$. We know that

$$P(L \cup M) = P(L) + P(M) - P(L \cap M)$$
$$= 0.80 + .50 - .40$$
$$= 0.90.$$

Solution. (b) We first notice that the event, $L \cup M$, means that the person subscribes to at least one paper. This can be represented by the union of the event, E, that the person subscribes to *exactly* one paper, and the event $L \cap M$, that the person subscribes to both papers. And since E and $L \cap M$ are mutually exclusive, we have that,

$$P(L \cup M) = P(E) + P(L \cap M).$$

Substituting the values we know in this equation, we get

$0.9 = P(E) + 0.4.$

From which it follows that $P(E)$, the probability that the person subscribes to exactly one paper is 0.5.

Student Learning Opportunities

1 (C) You asked your students whether they thought there was a higher percentage of people who were 55 or older and had had a heart attack or people who only had had a heart attack. They chose the first group. Were they correct? Why or why not?

2 (C) You ask one of your students, Julie, to come to the front of the room and randomly pick a card from an ordinary deck of playing cards. Now, you ask the class what the probability is that Julie picked a card that is either red or has an even number (10 or less). Ricky gives the following answer: Since half the deck has red cards, the probability that it is red is $\frac{26}{52}$. Also, since there are only 5 possible even numbers in the deck (2, 4, 6, 8, 10) and there are 4 suits, that makes 4 times 5, or 20 cards with even numbers. Therefore, the probability of picking a card with an even number is $\frac{20}{52}$. So, the probability is $\frac{26}{52} + \frac{20}{52} = \frac{46}{52}$. Is Ricky correct? Why or why not?

3 A construction firm has bid on two projects. The probability of getting the first project is 0.35 and the probability of getting the second project is 0.40. The probability that the construction firm gets both projects is 0.22. Find:

(a) The probability that the construction firm gets one or the other or both projects.
(b) The probability that the construction firm doesn't get the first project.
(c) The probability that the construction firm gets the first project but not both.
(d) The probability that the construction firm gets neither project.
(e) Which of the above 4 probabilities is the greatest and explain why this makes sense?

4 Ed, the painter, has two paint jobs for this week that must be completed. He estimates that there is a 95% chance he can finish the first job on time, a 70% chance he can finish the second job on time, and a 99% chance that he can finish one or the other on time. Assuming that these subjective probabilities are correct, and that the Kolmogorov Axioms apply, what is the probability that:

(a) he will finish both on time?
(b) he will finish neither on time?
(c) he will finish just one of the projects on time?
(d) Which of the above 3 probabilities is the greatest and explain why this makes sense?

5 Show that, if the sample space in an experiment is $S = \{a_1, a_2, a_3, \ldots, a_n\}$, that $P(\{a_1\}) + P(\{a_2\}) + \ldots + P(\{a_n\}) = 1$. Show also that it follows from this that, if all the singleton events are equally likely, that each has probability $\frac{1}{n}$ and that any event with m elements in it has probability $\frac{m}{n}$. Explain why this means that the Kolmogorov Axioms imply the classical approach.

6 **(C)** One of your students has calculated the probability of three mutually exclusive events A, B, and C. She has arrived at the solution that $P(A) = 0.5$ and $P(B) = 0.3$ and $P(C) = 0.4$. She asks you if she is correct. Can you tell? How?

7 An unfair die has the probability that $P(1) = P(2) = x$ and that $P(3) = P(4) = 2x$ and that $P(5) = P(6) = \frac{x}{2}$. Find x and then find the probability that a number divisible by 3 comes up on a roll of this particular die.

8 Show, using a Venn Diagram that if E and F are sets that $E - F$ (the set of those things in E but not in F) is the same as $E - E \cap F$. From this result, conclude that $P(E - F) = P(E) - P(F)$.

9 In the newspaper Example, 12.9, show that the event E can be described as $(L - L \cap M) \cup (M - L \cap M)$. Use this and the theorems in this section to compute $P(E)$ another way than was shown in the example.

10 Show that, if $A = \phi$ (where ϕ is the empty set), then $P(A) = 0$. [Hint: $A = A \cup \phi$.]

11 **(C)** We indicated that it is reasonable to assume that the probability of getting heads facing up when a coin is tossed is taken as $1/2$, since it has been shown experimentally many times that when a coin is, in fact, tossed repeatedly, the ratio of the number of heads to the total number of flips is close to 50%. But, your astute student argues, "Well, 49.98% is close to

50% and if we flipped a coin millions and millions of times, we really don't expect that we will get exactly half of the flips heads, so how do we know that the probability of getting heads isn't 49.98%? Furthermore, if we take the probability to be 50% when, in fact, it is less, aren't we opening ourselves up to accumulated error if we compute using a wrong probability?" What is your answer?

12.4 Elementary Counting

 **LAUNCH**

You have rented a room for a family birthday party that you are hosting, in which there are 20 people in attendance (including you). You have been told that you must vacate the room at exactly 3pm, since the room will be needed at that time for another event. Knowing that each of your family members must hug each other goodbye, and knowing that it takes each family member approximately 6 seconds per hug, at approximately what time should you announce to your family members that they must begin their goodbye hugs?

One of the first things you learned how to do as a very young child was to count. However, as you discovered in secondary school, simple counting is often not sufficient when it comes to problems such as the one you encountered in the launch. For problems such as this, straightforward counting takes far too long, so short cuts are needed that require a far more sophisticated approach. Furthermore, since probability problems often require knowing the number of elements in a sample space, accurate counting methods are essential. That is why such topics as permutations and combinations are included in the secondary school curriculum. We begin with the **counting principle**.

Suppose we have k tasks $T_1, T_2, T_3, \ldots, T_k$. If a task T_1 can be done in n_1 ways, and, once T_1 is done, T_2 can be done in n_2 ways, and, once these are done, T_3 can be done in n_3 ways, and so on, then the number of ways of performing the tasks $T_1, T_2, \ldots, T_k$ in succession is given by $n_1 n_2, \ldots, n_k$.

Example 12.10 *How many code words can be formed from 2 letters of the alphabet followed by a single digit, if the letters chosen must be different.*

Solution. Let task 1 be the task of choosing the first letter, task 2 the task of choosing the second letter, and task 3 the task of choosing the single digit. Task 1 can be done in any of 26 ways since we have 26 letters of our alphabet to choose from. Task 2 can then be done in 25 ways, since we have 25 letters to choose from, since we have already chosen one letter. Finally, task 3 can be done in any of 10 ways, since we can pick any digit from 0 to 9. Thus, the number of code words is $26 \cdot 25 \cdot 10$ or 6500.

Example 12.11 *A photographer wants to take a picture of 5 children in a family. He lines up 5 chairs in a row in which the children will sit.* (a) *How many different seating arrangements can he make with the 5 children?* (b) *Next the photographer wants to take some pictures using only 3 of the 5 children. So he removes two of the chairs. How many different seating arrangements can he make using only 3 of the 5 children.*

Solution. (a) Let T_1 be the task of putting a child in the first seat. Since the photgrapher has 5 children from which to choose, this task can be done in 5 ways. Let T_2 be the task of filling the second seat. Since 4 children are left to be placed in the second seat, this task can be done in 4 ways. Letting T_3, T_4, and T_5 represent the tasks of filling the third, fourth, and fifth seat, respectively, these tasks can be done in 3, 2, and 1 way, respectively. Thus, by the counting principle, we get that the total number of seating arrangements is $5 \cdot 4 \cdot 3 \cdot 2 \cdot 1$.

Solution. (b) Now we have only 3 seats. Proceeding as we did above, we let T_1, T_2, and T_3 represent the tasks of filling the first, second, and third seat. These can be done in 5, 4, and 3 ways, respectively, so by the counting principle, now the number of seating arrangements is $5 \cdot 4 \cdot 3$.

An arrangement of objects in which the order counts is called a **permutation**. In part (*a*) of the above example, we asked for the number of different seating arrangments of the 5 children. Each different seating arrangement of the children represents a different permutation of the children in the seats. In general, when we arrange r distinguishable objects taken from a set of n objects, the number of permutations we get is known as *nPr*. Using the counting principle and arguing as above, we can compute *nPr* as follows:

$$nPr = n(n-1)(n-2)\ldots(n-r+1). \tag{12.8}$$

Thus, in part (a) of the above example, we were arranging $r = 5$ children from a set of $n = 5$ children in seats, and this can be done in $5P5$ ways. By equation (12.8) this evaluates to $(5)(4)(3)(2)(1)$, which as most of you probably know is abbreviated as 5! (five factorial). Notice that the final 1 in this product is $n - r + 1$, since n and r are both 5. In part (b) of the above example, we were arranging $r = 3$ children taken from a set of $n = 5$ children. This can be done in $5P3$ ways. Again, by equation (12.8), this evaluates to $5(4)(3)$ ways. Notice the final 3 in the product is $n - r + 1$, since $n = 5$ and $r = 3$.

We also review counting combinations. We learn in discrete mathematics or in the beginning of a probability course that, if we have n objects, and we wish to choose a set of r of them, the number of sets of r objects where the order of the r objects is irrelevant is given by nCr, alternately denoted by $\binom{n}{r}$, where either one of these means

$$\frac{n!}{r!(n-r)!}.$$

Here are some typical secondary school problems:

Example 12.12 *How many committees of 3 people can be chosen from 10 people?*

Solution. Here we are looking for committees of size 3 from a set of 10 people. The number of ways this can be done is $\binom{10}{3} = \frac{10!}{3!(10-3)!} = \frac{10!}{3!7!} = \frac{10 \cdot 9 \cdot 8}{3!} = 120$. Thus, there are 120 such committees.

Example 12.13 *You visit an antique dealer and decide to buy 5 of the 14 antiques available. Two of the items make a pair, and the dealer will not sell them separately. So, you either buy them both, or buy neither. How many sets of 5 antiques can you buy?*

Solution. You either buy the pair that must be sold together or you don't. If you buy the pair, then you need to choose 3 more antiques from the remaining 12 antiques. If you don't buy the pair, then all 5 of your antiques must be chosen from the 12 antiques remaining, not counting the pair. Thus, the total number of sets of 5 antiques you can buy is

$$\binom{12}{3} + \binom{12}{5} = 1012.$$

Example 12.14 *We wish to choose a team consisting of 5 boys and 3 girls. The boys are to be chosen from a set of 10 boys and the girls from a set of 6. (a) How many such teams can we form? (b) What is the probability that, if we choose a team of 8 people at random from this group of 10 boys and 6 girls, the team will consist of 5 boys and 3 girls?*

Solution. (a) Let task 1 be the task of choosing the boys and task 2 the task of choosing the girls. Then task 1 can be done in $\binom{10}{5}$ or 252 ways and task two can be done in any of $\binom{6}{3}$ or 20 ways. Thus, the number of teams we can choose consisting of 5 boys and 3 girls is therefore

$$\binom{10}{5} \cdot \binom{6}{3} = 5040$$

and that is a lot of teams!

Solution. (b) The number of teams consisting of 8 people chosen from the 10 boys and 6 girls is $\binom{16}{8} = 12870$. Thus, the probability of getting a team consisting of 5 boys and 3 girls when a team of 8 is drawn at random is $\frac{5040}{12870} = 0.39161$.

Student Learning Opportunities

1 (a) How many license plates can one form using the digits from 1 to 9, if the license plates only consist of 6 digits?
  (b) How many license plates can be formed from any 3 letters followed by any 3 single digits from 1 to 9?
  (c) How many license plates can be formed from any three letters followed by any 3 digits from 1 to 9, if none of the letters are the same and none of the digits are the same?
  (d) Compare your answers to parts (b) and (c). Is one greater than the other? Was this predictable? Explain.

2 Dinner at your local restaurant consists of an entree, dessert, and a drink. If one must choose one of each and there are 5 entrees, 6 desserts, and 3 drinks to choose from, how many different dinner combinations can one have?

3 We have six switches right next to each other and each could be on or off. How many different on–off arrangements of the switches are possible? Explain.

4 In how many ways can you arrange the first 8 letters of the alphabet? Did you use permutations or combinations to answer this question? Explain.

5 What is the number of permutations of the first 8 letters of the alphabet taken 5 at a time? Explain what you did.

6 It is a long way to Tipperary. From your hometown you can get to town A by any of three roads, and you can get from there to town B by any of five roads, and finally from there you can get to Tipperary by any of seven roads. How many different itineraries are there from your home to Tipperary passing through town B?

7 (C) One of your students was reading about permutations and came across the formula: $nPr = \frac{n!}{(n-r)!}$. She is now confused since she learned that the formula was $nPr = n(n-1)(n-2)\ldots(n-r+1)$. How can you help her resolve her confusion?

8 (C) There are 5! ways of arranging the letters *abcde*. But if we had to arrange the following string of letters, *aaabb*, then the number of ways of arranging them would no longer be 5!, since we cannot distinguish one *a* from the next. Thus, if we represent the first *a* by a_1, the second by a_2, and the third by a_3, with similar defintions for b_1 and b_2, then $a_1 a_2 a_3 b_1 b_2$ and $a_3 a_2 a_1 b_2 b_1$ are the same arrangment, since our eyes cannot distinguish between the two. All we see is *aaabb*. It is a fact that, if we arrange n items where n_1 are of one type and indistinguishable, n_2 are of a second type and indistinguishable, ... n_k are a kth type and indistinguishable, then the number of distinguishable arrangements of the n objects is $\frac{n!}{n_1! n_2! \ldots n_k!}$. Using this fact,
(a) find the number of distinguishable arrangements of 5 beads, 3 of which are red and identical, and 2 of which are blue and identical and then list the arrangements.
(b) find the number of distinct arrangements of the word Mississippi.
(c) find how many ways can you arrange 9 othewise identical flags on a flag pole if 3 are red, 2 are blue, and 4 are green.

9 (C)
(a) You give your students the following problem: "You just bought 30 beads of which 18 are gold and identical and 12 are silver and identical. You want to line them up in a row. How many different arrangements of the beads can you make?" Elizabeth solves the problem by finding the number of distinguishable permutations of 30 objects, 18 alike of one kind and 12 alike of another kind. That is, she uses the result of the previous problem. Natalie solves the problem by finding the number of combinations of 30 different objects taken 18 at a time. Who is correct?
(b) Does the number of permutations of n objects, r alike of one kind and $n-r$ alike of another kind, always equal the combinations of n different objects taken r at a time? Explain.

10 (C) You tell your 10 Mathletes that you will be forming committees to work on different problems. You ask them if there would be more possible committees consisting of 8 people, or more possible committees consisting of 2 people. Envisioning all the different pairs they can form, most of your students say that there will be more committees of 2 people.

Are they correct? If not, what is the correct answer? How can you help your students understand it?

11 We have a group of 12 people and two of them, Abe and Carol, refuse to work together. How many 5 member committees can we make taking this into account? [Hint: Count how many committees contain Abe but not Carol. Count how many contain Carol but not Abe. Count how many contain neither.]

12 Show that $\binom{n}{n-r} = \binom{n}{r}$ and that $\binom{n}{n-2} = \frac{n(n-1)}{2}$. What relation does this problem bear to Student Learning Opportunity 10?

13 (C) One of your students was reading about combinations and came across the formula: $nCr = \frac{nPr}{r!}$. He is now confused since he learned that the formula was $nCr = \frac{n!}{r!(n-r)!}$. Show that the two formulas are really the same and try to explain the concept behind the formula $nCr = \frac{nPr}{r!}$.

14 Suppose that, in a lottery, you had to choose 3 numbers from the numbers 1-20. Suppose the winning numbers were 7, 11, and 19. If you chose your three numbers at random, what is the probability that you would have won this lottery according to the classical approach?

15 I have 10 disks in my car, 3 of which are rock music. If I pick 2 disks at random, what is the probability that (a) both are rock disks (b) at least one of them is not a rock disk?

16 What is the probability of choosing 13 cards from a deck of cards and having them all be spades?

17 A card is drawn at random from an ordinary deck of 52 playing cards. Determine the probability of getting the ace of spades when you pick a card from an ordinary deck of cards. Determine the probability of getting a picture card. Determine the probability of getting either an ace or a spade.

18 Eight different books are placed on a shelf at random. Three of them are math books and 5 of them are chemistry books. What is the probability that the 3 math books are together?

19 If one rolls a fair 6–sided die 3 times, what is the probability that the second and third rolls are greater than the first?

20 If we roll 3 fair 6-sided dice, what is the probability that they don't all show the same number?

21 A shipment of 100 porcelain figures is made. In that shipment, 13 have minor defects, and 5 have major defects, the rest are perfect. If we pick two of them at random, what is the probability that:

(a) both have minor defects?
(b) both of them have some defect?
(c) at least one of them has some defect?
(d) Compare your answers in (a)–(c). Which one was biggest and was this predictable? Explain.

22 A box contains 6 black balls and 5 white balls. Two balls are simultaneously drawn from the box after the balls have been thoroughly mixed. What is the probability that both balls are black? How did you arrive at your conclusion?

23 (C) Your student gives the following argument: We have 4 people, Ricky, Braha, Lakeisha, and David. We have to choose a president, secretary, and treasurer from among these people, only Ricky can't be president, and either Lakeisha or David must be the secretary. To count how many ways the officers can be chosen, we observe that there are 3 choices for president, three for treasurer (all except the one chosen as president) and 2 choices for the secretary. Thus, there are $3 \cdot 3 \cdot 2$ or 18 ways to choose the officers. How can you help your student understand what is wrong with this argument?

24 (C) Your student asks why the definition of 0! is 1. What possible reasons can you give for this defnition?

12.5 Conditional Probability and Independence

LAUNCH

Heidi is planning a trip to Europe to visit her grandparents who live up in the mountains. In Europe, 90% of all households have a television. 50% of all households have a television and a DVD player. Heidi knows that her grandparents have a television, but she forgot to ask if they also have a DVD player. She will only bring her DVDs with her if there is more than a 40% probability that they also have a DVD player. What is the probability that they have a DVD player? Should she bring her DVDs with her?

You might have recognized that the launch problem was a bit more complex than the ones you have examined in the previous sections. That is, in order to determine the probability that Heidi's grandparents had a DVD player, you had to take into account the fact that they had a television. That is, a previous condition was given to you that had a direct effect on the probability. As you are probably aware, this problem required that you understood notions of conditional probability.

Now that you are able to employ some sophisticated counting methods, we can investigate these more complex probability problems that are included in upper level secondary school mathematics classes. We will begin by developing the important concepts of conditional probability and independence and then highlight some very common misconceptions.

Suppose we have a jar containing 3 white balls and 4 red balls. If we draw a ball only once, we know that the probability of picking a white ball is $\frac{3}{7}$ and the probability of picking a red ball is $\frac{4}{7}$. Now suppose we wish to pick another ball, without replacing the first ball and we ask "What is the probability that it is white?" This is a question whose answer depends on what was picked first. Assuming the original ball was not put back, if the first ball picked was a white ball, then the probability that the next ball picked is a white ball is 2 out of the remaining 6 or $\frac{2}{6}$. If the first ball picked was a red ball, then the probability of picking a red ball on the second draw is $\frac{3}{6}$. This example leads to the notion of **conditional probability**. Specifically, if A and B are events and $P(A)$ represents the probability of A occurring and $P(B)$ represents the probability of B occurring, then the probability of B occurring given that A has occurred is denoted by $P(B|A)$. So, if in this jar problem we let A be the event of picking a white ball on the first pick, then we know that $P(A) = \frac{3}{7}$. If we let B be the event of picking a white ball on the second pick, then $P(B|A)$ (the probability that the second ball picked is a white ball given that the first ball picked is a white ball), is $\frac{2}{6}$. Now, suppose we want the probability that both the first ball picked is white and the second ball picked is white, which we denote by $P(A \cap B)$. Since ball 1 can be chosen in 7 ways and ball 2 in 6 ways, there are a total of 42 ordered pairs (ball 1, ball 2). If both balls are to be white, the first ball can be chosen from any of 3 white balls and the second, any of the two remaining white balls after the first is picked. Thus, there are 3×2 or 6 possibilities for the event $A \cap B$ to occur. Thus, the probability of getting a white ball on the first draw followed by a white ball on the second draw is $\frac{6}{42}$ or $\frac{1}{7}$. We observe that $P(A \cap B) = P(A) \cdot P(B|A) = \frac{3}{7} \cdot \frac{2}{6}$. In fact, this statement

$$P(A \cap B) = P(A) \cdot P(B|A) \qquad (12.9)$$

is always true and is used to define conditional probability. That is, by dividing both sides by $P(A)$, $P(B|A)$ **is defined to be**

$$P(A \cap B)/P(A) \qquad (12.10)$$

using equation (12.9) assuming that $P(A)$ is not zero.

One helpful way of thinking about conditional probability is that we are dealing with a smaller sample space than we were originally. When we ask for the probability $P(B|A)$, we are only looking at outcomes where A has occurred (hence the words probability of B given that A occurred). We are then taking the part of that sample space where B also occurs. (12.10) is really expressing that in mathematical terms. Here is an example which illustrates conditional probability.

Example 12.15 *We toss a pair of dice. (a) What is the probability of getting a sum of 5? (b) What is the probability of getting a sum of 5, given that one of the dice had a 4 facing up?*

Solution. (a) There are 36 possible outcomes when we toss a pair of dice. They are

(1, 1), (1, 2), ..., (1, 6), (2, 1), (2, 2), ..., (2, 6), ..., (6, 1), (6, 2), ..., (6, 6)

where the first number in each of the ordered pairs represents the outcome of the first die falling and the second number the outcome of the second die falling. Of these, the 4 pairs, (1, 4), (2, 3), (3, 2), and (4, 1) represent all pairs where the sum of the rolls is 5. Thus, the probability of getting a sum of 5 is $\frac{4}{36}$ or $\frac{1}{9}$.

Solution. (b) If we are given that one of the dice had a 4 facing up, our reduced sample space consists of the outcomes (4, 1), (1, 4), (4, 2), (2, 4), ..., (4, 6), (6, 4) which consists of 12 outcomes. Of these, only 2 outcomes, (4, 1) and (1, 4) give a sum of 5. Thus, the probability of getting a sum of 5, given that a 4 came up is $\frac{2}{12}$ or $\frac{1}{6}$.

To see how we use formula (12.10) to compute this, we can let A be the event when we get a sum of 5 and B be the event that one of the dice falls with a 4 facing up. Then $A \cap B$ is the event when we get a sum of 5 and one of the dice falls with a 4 facing up. This event is the set

$A \cap B = \{(1, 4), (4, 1)\}$. Since there are 36 outcomes and $A \cap B$ is represented by two of them, $P(A \cap B) = \frac{2}{36}$. The event B, on the other hand, is the set $\{(4, 1), (1, 4), (4, 2), (2, 4) \ldots (4, 6), (6, 4)\}$. This set has 12 of the 36 possible outcomes, so $P(B) = \frac{12}{36}$. Thus, $P(A|B)$, the event that we want, is computed by

$$P(A|B) = \frac{P(A \cap B)}{P(B)} = \frac{\frac{2}{36}}{\frac{12}{36}} = \frac{1}{6}.$$

The reduced sample space approach seems so much simpler.

When a person takes a blood test to test for a serious disease, there is always the possibility of the test yielding what is called a false positive. This is when the person is told that he or she has the disease, when in fact he or she doesn't. This information can be devastating to the person. What this next example shows is that, even when tests have what might be considered high reliability, there is still a reasonable chance of having a false positive result. That is why to be more certain, it is always wise to repeat such a test or take other tests to corroborate the results. There is also the possibility that a test may yield a false negative. That is, the person's test result can show that the person doesn't have the disease, when in fact he or she does have it. In some cases the latter is more insidious than the former, for the disease doesn't get treatment and may advance to a point where treatment is impossible or no longer effective.

Example 12.16 *Suppose that we currently have a test for a serious disease, say tuberculosis, which has 100% reliability, but is very expensive, to perform. A new, and much less expensive, test comes along and we want to determine how effective it is in determining if a person has tuberculosis. One way of doing this is to test, say, 1000 people with the more expensive test to determine how many of the people have the disease. Then test these same people with the new method and see in how many cases it properly predicts the disease. Let us imagine that we have done this with the following results. According to the completely reliable expensive test 8% of the 1000 people have the disease. Of those who had the disease (event A), the new test indicated such in 98% of the cases. Of those who didn't have the disease (event B), the new test indicated such in 98% of the cases. Thus, the test is what we call 98% accurate. (a) What is the probability that a person chosen at random from this 1000 people test positive? (b) What is the probability that a person will test negative? (c) What is the probability that a person who tested positive actually had the disease? Use this to find the probability of a person having a false positive (d) What is the probability that a person who tested negative actually did have the disease? (That is, what is the probability of a false negative?)*

Solution. (a) Let the event where a person tested positive be represented by T. This event, T, is the union of two mutually exclusive events, $A \cap T$ and $B \cap T$, where $A \cap T$ represents those who have the disease and test positive and $B \cap T$ represents those who don't have the disease but who nevertheless test positive. In symbols,

$T = (A \cap T) \cup (B \cap T)$.

Thus, $P(T) = P(A \cap T) + P(B \cap T)$. We are told that 98% of those who test positive actually have the disease. This translates to $P(T|A) = 0.98$. This means that the remaining 2% who don't have the disease test positive. That is, $P(T|B) = 0.02$. We now have

$$P(T) = P(A \cap T) + P(B \cap T) \tag{12.11}$$
$$= P(A) \cdot P(T|A) + P(B) \cdot P(T|B)$$
$$= (0.08)(0.98) + (0.92)(0.02)$$
$$= 0.0968.$$

Solution. (b) Let N be the event that a person tests negative. This too is the union of two mutually exclusive events: That the person does have the disease and tests negative and that the person doesn't have the disease and tests negative or that in symbols, $N = (A \cap N) + (B \cap N)$. Thus, we have

$$P(N) = P(A \cap N) + P(B \cap N) \tag{12.12}$$
$$= P(A) \cdot P(N|A) + P(B) \cdot P(N|B)$$
$$= 0.08 \cdot 0.02 + 0.92 \cdot 0.98$$
$$0.9032.$$

Solution. (c) We are looking for the probability $P(A|T)$. However, this is $P(A \cap T)/P(T) = (0.08)(0.98)/0.0968 = 0.80992$ from part (a). What this is saying is that, even with a test that is 98% reliable, there is roughly an 81% chance of a person who has the disease being correctly diagnosed, which means *there is a 19% chance that he won't have the disease even though he tested positive*. This is unacceptable, suggesting that it is always advisable that, if you test positive, you repeat the test, or do other tests to corroborate it.

Solution. (d) In this case we are really asking for the probability of a false negative, or in symbols, $P(A|N)$. This is equal to $P(A \cap N)/P(N) = (0.08 \cdot 0.02)/0.9032 = 0.001$ using the results from part (b). This small probability of a false negative is good.

This last kind of problem can be hard to keep track of. Sometimes, doing a tree diagram helps. Here in Figure 12.4 we show how such a diagram would look in this example:

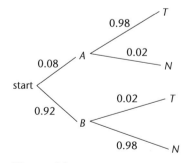

Figure 12.4

This diagram is read as follows. There are two branches from "Start" to A and B. These represent the events A and B happening and on the branch we put the probability of this event happening. From A there are two other branches to T and N. These represent the events T and N occurring, given that A has occurred. On these branches we put the conditional probabilities. The probability of having the disease (Event A), is 0.08 and the probability of testing positive, given that you have the disease, $P(T|A)$, is 0.98. Thus, the branches from "Start" to A to T are labeled 0.08 and 0.98. When we multiply these probabilities, we get $P(A \cap T)$ since $P(A \cap T) = P(A) \cdot P(T|A)$. Thus, we can think of the path "Start"–A–T as the path representing the simultaneous occurrence of A and T, that is, $A \cap T$. Similarly, when we multiply the probabilities along the path "Start"–A–N, we are getting $P(A \cap N)$. Looking at the above tree diagram, we see that the only way a person can test positive is to follow the path, "Start"–A–T **or** "Start"–B–T. Thus, the equations in (12.11) and (12.12) can be read immediately from the diagram.

Notice in part (c), when we asked what is $P(A|T)$, we were asking what part of T came from the path "Start"–A–T. We get this right away from the diagram by multiplying the probabilities along the path "Start"–A–T and dividing by the sum of the products of the probabilities on all paths from "Start" to T. Similarly, in part (d), when we asked what is $P(A|N)$ we were asking what part of T came from the path "Start"–A–N. We get this right away from the diagram by multiplying the probabilities along the path "Start"–A–N and dividing by the sum of the products of the probabilities on all paths from start to N.

We need one last definition, that of independent events. Two events, A and B, are **independent events** if $P(A \cap B) = P(A) \cdot P(B)$. This is equivalent to saying that $P(B|A) = P(B)$ and $P(A|B) = P(A)$ or in more intuitive terms that the probability of B occurring does not depend on whether or not A occurred and vice versa.

Suppose that, in the jar example that began this subsection, we picked a ball and then replaced it. Now, if we let A be the event that the first ball was white and B be the event that the second ball was white, $P(B) = \frac{3}{7}$, whether or not A occurred since the ball was replaced. Thus, the event B occurring does not have any relation to the event A occurring. If one were to ask what is the probability of both balls being white in this case since the ball was replaced, it would be $P(A) \cdot P(B) = \frac{3}{7} \cdot \frac{3}{7} = \frac{9}{49}$.

In a similar manner, suppose that in your right hand you have a coin and in your left hand you have a die and you toss the coin in your right hand and toss the die in your left hand. The outcomes of the coin and the die presumably have nothing to do with one another and so are independent events. Thus, if one asks what is the probability that the coin falls on heads, the answer is $\frac{1}{2}$. If one asks what is the probability that the coin falls on heads given that the die fell with a 6 facing up, the answer is still $\frac{1}{2}$. The roll of the die does not affect the fall of the coin and thus, these events are independent.

If we have several *independent* events (but a finite number of them) say A, B, C, and so forth, then the probability of *all* occurring, that is, $P(A \cap B \cap C \ldots)$ is $P(A) \cdot P(B) \cdot P(C) \ldots$ Thus, if A is the event where you get a head when you toss a coin and B is the event where you get a 6 when you roll a die, and C is the event where your birthday is in December, then the probability of all three occurring simultaneously is $P(A) \cdot P(B) \cdot P(C) = \frac{1}{2} \cdot \frac{1}{6} \cdot \frac{1}{12}$.

There is a very famous case that almost every law student studies which is called "Trial by Mathematics." In this case a couple was seen fleeing from a mugging. The girl was blonde with a pony tail, the car they drove was yellow, and the man was black with a beard and mustache. An interracial couple having these characteristics was picked up and tried for the mugging. The prosecution brought in a professor to assess the probability that a couple chosen at random would have the characteristics: (1) Black man with a beard (2) Man with a mustache (3) White woman with blonde hair (4) Woman with a ponytail (5) Interracial couple in a car (6) Yellow car. Conservative estimates were made of each of these probabilities and the probability of all six events was computed by multiplying the probabilities. The result was that the chance that this couple picked up was innocent was 1 in 12 million. They were convicted on that basis. The conviction was overturned when it was brought to the court's attention that the events (1)–(6) were not independent, and multiplying the probabilities was not a valid way to compute the probability of innocence. In fact, (1) and (2) overlap. Furthermore, there is a difference between the probability of a couple's matching the characteristics and the probability of innocence. By revising the argument and using another probabilistic argument, it was determined that there was a 40% chance that another couple lived in that town with the same characteristics and this was not enough to convict. This is a classic case of misuse of mathematics.

12.5.1 Some Misconceptions in Probability

As pointed out earlier, probabilistic issues pervade our everyday life. That being the case, it is important that we don't fall prey to some very tempting misconceptions that will be described in this section.

The first misconception is that probability has some definitive predictive value, the way some of the models we discussed in Chapter 9 had. A probability statement will never tell you what will happen, just what is likely to happen. Thus, saying the chance of rain is 95% today does not mean that it will rain today, though we have a lot of confidence that it will.

Another common misconception is something that goes like this: My mom had 10 boys in a row. There is a high probability that her next child will be a girl. That is not true. If the probability of getting a girl is 1/2, then the probability that her next child is a girl is still 1/2. That is, the results of getting a girl on the eleventh pregnancy is independent of the number of boys your mother had before. The next child is just as likely to be a boy as a girl if the probability of getting a girl is 50%. However, the probability that she will get 11 boys *in a row* is small, even though the probability of getting a male child is 50%. Since the events of having a male child on any pregnancy are independent of one another, we can calculate the probability of getting 11 boys in a row. If we let B_1 be the event where your mother had a boy on her first pregnancy, and B_2 be the event where your mother had a boy on her second pregnancy, and so on, then the probability that she got all boys on the first 11 pregnancies is expressed by $P(B_1 \cap B_2 \ldots B_{11})$. And since the events are independent, this is equal to

$$P(B_1) \cdot P(B_2) \ldots \ldots \cdot P(B_{11}) = \underbrace{\frac{1}{2} \cdot \frac{1}{2} \cdot \frac{1}{2} \cdot \ldots \cdot \frac{1}{2}}_{11 \text{ times}} = \frac{1}{2^{11}} = \frac{1}{2056}.$$

We have been assuming that the probability of getting a boy is $\frac{1}{2}$. After all, one either gets a boy or a girl and there is no reason for one to be more prevalent than another. So classical probability theory tells us that each should be considered equally likely and each has probability 1/2. Actually, the probability of getting a female is a bit more than 50% as real data show, which is why we have to be careful with what "seems" to be true, but we will ignore this and continue to take the probability of getting a boy and girl to be each equal to 1/2 for our examples.

Another misconception in probability involves the notions of certainty that an event will, or will not happen. It is true, that if the sample space is finite and all events in the sample space are equally likely, then an event having probability zero is equivalent to its being impossible, and having probability 1 is equivalent to its being certain. However, in the case where the sample space is infinite or the events not equally likely, then it is not always true that, if the probability of an event is zero, that the event cannot occur. We will have to wait until the section on Geometric Probability to explain this further. In a similar manner, we teach that, if an event is certain to happen, then the probability that it happens is 1. That is true. But if the probability of an event is 1, it is *not* certain that it will happen. This is quite subtle. The example we present later uses the notion of countability, which we discussed in Chapter 6 and therefore is somewhat more advanced than the secondary school level. Perhaps the best way of thinking of an event with probability zero is that it is so rare that, for all practical purposes, we don't expect it to occur. And if an event has probability one, it is *almost certain* to occur, though it might not.

Student Learning Opportunities

1 (C) Several of your students want to know what the difference is between mutually exclusive events and independent events. How do you clarify the difference?

2 Two defective keyboards have been mixed up with three good ones. To find the defective ones, the keyboards must be tested one by one. If we select our keyboards at random and test them, what is the probability that we find our two defective keyboards in the first two picks?

3 John feels pretty good about his recent LSAT exam. He assesses that there is a 70% chance that he did well, and that there is a 90% chance that a law school will accept him if he did well. Assuming these chances are accurate, calculate the probability that John will score high on the LSAT and be accepted to law school.

4 From an ordinary deck of cards, you pull out the ace of spades, the ace of hearts, the two of clubs, and the jack of diamonds. You now shuffle these 4 cards and randomly pick 2 cards from these 4. What is the probability of getting a hand with an ace? If you have the ace of spades, what is the probability that you have the second ace?

5 You roll a die until a "1" shows up, at which point the game ends. What is the probability that the game ends in 3 or fewer throws?

6 If a box contains 3 dimes and 4 nickels and you pick a coin and then return it to the box, mix the coins well, and then pick another coin: (a) What is the probability that both coins are dimes? (b) How does the answer change if you don't replace the coin?

7 Mary is a modern woman. She is trying to decide whether to call Jack or Larry for a date. She feels that, if she calls Jack, there is a 30% chance he will say "Yes" and, if she calls Larry, there is a 50% chance he will say "Yes." She decides to flip a coin to decide who to call—heads she calls Jack, tails she calls Larry.

   (a) What is the probability she will get a date with Jack?
   (b) What is the probability that, if Mary makes only one call to one of these guys, she will get a date with one of them?

8 A medical researcher estimates that 2% of a population is infected with a rare disease, which for this problem we will assume is accurate. The procedure used for for testing for this disease is 90% accurate. So, if one has the disease, the test will indicate that 90% of the time, and if the person doesn't have the disease, the test will indicate that 90% of the time. Of course, what this means is that the test will give a false positive result 10% of the time. If a person is picked at random and tested:

(a) What is the probability that the person tests positive?

(b) What is the probability that, if a person tested positive for the disease, he or she actually had the disease?

9 Mr. Krule, Mr. Vitious, and Ms. Meany are auditors for the tax department. They, respectively, handle 35%, 45%, and 20% of the tax returns that come in during the week. The two men are so experienced that they make errors, in fact, only 1% of the time. Ms. Meany on the other hand is new and she makes errors 10% of the time, which is why she does so few audits. If a return from the returns processed by these people is picked at random and an error is found in it, what is the probability it came from Ms. Meany?

10 A company produces stoves at 3 different factories. The probability of a stove being made at factories A, B, and C, respectively are 0.35, 0.25 and 0.40. Production records indicate that in the past, 5% of the stoves made at factory A are defective, 3% of those made at factory B are defective, and 7% of those made at factory C are defective. All the stoves are shipped to a central warehouse before being sent to stores.

(a) Draw a tree diagram which has all these facts. (Don't forget to put in the probabilities of a stove not being defective.)

(b) Find the probability a stove picked at random comes from factory A and is defective.

(c) Find the probability a stove picked at random comes from factory A and is not defective.

(d) Find the probability that, if a stove picked at random is defective, the stove came from factory A.

11 (C) Your students make the following computations. Explain what is wrong with each of them and then correct it:

(a) If the probability of getting a 1 on a roll of one die is 1/6, then the probability of getting two 1's on one roll of a pair of dice is 2/12, since each die can fall 6 ways for a total of 12 ways and 2 of them are successes.

(b) Two cards are picked from an ordinary deck of cards. The probability that the first card is a spade (event A) and that the second card is a diamond (event B) is $\frac{1}{16}$, figured out as follows: $P(A) = \frac{13}{52} = \frac{1}{4}$, since there are 13 spades in the deck of 52 cards. Similarly, $P(B) = \frac{13}{52} = \frac{1}{4}$, so $P(AB) = P(A)P(B) = \frac{1}{4} \cdot \frac{1}{4} = \frac{1}{16}$, since A and B are independent.

12 (C) Your student, John tells you the following: Although he likes to fly, he has this terrible fear that someone is going to bring a bomb on the plane and blow it up. He knows that the chances of this happening are small. But he also knows from probability, that the chances of two bombs being brought on board by two different people who don't even know each other is even smaller. So, as an insurance policy, on each flight, he brings along a fake bomb with him, feeling that this makes the chances of another bomb on board tiny. Comment on his thinking.

13 Show that, if A and B are mutually exclusive and if $P(A)$ and $P(B)$ are both positive, then A and B cannot be independent. [Hint: What is $P(B|A)$?]

14 (C) Assuming that B represents boy and G represents girl, which sequence of children is more probable: (a) *BGBGB* or (b) *BGGGG*? (Assume, though it is not really true, that the probability of getting a boy is the same as the probability of getting a girl and that both are 1/2.) Your students, Mal and John are having an argument. Mal claims (a) is more probable, since the pattern is more regular and there are an equal number of boys and girls. John claims that the two sequences are equally likely. Who is correct, Mal or John? Why?

15 A gambler has watched the roulette wheel come up black 40 times in a row. He knows it is due to come up red soon. So he starts betting big. If you were his friend, what would you tell him?

16 (C) Comment on the following statements made by students:

(a) Toss 3 coins. We are interested in the probability that all three will match. We know for sure that at least two will match. Now either the third will match or it won't, and both have equal probability. So the probability of a match is 50%.

(b) There is a higher probability of getting either 3 boys or 3 girls in a set of 4 children than getting 2 boys and 2 girls.

12.6 Bernoulli Trials

LAUNCH

37% of people over age 80 will not be around within the next 10 years. Terry is happy about this report, since he wants his grandfather, who is over 80 to live for a long time. But, his grandfather lives in a community with 110 other senior citizens, and he doesn't care to live if most of them die. Find the probability that in the next 10 years, at least 75 of the other 110 will die.

Problems such as the one you tried solving in the launch are characteristic of the more advanced problems you might have encountered in secondary school.

By using the counting techniques developed in the previous section, we can now work out a very important formula, which is used in many probability applications and which is needed to solve the launch problem. We begin with something called a **Bernoulli trial**, where there can only be two outcomes from an experiment, success and failure. If success has probability p, then failure has probability q where $q = 1 - p$, as we have seen earlier in the chapter. In Bernoulli trials, we assume that the probabilities of success and failure stay the same from trial to trial and the trials are independent. For example, we may toss a die and be interested in the event of getting a 6. Then the probability of success is $p = \frac{1}{6}$ and the probability of failure is $q = 1 - p = \frac{5}{6}$. Each time we toss the die, these probabilities don't change. Furthermore, and this is key, the trials are independent.

Now, suppose that we tossed the die 10 times and were interested in the probability of getting exactly four 6's, or put another way, 4 successes in 10 trials. Let us analyze the specific case where the first 4 trials are successes and the remaining trials are failures. Since the trials are independent, the probability that the first 4 trials result in success is $\frac{1}{6} \cdot \frac{1}{6} \cdot \frac{1}{6} \cdot \frac{1}{6} = \left(\frac{1}{6}\right)^4$. The probability that the remaining 6 trials result in failure, is similarly $\left(\frac{5}{6}\right)^6$. Thus, the probability that only the first 4 trials are successful and the remaining 6 trials are not, is

$$\left(\frac{1}{6}\right)^4 \cdot \left(\frac{5}{6}\right)^6. \tag{12.13}$$

The same analysis works if the order of the 4 successes and 6 failures is different. So if we had two successes, followed by two failures, followed by another two successes, followed by four failures, the probability of this happening is also $\left(\frac{1}{6}\right)^4 \cdot \left(\frac{5}{6}\right)^6$. (Why?) Thus, the probability of *any* set of 4 successes and 6 failures is given by expression (12.13). But there are $\binom{10}{4}$ sets of 4 successes and 6 failures, each having probability given by expression (12.13) and therefore the probability that one of them occurs is the sum of the individual probabilities, since the events are mutually exclusive. Thus, the total probability of getting 4 successes out of 10 trials is

$$\binom{10}{4}\left(\frac{1}{6}\right)^4\left(\frac{5}{6}\right)^6.$$

Now, suppose that we perform *n* trials of the experiment, and seek the probability that *exactly k* of the trials result in successes. Doing a similar analysis we get that, if we perform *n* trials of a Bernoulli experiment, then the probability of *exactly k* successes where the probability of success is *p* and the probability of failure is *q* is

$$\binom{n}{k}(p)^k(q)^{n-k}. \tag{12.14}$$

Example 12.17 *A company that manufactures computer chips finds that approximately* 1% *of the chips they make are defective. Which of the cases that follow do you think is most probable if* 100 *chips are made* (a) *exactly 3 defective chips* (b) *less than three defective chips.* (c) *at least 3 defective chips? Calculate the probabilities in each case.*

Solution. (a) We will calculate the probabilities in this example, where getting a defective chip is a "success." The probability of success according to the problem is 0.01. The probability of getting exactly 3 defective chips in a lot of 100 is the same as getting 3 successes. The probability of this happening according to expression (12.14) is:

$$\binom{100}{3}(0.01)^3(0.99)^{100-3} \approx 0.006.$$

This is quite unlikely. Is that what you would have expected?

Solution. (b) If we get less than 3 defectives, then we get 0, 1, or 2 defectives. The probability of this happening is:

$$\binom{100}{0}(0.01)^0(0.99)^{100} + \binom{100}{1}(0.01)^1(0.99)^{99} + \binom{100}{2}(0.01)^2(0.99)^{98} \approx 0.92.$$

Solution. (c) If we approach this as in the previous example, we would have to find the probabilities of 3 or 4 or 5 and so on up to 100 defective chips, which is too cumbersome. Therefore, we find an equivalent value, which is 1 minus the probability of less than three defective chips.

628 Data Analysis and Probability

The probability of at least 3 defective chips is $1 - P(\text{less than 3 defective chips}) = 1 - 0.92 = 0.08$. By ordinary industrial standards, this figure is considered pretty high and thus, a company that has a 1% defective rate in manufacturing should work to greatly improve it or they may find themselves out of business.

> **Example 12.18** *In a management training program, it has been established from past data, the probability that a new trainee will drop out of the training program before it is over is 0.10. Suppose that the company brings in 40 new trainees. What is the probability that no more than 3 trainees will drop out?*

Solution. Here "success" means dropping out. Thus, the probability of "success" is 0.10.

If no more than 3 drop out, then either none dropped out, 1 dropped out, 2 dropped out, or 3 dropped out. Thus, the probability of this happening is

$$\binom{40}{0}(0.10)^0(0.90)^{40} + \binom{40}{1}(0.10)^1(0.90)^{39} + \binom{40}{2}(0.10)^2(0.90)^{38} + \binom{40}{3}(0.10)^3(0.90)^{37} \approx 0.42.$$

Therefore, there is a 42% chance that no more than 3 will drop out and the company can plan accordingly.

In the case of the launch problem, p would be the probability that a person of 80 years of age would die within the next 10 years. So, $p = 0.37$. q would be the probability that the person would not die within the next 10 years. In this case, q would be $1 - 0.37$, or 0.63. We leave it to you to solve the rest of the problem. The next section will make the computation easier.

Student Learning Opportunities

1 **(C)** You gave your students the following problem: The Yummy Cereal company puts a shiny sticker in 4 out of every 10 boxes. What is the probability that Mrs. Cheerio will find 3 stickers in the next 5 boxes of cereal that she buys? One of your students says the following: "The probability of a success (finding a sticker) is $\frac{4}{10}$. Therefore, the probability of a failure (not finding a sticker) is $\frac{6}{10}$. Since we want to find 3 stickers in the next 5 boxes, that means there will be 3 successes and 2 failures. The probability of that is $\frac{4}{10} \cdot \frac{4}{10} \cdot \frac{4}{10} \cdot \frac{6}{10} \cdot \frac{6}{10} = \left(\frac{4}{10}\right)^3 \left(\frac{6}{10}\right)^2$." What has your student forgotten to take into account? Explain.

2 A die is tossed 5 times. What is the probability that one gets exactly three 6's?

3 The probability that children in a certain family have brown hair is $\frac{5}{6}$. What is the probability that out of 8 children:

   (a) exactly 2 will have brown hair?
   (b) none will have brown hair?
   (c) at least one will have brown hair?
   (d) Which of the above cases was most probable? Explain why that makes sense.

4 There is an 85% chance that a graduate from King's College in Sante Fe will have a job within 3 months of graduating. If 100 people graduate from King's College this year, what is the probability that at least 98% of them have a job within 3 months of graduation?

5 The Cheetahs are a local baseball team. They will play a series against the Lions, another baseball team. The series will end when one of the teams has won 4 games. What is the probability that the series ends in 4 games if the probability of the Cheetahs winning the game is $\frac{3}{5}$ and the Lions winning the game is $\frac{2}{5}$? Explain.

6 Anton has this strange habit of dropping paper clips out of the window at passers by below. His data from the past indicate that he has hit his victims 10% of the time. He wants to be at least 80% certain to hit at least one person today. What is the minimum number of clips he must drop to achieve this? Explain.

7 Recent data have shown that 3% of people taking the new drug hexatrine suffer from side effects. What is the probability that, out of a sample of 50 people who take the drug, exactly 4 suffer side effects?

8 There is a 20% chance that a person with a heart transplant will reject the new heart.
  (a) Find the probability that, in 5 such operations, none of the patients reject the heart.
  (b) Find the probability that at least one person rejects the heart.
  (c) Find the probability that all 5 reject the new heart.
  (d) Which of the above had the highest probability? Explain why this makes sense.

9 The probability of a male being color-blind is 0.042. Find the probability that, in a sample of 120 males, exactly 2 are color-blind.

10 The local hospital in your community is very concerned about accurate diagnoses of cancer conditions. Currently, they have on staff a specialist who reads CAT scans. Past performance shows that he has a 98% chance of reading the CAT scan correctly and diagnosing the cancer when a patient has it. But, of course, there is always the 2% chance that the specialist will say that there is no cancer when there is. This is a very serious mistake. What is the probability that, if 10 CAT scans from patients with lung cancer are given to this specialist, he will misread at least 1 and claim there is no cancer?

12.7 The Normal Distribution

LAUNCH

Consider the following sets of data: (1) the IQ scores of 1000 people; (2) the diameter of 1000 maple trees; (3) the hourly wage rates of 150,000 sanitation workers; (4) the heights of 3000 women of age 26. Clearly, these data come from very diverse fields, yet there is something that is quite similar about them. What is it? If you are beginning to consider the way the data in each of these cases is distributed, you are on the right track. As you read this section, you will become informed about the normal distribution, which is the underlying concept that unifies the four examples above.

In the previous section we discussed Bernoulli trials and found that, in an experiment, if the probability of success on a single trial is p, then the probability of getting k successes in n trials is given by

$$\binom{n}{k} (p)^k (q)^{n-k}. \tag{12.15}$$

We can actually create a graph that shows these individual probabilities. Such a graph is called a probability histogram. We label the x-axis with the number of successes, and on each tick mark, we construct a rectangle whose area is the probability that we get that number of successes. We take the base of each rectangle to be 1. Let us illustrate with an example.

Example 12.19 *We have a crooked coin. The probability of getting a head with this coin is $\frac{1}{3}$. We toss the coin 8 times. Draw the probability histogram that represents the probability of getting x heads.*

Solution. We know from equation (12.15) that $P(x = k) = \binom{8}{k} \left(\frac{1}{3}\right)^k \left(\frac{2}{3}\right)^{8-k}$. Substituting in $k = 0, 1, 2, \ldots 8$, we obtain the following probabilities: $P(x = 0) = \frac{256}{6561}$, $P(x = 1) = \frac{1024}{6561}$, $P(x = 2) = \frac{1792}{6561}$, $P(x = 3) = \frac{1792}{6561}$, $P(x = 4) = \frac{1120}{6561}$, $P(x = 5) = \frac{448}{6561}$, $P(x = 6) = \frac{112}{6561}$, $P(x = 7) = \frac{16}{6561}$ and $P(x = 8) = \frac{1}{6561}$. The probability histogram is given in Figure 12.5 below. As pointed out, each rectangle has a base of length one and a height equal to the probability that x occurs where x is the value in the center of the rectangle.

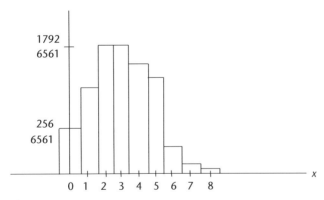

Figure 12.5

Notice what this graph does. It allows us to envision probabilities as areas of rectangles. In fact, from the graph, we can easily compute more complex probabilities. For example, to compute the probability that $x = 3$ or $x = 4$, we simply add the areas of the rectangles surrounding $x = 3$ and $x = 4$. To compute the probability that $x > 5$, we add the areas of the rectangles that correspond to $x = 6, 7,$ and 8. So, having the graph enables us to figure out many different probabilities rather easily.

This last problem was an example of what is referred to as a discrete probability histogram. The word "discrete" in this instance means that we can only have a finite number of outcomes, as was the case with our example. However, many times in probability we are interested in the probabilities of events occurring where the number of outcomes can be infinite. For example, the life of a light bulb. A light bulb can last from no time at all, to forever. A manufacturer would be very interested in finding out the probability that the light bulbs he manufactures would last more that 3000 hours. This way he could decide on an effective pricing strategy.

A function whose graph enables us to find probabilities of events in an experiment where there are an infinite number of outcomes is known as a **probability density function**. These are discussed extensively in college-level probability courses. However, on the secondary school level, there is one very important graph that is discussed and that is the **normal distribution**. The normal distribution with parameters u and σ is given by $f(x) = \frac{1}{\sqrt{2\pi}\sigma} e^{\left(-\frac{(x-u)^2}{2\sigma^2}\right)}$. ($\mu$ and σ are the mean and standard deviation of the population which you may have studied in secondary school. We discuss this a bit further at the end of this section.) When $\mu = 0$ and $\sigma = 1$, we get the graph shown in Figure 12.6

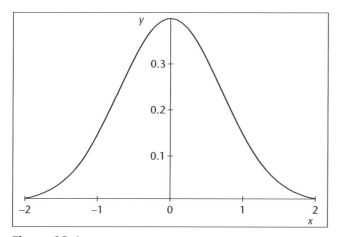

Figure 12.6

where the "hump" occurs at $x = 0$. In general, the graph of $f(x)$ for any μ and σ looks just like this, only the hump is at $x = \mu$, and the width varies with σ. The inflection points are at $x = \mu \pm \sigma$. Below in Figures 12.7 and 12.8 are the graphs of normal density functions with $\mu = 5$, $\sigma = 2$ and $\mu = -2$, $\sigma = 5$, respectively.

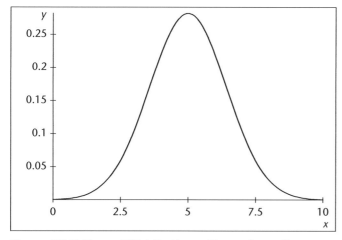

Figure 12.7 Normal Distribution with $\mu = 5$, $\sigma = 2$.

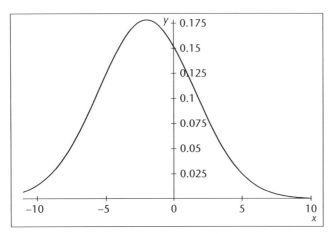

Figure 12.8 Normal Distribution with $\mu = -2$, $\sigma = 5$.

Many practical examples have probability histograms that appear to be normal or approximately normal. For example, variables like SAT scores, shoes sizes, people's heights, IQ scores, and diameters of tree trunks of a certain species of trees are usually normally distributed.

The key point about events having (continuous) probability densities is that probabilities associated with events can be obtained by computing *areas* under the curve. (Recall these are done by computing integrals.) While in the past there were tables that were used to do this (and the tables were constructed by using a great deal of calculus and numerical methods), many of the current graphing calculators have the capability of computing these probabilities with ease. We illustrate the syntax here for the TI series. Suppose that some variable x had a normal probability density function, $f(x)$ with parameters μ and σ and we wish to compute the probability that $a < x < b$. (This is given by $\int_a^b f(x)dx$.) On these machines we would access the "normalcdf" function (usually obtainable from the catalog capability of the calculator.) The correct syntax for our computation is normalcdf (a, b, μ, σ). Let us illustrate.

Example 12.20 *Suppose that IQs in a town are normally distributed with parameters $\mu = 100$ and $\sigma = 10$. Find the probability that a person in the town has (a) an IQ score between 100 and 125 and (b) an IQ score more than 140.*

Solution. (a) If we let x represent the IQ score, we are interested in the probability that x is between 100 and 125. Our probability is therefore, normalcdf (100, 125, 100, 10) = 0.4937.

Solution. (b) Here we are interested in the probability that $x > 140$. To figure out this part, we simply replace b by a large number. Thus, normalcdf (100, 1000000, 100, 10) would do the trick. Our calculator tells us this is equal to 0.00003, which is rather small. But that is no surprise. Few people have IQs above 140 .

Example 12.21 *The Bright Light Company manufactures light bulbs whose lives are normally distributed with parameters $\mu = 1000$ and $\sigma = 300$. (a) Find the probability that a light bulb will last for more than 1500 hours and (b) find the probability it will last less than 800 hours.*

Solution. (a) We calculate normalcdf (1500, 1000000, 1000, 300) and get 0.0477.

Solution. (b) Here we calculate normalcdf (0, 800, 1000, 300) which equals approximately 0.2520.

Probabilities that are computed using Bernoulli distributions are important in applications, but working with them at times can become very cumbersome, as is illustrated by the example below.

> **Example 12.22** *It is known that, in a 24 hour period, the probability that a given type of atom will split is $\frac{1}{10^{10}}$. Suppose that we have a sample of 10^{50} atoms. Compute the probability that, in a 24 hour period, 3 or fewer atoms split.*

Solution. Using the theory that we have presented in the last section, we would get that our probability is given by

$$\binom{10^{50}}{0}\left(\frac{1}{10^{10}}\right)^0\left(1-\frac{1}{10^{10}}\right)^{10^{50}} + \binom{10^{50}}{1}\left(\frac{1}{10^{10}}\right)^1\left(1-\frac{1}{10^{10}}\right)^{10^{50}-1}$$
$$+\binom{10^{50}}{2}\left(\frac{1}{10^{10}}\right)^2\left(1-\frac{1}{10^{10}}\right)^{10^{50}-2}.$$

This is a rather daunting calculation that our hand calculators would most likely have trouble with, since the numbers have more decimal places than the calculator can handle. In problems when n is large, and the computations are cumbersome, probabilists often use the following theorem that helps them do the necessary computations. Note that the theorem gives us an exceptionally efficient way to compute certain probabilities. The proof is beyond the scope of this book.

> **Theorem 12.23** *If x represents the number of successes in an experiment consisting of n Bernoulli trials, where the probability of success is p, then the probability that $a < x < b \approx$ normalcdf $(a, b, np, \sqrt{npq})$.*

> **Example 12.24** *Suppose that we toss a coin 324 times. Determine, approximately, the probability that we get between 150 and 180 heads.*

Solution. Few would want to do this problem using the Bernoulli probabilities discussed earlier. If we let x be the number of heads, we want the probability that $150 < x < 180$. However, by the above theorem, this computation becomes much easier. First, we observe that $np = (324)(\frac{1}{2}) = 162$ and that $\sigma = \sqrt{npq} = \sqrt{324 \cdot \frac{1}{2} \cdot \frac{1}{2}} = 9$. Thus, our probability simplifies to: normalcdf $(150, 180, 162, 9) \approx 0.8860$.

> **Example 12.25** *Suppose that 2% of people who take a penicillin shot have an allergic reaction. Now suppose that we inject 8000 people with penicillin. Find the approximate probability that more than 200 people will have a reaction to the shot.*

Solution. If we interpret success to mean that a person has an allergic reaction to penicillin, then $p = 0.02$. If x represents the number of allergic reactions, we need to compute the probability that $x > 200$. Computing $np = 8000 \cdot 0.02 = 160$ and $\sigma = \sqrt{8000 \cdot 0.02 \cdot 0.98} \approx 12.5220$, our probability is approximately normalcdf(200, 1000000, 160, 12.552) ≈ 0.0007.

A natural question is, "Suppose it is reasonable to assume that the values of x are normally distributed. How do people decide what the right μ and σ are for this distribution?" The answer is, this is done by experimentation. To find μ one takes the average of a large sample of values of x, say n values. This is what is taken for μ. Of course, once we have μ, if the variable x arises from a Bernoulli trial, it is easy to compute σ, since σ in that case is $\sqrt{npq}$. When x doesn't arise from a Bernoulli trial, σ is computed as follows:

$$\sigma = \sqrt{\sum_{i=1}^{n} \frac{(x_i - \mu)^2}{n-1}} \tag{12.16}$$

where the x_i's are the values of x in the sample taken. We assume that this sample is large enough to make this a viable estimation for σ. That means that we have at least 30 sample points. Calculators automatically compute this value of σ. On the TI calculator, when you put the sample measurements in a list and ask the calculator to do one variable statistics, the calculator gives you a value called S_X. This is called the sample standard deviation. This is precisely what we computed in equation (12.16). S_X gives us a measure of how spread out the data are. Small values of S_x mean the data are close to μ, while a large value of S_X means the data points are far from the mean.

Student Learning Opportunities

1 On the secondary school level, students are often given the following facts about a variable x having a normal distribution: Approximately 68% of the values of x lie between $\mu - \sigma$ and $\mu + \sigma$, approximately 95% lie between $\mu - 2\sigma$, and $\mu + 2\sigma$ and 99% between $\mu - 3\sigma$ and $\mu + 3\sigma$. See Figure 12.9 below:

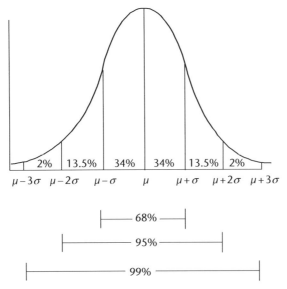

Figure 12.9

Using these facts, answer the following "typical" secondary school questions, being sure to explain how you arrived at your answers.

(a) At Bay Water secondary school, the student SAT scores are approximately normally distributed with parameters $\mu = 550$ and $\sigma = 75$. If 600 students took the SAT, approximately how many of them got scores between 475 and 625?

(b) Using the scenario from part *a*, what percent of students got scores between 625 and 700?

(c) If we use the normal distribution, what is the approximate probability that $x > \mu$?

(d) True or False: If a variable x is truly normally distributed, then the probability that $x > \mu$ and the probability that $x \geq \mu$ are the same. Explain your answer.

2 (C) When you discuss distributions of real data, you are always careful about specifying that the data are "approximately" normally distributed. Your students are curious why you don't just say the data are normally distributed. How do you explain your wording?

3 (C) One of your students, Matthew, is curious to know under what conditions would $\sigma = 0$. Using equation (12.16) as a guide, what do you say?

4 A bag containing 300 ordinary pennies is dropped on a table. What is the approximate probability that between 200 and 230 coins fall heads up?

5 The quality control at a refrigerator factory has shown that approximately 3% of the refrigerators that come off the assembly line are defective. If we make 1000 refrigerators, what is the approximate probability that more than 30 refrigerators are defective?

6 Of people with stage 3 breast cancer, 80% survive after 5 years. If 100 such people are given treatment, what is the probability that between 90 and 95 of these people will live?

7 A new material has been created and needs to be tested to see if it is reliable or not. The breaking strength is normally distributed with $\mu = 180$ and $\sigma = 5$. Any breaking strength less than 175 is considered defective. What is the probability that a sample of this material is defective?

8 The values of x are normally distributed, and in a random sample of 35 values of x we got 3 fives, 4 sevens, 8 nines, 13 ones, and 11 sevens. Key the values into the calculator and find the approximate values of μ and σ for this normal distribution. Then find the probability that x is between 7 and 8.

12.8 Classic Problems: Counterintuitive Results in Probability

LAUNCH

If you were in a room with 35 people, and you were asked to bet 100 dollars that at least two people in the room had the same birthday, would you be willing to make that bet?

If you are like most people, your reaction to the launch problem would be, "There is a very small chance that I would win, so this is not a bet I should make." Would you be surprised to find out that, if you took this bet, you would have better than an 80% chance of winning? It sounds unbelievable, but if you randomly pick samples of 35 people and ask them their birthdays, you would expect that about 8 times out of 10, you would get a match. If you don't believe it, try it! This very counter-intuitive result surely needs explanation. In this section we will investigate this classic problem, called "The Birthday Problem," as well as several other most interesting problems that also have counterintuitive results. We will also discuss how to better convince yourselves of the results you calculate, by using simulation.

12.8.1 The Birthday Problem

Let us now take a better look at this "birthday problem" that you confronted in the launch question. We have already seen that, if the probability of an event not occurring is p, then the probability that it does occur will be $1 - p$. Thus, if we can show that the probability is 0.19 that in a group of 35 people, all people have *different* birthdays, then it will follow that the probability that this is not true, that is, the probability that at least two people have the same birthday is $1 - 0.19 = 0.81$. We will approach this problem by computing the probability that all the people in the group have different birthdays. Since the birthdays of the people are independent events, to find the probability that all the people have different birthdays, we will multiply the probabilities together.

Let us use the problem-solving approach of starting with a simpler problem and working our way up to the problem at hand. Suppose we pick two people at random. We will be assuming that one year has 365 days. The first person has some birthday, and what that day is doesn't matter. So the probability that that person has a birthday is 1. The probability that the second person has a birthday different from the first is $\frac{364}{365}$. Thus, the probability that the two have different birthdays is $1 \cdot \frac{364}{365}$. Now, suppose we pick a third person. The probability that his or her birthday is different from the other 2 is $\frac{363}{365} \approx 0.99452$, since there are only 363 days left that will give a different birthday from the other two. Thus, the probability that all three have different birthdays is $1 \cdot \frac{364}{365} \cdot \frac{363}{365}$, or approximately 0.99180. Now let us continue by picking a fourth person. In order for his or her birthday to be different from the others, his or her birthday must occur on any of the remaining 362 days in the year and thus the probability of having a different birthday is $\frac{362}{365}$. Thus, the probability that all four have different birthdays is $1 \cdot \frac{364}{365} \cdot \frac{363}{365} \cdot \frac{362}{365}$, which is approximately 0.98364. We continue in this manner to find that the probability of randomly picking 35 people with different birthdays is $1 \cdot \frac{364}{365} \cdot \frac{363}{365} \cdot \frac{362}{365} \cdot \ldots \cdot \frac{331}{365} \approx 0.1857$. Therefore, the probability of two or more having the same birthday is $1 - 0.1857 \approx 0.8143$, which is better than an 80% chance! Remarkable!

12.8.2 The Monty Hall Problem

Another famous problem that is counterintuitive is the so-called "Monty Hall Problem", based on the show "Let's Make a Deal," whose emcee was named Monty Hall. The game worked as follows. The contestant was shown three doors. Behind one of the doors was a car, and behind the other two doors were goats. The contestant would pick a door, and the emcee, who knew what was behind the other two doors, would open one of the other doors, behind which he knew there was a goat. Then he would ask the contestant if he would like to switch his choice of doors. (The contestant received the prize behind whatever door he ultimately chose.) The question is, does it pay for the contestant to switch?

Most people who hear this problem think, "Having seen that there is a goat behind one of the doors, there are only two doors left, and one has a car and the other a goat. So switching, should make no difference, since there is a 50% chance he has the car, and there is a 50% chance that, if switched, he would have the car." Even professional mathematicians would argue this way. The surprise is, that it really does pay to switch. In fact, you have a better chance of winning if you switch. This requires explanation.

The picture in Figure 12.10 below shows the situation when the car is behind door 1.

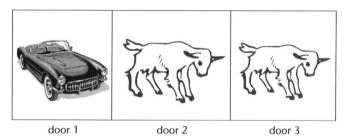

door 1   door 2   door 3

Figure 12.10

Let us assume that the contestant originally chose door 1. Then if he switched, he would lose. If he had chosen door 2, then the emcee would have opened door 3, and if the contestant switched to the only unopened door left, door 1, he would win. Similarly, had he picked door 3, the emcee would have opened door 2 and again switching to the only door left, door 1, the contestant would again win. Thus, in two out of the three cases, when the car is behind door 1, the contestant would win by switching. The same is true when the car is behind door 2 or 3. So in all cases, 2 out of 3 times the contestant would win by switching. This is quite a surprise!

12.8.3 The Gunfight

After a huge bar room brawl, three men, A, B, and C "settle it" outside. They take their guns and stand in a triangle but, before shooting, decide the rules of the gunfight. A is to shoot first, followed by B, followed by C, followed by A, and so on, round-robin style. The probability of A hitting his target and killing him is 0.3. B never misses, and C's chance of hitting his target and killing him is 0.5. A will shoot first. Who should he point his gun at and shoot?

Logic tells us that, if he kills C, then he is a goner, since B will shoot him next and will surely kill him. So he must shoot B. One of two things will happen.

Case 1: He will kill B and then it will be a fight to the end with C, with C shooting first.

Case 2: He will miss B and then it will be B's turn. Since C is a better shot than A, B will kill C. Then A has only one chance to shoot at B since, if he misses, B will kill him. He will hit B with probability 0.3. In summary, if A misses B, his chance of surviving is 0.3.

What comes next is very surprising. Let us return to Case 1. What if he kills B? What are his chances of surviving then? Our intuition tells us, "at least 0.3 since he has gotten rid of the most dangerous guy." Probability tells us otherwise. If he kills B, then the gunfight continues with him and C. What is A's chance of winning now? Well, A will win if C keeps missing and A hits C eventually. That is, C must continually miss on each turn with probability 0.5, while A must succeed in hitting C either on the first chance he gets with probability 0.3, or he has to shoot C on the second chance he gets, or third chance, and so on. The probability that A wins on his first shot is $(0.5)(0.3)$ (After A shot B, C missed A with probability 0.5 and A hit C with probability 0.3. We multiply the probabilities since the events are independent.) The probability that A hits C on the second round is $(0.5)^2(0.7)(0.3)$ which is figured as follows. C has missed A on his first shot, then A has missed C, then C has missed A again, and finally A hits C. These events occur with respective probabilities, 0.5, 0.7, 0.5, and 0.3, yielding a combined probability of $(0.5)^2(0.7)(0.3)$. Continuing this way, the probability that A hits C on the third round is $(0.5)^3(0.7)^2(0.3)$. You try to explain why. Similarly, the probability that A wins on the fourth, fifth, and sixth round are, respectively, $(0.5)^4(0.7)^3(0.3)$, $(0.5)^5(0.7)^4(0.3)$, and $(0.5)^6(0.7)^5(0.3)$.

We know that A will survive if he kills C on either the first round, the second round, the third round, and so on. Thus, using the countable additivity axiom of probability (Axiom (3)' of the Kolmogorov axioms) the probability of this happening is

$$(0.5)(0.3) + (0.5)^2(0.7)(0.3) + (0.5)^3(0.7)^2(0.3) + (0.5)^4(0.7)^3(0.3) + \ldots .$$

This is a geometric series with constant $r = (0.5)(0.7) < 1$ and so by the results of Chapter 6 Section 10 Theorem 6.47, this series converges to

$$\frac{a}{1-r} = \frac{(0.5)(0.3)}{1-(0.5)(0.7)} = \frac{3}{13} = 0.23077.$$

In summary, if A hits B, his chances of surviving are $\frac{3}{13}$. So what? Well, this is less than the probability 0.3 (computed earlier) of his surviving if he misses B. Thus, strangely, A is better off missing B than hitting him.

So what should A do on his first shot, since hitting C will make him a dead man?

Answer: Fire his gun into the air or do anything that will ensure that he will miss B! He has a better chance of surviving! Now this is really counterintuitive!

12.8.4 Simulation

In the previous section we presented some problems where the solutions seemed counterintuitive. Many people who see these problems don't believe that the solutions are correct. They ask for some kind of corroborating evidence. One particularly nice way of verifying solutions is by a procedure known as **simulation**. In a simulation, we try to imitate what we are describing. For example, in the birthday problem we were interested in the probability that, if we took 35 people at random, at least two would have birthdays that match. One way of simulating the problem is to randomly stop people on the street and ask them their birthdays. Once we get 35 people, we stop and see if there is a match. Now we pick another 35 and do the same thing. We can do this over and over and over and then tabulate how many times we got a match. Theoretically, we should get a match about 80% of the time. The only problem with this approach is that the people you stop on the street may get annoyed, and you may find yourself in a tiff with someone. A less intrusive way of simulating the birthday problem would be to use a manipulative that could represent random birthdays. For example, we could use two spinners; one broken into 12 equal sectors, each sector representing a different month and another spinner divided into 31 equal sectors numbered 1, 2, 3, and so on for the days of the month. Then spin the first spinner. It will land on a month. Now spin the second spinner. It will land on a date. Put the results of the two spins together to get a birthday. Thus, if the first spinner landed on December and the second on 5, that represents December 5. Impossible Days like February 31 we ignore. Now, we do this 35 times to generate 35 legitimate birthdays and then see if there is a match. We do this over and over with sets of 35 dates and then tally how many times we got a match. You very likely will get matches 80% of the time.

Another way to simulate this birthday problem is to use what is known as a random number generator. These are usually built into calculators and you can instruct the calculator to generate ordered pairs of integers (x, y) where x is between 1 and 12 and y is between 1 and 31. The x represents the month and the y the date. Again, we ignore impossible dates like February 31. The manual that came with your calculator will usually detail how to do this. When you are done, you tally your results. Try it! See if you get a match more often than not.

Student Learning Opportunities

1 Suppose you have 4 strangers in a room. What is the probability that they all have different zodiac signs? There are 12 zodiac signs.

2 (C) Your student Kaylee would like to know how many people she must pick at random before she would have at least a 50% chance of finding someone who has her birthday. What is your guess? Now analyze the problem as follows: The probability that someone misses your birthday is $\frac{364}{365}$. The probability that each person you pick misses your birthday remains the same from person to person. Thus, if you pick n people, the probability of no one matching your birthday is $\left(\frac{364}{365}\right)^n$. Thus, the probability that at least one person matched your birthday is $1 - \left(\frac{364}{365}\right)^n$. Take it from there. Find the smallest n that makes this probability greater than or equal to 50%. Does your answer surprise you? Why or why not?

3 (C) You ask your students to simulate the following problem: "Allison has been practicing her shots in basketball and now, when she stands at the foul line, she usually gets the ball in the basket 75% of the time. Her boyfriend, Greg tells her that if she makes exactly 4 out of her next 7 shots he will take her to dinner. What is the probability she will make exactly 4 out of the next 7 shots?" Your students design a spinner that is divided into four equal sections, of which three of these sections are shaded in. If the spinner lands in a shaded section, they count it as a hit. If it lands outside the shaded section, they count it as a miss. They then begin to spin the spinner and record their results. In this simulation:

(a) What will constitute a trial?
(b) What will constitute a successful trial?
(c) Approximately how many trials should they conduct?
(d) Based on these trials, how can they figure out the probability that Allison will make exactly 4 out of the 7 shots?
(e) What is another instrument they could have used to simulate this experiment? Explain exactly how it would be set up.

(f) Use the Bernoulli formula to calculate the probability that Allison will make exactly 4 out of the next 7 shots, given her shooting percentage of 75%.

(g) Is it likely that Greg will have to take her to dinner?

4 **(C)** You are having a school carnival and you create a game which involves two spinners that are the same. Each has one half colored yellow and the other half colored green. The player will win a prize only when both spinners land on green after each spinner has been spun once. Toula thinks there is a 50–50 chance of winning. Devise a simulation not using spinners that she could use to convince her that the chance of winning is not 1/2. What is the correct probability of this happening? If you run this simulation yourself, say 60 times, what estimate do you get for approximate probability of winning?

5 Your student Neil wants to estimate the probability that, in a family of two children, at least one is a girl. At first he thinks he should flip two coins simultaneously and if a head comes up, it means a girl is born. Then after some thought he changes his mind and thinks he should flip two separate coins and, if a head comes up, it means a girl is born. Comment on the two approaches to simulation. Are they both correct?

6 **(C)** Describe how you might simulate the Monty Hall problem in the classroom. (To actually do a simulation online, go to http://math.ucsd.edu/~anistat/chi-an/MonteHallParadox.html.)

7 **(C)** You give your students the very classic "urn" problem: "An urn contains two red balls and two green balls. Two balls are drawn out without replacing the first ball."

(a) What is the probability that the second ball is red, given that the first ball was red?

(b) One can show that the probability that the first ball was red, given that the second ball is red, is 1/3. Most students don't believe this. In fact, they say that it is impossible to compute such a probability, since the outcome on the first ball cannot depend on the outcome on the second ball, since the second ball has not yet been picked. Devise a way to simulate this process and then run your simulation to see if this probability asked for here is close to 1/3.

(c) When a student claims that it is impossible to compute the probability in (b) since the outcome on the first ball cannot depend on the outcome on the second ball, what error is the student making in his thinking?

8 Jake has been pretty lucky. He hasn't studied much and has gotten good grades. He now has a new teacher with a reputation for giving quizzes with 10 true–false questions. Jake wants to figure out what his chances are of getting at least 7 of these questions right on a quiz by simply guessing, so that he can continue his reputation as a slouch. How can he simulate the results of a quiz using a coin? What would be the probability of this happening had we computed it using our knowledge of probability, assuming that the probability of guessing true and false is the same, namely 1/2?

9 **(C)** You did the following activity with your students: You showed them three cards which you placed in a bag. One card had both sides red, one card had both sides black, and the third card had a red side and a black side. You pulled a card out, and showed the class one of the sides, which was black. You asked them what the probability was that the other side was also black. Your students met in groups and all agreed that the probability that the second side was black was 1/2. They claimed that there are two possible cards to consider after the first one is shown to have a black side, the BR card and the BB card. Are your students correct? Devise a simulation to see if the students are right and then run it. Do you still feel the same way?

10. Millie is a card shark. She is working on a new scheme and needs to know the probability that, of 4 cards picked from an ordinary deck of cards, 2 of them are kings. Help her devise a simulation to do that. Also, tell her what probability she should expect to get.

11. (C) Your very astute student claims that simulating the Monty Hall problem means nothing. The contestant is given one chance and only one chance to win. Simulation takes the frequency approach to probability, and since the player cannot play over and over, whatever results you get don't apply. How do you answer her?

12.9 Fair and Unfair Games

LAUNCH

Today we are going to play a game in which there will be two players, Player Brilliant and Player Genius. Here are the rules of the game. Two fair die will be rolled. Each time the sum of the numbers on the face of the die is either 5, 6, 7, 8, or 9, Player Brilliant gets one point. Each time the sum of the numbers on the face of the die is either 2, 3, 4, 10, 11, or 12 Player Genius gets one point. The first player who gets 20 points wins the game. Which player would you like to be? Is this a fair game? Why or why not?

We hope that you had the chance to actually play this game with your class, so that you could enjoy the surprise. At first glance, you might have thought that, since Player Genius got a point when one of 6 sums was rolled and Player Brilliant only got a point when one of 5 sums was rolled, that Player Genius was clearly at an advantage. We hope that, after playing the game and thinking it through, you realized that there was far more involved here than just the number of possible outcomes for each player. The number of ways to get each of these outcomes accounted for the reason that, in fact, Player Brilliant had an advantage, and indeed this was not a fair game. In this section, we will discuss more about fair and unfair games and examine what happens when money is involved in the playing of the game.

12.9.1 Games Where No Money is Involved

One of the applications of probability that interests secondary school students the most is examining the fairness of games. When we play a game, we have a sense of whether or not it is fair. If we can't win, or win rarely, our tendency would be to say that the game is not fair. If the chances of winning and losing were the same, we would probably say the game is **fair**. We begin by examining games where the rewards are simply the satisfaction of winning.

In what follows, we give several games and discuss whether or not they are fair.

Example 12.26 *In the first game we put 3 red marbles and 1 blue marble in one bag and 2 red and 2 blue marbles in the second bag. Now we pick 1 marble from each bag and if they match, we win; if they don't, we lose. Is this a fair game in the sense that we have the same chance to win as to lose?*

Before we proceed with the solution, why don't you think about it and make your decision?

Solution. We can put all the information in a table. In the left column we list the balls in the first bag and across the top row we list the balls in the second bag. We obtain the following table where "R" stands for red and "B" stands for blue and where an x represents a match.

	R	R	B	B
R	x	x		
R	x	x		
R	x	x		
B			x	x

We can see very clearly that this is a fair game, since 8 of the 16 equally likely outcomes result in a match. Was this what you expected?

A simplistic approach to this problem is to say there are four possibilities here, RR, RB, BR, BB, and two of the four result in a match and the other two don't, so the game is fair. The error in that reasoning is that these outcomes are not equally likely as the above table shows. Change the number of marbles of each type in each bag and there are still four possible outcomes, but some games where we change things will be fair and some won't. You will work out some of these in the Student Learning Opportunities.

Example 12.27 *A fishbowl has 10 red marbles and 10 blue marbles in which they are thoroughly mixed. You play a game with your friend. He picks two marbles with his eyes closed. If they match he wins. If they don't you win. The balls are returned and the game continues. Your friend is happy to play this game since he thinks he will win 2/3 of the time. "There are three outcomes" he says. "Either they are both red, or they are both blue, or they are different colors. I have a 2/3 chance of winning." You are smiling at his naiveté. "There are really four outcomes" you reason. "They are both red, they are both blue or the first is red the second is blue, or the first is blue and the second is red. So in only 1/2 the cases will he win." So you see this as a fair game. Who is right? (You might want to try simulating this by putting ten pieces of paper numbered with the number 1 and ten pieces of paper numbered with the number 2 in a bag and mixing them thoroughly and then picking as described and listing your outcomes.)*

Solution. Neither of you is right. There are 10 balls of each type. The number of pairs of balls both blue is $\binom{10}{2}$ and the same is true for the number of pairs of balls that are both red. There are

$\binom{20}{2}$ pairs of balls that can be picked. Thus, the probability of a match is

$$\frac{\binom{10}{2} + \binom{10}{2}}{\binom{20}{2}} = 0.473\,68.$$

Therefore, since the probability of winning is not 0.5, this is not a fair game. Is this what you expected?

12.9.2 Games Where Money is Involved

When money is involved in game playing, the issues become a bit more complicated. When playing games which involve money, a **fair game** is one in which you can expect to win nothing in the long run. That is, the wins and losses will balance out.

> **Example 12.28** *Suppose that you play the following game: You take a 12 sided die. If when you roll the die you get a number from 1 to 6, you win a dollar. If you get a number from 7 to 10, you lose 2 dollars, and if you get 11 or 12, you win 3 dollars. If you play this game a large number of times, what, approximately, will be your average gain per game?*

Solution. In this game any outcome from 1 to 6 inclusive results in a win of 1 dollar. Thus the probability of winning a dollar is $\frac{6}{12}$ or $\frac{1}{2}$. Any outcome from 7 to 10 inclusive results in a loss of 2 dollars. Thus the probability that you will lose 2 dollars is $\frac{4}{12}$ or $\frac{1}{3}$. The outcomes of 11 and 12 result in a gain of 3 dollars, and the probability of getting these outcomes is $\frac{2}{12}$ or $\frac{1}{6}$.

Let us assume that we play a large number of games, say N games. Then, using the frequency approach to probability, we should win a dollar about $\frac{1}{2}N$ times, putting $1 \cdot \frac{1}{2}N$ dollars in our pocket. We will lose two dollars about $\frac{1}{3}N$ times yielding a gain of $-2 \cdot \frac{1}{3}N$ dollars (a negative gain means a loss). We will win three dollars about $\frac{1}{6}N$ times, yielding a gain of $3 \cdot \frac{1}{6}N$ dollars Our net gain is given by

$$1 \cdot \frac{1}{2}N - 2 \cdot \frac{1}{3}N + 3 \cdot \frac{1}{6}N.$$

To find our average gain per game, we divide this by N, since we played N games to get

$$1 \cdot \frac{1}{2} - 2 \cdot \frac{1}{3} + 3 \cdot \frac{1}{6} \approx \$0.33/\text{game}. \tag{12.17}$$

The average gain per game is given a name. It is called our **expected gain**. Expression (12.17) tells us how to compute it. We *multiply each payoff by the probability of the payoff*, and that is our expected gain. Here are some examples to illustrate this.

> **Example 12.29** *You toss a pair of dice. It costs you 5 dollars to play. If the sum of the die rolled is from 2 to 6, you will keep your 5 dollars and win 8 dollars. Otherwise, you will lose your 5 dollars. Is this a fair game? If it is not fair, what must you pay when you bet to make it a fair game?*

Solution. The probability of getting a sum of 2 to 6 inclusive is $\frac{15}{36}$, while the probability of getting a sum from 7 to 12 is $\frac{21}{36}$. Your expected gain for the game is $8(\frac{15}{36}) - 5(\frac{21}{36}) = 0.41\overline{6}$ or 41 and 2/3 cents per game. Since your expected gain is not zero, this is not a fair game, though it is a good game for you to play, as on the average you will win about 41 cents per game. But you must remember this is probability. Our conclusions are for long stretches of play. It is very possible that you will lose several times in a row and use up all your money and then you won't be able to play the game long term. In short, you may go broke before you get a chance to make any money. Indeed, there is something known as Gambler's Ruin which says there is a very high probability that, if you play long enough, you *will* go broke since there is always a chance, though small, of a long stretch of losses in a row. Of course, how long is "long enough" for you to play before that happens? One never knows. That is why it is called gambling.

We still need to answer the question of what you must pay to make it fair. There are two answers to this. One can say that you can pay $41\frac{2}{3}$ cents, on the average, to play this game to make it fair. However, you might also argue that, since you cannot pay that amount per game, and since it is impossible to predict how many games you will play before you are playing "long run," you will never be able to pay this amount on average, and hence never make this fair. We leave that debate to you. This a good question to pose to your secondary school students since it is open ended and will promote interesting debate.

12.9.3 The General Notion of Expectation

We spoke about expected gain in the context of games, but the word expectation is ubiquitous in probability. For example, we can talk about the expected number of people who will survive an illness, the expected number of games played before a team wins, the expected life of a light bulb, and so forth. All of these events, number of people that survive an illness, number of games a team plays before they win, number of hours a light bulb runs before it fails, are unpredictable and hence are called **random variables**. The definition of expected value of any random variable is always defined to be the sum of the values that the random variable can take on times the probability of it taking on that value when the sample space is finite or countable. When the sample space is not finite and not countable, it is defined to be a certain improper integral. To keep things on the secondary school level, we only deal with finite and countable sample spaces. Here is a typical example.

> **Example 12.30** *If we toss a die, what is the expected value of the roll?*

Solution. The outcomes are 1, 2, 3, 4, 5, 6 and all occur with probability 1/6. Thus, the expected value of the roll is

$$1\left(\frac{1}{6}\right) + 2\left(\frac{1}{6}\right) + \ldots + 6\left(\frac{1}{6}\right) = 3.5.$$

The word "expected" here is problematical for a lot of people. The expected roll is simply an (approximate) average of the numbers coming up on the various rolls, assuming you play many times. You may find it strange that the expected value is 3.5, when it can never occur on any particular roll and therefore certainly cannot be expected to occur. Nevertheless, the word "expected" is used.

Here are some other typical examples:

Example 12.31 *Your town has a weekly lottery. Each week there is a grand prize of* 10, 000 *dollars. You must pay one dollar per ticket. To win the contest, you must pick* 3 *numbers from the numbers* 1 *to* 40. *Then, three numbers from* 1–40 *are drawn at random. If your numbers match, you win. (a) What is your expected gain for this game assuming you bet a dollar per week? (b) Do you think you should enter this lottery? Why or why not?*

Solution. Since you are betting a dollar each time, if you win, you net $10,000 - 1 = 9999$ dollars. If you lose, you lose 1 dollar. The probability of winning is $\frac{1}{\binom{40}{3}}$ since there are $\binom{40}{3}$ ways of picking three winning numbers and only one of them is the real winner, while the probability of losing is $1 - \frac{1}{\binom{40}{3}}$. Our expected gain is

$$9999 \cdot \left(\frac{1}{\binom{40}{3}}\right) + (-1)\left(1 - \frac{1}{\binom{40}{3}}\right) \approx 0.01.$$

So, if you play this lottery every week, in the long run, you can look forward to your average winnings being about 1 cent per game. So, in answer to (b), unless you get a thrill out of playing, think twice about playing this lottery.

Although expected value is a long-run phenomena, many people use it to make decisions that are one- time events or only short-run events. Consider the following business situation where subjective probabilities and expected value come together, but only for a one-time decision.

Example 12.32 *You are a toy consultant and are bidding on a consulting contract with firms Tillie Toys and Silly Toys. Tillie Toys requires a lot of personal information about your history as a consultant before it will allow you to enter the bidding process. To get this done, you will have to pay your lawyer about* 300 *dollars to prepare the documents. However, your profit from getting the contract will be* 5000 *dollars. Silly Toys' requirements for information are much less. To prepare the documents for Silly Toys' consideration of your bid, it will only cost* 100 *dollars. However, if you get that contract, you will only get a profit of* 3500 *dollars. Your gut tells you that you have a* 25% *chance of winning the contract from Tillie Toys and a* 30% *chance of winning the contract from Silly Toys. You decide to use expected value to make your decision. Both Tillie Toys and Silly Toys are subsidiaries of the same company, Great Toys, and the rules for Great Toys require that you can only bid on one contract. Which contract should you bid on?*

Solution. If you bid on the Tillie Toys (T) contract, your expected gain is

$E_T = 5000(0.25) + (-300)(0.75) = 1025$

If you bid on Silly Toys (S) your expected gain will be

$$E_S = 3500(0.30) + (-100)(0.70) = 980.$$

Although the 45 dollar difference in expected gain is not much, you are a business person, and since your goal is to make more money, you decide to bid on Tillie Toys.

12.9.4 The Cereal Box Problem

Toasty Flakes is a cereal company, which is putting prizes in their boxes in an attempt to lure customers into buying their products. They put one of six different prizes in each of their boxes and your cute little niece really wants those prizes. So, of course, you are going to try to make your niece happy and buy some boxes of that cereal. A natural question is to ask how many boxes you will have to buy before your niece gets all six prizes. What would you guess is the answer? The result might surprise you. This is known as the "cereal box problem" whose solution we will now investigate.

To get a better feel for the problem and an inkling of the solution, let us first think of an appropriate simulation. What device can you use to simulate picking six different equally likely prizes? You've got it! A fair six-sided die or a spinner with six equal sections. Using a fair die, toss it as many times as it takes so that each number turns up at least once. This represents 1 trial. Perform many trials and average your results to get an estimate of the solution: the average number of boxes you would have to buy to get all six prizes. Was your answer close to your original guess? Having done this simulation, you will be able to appreciate better the theoretical analysis that follows.

We know that the probability of getting a "1" when tossing a die is 1/6. You ask "Suppose we toss the die over and over. What is the expected *number of tosses* one must perform to get a success, where success means getting a 1?" Our intuition tells us that, if the probability of getting a "1" is 1/6, then it should take on *average* about 6 throws to get one success. Showing it, however, requires more work. The possibilities are, you get a success on the first toss, or you fail at the first toss, and get a success on the second toss, or you fail on the first two tosses and get a success on the third toss, and so on. These probabilities are summarized in the table below, where we use the fact that the probability of success on any roll of the die is independent of the probability of success on any other roll of the die. Thus, to compute the probability of say, getting 2 failures followed by a success, we multiply the probabilities of failure on each of the first two rolls, by the probability of success on the third roll which gives us (5/6)(5/6)(1/6) or $(5/6)^2(1/6)$.

Number of tosses till you get a 1	Probability that this happens
1	1/6
2	(5/6)(1/6)
3	$(5/6)^2(1/6)$
4	$(5/6)^3(1/6)$
5	$(5/6)^4(1/6)$
etc.	etc.

Now using the fact that the expected number, E, of tosses till we get a success is the sum of the (number of tosses till success)× (the probability of that happening), we have, using the table above, that E is

$$E = 1 \cdot 1/6 + 2 \cdot (5/6)(1/6) + 3 \cdot (5/6)^2(1/6) + 4 \cdot (5/6)^3(1/6) + \ldots. \tag{12.18}$$

Now to find this sum, we multiply the above equation (12.18) by 5/6 to get

$$(5/6)E = 1 \cdot (5/6)(1/6) + 2 \cdot (5/6)^2(1/6) + 3 \cdot (5/6)^3(1/6) + 4 \cdot (5/6)^4(1/6) + \ldots. \tag{12.19}$$

Subtracting equation (12.19) from equation (12.18) and subtracting like terms from like terms we get that

$$(1/6)E = 1 \cdot 1/6 + 1 \cdot (5/6)(1/6) + 1 \cdot (5/6)^2(1/6) + 1 \cdot (5/6)^3(1/6) + \ldots. \tag{12.20}$$

which is a geometric series with $|r| < 1$. So it converges to $\frac{1 \cdot 1/6}{1 - 5/6} = 1$. So equation (12.20) becomes

$$(1/6)E = 1$$

or $E = 6$.

The identical proof shows that, if the probability of success is p, then $1/p$ trials is the expected number of trials before we get a success. We just replace $1/6$ by p and $5/6$ by q where $q = 1 - p$ and proceed as above to get

$$pE = 1$$

or

$$E = \frac{1}{p}. \tag{12.21}$$

This brings us back to our original cereal box problem: That is, given that Toasty Flakes has put six different prizes in their boxes and has put one prize per cereal box, what is the expected number of boxes one must buy to get all six prizes?

Solution. The probability of getting some prize in the first box is $p = \frac{6}{6}$ or 1. Thus, the expected number of boxes one must buy to get the first prize is, by equation (12.21), $\frac{1}{p} = \frac{6}{6}$. The probability of getting a second prize different from the first is $p = \frac{5}{6}$. Thus, again by equation (12.21), one would have to buy an additional $\frac{1}{p} = \frac{6}{5}$ boxes, on average, to get a different second prize. The probability of getting a third prize different from the first two is now $p = \frac{4}{6}$, so we would need to buy an additional $\frac{1}{p} = \frac{6}{4}$ boxes to get that prize, and so on. So, the expected number of boxes you will have to buy to get all six prizes is,

$$\frac{6}{6} + \frac{6}{5} + \frac{6}{4} + \frac{6}{3} + \frac{6}{2} + \frac{6}{1} = 14.7.$$

Thus, surprisingly, on the average, you wouldn't have to buy that many boxes to get all six prizes.

You can actually run a simulation of the cereal box problem online, if you find this hard to believe. Visit http://www.mste.uiuc.edu/reese/cereal/cereal.html.

Student Learning Opportunities

1. In Example 12.26, suppose that there were 1 red marble and 3 blue marbles in the first bag and 2 red marbles and 2 blue marbles in the second bag. The rules of the game are the same. Is this a fair game? Explain.

2. You and your friend are playing a game. You toss a pair of dice and subtract the smaller number that comes up from the larger number. (If the numbers are the same, the difference is 0.)
   (a) What is the probability of your computing the above difference and getting a difference of 0?
   (b) What is the probability of getting a difference of 1?
   (c) Now we give the rules of the game: If the difference is 1, 2, or 3, you win. If not, your friend wins. Is this a fair game to you?
   (d) How can you modify the game to make it fair?

3. Suppose that there are $2n$ balls in the jar of Example 12.27, n of which are red and n of which are blue. Show that the probability of getting a match is $\frac{n-2}{2n-2}$, which obviously depends on n. Compute this value for different values of n and tell what this probability approaches as n gets very large. Since the proportion of red and blue are the same in both cases, does it surprise you that the probability of a match for 10 balls, half of which are red and half of which are blue, is different from the probability of this happening when there are 20 balls, half of which are red and half of which are blue?

4. The first of four 6-sided dice has four 4's and two 0's. The second has 3's on all of its sides. The third has four 2's and two 6's, while the fourth has three 1's and three 5's. Play the following game with a friend. Let her choose a die and roll and then you do the same. The person who rolls the higher score wins. Explain why rolling the first die is preferable to rolling the second, that rolling the second is preferable to rolling the third, that rolling the third die is preferable to rolling the fourth, but paradoxically, that rolling the fourth die is preferable to rolling the first. Make a convincing argument that whoever goes first is at a disadvantage. [Hint: Show that no matter what die that person picks, you can always pick a die that gives you a better chance at winning.]

5. On a piece of paper write the letter "X" on both sides. This is an X–X card. On a second piece of paper write the letter "X" on one side and the letter "Y" on the other side. This is an X–Y card. On a third piece of paper, write the letter "Y" on both sides. This is a Y–Y card. Put all three papers in a hat and let your friend pick one with his eyes closed and put that on the table. Now, bet him 10 dollars that the side of the paper on the table, which is facing up, matches the letter on the other side of the paper. You argue that this is a fair game since if the letter "X" is facing up, you know it is not a Y–Y card, so it is either an X–X card or an X–Y card. So you have a 50% chance of winning and a 50% of losing. You use a similar argument if the letter "Y" is facing up. Is this a fair game? Explain.

6. (C) Describe a childhood game you played and explain why it was or was not fair or make up a game using two dice and tell whether or not it is fair. (This is a good question to ask your own students.)

7 A used car dealer is accustomed to getting complaints. Over the past few months he has determined the following data about the number of complaints and the probability of getting that number of complaints per day.

Number of complaints	0	1	2	3	4	5
Probability of complaints	0.01	0.15	0.35	0.40	0.05	0.04

Find the expected number of complaints per day.

8 (C) Your student figures out that the expected number of girls in a family with 3 children is 1.5, and thinks he has made a mistake. He claims, how could one expect to have 1.5 children in a family when that is impossible? Did your student make a mistake. Verify his calculations and explain the meaning of the result.

9 Each week your local supermarket has a contest. There is a 30% chance you will win a 500 dollar prize in the contest if you pay 50 dollars to enter. You can only buy one ticket each week. What is your expected gain?

10 The Munchy And Crunchy company is introducing a new cereal. To get people to buy it, they put one model car in each box. A complete set consists of 8 model cars. How many boxes should you expect to buy if you want all 8 cars? Suppose they put N cars in their boxes. What is the expected number of boxes needed to be bought (in terms of N) before all these cars are obtained?

11 (C) One of your students, Michele comes to you for advice. She claims that her friend Kurt has asked her to play a dice game with him and, since he is so crafty, she is concerned that he might be tricking her into playing a game in which she can expect to lose money. Here is the game Kurt has proposed. Kurt tosses a fair die. If a 1 or 2 comes up he will pay her one dollar. But if a 3, 4, 5, or 6 comes up, Michele will have to pay Kurt 50 cents. How do you advise Michele? Is this a fair game?

12 You roll a pair of dice. If both die turn up 6, you win $5.00. Any other outcome results in a loss of 50 cents.

(a) What is your expected gain for this game?
(b) Will you ever get your expected gain on any play of the game?

13 You give your students the following problem: Your school is having a carnival with a game called the Wheel of Luck, which consists of a spinning wheel and a pointer. The player spins the wheel and whichever section the pointer lands on determines if the player will win or lose that amount of money. You pay $2.00 to play this game. The wheel is cut into 8 equal sections. There are 5 sections where the player can land and lose $2.00, 2 sections where the player can land and win $4.00, and 1 section where the player can land and win $5.00. Julia is asked to decide if she would play the game and why. Julia figures out her expected gain as follows.

$$\frac{5}{8}(-2) + \frac{2}{8}(4) + \frac{1}{8}(5) = \frac{-10}{8} + 1 + \frac{5}{8} = \frac{-2}{8} + \frac{5}{8} = \frac{3}{8} = 0.375.$$

She then says, "This game is not fair since the expected value is not zero. However, I would like to play this game since, on average, each time I play this game I can expect to win about

37.5 cents." Comment on her response and determine if she computed her expected gain correctly. If she hasn't, compute the correct expected gain.

14 Suppose the game in Example 12.27 has the following payoffs: If your friend wins, he gets 10 dollars. Otherwise, he loses the 10 dollars. What is his expected gain for that game?

15 In Example 12.26, we add some excitement to the game. You play, and if you get a match you win 2 dollars, otherwise you lose 3 dollars. What is your expected gain? Is the game fair? Explain.

16 (C) Your students have been made aware that gambling casinos don't charge a fair price for playing their games. They ask you why people gamble when the money it costs to play the game is always more than the expected value of the play? How do you respond?

12.10 Geometric Probability

LAUNCH

Jackie went to the carnival at her school last weekend and played an unusual dart game. There was a circular target that had a radius of 24 in. The inner concentric circle of the target had a radius of 12 in. Every time she threw the dart, it landed on a random point on the target. Do you think that it was likely that it landed on the circular region between the two circles? What was the probability that it landed on the circular region between the two circles?

In doing this launch problem, you probably noticed that it was impossible to use the classical approach and count the number of outcomes. In cases such as this, we can sometimes use another clever method called geometric probability, which has some interesting applications and mind-boggling consequences. After reading this section, you will have a clear idea of how to solve the launch question. When the occurrence of an event E can be described as part A, of a geometric figure B, we have another way of calculating the probability of the event occurring, which can often be simpler than other methods.

$$P(E) = \frac{\text{Size}(A)}{\text{Size}(B)}$$

where the word "size" means either length, area, or volume depending on the dimension (1, 2, or 3) that describes the region we are talking about. Let us illustrate with some examples,

Example 12.33 *Suppose we pick a point, (x,y), at random from the square $-2 \leq x \leq 2$, $-2 \leq y \leq 2$. Find the probability that $x^2 + y^2 \leq 4$.*

Solution. Below in Figure 12.11, we draw the region $-2 \leq x \leq 2$, $-2 \leq y \leq 2$ as well as the region $x^2 + y^2 \leq 4$.

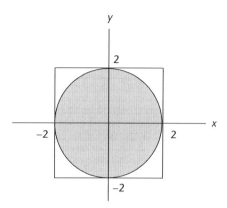

Figure 12.11

The reader needs to recall that the latter region is a circle with the center at the origin and radius 2. If we call the event that $x^2 + y^2 \leq 4$, E, then

$$P(E) = \frac{\text{area of the circle}}{\text{area of the square}} = \frac{4\pi}{16} = \frac{\pi}{4} \approx 0.785.$$

So it is likely that, if we pick a point at random in the given region, it will likely fall in the circle.

Example 12.34 *I arrive to meet my friend for lunch at some time between 12 noon and 1 pm, and he arrives at some time between 12 noon and 1 pm. Our agreement is that each of us will wait 15 minutes for the other and, if the other does not arrive within that time frame, then we will leave and the lunch date is off. What is the probability that we will have lunch together?*

Solution. Let the number of minutes that I arrive after 12 be x and the number of minutes after 12 that my friend arrives be y. Then $0 \leq x \leq 60$ minutes and similarly for y. The (nonnegative) difference in time between my arrival time x, and my friend's arrival time y is given by $|x - y|$. We will meet if $|x - y| \leq 15$. In secondary school we learn that this inequality is equivalent to

$$-15 \leq x - y \leq 15.$$

Solving for y, we have that this is the same as the two inequalities

$$y \geq x - 15 \quad \text{and} \quad y \leq x + 15 \tag{12.22}$$

and these two inequalities represent the event whose probability we wish to find. We graph our arrival times and the inequalities of (12.22) on the same set of axes and we get the picture shown in Figure 12.12 below.

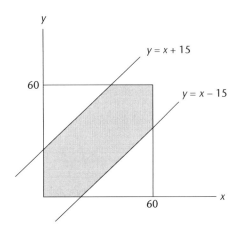

Figure 12.12

The shaded region represents the region where we will meet. If we call the event of our times, (x, y) landing in this region, E, then

$$P(E) = \frac{\text{Area of the shaded region}}{\text{Area of the square}} = \frac{1575}{3600} \approx 0.4375.$$

In the Student Learning Opportunities you will show how we arrived at 1575 for the numerator of the above fraction.

What does our answer mean? It means that there is less than a 50% chance that my friend and I will have lunch together.

12.10.1 Some Surprising Consequences

We mentioned earlier in the chapter a misconception concerning an event that has probability zero. We said that such an event can, in fact, occur. The following example shows why.

Example 12.35 *Suppose that a number is chosen at random from the interval [0.1]. (a) What is the probability that the number we pick is $\frac{1}{2}$? (b) What is the probability of picking a rational number? (c) What is the probability of picking an irrational number?*

Solution. (a) The event where we get $\frac{1}{2}$ is $C = \{\frac{1}{2}\}$. Now let A_1 be the event where the number picked is in the tiny interval $[\frac{1}{2}, \frac{1}{2} + \frac{1}{1\,\text{million}}]$. Now C is a subset of A_1, so by Corollary 12.6, $P(C) \leq P(A_1) = \frac{1}{1\,\text{million}}$. C is also a subset of $A_2 = [\frac{1}{2}, \frac{1}{2} + \frac{1}{1\,\text{billion}}]$. So $P(C) \leq P(A_2) = \frac{1}{1\,\text{billion}}$. Continuing in this way, we can make $P(C) \leq \frac{1}{1\,\text{trillion}}$, $P(C) \leq \frac{1}{1\,\text{quintillion}}$, and so on. The point is, $P(C)$ can be made smaller and smaller than a sequence of positive numbers that go to 0. The only nonnegative number that has that property is 0. That is, $P(C) = 0$.

It surprises people that the probability of picking the number $\frac{1}{2}$ is zero. But surely it is possible to pick the number $\frac{1}{2}$! Thus, just because the probability of a number is 0, that does not mean the event cannot occur. It can. In fact, we can show, and you will in the Student Learning Opportunities, that the probability of choosing any fixed rational number is 0.

Solution. (b) Let us call the event of picking a rational number in [0, 1], R. Then the event R is the set of rational numbers. From Chapter 6, we can enumerate R, since it is countable. That is, we can write $R = \{r_1, r_2, r_3, \ldots\}$. But by the countable additivity property (3)', of probability, $P(R) = P(r_1) + P(r_2) + P(r_3) \ldots = 0 + 0 + \ldots = 0$. Thus, the probability of choosing *any* rational number is 0. This is mind boggling and certainly counterintuitive.

Solution. (c) Call the event of picking an irrational number in [0, 1], I. Then $R \cup I = [0, 1]$. Thus,

$$P(R \cup I) = P([0, 1]) = 1. \tag{12.23}$$

By axiom (2) of the probability axioms $P(R \cup I) = P(R) + P(I)$, since the events R and I are mutually exclusive. (You can't pick a rational number and an irrational number at the same time.) So equation (12.23) becomes

$$P(R) + P(I) = 1$$
$$0 + P(I) = 1$$

or just, $P(I) = 1$. Thus, the probability of picking an irrational number is 1, yet, we don't have to pick an irrational number if we choose a number from [0, 1] at random. Therefore, here we have an event with probability 1 (picking an irrational number), which is *not* certain to occur.

This example illustrates that the probability of an event being 0 does not mean it is impossible, and the probability of it being 1 does not mean it is certain. However, for finite sample spaces, where all the events are equally likely, a probability of zero does mean the event cannot happen and a probability of 1 does mean that the event is certain.

We end with one final example that ties together first year calculus and some of the concepts studied in secondary school.

> **Example 12.36** *Find the probability that the roots of the quadratic equation*
>
> $$x^2 + 2bx + c = 0$$
>
> *are real if b and c come from the interval* $-4 \leq b \leq 4$ *and* $-4 \leq c \leq 4$.

Solution. We learn that roots are complex if the discriminant $(2b)^2 - 4c < 0$, which implies that $c > b^2$. If we replace the x-axis by the b-axis and the y-axis by the c-axis, our region for b and c is a square having side 8. We draw the parabola $c = b^2$ (which is like drawing $y = x^2$ only y is c and x is b). See Figure 12.13.

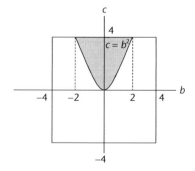

Figure 12.13

We call the shaded region $c > b^2$, R. The probability that the roots are complex is the probability that $c > b^2$ and this is:

$$\frac{\text{Area of } R}{\text{Area of the square}}. \tag{12.24}$$

To find the area of R, we use calculus. We find the intersection of the parabola with the line $c = 4$, which is at the points where $b = -2$ and $b = 2$. Recalling that the area between two curves is the integral of the "higher minus the lower curve," we get that our probability of the roots being complex, from equation (9.5), is

$$\frac{\int_{-2}^{2}(4 - b^2)db}{64} = \frac{\frac{32}{3}}{64} = \frac{1}{6}.$$

Thus, the probability of getting real roots is $1 - \frac{1}{6}$ or $\frac{5}{6}$, which is pretty likely.

12.10.2 Monte Carlo Revisited

It is often said that "necessity is the mother of invention." This is certainly true for many mathematical techniques, one being the Monte Carlo Method. Specifically, have you ever played the game of solitaire and wondered what the probability of winning was? The well-respected and well-known mathematician, Stanley Ulam (1909–1984) wondered about this problem. He was trying to determine the fraction of all games of the game of solitaire that could be completed satisfactorily to the last card. He thought that, if he had a computer play many games and studied the proportion of times the computer could successfully complete the game, he would have a sense of the answer to his problem. This is how the Monte Carlo Method was born. Ulam used this method in his research studies while he was working with other famous mathematicians at Los Alamos during World War 2.

In Chapter 4 we used the Monte Carlo Method to estimate π. (You might wish to review that before reading further.) The Monte Carlo Method is based on the frequency approach to probability. The idea is that, if we know that p is the probability of an event E occurring in some experiment, we can try to repeat the experiment over and over, and by taking the ratio of the number of successes to the total number of trials, we can estimate the probability, p of the event E. While this section could have gone in Chapter 4 with the other discussion of Monte Carlo, in Chapter 4 our goal was to study the circle and π. Now we go beyond that.

We begin with a function $f(x) \geq 0$ on $[a, b]$, which is the graph of a function that is above the x-axis and can possibly touch it. Then we know from calculus that $\int_a^b f(x)dx$ gives the area under the graph of $f(x)$ and above the x-axis. We also know how to evaluate such an integral when $f(x)$ is continuous on $[a, b]$. We compute $F(b) - F(a)$, where $F(x)$ is any antiderivative of $f(x)$ on $[a, b]$. However, suppose that an integral occurs in a practical application, and suppose that finding an antiderivative for $f(x)$ is difficult or even impossible, which happens quite often. For example, what if we want to compute a complicated double or triple or 12-fold integral, the kind that often occurs in physical applications? How can we do it if antiderivatives with respect to the variables in the problem can't be found? The answer is, we can use the Monte Carlo method, which is a clever way of dodging difficult mathematics and arriving at difficult results easily. Let's illustrate this by working a difficult integral with one variable.

Suppose we wish to compute $\int_0^1 e^{-x^2} dx$. This is not an easy integral to do without the use of power series. If you were given this problem in your first course in calculus, you would not be able to do it, as there is no closed form antiderivative for e^{-x^2}. But, watch how we can solve it with the Monte Carlo method. We enclose the curve in rectangle, R, whose base is [0, 1] (the interval over which we are integrating) and whose height is such that the area under the curve in the interval of integration (in this case [0, 1]) is contained in the rectangle. In this case, a height of 1 will work. (See Figure 12.14 below.)

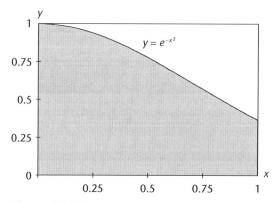

Figure 12.14

Now using, say, a random number generator, we generate ordered pairs, (x, y) of numbers where x and y are between 0 and 1. These ordered pairs, (x, y), are in the square. To estimate the probability of the point being in the region under the curve, we generate many points and find the ratio of the number of points under the curve to the number of points generated. That is,

$$P(\text{being under the curve}) \approx \frac{\text{number of points generated under the curve}}{\text{total number of points generated}}. \qquad (12.25)$$

However, this probability is

$$P(\text{being under the curve}) = \frac{\text{area under the curve}}{\text{area of the square}} = \frac{\int_0^1 e^{-x^2} dx}{1} = \int_0^1 e^{-x^2} dx. \qquad (12.26)$$

From equations (12.25) and (12.26) we get that

$$\frac{\text{number of points generated under the curve}}{\text{total number of points generated}} \approx \int_0^1 e^{-x^2} dx. \qquad (12.27)$$

Look at approximation (12.27). It is telling us that this difficult problem is easy to solve! We just take a ratio! We actually did this on the TI83 using the program given below:

1:	$0 \rightarrow T : 0 \rightarrow D$	(Initialize the values of the total number of points generated, and the number which lie under the curve.)
2:	FOR $(I, 1, 1000)$	(We are about to generate 1000 sets of random numbers, (x, y).)
3:	rand $\rightarrow x$	(Generate 1 random number for x between 0 and 1.)
4:	rand $\rightarrow y$	(Generate 1 random number for y between 0 and 1.)
5:	$T + 1 \rightarrow T$	(Each time we generate a new ordered pair, we increase the count by of T by 1.)
6:	If $y \leq e^{-x^2}$: $D + 1 \rightarrow D$	(If the point lies under or on the curve, increase the count of D by 1.)

7: END	(This signals the end of the generation of our pairs of numbers.)
8: "Our estimate for the integral is"	(We are telling the machine to write the words on the screen "The value of the integral is".
9: Display $\dfrac{D}{T}$	The machine displays our estimate of the integral.)

$$\int_0^1 e^{-x^2} dx \approx 0.763.$$

When we asked the program with which this book was written to evaluate the integral using its own techniques we got

$$\int_0^1 e^{-x^2} dx \approx 0.74682.$$

We didn't do badly at all with Monte Carlo. Of course, the more points we generate using the Monte Carlo method, the better we expect our answer to be.

Student Learning Opportunities

In all of the following problems, show and explain your work and whenever possible, discuss the reasonableness of your results.

1 A dart board is circular with radius 12 inches. The "bull's-eye" is at the center and consists of a circle of radius 1 inch, whose center is the center of the dart board. What is the probability that a dart, which hits the dartboard, hits the bull's-eye?

2 The telephone company is installing a telephone line that is 60 meters long and is suspended between two poles, one of which contains a transformer. The company is afraid that, in a storm, a break will occur at a random point in the line and they don't want that to happen close to the transformer. What is the probability that the break will be at a distance no less than 15 meters from the transformer?

3 A piece of spaghetti 10 inches long is dropped and breaks into two pieces. What is the probability that one of the pieces is longer than 8 inches?

4 A circular disk with radius 1 foot is placed so that its center is somewhere on a square table 6 feet by 6 feet. What is the probability that the circle lies totally on the table?

5 (C) One of your students, Jason, tells you that he is planning on going to his local county fair on the weekend, and he knows from past experience that there will be coin tossing game there. It always looks like fun to him, but he has a feeling that there is a low probability of winning anything and he is therefore, reluctant to play. (You win if you toss a coin and it lands entirely within one square.) He has looked into the game and this is what he knows: You toss the coin onto a large table ruled into congruent squares that have sides of 5 centimeters. The coin's diameter is 2 centimeters. (Assume that the markings on the table have no thickness.) What is the probability of Jason winning? Was he right about there being a low probability of winning?

6 Suppose that 2 numbers x and y are chosen so that both are between 0 and 1 inclusive.
   (a) Generate 20 such pairs of numbers using the random number generator capability of your calculator.
   (b) Find the ratio of the number of pairs of points, which satisfy $x + y \leq \frac{1}{2}$ to the total number of points generated.
   (c) What is the probability computed geometrically that $x + y \leq \frac{1}{2}$?
   (d) Is your answer from part (b) close to your answer to part (c)? Explain.

7 A point is randomly chosen on the line segment joining (0, 0) to (10, 20).
   (a) Generate 20 such pairs of numbers using the random number generator capability of your calculator.
   (b) Find the ratio of the number of pairs of points, which satisfy $y \geq 8$ to the total number of points generated.
   (c) What is the probability computed geometrically that $y \geq 8$?
   (d) Is your answer from part (b) close to your answer to part (c)? Explain.

8 Suppose that a and b are two numbers chosen at random and that $-4 \leq a \leq 1$ and that $-2 \leq b \leq 4$.
   (a) Generate 20 such pairs of numbers using the random number generator capability of your calculator.
   (b) Find the ratio of the number of pairs of points, which satisfy $ab > 0$ to the total number of points generated.
   (c) What is the probability computed geometrically that $ab > 0$?
   (d) Is your answer from part (b) close to your answer to part (c)? Explain.

9 Given right triangle ABC with right angle at C. P is chosen inside the triangle. What is the probability that triangle PBC has area less than, or equal to, 1/2 the area of triangle ABC?

10 (C) Your students are really bewildered by the fact that you can have an event be possible and yet have zero probability of happening. They insist that if you pick any rational number r in the interval from [0, 1] that you can figure out a non-zero probability of its occurring. They want you to prove to them that the probability of picking r is really 0. How do you do it?

11 Show that, if we pick a point at random inside the square $-5 \leq x \leq 5$ and $-5 \leq y \leq 5$, then the probability that it lies on the portion of the line $y = 3$, which lies inside the square, is 0. Show that the probability of choosing the point along any horizontal line segment in the square is 0. This is yet another example of an event whose probability is 0 but which can happen.

12 When we proved that the probability of picking a number at random in [0, 1] and getting a rational number was zero, we enumerated the rational numbers, r_1, r_2, and so on, and then reasoned that P(picking a rational) $= P(r_1) + P(r_2) + \ldots = 0 + 0 + \ldots = 0$. Can't we do a similar thing with all the real numbers in the interval [0, 1]? (That is, call the real numbers real number 1, real number 2, and so on and then P (picking a number in [0, 1] and getting a real number) $= P$ (picking real number 1) $+ P$ (picking real number 2)$\ldots = 0 + 0 + \ldots$ and thereby conclude the rather strange result that, if we pick a number at random in [0, 1], the chances that it is a real number is 0.

13 Redo Example 12.36 for $-100 \leq b \leq 100$ and $-100 \leq c \leq 100$, then for $-10^6 \leq b \leq 10^6$ and $-10^6 \leq c \leq 10^6$. Show that the larger the square gets, the closer the probability is to 1 that the roots are real if b and c are in the square. Conclude that, if b, c are any real numbers, the probability of getting real roots to the quadratic equation is 1. Here is yet another example where the probability of an event is 1, but the event is not certain to happen.

14 George has a girlfriend in the north part of town and another in the south part of town. The buses to each part of town stop at the same bus stop and both stop every 10 minutes. George arrives at the bus stop at a random time each day and takes whichever bus comes first. Here is the bus schedule:

Northern Bus	Southern Bus
12:00	12:01
12:10	12:11
12:20	12:21
etc.	etc.

The girl in the north really likes the way things are going because she sees George 90% of the time that he says he might come. The girl in the south is really unhappy, because she hardly sees George. Can you explain why? Does this have anything to do with geometric probability?

15 Use the Monte Carlo technique to evaluate each of the following integrals and then compare the values to the value the calculator gives.

(a) $\int_0^1 e^{-x^3} dx$

(b) $\int_0^{\sqrt{\pi}} \sin(x^2) dx$

(c) $\int_1^2 \frac{x}{x^4+1} dx$

12.11 Data Analysis

Below is a chart documenting the number of traffic fatalities per 100 million vehicle miles from the year 2003. From the data answer the following questions:

1 Which states have the least number of traffic fatalities? the greatest number of traffic fatalities?
2 Where does the state you live in rank in terms of number of traffic fatalities throughout the United States?
3 In the United States, what is the approximate number of traffic fatalities per 100 million vehicle miles?

Traffic fatalities per 100 Million vehicle miles by states (2003)

AL	1.71	HI	1.45	MA	0.86	NM	1.92	SD	2.38
AK	1.92	IO	2.05	MI	1.27	NY	1.10	TN	1.73
AZ	2.08	IL	1.36	MN	1.19	NC	1.63	TX	1.64
AR	2.05	IN	1.15	MS	2.32	ND	1.41	UT	1.29
CA	1.30	IA	1.42	MO	1.81	OH	1.17	VT	0.83
CO	1.46	KS	1.64	MT	2.41	OK	1.46	VA	1.23
CT	0.94	KY	1.99	NE	1.54	OR	1.46	WA	1.09
DE	1.57	LA	2.02	NV	1.91	PA	1.48	WV	1.96
FL	1.71	ME	1.39	NH	0.96	RI	1.24	WI	1.42
GA	1.47	MD	1.19	NJ	1.07	SC	2.01	WY	1.79

Today is known as the information age. Never before has there been such an explosion of available data. On a daily basis, we are bombarded with statistical information that ordinary citizens must be able to understand. This is one of the reasons that statistics now takes an important place in the secondary mathematics curriculum. Statistics is essentially the study of data and making conclusions from this study. One of the first things one examines in statistics is how to organize what seems to be random information and put it in some form from which we can note patterns or a lack of patterns and then draw conclusions. We hope that you were able to organize the data in the launch question in a way that helped you answer the questions and thereby find out more about traffic fatalities in your own and in other states. After reading this section, you will be reminded of other methods you could have used to plot and analyze the given data.

12.11.1 Plotting Data

Histograms

As teachers, after testing, we often wish to get information about the distribution of scores. Students are also interested in how the rest of the class does on an exam. Early in their school careers students are exposed to picturing the data in a way that is known as a **histogram**. In a histogram the data are divided into class intervals and the frequency with which data occurs in the intervals is plotted.

> **Example 12.37** *Given the following test scores on a recent test:*
>
> 32 63 72 49 85 34 14 86 56 65 72 78 23 75 86 95 100 22 49 68
>
> *Draw a histogram for this data.*

Solution. After we sort the data we get the following:

14 22 23 32 34 49 49 56 63 65 68 72 72 75 78 85 86 86 95 100

We notice that the data range from 14 to 100. We could, if we wish, divide the range from 14 to 100 into parts, but we could just as well divide the slightly larger interval from 10 to 101, which contains these data, into parts. These parts, are called **class intervals**. We then count how many scores occur within each class interval and make a plot which indicates this. Let us illustrate below. Suppose we wish to know how many students got scores below 25, how many from 25–50 (but not including 50), how many from 50–75 (but not including 75), and how many from 75–101. We could divide the data into these intervals. Suppose we did this. Our histogram would now look like the one shown in Figure 12.15.

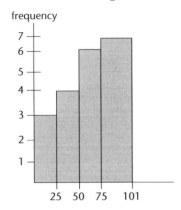

Figure 12.15

This is read as follows. The number of scores from 0 to 25 but not including 25 is 3. The number of scores from 25 to 50 but not including 50 is 4, and so on.

It is not as if we cannot get this information from the sorted data. We can just count. But the histogram gives a picture which makes an impression. Furthermore, if there were thousands of scores, the histogram would be a nice way of summarizing the data.

If we wanted more detail about the above distribution of the scores, we could refine the intervals. For example, a convenient division would be to divide this interval from 10 to 101 into 9 parts. Thus, our first class interval will be the scores from 10 to 20, including 10 but excluding 20. Our second class interval of data would be from 20 to 30, including 20 but not including 30, and so on. Our histogram would look like the one shown in Figure 12.16.

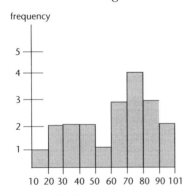

Figure 12.16

There is no "right" size for the class interval. We can make it as big or small as we want, and the class intervals do not have to be equal in length. We take what we feel gives an impression we want to give. While a histogram is very useful when dealing with very large data sets, without the use of technology, it is unnecessarily cumbersome when a teacher wants to examine a class of test scores, such as that described here. Watch how much simpler it will be to use a newer kind of plot that we will now describe.

Stem and Leaf Plots

Notice that to draw a histogram, we must first sort the data and then we must decide on the appropriate size for the class interval. When we are finished, while the histogram does give us a picture of the data, certain information is lost. We cannot tell exactly what the values of the data points are, since the histogram counts only how many data points are in the intervals. Wouldn't it be nice if we didn't have to sort the data and we didn't have to determine class intervals, and we could get a picture of how many data points are in the interval and what their values are. Well there is such a picture that does this and it is known as a **stem and leaf plot** which is a relatively new invention created by John Tukey in the 1970s and is now part of the school curriculum.

Watch how simple it is to plot the data we used from the last example. We notice that the test scores range from 14 to 100. When constructing a stem and leaf plot for this set of data, we have two columns. The first column is for the stem and, in this case, would have numbers 1 through 10. The "1" stands for scores in the 10s, the "2" for scores in the 20s, the 3 for scores in the 30s, and so on. The second column is for the leaves, which essentially tell us what the actual scores are. Thus, before we start filling in the table it looks like:

Stem	Leaf
1	
2	
3	
4	
5	
6	
7	
8	
9	
10	

Now, we start looking at the data and filling in the table. Our first score is 32. We go to the stem marked "3" and in the leaf column put in the number 2 representing the number 32. Our next score is 63. We go to the stem row with the 6 and put in the leaf column the number 3 representing the score 63. We continue in this manner and generate the following table:

Stem	Leaf
1	4
2	3, 2
3	2, 4
4	9, 9
5	6
6	3, 5, 8
7	2, 2, 8, 5
8	5, 6, 6,
9	5
10	0

This table shows us at a glance the distribution of the scores together with the actual scores. We can also count how many scores are in each row, so in a sense this is like a histogram with more detail. In fact, if we rotated the stem and leaf plot 90 degrees counterclockwise, we would get a histogram where the class intervals are 10–20, 20–30, and so on. One big advantage of this type of plot is that the data need not be sorted before drawing the plot.

When it is necessary to compare two sets of data, it is possible to graph side-by-side stem and leaf plots. Here is an example which demonstrates that idea.

Example 12.38 *The following represents the sales of computers for two groups of salesmen during the month of January to April. The first group is on the right, the second group is on the left. Thus, the number of computers sold by the 4 salespeople in the first group represented by the first line of numbers is 54, 56, 57, and 59 computers, respectively. For those 3 salespeople in the second group whose leaves are on the left, the numbers are 51, 53, and 57. Looking at this double stem and leaf plot, what can you say about the sales of the two groups?*

Second group	Stem	First group
7, 3, 1	5	4, 6, 7, 9
7	6	2, 5, 7, 8, 8
5, 3, 3, 0	7	2, 2, 5, 6
6, 2	8	3, 5, 7, 9
7, 1	9	
3	10	6

Solution. Overall, it looks like the first group sold more computers in the 50–70 range than the second group did. In certain ranges the second group did better. For example, the first group had no sales in the 90s while the second group did.

When there are a lot of data points, or if the data values vary widely, representing the data through a stem and leaf plot is probably not a good idea. In that case the histogram is a better choice.

Box and Whisker Plots

Another relatively new way of organizing data is the box and whisker plot. It was invented within the last 40 years and is an effective way of representing the spread of data. It is now included as part of the secondary school mathematics curriculum.

Here is how we create a box and whisker plot. First, we find the median of the data. Then we find the median of the first half of the data which we call the **first quartile**, and then the median of the second half of the data which we call the **third quartile**. (For a review of how to find the median, read the next section.) On a number line containing the data points, we draw a box where the left edge is at the first quartile, the right edge is at the third quartile, and where a vertical line is drawn at the median of the full set of data. This is our box. We draw lines from the left edge of the box to the smallest data point and from the right edge of the box to the largest data point. These are our whiskers. And, we now have our **box and whisker plot**. Let us illustrate this using data from an example which occurs in the next section. These data represent the number of years 30 people diagnosed with stage 4 breast cancer lived beyond their initial diagnosis.

2.1	2.3	3.1	3.2	4.2
4.4	4.6	4.7	4.8	5.1
5.7	5.9	6.2	6.2	6.2
6.6	6.6	7.3	7.4	7.5
7.7	8.2	8.3	8.4	8.6
9.1	9.7	10.5	12.5	15.5

(12.28)

There are 30 numbers here so the median is the average of the 15th and 16th number. That yields $\frac{6.2+6.6}{2} = 6.4$. The median of the first half of the data which consists of 15 data points is the 8th number which is 4.7. This is our first quartile. The median of our second half of the data is the 23rd number or 8.3. This is our third quartile. Our box and whisker plot is shown in Figure 12.17.

Figure 12.17

The box and whisker plot is also called the 5 number summary plot, since there are 5 key numbers listed. The lowest value, the highest value, the median, the first quartile, and the third quartile. It is important to note that, in this example, the median, 6.4 does not lie in the middle of the box. Don't be fooled by this, since there are the same *number* of data points between 4.7 and 6.4 as there are between 6.4 and 8.3. The median can lie anywhere in the box depending on the data.

The **interquartile range** in a set of data is the positive difference between the first and third quartiles. In this case it is $8.3 - 4.7 = 3.6$. An important number in the box and whisker plot is one and a half times the interquartile range. In this case this number is $1.5(8.3 - 4.7) = 5.4$. Any number more than this number to the left of the first quartile or the right of the third quartile is called an outlier. Thus, any data point more than $8.3 + 5.4 = 13.7$ or less than $4.7 - 5.4$ is called an **outlier**. This data set has only one outlier, 15.5. An outlier is considered to be an unusual data point. While survival rates for early stages of cancer can be quite long, the woman who lived 15.5 years after the diagnosis of such a late stage cancer was exceptional. There are many definitions of outlier, but the one we presented here is a very popular one and is due to John Tukey, a famous statistician. He is also responsible for the box and whisker plot.

There is another kind of box and whisker plot, where outliers are plotted separately, and the right whisker of the plot goes only till the largest data point, which is not an outlier. Outliers are represented by circles. Thus, our box and whisker plot that indicates outliers by circles will look like that shown in Figure 12.18 for the above example:

Figure 12.18

Notice that the box and whisker plot clearly indicates where the middle half of the data is, where the outliers are, and how the data are spread.

The box and whisker plot is particularly useful for comparing two or more sets of data. This is done by drawing them parallel to one another, as is shown in the next example.

Example 12.39 *Below in Figure* 12.19, *we have drawn the box plot of the heart rates of* 30 *sixth graders before and after exercise.*

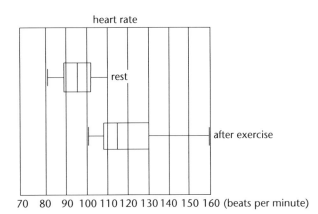

Figure 12.19

What conclusions can you make?

Solution. The obvious, and expected conclusion, is that the heart rates would increase after exercise. The lowest increase in heart rate was at least 20 beats per minute. After exercise, we see that at least one person's heart rate went up at least 50 points (look at the right whisker of each plot). Before the exercise, most of the heart rates were close to one another. After exercise, there was much greater variability in the heart rates. You might be tempted to think that more people have heart rates between 115 and 130 beats per minutes than between 108 and 115 beats per minute, but this is NOT the case. What is true, is that there is a much greater variation in the number of beats per minute of those people who have above 115 beats per minute, some even reaching dangerous levels. This might be important information for doctors to know when recommending exercise routines for their patients. Also note that the median heart rate after exercise is not in the middle of the box.

Note that one of the greatest advantages of the box and whisker plot is that any number of data sets can be compared at one time. Therefore, in the example above, we could examine the heart rates of these sixth graders when they get up in the morning, immediately after they exercise, right before they go to sleep at night, and so on. Unlike the histogram and stem and leaf plots, we are not limited to comparing only two data sets.

12.11.2 Mean, Median, Mode

Mean, median, and mode are called measures of central tendency and can play an essential role in real- life decision making . Not only is it important for students to know the meanings of these

Data Analysis and Probability   665

terms, but it is critical that they understand when it is most appropriate to use each one. For example, mean is useful when reporting things like the average price of a movie in a certain city. Mode might be useful for department stores buyers who need to know which dress size sells best. Medians are useful when one wants to get an idea of the incomes of people living in a certain neighborhood. We will now address some of these concepts using real-life situations.

> **Example 12.40** *The following data from a long range study represent the number of years each of 30 patients survived after being diagnosed with early stage 4 breast cancer. All recorded deaths are those that were from the cancer.*
>
> | | | | | |
> |---|---|---|---|---|
> | 5.7 | 4.2 | 6.6 | 12.5 | 3.1 |
> | 8.6 | 2.1 | 7.3 | 7.5 | 10.5 |
> | 4.4 | 6.2 | 7.7 | 8.2 | 4.8 |
> | 5.9 | 6.6 | 3.2 | 9.1 | 15.5 |
> | 7.4 | 2.3 | 6.2 | 9.7 | 8.4 |
> | 4.6 | 4.7 | 6.2 | 5.1 | 8.3 |
>
> *Looking at the data, it is hard to draw any conclusions. This is to be expected, since we are talking about length of life and that depends on many different factors. It is considered a random event by many. But can we make any conclusions from the data?*

Solution. In middle and secondary schools we teach that the first way to study data is to organize it unless we are using the stem and leaf plot). And one way to organize it is to put it in increasing order. Here are the same data put in increasing order.

2.1	2.3	3.1	3.2	4.2
4.4	4.6	4.7	4.8	5.1
5.7	5.9	6.2	6.2	6.2
6.6	6.6	7.3	7.4	7.5
7.7	8.2	8.3	8.4	8.6
9.1	9.7	10.5	12.5	15.5

Now, we can clearly see what the shortest life span was and what the longest life span was, and we can also see the **range** of the data, which is defined to be the difference between the smallest and the largest data points. So in this case, the range is $15.5 - 2.1 = 13.4$. What else do these data tell us?

In schools we ask students to learn three basic measures, the mean, the median, and the mode. The **mean** is just the average of the data, which in this case is ≈ 6.75. The **mode** is the number that occurs most often, which in this case is 6.2. A set of data can have many modes. Finally, the median is the middle number.

While everyone seems to agree on the definition of mean and mode, when it comes to median, one will see different definitions. It seems that the easiest and least problem prone definition is that given a sample of data, a **median** is a number which separates the higher half of the data from the lower half *when the data points are in order of size*. Notice the words "a number which separates" in the previous sentence. That seems to imply that there can be more than one median, and there can be for a set of data. For example, when we examine the data 1, 2, 3, 4, we notice that the number 2.1 separates the higher half from the lower half in the sense that half the numbers are less than 2.1 and half are greater. The same is true if we take the number 2.3 or 2.5. Thus, all of these can be considered medians. However, to avoid confusion when talking about medians, and to ensure that the median is unique, we agree to compute it as follows: When there is an odd number of data points, the median is the middle one which is the $\frac{n+1}{2}$ number in the list, where n is the total number of data points. Thus, if there are $n = 5$ data points, the $\frac{5+1}{2} = $ 3rd number is the median. If there are an even number of data points, then the median is the average of the two middle numbers, which are the $\frac{n}{2}$ and $\frac{n}{2} + 1$ numbers where n is the number of data points. Thus, if there are $n = 4$ data points, the average of the $\frac{4}{2} = $ 2nd and 3rd numbers is the median. So, if our data points are 1, 2, 3, 4, 5, we have an odd number of data points and our middle number is 3. This is our median. If our data points are 1, 2, 3, 4, we now have an even number of data points and we take the median to be the average of the two middle numbers, 2 and 3. So the median in this case is 2.5. When we speak of half the data, each data point is considered separate even if they have the same value. Thus, in our example above, we saw three 6.2s. These are considered as 3 separate 6.2s even though they have the same value.

One often hears teachers saying that "Half the numbers are below the median and half are above." The words can be confusing. For example, suppose that we have the data values, 2, 2, 2, and 3. The median here is 2 and it is very likely that someone will say, "But 3 out of the 4 numbers are equal to the median. So how could half be above and half below?" So now we have a problem. A better way to describe median would be to say that half the *data points* (not their numerical values) are to the left of the median and half to the right when the numbers are strung out in a line in numerical order and each data point is considered a separate entity, even if the values are the same. Thus,

$$2, 2, \overset{(2)}{|} 2, 3$$

shows that 2 is the median of the data set 2, 2, 2, 3. We have put the bolded 2 in parentheses to indicate that, while the numerical value of the median is 2, that 2 above the vertical line is not one of the data points, rather it is only a number computed using the two middle data points and taking an average. This picture

$$1, 2, \overset{3}{|} 4, 5$$

indicates that the 3 is the median of the set of data 1, 2, 3, 4, 5. We didn't put the three in parentheses, since it now *is* part of the data. This device is our own concoction, just to clarify what one means by median. But if we are interested in the actual numerical value of the median, which we often are, then a correct definition of a median is as follows:

"A **median** is a number where at most half of the number of data points have values less than the median and at most half of the data points have values greater than the median." Notice the words "have values less than" in the previous definition. We ask you in the Student Learning Opportunities to show why the words "less than or equal to" would not work. While this is the correct definition, most secondary school students would prefer the intuitive "middle number" of the data and we feel this is enough. We point out this correct definition, since you are likely to run into this problem with a more discriminating student.

What can we learn from the mean? What can we learn from the median?

One of the first things that we teach regarding the mean is that while it is important in many statistical analyses, it can severely distort our impression of the data. For example, suppose that the scores on an exam were 1 and 99. The average of the scores is 50, but neither score is close to 50. The median here is also 50. Again, it tells us little about the scores, except half of the scores are below the median and half are above. When there are many data points there may be large values that are not consistent with the rest of the values, known as outliers, and these can give us a false impression of what the data represent when we take an average. For example, suppose that a company has ten employees plus the president. Each of the ten employees earns 10,000 dollars a year, and the president earns 1 million dollars a year. The average of the numbers is $\frac{1100000}{11} = 100,000$ and, of course, no one earns over 100,000 except the president. Now let us compute the median of the data. When the data points are listed in order, the two middle numbers are both 10,000 and the average of these is 10,000 and therefore the median of this data is 10,000. Now, whether the president earned 1 million dollars a year or one billion dollars a year, the median would still be the same. And so we see that the median is unaffected by changes in the data at the very beginning and very end of the data set assuming the data set has more than two elements.

Student Learning Opportunities

1 (C)
   (a) Some of your students are confused by what the quartiles tell you. They ask you to explain in words what the lower and upper quartiles mean. What do you say?
   (b) Some of your students are reluctant to use the box and whisker plot and want to know what it tells you that a stem and leaf plot doesn't. How do you respond?
   (c) One of your students, Claudia asks why statisticians use three different types of measures of central tendency (mean, median, mode). She thinks that using the mean alone would be good enough to summarize data. How do you respond?

2 (C) Some of your students insist that, in a box and whisker plot, the median of the data falls exactly in the middle of the box. How do you help them understand why this is not necessarily the case?

3 Consider the following data for the heights (in centimeters) of boys and girls in a secondary school class.

| Boys | 156 | 141 | 163 | 151 | 147 | 146 | 178 | 132 | 152 | 173 | 174 |
| Girls | 163 | 128 | 135 | 144 | 131 | 132 | 139 | 147 | 163 | 155 | 150 | 128 |

   (a) Draw separate histograms for the boys and girls. Use 6 class intervals.
   (b) Draw a side by side stem and leaf plot for this data.
   (c) Draw box plots for the boys and girls on the sheet and determine from the plots if the following statements are true or false:
      (i) The interquartile range for the girls is 3 more than the boys.
      (ii) The median line for the girls is in the middle of the box from the box and whisker plot, but that is not true for the boys.
      (iii) The boys appeared to be taller overall.

(iv) The shortest boy is slightly taller than the shortest girl.

(v) The tallest girl is shorter than about 25% of the boys.

4 Samantha and Ally both are camera salespeople. The number of cameras each of them has sold each month over the last 12 months are given below.

Samantha	13	17	25	39	7	49	62	20	41	51	43	35
Ally	24	13	57	1	37	50	20	47	15	34	19	28

Draw box and whisker plots for each of these data sets and place one under the other. Try to make some conclusions about the data sets. Would it be fair to say that one person is a more successful camera salesperson than the other? If so, who?

5 Brand A and Brand B are two brands of licorice. James doesn't care that both boxes have the same weight. He is interested in buying the brand that has the most pieces of licorice. He has purchased 12 eight ounce boxes of each and has found the following number of pieces of licorice in each box. In your opinion, which brand should he buy using this criteria? Why?

Brand A	26	32	45	26	29	38	32	22	26	31	40	25
Brand B	34	23	33	26	19	44	29	32	35	34	23	27

6 In Figure 12.20 we see a box and whisker plot for the number of appendectomies performed by male doctors and female doctors at Mercy Hospital in the past year.

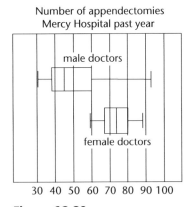

Figure 12.20

(a) In the top plot, what does it mean that the right whisker is longer than the left whisker?

(b) In the top plot, why isn't the median in the middle of the box?

(c) In the top plot, which measure of central tendency do you think might be most representative of the data, the mean or the median? Explain.

(d) In the bottom plot, which measure of central tendency do you think might be most representative of the data, the mean or the median? Explain.

(e) What can you conclude about the number of operations performed by women and men doctors in Mercy hospital during the last year?

7 (C) Your astute student, Johanna, questions the definition of median and asks, "Why do we use the words 'have values less than' rather than 'less than or equal to?' How do you respond?

8 (C) Hudson, one of your clever students claims to have found a new way to find the mean of two numbers. He shows you an example of finding the average of 14 and 20. He explains that he subtracts 14 from 20 and gets 6. Then he divides this difference (6) by 2 and gets 3. He adds 3 to 14 and gets 17, which he says is the average. Is Hudson right? How could you prove it?

9 (C) One of your students, Ann wants to know what how the mean of a set of data changes if you add or subtract the same number to each of the data points. She also wants to know how the mean will be affected if each of the data points are multiplied or divided by the same nonzero number. How can you respond to Ann and how could you prove your answer?

12.12 Lying with Statistics

**LAUNCH**

On February 5th, 2001, the following data were published regarding complaints about airlines in US News and World Report.

Most complaints (November 2000)
United Airlines 252
American Airlines 162
Delta Air Lines 119

Fewest complaints (November 2000)
Alaska Airlines 13
Southwest Airlines 22
Continental Airlines 60

Is it correct to conclude that United, American, and Delta were the worst airlines and Alaska, Southwest, and Continental were the best? Why or why not?

In case you are unaware, there are many disparaging quotes about statistics such as "There are three kinds of lies: lies, damned lies, and statistics," attributed to Disraeli and popularized in the United States by Mark Twain. We hope that you were able to point out the fact that, without knowing the numbers of passengers who flew these airlines, the data in the launch question are completely meaningless, or better yet, misleading.

Today, whenever we open a newspaper we are bombarded with statistics and graphs. Although graphs are wonderful visuals, they are often used to fool the public, as are the presentation of the numbers which relate to the statistics. We show here a few cases of how this works.

670 Data Analysis and Probability

Example 12.41 *Consider a study of 100 people who were at some risk of having a heart attack. Under normal circumstances, 2 out of these 100 people would have a heart attack. In a drug trial all 100 were given a drug and only one person had a heart attack. When the data come out, the drug company calls the newspaper and declares that their "New Drug Decreases Risk of Heart Attack By 50%." Are they lying?*

Solution. They are not lying, rather they are making a misleading claim. Reporting data with percents can give the wrong impression. Fifty percent seems like a terrific decrease. But suppose that the Wen Ching Tea company had given a similar group of one hundred people green tea to drink, and they too saw that only 1 person died of a heart attack. The results are the same, but the drug company could now have made the headlines even more deceptive by saying "New Drug Decreases Risk of Heart Attack by 50% While Green Tea (with the same exact results on another 100 people) Decreases Heart Attack in Only One Out of 100!"

Example 12.42 *Another instance of this type of deception is the following which appeared in Reuters concerning the new tax laws:*
"Those making between $30,000 and $40,000 per year will have a federal income tax cut averaging 38.3 percent, against 8.7 percent for people making more than $200,000 per year, the figures compiled by the treasury department showed."
This looks like a very big break for the common man. But is it?

Solution. Let's see. The single person who makes $30,000 with no deductions pays roughly $2994 in federal tax. If we cut this by 38.3% the person now pays 62.7% of what he paid before, which amounts to approximately: $1877, a savings of a bit over $1000. Divide that by 52 paychecks, and how much extra is this person getting a week? Now let us look at the same person earning 200,000 dollars a year. This person's federal tax will amount to approximately $52,000 and the reduction, 8.7% of this, amounts to roughly $4524. Who is really getting the better deal, the rich guy or the poor one?

Statistics can easily distort and give a wrong impression of events. Consider the following example:

Example 12.43 *The workers at the local tin factory are disgusted. It is has been ten years, and their salaries have not risen much. So, they decide to strike and get the media's attention. They show the media a graph of the rise in their salaries over the last 10 years. (See Figure 10.21.)*

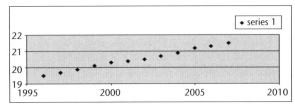

Figure 12.21

The management of the company is quite concerned about this bad press and presents its own graph of their salaries over this 10 year period. In Figure 12.22 you see their graph.

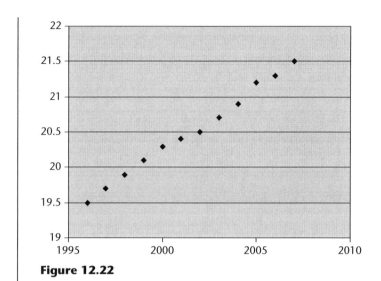

Figure 12.22

How much better things are looking now! Yet the data are the same. Can you see what happened?

Solution. We have just changed the scale on the *y*-axis. This just shows how by proper scaling, you can make data look much better or worse than it actually is.

As another example, consider the following:

Example 12.44 *In the student newspaper the following alarming headline appeared, "Number of Drunk Student Arrests on Campus Increases by 300%." To make matters worse, they presented the picture shown in Figure 12.23 to show this.*

Figure 12.23

What impression do you get? The figure on the right is the figure on the left scaled by a factor of 3. Does it appear 3 times as big?

Solution. The figure is two dimensional and scaling by a factor of 3 multiplies the area by 9. (And if you picture this in 3 dimensions, the volume is multiplied by 27. So, it is even worse!) This drawing gives the mis-impression that campus arrests really were much more than they were the year before. This kind of distortion is not uncommon in the newspapers.

Another type of distortion that is unfortunately common is the wrong conclusion made from a sampling procedure that has some kind of built-in bias in it. To get any kind of real information from statistics, one has to do what is known as a good sampling. This can mean many things to many people. For example, you might want to know the average American man's opinion in the United States of New Yorkers. So, you walk along the street and ask as many people as you meet, "What do you think of New Yorkers?" Now, if you are asking this question in New York, and you are asking the question only to New Yorkers, how likely are you to get a negative answer? Is your sample really a representative sample of what the average American thinks of New Yorkers? No!

Marketers use a similar strategy to mislead the public when they report the results of surveys. Typical headlines might be "Nine out of ten doctors recommend Brand X Aspirin." Nine out of which ten doctors? Were only ten doctors interviewed? Were the doctors on retainer by the Aspirin company? Was a legitimate cross-section of the doctors made? What did they recommend about Brand X aspirin? That it be thrown out? That it be banned?

Magazines often send surveys to their readers. But sending surveys to their readers builds in a bias since their readers are not a representative sample of the entire population. For example, does the average reader read "Home Handyman?" Would a survey of the readers of this magazine represent what people in general think?

One of the most notorious magazine surveys was done by the now defunct *Literary Digest*. On the basis of a large number of phone calls to what they felt was a random sample of the population they predicted that Franklin Delano Roosevelt would lose the election to Alfred Landon. That didn't happen. In fact, Roosevelt won by a large margin. How could they have been so mistaken, since they called over 3 million people from all over the country to get their results? The answer is that in those days, (1936) only the affluent had phones, and most of them were Republican! So their sample, though large, had a built in bias.

Even if a survey is sent to a representative portion of the population, many will choose not to respond. So, if a person running for president wants to get a sense of how popular he or she is, it might be wise to send out a survey. People who don't like the candidate may not even bother answering the questionnaire, while those who like the candidate may be happy to respond. So, the candidate may get responses only from his or her supporters which may only be a small portion of the population. So, to think that these responses are a representative sample and make conclusions on the basis of them can easily lead to false conclusions. Related to this issue is how much you can trust the answers people give. If you ask a representative group of people how often they bathe, is a person who bathes once a week likely to give an honest response? If you ask a person how much money he makes, is the low earner likely to tell you the truth? Sometimes even the high earner feels that what he is making is not enough and so might embellish his salary. On the other hand, he might not want you to think he is a high earner, for all sorts of reasons, so he may give you a substantially lower figure. Thus, one must be critical about the possibility that the answers reported in a survey are not accurate.

Even professional outfits that are aware of the problems with sampling and choosing representative samples make mistakes . For example, as recently as April 2007 the Gallup Organization, the premier sampling organization, did a study on how well integrated Muslims were in Britain, France, and Germany, and how much they identified with their nations, their faith, and their ethnicity. But afterwards, the Gallup Organization admitted that their study may have been biased for they realized that the Muslims in London, who were the main interviewees, were somewhat different in attitudes from Muslims who lived in England but who lived outside of London.

12.12.1 What Can you Do to Talk Back to Statistics?

The reporting of data is often meant not to be objective, but to influence how you think. Surely, there are those organizations that present statistics because they truly are interested in what the results mean. Nevertheless, you need to be skeptical. So what should you be careful of when reading statistics? Here we give some tips to watch out for.

Percentages are used often in statistics and can easily distort the facts. For example, if in a city of 15 million people, one person has contracted the West Nile Virus, and next year two people contract it, what does this really mean? While it is true that the percentage increase in the contraction of the disease is 100%, does this warrant the alarming headline in 96 point type that there was a 100% increase in the number of contracted cases? Percentages without the clarifying data are often misleading. So, don't look at percentages alone. Check the data!

Similarly, sometimes percentages overlap and companies mistakenly add the percentages to purposely distort the data. For example one might report that wages went up 20%, and the total manufacturing cost went up 10% to make it seem like a 30% increase overall. But, here you are counting things twice. The wages are part of the 10% manufacturing cost. This is one of the ways companies try to justify unjustifiable raises in prices!

A common ploy in department stores is to advertise that you will get 50% off, followed by another 15% off. So are you getting 65% off? That is, do you just add the percentages? The answer is "No," though many people will think otherwise. If your item costs 100 dollars to begin with, then after a 50% discount you pay 50 dollars. The 15% is now applied to the 50 dollars, not the original amount. This gives you an extra 7.50 discount. Your total discount is 57.50 which is a 57.50% decrease. Why don't they just say you get a discount of 57.50%? Because they know you will think you are getting a 65% discount if you say it the other way.

Car salesmen do something very similar and are particularly dangerous in this respect. Consider the following scenario. You buy a used car for $1200. The dealer gives you a loan for the $1200 which you will pay back in equal amounts of 106 dollars per month over a year, starting on the day you receive the loan. The dealer says, "You are only paying a total of $1272. So, your interest is only $72. That means you are paying only 6% interest." Are you? Let's see.

Figure out exactly what 6% interest is on the amount you owe each month. Remember though, 6% is the annual rate. So, if you are truly paying 6% interest, then you are paying interest of $0.06/12 = 0.005$ per month. Your first payment is 106 dollars paid as soon as you get the loan. So you are really only being loaned $1094. Your interest for the next month should be $1094 \times 0.005 = 5.47$. At the beginning of the month after that, you pay your next monthly payment of $106 and then you only owe $988. Your interest on that for that month is $988 \times 0.005 = 4.94$ and so on. The following table summarizes this:

You owe	1094	988	882	776	670	564	458	352	246	140	34
Interest at 0.005 per month	5.47	4.94	4.41	3.88	3.35	2.82	2.29	1.76	1.23	0.70	0.17

By the beginning of your last month, you only owe $34 dollars on the loan at an interest rate of 0.005 which yields interest of 17 cents. The total you would owe, if you were truly paying 6% interest, is the sum of the entries in the second row of the above table which is:

$$\$(5.47 + 4.94 + 4.41 + 3.88 + 3.35 + 2.82 + 2.29 + 1.76 + 1.23 + 0.70 + 0.17) = \$31.02$$

But you paid nearly $2\frac{1}{2}$ times that much! Thus, you paid almost 15% interest, not 6% as the dealer said. So, next time your hear a car salesman telling you about the great deal he or she is offering, be skeptical.

Many times people make up their own figures and then compute accordingly. Someone may begin an argument with, "Let's assume that the probability of slipping on the ice is 0.2, and that the probability of tripping over a curve is 0.5 etc., etc." Let's assume? Why? Do the numbers make sense? You can't just make assumptions and proceed from there without a basis for the numbers.

Often people use numbers to make their arguments more convincing. But beware of whether there is truth in the numbers. Consider the following comment made in a book on the history of an oil company: "Price cutting in the southwest... ranged from 14% to 220%." Come on now! How could the price decrease by 220%? Yet, this comment passed through the editors and proofreaders. It was, after all mathematics, and like many, they were too intimidated or unfamiliar with the mathematics to question it.

Percentages must be read with a critical eye. Also, beware of the difference between percentage and percentage points. For example, suppose you sold 100 dollars worth of goods and made a 5 dollar profit. The next year you sold the same 100 dollars worth of goods and made a 10 dollar profit. Was your increase 5 percentage points, or 100%? Actually, both are correct, but by using percentage instead of percentage points, one can easily make the profit appear much greater.

Sometimes people report their conclusions on the basis of an inadequate sample, or a defective sampling procedure. So, if a person tosses a coin 10 times he might get 8 heads. Is this enough to conclude that the probability of getting heads is 0.8? The same kind of error can happen in a survey of 200 people which might not be sufficient to make a valid conclusion. The issues here are subtle and they are discussed in most statistics courses. Suffice it to say that when a conclusion is made based on a limited sample, you want to look for a figure called the reliability of the study. If the figure is 95% reliable or better, you may be able to trust the conclusions. Notice the word "may." A defective experiment with defective numbers can be highly reliable based on the numbers, but it is highly reliable junk!

You may be thinking that we have raised more questions in this chapter than we have answered, and that is probably correct. Probability and statistics have many subtle issues. However, on a secondary school level, we expect students to be aware of the issues that arise in the media and be critical judges of what they hear and what they see.

Student Learning Opportunities

(C) Your students are first learning about how to be careful interpreters of statistical and probabalistic statements. They want your help in criticizing the claims made in the problems that follow. In each case explain what errors are being made

1 The average income per person in the United States in the year 2006 was roughly 22,000 dollars per year. Thus, families of 4 are, on average, twice as wealthy as families of 2.

2 A survey was taken with 50 people. They were asked the number of hours they slept last night. The average was taken and it was concluded that the group average for sleep last night was 7.8 hours.

3 Last year workers took a pay cut of 20%. But this year they got a raise of 5% recovering $\frac{1}{4}$ of what they had lost. What kind of pay increase must they get to restore a pay cut of 50%?

4 When working out the average hourly wage at the plant, the boss took the 5 dollar wage per hour, added the 10 dollar overtime wage per hour and added the 8 dollar time and a half wage and divided the sum by 3 to conclude that the average worker made about 7.66 an hour!

5 Some years ago it was reported that $33\frac{1}{3}$% of the women attending a major university married faculty members.

6 Mack was complaining to the truant officer that he had no time to go to school. "There are 365 days a year and I spend 8 hours a day sleeping which uses up 1/3 of them. That leaves me with roughly 243 days. I spend about 2 hours a day eating with my family, which takes up another 30 days. I now have 213 days left. Subtract from that the weekends which account for 104 days, and I now have 109 days left. Of course, there is no school on Christmas and Easter, so subtract another 20 days. That leaves me with 89 days which is the length of my summer vacation. So I have no time to go to school!"

7 If you buy 20 items today and they increase in price by 5% tomorrow, the price of the items has gone up by 100%.

8 A newspaper sent out 1200 questionnaires, and asked companies about their price gouging policies. On the basis of the returns, they concluded that companies were not gouging as American's thought.

9 During a recent week of storms the death rate jumped up quite a bit. The storms were blamed for the deaths.

10 A survey of college students asked if they were in a relationship but were seeing someone "on the side." On the basis of the results, the surveyor concluded that since no one answered that they were seeing someone on the side, that all (or at the very least, most) college relationships on that campus were honest and monogamous.

11 The average SAT score in mathematics for all high school students in Eversosmart School District is known to be 700. You pick a random sample of 10 students and the first student you pick has an SAT score in mathematics of 500. Even though this score is lower than 700, the other 9 scores will make up for it so that the average SAT score for the 10 students will be 700.

CHAPTER 13

INTRODUCTION TO NON-EUCLIDEAN GEOMETRY

13.1 Introduction

Children like to play in the sand. You often see a child using a shovel to draw a line in the sand. One is inclined to think, "How long will the child make this line?" But a child's world is very limited, and if you asked the child, "Can you make the line longer? How long can the line be made?" you are likely to get an answer like "Sure I can make the line longer. It can go on forever."

But can it?

If you are thinking, "Yes" then you are forgetting that we live on a spherical planet. And if we continued our line, assuming we could continue drawing for a long time, we would eventually come back to our starting point. Our "line" is really going around the earth and, in fact, is really circular! Put another way, there is no such thing as drawing a line on earth. We can only draw something that looks like a line, but is, in fact, part of the arc of a circle.

Our point is that the concept of a straight line is an abstraction. However, the "lines" that we do draw on earth take up such a small part of the earth's surface that the curvature is indiscernible. They appear straight and flat, as if they were drawn on a plane. And that, of course, is because the earth to us does appear to be a plane in our immediate neighborhood. Even the notion of a plane is an abstraction of what we see around us. There is no physical object that is a plane as far as we know, as a plane is flat and goes on in all directions, forever. Even though there is no such thing as a plane, near us the earth does seem planar and we use plane geometry as an excellent model of our world.

Geometry comes from the words "geo" and "metry" which mean to measure the earth. To ancient civilizations, geometry was a practical subject. These civilizations had made all different kinds of geometric observations, like what similar triangles were and their properties, and used them for measurement and construction. Some of the relationships people discovered, and believed, were correct. Some were not. For example, it was believed by the Egyptians for some time, that the area of a parallelogram is the product of the lengths of the sides. Of course, we now know that is not true. Thanks to the Greeks who put the subject of geometry on a firm foundation, we now know how to find the area of a parallelogram and so many other figures.

The Greeks also took geometry to a much higher level. It was all part of the Greek tradition of seeking truth through logic and analysis. One wouldn't accept anything without proof that it was true. But, how could one be sure that something has been proven? The question is not easy to answer, and we show you, in the next section, why extreme care is needed in trying to prove anything lest we make assumptions that appear to be true, but aren't. Questions like

678 Introduction to Non-Euclidean Geometry

this helped convince mathematicians of the need to put geometry on an even more rigorous foundation than the Greeks had, and this led to new geometries, which will be described in this chapter.

13.2 Can We Believe Our Eyes?

 **LAUNCH**

Carefully examine the following proof that, through a point, *D* outside a line, *CE*, one can draw two perpendicular lines to *CE*.

Proof. We begin by drawing both *CD* and *DE* as shown in Figure 13.1. We then draw circles using *CD* and *DE* as diameters.

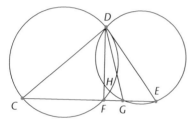

Figure 13.1

Our circles will hit our line, *CE*, in two points, *F* and *G*, as shown in the figure above. Angle *DFC* being inscribed in a semicircle with arc *DFC* is a right angle and angle *DGE* being inscribed in the semicircle with arc *DGE* is also a right angle. (See Chapter 1 Theorem 1.2.) Thus we have two perpendiculars from *D* to *CE*, namely, *DF* and *DG*. ∎

In secondary school we learned that there is only one perpendicular that can be drawn from a point outside a line to the line. So which is it, one, or two, as our proof just now showed? Explain your reasoning.

In your past studies in geometry, you most likely examined many proofs of theorems. Often, these proofs were supported by diagrams which you believed were correctly drawn. Did you notice anything that might have been incorrectly drawn in the launch problem? Now, consider the following proof, which you probably saw in your secondary school geometry class.

Theorem 13.1 *The area of a parallelogram is base times height.*

Proof. We begin with parallelogram *ABCD* shown in Figure 13.2 below, where the base has length *b*. The height, *BE*, of length *h*, is drawn.

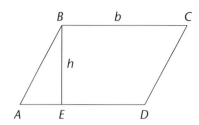

Figure 13.2

Now extend AD and draw altitude CF as shown in Figure 13.3 below.

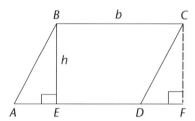

Figure 13.3

Since parallel lines are everywhere equidistant, $BE = CF$. Since opposite sides of a parallelogram are equal, $AB = CD$. It follows that right triangle ABE is congruent to right triangle DCF by $HL = HL$ (See Chapter 5 Theorem 5.3.) Thus,

Area of ABE = Area of CDF. (13.1)

It follows that

$$\begin{aligned}\text{Area of parallelogram } ABCD &= \text{Area } ABE + \text{Area } BCDE \\ &= \text{Area } CDF + \text{Area } BCDE \quad \text{(by equation 13.1)} \\ &= \text{Area of rectangle } BCFE \\ &= bh.\end{aligned}$$ (13.2)

The proof is hard to argue with. But let us take a second look at the diagram. Suppose our parallelogram looked like that shown in Figure 13.4.

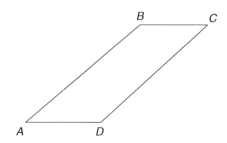

Figure 13.4

Then, when we drew BE and CF our figure would look like that shown in Figure 13.5.

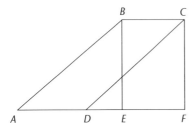

Figure 13.5

This would mean that the statement of equation (13.2) concerning how the areas add up to the area of the parallelogram wouldn't be correct, and the whole proof would collapse. ∎

We used the picture to guide our original proof and in secondary school geometry, we assume these pictures are correct. But is this assumption correct? From what we have just demonstrated, it appears that we have to be much more careful about our assumptions.

We hope that you noticed that, in the launch problem, we presented a fallacious proof that was a result of a flawed picture. In what follows, we present another fallacious proof. See if you can find out what is wrong. We will call the result a "Theorem" in quotes because, although the logic seems sound, the proof is not. It was instances like this that led mathematicians to look much more critically at the assumptions they made, which eventually led to new geometries.

"**Theorem 1**" *All triangles are isosceles.*

Proof. We start with a non-isosceles triangle ABC, and draw the angle bisector of angle B, dividing angle B into equal angles 1 and 2. We also draw the perpendicular bisector of AC and let the angle bisector of ∠B and the perpendicular bisector of AC meet at D as shown in Figure 13.6 below.

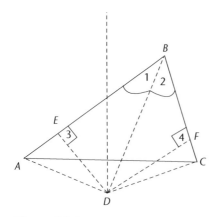

Figure 13.6

Now draw DE perpendicular to AB and DF perpendicular to BC. Since all right angles are equal, ∠3 = ∠4. Since BD is common to both triangles EBD and FBD, we have triangle FBD is congruent to triangle CBD by AAS = AAS. (See Chapter 5 Section 2.2 Exercise 7.) It follows that

$$BE = BF. \tag{13.3}$$

It also follows from this that

$$DE = DF. \tag{13.4}$$

Now we have to recall a fact from geometry that you will prove in the Student Learning Opportunities: Any point on the perpendicular bisector is the same distance from the endpoints. From this, it follows, that

$$AD = DC. \tag{13.5}$$

Since in right triangles EDA and FDC, the hypotenuses and legs are equal (see equations (13.4) and (13.5), it follows that triangles EDA and FDC are congruent by $HL = HL$. So

$$EA = FC. \tag{13.6}$$

Adding equations (13.3) and (13.6), we get that

$$BA = BC. \tag{13.7}$$

This tells us that triangle ABC is isosceles. Of course, ABC was any non-isosceles triangle. So all triangles, isosceles or not, are isosceles! ∎

We will prove one last fallacious theorem before we proceed, but we will need a lemma.

Lemma 13.2 *If $\frac{p}{q} = \frac{a}{b} = \frac{c}{d}$ then $\frac{p}{q} = \frac{a-c}{b-d}$ if $b - d$ is not zero.*

Proof. Call the common value of the fractions $\frac{p}{q}$, $\frac{a}{b}$, and $\frac{c}{d}$, m. Since $\frac{a}{b} = m$, we have upon cross-multiplying that

$$a = bm. \tag{13.8}$$

Since $\frac{c}{d} = m$, we similarly have

$$c = dm. \tag{13.9}$$

Subtracting equation (13.9) from equation (13.8) and factoring out m, we have that

$$a - c = (b - d)m.$$

Dividing both sides of this by $(b - d)$ we get

$$\frac{a - c}{b - d} = m. \tag{13.10}$$

Since m, is also $\frac{p}{q}$, we have, substituting into equation (13.10) and switching sides that

$$\frac{p}{q} = \frac{a - c}{b - d}.$$

∎

Let us illustrate this lemma. We know that $\frac{1}{2} = \frac{2}{4} = \frac{3}{6}$. Thus $\frac{1}{2} = \frac{2-3}{4-6}$, which is true!

682 Introduction to Non-Euclidean Geometry

We now come to our next erroneous proof. It requires a more careful reading than the last results, and uses similar triangles. This time the error is a bit more subtle.

"Theorem 2" *Opposite sides of a trapezoid have the same length!*

Proof. We begin with trapezoid *ABCD* as shown in Figure 13.7 below.

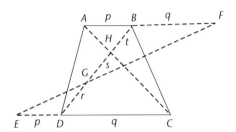

Figure 13.7

We suppose that the length of *AB* is p and that the length of *CD* is q. Extend *CD* a distance p to *E*, and extend the length of *AB* distance q to *F*. Since *AB* and *CD* are parallel, $\angle ABH = \angle CDH$. Also $\angle BHA = \angle CHD$ since they are vertical angles. Thus, triangles *BHA* and *DHC* are similar by *AA = AA*. In a similar manner, since $\angle E = \angle F$ and $\angle FGB = \angle EGD$ triangles *DEG* and *BFG* are similar. From the similar triangles *BHA* and *DHC* we have

$$\frac{AB}{CD} = \frac{BH}{DH}$$

or,

$$\frac{p}{q} = \frac{t}{r+s}. \tag{13.11}$$

From the second set of similar triangles *DEG* and *BFG* we have

$$\frac{ED}{FB} = \frac{DG}{BG}$$

or,

$$\frac{p}{q} = \frac{r}{s+t}. \tag{13.12}$$

Hence from equations (13.11) and (13.12) we have

$$\frac{p}{q} = \frac{t}{r+s} = \frac{r}{s+t}.$$

By our lemma we have that,

$$\frac{p}{q} = \frac{t-r}{(r+s)-(s+t)} = \frac{t-r}{r-t} = -1 \tag{13.13}$$

Taking the absolute value of both sides we get that

$$\frac{|p|}{|q|} = 1$$

and it follows from this that $|p| = |q|$. But the length of $AB = |p|$ and the length of $CD = |q|$. Thus, the length of AB is equal to the length of CD. ∎

As an immediate corollary of this we get:

"Corollary" *Any two line segments have the same length.*

Proof. Take any two line segments, AB and CD with different lengths, and place one above the other and parallel to it. Then draw trapezoid $ABDC$ as shown in Figure 13.8 below.

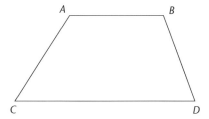

Figure 13.8

By our theorem, $AB = CD$, and thus, any two lines have the same length. ∎

Wasn't that simple?

13.2.1 What Are the Errors in the Proofs?

You are probably thinking "What is going on with these proofs and where are the flaws?" Here are the answers. In the proof in the launch the points, F, G, and H are all the same. In the Student Learning Opportunities, we will guide you through a proof of how we know this. Our picture, therefore, was wrong, and the two perpendiculars we got are really one. If you try drawing the figure, you will see this very clearly.

In "Theorem" 1 the picture again is wrong. The two lines drawn, DE and DF do not lie where we drew them. Rather, they look something like the picture shown in Figure 13.9, where one of the points is inside of the triangle, and the other is outside of the triangle.

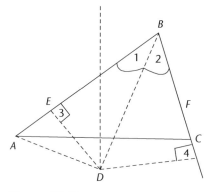

Figure 13.9

Again, we will guide you through a proof of this in the Student Learning Opportunities. Thus, adding equations (13.3) and (13.6) did *not yield equation* (13.7) *as we said*. Hence, our proof was wrong.

684 Introduction to Non-Euclidean Geometry

Finally, in the proof of "Theorem 2", r turns out to be equal to t from the figure in that problem, as you will see in the Student Learning Opportunities, and so equation (13.13) does not follow from the lemma, since we would then be dividing by 0.

As you have seen, in geometry we must be exceedingly careful. Misleading diagrams and faulty logic or faulty assumptions can lead us to believe things that are not true. Errors such as these, together with a desire to prove the parallel postulate from the other axioms (see next section), led mathematicians of the 19th and 20th century to re-examine very closely the assumptions that Euclid made in his treatment of Geometry in *The Elements*. And thus, new geometries were born.

Student Learning Opportunities

1 When we resolved the launch problem we said that the points F, G, and H were all the same. Prove that. [Hint: Draw CH and HE and show that angles CHD and DHE are adjacent right angles so that CHE is a straight line. But there is only one straight line from C to E. Finish it.]

2 Prove that every point, P, on the angle bisector of angle BAC is the same distance from AB and AC. [Hint: Form congruent triangles.]

3 Prove that every point, P, on the perpendicular bisector of line segment AB is the same distance from A and B. We used this fact in our "proof" that all triangles are isosceles.

4 (C) You have asked your students to use a geometric dynamic software program to create 3 triangles and then try to inscribe each of them in a different circle. All have been successful each time and are convinced that every triangle can be inscribed in a circle. You have taught them the importance of proof and have encouraged them to prove their conjectures. What proof of this can you give before guiding them to discover the proof themselves? [Hint: Use the result of the previous problem.]

5 Prove that, if angle ABC is an angle inscribed in a circle, then the angle bisector of angle B bisects the arc AC. Then show that the perpendicular bisector of line segment AC also bisects that same arc. Thus, the perpendicular bisector of AC and the angle bisector of B always meet at some point outside of the triangle.

6 Here is an outline of how we resolve the flaw in "Theorem 1." Using the previous problem and Figure 13.10 below, where GD is the perpendicular bisector of AC and BD is the angle bisector of angle B,

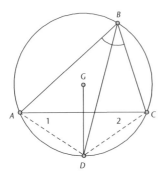

Figure 13.10

(a) Show that, ∡BAD and ∡BCD are supplementary.
(b) Show that ∡1 = ∡2.
(c) Show that, if the triangle ABC is not isosceles, then angles BAD and BCD cannot both be right angles. Hence, one of them is obtuse and the other is acute.
(d) Using the fact that one of the angles BAD and BCD is acute and the other is obtuse, show that one of the perpendiculars drawn from D to sides AB and AC is inside the triangle and the other is outside the triangle.
(e) Resolve the fallacy in "Theorem" 1.

7 Find the error in the following proof that every right angle is obtuse. Begin with the rectangle ABCD shown below in Figure 13.11.

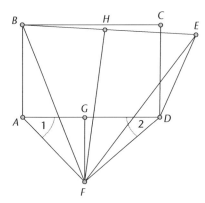

Figure 13.11

Draw line segment DE with length equal to CD. Draw the perpendicular bisectors of BE and AD and let them intersect at F. Then BF = EF and AF = DF, since any point on the perpendicular bisector is equidistant from the endpoints. Since triangle AFD is isosceles, angle 1 is congruent to angle 2. Also triangles AFB and DFE are congruent by SSS = SSS. Hence ∡BAF = ∡EDF. Using this last statement, together with ∡1 = ∡2, show, by subtraction, that ∡ABC, which is a right angle, equals ∡EDA which is obtuse.

8 Prove that r = t in the fallacy that opposite sides of a trapezoid are equal, thus explaining the fallacy.

13.3 The Parallel Postulate

LAUNCH

Cut out a large triangle from a piece of construction paper and label the angles A, B, and C. Then snip off each of the three angles and rearrange them so that angles A, B, and C are adjacent. What is the measure of the sum of angles A, B, and C? What have you shown? Was this a proof? Did the parallel postulate play a part in this activity? Explain.

If you did the launch activity correctly, you will have given an informal demonstration that the sum of the angles of a triangle is 180 degrees. But, you were probably aware that this did not constitute a proof. It was also not obvious what role, if any, the parallel postulate played in your demonstration. After reading this section, you will get a better idea of the critical role of the parallel postulate in the different geometries you will encounter.

13.3.1 What Can We Prove with the Parallel Postulate?

We began Chapter 4 with the simple result that the area of a right triangle was one half base times height. If you follow the development in Chapters 4 and 5, we used this result to derive all the area formulas for polygons, circles, the Pythagorean Theorem, all the laws of similarity, the trigonometric laws, laws about secants and tangents, and many more. Thus, literally, almost the entire secondary school curriculum was developed from the area of a right triangle! Isn't that amazing? Who would have thought that the formula for the area of a right triangle was so important? The proof of that formula depended on the fact that every right triangle is half a rectangle. But how do we really know that every right triangle is half a rectangle?

You might smile, and say, "Here is how. If we are given right triangle *ABC* as shown in Figure 13.12 below,

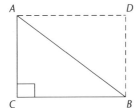

Figure 13.12

we can simply draw a line through *A* parallel to *BC* and another through *B* parallel to *AC* and form quadrilateral, *ADBC*, which is a rectangle."

But, we now ask, "How do we know that we can draw the parallel lines that we said we could draw?"

At this point you may be thinking this is just pedantic on our part. "Of course, you can draw a line parallel to a given line! Just use the parallel postulate!" Ah yes! The parallel postulate! That would do it ! (See below.)

When Euclid wrote his *Elements*, he began with certain assumptions that he felt were obvious. These were:

A(1) For any two distinct points *P* and *Q*, there is a unique line containing them.
A(2) Given a line segment *AB* and another line segment *DE*, we can always extend *AB* to *AC* where *BC* is congruent to *DE*.
A(3) For every point *O* and every point *A* not equal to *O*, there is a circle with center *O* and radius of length *OA*.
A(4) All right angles are congruent to one another.
A(5) (**Parallel postulate**) From a point P outside a line *l*, one can draw a unique line *m* parallel to *l*. (Parallel lines in Euclidean geometry are defined as those which never meet no matter how far extended. This is not the form in which Euclid stated it, but is equivalent to Euclid's statement.)

These assumptions, accepted without proof, were called **axioms** or **postulates**.

Axioms $A(1)$–$A(4)$ were simple, clear, and acceptable for most people. But axiom $A(5)$, the parallel postulate, was not readily accepted. People felt that it was much more complex than the other axioms and should be provable from them. Euclid himself seemed somewhat unhappy about $A(5)$ and thus, tried to prove as many theorems as possible without using $A(5)$. Indeed, he was able to prove 28 theorems using only $A(1)$–$A(4)$. But eventually he had to use it.

What happened next was interesting. For over 2000 years mathematicians tried to prove $A(5)$ from the other axioms. No one succeeded. Many false proofs were proposed. Some tried replacing $A(5)$ by a simpler axiom, which would allow them to prove $A(5)$. But these simpler axioms turned out to be equivalent to $A(5)$ which, for all practical purposes meant that they were really assuming $A(5)$ without knowing it. So proving $A(5)$ was difficult to do. Eventually, it was proved that $A(5)$ cannot be proved from the other axioms. That is, it is completely independent of them. And this put an end to 2000 years of hard work, but not before some new geometries were discovered.

One of the most important theorems, with which we are very familiar, is that the sum of the angles of a triangle is 180 degrees. Let us review the proof. Take note of the heavy use of the parallel postulate.

Theorem 13.3 *The sum of the angles of a triangle is* 180 *degrees.*

Proof. Given triangle ABC. Through A draw a line parallel to BC. (See Figure 13.13 below.)

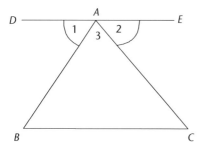

Figure 13.13

We know that

$$\angle 1 + \angle 2 + \angle 3 = 180° \tag{13.14}$$

since a straight angle has 180 degrees. Since DE is parallel to BC, angle 1 equals angle B and angle 2 equals angle C. (When two parallel lines are cut by a transversal, the alternate interior angles are equal.) Replacing $\angle 1$ in equation (13.14) by $\angle B$ and $\angle 2$ by C equation (13.14) becomes

$$\angle B + \angle C + \angle A = 180°.$$

∎

Let us reflect on this proof. To prove this theorem, we needed to draw the line parallel to the base BC and then use facts about alternate interior angles being equal when two parallel lines are cut by a transversal. Thus, it is clear that this proof depends very heavily on the parallel postulate.

We will return to this shortly.

13.3.2 What Can We Prove Without the Parallel Postulate?

It is natural to ask, "Which are the theorems that we can prove without the parallel postulate?" There are many. Here are some of them. Take particular note of (T10). That one is disturbing.

(T1) If two lines cut by a transversal have alternate interior angles equal, then the lines must be parallel. (To prove the converse we need the parallel postulate.)

(T2) Two lines perpendicular to the same line are parallel to each other.

(T3) If P is any point not on a line l, then there exists *at least* one line through P parallel to l. (Remember we are *not* assuming $A(5)$!)

(T4) An exterior angle of a triangle is greater than either of the remote interior angles.

(T5) Two triangles are congruent if two angles and a side of one are equal to two angles and the side of the other.

(T6) Two right triangles are congruent if the hypotenuse and leg of one are equal to the hypotenuse and leg of the other.

(We proved this in Chapter 5 using the Pythagorean Theorem, but the Pythagorean Theorem is not one of those theorems that can be proved without $A(5)$! So the proof we refer to here does not use the Pythagorean Theorem.)

(T7) Every angle has a unique bisector.

(T8) In a triangle, the greater angle lies opposite the greater side and vice versa.

(T9) If A, B, and C are three non-collinear points, then $AC < AB + BC$ where AB, AC and BC represent the lengths of the sides of the triangle. (Another way of saying this is that the shortest distance between two points is a straight line.)

(T10) The sum of the degree measures of a triangle is *less than or equal to* $180°$.

This last result may trouble you. For we "know" that the sum of the angles of a triangle is equal to $180°$. Actually, we don't know this without the parallel postulate. Remember that our proof that the sum of the angles of a triangle being 180 degrees used the parallel postulate.

You may be thinking "Surely it must be true, even without the parallel postulate. There must be another way to prove it without the parallel postulate."

The mathematician Gauss (1777–1855), who was the foremost mathematician of his day and certainly considered by many to be one of the greatest mathematicians who ever lived, decided to take up this problem of proving that the sum of the angles of a triangle had to be 180 degrees but without the parallel postulate. He too couldn't believe that the sum of the angles was anything other than 180 degrees. And he couldn't believe that the only way to prove it was to assume the parallel postulate, $A(5)$. So he didn't assume it.

He said there could only be two choices. Either the sum was less than 180 degrees or it was greater. Using the fact that lines are *infinitely long*, he showed that the sum of the angles of the triangle couldn't be more than 180 degrees. But when he considered the case of the sum of the angles being less than 180 degrees, he could find no contradiction. He soon became convinced that there existed a new kind of geometry, completely consistent with axioms $A(1)$–$A(4)$ where the sum of the angles of a triangle could be less than 180 degrees!

This was mind boggling. He decided not to publish this result for in his words, "I fear the howl of the Boetians if I speak my opinion." That is, his idea was so radical that he feared people would think he was crazy. So he let it lie and kept his thoughts on paper, in a dresser drawer.

Then Janos Bolyai (1802–1860), a Hungarian mathematician entered the scene. His father, Farkas, had made a career out of trying to prove the parallel postulate, and his son Janos, like him, decided to enter the game, even though his father warned him not to. He told his son, "I have traversed this bottomless night which extinguished all light and joy of my life... I entreat you, stay away from the science of parallels."

His son would not hear of it. With typical youthful bravado, he was going to try. And try he did, but with no success. However, he did come to the conclusion that he had, in fact, discovered this strange new geometry where the parallel postulate did not hold. He published his results in an appendix to an 1832 work by his father Farkas. Farkas, years earlier, seeing his son's talents, asked Gauss via a letter, to let Janos be his apprentice. Gauss never responded to his letter, perhaps because at the time he was having trouble with his own son who had run away. So when Janos discovered this new geometry, his father was quick to send this appendix to Gauss. It was probably his way of saying to Gauss "You see, I told you he was good."

Gauss's response devastated Janos. He told him that he had discovered the results decades before, but at the same time told him how "overjoyed" he was that this was the discovery of the son of an old friend "who outstrips me in such a remarkable way." In fact, Gauss described Janos as a genius to one of his associates. Janos, who was known as a man of fiery temperament, was angry with Gauss and felt that Gauss was trying to take credit for his (Janos') results. Gauss' response was so upsetting to Janos, that he went into a deep depression and never published anything again.

So what is this strange geometry that we are talking about that both Gauss and (Janos) Bolyai discovered? It is called **hyperbolic geometry**. In fact, there is another Geometry that we will present too called **spherical geometry**. These new geometries opened up new worlds and both have found uses in the sciences, one in Einstein's Theory of Relativity.

Let us look again at the facts we can prove without the parallel postulate. In particular, let's look at (T3). That statement tells us that we can prove that there is at least one line parallel to a given line from a point outside. Does that mean that there might be more than one parallel line? Well, that too turned out to be the case.

Before we get into a discussion of hyperbolic geometry and spherical geometry, we need to take one last digression and discuss some issues in basic geometry—describing a line and a point, which brings us to the section on undefined terms. This will be the last section before we investigate the Hyperbolic world.

Student Learning Opportunities

1 In exercise 1(b). of the second section of Chapter 1, we asked you to prove (T4). If you haven't already done so, prove (T4).

2 Using (T4), prove (T1).

3 Using (T4), prove (T2).

4 Beginning with T(4), prove T(10).

5 Try proving T(6) without the Pythagorean Theorem. [Hint: Place the triangles so that their equal legs coincide and their right angles are adjacent. This will form a large isosceles triangle. Use this to show that the given triangles are congruent by AAS.]

6 (C) You have asked your students to use a dynamic geometry software program to investigate the relationships between the sides and angles within triangles. They have arrived at the conjecture that, in a triangle, the greater angle lies opposite the greater side (part of T(8) in this chapter), but they are having trouble devising a proof. Devise a proof for them.

[Hint: Begin with triangle ABC, and assume that AC is greater than AB. First show that angle B can't equal angle C. After doing that go back to the original triangle ABC. Mark off on AC a distance AD equal to AB and draw BD. Now use T(4) to show that the measure of angle ADB is greater than angle DCB. Finish it.]

7 Using the previous problem as a tool, show that, if in triangle ABC the measure of angle A is greater than the measure of angle B, then BC is greater than AC.

13.4 Undefined Terms

LAUNCH

Suppose your English is weak and you are looking up the word: "chair" in the dictionary. According to 'Webster's dictionary, a chair is "a seat with a back." But, since you don't know what "seat" means, you look that up. According to Webster, a seat is "a place to sit." Now since you don't know what "sit" means, you need to look up the word "sit." You find, "take a seat." But you are still trying to find out what the word "seat" means. How can you ever learn what the word "chair" means? What does this example have to do with learning terms in geometry?

Are you wondering what the point was of our launch question? Our point is, that there are only a finite number of words in the English language, and if we look up any word, it will be defined in terms of other words, which in turn will be defined in terms of other words, and so on. Eventually, we will come back to some of the same words we came across when we tried to define our original term. What this illustrates, is that it is impossible to define every single word. Some words are intuitive and we must just leave them as undefined.

This is no different in mathematics. We cannot define all our terms. And some concepts are so abstract that trying to define them clearly can actually lead to all sorts of confusion. Euclid attempted to put geometry on a firm foundation by defining all terms. He began with the definition of a point. A point is "that which has no part" and a straight line "that which lies evenly with the points on itself."

If you are thinking, "What does this mean?" you are not alone. Although Euclid tried to express our intuitive notions of a point as a "dot" on a page and a line as the result of penciling along a straight edge on a page, his resulting definitions don't make sense. In fact, mathematicians of the 19th and 20th century felt that the words "point" and "line" should be left undefined. Rather, one should describe the relationships between words, that is that lines should consist of points. This led to the following definition.

A **geometry** is a set S, whose elements we call points. Certain subsets of S are called lines. And so lines, being subsets of S, contain points.

This is the broadest definition of a geometry. But without any axioms to work with, this leads to little, if any, consequences. A structure of relationships is needed, and this can be done in many ways.

Here is a possible set of relationships that can be required of points and lines, which certainly hold in Euclidean Geometry.

(*I*1) *For any two distinct points P and Q in S there is a unique line containing them.*
(*I*2) *Every line contains at least two points.*
(*I*3) *There exist at least three non-collinear points. That is, there are three points which don't all lie on the same line.*

Any system in which axioms $I(1)$–$I(3)$ holds is known as an **incidence geometry**. Here is an example of an incidence geometry with 3 points that shows us that we have not fully captured Euclidean Geometry with axioms $I(1)$–$I(3)$.

Example 13.4 *Let us consider the set S = {a, b, c} and call the elements of S "points." Call a subset consisting of two distinct elements a "line." Show that S is an incidence geometry. In this geometry, parallel lines are lines that have no points in common. What unexpected things can we say about parallel lines in this geometry?*

Solution. To show that this is an incidence geometry, observe that the lines in this geometry are the sets {a, b}, {a, c}, and {b, c}. It is clear that, if we pick any two "points", there is a line containing them. For example, if we pick the points *a* and *b*, then {a, b} is the line containing them, since it has both of these as elements. Thus $I(1)$ holds. That every "line" contains at least two "points" is obvious since every line consists of precisely two points. $I(3)$ requires that there exist 3 points that don't lie on any line. The three points are *a*, *b*, and *c*. No "line" contains all three of them, since lines only contain two "points."

We observe that there are only three lines in this geometry, {a, b}, {a, c} and {b, c}, and any two of them have a point in common. Thus, in this geometry there are *no parallel lines*.

Here is another example of an incidence geometry with 4 points, where the results about parallel lines are similar to those in Euclidean geometry.

Example 13.5 *Let the points be the four letters, a, b, c and d. Let the lines in this geometry be the sets {a, b}, {a, c}, {a, d}, {b, c}, {b, d}, and {c, d}. This geometry satisfies I(1)–I(3) as you will verify in the Student Learning Opportunities. But in this geometry, it is true that from a point outside a given line, l, there is one and only one line parallel to the given line. Illustrate this last statement.*

Solution. Let us illustrate this with one example. Consider the line {a, b}. *c* is a point not on that line ("not on" means not contained in the set we are calling a line). The only line parallel to {a, b} (not having anything in common with *ab*) containing *c* is {c, d}. Thus, there is only one line through *c* which is parallel to *ab*.

To finish this example, we have to consider all cases of lines and points not on the line and show in each case that there is one and only one parallel line to it. We leave this as a Student Learning Opportunity.

In the Student Learning Opportunities we will give an incidence geometry where strangely, there are *more than two parallel lines* through a point not on the line. Of course, the word "line" now means just a set of points and has no visual representation that we are accustomed to. In summary, we see that, with the incidence axioms, we can have one line, no lines, or many lines parallel to a given line.

692 Introduction to Non-Euclidean Geometry

Do these incidence geometries we discussed satisfy Euclid's axioms $A(1)$–$A(4)$? Well, they really can't, since there is no notion of distance. So, we can't even talk about circles with radius r, which is what axiom $A(2)$ deals with.

Our goal in the next section is to develop a geometry that satisfies Euclid's first four axioms, but not the fifth. This will lead us directly to hyperbolic geometry. If, in our geometry, we wish to discuss such concepts as congruence and area, we will need to define the notion of distance and its properties. The properties of distance we give you in the next section will be familiar. The consequences will most likely surprise you.

Student Learning Opportunities

1 **(C)** One of your students, Maxwell, is very confused about why you have to leave the terms "point" and "line" undefined in Euclidean geometry. He claims to know what they mean, so doesn't that mean they have a definition? How can you help Maxwell understand this situation?

2 Verify the claim made in Example 13.5 that there is only one line parallel to the given line through a point not on the line.

3 Consider a geometry with five points a, b, c, d, and e. Let the lines consist of sets of two points. There are ten lines in this geometry. Show that, for every point P not on a line l, there are at least two lines parallel to l.

4 Prove the following theorems in an incidence geometry:
   (a) If l and m are distinct lines that are not parallel, then l and m have a unique point in common.
   (b) There exist three different lines that don't intersect in the same point.
   (c) For every line, there is at least one point not on it.
   (d) For every point, there is at least one line not passing through it.
   (e) For every point P, there exist at least two lines through P.

13.5 Strange Geometries

LAUNCH

For this launch you will need to have a large rubber ball, the size of a dodge ball, as well as a marker. Imagining the ball as a model of the earth, draw a line around the equator. From the tip of the north pole, draw a line perpendicular to the equator. From the same point, draw a second line, different from the first, that is also perpendicular to the equator. You have now drawn two lines perpendicular to the one line, the equator. Didn't you learn that, from a point outside a line, only one perpendicular line could be drawn? What seems to be the problem?

We hope that this launch activity has you completely baffled. Read on to get more insight into what is happening.

We said that a geometry would consist of a set of objects called points and lines were simply sets of points. We also saw in the last section that geometries can exist where for a given line there are (a) no lines parallel, (b) one line parallel, or (c) many lines parallel to it. Thus, we can expect strange things to happen when we abstract things. This ability to freely abstract leads us to our first geometry that satisfies Euclid's axioms.

13.5.1 Hyperbolic Geometry

There are many different models for hyperbolic geometry and all are essentially the same. The one we discuss here is called the **Poincaré model**. We will refer to this as **hyperbolic space**, or the **hyperbolic plane**, or when we want to be more informal, the **hyperbolic world**. In this model our geometry consists of the set of "points" in the interior of a circle C of radius 1. Our "lines" consist of arcs of circles that intersect our circle at an angle of 90 degrees together with diameters of C. (See Figure 13.14.)

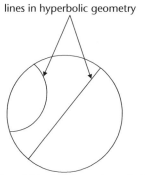

Figure 13.14

What does it mean to intersect the circle at 90 degrees? More generally, what do we mean by the angle between two curves? The angle between two curves at an intersection point is defined to be the angle between the tangent lines drawn to the curves at the point of intersection. So, for two curves to meet at right angles at a point, we mean that, if we draw the usual Euclidean tangent lines to the curves at that point, they meet at right angles.

In Figure 13.15 below we see two circles intersecting at right angles at A as measured by the tangent lines drawn at A.

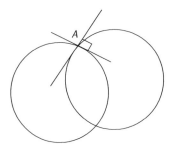

Figure 13.15

694 Introduction to Non-Euclidean Geometry

In Figure 13.16 below, C represents our hyperbolic world. Lines are defined to be arcs of circles intersecting C at right angles, together with diameters of C. Thus, the arc AB is a line, as is the diameter $A'B'$.

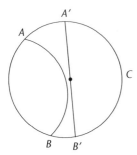

Figure 13.16

Since these lines don't intersect, AB and $A'B'$ are parallel in this geometry (parallel meaning having no points in common). In this geometry, there are many lines parallel to AB. Any diameter of C that doesn't intersect AB is a line parallel to AB. *Thus in this geometry, there are infinitely many lines parallel to AB.*

Given a line in hyperbolic space, if we pick two points on that line and connect them, we get what is called a **line segment**. Thus, in the figure below, we show three line segments, AB, BC, and CA. What do we mean by a **triangle** in this geometry? Well, it is a closed figure formed by three line segments. Figure 13.17 below gives a picture of a hyperbolic triangle. The line segments AB, BC, and CA are called the **sides of the triangle**.

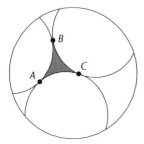

Figure 13.17

The **angle** between two sides of a triangle in this world is defined to be the (usual Euclidean) angle between the tangents to the arcs making up these angles. In the Figure 13.18 below, using dotted lines, we have drawn the Euclidean triangle ABC from the picture above as well as the hyperbolic triangle. We have drawn the hyperbolic angle A, which is measured by angle DAE between the tangent lines to the arcs AB and AC.

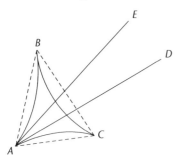

Figure 13.18

We can see that the hyperbolic angle at *A* has a measure smaller than the angle *A* in Euclidean triangle *ABC*, since it is measured by the tangent lines *AE* and *AD* to arcs *AB* and *AC*, respectively, which are between the sides *AB* and *AC* of the Euclidean triangle. Similarly, the measure of the hyperbolic angle at *B* is less than the angle *B* in Euclidean triangle *ABC*, and the measure of hyperbolic angle *C* is less than the Euclidean angle *C*. Since in the Euclidean triangle, the sum of the angles is 180 degrees, it follows that, in this hyperbolic geometry, the sum of the hyperbolic angles of the hyperbolic triangle is less than 180°. Thus we have

> **Theorem 13.6** *In hyperbolic space, the sum of the angles of a triangle is less than* 180 *degrees.*

13.5.2 Euclid's Axioms in the Hyperbolic World

Euclid was able to arrive at the results on congruence just from his first four axioms without even discussing parallel lines. One could follow his proofs verbatim in this hyperbolic world, and prove theorems such as two triangles are congruent if it was true that *SAS = SAS* or *ASA = ASA* and so on. One simply has to think of a "line" and a "triangle" in the context given in the previous section. But, we need axioms *A*(1)–*A*(4) to be able to do this. Let us examine *A*(1)–*A*(4) in this new Hyperbolic world.

A(1): If *P* and *Q* are two points in our hyperbolic world, then it can be shown that there is one and only one "line" that passes through them. This line hits circle C in two points *R* and *S* as shown in Figure 13.19 below.

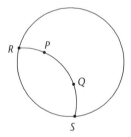

Figure 13.19

A(2) (Extending a line segment any desired distance.) Given your previous notion of Euclidean geometry, you might think this axiom presents a problem. For, if you think in terms of ordinary distance, if we extended the line far enough, we would eventually go outside our circle. Since the circle is our world, this would not be allowed. So, if there is any hope for this axiom to hold in this world, we must create a new notion of distance so that this is true.

In the hyperbolic world, there will have to be some kind of distortion of distances or else how could we extend the line segment any distance without leaving the circle?

Furthermore, if we want to mimic some of the theorems we have in Euclidean geometry, the definition of distance in the hyperbolic plane will have to satisfy some of the conditions that we ordinarily associate with distance in the Euclidean plane.

What are some of these properties that ordinary distance satisfies? Well, if *P* and *Q* are points, the distance from *P* to *Q* should be some nonnegative quantity and be zero if and only if *P* = *Q*. Another property that the distance function must satisfy is that the distance from *P* to *Q* be the same as the distance from Q to P. A third important characteristic of Euclidean geometry is that the shortest distance between two points is a line. So, if you travel from P to Q via some point R not on the line, then you will have traveled a greater distance than if you had gone directly from P to Q. Stated mathematically, if P, Q, and R are three points, the distance from P to Q is less than or equal to the distance from P to R added to the distance from R to Q.

To summarize, if we denote the distance between two points P and Q in this world as $d(P, Q)$, then $d(P, Q)$ must satisfy the following **distance properties**:

(D1) $d(P, Q) \geq 0$
(D2) $d(P, Q)$ equals zero if and only if $P = Q$
(D3) $d(P, Q) = d(Q, P)$
(D4) If P, Q and R are 3 points then $d(P, Q) \leq d(P, R) + d(R, Q)$

It was discovered by Poincaré that the following hyperbolic distance formula for the distance. $d(P, Q)$ between two points in our hyperbolic world satisfies $D(1)$–$D(4)$ and makes $A(2)$ true:

$$d(P, Q) = \left| \ln \left(\frac{PR}{PS} \div \frac{QR}{QS} \right) \right|$$

where PR, PS, QR, and QS are the ordinary distances in the plane and we are using Figure 13.20.

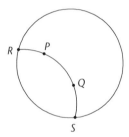

Figure 13.20

(The motivation for this definition is beyond the scope of this book.)

Because of this distortion of distances, lines that have the same length in hyperbolic geometry may appear to have different lengths. Below, in Figure 13.21 we show three lines each of length 1 in the Hyperbolic world according to this definition of distance. Notice that the closer the points are to the boundary of our circle, the smaller the length 1 appears to our eyes.

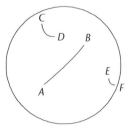

Figure 13.21

Because of this distortion, triangles that look different to our eyes may, in fact, be congruent. Below in Figure 13.22 we show you a figure of two congruent triangles in hyperbolic space.

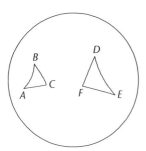

Figure 13.22

Also, because of this distortion of distance, the midpoint of a line segment (that point at equal distance from the endpoints) may not be at what appears, to our eyes, to be the midpoint. For example, in Figure 13.23 below, C is the midpoint of AB. One must remember that we are using the distance function defined above to measure distances in this world in a circle.

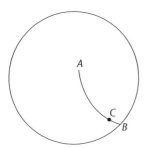

Figure 13.23

No doubt, this is all feeling very strange to you. So, to get a better feel for this, we recommend that you go to this wonderful website where you can experience the Hyperbolic world: http://www.cs.unm.edu/~joel/NonEuclid/NonEuclid.html. It is worth the trip! You can draw figures, measure angles and distances and it behaves a lot like Geometer's Sketchpad.

For now though, we continue summarizing some features of the hyperbolic world.

Since we have a notion of distance, we can now define a circle with center at any point in the hyperbolic world. A **circle** is simply the set of all points at a fixed distance from a center O. So $A(3)$ now automatically holds. Hyperbolic circles look no different from ordinary circles, only the center of the circle is not at the visual center of the circle because of the strange notion of distance we have. Below in Figure 13.24 we have drawn a hyperbolic circle together with its center:

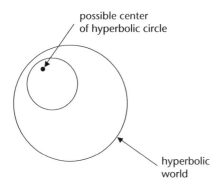

Figure 13.24

698 Introduction to Non-Euclidean Geometry

Being that A(4) holds in the Euclidean plane, and since angles are measured in the usual Euclidean way, between tangent lines to curves, A(4), will hold in the hyperbolic plane.

So having established that A(1)–A(4) hold, we can now prove many of the usual theorems proved in geometry. All of T(1)–T(10) for example can be proven. Also, we can prove that, if two lines in this world intersect, then vertical angles are equal and that any straight angle has 180 degrees. We can also prove the congruence theorems. Thus, if two hyperbolic angles of one triangle and the included side are equal in measure, respectively, to two hyperbolic angles and the included side of the other triangle, then the triangles are congruent. But, while we can prove these ordinary theorems, there are several extraordinary results. The first one we saw was that the sum of the angles of a triangle was less than 180 degrees. Here is an even more surprising result.

> **Theorem 13.7** *In hyperbolic space, if three angles of one triangle are equal in hyperbolic measure to three angles of another triangle, then the two triangles are **congruent**. (Here congruence takes on its usual meaning, namely the lengths of the three sides of one triangle match those of the other triangle and the measures of the corresponding angles in the two triangles are the same.)*

Proof. Suppose that we are given triangles ABC and DEF shown in Figure 13.25 below, and suppose that $m\angle A = m\angle D$, $m\angle B = m\angle E$ and $m\angle C = m\angle F$.

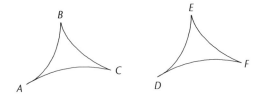

Figure 13.25

We will ultimately show that $AB = DE$ and it will follow that triangles ABC and DEF are congruent by AAS. The proof is a bit circuitous, so it requires a careful read. It is a proof by contradiction, but establishing the contradiction is a lot of work.

So, suppose that $AB < DE$. (If $DE < AB$ a similar proof works.) Then we can copy the length of AB to DE as shown below. (That we can do this is more sophisticated than you might think.) We call that copy GE. So $AB = GE$. (See Figure 13.26 below.)

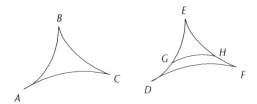

Figure 13.26

Also, copy angle A to that picture at G and call that angle EGH. (See Chapter 14 to see how that is done in the Euclidean plane. A similar construction can be done here.) So $\angle A = \angle EGH$. Since we are given that $\angle A = \angle D$ we have, using the last equation that

$\angle EGH = \angle D.$ \hfill (13.15)

Now since triangles *ABC* and *GEH* have two pairs of congruent angles and the included sides *AB* and *GE* are congruent, the triangles *ABC* and *GEH* must be congruent by $ASA = ASA$. Thus, $\angle C = \angle EHG$. Since $\angle C$ was given to be equal to $\angle F$ we have

$$\angle EHG = \angle F. \tag{13.16}$$

Next we will show that quadrilateral *GHFD* has an angle sum of 360°. Since, as we will show, in hyperbolic geometry this is not the case, we will arrive at our contradiction.

Since angles *EGH* and *HGD* are supplementary,

$$\angle HGD = 180 - \angle EGH = 180 - \angle D \tag{13.17}$$

by equation (13.15). In a similar manner,

$$\angle GHF = 180 - \angle EHG = 180 - \angle C. \tag{13.18}$$

If we add the angles of quadrilateral *GHFD* we get

$\angle GHF + \angle F + \angle D + \angle DGH$

$= \angle GHF + \angle C + \angle D + \angle DGH$ (Since we were given $\angle F = \angle C$)

$= 180 - \angle C + \angle C + \angle D + 180 - \angle D$ (By equations (13.18) and (13.17))

360 degrees. (Simplifying)

Now we ready to get our contradiction. We break quadrilateral *GHFD* into two triangles by drawing a "line" from *D* to *H*. Since, by Theorem 13.6, each of the two triangles has the sum of its angles less than 180 degrees, the quadrilateral *GHFD* must have the sum of its angles less than 360 degrees. This contradicts what we already showed, namely that the quadrilateral had an angle sum of 360 degrees. ∎

Our contradiction arose from first assuming that *AB* was not equal to *DE*. It must follow that $AB = DE$ and therefore that triangle *ABC* is in fact, congruent to triangle *DEF* by $ASA = ASA$.

What the above theorem is saying is that, in the hyperbolic world, there is no difference between congruence and similarity. If two triangles are similar, they are congruent! Isn't this remarkable?

In the course of proving the above theorem, we broke the quadrilateral into two triangles, and since the sum of the angles of the two triangles was less than 360°, the sum of the angles of the quadrilateral would have to be less than 360°. This leads to our next result:

Corollary 13.8 *In hyperbolic space, there are no rectangles. (A rectangle is a 4 sided figure with 4 right angles.)*

The fact that there are no rectangles in hyperbolic space leads to some other interesting results. For example, if rectangles don't exist, then neither does the Pythagorean Theorem. At least it seems that way, since all the proofs that we have given for the Pythagorean Theorem made use of areas, which in turn made use of areas of right triangles, which in turn made use of the fact that a right triangle is half of a rectangle. Since there are no rectangles, none of the proofs we gave would have been possible. This seems to indicate that there is no Pythagorean Theorem in the hyperbolic world, unless, of course, someone could figure out a way to prove the theorem without the use of rectangles. However, this won't happen, for it can be shown that, *if* the Pythagorean Theorem holds in a geometry satisfying A(1)–A(4), that geometry *must be* Euclidean and A(5) *must* hold. So it is indeed true that there is no Pythagorean Theorem in the hyperbolic plane since it is not Euclidean as the last two theorems have indicated. A consequence is that a geometry satisfies the Pythagorean Theorem if, and only if, it satisfies the Parallel Postulate. Thus, the Pythagorean Theorem is, in fact, equivalent to the Parallel Postulate. How surprising!

There is an interesting corollary of this theorem:

> **Corollary 13.9** *The sum of the angles of a triangle in hyperbolic space is not constant. Different triangles have different angle sums.*

How remarkable! Here is the proof:

Proof. Suppose that the sum of the angles of any triangle in hyperbolic space is a constant K. Begin with triangle ABC. Make AB shorter than it is by moving A. Call the moved point A, A'. At A' copy angle A, and let the side of A' meet AC at C'. (See Figure 13.27 below where $\angle 1 = \angle 2$.)

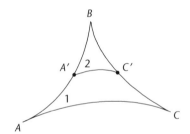

Figure 13.27

If the sum of the angles of a triangle were constant, then since two angles of triangle ABC and $A'BC'$ are the same ($\angle 1 = \angle 2$ and B is common), it must follow that $\angle C = \angle C'$. So, the three angles of triangle $A'BC'$ are the same as the three angles of ABC. It follows from Theorem (13.7) that triangles ABC and $A'BC'$ are congruent. But, how can this be since $A'B$ is shorter than AB? We have our contradiction.

Since our contradiction arose from assuming that the sum of the angles of a triangle in hyperbolic space is constant, it must be that the sum of the angles in hyperbolic space is not constant! ∎

A corollary of this is that, if we know two angles of a triangle in hyperbolic space, we cannot determine the third angle! We hope that by now you have noticed that the hyperbolic world is full of surprises!

13.5.3 Area in Hyperbolic Space

In Euclidean geometry, we develop the entire theory of area from the area of a square. After all, we always talk about square units. But in hyperbolic space, there are no squares, since there are no rectangles! So how does one find areas of figures in hyperbolic space? We can't go into too much detail here, since there is a great deal involved, but let us indicate what surely will be a major surprise to you: In the hyperbolic world the area of a triangle does *not* depend on the lengths of its sides and is *not* 1/2 base times height. Indeed, if you computed 1/2 base times its corresponding height for each of the three sides of a single hyperbolic triangle, you would not get the same answers. A visit to the website we mentioned earlier will clearly demonstrate this. So area can't be 1/2 base times height. This does not mean that the area is independent of the sides, only that the usual formula for the area of a triangle does not hold. We will show, however, that *any* notion of area in hyperbolic geometry cannot depend on the sides. This is striking!

So what is it that we require of an area function? Well, any definition of area should satisfy certain rules that area satisfies in Euclidean space. For example, with each region, we associate a nonnegative area. If we denote the area of a region R by $A(R)$, then

$A(R) \geq 0.$

Another property that should hold is that our definition of area should be additive. Thus, if a polygon, say $ABCD$ can be decomposed into non-overlapping triangles (triangles that have only edges and vertices in common, as in the Figure 13.28 below),

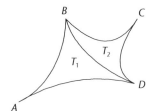

Figure 13.28

the sum of the areas of those triangles add up to the area of the whole polygon. Thus, in the above figure the area of the polygon $ABCD$ should be the sum of the areas of triangles T_1 and T_2.

Finally, we ask that congruent triangles have the same area.

We need one last definition before we get to the big surprise. We saw that the sum of the angles of a triangle in hyperbolic space was always less than $180°$ and was not constant. Different triangles can have different angle sums. We define the **defect** of a triangle to be how much the sum of the angles differs from $180°$. Thus, if the sum of the angles of the triangle was $170°$, the defect would be $10°$. When a polygon is decomposed into non-overlapping triangles, the defect of the polygon is defined to be the sum of the defects of the non-overlapping triangles.

The defect is a function that satisfies the rules that we want to hold for an area function. That is, the defect of a triangle in Hyperbolic space is always ≥ 0. It can be show that the defect of a polygon is the sum of the defects of the triangles in its decomposition and, if two triangles are congruent, the defects are the same. Thus, we could define the defect of a polygon to be it's area. But that seems unreasonable. The following theorem is stunning!

Theorem 13.10 *Any area function in hyperbolic space that satisfies the minimal conditions for area we required above depends only on the defect and is, in fact, a constant times the defect. Thus in any model of hyperbolic space, area will depend only on the defect.*

What the theorem says is that, in hyperbolic space, it is the angles and only the angles of a triangle that determine the area of a triangle. This has to leave you with an utter sense of disbelief!

A consequence is that the area of a triangle cannot get arbitrarily large (since the defect is limited to less than 180°), even though the sides can! The proof of this theorem is quite complex and will not be given here but can be found in many books on hyperbolic geometry.

So you may be wondering of what possible use could this geometry be? Well, Einstein used it in his special theory of relativity with significant results. In fact, many scientists believe that hyperbolic geometry might be the appropriate geometry with which to study the universe. Time will tell whether this true.

We now give a brief overview of a different geometry which is the geometry of the earth: spherical geometry. There are many surprises here also.

13.5.4 Spherical Geometry

There is a famous recreational problem that one often sees in puzzle books, which goes something like this: Jack left his house and took a morning stroll. He first walked 10 meters South, then 10 meters West, and then 10 meters North and found himself back at his house door but facing a bear. What color was the bear?

When most people hear this problem, they think this stroll is impossible and believe that the reference to the bear is just to fool them. But, if Jack lived at the north pole and did exactly what we said above, he indeed would find himself back at home. Things happen on the earth that cannot happen in Euclidean (plane) geometry. And as for the question about the bear's color, well it has to be a polar bear, so it appears (according to the puzzle books) that it must be white!

One other type of geometry that one studies is the geometry on the sphere. It is very different from hyperbolic geometry we presented (geometry in the circle) but is, after all, what happens in our world. The points on the sphere are the "points" in this geometry, and the great circles are our "lines." Recall that a great circle is a circle on the earth's surface that results from cutting the earth with a plane that goes through the center. (We are assuming the earth is spherical. This is not quite true, so if you are unhappy with this, just imagine a genuine sphere instead.) The equator is one great circle, but so is any circle on the surface of the earth that goes through the north and south pole. The earth with this definition of "point" and "line" is an example of something called **spherical geometry**.

It can be shown that any two points on the sphere lie on a great circle. This allows us to define the distance between two points P and Q on this sphere to be the arc distance between them along the great circle. It can be shown that this distance function has all the properties that we talked about in the previous section on hyperbolic geometry. It follows that the shortest distance between two points on the sphere is the "great circle" distance. Consequently, one travels *north* from Florida to Alaska to reach the Philippines (even though Philippines are south of Florida). This is because Florida, Alaska, and the Philippines all lie on a great circle and this gives us the shortest distance to go from one to the other. Stated another way, in spherical geometry, Alaska, Florida, and the Phillipines are all collinear.

Since any two great circles intersect, and the great circles are our lines in this geometry, there are *no parallel lines*. Furthermore, in this geometry, the sum of the angles of a triangle is *more* than 180 degrees as we shall soon see.

Earlier in the chapter in Section 3 we said that, when Gauss began his study of hyperbolic geometry, he ruled out the case that the sum of the angles of a triangle is more than 180 degrees by using the fact that lines were infinitely long. What he failed to note was that in this elliptic geometry, lines would never end since they were circles, but were not infinitely long. That is, he was biased by his Euclidean notion of line. In spherical geometry, the sum of the angles of a triangle can be (and is!) more than 180 degrees.

On the sphere, a triangle, *PCB*, is the intersection of three great circles that don't intersect in the same points. We illustrate such a circle in Figure 13.29 below, where we show one such triangle formed by the equator and two other great circles. Here we see each of the base angles is 90 degrees.

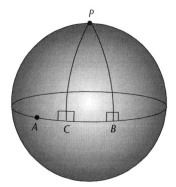

Figure 13.29

If you have been reading very carefully, you might be thinking, "Wait a minute. We pointed out that, if Euclid's first four axioms, $A(1)$–$A(4)$ hold, then there must be at least one parallel line to a given line through a point outside the line." Indeed, we did state that as Theorem T(3). But remember, that was *if* all of Euclid's first four postulates hold. In this geometry, the first postulate does not hold. Given any two points, there is *not* a unique "line" which passes through them. For example, take as the two points the north pole and south pole. Through these points, there are infinitely many great circles ("lines") passing through them. Because not all of Euclid's postulates hold in this geometry, some of the ten theorems we spoke about won't hold either.

Let's examine the theorem for Euclidean geometry that states that the exterior angle of a triangle is greater than either of the two remote interior angles. (This was Theorem (T4).) Using the same picture above, we notice that angle *PCA* is 90 degrees, as are the interior angles *PBC* and *PCB* of the triangle. Thus in this geometry, it is possible for the exterior angle of a triangle to equal one of the remote interior angles. This just corroborates what we said earlier, that some of the theorems of Euclidean geometry simply will not be true.

As you can imagine, there is far more that can be said about these interesting geometries. Since most prospective secondary school math teachers are required to take a course that discusses non-Euclidean geometries, we end this chapter expecting and hoping that many of you will continue your studies of this fascinating subject.

Student Learning Opportunities

1 Show that in hyperbolic space, if a quadrilateral has three right angles, the fourth *must* be acute.

2 Go to the website http://www.cs.unm.edu/~joel/NonEuclid/NonEuclid.html and construct each of the following in Hyperbolic space.
   (a) An isosceles triangle
   (b) An equilateral triangle

(c) An angle bisector in the vertex angle of an isosceles triangle
(d) A right triangle
(e) A parallelogram

3 **(C)** Your students have become very intrigued by hyperbolic geometry and have made the following statements. They want to know whether what they have claimed is true or not. To help you figure out which of the following are true, use the website from the previous problem to draw some figures, then answer their questions about the truth of each statement.

(a) The angles of an equilateral triangle are equal, but they are not each 60°.
(b) The angle bisector of the vertex angle of an isosceles triangle bisects the base.
(c) The Pythagorean Theorem does not hold.
(d) If two parallel lines are cut by a transversal, then alternate interior angles are equal.
(e) Any two equilateral triangles are congruent.
(f) Opposite sides of a parallelogram are equal.
(g) If opposite sides of a quadrilateral are equal, the quadrilateral is a parallelogram.
(h) Big triangles (to our eyes) have big areas while small triangles (to our eyes) have small areas.

4 What is the maximum distance between two points on a sphere of radius 1? How does your answer relate to the statement, (true in Euclidean geometry), that lines are infinitely long?

5 In Euclidean geometry one axiom is: If A, B, and C are three points lying on a line, then only one of them is between the other two. Show that if we work on a sphere and take our "lines" to be great circles, then this axiom fails.

6 After doing a lesson on spherical geometry, you are planning to give your students the following statements. They will have to decide whether they are true or false and give explanations. Decide which are true and which are false and give your own explanation.

(a) Any two lines intersect at a unique point.
(b) A triangle can have three right angles.
(c) The Pythagorean Theorem holds.
(d) A triangle can have more than one obtuse angle.
(e) Through a point outside a line, one can draw one and only one parallel line.

7 Look up Girard's theorem on how to find the area of a triangle on a sphere and write at least two things you find interesting about it. (It involves the angles of the triangle and tells exactly how much the sum differs from 180.)

8 Look up the Law of Sines and Cosines on the sphere. How are they similar to the Euclidean Laws? How are they different?

9 **(C)** After learning about spherical geometry, your students are really confused about why they spent all of their time in school learning about plane geometry, when in reality they live on a sphere and that is the geometry that should most concern us. Why do we study Euclidean geometry?

CHAPTER 14

THREE PROBLEMS OF ANTIQUITY

14.1 Introduction

For many years, geometric constructions were part of every secondary school mathematics curriculum. Students were required to know how to use a straightedge (a ruler without markings) and a compass to construct certain geometric figures. These kinds of problems occupied the Greeks for many years. Some of them have historical significance, intriguing many mathematicians and taking almost 2000 years before it was shown that they could not be done. We will begin with some of the elementary constructions and then proceed to a discussion of these more complex constructions. We tie together many of the concepts discussed in this book and in the secondary school curricula to show, informally, how some of these issues were resolved.

14.2 Some Basic Constructions

LAUNCH

Using just a straight edge (with no ruler markings), draw a line ℓ on a piece of paper. About two inches above the line, mark a point, P. It should look similar to the picture in Figure 14.1 below. Now, using a compass and your straight edge, try to construct a line that goes through point P that is perpendicular to line ℓ. Explain what you did. Prove why it works.

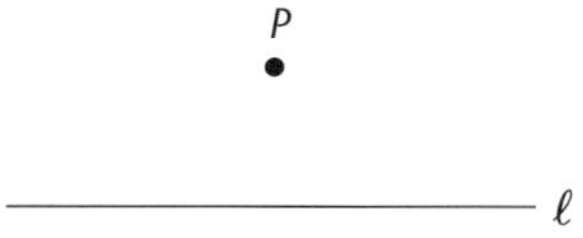

P
•

———————————— ℓ

Figure 14.1

If you are like most math majors, you probably recalled something from your secondary school mathematics education about geometric constructions. Hopefully, you recalled how to construct a line perpendicular to line ℓ from a point, P, outside line ℓ, using just a straight edge and compass. But, did you ever learn why the construction works? Perhaps as you worked on the launch problem, you figured it out. But, in case you didn't we will explain the construction and the proof of why it works now.

We begin by placing one leg of our compass at P and then draw a circle with radius more than the distance from P to the line. This circle will intersect the line in two points A and B. See figure below. Now starting with A as the center, construct two circles one centered at A and the other centered at B with the same radius. (Taking radius AB would certainly be fine.) These two circles will intersect at Q as shown in Figure 14.2 below.

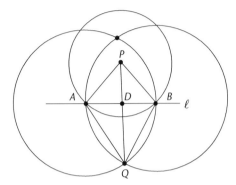

Figure 14.2

Draw the line PQ intersecting the line ℓ at D. We claim that PD is perpendicular to ℓ.

Proof. PA and PB are equal, being radii of the same circle. AQ and BQ, also being radii of congruent circles, are equal, and of course $PQ = PQ$. Thus, triangles APQ and BPQ are congruent by SSS. It follows that their corresponding parts, angles APQ and BPQ are equal. Since $PD = PD$ and $PA = PB$, it follows that triangles APD and BPD are congruent by SAS. It follows that the corresponding angles PDA and PDB are equal, and since they add up to 180, both are right angles. So, line PDQ is perpendicular to line ℓ.

Note that, in the above figure, we drew full circles, but one needs only draw the relevant arcs. Here is the shortened version of our construction. First we draw an arc intersecting the line ℓ at A and B. Then we draw arcs of congruent circles with centers at A and B that intersect at Q. Then we draw PQ using our straightedge as shown in Figure 14.3 below.

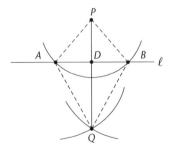

Figure 14.3

Notice that, in this construction, we first found the intersection of the arc AB of the circle with the line ℓ. Then we found the intersection of the two circles with centers at A and B. They intersected at Q. We then used our straightedge to draw the line between P and Q and found the intersection, D of that line with line ℓ. It is these four operations: (1) intersecting circles;

(2) drawing lines; (3) intersecting lines with circles; and (4) intersecting lines with lines that are the building blocks of all constructions.

Let us turn to another construction. This time, instead of using point P outside a line, let us pick P on the line and draw a perpendicular from there. So we begin with a point P on a line ℓ as shown in Figure 14.4 below,

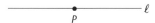

Figure 14.4

and try to draw a perpendicular that passes through P but which is perpendicular to the line.

We begin by drawing a circle with center at P that intersects ℓ in points A and B. Now from A and B draw circles with equal radii greater than the distance from P to A. Let them intersect at Q. Draw PQ. See Figure 14.5 below. We claim that PQ is the required perpendicular.

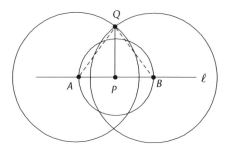

Figure 14.5

A shorter version of this construction is given in Figure 14.6.

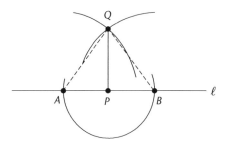

Figure 14.6

Here is the proof that PQ is perpendicular to ℓ: $PA = PB$, since they are radii of the same circle. $AQ = BQ$ since they are radii of congruent circles. Of course $PQ = PQ$. Thus, triangles AQP and BQP are congruent by *SSS*. It follows that angles APQ and BPQ are equal and hence both are 90 degrees since together they make a straight angle.

Let us now examine how to copy an angle. We begin with angle A and line ℓ shown in Figure 14.7. How can we copy angle A to line ℓ?

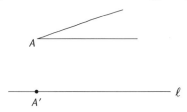

Figure 14.7

We begin by swinging an arc from A so that it intersects the sides of angle A at points B and C. Starting at point A' on ℓ we swing an arc of the same radius which intersects line ℓ at B'. (See Figure 14.8 below.)

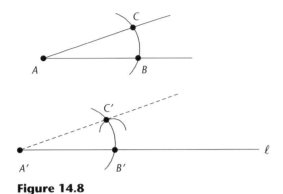

Figure 14.8

We now place the point of our compass on B and open it a distance of BC. Now, starting at B' we swing an arc with radius the length of the line segment BC. This intersects our previous arc in a point C'. (Again, see figure above.) Now draw A'C'. Do you know why angle A' has the same measure as angle A? Here is the reason: By construction, AB = AB', and AC = AC' since they are radii of congruent circles whose centers are at A and A', respectively. Also, by the way we set our compass, BC = BC'. So triangles CAB and C'A'B' are congruent by SSS, from which it follows that angle A = angle A'. So, we have copied angle A.

Now that we have the ability to copy an angle, we can construct a line parallel to a given line from a point outside the line. Here are the steps: (1) Begin with line *l* and point *P* outside *l*. (2) Draw any line from *P* intersecting *l* at *D* and call the angle at *D* angle 1, as indicated in Figure 14.9 below. (3) Copy angle 1 to line PD using *P* as a vertex. Call it angle 2, as indicated in the diagram below. Since the alternate interior angles 1 and 2 are equal, lines *l* and *m* are parallel.

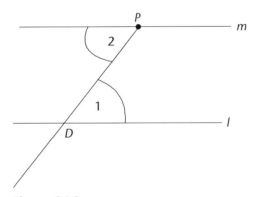

Figure 14.9

Finally, let us investigate how to bisect an angle. We begin with angle ABC shown below. Placing our compass at B, draw an arc intersecting sides BA and BC at D and E, respectively. Now, putting the point of the compass at D we draw another arc, and keeping the legs of the compass at the same distance apart, we put the point of our compass at E and swing an arc of the same radius, intersecting our previous arc at F as shown in Figure 14.10 below.

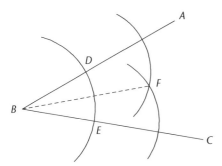

Figure 14.10

Draw BF. Why is BF an angle bisector of angle B? The proof is not hard. Referring to the Figure 14.11 below,

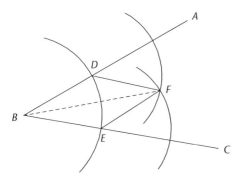

Figure 14.11

BD and BE are equal, being radii of the same circle. So DF and EF are equal, since they are radii of congruent circles. Of course $BF = BF$. So triangle BDF is congruent to triangle BEF and the corresponding angles DBF and EBF are equal, which means that BF is the angle bisector of angle ABC.

Student Learning Opportunities

1 **(C)** You asked your students to use a dynamic geometric software program to draw any triangle and then construct perpendicular bisectors on two of the sides. You then asked them to circumscribe a circle around the triangle, and they all noticed that the center of the circumscribed circle is the intersection of the perpendicular bisectors of the two sides. They are curious about why this is the case. Clarify why, by actually constructing the perpendicular bisectors of two sides of a triangle and then showing that the point of intersection is equidistant from the three vertices. Provide your students with a proof that this always works.

2 Show that, if we construct two angle bisectors in a triangle, that the point, P, of intersection is the center of a circle that can be inscribed in the triangle. The radius of that circle is the perpendicular distance from P to the side of a triangle.

3 Construct two congruent tangent circles using only straight edge and compass. Show why the construction works.

4 Construct three congruent tangent circles using only straight edge and compass. Show why the construction works.

5 Construct an equilateral triangle using only straight edge and compass. Show why the construction works.

6 Construct a line segment of length $\sqrt{ab}$ where $a, b > 0$. Show why the construction works.

7 In each of the following, a construction is described. Show why the construction works.

  (a) Constructing a circle through three given points A, B and C: Construct the perpendicular bisectors of line segments AB and BC and suppose that they intersect at P. Then, the circle with center P and radius PA also passes through points B and C.

  (b) Constructing a line through a given point P outside a circle with center at C, which is tangent to the circle: Draw PC and then construct its perpendicular bisector, giving you the midpoint M of PC. Construct a circle with center M and radius MC, intersecting the original circle at points R and S. Show that PR and PS are tangents to the circle.

  (c) Divide a segment AB into n congruent segments: Draw a different line through A and on it construct equal segments with a compass. Let the last segment intersect the line at F. Draw an arc with center A and with radius BF. Draw another arc with center B and radius AF. Suppose these two arcs intersect at D. Show that $ADBF$ is a parallelogram. Now along DB mark off the same segments you marked off along AF and connect the endpoints of the corresponding segments. These new line segments will divide AB into n congruent segments.

14.3 Three Problems of Antiquity and Constructible Numbers

LAUNCH

Using a straight edge, draw an angle that measures approximately 60 degrees. Using what you learned from the previous sections, construct the following, using only a straight edge and compass.

1 Divide the angle into four equal angles.
2 Divide the angle into three equal angles.

You probably had no difficulty drawing the first construction of the launch, since all you had to do was to bisect the first angle and then bisect each of the two smaller angles. But, surely you encountered difficulty drawing the second construction. As you read this section, you will find out why it was so difficult!

We have seen how to construct a perpendicular to a line segment from a point outside the line segment, how to draw a line parallel to a given line, how to bisect an angle, and how to copy an angle. There are many other constructions that one can do, and among geometers, determining which other constructions could be done with straightedge and compass became a bit of a game.

Many problems were posed, asking people to construct things with only straightedge and compass, and most were either solved or shown to be not solvable. However, there were three constructions that were posed whose answers were simply not forthcoming. These were:

Problem 1: Is it possible to double the cube, that is to construct using only straightedge and compass, the side of a cube whose volume is twice that of a given cube?

Problem 2: Is it possible to trisect an angle, using only straightedge and compass?

Problem 3: Is it possible to square the circle, that is, construct a square, using only straightedge and compass, whose area is the same as the area of a given circle?

For many centuries mathematicians tried unsuccessfully to solve these problems. It wasn't until the 18th century that they discovered the root of their difficulties. The fact was, these constructions were impossible and they proved it!

Although proving that something is impossible to do is quite difficult, we believe that it is worthwhile for current and future teachers of geometry, to get a sense of what these types of proofs are like. Therefore, we will give an informal presentation of the proofs. You will notice that they connect many of the secondary school topics that one usually discusses, such as solving simultaneous equations, finding roots of polynomials, finding equations of lines and circles, and finally, solving trigonometric identities. The key idea in each of these proofs is the notion of constructible numbers.

14.3.1 Constructible Numbers

Any number that can be formed from the number 1 by using the operations of addition, subtraction, multiplication, division, or extracting square roots of nonnegative numbers, is called a constructible number. Thus, 2 is constructible since it can be written as $1 + 1$. Similarly, 3, 4, 5, and so on are constructible, as are the numbers 0 (which is $1 - 1$), -1, -2, and so on. All rational numbers p/q are constructible, since both the integers p and q are, and according to the definition of constructible number, we are allowed to divide constructible numbers to get new construtible numbers. Numbers like $\sqrt{3}$ are constructible, since 3 is constructible and we are allowed to take roots of constructible numbers to get new constructible numbers. Since we can add constructible numbers to themselves and get other constructible numbers, and since both -1 and $\sqrt{3}$ are constructible, so are $-1 + \sqrt{3}$ and, since we can take square roots, $\sqrt{-1 + \sqrt{3}}$ is constructible, as well as $\sqrt{1 + \sqrt{-1 + \sqrt{3}}}$ and $\sqrt{\sqrt{1 + \sqrt{-1 + \sqrt{3}}}}$. Since we can also divide the constructible numbers that we have created, the number $\dfrac{\sqrt{-1 + \sqrt{3}}}{\sqrt{1 + \sqrt{-1 + \sqrt{3}}}}$ is also constructible. We can continue this way creating all kinds of constructible numbers and then add, subtract, multiply, divide, or take square roots of any of them in any combination to generate new constructible numbers. You may wonder why these numbers are called constructible. Well, read on.

14.3.2 Geometrically Constructible Numbers

The first thing we wish to show, which will take a few paragraphs, is that every constructible number, as defined above, can be constructed with a straightedge and compass, which is why we call them constructible. In what follows, any number constructed with a compass and a straightedge will be called **geometrically constructible**. We will always assume that, if needed, we have available a line segment of length 1 to be used in any geometric construction.

To begin this process, we first observe that, if we have a line segment of length 1, then it is easy to construct a line segment of length 2 using our line segment of length 1. We set our compass legs to a distance of 1, using our line segment of length 1. We draw any line segment. Starting with one leg of our compass at the left endpoint of the line segment, which we call A, we strike off a distance of one. (If needed, we extend the segment with our straightedge.) Say this intersects the line segment at B, then AB has length 1. Now, keeping the legs of the compass exactly the same distance of 1, we put one leg of our compass at B and strike off another arc of length 1, intersecting the line segment at C. Then AC has length 2. Continuing, we can construct a length of 3, 4, and so on. (See Figure 14.12 below.)

Figure 14.12

Thus, all positive integer lengths are constructible.

Next, consider how we might construct a length of $\frac{1}{3}$. Begin with segment OA of length 1. Through O draw line ℓ as shown, and mark off 3 segments of length 1 on line ℓ, calling the first segment OC, the second CD, and the third DE. Connect E to A as shown in Figure 14.13 below,

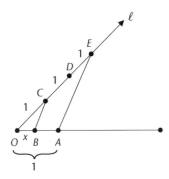

Figure 14.13

and through C draw a line parallel to EA, hitting OA at B. We claim that OB has length $\frac{1}{3}$. This follows, since triangles OCB and OEA are similar (why?), so

$$\frac{OC}{OE} = \frac{OB}{OA}$$

or, put another way,

$$\frac{1}{3} = \frac{x}{1}.$$

Solving this equation for x, we get $x = \frac{1}{3}$. In a similar manner we construct $\frac{1}{q}$ by marking off q segments of length 1 on ℓ and calling the first OC, and proceeding as before. Having constructed a length of $\frac{1}{q}$, we can mark p of these segments off on a line to get a line segment of length $\frac{p}{q}$. Thus, we can construct any rational number. We will show in the Student Learning Opportunities that one can also construct a length a/b, when a and b have already been constructed. Here a and b don't have to be rational.

Finally, if we are given a length of a, do you think we can construct a length of $\sqrt{a}$? If so, how? Here is how to do it. We mark off a length of $AB = a$ and $BC = 1$ on a line segment. (See Figure 14.14 below.) We construct the midpoint O of AC (not shown in the diagram) and then construct a semicircle with center O and radius OA. Finally, through B we construct a perpendicular BD, intersecting the semicircle drawn at D as shown below. We will show that $BD = \sqrt{a}$.

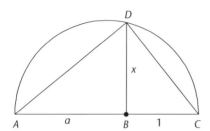

Figure 14.14

Being inscribed in a semicircle, angle ADC is a right angle. Since segment DB is perpendicular to segment AC, angle DBA is a right angle. So both triangles ADC and DBA have a right angle. Also, both triangles have angle DAB in common, so we have that triangles ABD and ADC are similar by AA. By a similar argument, we have that triangle ADC is similar to triangle DBC, and by transitivity, it follows that triangle ABD is similar to triangle DBC. It follows that

$$\frac{AB}{BD} = \frac{BD}{BC}$$

or,

$$\frac{a}{x} = \frac{x}{1}.$$

It follows, by cross-multiplying and taking square roots, that $x = \sqrt{a}$.

In the Student Learning Opportunities we will outline how to construct lengths ab, a/b, $a \pm b$ once we have lengths a and b constructed. Thus, from those examples and what we have shown in the last few paragraphs, if a and b are (geometrically) constructible, then so are $a \pm b$, ab, $\frac{a}{b}$, and $\sqrt{a}$. Thus, since 1, 2, and 3 are (geometrically) constructible, so are $1 + \sqrt{2}$, $\sqrt{1 + \sqrt{2}}$, $\sqrt{2 + \sqrt{3}}$, $\frac{\sqrt{1+\sqrt{2}}}{\sqrt{2+\sqrt{3}}}$, and so on. That is, all constructible numbers are geometrically constructible. Any positive geometrically constructible number has a corresponding negative geometrically constructible number. We simply mark off the positive constructible number to the left of the origin with our compass, and this will represent the corresponding negative constructible number.

14.3.3 The Constructible Plane

For what follows, we imagine that everything is taking place on a coordinate plane. This plane will be somewhat special, and will consist *only* of points (a, b), where a and b are both constructible numbers. This plane will be known as the **constructible plane** and all points in this plane will be referred to as **constructible points**.

We now ask a question: If we work in the constructible plane, and start with constructible points and use circles with centers at constructible points and radii which are constructible numbers, and lines joining constructible points, then will the intersections of any two such circles, or any two such lines, or any such circle and line be constructible points? It turns out that they will! And, we will now explain why.

First we show that every line joining two points whose coordinates are constructible has the form $Ax + By = C$, where A, B, and C are constructible. We suppose that the line connects two points $P = (a, b)$ and $Q = (c, d)$, where a, b, c, and d are constructible. Then the slope of the line is $m = \frac{d-b}{c-a}$ which, being the quotient of constructible numbers, is constructible. Now, the equation of the line joining P and Q (by the point slope formula) is: $y - d = \frac{d-b}{c-a}(x - c)$. Multiplying both sides by $c - a$ and getting all the x's and y's on one side we have

$$(d - b)x - (c - a)y = ad - bc$$

which is of the form $Ax + By = C$, where $A = d - b$, $B = -(c - a)$ and $C = ad - bc$. Of course, A, B, and C are constructible numbers, since they are formed by the operations of addition, subtraction, and multiplication of constructible numbers.

Now, suppose we start with two non parallel lines $ax + by = e$ and $cx + dy = f$ obtained by connecting points with constructible coordinates. Then, of course, their coefficients, a, b, c, d, e, and f will be constructible. If we try to solve this system

$$\begin{cases} ax + by = e \\ cx + dy = f \end{cases} \tag{14.1}$$

for x and y, we can do it by multiplying the top equation by $-d$ and the bottom by b and then add the equations. We we will get the following solutions for x and y which you can verify.

$$x = \frac{(bf - de)}{bc - ad} \tag{14.2}$$

$$y = \frac{(ce - af)}{bc - ad}. \tag{14.3}$$

Now, since coefficients a, b, c, d, e, and f in system (14.1) are constructible numbers, we see that the values of x and y given in equations (14.2) and (14.3) are constructible numbers as well, since they are being formed by the allowed operations of addition, subtraction, multiplication, and division of constructible numbers. So, when solving linear equations with constructible coefficients, we get constructible intersection points (x, y). (The only issue is if $bc - ad = 0$, but this will happen only when the lines are parallel and so there won't be an intersection in that case.)

Let us continue. Suppose we wish to find the intersection of the line $ax + by = c$ and the circle $(x - h)^2 + (y - k)^2 = d^2$, where a, b, c, h, k, and d are constructible numbers. That is equivalent to solving the system of equations

$$\begin{cases} ax + by = c \\ (x - h)^2 + (y - k)^2 = d^2 \end{cases}$$

simultaneously. We can solve for y in terms of x in the linear equation and substitute into the equation of the circle and solve the resulting equation. The solution (obtained by computer), is,
$$x = \frac{-ac + abk - b^2h}{-a^2 - b^2} \pm$$
$$\frac{1}{-a^2 - b^2}\sqrt{2b^3ck + 2ab^2ch - 2ab^3hk - b^2c^2 - b^4k^2 + b^4d^2 - a^2b^2h^2 + a^2b^2d^2}$$ with a similarly looking expression for y. What we wish to notice here, is that if the values of $a, b, c, h, k,$ and d are constructible, so are the values of x and y since they are being formed from the allowable operations of addition, subtraction, multiplication, division, and extracting square roots, all of which are geometrically constructible.

Finally, let us find the intersection of two circles. For example, suppose we wanted to find the intersection of the circles

$$\begin{cases} (x-1)^2 + (y-2)^2 = 25 \\ (x+2)^2 + (y-3)^2 = 9. \end{cases} \tag{14.4}$$

We notice that, if we expand the left side of each equation and subtract the resulting equations, then the x^2 and y^2 terms are eliminated, and we are left with a linear equation which we can solve for y in terms of x and substitute into either equation in (14.4). We are in the same boat as we were before, solving a linear and quadratic equation and our solutions will again be of the form $p \pm q\sqrt{r}$, where $p, q,$ and r are constructible. Thus, our intersection points are constructible.

We summarize our findings in the following theorem where the word "line" means line joining points with constructible coordinates, and "circle" means circle with center at a point that is constructible, and whose radius is a constructible number.

Theorem 14.1 *When we find the intersection of two lines, a line and a circle, or two circles, that is, when we do geometric constructions, the coordinates of the points of intersection are constructible lengths.*

We now come to a key step in our goal: the observation that constructible numbers form a hierarchy. Each constructible number is at a certain level, which intuitively is the number of square roots we have taken to form the number. Thus, rational numbers have no square roots and are at level 0. Numbers like $1 + \sqrt{2}$ have one square root and thus are at level 1, while a number like $\sqrt{1+\sqrt{2}}$ or $\sqrt{2} + \sqrt{3}$ has two root extractions and are thus at level two. The real definition of level requires a knowledge of field extensions and is a bit more complex than we have presented. But, what we have done is sufficient for our purposes, which is to present a fascinating relationship between the level of a number and the degree of an equation for which it is a solution.

Observe that every rational number satisfies a first degree equation with integer coefficients. For example, the number $x = 2/3$ satisfies the equation $3x - 2 = 0$, while the rational number $x = -4/5$ satisfies the equation $5x + 4 = 0$. Thus, all rational numbers (all level 0 numbers), satisfy equations of degree 2^0 or 1.

A level one number is of the form $a + b\sqrt{k}$, where a and b are rational numbers and $\sqrt{k}$ is not. We already know these are constructible, since they are formed by the operations needed to form constructible numbers. Can we find an equation with integer coefficients with these numbers as roots, and if so, what is their degree? To answer this, we first show it for a specific case. Suppose we had the number $x = 3/2 + 5\sqrt{2}$. We can subtract $3/2$ from both sides to get $x - 3/2 = 5\sqrt{2}$, then square both sides to get $x^2 - 3x + 9/4 = 50$, and then multiply by 4 to get an equation of degree 2 with integral coefficients, which is satisfied by $x = 3/2 + 5\sqrt{2}$. A similar argument shows that any number of the form $a + b\sqrt{k}$ satisfies a quadratic equation with integral coefficients, as long as k is not a perfect square.

A typical level 2 number might be $\sqrt{1 + \sqrt{2}}$ or $\sqrt{2} + \sqrt{5}$. We will show that both of these level two numbers satisfy an equation of degree 4. Here is how. If $x = \sqrt{1 + \sqrt{2}}$, then squaring both sides we get $x^2 = 1 + \sqrt{2}$. Rewriting this as $x^2 - 1 = \sqrt{2}$ and squaring again, we get $(x^2 - 1)^2 = 2$, which is a fourth degree polynomial with integral coefficients that is satisfied by $x = \sqrt{1 + \sqrt{2}}$.

Similarly, if $x = \sqrt{2} + \sqrt{3}$, then $x - \sqrt{2} = \sqrt{3}$. Squaring both sides we get $x^2 - 2\sqrt{2}x + 2 = 3$, which can be rewritten as $x^2 - 1 = 2\sqrt{2}x$. Squaring both sides again, we get $x^4 - 2x^2 + 1 = 8x$, which is a fourth degree polynomial with integral coefficients, which is satisfied by $x = \sqrt{2} + \sqrt{3}$.

Although we have not been overly formal, we have shown, by example, that level 0 numbers satisfy equations with integral coefficients of degree 2^0 or 1, level 1 numbers satisfy equations with integral equations of degree 2^1 or 2, level 2 numbers satisfy equations with integral coefficients of degree 2^2 or 4, and so on. We seem to have established a pattern here. In fact, we have observed the general principle, which can be proved by induction, but which for now, we hope the reader will accept.

Theorem 14.2 *Every geometrically constructible number c, is at some level k, and satisfies an equation with integral coefficients of degree a power of 2^k and no lower degree. Thus, all constructible numbers are algebraic.*

The polynomial of minimal degree which c satisfies is called its **minimal polynomial**.

We are now ready to turn to three problems that occupied mathematicians for thousands of years and which were solved using the ideas presented in this section.

14.3.4 Solving the Three Problems of Antiquity

We recall the three problems of antiquity we are discussing.

Problem 1: Is it possible to double the cube, that is to construct using only straightedge and compass, the side of a cube whose volume is twice that of a given cube?

Problem 2: Is it possible to trisect an angle, using only straightedge and compass?

Problem 3: Is it possible to square the circle, that is, construct a square, using only straightedge and compass whose area is the same as the area of a given circle?

Let us consider problem 1 by beginning with a cube whose volume is 1. To double the cube requires constructing a cube whose volume is 2, which is equivalent to constructing a side whose length, $s = \sqrt[3]{2}$. We will show that it is impossible to construct a length of $\sqrt[3]{2}$.

Now $s = \sqrt[3]{2}$ satisfies the cubic equation

$$s^3 - 2 = 0. \tag{14.5}$$

If s is in fact, constructible and has level n, by Theorem 14.2 its minimal polynomial is of degree 2^n, and s satisfies no polynomial of smaller degree with integral coefficients. Since s satisfies the polynomial in equation (14.5), 2^n cannot be > 3 (because then it would satisfy the polynomial in equation (14.5) which would have smaller degree). So, $2^n < 3$. It follows that n can only be 0 or 1, which means that the polynomial that $\sqrt[3]{2}$ satisfies must be of degree 1 or 2. If $\sqrt[3]{2}$ satisfies an equation of degree 1 with integer coefficients, then it would satisfy $px + q = 0$ and x would be $-q/p$ making it rational. But, we have already shown in Chapter 3 that $\sqrt[3]{2}$ is not rational. Thus, if $\sqrt[3]{2}$ is constructible, the only other possibility is that it satisfies a polynomial of degree two with integer coefficients, say $px^2 + qx + r = 0$. Using the quadratic formula, we see that x must be of the form $a + b\sqrt{k}$ where a and b are rational, $b \neq 0$, and $k > 0$ is not a perfect square. (k can't be a perfect square for if it were, then $a + b\sqrt{k}$ would be rational, which we have already indicated is not the case.) But, we have already shown in Chapter 7 Theorem 7.26 that, if $a + b\sqrt{k}$ is a root of a polynomial with integral coefficients, specifically equation (14.5), then so must $a - b\sqrt{k}$ be a root of that equation. Thus, equation (14.5) has at least two real (constructible roots). But, if we graph $f(s) = s^3 - 2$, we see it crosses the x-axis only once. Thus, it has only one real root. This contradicts what we have said earlier about there being two real roots, $a + b\sqrt{k}$ and $a - b\sqrt{k}$. Thus, $\sqrt[3]{2}$ is not constructible, and we cannot double the cube.

We now examine Problem 2. We want to show that it is impossible to find a general method that will trisect any angle using compass and straight edge. We begin by showing this for the special case of a 60 degree angle. Interestingly, showing that we can't trisect a 60 degree angle will prove that there is no general method that will trisect every angle.

We assume that the 60° angle is part of an equilateral triangle with sides of length 1. By doing this, we can place all its vertices at points whose coordinates are constructible numbers. (See Figure 14.15 below.) Now, if there was a method of trisecting an angle, we should be able to apply this method to trisect the 60° angle at the origin O.

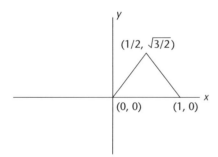

Figure 14.15

Suppose that the last line in the construction process is the line joining the vertex at O to some previously constructible point P and OP is the trisector. (See Figure 14.16 below.)

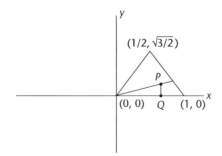

Figure 14.16

Of course, the x and y coordinates of P are also constructible numbers, as we have pointed out in Theorem 14.1. We know, using the distance formula that the length OP is a constructible number, since it is computed by using the operations of addition, subtraction, raising to powers, and extracting square roots. Now, from P we drop a perpendicular, and the point Q where it intersects the x-axis, has a constructible x coordinate since this is the same x coordinate as P, which we already know is constructible. Also, since POQ is the trisector, angle POQ is 20°. So, referring to the above figure, $\cos 20° = \dfrac{OQ}{OP}$ where O is the origin. Thus, $\cos 20°$, being the quotient of two constructible numbers, is a constructible number.

Our plan is to show that $\cos 20°$ satisfies a cubic equation and no smaller degree equation. It will follow that $\cos 20°$ can't be constructible by Theorem 14.2, since 3 is not a power of 2. If $\cos 20°$ can't be constructed, then it follows that P couldn't have been constructed. Thus, the angle POQ couldn't be trisected with ruler and straightedge. The ideas are simply ingenious! Let us proceed to show that $\cos 20°$ satisfies a cubic equation.

To show that $\cos 20°$ is not constructible, we need to use a result from Chapter 7. There we showed that $\cos 3\theta = 4\cos^3\theta - 3\cos\theta$. If we let $\theta = 20°$, we get the equation

$$\cos 60° = 4\cos^3 20° - 3\cos 20°$$

or just

$$\frac{1}{2} = 4\cos^3 20° - 3\cos 20° \tag{14.6}$$

since $\cos 60°$ is $\dfrac{1}{2}$. Now call $\cos 20° = y$, and equation (14.6) becomes

$$\frac{1}{2} = 4y^3 - 3y.$$

We multiply this by 2 and bring all the terms over to one side to get

$$8y^3 - 6y - 1 = 0. \tag{14.7}$$

Now, if y (which is $\cos 20$), is constructible, y satisfies a polynomial with integral coefficients of degree 1 or 2, just as we saw in the proof of the impossibility of doubling the cube. If y satisfies a polynomial of degree 1 with integral coefficients, it is a rational number. But, by using the rational root theorem, you can show that equation (14.7) has no rational roots. Thus, y can't be rational. This contradiction shows that y cannot satisfy a first degree polynomial.

Now, by the same argument used in solving the doubling the cube problem, if $\cos 20°$ is constructible, then it satisfies a quadratic polynomial with integral coefficients. Hence by the quadratic formula, $y = \cos 20 = a + b\sqrt{k}$, where a and b are rational. So $a - b\sqrt{k}$ is a root of equation (14.7) also. As a result, we find that equation (14.7) has two real (constructible roots), $r_1 = a + b\sqrt{k}$ and $r_2 = a - b\sqrt{k}$. Now the argument becomes more subtle. We know that the sum of the roots of the cubic equation (14.7) is the coefficient of y^2 or 0. (See Chapter 3 Section 6 Exercise 12). Since the sum of the two roots r_1 and r_2 is $2a$, the third root must be $-2a$ for the sum of the roots to be 0. But a is rational! *Thus the third root of equation (14.7) must be rational.* We have our contradiction, because we have already indicated that equation (14.7) has no rational roots.

Our contradiction arose from assuming that cos 20° was constructible. Thus, cos 20° is not constructible, and we cannot trisect the angle using straightedge and compass.

What a clever proof! And yes, what a difficult proof!

You may need to read it over a few times to convince yourself of its truth. It is the way it connects so many mathematical concepts that makes it so interesting and probably explains why the proof took so long to come about.

Note: While we cannot trisect angles with straightedge and compass, it is possible to trisect an angle with a marked ruler and compass. We refer the reader to http://www.uwgb.edu/dutchs/PSEUDOSC/trisect.HTM to see how this can be done.

Problem 3, the problem of squaring the circle, is a lot easier to solve, once we know that π is transcendental, which we stated in Chapter 3. (The proof that π is transcendental is quite difficult, however.) Now, consider a circle of radius 1. If we could construct a square with side s equal to the area of a circle, then $s^2 = \pi$ and hence $s = \sqrt{\pi}$ would need to be constructed. But, if a number is constructible, its square is also constructible. Thus, π must be constructible. But, if π is constructible, by Theorem 14.2, it must be algebraic, and that can't be since it is transcendental. So, we cannot square the circle.

Student Learning Opportunities

1 Show that, if a and b are geometrically constructible, then so are $a + b$ and $a - b$. (Here $a > b$.)

2 Show how to geometrically construct $1/a$ if a is geometrically constructible.

3 Show how to construct $\sqrt{a^2 + b^2}$ geometrically if a and b are constructible.

4 Follow the steps outlined below to show that, given segments of length a, b, and 1, the given constructions work.

(a) Mark off a length of $OD = 1$ and then a length of $OB = b$. Draw any ray OQ and mark off a length of $OA = a$ as shown in Figure 14.17 below. Draw AB and through D draw a line parallel to AB intersecting ray OQ at C. Show that $OC = a/b$.

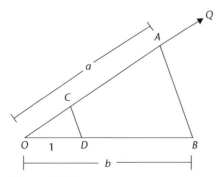

Figure 14.17

(b) Mark off line segments $OC = 1$ and $OA = a$, and $OD = b$. Draw CD and through A draw AB parallel to CD as shown in Figure 14.18. Show that $OB = ab$.

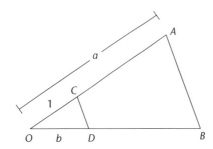

Figure 14.18

5 Find a polynomial of smallest degree with integer coefficients and one of whose roots is $1 - \sqrt{5}$. At what level is $1 - \sqrt{5}$?

6 Find a polynomial of smallest degree with integer coefficients, of which $\sqrt{2} + \sqrt{7}$ is a root. At what level is $\sqrt{2} + \sqrt{7}$?

7 What is the degree of the smallest polynomial that $\sqrt{2} + \sqrt{3} + \sqrt{7}$ satisfies?

8 What is the degree of the smallest polynomial that $\sqrt{2} + \sqrt{3} + \sqrt{6}$ satisfies?

BIBLIOGRAPHY

Anton, H.A. & Kolman, B. (1974). *Applied Fine Mathematics.* NY: Academic Press.
Bell, E.T. (1986). *Men of Mathematics.* NY: Simon and Shuster.
Boyer, C. (1968). *A History of Mathematics,* 2nd edn. NY: Wiley.
Brown, R. (1987). *Advanced Mathematics.* NY: Hougton Mifflin.
Brown, J.W. & Churchill, R.V. (2004). *Complex Variables and Applications,* 7th edn. NY: McGraw Hill.
Courant, R., Robbins, H., & Stewart, I. (1996). *What is Mathematics.* NY: Oxford University Press.
Devlin, K. (2002). *The Language of Mathematics.* NY: Owl Books.
Dunham, W. (1994). *The Mathematical Universe.* NY: John Wiley & Sons.
Feller, W. (1968). *An Introduction to Probability Theory and Its Applications.* 3rd edn. NY: John Wiley & Sons.
Fey, J.T., Hirsch, C.R., Hart, E.W., Schoen, H.L. & Watkins, A. E. (2009). *Core-Plus Mathematics: Contemporary Mathematics in Context Course 3.* 2nd Edn. NY: McGraw Hill.
Fomin, D., Genkin, S., & Itenberg, I. (1993). *Mathematical Circles.* Providence, RI: American Mathematical Society.
Gamow, G. (1988 Reprint). *One Two Three Infinity.* NY: Dover Publications.
Hadar, N. M. & Webb, J. (1998). *One Equals Zero.* Emeryville, CA: Key Curriculum Press.
Heath, T.L. (Ed.) (2002 reprinted). *The Works of Archimedes and the Methods of Archimedes.* NY: Dover Publications.
Hirsch, C.R., Fey, J.T., Hart, E.W., Schoen, H.L., & Watkins, A. E. (2008). *Core-Plus Mathematics: Contemporary Mathematics in Context Course 2.* 2nd edn. NY: McGraw Hill.
Hirsch, C.R., Fey, J.T., Hart, E.W., Schoen, H.L., & Watkins, A. E. (2008). *Core-Plus Mathematics: Contemporary Mathematics in Context Course 1.* 2nd Edn. NY: McGraw Hill.
Jones, G. A., Langrall, C.W., & Mooney, E.S. (2007). Research in probability: Responding to classroom realities. In F.K. Lester, Jr. (Ed.), *Second Handbook of Research on Mathematics Teaching and Learning* (pp. 909–955). Reston, VA: National Council of Teachers of Mathematics.
Niven, I. (1961). *Numbers, Rational and Irrational.* Washington, D.C.: New Mathematical Library, Mathematical Association of America.
Kline, M. (1985 Reprint). *Mathematics for the Nonmathematician.* NY: Dover Publications.
Lehocsky, S. & Rusczyk, R. (2003). *The Art of Problem Solving.* 5th edn. Alpine, CA: AOPS.
Maor, E. (1998). *Trigonometric Delights.* Princeton, NJ: Princeton University Press.
Mendelson, E. (1985). *Number Systems and the Foundations of Analysis.* Malabar, FL: Krieger Publishing Company.
Moise, E. (1990). *Geometry From An Advanced Standpoint,* 3rd edn. Reading, MA: Addison Wesley.
Nahin, P.J. (2006).*The Story of i.* Princeton, NJ: Princeton University Press.
Posamentier, A. (2003). *Math Charmers.* NY: Prometheus Books.
Posamentier, A. (1998). *Problem Solving Strategies for Efficient and Elegant Solutions.* NY: Corwin Press.
Shroeder, M. R. (1988). *Number Theory in Science and Communication.* NY: Springer Verlag.

Shaughnessy, J.M. (1981). Misconceptions of probability: From systematic errors to systematic experiments and decisions. In A.P. Shulte (ed.), *Teaching Statistics and Probability, 1981 Yearbook*. Reston, VA: National Council of Teachers of Mathematics.

Sloyer, C. W. & Crouse, R.J. (1977). *Mathematical Questions from the Classroom.* Boston, MA: Prindle, Weber & Schmidt, Inc.

Sproull, R.F. & Newman, W.M. (1979). *Principles of Interactive Computer Graphics*. NY: McGraw Hill.

Toussaint, G. (July 31–August 3, 2005). The Euclidean Algorithm generates traditional musical rhythms. *Proceedings of the 9th Annual Conference of BRIDGES: Mathematical Connections in Art, Music and Science*. Banff, Alberta, Canada.

Usiskin, Z., Perressini, A., Narchisotto, E., & Stanley, L. (2003). *Mathematics for High School Teachers, An Advanced Perspective*. Upper Saddle River, NJ: Prentice Hall.

APPENDIX

An $m \times n$ matrix is a rectangular array consisting of m horizontal rows and n vertical columns. We say that the size of the matrix is m by n or that its size is $m \times n$. Consider the matrix $A = \begin{pmatrix} 2 & -1 & 4 \\ 0 & 5 & 1 \end{pmatrix}$. This has two horizontal rows and three vertical columns, so this is a 2×3 matrix. The entries in the matrix are called its components. To add or subtract matrices, they must be the same size. In this case we add or subtract them component-wise (that is, entry by entry). Thus, to add $A = \begin{pmatrix} 2 & -1 & 4 \\ 0 & 5 & 1 \end{pmatrix}$ and $B = \begin{pmatrix} 5 & 2 & -3 \\ 1 & 2 & 0 \end{pmatrix}$ we get $A + B = \begin{pmatrix} 2+5 & -1+2 & 4+-3 \\ 0+1 & 5+2 & 1+0 \end{pmatrix}$ or just $\begin{pmatrix} 7 & 1 & 1 \\ 1 & 7 & 1 \end{pmatrix}$. Similarly, $A - B = \begin{pmatrix} 2-5 & -1-2 & 4--3 \\ 0-1 & 5-2 & 1-0 \end{pmatrix}$ or just $\begin{pmatrix} -3 & -3 & 7 \\ -1 & 3 & 1 \end{pmatrix}$. We can multiply a matrix by a constant k by multiplying each component by k. Thus $5A = \begin{pmatrix} 5 \cdot 2 & 5 \cdot -1 & 5 \cdot 4 \\ 5 \cdot 0 & 5 \cdot 5 & 5 \cdot 1 \end{pmatrix} = \begin{pmatrix} 10 & -5 & 20 \\ 0 & 25 & 5 \end{pmatrix}$. The following is true of matrices.

Theorem A.1 *If A, B, and C are matrices of the same size and k is a constant, then* (a) $A + B = B + A$, (b) $A + (B + C) = (A + B) + C$, (c) $k(A + B) = kA + kB$.

If $A = (a_1, a_2, \ldots, a_n)$ is a $1 \times n$ matrix and $B = \begin{pmatrix} b_1 \\ b_2 \\ \ldots \\ \ldots \\ b_n \end{pmatrix}$ is an $n \times 1$ matrix, then we can define the product of A by B as the *number* $a_1 b_1 + a_2 b_2 + a_3 b_3 + \ldots + a_n b_n$. Thus, to multiply $A = \begin{pmatrix} 4 & 2 & -1 \end{pmatrix}$ by $B = \begin{pmatrix} 1 \\ 2 \\ 3 \end{pmatrix}$ we get the number $4 \cdot 1 + 2 \cdot 2 + -1 \cdot 3$ or 5.

If A is an $m \times n$ matrix and B is an $n \times p$ matrix (that is, if the number of columns of A is equal to the number of rows of B), then we can multiply AB to get an $m \times p$ matrix. The entry in the ith row and jth column of AB is the product of the ith row of A multiplied by the jth column of B. Thus, to multiply the 2×3 matrix $A = \begin{pmatrix} 4 & -6 & 3 \\ 2 & 1 & 0 \end{pmatrix}$ by the 3×2 matrix $B = \begin{pmatrix} -1 & 6 \\ 5 & 2 \\ 7 & -3 \end{pmatrix}$, we get the 2×2 matrix

$$AB = \begin{pmatrix} 4 & -6 & 3 \\ 2 & 1 & 0 \end{pmatrix} \begin{pmatrix} -1 & 6 \\ 5 & 2 \\ 7 & -3 \end{pmatrix} = \begin{pmatrix} -13 & 3 \\ 3 & 14 \end{pmatrix}$$

obtained as follows: The entry in the first row first column of our answer, namely -13, is the product of the first row of A or $\begin{pmatrix} 4 & -6 & 3 \end{pmatrix}$ multiplied by the first column of B namely $\begin{pmatrix} -1 \\ 5 \\ 7 \end{pmatrix}$. This yields $4 \cdot -1 + -6 \cdot 5 + 3 \cdot 7$ or -13. Similarly, the entry 3 in the first row second column of AB is obtained from multiplying the first row of A by the second column of B. That is, $\begin{pmatrix} 4 & -6 & 3 \end{pmatrix} \cdot \begin{pmatrix} 6 \\ 2 \\ -3 \end{pmatrix}$ or just $4 \cdot 6 + -6 \cdot 2 + 3 \cdot -3 = 3$, and so on. The graphing calculators of today easily multiply matrices.

A square matrix is a matrix that has the same number of rows as columns. Square matrices that have 1's along the diagonals and 0 s everywhere else are called identity matrices. Below, we see two of them.

$$I_2 = \begin{pmatrix} 1 & 0 \\ 0 & 1 \end{pmatrix} \quad \text{and} \quad I_3 = \begin{pmatrix} 1 & 0 & 0 \\ 0 & 1 & 0 \\ 0 & 0 & 1 \end{pmatrix}.$$

If X is any matrix and, if we compute IX where I is the identity matrix that makes the multiplications defined, we get X. Thus, multiplying a matrix by an identity matrix does not change the matrix.

If A is a square matrix having n rows and n columns, then A is said to have an inverse if there is a square matrix B (necessarily n by n) such that $AB = BA = I_n$. B is called the inverse of A and is denoted by A^{-1}. There is only one inverse of a matrix and it can easily be found with most graphing calculators.

Here are some examples: If $A = \begin{pmatrix} 1 & 2 & 4 \\ 0 & 3 & 1 \\ 0 & 0 & 2 \end{pmatrix}$, then $A^{-1} = \begin{pmatrix} 1 & -\frac{2}{3} & -\frac{5}{3} \\ 0 & \frac{1}{3} & -\frac{1}{6} \\ 0 & 0 & \frac{1}{2} \end{pmatrix}$ and we can verify that $AA^{-1} = I_3$.

Some properties of matrix multiplication follow.

Theorem A.2 *If A, B, and C are matrices such that the matrix multiplications and additions are defined, and k is a constant, then* (a) $A(BC) = (AB)C$, (b) $A(B + C) = AB + AC$, (c) $A(kB) = k(AB)$, (d) $AI_n = A$.

If we have a system of linear equations, say

$$3x + 3y + 7z = 30$$

$$4x - y + z = 5$$

$$7x + y - z = 6,$$

we can write this system in matrix form as

$$\begin{pmatrix} 3 & 3 & 7 \\ 4 & -1 & 1 \\ 7 & 1 & -1 \end{pmatrix} \begin{pmatrix} x \\ y \\ z \end{pmatrix} = \begin{pmatrix} 30 \\ 5 \\ 6 \end{pmatrix} \tag{A.1}$$

or just $AX = B$ where $A = \begin{pmatrix} 3 & 3 & 7 \\ 4 & -1 & 1 \\ 7 & 1 & -1 \end{pmatrix}$ $X = \begin{pmatrix} x \\ y \\ z \end{pmatrix}$ and $B = \begin{pmatrix} 30 \\ 5 \\ 6 \end{pmatrix}$. The $AX = B$ form looks just like the simple equation $3x = 9$ that we solved in elementary algebra. There we solved by multiplying both sides of the equation by $1/3$ or 3^{-1}. This suggests that we solve $AX = B$ by multiplying both sides by A^{-1}. If we do this, we get

$$A^{-1}(AX) = A^{-1}B$$

which, by the associative law of multiplication of matrices, gives us

$$(A^{-1}A)X = A^{-1}B$$

or just $I_3 X = A^{-1}B$. Since $I_3 X = X$ this gives us the solution right away: $X = A^{-1}B$. A calculator easily computes A^{-1} for us, and using the calculated value we get

$$X = A^{-1}B$$
$$= \begin{pmatrix} 0 & 1/11 & 1/11 \\ 1/10 & -26/55 & 5/22 \\ 1/10 & 9/55 & -3/22 \end{pmatrix} \begin{pmatrix} 30 \\ 5 \\ 6 \end{pmatrix}$$
$$= \begin{pmatrix} 1 \\ 2 \\ 3 \end{pmatrix}.$$

Since $X = \begin{pmatrix} x \\ y \\ z \end{pmatrix} = \begin{pmatrix} 1 \\ 2 \\ 3 \end{pmatrix}$, $x = 1$, $y = 2$, and $z = 3$. We can check in (A.1) that this works.

For a 2×2 matrix $A = \begin{pmatrix} a & b \\ c & d \end{pmatrix}$, we define the determinant of A, denoted by det (A) or $|A|$ to be the number $ad - bc$, and the following is easy to verify by straight multiplication.

Theorem A.3 *If A is a 2×2 matrix and* det $(A) \neq 0$, *then A has an inverse. The inverse is given by*

$$A^{-1} = \begin{pmatrix} \frac{d}{ad-bc} & \frac{-b}{ad-bc} \\ \frac{-c}{ad-bc} & \frac{a}{ad-bc} \end{pmatrix} \text{ or just } \frac{1}{ad-bc} \begin{pmatrix} d & -b \\ -c & a \end{pmatrix}.$$

If A is a 3×3 matrix, say

$$A = \begin{pmatrix} a & b & c \\ d & e & f \\ g & h & i \end{pmatrix}$$

then we define the determinant of A to be the number $a \cdot \det \begin{pmatrix} e & f \\ h & i \end{pmatrix} - b \cdot \det \begin{pmatrix} d & f \\ g & i \end{pmatrix} + c \cdot \det \begin{pmatrix} d & e \\ g & h \end{pmatrix} =$
$a(ei - fh) - b(di - fg) + c(dh - ge)$ which in turn equals

$$aei + bfg + cdh - gec - hfa - idb. \tag{A.2}$$

This is easier to remember by lining up A next to itself as follows:

$$\begin{pmatrix} \mathbf{a} & b & c & a & b & c \\ d & \mathbf{e} & f & d & e & f \\ g & h & \mathbf{i} & g & h & i \end{pmatrix}$$

and multiplying the numbers diagonally, starting with a. On the first diagonal we have a, e, and i which we have bolded. We multiply them. On the second diagonal we have b, f, and g. We multiply them. On the third diagonal we have c, d, and h. We multiply them. We now add the three products we get. This gives us the first half of (A.2). The next three products are subtracted. They come from multiplying the diagonal entries starting at the lower left hand corner, g. On that first diagonal, you see g, e, and c. We multiply them.

On the second diagonal we see h, f, and a. We multiply them. On the third diagonal we see i, d, and b. We multiply them. We subtract these three products from the first 3 we added, and we get our determinant. This seems rather difficult. And the definition of determinants of larger matrices becomes even more complex. Whatever the size of a matrix, its determinant is denoted by det (A) or |A|. Most graphing calculators compute the determinant of a matrix for us.

There are some theorems about determinants that are useful to know. Here is one.

Theorem A.4 *If we interchange two rows of a matrix, the determinant of the resulting matrix will be the same as the original, except that the sign will change.*

So, for example, if we compute $\det \begin{pmatrix} 5 & 1 \\ 2 & -3 \end{pmatrix}$ we get -17, while if we compute the determinant of the matrix $\begin{pmatrix} 2 & -3 \\ 5 & 1 \end{pmatrix}$ where we have switched rows we get 17, the opposite. Similarly, $\det \begin{pmatrix} 3 & 3 & 7 \\ 4 & -1 & 1 \\ 7 & 1 & -1 \end{pmatrix} = 110$ while $\det \begin{pmatrix} 7 & 1 & -1 \\ 4 & -1 & 1 \\ 3 & 3 & 7 \end{pmatrix}$, where we have switched the first and third row of the original matrix, is -110 as you can easily check on your calculator. There is a similar theorem for switching two columns.

The following is a famous theorem.

Theorem A.5 (a) *If A is a square matrix and if* $\det(A) \neq 0$, *then A has an inverse.* (b) *If A and B are square matrices of the same size, then* $\det(AB) = \det(A)\det(B)$.

One last thing. Suppose that we have the system of equations

$$\begin{cases} ax + by = e \\ cx + dy = f \end{cases}$$

Then we can solve for x and y using determinants. Namely, $x = \dfrac{\begin{vmatrix} e & b \\ f & d \end{vmatrix}}{\begin{vmatrix} a & b \\ c & d \end{vmatrix}}$ and $y = \dfrac{\begin{vmatrix} a & e \\ c & f \end{vmatrix}}{\begin{vmatrix} a & b \\ c & d \end{vmatrix}}$. This is known as **Cramer's Rule**. To illustrate, the solution of $\begin{matrix} 3x + 2y = 7 \\ x - 2y = -3 \end{matrix}$ is, $x = \dfrac{\begin{vmatrix} 7 & 2 \\ -3 & -2 \end{vmatrix}}{\begin{vmatrix} 3 & 2 \\ 1 & -2 \end{vmatrix}} = 1$ and $y = \dfrac{\begin{vmatrix} 3 & 7 \\ 1 & -3 \end{vmatrix}}{\begin{vmatrix} 3 & 2 \\ 1 & -2 \end{vmatrix}} = 2$.

Student Learning Opportunities

1 Given $A = \begin{pmatrix} -1 & 2 & 3 \\ 4 & -5 & 6 \end{pmatrix}$ $B = \begin{pmatrix} 4 & 2 \\ 0 & 3 \\ 5 & 1 \end{pmatrix}$, $C = \begin{pmatrix} 2 & 1 \\ -3 & 2 \end{pmatrix}$, and $D = \begin{pmatrix} 1 & 5 \\ -3 & 2 \\ 4 & -1 \end{pmatrix}$. Perform each of the following operations or tell if an operation is undefined. Do it by hand and then check your answer with the calculator.

(a) AB
(b) BA
(c) $A + B$
(d) $B + D$
(e) $4B - 3D$

(f) $3A$
(g) $\det(A)$
(h) $\det(C)$
(i) $\det(DA)$
(j) $\det(AD)$
(k) A^2
(l) C^3

2 Verify the Associative Law $(AB)C = A(BC)$, for the matrices $A = \begin{pmatrix} 2 & 1 \\ -3 & 2 \end{pmatrix}$, $B = \begin{pmatrix} 4 & -3 \\ 2 & 1 \end{pmatrix}$, $C = \begin{pmatrix} 5 & 9 \\ 7 & -6 \end{pmatrix}$. Do it by hand and then check your answer using the calculator.

3 Verify that $\det(AB) = \det A \det B$ for the matrices from Question 2. Don't use the calculator.

4 Pick any two 3×3 matrices of your choice, A and B, and using the calculator show that $\det(AB) = \det A \det B$.

5 Solve the following systems of equations by using inverses of the matrices. Check your answer.

(a) $\begin{aligned} 2x + 3y &= 5 \\ 7x - 2y &= 7 \end{aligned}$

(b) $\begin{aligned} 4x + y - z &= 5 \\ 3x - 2y + 4z &= 2 \\ x + y + z &= 9 \end{aligned}$

6 Determine the inverse of each of the following matrices by hand or with the calculator.

(a) $\begin{pmatrix} 5 & 6 \\ -1 & 2 \end{pmatrix}$

(b) $\begin{pmatrix} 4 & 2 & -3 \\ 3 & 1 & 5 \\ 0 & 2 & 0 \end{pmatrix}$

7 Show that, if $AB = BA$, then both A and B must be square matrices of the same size.

8 Using matrices find the equation of the parabola passeing through the three points (1, 3), (3, 13), and (7, 57).

9 Suppose that we have a 3×3 matrix and switch rows 1 and 3 first and then rows 2 and 1 in the resulting matrix. How does the determinant of the final matrix compare with that of the original? Explain.

10 Solve the following systems of equations using Cramer's Rule

(a) $\begin{cases} 2x + 3y = 8 \\ 4x - y = 2 \end{cases}$

(b) $\begin{cases} -3x - 2y = -5 \\ 4x - y = -3 \end{cases}$

INDEX

Abel-Ruffini 102
addition and multiplication of signed Numbers (Definition) 211–22; of fractions (see fractions rules)
additive inverse 220, 235
additivity 607
algebraic number 88, 303
altitude(s) 115
angle positive and negative 524; quadrantal 526; inscribed in a semicircle; 10 hyperbolic 694
Anscombe's data 425
Ars Magna 307
Arc sine (sine inverse) 563
area (Euclidean) under a curve 140–2, circle 126–39, 496; ellipse 496; hyperbolic 700 irregular shapes 139; parallelogram 116; polygon 120; sector 539; trapezoid 117; triangle 116, 494 (see also Heron's theorem, Pick's Theorem); transforming 492
area in Hyperbolic space 701
Argand Diagram 316
arithmetic sequence 361
Associative Laws of Addition and Multiplication 217; for complex numbers 310; generalized 218
asymptotic 49
attractor (see Newton's basin)

base number systems 51–6
best fit (see regression)
Barnsley M. 501
Bernouli trial 626
Bolyai J. 688
Bombelli R. 100
Bowditch (see Lissajous)
best fit 420

binary operation 217; representation 53
Birthday Problem 636
Bisection method 108–9
Box and Whisker Plot 662

Cantor G. 298
Cardan G. 100–1
cardinality 199, 301–4
Cavalieri's Principle 153
central angle 180
Cereal Box Problem 646
Ceva's Theorem 195–8
cevian 196
Chaos Game 389
characteristic equation 366
cis θ 326–9, 345
classical approach to probability 602
class intervals 659–60
coefficient of determination 431
combination 614
Commutative Laws of Addition and Multiplication 217, 230
completing the square 90
complex number(s) conjugate 309, 319, 331; distance between 337; Distributive Law 310; division and multiplication 309 equality of 308; exponent 353; geometry of 335–8; logarithm of 352; magnitude 323; polar form 326; relation to fractals 347–52; roots of 330–2; subtraction 308; vector interpretation 318
complex Plane 316
components of force 507; of vector 587
composite number 31
composition (see functions)
congruence ASA 166; HL 163; in Hyperbolic world 698; mod m 57; SAS 163; SSS 162

conjecture 12
conjugate 304; surds 342
constructible 711; plane 713
constructions 705–8
cosine 159; of A-B 532
cosecant (*see* reciprocal trigonometric functions)
cotangent (*see* reciprocal trigonometric functions)
countable 297–300; algebraic numbers are 303
counting principle 613

data fit 425; issues with 431; exact (polynomials) 431–7
diagonal argument 300
decimal expansion 277–88; representation 293–6
defect 701
degree 410
dependent variable 399
DeMoivre's Theorem 328
dependent variable 398
Diophantine Analysis 61–7; equation 63
dilation 455–6; resulting from complex arithmetic 318, 335
distance properties 696
Distributive Law 217, 310
divisibility 19; relationships 17–25; rules 26–8
Division Algorithm 39; for polynomials 48–50
differences 433
divisor 20
domain 399–403; of inverse function 446

empirical probability 601
experimental probability *see* empirical probability
equivalent equations 272–4; issues with solving 268
Euler's conjecture 2
Eratosthenes 515
Escher 452–3
Euclidean Algorithm 43–6
Euler's amazing identity 344
even 18
event 606; complementary 608
expected gain 643; value 644
exponents, laws of 242–4
exponential function 409
extraneous solutions 270

factor 19
factor theorem 72
Fagnano's problem 509–11
fair game 641–4

fallacious proofs 678–84
Fibonacci sequence 376–7
fractal 247–352, 497, 501; dimension 390–3
fraction rules 224–30
fractional exponents 247–52
function 398–401; composition of 469–74; inverse 442–8; modeling with 70, 406–12; model to use 412–14; ways of representing 401–3
Fermat's conjecture 2
Ferro S. 95–7
Fundamental Thoerem of Algebra 81
Fundamental Theorem of Arithmetic 380
Fundamental Theorem of Calculus 140

Gauss C. F. 9, 36, 688
Geometric Probability 650
Golden Rule of Fractions 226–7
geometric series 277
geometry (definition of) 690; incidence 691
Great Survey of India 531
greatest common divisor 42–7
gunfight problem 637

Heron's theorem 205
histogram 659; discrete probability 630
homogeneous coordinates 498
horizontal line test 440
hotel infinity 298
Hyperbolic geometry 693–701

image 459
imaginary part 311
independent events 622; variable 398, 622
inequalities 253–8
inscribed angle 180–2
integers 221
Intermediate Value Theorem 98
Interquartile range 663
inverse function 442–8
invertible 495, 495, 506
irrational number(s): are nonrepeating decimals 288; between any two real numbers 232; e is 346; exponent 250; square root of 2 is 11, 73, 87
infinite complexity 361

Kepler's law 415
Koch curve 388
Kolmogorov axioms 607

Law of Cosines 160, 529, 537
Law of Sines 165–7, 531
linear correlation coefficient (*see* r)
curve of best fit 421
least common multiple 46–7
linear function 407
linear recurrence relations 366
Lissajous curves 578–81
logarithms 259; common and natural 260; of complex number (*see* complex number) principal 354; rules for 263–5

magnification factor 391–3
magnitude 323
mathematical model 411
matrix transformation 483–90
measures of central tendency (*see* mean median and mode)
mean 631, 634
median 666
minimal polynomial 716
mode 665
modeling (*see* functions)
modular arithmetic 59–60
Monte Carlo method 137, 454
Monty Hall problem 636
multiple 20
multiplication principle (see counting principle)
multiplicative identity 239; inverse 239
multiplicity 80
mutually exclusive 606

natural numbers 216
Newton's Method (Newton-Raphson method) 104–9, 351; basin 351; Law of Cooling 261
Normal distribution 623

odd number 18
one to one functions 435–42
outlier 663

Parallel postulate 686; what we can prove without 688–9; more than one 689, 691
partial sum 277
period 267, 546
periodic simple and delayed 290–1
peridocity 289–92
permutation 614

phase shift 549
pi (computation of) 133–6; is transcendental 88
Pick's Theorem 206–11, 382
polar form of angle 326
polynomial 71; functions 410
postulate (*see* axiom)
power function 411
Principle of Mathematical Induction 374; strong 379
principal square root 246, 330; n^{th} root 247
polynomiography 82
prime number 21, 31–5; Theorem 36
principal solution 354; logarithm 354
probability classical approach 602; conditional 618–22; counterintuitive results in 635–8; density function 631; elementary results 609–11; frequency approach 600; issues with 603; misconceptions 623–4; one 624; subjective approach 603; zero 623–4, 652
proof by contradiction (indirect proof)) 10–12; by counterexample 12; direct 9–11
Ptolemy's Theorem 185
Pythagorean theorem 117, 593; converse 119, 123; equivalent to parallel postulate 700 triple 200–2

Quadratic formula 93; function 407
quartile 662

r value 425
radian measure 537–8; area of sector and length of arc 539
Random Variable 644
Rational Root Theorem 85–8
range 400–3, 442 of inverse function 446
rational number between any two reals 232; repeating decimal 287; root theorem 86
real part 316
reciprocal trigonometric functions 534
recursive (recurrence) relations 359; solving 361–5
reference angle 563
reflections 317, 355, 453, 461, 467–8, 477, 479, 482, 485, 499–500, 504, 507–10
regression line or curve 413, 421
relatively prime 36
remainder 71–2
repeating decimals 287–8
root 71, 82; of complex number 330–1
rotations 317, 335, 454–7; matrix of 468
reflections 454, 467–8, 474, 477, 509

rotation 454, 463, 467, 479
RSA encryption 59–61

sample space 606
scatter plot 412
secant line 182–5; (*see* reciprocal trigonometric functions)
Sierpinski triangle 361; carpet 395
signed numbers 220–3
similar triangles 169–72
simple solid 146–55
simulation 638–9
sine 159; of A+B 176, 186
spherical geometry 702
square unit 114
statistics (lying with) 669–72
stem and leaf plot 661
synthetic division 75–9, 84
sum of the angles of a triangle is 180 degrees 687; is less than 180 degrees 695; is more than 180 degrees 703

tangent 159; to circle 184 graph 552–3
terminal side 524

terminating decimal (*see* decimal representation)
theoretical probability (*see* classical probability)
three problems of antiquity 716–19
Tower of Hanoi 369–70, 378
transcendental 88, 303
transformation 459; composition of 469; in 3 dimensions 504–6; matrix of 468–73; area under (*see* area); image of polygon under 485–9
translation 454, 459–60, 463; matrix of 498
trial 601
trigonometric identities 567–73; functions 159; graphs of 544–52
trial by mathematics 622

uncountable 299–300

vectors 581–94
vertical line test 440
volumes of solid 146–56

whole numbers 217

zero of function 71; to the zero power 244